D0875478

Quantitative Geography

Techniques and Theories in Geography

Quantitative Geography

Techniques and Theories in Geography

JOHN P. COLE
Lecturer in Geography,
Nottingham University

CUCHLAINE A. M. KING
Reader in Geography,
Nottingham University

JOHN WILEY & SONS LTD
LONDON NEW YORK SYDNEY

Library of Congress catalog card number 68-54882

SBN 471 16475 5 Cloth bound

SBN 471 16476 3 Paper bound

Printed in Great Britain by Robert MacLehose and Co. Ltd, The University Press, Glasgow

Preface

The purpose of this book is to present in one volume a large number of techniques and theories used in geography. For the most part it is fairly elementary in scope. Indeed it starts each topic assuming that the reader knows nothing or very little about it. In this way it is hoped that many people will be able to obtain a background that will enable them both to appreciate the so-called 'quantitative' papers now appearing in many geographical periodicals, and also to practise techniques and develop theories in their own work.

When using this book the reader is asked to abandon temporarily all his preconceived ideas about the scope and content of geography. He will then be more receptive to the mathematics and statistics introduced in chapters 2 and 3 as an essential basis for material discussed in the rest of the book. A large number of step by step worked examples are given, and numerous maps and diagrams have been included to help those who learn more quickly with the help of visual illustrations.

The order in which the topics are arranged was given much thought. It was based as far as possible on the idea that some topics have to be covered and learned before others can be started. There are however many ways in which the 15 chapters could have been arranged. Clearly therefore it will sometimes be desirable for the reader to go back and reconsider earlier chapters or sections after having read later ones. To some extent the chapters may be considered self-contained entities and read independently, and the book may prove useful as a kind of handbook to be referred to from time to time.

In connexion with the scope and contents of the book, the following points should be noted. Firstly, there is very little about the nature of geography, though some references are given on this matter in chapter 1. Secondly, obviously even in a book this size it has not been possible to cover all techniques and theories in geography. Some have been given considerable attention while others have been neglected. Thirdly, both the choice of techniques and theories, and the choice of examples, have been dictated by the limitations of knowledge and specialisms of the authors. Geomorphology and political geography are over represented at the expense of other branches. Fourthly, both physical and human geography are considered. Often similar methods of approach can be used. Cross-fertilization between these major branches of geography can greatly enrich the whole subject, and help to maintain the unity of geography at a time of increasing specialization. Fifthly, of the many examples in the book, a considerable number are fictitious; some are unrealistic in order to be simple.

Briefly, next, a few words on why a book of this kind is necessary. Awareness of place, relative location, distance and other underlying concepts in geography have presumably long been part of the experience of man. Geography as a discipline has however been widely regarded by people outside professional geography itself in a very limited sense. In particular, it is seen and referred to by many either as an inventory of locations or positions of things in the world, or as the physical background of areas being studied for particular purposes. It is to the credit of a limited number of geographers, mainly American, that an attempt has been made to turn geography into the science of areal or spatial relationships. There have been various isolated attempts to do this previously, but since the Second World War, and particularly since about the mid-1950s, there has been a more conscious movement to borrow techniques and theories from such diverse disciplines as statistics, psychology, physics and engineering, and to use them in geography.

One purpose of this book, then, is to bring together some of the things that have been done during the so-called quantitative revolution in geography, and to make them accessible to others. The authors feel that while geography is not a vocational subject, it has nevertheless a great deal to offer in an applied sense (see chapter 14) and as a way of seeing things. They are of the opinion that a quantitative approach is highly desirable in geography and beneficial to it.

The authors are grateful to many colleagues and other friends for ideas and advice of various kinds. Above all, they are greatly indebted to Professor C. W. J. Granger, of the Department of Mathematics, Nottingham University, for his continual encouragement and advice on statistical matters and his comments on geographical matters. Professor R. Emerson, of the University of East Anglia, gave much help and encouragement. Professor T. Hagerstrand of the University of Lund, Sweden, helped in the organization of the book and in many other ways. Mr. J. M. Bates of the Department of Economics, Nottingham University, very kindly read chapter 3 and Miss R. E. Gent similarly checked chapter 2. The authors also wish to thank Mr. G. A. Smith for agreeing to the inclusion of a great deal of material, in the collecting and processing of which he collaborated. It is perhaps unfair to select a few names from those of the many students at Nottingham who have become interested in quantitative methods, but discussion with the following has been most useful: Dr. John Andrews, Mrs. K. M. Bonniface, Mr. D. J. Blair, Mr. R. S. Andrew, Miss M. R. Tinning and Miss J. Brindley. Finally, the authors would like to thank Professor K. C. Edwards for the encouragement he has given to the development of quantitative techniques in the Geography Department at Nottingham University.

J. P. COLE
C. A. M. KING

Contents

Part 1

Introduction, mathematics and statistics

1 Introduction

'If the earth were uniform — well polished, like a billiard ball — there probably would not be any such science as geography . . .'

J. Gottmann (1951)

'Without the hard little bits of marble which are called "facts" or "data" one cannot compose a mosaic; what matters, however, are not so much the individual bits, but the successive patterns into which you arrange them, then break them up and rearrange them.'

A. Koestler (1964)

1.1 Introduction

The subject matter that comes within the field of geography is very diverse. This great diversity makes it essential that the geographer should be able to arrange his data systematically in order to make use of them in his own special way. The geographical way of looking at things differs from that of some other sciences, although the subject matter may be the same. The essential element in nearly all geographical work is spatial variation, as is so admirably suggested by Gottman's statement quoted above. The importance of this aspect will be stressed throughout the book.

The purpose of chapter 1 is to draw attention to some of the basic characteristics of geographical data and to consider ways of looking at this material. Section 1.2 discusses various definitions of the field of geography. It is not possible to set rigid bounds to this field, but suggestions can be made to help to delimit the very broad field of geography. Section 1.3 draws attention to some of the basic characteristics of geographical data. Attempts are made to systematize the data for ease of handling and to draw attention to similarities between apparently unlike situations in geography. Section 1.4 deals briefly with some of the criticisms levelled at traditional geography and with attempts to provide answers to them. In section 1.5 the need for a scientific approach is stressed. Section 1.6 outlines the sequence of topics adopted in this book to explore some of the ways in which geography can be studied.

1.2 The content of geography

The field of geography is the earth's surface*. This however is also the field of many other sciences. Geography is the science that is mainly concerned

* The term surface is used here not in a strictly mathematical sense, but synonymously with shell (see glossary).

with the distribution of elements that occur on the earth's surface and with the variations of the distributions through time and in space. Not all the distributions that occur however are necessarily within the field of geography. Moreover, geography does not have a particular set of objects, but shares those of other disciplines. Some limits can be set to this broad field, both by considering space and by considering the scope of the enquiry or study.

(1) The space limitations of geography can be fixed more easily than the scope of the enquiry. It has already been said that geography is concerned with the surface of the earth. The interior of the earth is therefore excluded. The core zone is the sphere of the geophysicist, and the geographer is only very indirectly concerned with this zone. The core is surrounded by the uniform mantle, and this in turn is separated from the crust by the Moho discontinuity. This boundary surface provides a reasonable lower limit to the earth's surface for most geographical purposes. The upper limit can be defined by the upper limit of the atmosphere, within which are confined the weather phenomena that have such an important effect on the earth's surface. With the wide occurrence of artificial satellites however it has now become important to reconsider the upper limit. Beyond this upper limit lie the solar system and the stars. Although the solar system has a great effect on the earth's surface, it does not lie directly in the geographical field. Indirectly its influence must often be taken into account in geographical problems. The stars lie beyond the scope of geography and can be left to the astronomers. In this sphere also, however, geographers may use the stars for surveying and fixing positions on the earth. Thus the solar system and the core of the earth have an effect on the earth's surface and from this point of view are relevant to geography, but in themselves they are not part of geography.

(2) It is more difficult to define the limitations of scope than of space. Geography is concerned with the distribution of different elements on the earth's surface. The larger elements, such as countries, towns or mountain ranges, usually have geographical relevance. The limit to the geographical study of these elements is usually reached where they become increasingly subdivided into smaller entities. As long as they are undivided units they often have geographical relevance. When however an object cannot be further subdivided into like units, its study often becomes more relevant to another science. Geography, therefore, deals mainly with whole objects and with organisms which exist as units and which, as such, are discrete and may be able to move or be moved as one body. This size limitation excludes nearly all objects that can only be studied with a microscope. Geography is therefore mainly concerned with objects of medium size or large size and not of very small or microscopic size.

The limit of size varies from one branch of geography to another. It is often related to the competence of the investigator. Some examples will illustrate how geographical data merge into those of other disciplines as the size of the object is reduced. Plant associations and entire single plants are studied in biogeography, but the dissection of plants is left to the botanists. The same applies to the animal in relation to the zoologist. Geographers are concerned with man as a whole body, but not with the way in which he is put together, this being the field of anatomy and physiology. Similarly towns form the basis of urban geography. They can be divided into buildings which are still whole units and can still be studied by geographers. However, the interior layout of the building is more the concern of the architect than of the geographer.

In a study of dynamic situations, processes, not discrete objects, are involved. Geomorphologists are interested in landscape evolution, but in some cases the study of the processes involved must be left to the physicist or the chemist. The study of climate is relevant to geography, but this leads to a study of meteorology, which when pursued in depth lies within the domain of physics. These limits are more often set by the competence of the worker. The transport network, another dynamic situation, is a legitimate field of study of the geographer, but he is not concerned with the workings of the vehicles that use the network, this aspect being the engineer's concern.

The indivisible unit may be taken as the smallest object that is really relevant to geography. The geographer may then study the interaction of independent units and their effect on each other. Each unit is normally a member of another more generalized class of units. The relationship between the units and their associated classes of units is considered in chapter 13, when classification is discussed. The lowest member of the system of classification is usually the indivisible unit.

(3) Distributions of individual units on the earth's surface can normally be considered to have geographical interest. Distributions within the units, for example the distribution of veins and arteries in the body, or certain other distributions, such as the distribution of stars in the sky, are not geographical. However such non-geographical distributions can often be made use of in geography. They can be used as an analogy to assist in the understanding of geographical distributions. J. Hutton (1795) for instance, used the analogy between the circulation of blood in the human body and the circulation of water on the earth. Other examples are discussed in chapter 11. It is also possible for the operation to be reversed. The geographical attitude to distributions could have useful analogies in non-geographical distributional problems. The principle of working with analogies from one discipline to another is one of the basic features of *General Systems Theory*, which is discussed in chapter 11.

1.3 The nature of geographical data

(*a*) *Spatial, temporal and other variations*

It was suggested in section 1.2 that one of the most important distinguishing features of geography is its concern with *spatial* variations. A distinction is sometimes made between spatial, temporal and other variations. This is illustrated by three examples in figure 1.1, each occurring on a line.

A spatial variation, at least from the point of view of the geographer, implies relative location of objects in space. Relative location may be expressed by distance along a line from a point of origin, by two co-ordinates on a surface, or by three co-ordinates in 3-dimensional space. A 2-dimensional example is the earth's surface, on which each place has a particular relationship to every other place. Thus each object and observation studied in geography has a unique location in space at any given time, and from its unique location each is in a special relationship to all other objects or observations in space.

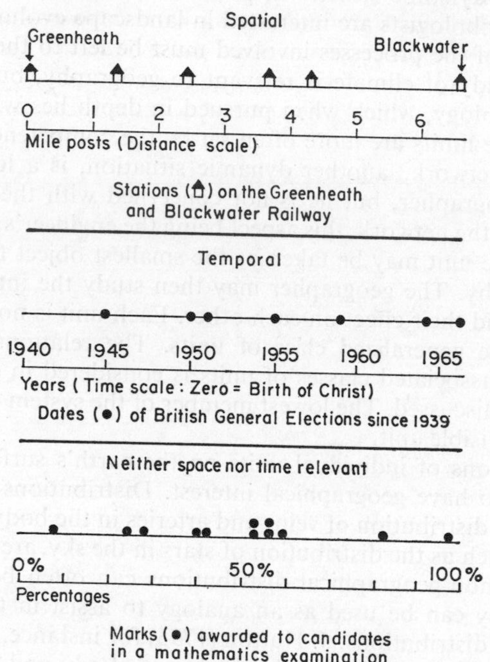

Figure 1.1 Spatial, temporal and other variations on a line.

Figure 1.1 shows simple examples of spatial and temporal variations and of a case of variation in which relative location is not relevant either in space or in time. On the imaginary railway, the stations are placed at clearly distinguishable points along a distance scale. On the temporal scale, the elections are similarly placed in order of occurrence in time. The marks in the mathematics examination were however simultaneously obtained by the candidates in a given room during a given period; in this case space and time are irrelevant for each candidate, although the room does of course occupy space and the exam covers a period of time.

To illustrate the nature of variations further, some information is given in the table below about ten fictitious school children who sit in the front row in a classroom. In practice it is more likely that one would try to obtain information about considerably more than ten items in a study of the kind shown, but to keep the example reasonably small, only ten have been used.

Moreover, the information is more appropriate to the field of education than to geography. The purpose of the example is however to show that some variations are essentially spatial in character and that others are not.

Table 1.1 Front row of class in Melford school

	I	II	III	IV	V	VI	VII	VIII
William Bacon	BR	E	3	M	5	108	8/3	950
Joseph Cunningham	BR	E	5	M	10	96	8/6	960
John Hemlock	BR	E	7	M	1	116	8/4	1110
Mavis Holt	BR	E	9	F	2	112	9/0	210
Edward Knight	BR	E	11	M	4	114	8/4	230
Joan Myers	BR	E	13	F	6	101	8/8	670
Jean Primrose	BR	E	15	F	7	95	8/2	400
Frank Smith	BR	E	17	M	8	93	8/11	820
Roger Treadgold	BR	E	19	M	3	114	8/11	1440
Anthony Yates	BR	E	21	M	9	102	8/10	90

I Nationality (BR=British)
II Native language (E=English)
III Distance in feet of seat in class from left hand wall of classroom
IV Sex (M=Male, F=Female)
V Teacher's ranking of the children according to ability
VI Result of IQ test adjusted for age but not yet revealed to teacher
VII Age in years and months to nearest month
VIII Distance of child's home in yards from school gate along shortest feasible route

Eight pieces of information are given about each child. I and II, nationality and native language do not vary; they are therefore 'constants', though potentially they could become variables if non-British or non-English speaking children happened to be in the set under consideration. III is a variable with spatial implications. Note, however, that as the desks in the class are spaced at regular intervals apart, the variation is *regular*. Moreover, the spatial arrangement of objects in a room (regular or irregular) is not generally considered by geographers to be of direct importance to them. IV is a biological variable (and therefore neither a spatial nor a temporal variable). It has only a yes/no answer. V and VI are educational (intelligence assessment) variables, and like IV have no spatial or temporal connotations. It may be noted in anticipation of the discussion of levels of data later in the book that V has its information ranked (with intervals of one unit), while VI has its information with different intervals between the marks. VII is of temporal interest since the ages of the children can be plotted on a time scale in relation to a base month, for example that in which the eldest was born. Of the eight columns of data, VIII is of particular interest to the geographer since it is of a spatial nature. The geographer might be interested solely in the distance each child lives from school (regardless of direction) as a measure of effort and/or time required to journey between school and home. He might however also be interested in the actual distribution of the homes of the ten children in area, not only in relation to the school, but also in relation to each other and even to all or some part of the general setting, say a town, in which the school is situated. Figure 1.2 compares variable VII, plotted on a time scale, with variable VIII plotted both on a distance line and in area.

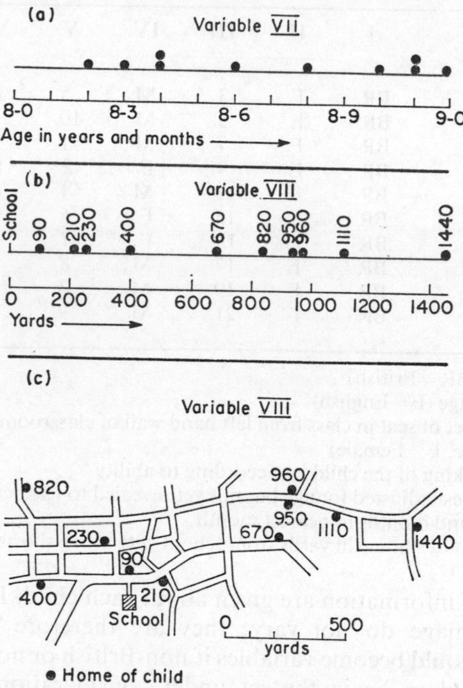

Figure 1.2 Examples of variations in time and in space. VII and VIII refer to information given about ten children in the table in the text. (*a*) Variations on a time scale. (*b*) Distances from a point (the school entrance). (*c*) Relative distances and directions from a point (school entrance) of ten children.

VIII is the only variable in the table that is of direct interest to the geographer. But in analysing the facts given by variable VIII the geographer might use some or all of the rest of the information available. Suppose, further, that several children had some other language, rather than English, for their native tongue. The urban geographer might then be interested to know if these non-English speaking children all lived in one part of the town, or if they were dispersed among the English speaking children. Thus the geographer can use non-spatial data in conjunction with spatial data.

In table 1.1 several kinds of information are given about the children. There is some confusion over the difference between quantitative and qualitative types of information. Levels or scales of data, nominal, ordinal, interval and ratio, are discussed in chapters 2 and 3. Here, attention is drawn to the possibility of converting apparently qualitative information into a quantitative, numerical form if only in terms of 1 to represent presence of some feature and 0 to represent its absence. Thus male and female can be expressed as 1 and 0, male or not male.

(*b*) *The objects and spaces studied by the geographer*

The geographer is interested in the arrangement of objects in space. The objects he studies are also studied in other disciplines (e.g. plants and animals in biology). Moreover, the geographer does not have a complete monopoly of the study of space. This interests equally for example the astronomer, geologist and architect. The geographer, however, handles both the objects and the spaces in a particular way, often making drastic simplifications and abstractions. Some of these must now be noted.

Both the spaces in which the objects are arranged, and the objects themselves exist in the 'real' world in 3-dimensional space. Although more sophisticated measures may be used, the shape of a given space or object may be described very approximately by a comparison of the length of its longest axis with the length of the other two axes at right angles to the longest axis. Very broadly, it is possible to have spaces or objects (which occupy spaces) that have all three axes roughly equal (ball or cube), two long and one short (paving slab) or one long and two short (pencil or rod). These are illustrated in figure 1.3*a*.

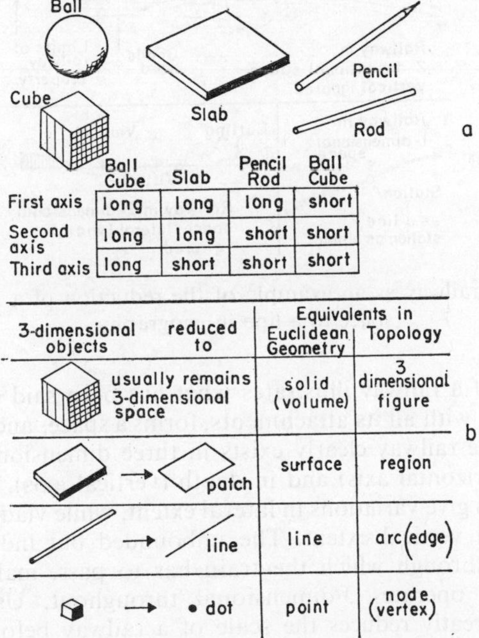

Figure 1.3 The shape of objects and spaces. (*a*) Long and short axes. (*b*) The elimination of dimensions.

Often in geography it is adequate and convenient to overlook one, two or even three of the dimensions (see figure 1.3*b*). Thus the short axis or axes may be overlooked. The slab may therefore be thought of as a 2-dimensional surface and the rod as a line. The small compact object may still be reduced to a dot of no dimensions by analogy with a point in geometry, if it is sufficiently small in relation to the space in which it is situated. Further aspects of 3-dimensional objects and space will be dealt with in chapter 9.

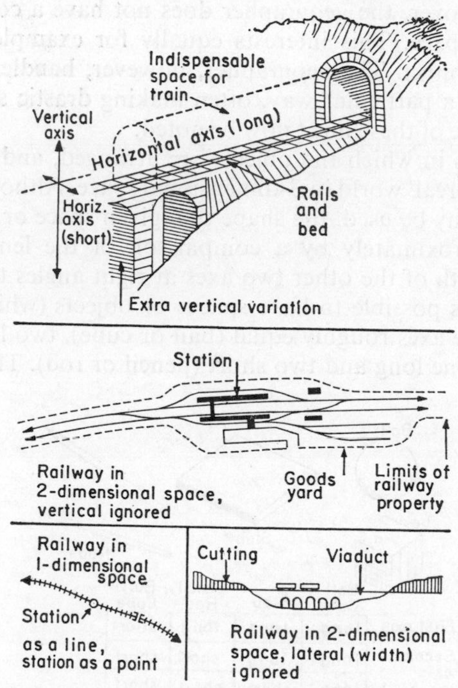

Figure 1.4 A railway as an example of the reduction of a 3-dimensional space to a line in geography.

The example of a railway illustrates what has been said so far (see figure 1.4). The railway, with all its attachments, forms a space, and the trains using it are objects. The railway clearly exists in three dimensions. It varies both in width (one horizontal axis) and in depth (vertical axis). Stations, sidings and other features give variations in lateral extent, while viaducts and cuttings give variations in vertical extent. The unbounded but indispensable space above the rails, through which the train has to pass, makes the space in which a railway operates 3-dimensional throughout. Usually, however, the geographer greatly reduces the scale of a railway before looking at it, and then loses sight of, or interest in, anything but its long axis or length. In other words, the railway, really a greatly elongated 3-dimensional space, can be thought of in geography as a line. Similarly, a train itself, really a 3-dimensional object, is usually much longer on one axis than the other two and may be represented as a line. On a small enough scale it can even be considered as a dot or point. Such drastic abstractions are commonly made in geography, and are an essential part of cartography.

To summarize, then, all objects and spaces in the real world however small exist in 3-dimensional space, but reductions in dimension may be made to produce *patches*, *lines*, or *dots*. In Euclidean geometry the rough equivalents are surfaces, lines and points, while in topology they are regions, arcs and nodes (see chapter 2). *Three real* dimensions are for convenience reduced to two, one, or none. If the shape is compact, the reduction, if any, is usually from 3 to none (dot). If it is flat, usually from 3 to 2 (patch) and if elongated, usually from 3 to 1 (line). Altogether there are ten possible combinations of dots, lines, patches and 3-dimensional spaces. Examples of these are given in fig. 1.5.

(*c*) *The character of the data*

Geography studies objects and areas, and their dimensions, distributions and attributes. These basic data can be studied as whole units. The geographer is often interested in the number of objects, because the best way to deal with the problems associated with the objects depends partly on their number. If the number is small enough for all the members of the set to be available for study, then the set can be called finite. This is in some ways not very helpful, as strictly all unit objects must be finite in number, especially as far as geographical data are concerned. Even the number of atoms in the universe is presumably finite, though very large. In many instances however, the number of objects with which geographers are dealing is in practice uncountable and can be considered as virtually infinite, although this is not strictly true in a mathematical sense. The sand grains on a large beach are an example of an uncountable set of objects. The important point concerning this type of object is that it must be studied by the statistical techniques of sampling. These are discussed in chapter 3. Separate objects, therefore, must be discrete units, and they must be finite in number, although for purposes of analysis they may be divided into countable and uncountable groups.

Complications arise when the character of the objects is being considered. An object must exist in space and time and it must have dimensions. Each dimension is itself finite (x), but it can be subdivided into an infinite number of points, as a point has no dimensions ($x/0=\infty$). This infinite number of points can be represented by the real number system (see section 2.5). A point can move along a line continuously, moving an infinitesimally small distance in each infinitesimally small time span. These distances can however be considered to be very small steps. In fact the limit is fixed by the accuracy of measurement. Thus the line can in practice be divided into a finite number of small steps. The same principle applies to surfaces and volumes. The line, surface or volume is itself a discrete unit, but, because it must have magnitude,

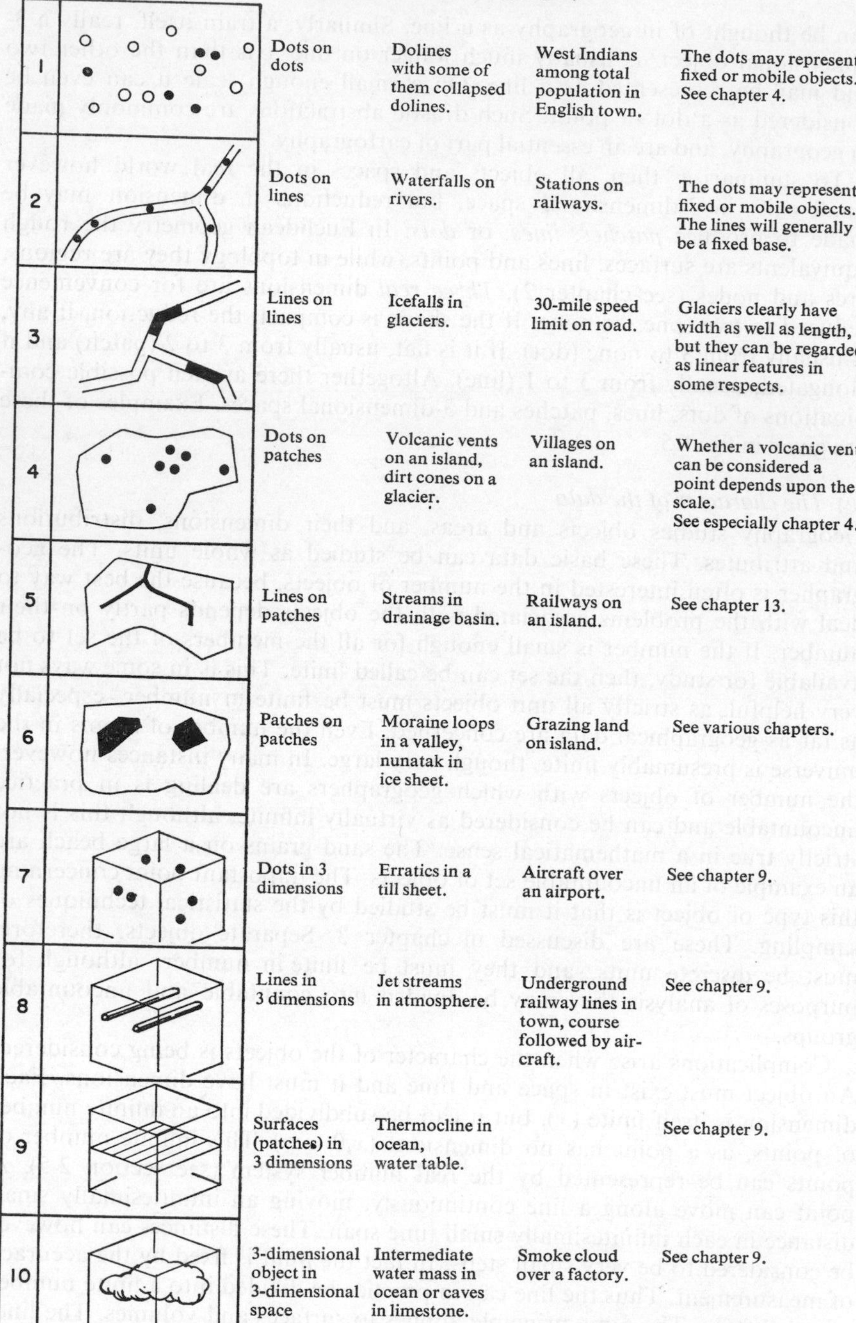

1
Dots on dots
Dolines with some of them collapsed dolines.
West Indians among total population in English town.
The dots may represent fixed or mobile objects. See chapter 4.
2
Dots on lines
Waterfalls on rivers.
Stations on railways.
The dots may represent fixed or mobile objects. The lines will generally be a fixed base.
3
Lines on lines
Icefalls in glaciers.
30-mile speed limit on road.
Glaciers clearly have width as well as length, but they can be regarded as linear features in some respects.
4
Dots on patches
Volcanic vents on an island, dirt cones on a glacier.
Villages on an island.
Whether a volcanic vent can be considered a point depends upon the scale. See especially chapter 4.
5
Lines on patches
Streams in drainage basin.
Railways on an island.
See chapter 13.
6
Patches on patches
Moraine loops in a valley, nunatak in ice sheet.
Grazing land on island.
See various chapters.
7
Dots in 3 dimensions
Erratics in a till sheet.
Aircraft over an airport.
See chapter 9.
8
Lines in 3 dimensions
Jet streams in atmosphere.
Underground railway lines in town, course followed by aircraft.
See chapter 9.
9
Surfaces (patches) in 3 dimensions
Thermocline in ocean, water table.
See chapter 9.
10
3-dimensional object in 3-dimensional space
Intermediate water mass in ocean or caves in limestone.
Smoke cloud over a factory.
See chapter 9.

Figure 1.5 Combinations of dots, lines, patches and 3-dimensional spaces and objects, with geographical examples.

and therefore be measureable, it must consist of one or more continuous distances, which consist of an infinite number of points.

Some things have no dimensions, such as temperature. In this instance the temperature is an attribute of a particular parcel of air at a fixed point, the thermometer bulb. It is a continuously varying attribute. For purposes of measurement temperature can be converted into a specified volume of mercury. This provides a continuous measureable dimension. Temperature, however, exists, even if no thermometer is there to measure it, as a continuously varying quantity that has finite limits.

The table 1.2 attempts to illustrate some of the possibilities that exist when the different dimensions, including time and space–time, are considered. It is worth noting that lines and surfaces can be finite and bounded, finite and unbounded or infinite and unbounded. The second possibility is of interest in considering movement around a great circle or parallel, for example the west wind drift in the southern ocean, which provides a finite unbounded route by which water can and does continuously flow around the earth. The third situation is less likely to be of geographical importance, but a line running through the centre of the earth to infinity at both ends illustrates the possibility.

1.4 Conventional geography and its critics

Many disciplines have undergone drastic reappraisals in the last two or three decades (e.g. mathematics, linguistics, psychology) or have them looming on the horizon (e.g. history*). It would be surprising if geography were to escape reappraisal. Given the broad scope of the subject and the intersection of the subject with many others, geography may indeed be faced not only with its own internal reappraisal but with repercussions from those of other disciplines. In this book great importance is attached to the repercussions of the so-called *New Mathematics* on geography, and chapter 2 is mainly devoted to this theme.

In this section it is proposed to note some current or recent adverse criticism of geography as a discipline, criticism made both by geographers and by non-geographers, and to suggest some remedies. Subsection (*a*) considers criticisms of geography concerning its lack of definition and unity. Subsection (*b*) considers criticisms levelled at geography regardless of whether it is satisfactorily defined or not.

(*a*) *Criticism of geography on account of its lack of definition and unity*

Geography is still far from being a field of study clearly defined to the satisfaction of those who practise it. The reason for this is that several definitions, all of which to some extent intersect, are widely accepted.

* K. Thomas (1966), whose views may not be shared by many historians, is certain of the need for a numerical and statistical approach in much historical research: 'A more self-conscious statistical approach will enable him [historian] to be more objective when the facts are available, and to refrain from unverifiable pronouncements when they are not.'

Table 1.2 Dimensions

		Finite: Countable	Finite: Uncountable	Infinite	Units under consideration
Space 1–D 2–D 3–D	*Discrete*	Small number of units Objects e.g. people in village Points representing objects on a map	Large number of units Objects e.g. Sand grains on large beach	Number of points on a line, surface or in a volume	Number of objects
	Continuous	*Bounded* Length, area or volume of one object e.g. Length of line — river Area of surface — Ice cap Volume of space — pebble	*Unbounded* Length of closed line — Great circle Area of closed surface — earth	Number of lines on a surface Number of lines and surfaces in a volume	Attributes of objects
Time	*Discrete*	Number of units e.g. seconds, years — A life time	Number of seconds in geological time	Eternity	Units of time — no movement
Space–Time	*Continuous*	Movement along a finite bounded route or line	Movement along an unbounded finite route or line, when the line is not of permanent duration	Movement along an eternal unbounded line or surface or in a volume	Movement in Space–Time
	Discrete and continuous	Acceleration under alternating flow e.g. waves			Intermittent movement
	Discrete	Number of objects created in a finite time and space	Number of objects created in finite space and time when 1 space is very large or 2 time is very long or 3 objects are very numerous or 4 when they are reproduced very fast	Number of objects created or reproduced either during eternity or in an infinite space	Objects in time
Zero–D	*Continuous*	Temperature Density		Heat and energy	Dimensionless items

D = Dimension

(*i*) *Geography deals with certain spatial aspects of the earth's surface* Not all spatial aspects of the real world, whether on the earth's surface or not are of geographical interest. The distribution of knots on floorboards in a room, the wiring of an aircraft carrier and the arrangement of stars in a galaxy are all spatial situations. None concerns the geographer directly. Clearly there is considerable scope for dispute as to *which* spatial aspects the geographer should actually deal with. Geography as the study of *spatial relationships* is a subset of all spatial relationships.

(*ii*) *Geography is concerned with areal differentiation* Geography should describe (and account for) differences between areas. As in (i) it is debatable as to which areas are the concern of the geographer. The problem of defining regions is closely connected with this definition of geography. The region itself is currently a highly controversial concept in geography, and at the moment regional and systematic approaches are unfortunately quite separate in the minds of many people.

(iii) *Geographers study the landscape* The definition of geography is related to the two previous definitions, but it may be criticized because it gives insufficient emphasis to spatial variations and too much to the objects themselves. At the same time it is difficult to decide which objects in the landscape should receive the attention of the geographer and which should not.

(iv) *Man and* (*in*) *his environment* This view of geography is widely held both by geographers and popularly. In a popular sense one may meet a statement such as: 'Siberia is a difficult area to develop because of its geography'. Geography here means physical conditions, such as cold and ruggedness, lack of certain resources, such as farmland, and perhaps even big distances. While the study of man and his environment can be of great interest, it has certain drawbacks as a definition of geography. Firstly, it has given rise to many unfortunate statements about deterministic relationships. Historians and others, as well as geographers, are guilty of this. Secondly, man and the physical environment is only one of several possible relationships. Human activities may be affected by physical conditions, but they may also affect them. Moreover, it is possible to consider exclusively the relationship between physical distributions or between man-made ones. Thus the study of the influence of the physical environment on man as a definition of geography does not allow the study of an entirely human environment (e.g. town centre) or a purely physical one (e.g. an uninhabited Arctic Island) by the geographer. The definition is, in fact, an important special case of spatial relationships.

(v) *and* (vi) Two other more popular views of geography regard the subject either as a gazetteer of place names and products, or as a subject dealing with position. One reads, popularly, statements such as 'Because of its *geographical* position, Singapore is a strategic locality'.

In view of these and other possible definitions or views of geography it is not easy to decide what the subject actually is. One is also left with this impression after reading the magnificent summaries of thought on this matter by R. Hartshorne (1939, 1959). W. Pattison (1963) rationalizes the field of geography in a number of traditions. D. Lowenthal (1961) may also be consulted on this matter. Perhaps, indeed the word geography might best be used to refer loosely to some undefined compartment in total knowledge. The centre of the compartment is clearly distinguishable and is agreed upon by most if not all geographers but the periphery is fuzzy and flexible. Indeed there seems to be a move away from rigid disciplines in education and knowledge at least at the higher (university) and lower (junior school) ends. Even former distinguished and apparently unassailable disciplines like physics and chemistry have now produced mergers: for example physical chemistry.

If knowledge is indivisible, then exact definitions of individual disciplines are not feasible. Knowledge is then merely divided into rough compartments for organizational purposes. Perhaps new compartments are centring on very different things from the old: logic, communication theory, spatial variations, temporal variations. The establishment of new approaches to geography might lead to chorometrics. In particular, there may at the moment be a shift in some fields of study from compartments that focus on discrete 'objects' (e.g. the element in chemistry, the cell in biology, man in anthropology) to relationships and patterns. It is not difficult to see that some spatial patterns or arrangements, such as the streets of a town, the linked cells in the brain, or the form of buildings, may persist after the objects composing them (building materials, cells and so on) have been replaced once or even several times.

At all events it seems reasonable to expect changes in approach to geography no less drastic in the next few decades than in the last few. What is to be regretted is the tendency of some geographers to equate their own particular approach and achievements with the whole of geography, whether they argue that it must be spatial theory, man and land, applied mathematics, or some other definite approach. Surely it is in the spirit of the subject that just as many things vary on the earth's surface, so different environments, experiences and attitudes make for different emphasis in approach and content. Perhaps for example the North American geographer, especially if he lives in the Middle West, is more ready to see spatial regularities in human geography than his colleague in Europe. The latter may feel that past patterns greatly influence present ones and that physical conditions which often intrude to complicate patterns and areal variations in cultural features, are not to be overlooked.

(*b*) *Other criticisms of geography*

Some of the criticisms of geography are discussed in this subsection. They have been given in list form for convenience. Several are in fact connected.

(i) *Geography is too easy* It merely involves the memorizing of facts. In learning place names, or in studying what other parts of the world are like, geographers are working at a very low intellectual level. Lewis Carroll suggests this in *Alice through the Looking Glass.*

> 'Of course the first thing to do was to make a grand survey of the country she was going to travel through. "It's something very like learning geography", thought Alice as she stood on tiptoe in hopes of being able to see a little further. "Principal rivers there *are* none. Principal mountains — I'm on the only one, but I don't think its got any name. Principal towns . . .".'

The desire to memorize large amounts of information is deep rooted, and in many disciplines credit is given for this achievement. The need to memorize a large amount of information to some extent must date from a time before the invention of printing, or at least before books became cheap enough to become widely available or easily carried about, either privately or in libraries. Chaucer's Scholar in *The Canterbury Tales* was greatly esteemed by his fellow pilgrims for his vast amount of knowledge. Goldsmith's 18th century schoolteacher in *The Deserted Village* was likewise admired for what he retained in 'one small head'. Presumably it was useful for these gentlemen to have so much information at their fingertips. It is still useful for a geographer to store information in his head. A sensible criterion for what should be memorized and what looked up when needed seems to depend on the amount of time required to do one or the other. But if fact learning is needed in geography, much of it could be done through studying problems and doing exercises, rather than for its own sake. It will be suggested in the next section that the acquisition of information is one of the steps in the scientific method. Awareness of the importance and possibilities of the scientific method for geography raises questions about the teaching of the subject. Outside university geography departments, and often, indeed, within them, emphasis has been on presenting a large number of facts to pupils and students with a view, presumably, to making them into their own private data stores and ending the procedure, to be discussed, at step 3. If appropriate material could be provided and exercises worked out it would be valuable to teach awareness of the whole series of steps, even to fairly young children. Attempts have been made to do this in exercises prepared for children of about ten years of age by J. P. Cole and N. J. Beynon (1966). A parallel may be seen in the study of languages. Both vocabulary and grammar forms are now to some extent absorbed by using the language being learnt in conversation from the start, rather than by memorizing grammar paradigms and word lists and translating utterly artificial sentences.

(ii) *Geography is hopelessly divided into physical geography and human geography* One of the chief objections to geography has been that it suffers from dealing with both physical and human aspects of reality. In particular there has recently been a bitter dispute among Soviet geographers on this matter, over the incompatibility of physical laws and Marxist–Leninist laws. There is a strong feeling that science, in dealing with physical objects, makes laws that are 100 per cent certain, or that at least would be if enough were known about the processes involved (see chapter 12). At the other extreme human beings are considered by many to be so utterly unpredictable that at best only very vague tendencies can be stated about their behaviour. There is unfortunately some truth in this dichotomy, at least in geography as it is currently studied. However in the physical sciences there has been a trend towards a more probabilistic view of the physical world, in place of a mechanistic one, while some social sciences, such as economics, have also been making statements on probabilistic lines with some success. The contradiction between studying both physical and human features is not peculiar to geography. Students of agriculture are dealing both with physical features such as soil and with man-made features such as prices in agricultural economics. Similarly, the biologist and the psychologist may be dealing with man both as a physical (biological) object, and as an object capable of making decisions. These disciplines do not seem to have suffered in the way that geography suffers, or is supposed to suffer, through including both physical and human objects.

(iii) *Geographers are good at collecting and describing data, but bad at interpreting it* Certainly much effort has been devoted to teaching undergraduate geographers how to collect data in the field and how to record it on maps. Even in the gathering of data, geographers have generally failed to appreciate the importance and requirements of *sampling*. Once the data have been collected, they have been processed only inadequately, or merely described.

(iv) *Geography is an academic subject and has little or no practical application* There is no reason why studies should not be made of topics with no conceivable application. The development of violin making in 16th century Tuscany is a worthy research topic (not presumably geographical) requiring both energy and intellectual discipline on the part of the researcher. On the whole, however, geographers seem to feel frustrated that the findings of their subject have not been more widely applied. Decisions about spatial matters such as the location of new railways, roads, industrial plant and so on, have usually been made by non-geographers, and often without the advice of geographers. Geographers ought to be able to make sensible recommendations about spatial aspects of planning decisions, but in planning they are often thought of merely as the people who compile the maps. They tend to complain about this situation, but the fault may partly be their own. They have failed to learn the so-called *scientific method*, which should help them to predict and therefore to make recommendations. This method will be discussed in section 1.5. This is not to say that everything done by geographers must have a practical application in view or that there is no place for description alone. It has often happened in mathematics that some remote development in pure mathematics has later turned out to be vitally useful in applied mathematics. It is to be hoped that the same will happen in geography. It does seem desirable that geographers should be a little more aggressive about trying to find practical uses for what they do, and much more conscious about the methods that will *enable* them to offer useful advice about spatial problems.

Certainly there have been many attempts since the Second World War to clarify the aims of geographers, and to propose ways of making geography more rigorous, more scientific, and more predictive. R. Hartshorne (1959) has reconsidered some of his views. W. Bunge (1962 and 1966) provides a great deal of thought provoking material in his *Theoretical Geography* and the authors are indebted to his work for many ideas even if they have not always acknowledged them in this book. P. Haggett (1965) in his *Locational Analysis in Human Geography* deals rather with the findings of various new techniques and theories than with their precise application but in so doing provides a framework for human geography. The appearance in 1967 of *Models in Geography* edited by R. Chorley and R. Haggett is further massive evidence of the interest in Britain in what may be called new geography. *The Science of Geography* (1965), is an attempt by U.S. geographers to present the field of geography to non-geographers. E. A. Ackerman, chairman of the body of geographers who prepared the document, himself set out clearly in an earlier paper (E. A. Ackerman 1958), the following six steps as an essential part of the framework of modern geography:

(1) Description of the earth as a site
(2) Identification of the specific phenomenal content of earth space
(3) Identification of generic relations
(4) Identification of genetic relations
(5) Determination of covariation among earth features
(6) Integration of data on site, phenomenon and process so as to reveal the full pattern of space relations

Other important U.S. sources of new ideas, theories, and particularly techniques and findings in geography include the *Papers of the Regional Science Association*, the *Journal of Regional Science* and the (duplicated) *Discussion Papers* of the Michigan Inter-University Community of Mathematical Geographers. The U.S.A. does not however have a monopoly of new ideas as to what geography is. Sweden, the U.S.S.R., Canada, Poland, and India are some of the countries from which new ideas and techniques have come. Professor T. Hagerstrand, of the University of Lund, Sweden, has been responsible both for many new ideas of his own and for inspiring others to think on new lines. I. M. Mayergoyz's *Quantitative Methods of Research in Economic Geography* (1964) is one of many instances of the growing interest of new methods in the U.S.S.R. Fortunately, much is now available in English on the views of Soviet geographers thanks to the publication by the American Geographical Society of *Soviet Geography Review and Translation.*

1.5 The scientific analysis of data in geography

The term scientific has at least two very important, widely used, but fundamentally different meanings. The confusion over these must briefly be pointed out. Science may refer either to a certain set of phenomena, or to a method of study. In the first sense, it implies very broadly natural phenomena as opposed to man made phenomena; man himself falls perhaps into both categories. Educationally and academically, the sciences and the humanities, rightly or wrongly, provide the first subdivision of total knowledge both in many people's minds, and in the organization of educational establishments in many countries, regardless of their level of development, political doctrines, and other matters. In the second sense, science implies a method of study, a series of steps to be followed in research, whether the study involves natural phenomena, man himself, man made things, or conceivably combinations of any of the three. Both uses of the word science are legitimate. The discussion of science in this section concerns however the second meaning of science, not the first. M. Korach (1966) puts very concisely a view of the scientific method now widely accepted: 'In our opinion too there is no doubt that *everything* can be an object of scientific investigation. The possibility of a science existing is determined not by the *what* but by the *how*; not by the subject but by the method.'

This section is subdivided into (*a*) a consideration of the sequence of steps that might be followed in geographical work in order for the approach to be scientific, and (b) some further comments on the scientific method of approach.

(*a*) *Steps in scientific method*

The following sequence of steps is not relevant to every situation in geography. It is intended rather to suggest to the geographer the things he ought to be asking himself when working in geography. In most tasks, however, some things have to be completed before others can be started. Critical path analysis, discussed in chapter 14, is a method of catering formally for this fact, itself long a part of human experience.

(1) It is desirable, if not essential, to have in mind some objective in a given study, rather than to collect material in the hope of finding an objective. The objective may be more specifically some problem to solve, some hypothesis to test, or some theoretical model, borrowed even from another discipline, to try against a given situation.

(2) Once an objective has been worked out, it is usually necessary to collect information, or data. In geography, this may be obtained from field work, from published material in verbal or numerical form, or from sources such as aerial photographs or existing maps. Sampling techniques are often relevant at the stage of data collecting, but geography lags behind many other disciplines in the use of sampling.

(3) Data must then often be prepared and stored in some way. When data are recorded on a map, the map becomes a special kind of store, in which the relative location of things in space is given prominence. Data of geographical interest can however also be stored in tables, and particularly, in matrices (see section 2.8). The physical form of the store need not be a printed page. Modern methods of data storage allow very large amounts of data to be accommodated on punched cards or

punched tapes. If the data can be satisfactorily recorded and retrieved, then, as will be shown in chapter 14, modern geography can benefit greatly from new developments in data handling.

(4) After data have been prepared and stored they often have to be processed in some way. Increasingly, mathematical methods are being used in geography and statistical inferences are being derived from the mathematical procedures. Many techniques require computations that would be impossibly long with paper and pencil methods or even with the help of mechanical or electronic calculating machines operated manually. The development of electronic computers since the Second World War has revolutionized the possibilities of data processing.

(5) From the steps outlined so far, some results or output are produced. These constitute the findings of a particular piece of work or project. The interpretation of the findings will as far as possible be given in a form, verbal, numerical or in some other way, that is comprehensible, in the accepted language of science, to other workers, at least in the same field. The interpretation and emphasis is, nevertheless, up to a point bound at this stage to be subjective and personal. The findings may lead in several directions. Firstly, they may be so unsatisfactory that a fresh start from step 1 or step 2 may be needed in a particular piece of work. Secondly, they may need substantiation by supplementary work, or by further work beyond that so far carried out. Thirdly, they may be presented as a completed study, the possible forms of which are discussed in the next subsection.

(*b*) *Further comments on the scientific method*

Results expressed in numerical form have advantages over those expressed in verbal form. It is generally more concise, more precise, and more universally comprehensible. Thus a correlation index r of $+0.75$ means the same to a scientist anywhere in the world so long as he is acquainted with the product moment correlation technique. A verbal statement such as 'very strong', made about the same correlation, is less precise even within its own verbal language and possibly even ambiguous outside it. It must be stressed, however, that the further interpretation of numerical statements must usually be verbal. Moreover, the handling of numerical data can easily become mechanical and unimaginative, and their very precision and conciseness must not be an excuse for forgetting what the numerical data are actually describing.

The results, whether numerical or verbal, may be descriptive in character and could possibly form the raw material awaiting the work of someone else. They may result in some specific recommendation or prescription. They may, however, form the basis for some generalization, around which statements may be made about patterns, regularities or tendencies. Enough repetitions of the same patterns or tendencies may lead to the formulation of laws. Prediction with a given degree of confidence, if not with certainty, is then possible. The ability of a discipline to make predictions is in the view of many the principal criterion of success in a scientific sense. Prediction in general can be forwards or backwards in time, or sideways at the present. Geographers often use the space dimension for sideways prediction. This prediction can be from one area to another.

The appropriateness of the scientific method for a given problem or hypothesis depends on the number of things being considered. Thus, for example, whereas it is impossible to make any prediction about the location of capital cities in their territories simply from studying Paris, it is perfectly feasible to make predictions with considerable confidence about the voting or smoking habits of the population of France, even from quite a small sample. Paris is unique as the capital of France. Some 50 million French, on the other hand, are numerous enough to form the basis for many generalizations, even if these are always probabilistic rather than exact. Somewhere between one and many objects of study is a critical zone into which the scientific method may be introduced with caution. Perhaps this lies somewhere between a few and a few hundred. R. J. Chorley and P. Haggett (1967, p. 21) state 'Today the distinction is made commonly between the "humanities" which are primarily concerned with the unique and non-recurrent, and the "sciences" which seek to establish general statements for repeatable events and processes. Contemporary geography obviously lies athwart this apparent gulf . . . which must either be bridged or must lead to the dismemberment of the existing discipline'. The transference of geography wholly to the science side, depends upon the use of the scientific method.

Acceptance of the scientific method in geography may perhaps prove a blessing to those geographers who regret the dichotomy between physical and human geography. The reader will see from what was said at the beginning of this section that the first view of science, science as the study of natural phenomena, splits geography, whereas the second at least permits physical and human geography to share the same basic scientific method. In passing, it is worth drawing attention to the fact that both North American and Soviet geographers, though doubtless for different reasons, have tended to hold the view that physical and human geography are separate disciplines, whereas their West European counterparts have often stressed the unity of geography.

Geographical data have the characteristics of data used both by the physical and biological sciences, on the one hand, and the social sciences, on the other. Geography is, however, essentially a field science rather than a laboratory one, so that geographical phenomena are not often susceptible to controlled laboratory experiments as are those in some of the physical and biological sciences. As with many situations in the social sciences, geography is usually concerned with many uncontrolled variables. For this reason relationships are often rather weak.

The map is often thought of as the special tool of the geographer. Its place in the scientific method is, therefore, worthy of comment. In stage 1 the data portrayed on a map may provide an objective for study. The peculiarities of a drainage pattern, which are best shown on a map, may for instance suggest a problem. In stage 2, maps can provide valuable data. Thus drainage basin morphometric data can most readily be obtained from a map. Maps may also be used for storage of data collected from other sources in stage 3. This use, however, is probably not as efficient as numerical methods of storage. Maps do not often play a large part in stage 4, but in stage 5 they frequently provide a useful and concise method of displaying conclusions, which can readily be appreciated by other geographers. In all stages the map is a tool, a means to an end, and not often is it the end in itself.

1.6 The plan of the book

The order in which the chapters of the book have been arranged is considered next. Part I, consisting of chapters 2 and 3, is concerned with some of the fundamental mathematical ideas that can be applied in many fields of geography. The new mathematics is spreading fast and soon may be universally taught, so that many of the new ideas will soon become common knowledge. Their application to geographical situations is therefore very relevant. The situations in geography are usually complex and require the sorting of a large amount of data. Statistical techniques are essential for dealing with this material. These techniques are discussed in chapter 3 and examples of the application of the methods are given. This part of the book comes first as it provides the basic methods that are necessary throughout the rest of the book.

Part II is a detailed study of distributions, going from the relatively simple situations to the more complex situations that can only be dealt with adequately by the use of more advanced mathematical models, such as factor analysis. The four situations that are considered in these four chapters are (1) a single distribution in a single area (chapter 4); (2) the comparison of a single distribution in one area with a similar single distribution in one or more other areas (chapter 5); (3) the correlation between two or more distributions in a single area (chapter 6); (4) the comparison of the correlation between two or more distributions in one area with the correlation between the same two or more distributions in one or more different areas (chapter 7).

Part III of the book considers special situations that are particularly relevant to geographical data. The main field of geographical study is the surface of the earth. In chapter 8 situations are considered in which it is necessary to take the earth as a whole sphere or as a 2-dimensional spherical surface. In part II the distributions that were discussed were those in which the objects could be considered as points or attributes of areas. This type of simplification is not always possible, so in chapter 9 those situations that require a consideration of the three space dimensions are discussed. Some of the situations in which time plays an important part as well as space are reviewed in chapter 10.

Part IV is concerned with different stages of investigation. In chapter 11 some of the very numerous types of model are considered. These models provide a method of investigating the data or a theoretical framework which can direct an enquiry. The techniques stage was dealt with in part I, as it provides the methods that are needed for parts II and III. Chapter 12 considers in turn some of the theories, tendencies and laws that are relevant in geography. In chapter 13 classification of geographical data is discussed and this leads to the organization of the different types of network that involve movement. Chapter 14 draws on all the previous ones to mention briefly some of the applications of geography. In chapter 15, there are comments on the organization of selected branches of geography.

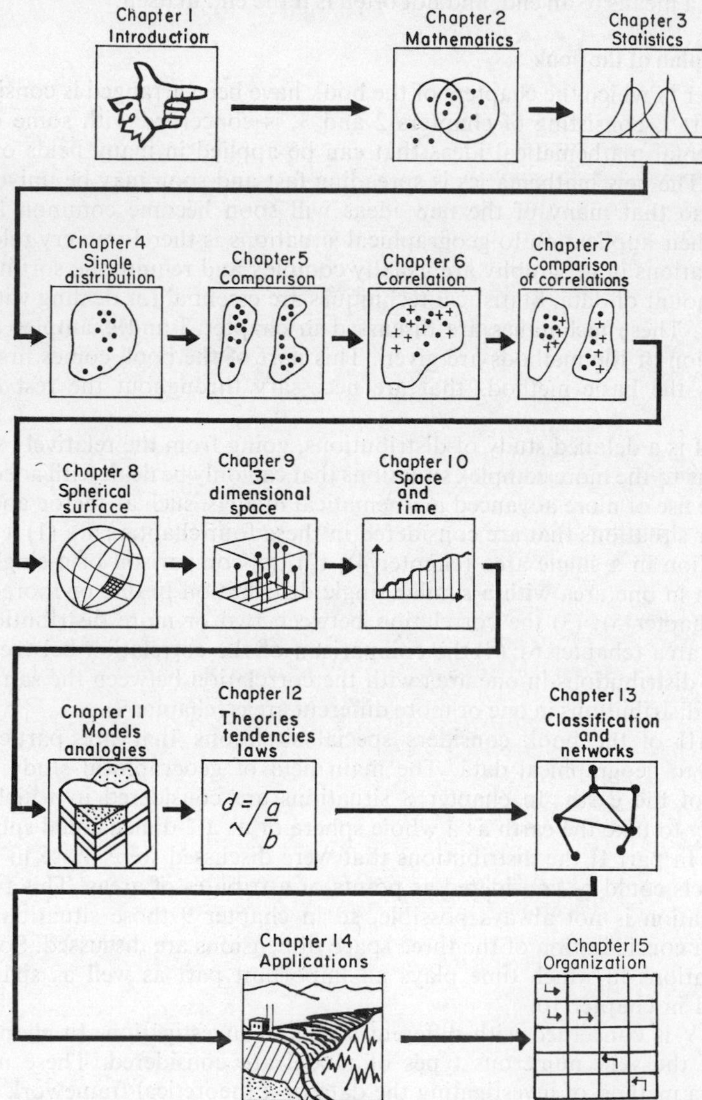

Figure 1.6 Flow chart showing the sequence of topics in the book by chapters.

Parts II and III exemplify techniques and theory in geography. They represent ways of looking at and studying the subject. As they precede the theoretical part, they provide examples of real situations but they can then be looked at from the more theoretical point of view in part IV. Geography aims to study reality, but theoretical views are necessary to accomplish this aim.

The reasons why the sections of the book have been placed in the specified order has now been given. They need not however be read in this order. In some instances for example it might be desirable to read part IV between parts I and II, as part IV provides some of the theoretical framework. On the other hand it may be easier to appreciate this framework after some simpler concrete examples have been discussed. The parts and the chapters of the book have to some extent been arranged in order from simple to complex, and from 1-dimensional cases to those with increasing numbers of dimensions. Figure 1.6 provides a diagrammatic idea of the arrangement of the book.

References

Ackerman, E. A. (1958). *Geography as a Fundamental Research Discipline*. University of Chicago, Dept. of Geography, Research Paper No. 53.

Berry, B. J. L. and Marble, D. F. (Edrs.) (1967). *Spatial Analysis: A Reader in Statistical Geography*, Prentice Hall, New Jersey.

Bunge, W. (1962). *Theoretical Geography*, Lund Studies in Geography (Ser. C. General and Mathematical Geography No. 1). C.W.K. Gleerup, Lund, Sweden.

Bunge, W. (1966). *Theoretical Geography*, revised and enlarged edition. Lund Studies in Geography, Lund, Sweden.

Burton, I. (1963). The quantitative revolution, and theoretical geography, *Canadian Geographer*, No. **7,** 151–162.

Chorley, R. J. and Haggett, P. (Eds.) (1967). *Models in Geography*, The second Madingley Lectures, Methuen, London.

Cole, J. P. and Beynon, N. J. (1966). Seven Plus Junior School Geography Project, *Bulletin of Quantitative Data for Geographers*, University of Nottingham. Material used in this project as well as new material are to be published by Basil Blackwell, Oxford.

Gottmann, J. (1951). Geography and International Relations, *World Politics*, **III,** ii, pp. 153–173.

Haggett, P. (1965). *Locational Analysis in Human Geography*, Arnold, London.

Hartshorne, R. (1939). *The Nature of Geography, A Critical Survey of Current Thought in the Light of the Past*, *Assoc. Am. Geog.*, **XXIX,** Nos. 3 and 4, Lancaster, Pennsylvania.

Hartshorne, R. (1959). Perspective on the nature of geography, *Assoc. Am. Geog.*, Rand McNally, Chicago.

Hutton, James (1795). *Theory of The Earth, with Proofs and Illustrations*, Edinburgh. 2 vols.

King, C. A. M. (1966). *Techniques in Geomorphology*, Arnold, London.

Koestler, A. (1964). *The Act of Creation*, Pan Books, London, p. 236.

Korach, M. (1966). The Science of Industry in *The Science of Science* (Eds. M. Goldsmith and A. Mackay), Penguin Books, Harmondsworth.

Lowenthal, D. (1961). Geography, experience and imagination: towards a geographical epistemology *Ann. Assoc. Am. Geog.*, **51**, No. 3, pp. 241–260.

Lukermann, F. (1964). Geography as a formal intellectual discipline and the way in which it contributes to human knowledge, *Can. Geog*. **VIII**, 4, p. 167.

Mayergoyz, I. M. (1964). *Kolichestvennyye metody issledovaniya v ekonomicheskoy geografii*, Moscow (Vsesoyuznyy institut nauchnoy i tekhnicheskoy informatsii).

Pattison, W. (1963). *The Four Traditions of Geography*. Paper presented at National Council for Geographic Education, Columbus, Ohio.

Spate, O. H. K. (1960). Quantity and quality in geography. *Ann. Assoc. Am. Geog.*, **50,** No. 1, pp. 377–394.

The Science of Geography, National Academy of Sciences — National Research Council, Washington D.C. Publication 1277.

Thomas, K. (1966). The tools and the job. *Times Lit. Suppl.*, April 7, 1966, pp. 275–276.

2

Mathematics

'The question of what properties, such as angle or area, are reproduced on a map without distortion is of prime interest to mathematicians. The question extends far beyond the confines of geometry, for all mathematics can be considered broadly as a study of maps and mapping.'
G. A. W. Boehm (1959)

'The book of nature is written in the mathematical language. Without its help it is impossible to comprehend a single word of it.' *Galileo*

2.1 Introduction

There is a wide divergence of views among geographers as to how important mathematics is in geography. At one extreme are those that consider that anything quantitative is suspect if not undesirable. At the other are those that would even go as far as to say that potentially anything of a spatial nature in mathematics is relevant to geography. Unfortunately the dispute is made more complex because the content of mathematics, for long, apparently, one of the most clearly defined and cohesive disciplines, at least in the eyes of the layman, is itself uncertain now. According to R. Courant (1964) 'It becomes the urgent duty of mathematicians therefore, to meditate about the essence of mathematics, its motivations and goals, and the ideas that must bind divergent interests together'. The aims and future development of mathematics are far from clear.

It is assumed by the authors of this book that at least some aspects of geography, both physical and human, have benefited and will continue to benefit from the introduction of mathematical concepts and techniques. In this period of experimentation, it is the duty of geographers to introduce into their subject the new ideas from mathematics. Most mathematicians, pure and applied, are no doubt too concerned with the rapid development of their own subject or with its progress in disciplines such as physics and engineering, which have depended heavily on it for a long time, to bother with geography. W. W. Sawyer (1964) points out: 'The scientist who uses mathematics should be aware that much new mathematical knowledge is being discovered. Nearly all of it will be irrelevant to his research, but he should keep his eyes open for the small piece that may be of value to him.'

It is not immediately obvious whether a new piece of mathematical knowledge will be beneficial to geography or not. The criterion of its usefulness and success will be its ability to promote in some way the progress of geography.

Unfortunately the geographer can offer little to mathematics to repay for what he acquires, since he is mainly involved in applying mathematics in his own subject rather than advancing mathematics itself. Indeed, much of the mathematics needed in geography seems simple or trivial to mathematicians. Similarly, many problems that occupy and in some cases still confound mathematicians (e.g. whether some digits occur more frequently than others in pi, and why in topology only four colours are needed in the map problem (see section 2.10)), may seem trivial to the geographer.

The geographer who takes up and applies mathematics in geography is liable to be criticized both by mathematicians, who will accuse him of over-simplifying or wrongly using what he has borrowed, and by fellow geographers who resent 'gimmicks', or who even feel that their own way of thinking and their own work may be threatened. Nevertheless, since the Second World War, many innovations have appeared in mathematics. Both old and new branches of the subject that were once studied only in higher educational establishments are now filtering down or deliberately being pushed down to school children of younger and younger ages. New methods of teaching mathematics are also being applied widely. Not surprisingly, therefore, the so-called New Mathematics has appeared in both North America and Western Europe in various forms and places.

Geometry, arithmetic and algebra, at least as taught widely in British schools until the 1960s, are hardly recognizable to the layman in the New Mathematics. The theorems of Euclid are given much less prominence than before. According to M. Kline (1964), there are several geometries, not just one. In arithmetic, less credit is now given to children for working out improbably complex money sums. A desk calculator handles these.

> 'People suddenly awoke to the fact that like it or not, today's world rests on science, and that science in turn rests on mathematics. The urgent need is no longer for clerks and navigators but for men able to describe scientific findings accurately. . . .'
>
> *Mathematics*, Life Science Library (1965, p. 193)

Basic laws of arithmetic are being introduced to children, rather than limited algorithms (see glossary) for calculations such as addition and multiplication. The question is not: 'what is 7×9?' but why does $7 \times 9 = 63$?' There are many arithmetics (see section 2.4) and arithmetic can employ numbers with many different bases in addition to what is commonly called decimal base (better denary or to the base 10), the only one with which most of us are familiar. As for algebra, in the last decades, this branch of mathematics has flourished to the extent that we are told that there are now at least several hundred different recognized systems of algebra, not just one.

Some appreciation of both the content and the new way of teaching mathematics, old and new, is more urgently needed now by geographers than ever before. The New Mathematics is taken here to mean, very broadly, developments in mathematics since the 17th century. For some strange reason, these were able to remain out of schools almost entirely. According to E. B. Rosenthal (1966) 'A traditional (United States) high school course contains nothing that was not known at the time of Isaac Newton; anything discovered or created after the middle of the seventeenth century is not included'. In the last decade or so, developments of the 18th and 19th centuries, such as probability, topology and set theory, have entered school curricula alongside such branches as game theory, linear programming, and computer programming, developments of the mid-20th century. As a result, around 1965, there were many schools with pupils leaving the top end versed largely in pre-Newtonian mathematics, and children at the lower end introduced to many post-Newtonian branches. By 1970, many students of geography entering university might be expected to have some notion of the New Mathematics. Whether they would think of linking this with geography without being shown the way seems doubtful in most cases. To progress and to communicate with their students geography teachers will clearly have to learn at least a little about New Mathematics. It would be better still if they could make an effort to find which new concepts and techniques are useful in geography. Among geographers who have made important contributions already in this direction may be mentioned B. J. L. Berry, W. Bunge, W. L. Garrison, P. R. Gould, W. R. Tobler and W. Warntz.

It seems important also for geographers to give more consideration not only to what mathematics children are taught at school but also to what they are able to absorb and how they see mathematics. There is a considerable literature on this subject. In particular, J. Piaget (1960) has done a great deal of work on the child's conception of geometry. The work is at least relevant to the geographer in the context of local studies in geography. Interestingly, Piaget (1953) notes that children appreciate topology (see section 2.11) before Euclidean geometry. I. Macfarlane Smith (1964) has assembled a great deal of material on spatial ability. Again this is somewhat marginal to geography but should not be overlooked. E. E. Biggs (1965) reviews mathematics in primary schools and notes in a limited section specifically on geography: 'Geography and mathematics have many important points of contact.'

Whether the reader is a teacher or a student of geography he is advised to consult mathematicians whenever possible on the use of mathematics in geography. Often out of interest or simply out of curiosity mathematicians will contribute in various ways. Indeed teachers of New Mathematics may be glad of geographical examples in their own work.

This chapter reviews some of the branches of mathematics that are of use or may prove to be of use in geography. The following chapter considers statistics. Mathematics and statistics have been kept apart because there is recognized to be a fundamental difference between the two disciplines. Mathematics builds on its own laws and is infallible or anti-exceptional. Statistics, supported by probability theory, deals with uncertainty. Mathematical techniques are however indispensable for the operation of statistics. In this sense, therefore, statistics may be considered to be a branch of applied mathematics.

Some branches of mathematics are accepted in geography, tacitly if not consciously, and are respectable and of long standing. Some have been introduced recently, and are a matter of controversy. Some that appear to be of considerable usefulness are hardly known at all by geographers. All three kinds will be introduced in this chapter. Among the oldest are measuring, map projections (using several kinds of geometry), surveying (using trigonometry), the idea of rectangular or Cartesian co-ordinates (indispensable in the making of many maps), and numbers for ordering and computations. In the last decade or two, the attention of geographers has been focused on statistical techniques of various kinds (with probability), multivariate analysis, the algebra of matrices, topology (particularly graph theory), and linear programming and game theory. Examples of branches that have hitherto largely been ignored by geographers are set theory, logic, switching circuits, affine and projective geometry (except in the special case of map projections), solid geometry, point set theory, mapping, packing, routing and erasing, to name a few.

The preceding tentative list of branches of mathematics based on their penetration of and standing in geography may usefully be supplemented by a list based on the kind of application that might be made of each particular technique in geography. Five broad classes, not by any means exhaustive or mutually exclusive, are suggested.

(i) Data gathering and processing (descriptive statistics)
(ii) Mapping (surveying, projections, transformations)
(iii) The study of distributions (geometries, descriptive and inferential statistics)
(iv) The study of relationships (correlations)
(v) Thinking and organization (set theory, logic, Boolean algebra)

Some of the topics listed above are very obviously techniques, some form the basis for models and theories in geography, while some do not clearly fall into either category, or serve some other purpose still.

At this stage a few words of introduction about some of the relatively new (to geography) branches of mathematics seem appropriate. Somewhere at or near the base of the massive structure of mathematics, where various branches have their roots, common features are found. The fundamental similarities between many branches of mathematics are naturally understood implicitly by mathematicians but may not be obvious to geographers.

> 'To explore the labyrinths where future giants of knowledge may crouch, modern mathematicians light their way by . . . set theory, group theory and symbolic logic . . . Group and set theory are both used to compare the tools of different branches of mathematics and to make them as freely interchangeable as possible.'
>
> *Mathematics*, Life Science Library (1965, p. 172)

Thus for example maps (of the geographer), graphs and matrices have certain features in common. Similarly set theory, logic, Boolean algebra, binary numbers and even matrices are connected, and all have connexions with well known operations such as addition and multiplication.

Sets and their visual expression in Venn diagrams (see next section) cover a number of very basic situations, relationships and operations. To many readers they will represent an obvious formalization of already familiar concepts. The need for using set vocabulary in geography is less obvious, but there is evidence that it may be entering the subject now in papers by M. B. Teitz (1963), R. G. Golledge and D. M. Amadeo (1966), and J. A. Shear (1966). L. Hamill (1966), on the other hand, expresses doubt about its usefulness. Set vocabulary and Venn diagrams may prove useful in the vocabulary of the geographer at least in the description of distributions (using particularly the concept of *subsets*), in relationships (*intersection* and *union*) and in comparisons (*correspondences*).

Our concept of numbers and counting is very firmly based on the use of ten digits one of which is 0 (zero), a decimal (better *denary*) based system. A binary based system will be introduced in section 2.5. This uses two digits, 1 and 0 (zero). There are several reasons for using binary based arithmetic as well as denary in geography, or at least for knowing something about it. Computers calculate in binary. Logarithms with a binary base (or with other bases, as in natural logarithms) give different views of distributions from those plotted on denary logarithm scales. Classification and logic can be learnt and manipulated more easily in some cases with the use of 1 and 0.

Combinations and permutations merit attention since they are suitable for handling many situations facing the geographer. They are vital, also, for an appreciation of probability and statistics. An appreciation of the large numbers of combinations or permutations possible even in many apparently simple geographical situations is salutary for the geographer, who might sometimes benefit by considering other possible arrangements of things he is dealing with as well as the one actually confronting him.

Matrix algebra is a branch of mathematics largely unknown to older geographers but now widely taught in schools. Matrices are useful for storing some kinds of geographical data. Matrix operations are also vital for computers to handle some kinds of data and to make calculations. Some matrix algebra is introduced in section 2.8.

Of many kinds of geometry, topology deserves more attention than it has hitherto received in geography. It is more flexible than Euclidean geometry and can help in the study and discovery of situations not satisfactorily handled by Euclidean geometry. One branch of topology, graph theory, is particularly relevant to networks in geography.

Mathematics disturbs many people on account of its widespread use of symbols. They are often a shorthand way of expressing some instruction or relationship that can quite legitimately be expressed verbally. Thus $\sum$ generally means 'sum (or add) all the values (numbers) that immediately follow'. Similarly, 10 sheep (denary base) is a convenient way of writing the word sheep ten times or drawing ten sheep. Unfortunately, mathematicians are not always consistent in the use of symbols. Thus the operation multiplication can be indicated by $\times$, $\cdot$ or no symbol at all. Sometimes, moreover, the same symbol can have different meanings. It is essential, therefore, to appreciate the particular meaning of each symbol used *in a given context*. It is assumed that here a mathematician will be consistent in his use of symbols, just as the author of a verbal work will be consistent in the meaning of his technical terms, at least within that work.

The list of symbols at the end of the book is far from complete, but has been included to help the geographical reader unfamiliar or out of practice with mathematics to make a start. Most symbols included occur at some point in the book and all appear with increasing frequency in the work of quantitative geographers.

A list of general mathematical works has been included at the end of this chapter to give the reader a guide to mathematical material which is reasonably elementary. It has not been convenient to list the contents of each book, but most contain a fairly wide cover of new topics. The works have been subdivided on the basis of the kinds of user to whom they are aimed. In general they increase in difficulty and complexity through the list.

2.2 Introduction to set theory

The use of the vocabulary and operations of sets is becoming very widespread both in school mathematics texts and in applied mathematics. Similarly Venn diagrams (see figure 2.1), which give a diagrammatic or visual view of simple set theory, are often seen in scientific material. Terms and operations of set theory will be described with the help of a simple example in subsection 2.2*a*. The possibilities of using set vocabulary in geography will be discussed in subsection 2.2*b*. Sets will also appear later in the book.

(*a*) *The vocabulary of set theory and Venn diagrams*

Set theory has several functions. Firstly, as a basic aspect of mathematics* it is introduced fairly early in many of the new mathematics projects. 'Modern mathematics is based [*sic*] on the idea of *sets*. It is, therefore, appropriate that we begin with a study of sets. We will also find that many of the problems of life in general are associated with the idea of sets', (Carter 1965). Secondly, set theory helps to provide a common vocabulary for various branches of mathematics and therefore helps to link them. If one aim of geographical research is to make generalizations, then the linking aspect of sets may be particularly useful. Generalization depends on the discovery of analogous structures in dissimilar things. Thirdly, set theory was developed to handle the question of infinite sets of numbers, and still serves this purpose.

Simple set theory and operations are described under a number of headings (i–ix). The topics discussed under these headings appear in many different introductory courses to set theory, but are not always presented in the same order. Further, some of the terms and operations are not always represented by the same symbols in different sources. This unfortunately causes confusion.

* Set theory is *not*, however, the basis of mathematics, classical or modern or any other. Soon after the development of set theory towards the end of the 19th century, paradoxes were revealed, including one by Bertrand Russell in 1901. The paradoxes of set theory are briefly discussed in *Scientific American* (Sept. 1964, pp. 126–7).

Certain set concepts including *superset* and *difference* are not mentioned in this section.

What is a set? Somewhat different wording is used by mathematicians to introduce the concept of a set and to say what it is. The following conveys adequately the general idea: 'The word *set* is used to refer to a quite arbitrary but well defined collection of objects; these objects are called the *elements* of the set' (Carter, 1965, p. 1). The elements (sometimes called members) of a set may be numbers (e.g. the positive integers 5, 6, 7), letters (e.g. the letters *c*, *a*, and *t* in the word *cat*), people (e.g. the first year Honours geography students at Nottingham University as on May 31st 1968). They may also be numerical observations, such as the rainfall data for weather stations in the U.S.S.R. A continuous feature such as a line, surface or 3–dimensional object may also be called a set. Sets can be infinite or finite. An example of an infinite set is the numbers 1, 2, 3 and so on, while a finite set would be the children in a classroom at a given time. An important thing is to be clear what the elements of the set are. It is essential to have a rule to determine this in the case of each set. In other words, while sets are very flexible as to what they can contain, they must be defined and used according to certain rigorous rules.

The basic notions about set theory will be illustrated with the help of information about a small finite set of ten villages. The villages are listed in alphabetical order and each has a reference number (1–10), which will be used in the examples to save writing the names repeatedly.

Table 2.1 Village information

		Church	Public Telephone	Post Office	Filling Station	Library	Cinema
1	Ashton	1	1	1	0	1	0
2	Dunnett	1	1	1	0	0	0
3	Ersham	1	1	1	1	0	0
4	Langham	1	1	1	1	0	0
5	North Biggs	1	1	0	1	0	0
6	Orford	1	1	1	1	0	0
7	Oxton	1	1	0	0	0	0
8	Pendleton	1	1	1	0	1	0
9	Sopton	1	1	1	0	0	0
10	Springwell	1	0	0	1	0	0

Attributes: 1 = does have; 0 = does not have.

Note that in this example the relative location of the village is *not* being considered.

(i) *Membership* There are ten elements or members in the set of villages under consideration. The members of a given set are usually enclosed in a special bracket, called braces (see figure 2.1). The members may be listed individually, or defined collectively. When it is infinite, the latter is obviously essential. In the given example, the set of villages could be shown as a list

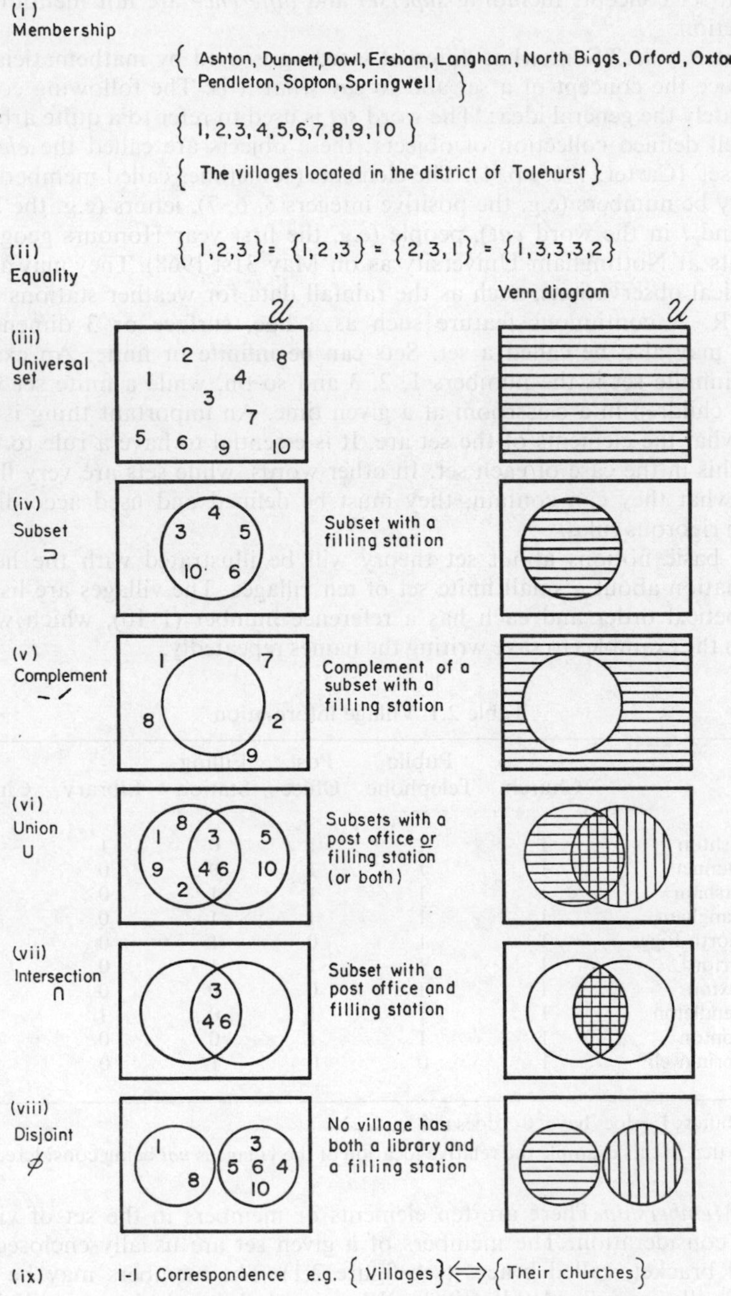

Figure 2.1 Nine situations and operations in simple set theory.

of their names, or of the reference numbers given to them, or could be defined in words. Sets of numbers may be defined entirely in terms of conventional mathematical symbols. For example, the quantity of *real* numbers (see section 2.5) between 1 and 2 is infinite, because it is possible to go on indefinitely finding more and more decimals between the two. It is thus impossible to list them. They would be expressed as a set in the following way: The set of all real numbers would be called R. The set of real numbers between 1 and 2 would then be $\{x|x \in R \text{ and } 1<x<2\}$ or in words, 'the set of all x such that x is an element of the set of real numbers and x is greater than 1, less than 2'. The vertical stroke (sometimes a colon) is read as 'such that'.

Membership of a set is often indicated by the symbol $\in$, read 'is an element of'. If the villages under consideration are called set T (short for the villages in Tolehurst district), then we can say: Ersham $\in T$.

The number of members in a set is often referred to by n. In our example, $n=10$.

A set may be *empty* (or *null*). Such a set has no members. An empty set is indicated by ϕ or { } (empty braces). A set with only one member is sometimes referred to as a *unit* set.

(ii) *Equal sets* Two or more sets are equal if the elements in them are identical. The order in which the elements are listed is irrelevant. Moreover, the fact that the same element recurs in one of the sets but not the other is also irrelevant, at least according to some sources: e.g. $\{1, 2, 3\}=\{2, 1, 1, 1, 3\}$.

(iii) *Universal set or universe of discourse* This concept refers to all the elements of a set that is under consideration at a given time for a given purpose. For example, the set of ten villages may be under consideration as part of a rural development project. The universal set is often referred to by $\mathcal{U}$, $\mathcal{E}$, or I.

(iv) *Subset* In the universal set there are many subsets. All the elements of a given subset must be elements of the set of which it is a subset. In the example, one subset is the villages having a library, another, the villages having a post office. Special subsets are the villages having a church and the villages having a cinema. In the first case the subset is equal to the original (universal) set. Such a subset is permissible and in fact essential in the operation of sets, but it is not called a *proper* subset, as all the others are. In view of this important feature of set theory, set and subset may in certain circumstances be interchangeable terms. In the second case, the subset is an empty set (since no villages have a cinema). This concept is also essential in set theory.

Two further points must be noted about subsets. Firstly, it can be seen in the information about the villages that there is a subset (with nine elements) of villages with telephones, and that the seven villages with a post office are a (proper) subset of the villages with a telephone. Thus it is possible to have different levels of subsets. It would be permissible, if desired, to upgrade the nine villages with public telephones to a new universal set, with the villages with a post office as a subset of it. Secondly, it may be noted, in anticipation of the sections on combinations (2.11) and on probability (3.1), that there are 2^{10} possible subsets (combinations) of the ten villages. 2^{10} is derived from the formula 2^n ,where n is the number of elements in the set. 2^{10} is 1024.

(v) *The complement of a set* Let us consider the villages with a filling station (3, 4, 5, 6, 10) as a subset, or new set in its own right, within the universal set, and call it set F. Then the complement of the set F is all the villages not in the set (i.e. 1, 2, 7, 8, 9). This can be represented as F' or $\bar{F}$ or 'not F'. The union (shown by a $\cup$ sign) of F with F' must make up the universal set. Thus $F \cup F' = \mathscr{U}$.

(vi) *The union of sets* The union of two or more sets consists of all the members of both or all sets. For example, which villages have a post office or a filling station (or both)? The key word in union is a special kind of *OR*, which may be thought of as *AND/OR*. In the example, we are interested in knowing whether a village has (at least) a post office *or* a filling station. It does not matter if the village happens to have both. In the example, all villages but Oxton belong to the union of the two sets. Only Oxton is without both. But there are four kinds of village, e.g. Ersham (3) has both, Ashton (1) has only a post office, Springwell (10) has only a filling station, Oxton (7) has neither. Union in set theory is associated with addition (+) in arithmetic.

(vii) *The intersection of sets* The *intersection* of two or more sets consists *only* of the members that are common to both sets. For example only three villages (3, 4, 6), have *both* a post office *and* a filling station. The key word in intersection is AND. In the example, we are interested in knowing which villages have both features. A comparison of the diagrams for union and intersection shows the basic resemblance between the two operations, but reveals that intersection is more restricted or particular in what it includes. Intersection in set theory is associated with multiplication ($\times$) in arithmetic.

(viii) *Disjoint sets* Two or more sets are called disjoint if they have no members in common. Thus, in the example, two villages have a library, five have a filling station, but no villages have both features. One way of expressing disjointness is to consider it as a special case of intersection, an intersection of two sets that produces an empty set. In this example, the intersection of F and L gives ϕ or $F \cap L = \phi$.

(ix) *One–one correspondence* ($\Leftrightarrow$) One–one correspondence is a concept in set theory somewhat distinct, at least to the layman, from the operations outlined above. It was in establishing correspondences, however, the mathematician G. Cantor developed sophisticated set theory. Very broadly, a one–one correspondence between two sets implies that a member of one of the sets can meaningfully be paired off with a corresponding member in the other set in such a way that the members of both sets are used up. Thus the two sets have to have the same number of members, but they are called equivalent sets (⇔indicates equivalence), not equal sets. In the example used in this section, each village has a church. There is a one–one correspondence, therefore, between the set of villages and the set of churches. Similarly, each village name has a corresponding counting number (positive integer). Further, each village *name* is no more than a *label* corresponding to the real object, the village, an agglomeration of buildings defined as such by general agreement. Counting may (naively) be considered as taking from the (infinite) set of counting numbers a unique token and putting this in a one–one correspondence with each object counted. Correspondences need not be one–one. They may be one–many; for example, a bus and the seats in it. In a general way, map-making in geography may be thought of as putting symbols on a piece of paper in a one–one correspondence with selected objects from the real world. Correspondences will be referred to in chapter 5 when comparison in geography is discussed.

(*b*) *The application of sets in geography*

Set vocabulary has penetrated modern geometries. Points, lines, surfaces and 3-dimensional objects (e.g. a cube) may be thought of as infinite sets of points*. Both the objects studied by the geographer, and the spaces they occupy, may be treated in terms of set vocabulary. But the real 3-dimensional nature of the objects that for convenience the geographer collapses to patches, lines or dots, must never be forgotten after the process of simplification. Furthermore the rigorous nature of mathematical thinking may to some extent be modified as the geographer adapts the terms and operations of sets for his own uses.

It is not always convenient to consider the spaces of the geographer as infinite sets of points. Often it is desirable to divide them into compartments. It is also possible to represent them by a regularly spaced finite set of dots standing in for the infinite set of points of the space itself. A 1-dimensional (line) space, a 2-dimensional (patch) space, and 3-dimensional space, are illustrated in figure 2.2*a*. They are divided into compartments and are also represented as dots. The fineness of the compartments, which on a surface (patch) are usually squares, but could be hexagons, depends on the amount of detail to be studied on the surface.

The recent introduction of sets into geography was noted in section 2.1 of this chapter. The successful use of sets in geography requires that both the user and his readers know the meaning of the basic terms and operations. It also requires a somewhat different kind of abstraction on the part of the geographer from his customary one. There is nothing to be gained from

* An introduction to a new way of looking at geometry will be found in *Modern Elementary Mathematics*, M. Ward and C. E. Hardgrove, Addison-Wesley, Reading, Mass. (1964), chapter 8, Geometry. The same book discusses Sets in chapter 2. The vocabulary of sets can be used for Euclidean geometry, by using point sets. All Euclidean space is a universal set of elements called points. Lines (straight, curved, etc.), surfaces and solids are families of 'distinguished subsets'. Euclidean geometry is redefined so that the interior points as well as the boundaries of plane figures such as the triangle and circle form part of the figures. Thus a triangle has a boundary and an interior region. The concept of ***neighbourhoods*** in this context should also prove useful in geography.

Geometrical (Euclidean)	Number of space dimensions	Topological equivalent	Geographical abstraction	Visual impression
point	0	node(vertex)	dot (small object)	
line	1	arc(edge)	line (e.g. route)	
surface	2	region (face)	area patch	
3-dimensional space(solid)	3	figure		

One dimension

Line as an infinite set of points

Line divided into a finite set of discrete compartments

Line represented by a selected finite set of points spaced at regular distances apart

Two dimensions

Area as an infinite set of points

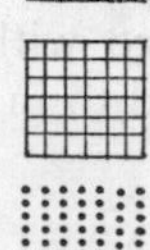

Area divided into a finite set of discrete compartments (a grid)

Area represented by a selected finite set of points spaced at regular distances apart

Three dimensions

3-dimensional space as an infinite set of points

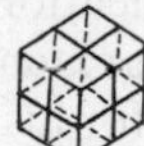

3-dimensional space as a finite set of discrete compartments

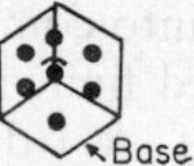

3-dimensional space represented by a selected finite set of points spaced at regular distances apart

a

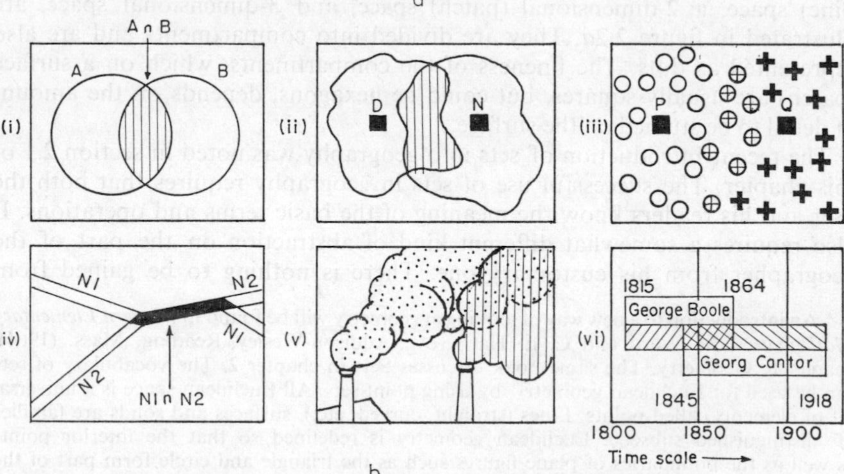

Figure 2.2 The application of sets in geometry and in real world situations. (*a*) The extension of sets to geometrical figures. (*b*) The extension of Venn diagrams from two dimensions to other dimensions.

introducing new vocabulary into geography unless the new vocabulary makes it easier to see situations that are otherwise missed or unless it makes thinking clearer and language more concise. In chapter 4 it will be shown how the use of the concept of *subsets* helps in the appreciation of some kinds of distribution. In chapter 5, the idea of *correspondences* will be introduced to help in the study of comparisons. In chapter 6, *intersection* of sets will be used to help to develop the idea of correlation. In this section, some of the features of a map will be described in set vocabulary.

To appreciate the application of set vocabulary in maps and in real world situations, the geographer must be prepared to make two drastic changes to the sets described in the previous section and illustrated in figure 2.1. Firstly *convert the conventionally tidy Venn diagrams* (as in figure 2.1), *into maps*, which usually have distributions of varying shape, and secondly, *allow the relationships portrayed by Venn diagrams in two dimensions to be extended to other dimensions.* When Venn diagrams are extended to other dimensions they appear different. They may apply however in one and three space dimensions and in time, as well as in the customary two dimensions. In figure 2.2*b*. the operation *intersection* (the overlapping and sharing of the same space or objects) is used to illustrate what has been said:

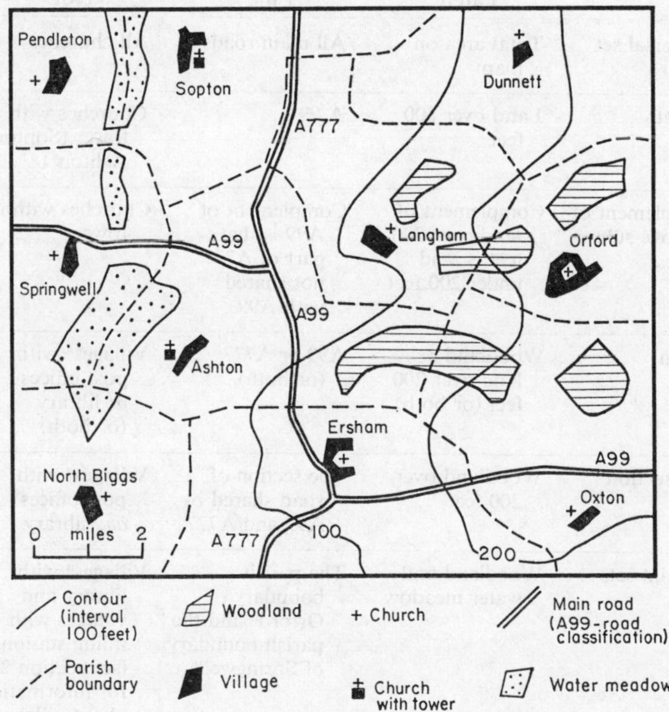

Figure 2.3 A fictitious area with ten villages to show the application of sets to a geographical situation (see table in text).

(i) A conventional Venn diagram, the intersection of sets *A* and *B*.
(ii) The universal set as a map which includes two towns, *D* and *N*, whose service areas overlap. The 2-dimensional service area of each is indicated. The inhabitants of a certain area lying between the two towns frequent both towns for services.
(iii) The intersection of two sets of dots, the people who use the services of towns *P* or *Q* (○ and +) and the people who use both (⊕).
(iv) The intersection of two line sets. For a stretch of road, N1 and N2 share the same line of road.
(v) The smoke from two chimneys mingles in 3-dimensional space.
(vi) Two mathematicians share some of the same time during their lives, (1845–1864).

Figure 2.3 shows a map of the ten villages already listed in section 2.1. Selected features have been shown on the map. The map is accompanied by a table showing examples of six set concepts: universal set, subset, complement, union, intersection and disjointness. Each of these is illustrated with patch (2-dimensional), line (1-dimensional) and dot (zero-dimensional) examples

Table 2.2 Sets on the map

	Patch	Line	Dot
Universal set	Total area on map	All main roads	All churches
Subset	Land over 200 feet	A 99	Churches with tower (Sopton, Ashton[a])
Complement of above subset	Complement of land over 200 feet is land under 200 feet	Complement of A99 is that part of A777 not shared with A99	Churches without tower
Union	Woodland *or* land over 200 feet (or both)	A99 *or* A777 (or both)	Villages[a] with post offices *or* library (or both)
Intersection	Woodland over 200 feet	The section of road shared by A99 and A777	Villages[a] with post offices *and* library
Disjoint sets	Woodland and water meadow	The parish boundary of Orford[a] and the parish boundary of Springwell[a]	Villages[a] with library and villages with filling stations (see section 2.1 for information about villages)

[a] Refer to table 2.1 earlier in this section for information about the villages.

which, as already stressed in chapter 1, are all really 3-dimensional objects but are reduced for convenience.

Often the geographer is concerned with the study of finite sets of objects that are small enough in a given area to be studied as points. He is particularly interested both in their arrangement in space and in the way they change in location in space during a given period of time. Point set theory, a very flexible geometry alluded to by R. Courant (1964), may prove to be of value in geography. 'In point-set theory . . . the order of points is not retained during the kind of transformation called "scattering". The scattered points do remain conumerous with the points in the original figure. Thus point set theory can be described as the study of the properties preserved under all one–one correspondences'. F. Waismann (1959) remarks: 'Hence in the series of geometries, metrical geometry stands at one end (or, strictly speaking, topography which regards every geometrical structure as an individual); at the other end, the theory of sets'. The scattering of points is now illustrated.

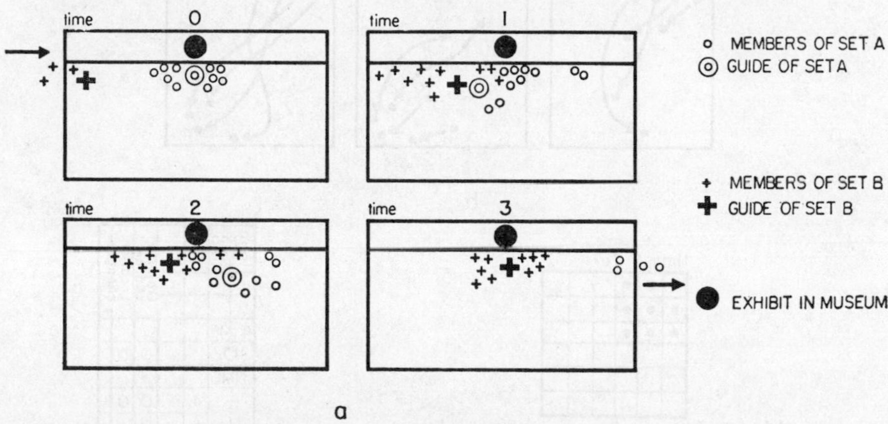

Figure 2.4 The movement of finite sets of objects (dots) over area during time. (*a*) Visitors to a Museum.

Figure 2.4*a* shows a situation in a Museum. Most geographers would not consider the situation a geographical one. It is spatial, however, and serves as a familiar and simple situation around which more complex geographical ones can be built. At time 0, a guide is talking to ten visitors (circles) about an exhibit. At time 1 a second party of visitors (crosses) with their guide has entered the same room, and the two guides are talking. At time 2 the first party is moving away. At time 3 only the second party is left facing the exhibit. Although the members of the two sets of visitors became mixed spatially for a while, everyone knows to which party he or she belongs. The purpose of this example is to suggest that although the situation illustrated must be familiar and obvious to the reader, it is difficult to describe to someone else. With the help of set vocabulary the description may be made easier. In practice a much more complex intermingling of sets of points often

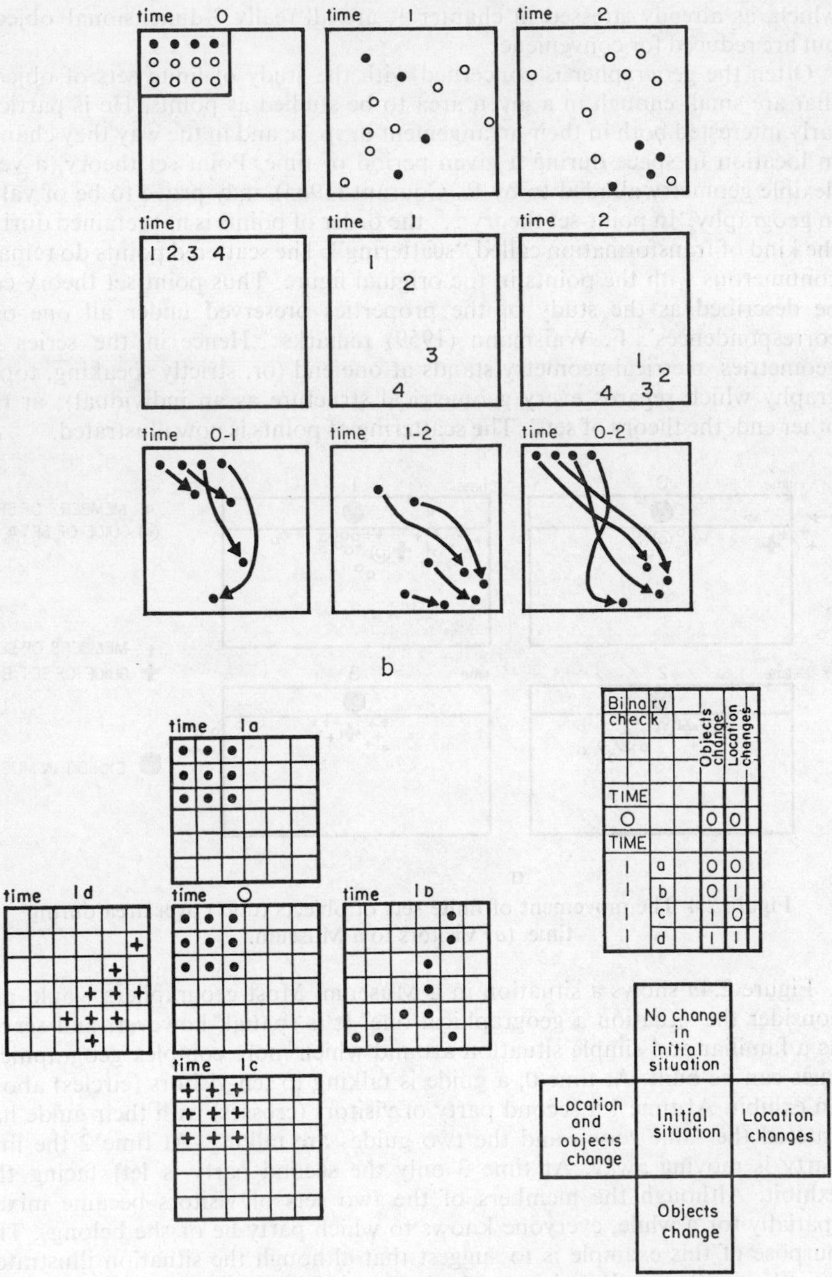

Figure 2.4 (*b*) The situation in more diagrammatical form. (*c*) The objects change in form as well as location.

occurs: for example many different species of plant or animal occupying the same area, different sets of people moving about in a town, different types of motor vehicle in a given stream of traffic.

Figure 2.4*b* traces the movements over time of a subset of four objects out of a set of twelve that find themselves in a particular location at time 0. The top row of three diagrams in figure 2.4*b* plots the objects at three instants in time [say time 0, time 1, (0 + 1 unit of time, say 1 hour) and time 2]. The second row of three diagrams picks out individually the four objects in the black subset. The third row of diagrams shows by lines courses followed by the four objects as they moved during specified periods of time, namely between times 0 and 1, 1 and 2, and 0 and 2.

The objects in a set may change during time not only in location but also in form. For example a commercial vehicle may be empty at some times, loaded at others. A fruit tree will be immature for some years, then mature (fruit bearing) for some more years. The diagrams in figure 2.4*c* illustrate what happens in a very simple situation when 9 objects (a finite set of dots) in a space change over time, (*b*) in location, (*c*) in form, (*d*) in both location and form, or (*a*) in neither way. A table with binary numbers serves to check that all possible situations have been considered.

In practice the situation described is too simple to occur often in the real world. The situation might be complicated at least in the following ways:

(i) There might be far more than nine objects.
(ii) There might be two or more sets of objects, rather than one.
(iii) Some of the objects might leave the area in question and/or others enter it during a period of time under consideration.
(iv) Some of the objects in a given set might move while others remained fixed.
(v) The size and/or shape of the base area on which the set of objects are located might itself change.
(vi) Time intervals might vary.

Finally, it must be appreciated that the time intervals, say between time 0 when the study starts and time 1, after the elapsing of one unit of time, may range between very short, say a second in studying traffic flows, and very long, say decades or even centuries for changes in patterns of distribution of plant species, towns and other objects that are conceived as actually fixed over short periods of time. The question of time will be reconsidered in chapter 10 and will recur in the chapters that follow 10.

An example illustrating the possibility of using set terminology to describe a specific situation in geomorphology will now be discussed. A river loaded with very mixed material, both in size and resistance, carries its load in an angular state to the sea. As the load reaches the sea its individual constituents arrive mixed together. The river has a seasonal flow, only bringing its load down to the sea after each annual period of flooding. All the load of one flood may be termed the universal set *A*. This set is made up of sand and angular rocks of varying sizes and resistance. For convenience the stones may be divided into sets according to size and resistance; for simplicity only two sets of each variable will be considered. There is thus a large resistant set, a large soft set, a small resistant set and a small soft set. The fine sediment (sand) will form a fifth set. This situation is shown in figure 2.5*a*. After an elapse of time, waves have moved the pebbles and sand along the beach and the intersecting sets have become sorted out into separate disjoint sets as far as size goes. The larger pebbles are at the top, the sand at the bottom of the beach. Time has not yet allowed rounding to occur. This situation is shown

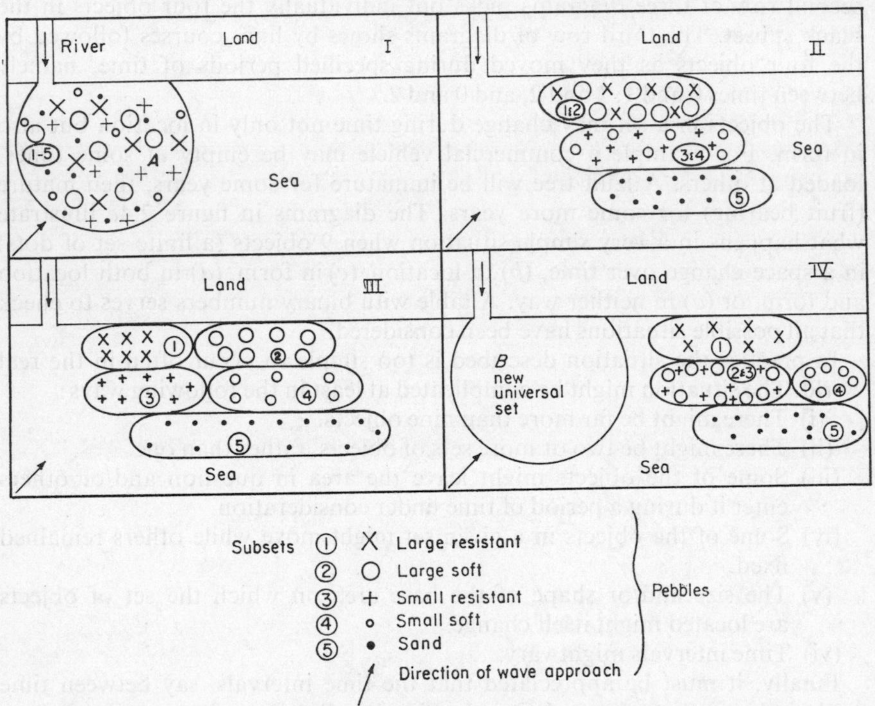

Figure 2.5 A hypothetical situation to show how sets could be used in physical geography to describe the sorting of different types of pebbles by marine action. Four stages of sorting are suggested at successive time intervals. This results in different subsets at different stages.

in figure 2.5*b*. After a longer period, figure 2.5*c*, the intersecting sets have become further separated and now the soft stones have become rounded and hence have moved further than the angular ones so that all five sets are now disjoint. Two of the sets have changed character, from angular to rounded. After a further period of time two of the sets merge again in figure 2.5*d* for the large soft stones become smaller by attrition and mingle with the small hard stones which they now resemble in size and shape. They, therefore, now accumulate on the same part of the beach. The sets in this example are separated spatially owing to their different response to wave action. The situation is simpler than a real one but it does indicate how sorting on a beach could be considered in terms of sets. After a year another universal set *B* of pebbles would be introduced into the sea and the process would begin again until some of the *A* subsets might intersect with the *B* subsets.

To conclude this section, the reader is asked to solve the following problem. 50 tourists leaving the state of Florida, U.S.A., were asked among other things, whether or not they had visited Miami and/or Tampa during their stay. The reply was:

Miami	23	Union
Tampa	19	
Neither	18	Complement

How many had visited both? The construction of a Venn diagram with two circles helps. The answer is the *intersection* of two sets.

2.3 Boolean algebra and logic circuits

It was pointed out in section 2.1 that the foundations of many aspects of mathematics are very similar and that some branches, including set theory, group theory and Boolean algebra help to link other branches. Several branches have some features in common with set theory. Boolean algebra and switching circuits are such branches. They are discussed very briefly. Binary numbers, which are introduced in section 2.5 are also closely connected.

(*a*) *Boolean algebra*

Boolean algebra is a system that operates with two constants, 0 and 1. A variable may therefore be either 0 or 1. There are three basic operations: *Inversion, and, or*. *Inversion* operates on a single constant or variable. When a constant is inverted it becomes equal to the other constant. Thus

$\bar{1}$ (not one) $= 0$ While $\bar{0}$ (not zero) $= 1$

This operation is analogous to *complement* in set vocabulary. *And* is an operation that connects two or more constants and/or variables. Thus $1 \cdot 0 = 0$. This operation is analogous to *intersection* in sets. The *and* operation is as follows:

$$0 \cdot 0 = 0 \qquad 1 \cdot 0 = 0$$
$$0 \cdot 1 = 0 \qquad 1 \cdot 1 = 1$$

The answer is 1 only if *both* constants are 1. *Or* is also an operation that connects two or more constants and/or variables. Thus $1 + 0 = 1$. This operation is analogous to *union* in set theory. The *or* operation works as follows:

$$0 + 0 = 0 \qquad 1 + 0 = 1$$
$$0 + 1 = 1 \qquad 1 + 1 = 1$$

The answer is 1 if either constant or if *both* constants are 1.

Variables are represented by letters, for example by A below. When a variable and a constant or when two variables are under consideration, the following situations occur:

$$\mathrm{A} \cdot 1 = \mathrm{A} \qquad \mathrm{A} + 0 = \mathrm{A}$$
$$\mathrm{A} \cdot 0 = 0 \qquad \mathrm{A} + 1 = 1$$
$$\mathrm{A} \cdot \mathrm{A} = \mathrm{A} \qquad \mathrm{A} + \mathrm{A} = \mathrm{A}$$
$$\mathrm{A} \cdot \bar{\mathrm{A}} = 0 \qquad \mathrm{A} + \bar{\mathrm{A}} = 1$$

Check these by replacing A by 0, then by 1 and verify in the previous tables. For example $A \cdot 1 = A$.

Substitute 0 for A then $0 \cdot 1 = 0$

Substitute 1 for A then $1 \cdot 1 = 1$ and so on.

(*b*) *Circuits*

According to D. C. Evans (1966): 'Electronic digital computers are made of two basic kinds of components: logic elements (often called switching elements) and memory elements. In virtually all modern computers these elements are binary, that is, the logic elements have two alternative pathways and the memory elements have two states'. Three relatively simple and two somewhat more complex binary logic elements are used in computers. They are represented by different symbols in different works. The following three symbols are equivalent to the three operations already briefly described in Boolean algebra: *Inversion, and, or*, and the symbols are I, ·, +.

The square box in figure 2.6*a* is a logic block or element. One or more inputs may enter the box. One output leaves it. The operations are illustrated

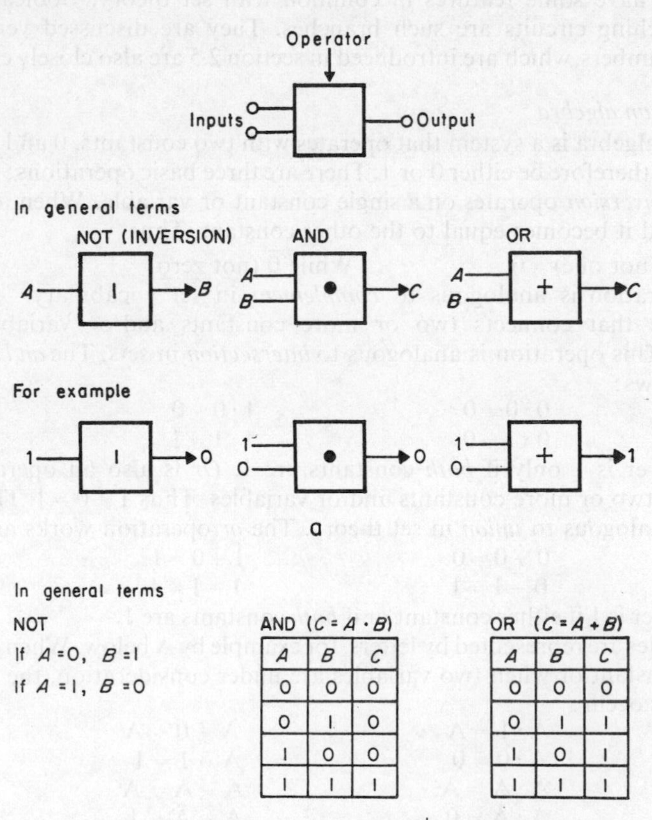

A	B	C
0	0	0
0	I	0
I	0	0
I	I	I

A	B	C
0	0	0
0	I	I
I	0	I
I	I	I

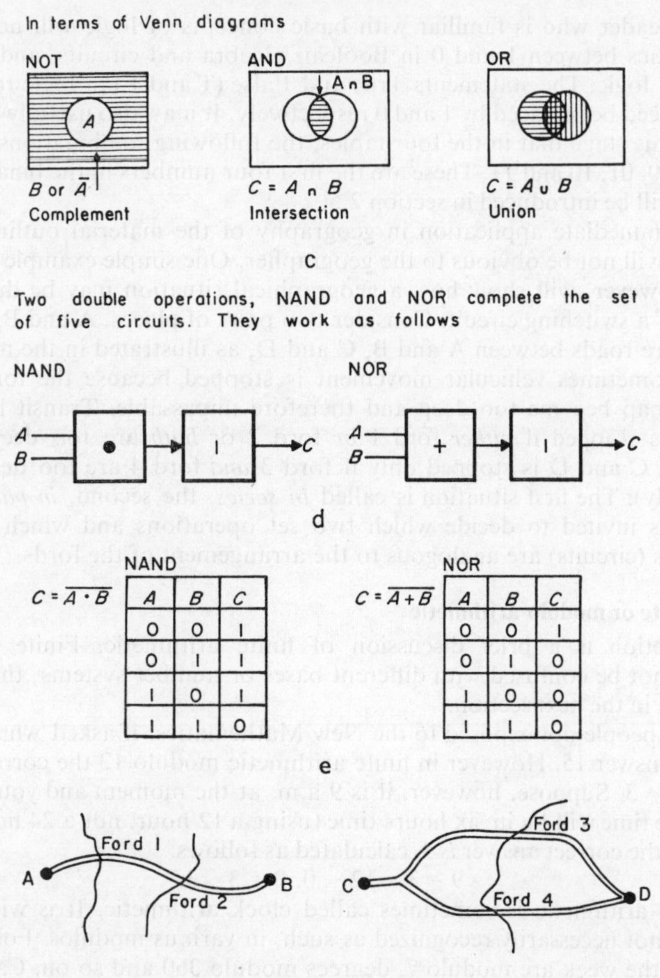

Figure 2.6 Logic circuits. (*a*) Three basic operations. (*b*) Outcomes of the three basic operations. (*c*) Basic operations in terms of Venn diagrams. (*d*) Two double operations. (*e*) Outcomes of two double operations. (*f*) Geographical examples of 'in series' and 'in parallel'.

first in general terms, then with an example. Combinations of inputs for A and B give appropriate outputs as C for the operations AND and OR. These are shown in the two tables in figure 2.6*b*. The three logic concepts are shown in Venn diagrams, figure 2.6*c*. A comparison of the C column in the AND and NAND tables and in the OR and NOR tables (figure 2.6*e*) respectively, shows that in each case C is the inverse because the outputs of the AND and OR boxes (figure 2.6*d*) have always been inverted.

The reader who is familiar with basic concepts of logic will notice many similarities between 1 and 0 in Boolean algebra and circuits, and True and False in logic. The statements True and False (T and F) in logic truth tables may indeed be replaced by 1 and 0 respectively. It may also usefully be pointed out at this stage that in the four tables, the following combinations of 0 and 1 occur, 00, 01, 10 and 11. These are the first four numbers in the binary system, which will be introduced in section 2.5.

The immediate application in geography of the material outlined in this section will not be obvious to the geographer. One simple example (see figure 2.6*f*) however, will show how a geographical situation may be discussed in terms of a switching circuit. Consider two pairs of places, A and B, C and D. There are roads between A and B, C and D, as illustrated in the map (figure 2.6*f*). Sometimes vehicular movement is stopped because the fords shown in the map become too deep and therefore impassable. Transit between A and B is stopped if *either* ford 1 *or* ford 2 or *both* are too deep. Transit between C and D is stopped only if ford 3 *and* ford 4 are too deep (simultaneously). The first situation is called *in series*, the second, *in parallel*. The reader is invited to decide which two set operations and which two logic elements (circuits) are analogous to the arrangement of the fords.

2.4 Finite or modulo arithmetic

This section is a brief discussion of finite arithmetic. Finite arithmetic should not be confused with different bases or number systems; these will be outlined in the next section.

Most people not exposed to the New Mathematics, if asked what $9 + 6 =$, would answer 15. However in finite arithmetic modulo 12 the correct answer is $9 + 6 = 3$. Suppose, however, it is 9 a.m. at the moment and you are asked what the time will be in six hours time (using a 12 hour, not a 24 hour clock). Clearly the correct answer *is* 3, calculated as follows.

$$9 + 3 = 12 = 0,\ 0 + 3 = 3$$

Modulo arithmetic is sometimes called clock arithmetic. It is widely used, though not necessarily recognized as such, in various modulos. For example, days of the week are modulo 7, degrees modulo 360 and so on. Calculations that involve cyclical or repetitive situations rather than cumulative ones use modulo calculations.

Figure 2.7 shows four 'clocks', Modulos 12, 7, 4 and 2. Each modulo has its own arithmetic. The operations of addition and multiplication for finite arithmetic modulo 4 are shown in the two tables. To find $3 + 2$ in modulo 4, read along row 3, then down column 2. The answer is where row and column intersect: $3 + 2 = 1$. In the multiplication table, $3 \times 2 = 2$. Reference to the binary 'clock' may be helpful in understanding binary base arithmetic described in the next section.

2.5 Numbers

In this section, numbers, the study of which has recently gained much prominence in mathematics, will be considered under three headings:

(a) Position and bases
(b) Number systems and their uses
(c) Numeracy

It is important that the reader should distinguish between different methods of representing numbers in subsection (a), and different *systems* of numbers themselves, beyond the peculiarities of their symbolic representation, in subsection (b).

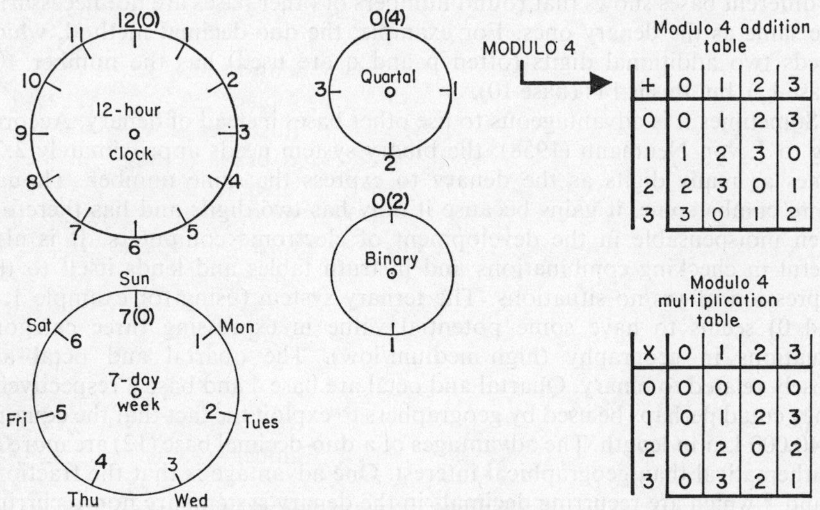

+	0	1	2	3
0	0	1	2	3
1	1	2	3	0
2	2	3	0	1
3	3	0	1	2

X	0	1	2	3
0	0	0	0	0
1	0	1	2	3
2	0	2	0	2
3	0	3	2	1

Figure 2.7 Modulo arithmetic in modulos 12, 7, 4 and 2.

(*a*) *Position and bases*

Very broadly, there are two fundamentally different ways of representing numbers: non-positional and positional. In the non-positional method, a given symbol represents a given quantity wherever it stands. For example, the X's in the Roman XXX are each equal to ten, and the total is 30 in our notation*. In contrast, in the positional method, the value of a *digit* in a sequence of numerals is decided by its position. Thus each of the 3s in 333 (three hundred and thirty three) represents a different quantity. Similarly 625 is different from 256. In the familiar denary base system that we use, 1, 2, 3, 4, 5, 6, 7, 8, 9 and 0 (zero) are the digits. The positional method, which requires a zero, is much more concise and viable than the non-positional method.

The denary positional method we use is one of a very large number of possible counting methods. Moreover, there is nothing special about having ten digits including zero, as we do†. It is possible to have fewer or more digits. Once two symbols (e.g. 1 and 0) are adopted it is possible to have a positional counting method (the binary). Only a method using a single

* Note that although the Roman method of representing numbers was basically non-positional, they did use some features of position: e.g. IV and VI are different.

The most widely accepted explanation seems to be that man has ten fingers.

symbol (for example, a 1) cannot be positional. Ticks are recorded in a one–one correspondence with the objects, such as goats, to be counted. Even the binary base method is much more sophisticated than this.

Attention is being drawn to bases because the denary base, while quite arbitrary, influences our thinking. For example, 100 years, 10 shillings, or 1000 miles assume a special place in our outlook on the things around us because they are tidy round numbers. The table comparing numbers expressed in different bases shows that round numbers of other bases are not necessarily the same as the denary ones. For example, the duo-decimal method, which needs two additional digits (often p and q are used) has the number 100 (base 12), but this is 144 (base 10).

Sometimes it is advantageous to use other bases instead of denary. According to J. von Neumann (1958), the binary system needs approximately 3.32 times as many digits as the denary to express the same number. Though more cumbersome, it gains because it only has two digits and has therefore been indispensable in the development of electronic computers. It is also useful in checking combinations and in truth tables and lends itself to the expression of yes/no situations. The ternary system (using for example 1, 2 and 0) seems to have some potential value in expressing three category situations in geography (high/medium/low). The quartal and octal are closely related to binary. Quartal and octal are base 4 and base 8 respectively. They could perhaps be used by geographers to exploit the fact that the equator is 40,000 km in length. The advantages of a duo-decimal base (12) are more of mathematical than geographical interest. One advantage is that the fractions $\frac{1}{3}$ and $\frac{2}{3}$ which are recurring decimals in the denary system, are non-recurring in duo-decimal. They are 0.4 and 0.8 respectively. Thus other bases have their 'decimals'. For example, 0.1 in binary is equivalent to 0.5 in denary.

It has been widely reported that school children find little difficulty in using various bases in the New Mathematics. Their parents, on the other hand, are not so fortunate. The reader who wishes to become proficient in the manipulation of other bases is advised to consider first the duo-decimal method which adds two digits to the familiar denary method, then to concentrate on binary. One way of practising binary is to calculate the weight of imaginary parcels using 1 oz, 2 oz, 4 oz and heavier weights, as shown at the bottom of table 2.3.

It is helpful to appreciate that many calculations are in fact made in modulos other than modulo 10 for base 10, but use base ten symbols. For example, 7935 seconds after mid-night uses base 24 for counting the hours and base 60 for counting the minutes and seconds. It would be expressed in denary as 02: 12: 15 hours, minutes and seconds respectively. The total is arrived at as follows: 15 seconds $+ 12 \times 60$ seconds $+ 2 \times 60 \times 60$ seconds. In passing, it is interesting to note that the Babylonians had a 60 base system as long ago as 1700 B.C. Various bases are shown in the table that follows. The lead-in zeros preceding the counting numbers in binary, ternary and quartal are not essential but are helpful in seeing the way bases work. Binary bicimals, addition, and multiplication are also introduced.

Table 2.3 Types of base

	Base 2	Base 3	Base 4	Base 10	Base 12
Unary	Binary	Ternary	Quartal	Denary (decimal)	Duodenary (duo-decimal)
	0000	000	000	0	0
1	0001	001	001	1	1
11	0010	002	002	2	2
111	0011	010	003	3	3
1111	0100	011	010	4	4
11111	0101	012	011	5	5
111111	0110	020	012	6	6
1111111	0111	021	013	7	7
11111111	1000	022	020	8	8
		100	021	9	9
			022	10	P
			023	11	Q
			030	12	10
			031	13	11
			032	14	12
			033	15	13
			100	16	14
			101	17	15
			102	18	16
			103	19	17
			110	20	18
			111	21	19
			112	22	1P
			113	23	1Q
			120	24	20
			121	25	21

Decimals (base 10) and Bicimals (base 2)

Base 2	Base 10
0.001	0.125
0.010	0.25
0.100	0.5
0.101	0.625
0.111	0.875

Binary addition

```
   1011  (11 base 10)
  +1101  (13 base 10)
  -----
  11000  (24 base 10)
  -----
```

Binary multiplication

```
1011×11  (11×3 base 10)
     11  (3 base 10)
  -----
   1011  (11 base 10)
  1011   (11 base 10)
 ------
 100001  (33 base 10)
```

A method of learning binary is to weigh parcels. Use only 1 each of 1 oz, 2 oz, 4 oz, etc. weights. Put a 1 in the table for each weight you use, a 0 for each weight you do not use:

Table 2.4 Binary by weights

Denary weight in oz	16 oz	8 oz	4 oz	2 oz	1 oz
3	0	0	0	1	1
4	0	0	1	0	0
7	0	0	1	1	1
17	1	0	0	0	1

The reader is invited to enlarge the table putting binary values for each parcel up to 32 oz.

(*b*) *Number systems*

Although there are different kinds of number, it is not widely appreciated yet by non-mathematicians that there are at least several different important kinds of number system. Very broadly, it is convenient for the layman to distinguish initially between whole numbers (integers) and 'decimal' numbers (real numbers). These are described below. Each of these may however be elaborated. The relationship of number systems to one another is difficult to appreciate at first. In some books it is explained in terms of sets and subsets. This approach is helpful but it overlooks the different nature and function of different number systems.

In figure 2.8, different number systems are arranged one below another in order of increase of the quantity of numbers they contain. The first system of wide application in geography is the system of *counting numbers*. This includes 0 (zero) because it is often necessary to note the absence of some object that might have been counted: e.g. there are no (= zero) glaciers in Texas. This system, basically, is used for counting discrete objects, items against which

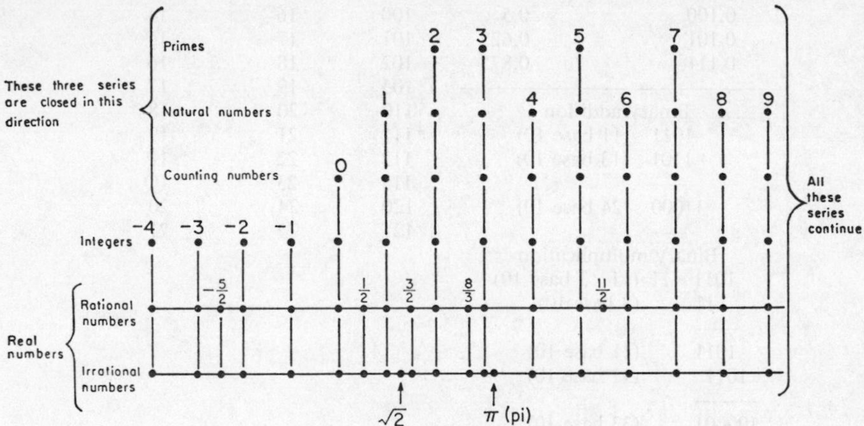

Figure 2.8 Number systems.

the counting numbers can be placed in a one–one correspondence. The so-called natural numbers are the same as the counting numbers except that they exclude zero. They start at 1 and are used for identification and ordering or ranking. Note that the natural numbers and the counting numbers are infinite sets because they continue endlessly off the page to the right in figure 2.8, though they are closed on the left because there are no negative numbers.

The integers are what are familiarly known as whole numbers, with negative as well as positive members. Each positive integer (e.g. 3) has a corresponding negative integer (− 3). Note however, that although the 'line' of integers is open off the page both to right and left, any number is greater than any number to its left. Thus 4 is greater than 2 but − 2 is greater than − 4.

Whole numbers or integers are not satisfactory, however, for representing any required point of an infinite set of points on a number line. In figure 2.8, the integers have no way of representing a point between 1 and 2. This is the function of the so-called real numbers. Real numbers are represented by fractions or 'decimals'. Thus the point on the real number line half way between 0 and 1 is $\frac{1}{2}$ or 0.5, that between 1 and 2 is $\frac{3}{2}$ or 1.5, and so on. However, the real number 1 itself is $\frac{1}{1}$ or 1.000 and is a different kind of number from the integer 1. Lack of space makes it impossible to go more deeply here into the difference between real numbers and integers, but there are now many references available for non-mathematicians. (e.g. I. Adler's *The New Mathematics* in *Signet Science Library* (1964); *Scientific American*, September (1964). The real numbers themselves are complicated by the fact that some are *rational*, and some *irrational*. Rational numbers are those that can be represented as a fraction or recurring decimal. Examples are $\frac{9}{8}$ or 1.12500, $\frac{1}{3}$ or 0.333 recurring. Irrational numbers are those that continue indefinitely without recurring. Examples are the square root of 2, and π (pi), the ratio of the length of the circumference of a circle to its diameter. Reputedly, one of the reasons why the Greek mathematicians developed geometry was that it enabled them to manipulate such irrational numbers as $\sqrt{2}$. This number can never be found, but it can be constructed geometrically as the hypotenuse of a right angle triangle with other sides each one unit in length.

In geography, zero and the positive integers are used for counting finite sets of discrete objects, but real numbers, taken to a required number of decimal places, are used to measure spaces, ratios, and many other features. In computer programming it is essential to distinguish between integers and real numbers.

There are other systems of numbers. The square root of -1 is an example of an '*imaginary*' number. An imaginary number may be combined with a real number to form a *complex* number. Complex numbers have a wide application in various fields including, for example, engineering. Unlike the real numbers, which can be conceived as an infinite set of points on a number line, complex numbers must be conceived on a plane with the real number along a horizontal axis and the imaginary along a vertical axis. They are not insuperably difficult for the non-mathematician, and are recommended, for example, for introduction in primary schools with the help of rotations and modulo arithmetic (E. E. Biggs (1965), Appendix 5).

Matrices, which will be introduced in section 2.8 are also considered by some mathematicians as a special kind of number (e.g. Schwartz (1961), p. 2). To conclude this subsection, according to Schwartz (p. 2): 'The hard thing is to invent useful kinds of numbers, and kinds of numbers that work . . . Nevertheless, a large variety of more or less successful new kinds of numbers have been invented by mathematicians'.

(*c*) *The appreciation of large quantities*

Large numbers, representing large quantities of objects, are often referred to in geographical material. That their magnitude is not always appreciated seems clear both from the frequency with which they are misprinted in texts and from the small upper limit many people seem to have when it comes to dealing with large quantities. An example of a misprint that was probably unnoticed by many readers was the quotation in the *Times* that the British Parliamentary Constituency of Sutherland is a difficult one for candidates to campaign in because of its size, 26,000 square miles. In fact, the area of Sutherland is roughly 2,600 square miles, which, by British standards *is* still quite large. The point is that the error of magnitude of 10 times is unnoticed because most people have no idea how many square miles to expect. One of the authors has asked many groups of students and others, how many possible ways there are of arranging (permuting) 10 things. Answers are usually 100, or 1000, rarely more. In fact, the answer is over 3 million. To take another example, suppose it is stated that by the mid 1960's the gross national product each year of the U.S.A. was approximately

$ 6,000,000,000
$ 60,000,000,000
$ 600,000,000,000
$6,000,000,000,000

which reader could tell the correct answer?

What remedies are there for the lack of ability to appreciate large numbers? The following may be suggested:

(i) Try to work out intelligently the approximate answer that might be expected. In the United States cxample above, the population of the U.S.A. is approximately 200,000,000. This gives a per capita income of 30, 300, 3000 or 30,000 dollars per capita. Commonsense or personal experience will find the right answer.

(ii) Memorize or have at hand some basic sets of figures against which others can be matched. For example, if the area of Yorkshire, Britain's largest county, is known to be approximately 6,000 square miles, then Sutherland cannot be 26,000 square miles, which is in fact roughly the area of the whole of Scotland.

(iii) Express numbers in a more satisfactory way. One aspect of the teaching of the New Mathematics is to put emphasis on what numbers really mean. Thus denary 345 (three hundred and forty five) is in full $3 \times 100 + 4 \times 10 + 5 \times 1$. Now 100 is itself 10×10, and can be expressed as 10^2. Similarly, 1000 is $10 \times 10 \times 10$, or 10^3. On this basis it is possible to express large numbers more concisely and at the same time to compare them more meaningfully. Thus 300 is 3×10^2; 3,000 is 3×10^3; 3,000,000,000 (approximately the total population of the world) is 3×10^9. This method is cumbersome for representing exact numbers (e.g. 7,219,424) but excellent for rounded numbers. It must always be remembered, however, that for example 10^3 is 10 times greater than 10^2, not $\frac{3}{2}$ times greater.

The following table shows some numbers of varying magnitudes with examples of numbers of people, distances, output figures and permutations and combinations.

Table 2.5 Large numbers

Quantity	Magnitude	Population	Distance (kilometres)
1	10^0	Individual	Within town
10	10^1	Big family	Across town
100	10^2	Hamlet	Regional
1,000	10^3	Village	National
1,024 (2^{10} in binary)			
10,000	10^4	Small town	Intercontinental
40,000	4×10^4		Equator
100,000	10^5	Medium town	
500,000	5×10^5		100 Transatlantic journeys
1,000,000	10^6	Large town	
3,628,800	$>3\times10^6$	Possible orders of 10 things	
10,000,000	10^7	Small country	
100,000,000	10^8	Fairly large country	
1,000,000,000	10^9	China fairly soon?	
3,000,000,000	3×10^9	*World*	
10,000,000,000	10^{10}		
87,178,000,000	9×10^{10}	Possible orders of 14 Chapters in a book	
100,000,000,000	10^{11}	Cells in a human brain	
507,000,000,000	5×10^{11}	Soviet electricity output in 1965 in kilowatt hours	
1,000,000,000,000	10^{12}		
6,000,000,000,000	6×10^{12}	U.S. gross national product in a decade in dollars	
	10^{14}	Cells in a human body	
	3×10^{64}	Possible journeys visiting each of the 50 U.S. state capitals in turn	
	10^{76}	Atomic particles in the universe according to Eddington	
	$10^{3,320}$	Possible combinations of students in a University with 10,000 students	
	$10^{3,000,000}$	Theoretically possible dynamic combinations and permutations of ensembles of homologous neurones to explain behaviour patterns of people	
	$10^{2,400,000,000}$	Theoretical possible combinations in DNA molecule which determines genetic endowments	

2.6 Dimensions, scales and precision

Geographers deal with a wide variety of phenomena which must be measured and then expressed in suitable terms. These require a consideration of dimensions, measurement scales, accuracy and precision of measurement. These aspects will be briefly commented upon.

(*a*) *Units of measurement*

Measureable parameters include length (L), mass (M), time (T) and temperature (t). These fundamental parameters may be used independently or they may be combined to form more complex measures, sometimes called derived units. Some of these complex measures are commonly used in geography. They include area (L^2), volume (L^3), and velocity (L/T). The last ratio is sometimes written as LT^{-1}. Acceleration, which can be written as L/T^2 or LT^{-2}, is given in terms such as feet per second per second. Density incorporates the two parameters L and M in the form M/L^3 or ML^{-3}, which is mass per volume, such as pounds per cubic foot. Other properties need three of the basic parameters, L, M and T. Viscosity is given as $ML^{-1}\,T^{-1}$, and energy is $ML^2\,T^{-2}$. All these properties can be expressed in specific units. Discharge, Q, can be given in cusecs (cubic feet per sec) or $L^3\,T^{-1}$. The parameters are some-sometimes combined in such a way that they cancel out to form dimensionless units. The simplest example is slope, which can be stated as a gradient in the form L/L. All angular measurements are dimensionless. The formulation of more complex dimensionless ratios is often convenient in scale model investigations where the problem of scale is particularly acute.

(*b*) *Scales of measurement*

Four scales of measurement have been recognized. It is useful to define these different methods of expressing the character of different objects, as the techniques of statistical analysis that can be applied to data depend in part on their scale of measurement. They are given in order of increasing exactitude or refinement.

(1) The *nominal* scale is only able to differentiate between objects by placing them into classes according to a recognizable attribute. A pebble sample from a particular locality might divide qualitatively into black stones and white stones. There is no continuum of characteristics in this example. The two subsets of stones come from two quite distinct sources. A nominal scale can be used to differentiate objects that have and do not have a particular property, or which differ in type. A village with or without church, a mining village or an agricultural one, a valley with or without lakes, and a glaciated valley or a fluvial one provide examples.

(2) The *ordinal* scale allows more refinement in that the values of the items under consideration can be ranked in order, even if absolute values are lacking, or would give a spurious impression of precision. It is possible to rank the towns of England in order according to population. This scale can be expressed in terms of positive integers, the natural numbers.

(3) The *interval* scale is the first that can be expressed in the real number system. One disadvantage is its lack of an absolute zero. Temperature for example is measured in °C or °F on the interval scale. It cannot be said that a temperature of 10°C is twice as hot as one of 5°C, since the zero 0°C is arbitrary. A good geographical example is given by altitude above present mean sea-level, which is itself a continually fluctuating zero. It cannot be said that a hill 200 ft high is twice as high as one 100 ft high because if sea-level were to change, this relationship of ratio would no longer hold.

(4) The *ratio* scale is the most precise and it incorporates equivalence of ratios and starts from an absolute zero. Temperature in °K can be measured

on this scale and a temperature of 10°K is twice as hot as 5°K. Where possible, measurements are made on the ratio scale, which uses the real number system and allows continuous measurement. Units can be converted directly from one system to another (e.g. metres to feet).

The major division is between the first two scales, which are non-parametric and the second two, which are parametric. H. R. Alker (1965, p. 19), goes so far as to say: 'Nominal and ordinal scales, the non-parametric scales, are usually considered examples of qualitative measurement.' Parametric scales provide values that can be used to calculate the mean and standard deviation, properties which are basic to all the parametric (see glossary) statistical tests discussed in chapter 3.

(*c*) *Accuracy and precision*

The aim of measurement is to record, as closely as required by the problem in hand, the true value of the object being measured. This aim involves the concept of accuracy and precision. These terms are clearly defined by C. Eisenhart (1952). Accuracy can be defined as the degree to which the measured value approximates to the true value. Accuracy includes the term precision, which is a measure of the closeness of the clustering of the values about the mean value measured. Precision is the degree of conformity of the values amongst themselves. The relationship between the two terms is shown most clearly in the form of a diagram (see figure 2.9) which illustrates the importance of bias or systematic error in the accuracy of a measurement.

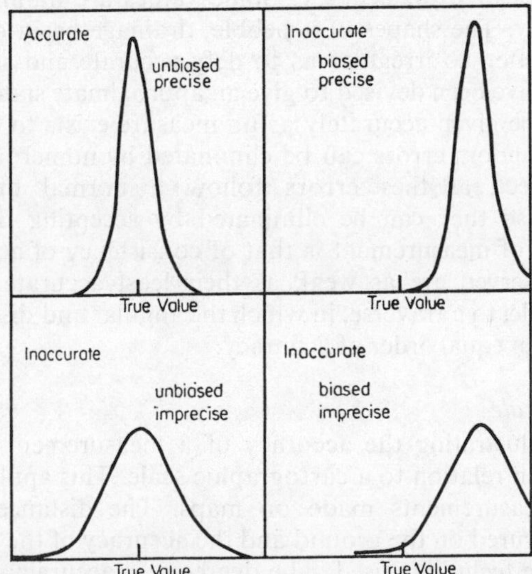

Figure 2.9 A diagram to illustrate the differences between precise and accurate observations. The standard deviation of the precise value is much smaller than that of the imprecise one as shown by the higher, narrower curve in the precise values. *After Eisenhart.*

To be accurate a measurement must be unbiased. A biased measurement may be precise but cannot be accurate. An accurate method of measurement must be both precise and unbiased. The values will then be closely clustered around the true value. The true value cannot however be known absolutely, owing to the limitations of the measuring technique being used. The accuracy is therefore impossible to assess. The true value is therefore defined for practical purposes as that value which is approached by the means of an indefinitely large number of observations. The true value defined thus is used in the theory of errors.

Neither accuracy nor bias can be known logically, but precision can be known. The precision of a measurement will depend on the measuring process and this affects the accuracy. The measuring process selected must therefore be chosen in relation to the required accuracy. All measurements must be made in terms of standard units, such as feet, seconds or degrees. The accuracy of measurement must depend on the purpose of the measurement. For example the level of a raised beach may be surveyed to the nearest $\frac{1}{100}$ ft but this accuracy is meaningless in view of the disturbance of the surface since the sea-level was at this height and also the lack of information concerning past tidal range and other matters. Nevertheless if the survey is accurate then at least one error or degree of uncertainty is reduced in amount.

All measurements are liable to error. These errors are of four types: gross errors, systematic errors, method errors and random errors. The first type can usually be identified fairly easily, but the systematic errors are less easy to deal with and lead to inaccuracy. Some values are infinitely difficult to measure accurately. The shapes of a pebble, drainage basin or political unit for instance are often so irregular as to defy accurate and simple measurement. Formulae have been devised to give an approximate statement of shape, but these cannot be given accurately as no measure exists to define irregular shapes exactly. Random errors can be eliminated by numerous observations of the same object, as these errors follow a normal distribution (see chapter 3.2) and so they can be eliminated by accepting the mean value. Another principle of measurement is that of consistency of accuracy. All the measurements observed are as weak as their least accurate member. This applies, for example, to a traverse, in which the angular and distance measurements must be of an equal order of accuracy.

(*d*) *Cartographic scale*

One method of illustrating the accuracy of a measurement is to consider the measurement in relation to a cartographic scale. This applies particularly to maps, and measurements made on maps. The distance between two points can be measured on the ground and the accuracy of the value obtained will depend on the technique used. The degree of inaccuracy becomes much greater when the distances are measured on maps, the values becoming increasingly uncertain and imprecise as the scale of map decreases.

The plottable error of the map provides a measure of the limit to which measurement can be made on maps of different scale. The plottable error is the smallest distance that can be recorded true to scale on a map, assuming the finest line on the map is $\frac{1}{100}$ in wide. On a scale of 200 ft to 1 in, the plottable error would be 2 ft.

On a 1 inch to 1 mile map the plottable error is 52.8 ft, so that any object of smaller dimensions must be shown symbolically with its size exaggerated. Some idea of the degree to which accuracy is lost and generalization increases, as the scale of the map diminishes, can be gained from considering the length of a stretch of irregular coastline, a meandering river or a tortuous road measured on different scale maps (see section e(iv)).

(e) Common methods of increasing precision of measurement

The precision of measurement of surveying instruments has been very greatly improved by the development of optical processes such as the micrometer, but these will not be considered in detail. However it is of interest to note that a theodolite circle of only 5 inches diameter can now be subdivided to give a direct reading to 1 second of arc, whereas about a century ago a circle 3 ft in diamcter was needed to give a similar or rather lower order of precision. Two simple methods of increasing precision that are of more direct interest to

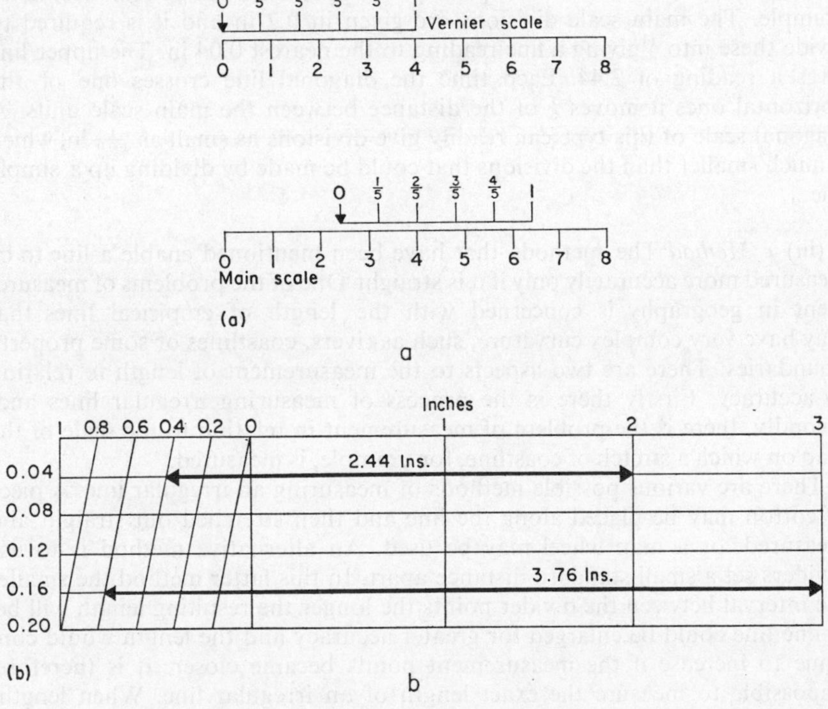

Figure 2.10 Diagrams to show (*a*) the construction and operation of vernier scales and (*b*) diagonal scales.

geographers will be mentioned. These are the vernier and diagonal scales.

(i) *Vernier Scale* A vernier scale is an optical device that depends upon the principle that it is easier to see the coincidence of two lines than to judge the distance between two lines. It requires that a small vernier scale shall be free to slide over the larger main scale and is used in instruments such as a barometer, a sextant and an Abney level. The scale is such that n divisions of the vernier scale are made equal in size to $n-1$ divisions of the main scale. If a scale reading $\frac{1}{5}$ of the main scale divisions is required then 4 of the main scale divisions are divided into 5 vernier divisions. In the situation in figure 2.10*a* the vernier index in the upper diagram is opposite 0 and the 1 is opposite 4 of the main scale. In the lower diagram the coincidence is at $\frac{2}{5}$ on the vernier and 4 on the main scale so that the reading is $2\frac{2}{5}$. Had the $\frac{3}{5}$ coincided with the 5 in the main scale the reading would have been $2\frac{3}{5}$. The vernier scale can very easily be adapted to give any required accuracy by making n larger and it can also be adapted for angular measurements.

(ii) *Diagonal Scale* The diagonal scale (see figure 2.10*b*) depends on the Euclidean geometrical properties of similar triangles and is a fixed scale. The method of construction and the principle of operation can be seen best in an example. The main scale divisions are given in 0.2 in and it is required to divide these into $\frac{1}{5}$ giving a fine reading to the nearest 0.04 in. The upper line gives a reading of 2.44. Each time the diagonal line crosses one of the horizontal ones it moves $\frac{1}{5}$ of the distance between the main scale units. A diagonal scale of this type can readily give divisions as small as $\frac{1}{100}$ in, which is much smaller than the divisions that could be made by dividing up a simple line.

(iii) ϵ *Method* The methods that have been mentioned enable a line to be measured more accurately only if it is straight. One of the problems of measurement in geography is concerned with the length of empirical lines that may have very complex curvature, such as rivers, coastlines or some property boundaries. There are two aspects to the measurement of length in relation to accuracy. Firstly there is the process of measuring irregular lines and, secondly, there is the problem of measurement in relation to the scale of the map on which a stretch of coastline, for example, is measured.

There are various possible methods of measuring an irregular line. A piece of cotton may be placed along the line and then stretched out straight and measured, or a map wheel may be used. An alternative method is to use dividers set a small standard distance apart. In this latter method the smaller the interval between the divider points the longer the resulting length will be. A fine line could be enlarged for greater accuracy and the length would continue to increase if the measurement points became closer. It is therefore, impossible to measure the exact length of an irregular line. When lengths are being compared it is necessary for the degree of accuracy of measurement to be stated if the results are to be comparable.

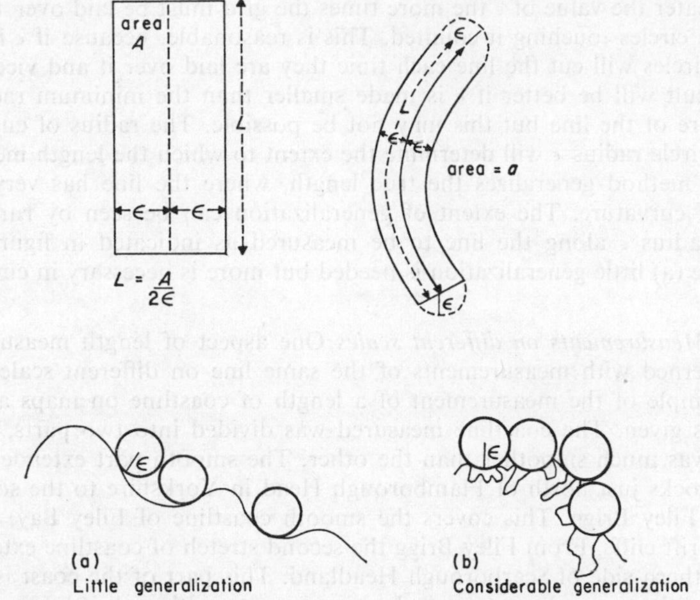

Figure 2.11 Diagram to illustrate a method of measuring length of lines, and the degree of generalization required for different outline complexity.

J. Perkal (1958, translated in Nystuen 1966) has suggested a method for measuring an irregular line, giving the ϵ length of the line. The basis of his method is that the length of a line is equal to an area parallel to the line and 2ϵ units wide divided by 2ϵ as shown on the diagram in figure 2.11 $L_\epsilon = \frac{A}{2\epsilon}$ in a rectangle and a similar formula gives $L\epsilon$ for an irregular curve. A correction must be made if the line is discontinuous by subtracting the end semi-circles. The formula is then

$$L_\epsilon(X) = \frac{A\epsilon\ (X) - \pi\epsilon^2}{2\epsilon}$$

The length of the line is directly proportional to the area A but inversely to the size of ϵ. ϵ/L should be a fairly small fraction $\frac{1}{10}$ or less. The problem then becomes one of measuring the area A. This can be done by using a lattice of points. A series of circles drawn round these points provides a better method. The circles can be arranged either in a square or triangular lattice. If the square lattice is used the formula is simple. Perkal has adjusted the constants in the formula to provide a very simple method of determining length by placing the lattice of circles K times over the line to be measured. The value of K is directly proportioned to ϵ, the circle size. The number n of circles cut by the line is counted. Then

$$L_\epsilon(X) = \frac{A^2}{2K\epsilon}\sum_{i=1}^{k} n - \frac{\pi}{2}\epsilon \quad \text{if} \quad K = 2\epsilon \quad \text{then} \quad L_\epsilon(X) = \sum_{i=1}^{k} n - \frac{\pi\epsilon}{2}.$$

The greater the value of ϵ the more times the grid must be laid over the line and the circles touching it counted. This is reasonable, because if ϵ is small many circles will cut the line each time they are laid over it and vice versa. The result will be better if ϵ is made smaller than the minimum radius of curvature of the line but this may not be possible. The radius of curvature of the circle radius ϵ will determine the extent to which the length measured by this method generalizes the true length, where the line has very small radii of curvature. The extent of generalization can be seen by running a circle radius ϵ along the line to be measured as indicated in figure 2.11. In curve (a) little generalization is needed but more is necessary in curve (b).

(iv) *Measurements on different scales* One aspect of length measurement is concerned with measurements of the same line on different scale maps. An example of the measurement of a length of coastline on maps at three scales is given. The coastline measured was divided into two parts, one of which was much smoother than the other. The smooth part extended from Crab Rocks just north of Flamborough Head in Yorkshire to the southern side of Filey Brigg. This covers the smooth coastline of Filey Bay, mainly cut in drift cliffs. From Filey Brigg the second stretch of coastline extends to the northern side of Scarborough Headland. This part of the coast is rocky and slightly to moderately crenulate in character although it has no major indentations. The lengths of these two stretches of coastline were measured on the 1:25,000 map, the 1:63,360 map and the 1:250,000 map which are respectively about $2\frac{1}{2}$ in to 1 mile, 1 in to 1 mile and about $\frac{1}{4}$ in to 1 mile. The results are given in the table.

(Distance in km)	1:250,000	1:63,360	1:25,000
Crab Rocks to Filey Brigg	12.5	13.2	13.375
Filey Brigg to Scarborough	15.0	16.2	21.75

The figures show that there is relatively little difference in the length of the smoother section of the coast where generalization as the scale decreases is not so necessary. The crenulate section on the other hand shows a marked increase on the largest scale, on which all the small indentations can be shown.

2.7 Calculations

The New Mathematics stresses among other things two aspects of computation, firstly the basic laws of arithmetic, and secondly the application of devices to cut short the time in making calculations. This section will be devoted largely to the second aspect. The attention of the non-mathematical reader is however drawn to the fact that the teaching of arithmetic until recently has largely been the teaching of certain tricks that could then conveniently be applied to everyday calculations such as adding and dividing. Underlying these are five fundamental laws for the arithmetic of integers*:

Commutative addition law: $a+b=b+a$

Associative addition law: $a+(b+c)=(a+b)+c$

Commutative multiplication law: $ab=ba$

Associative multiplication law: $a(bc)=(ab)c$

Distributive law: $a(b+c)=ab+ac$

The ordinary rules of arithmetic and algebra are based upon these laws and elaborations of them.

Much work in geography requires laborious arithmetic calculations or some kind of sorting. There seems no merit in using paper and pencil methods for this when various devices are available. It is desirable to use the quickest device for a given piece of work, so long as the particular device does not lose too much accuracy. Moreover, some calculations are so lengthy that they could not even be considered by pencil and paper methods.

Devices for making calculations will be outlined roughly in increasing order of complexity.

(*a*) *Simple 'manual' desk aids*

Among these are slide rules, logarithm tables, and various manually operated machines that are mostly cheap and portable. Addition and subtraction cannot be done on slide rules and logarithm tables, but can on hand calculating machines. The number of decimal places to which a calculation can be accurately made is usually limited, but in geography and in many social science subjects one or two decimal places are often adequate.

(*b*) *Mechanical and electronic calculating and sorting machines*

These use electric power but are still manual in the sense that number keys must be pressed by hand. Electronic calculating machines are considerably faster than mechanical ones. Both are much faster and more accurate than the simple desk aids, let alone pencil and paper. Card sorting equipment is also mechanical, but its function is to sort information rather than to perform arithmetical operations. Cards are punched at appropriate places on their margin or anywhere on the surface, and may then be selected mechanically as required.

(*c*) *Electronic computer*

The above devices may all broadly be called computers. The so-called electronic computer differs fundamentally from the others on account of both its structure and its great speed. The following points should be noted about it.

(i) Up to now electronic computers have largely been used either to perform arithmetical calculations or sorting tasks.

* The five laws are quoted from Luttewood, D. E., *The Skeleton Key of Mathematics*, Harper Torchbooks, New York (1960, p. 17).

(ii) A computer does what it is instructed to do by a program, which is as good or as bad as the person writing it. The output of a computer, therefore, depends on the program given to it, the data to be processed by the program, and the processing equipment of the computer.

(iii) A computer differs from other devices because of its great speed and the consequent low cost of its work. Breakthroughs in technology have revolutionized food production, transport speed and costs and so on. Usually, however, these have only increased capabilities a few times. For example, in 1910 it was possible to travel across the U.S.A. by train at something like 50 m.p.h. In 1960 it was possible to do so at 500 m.p.h. by jet aircraft, an increase in speed by a factor of 10^1 in half a century. A large electronic computer can calculate 1,000,000 times or 10^6 times faster than a mechanical calculating machine. It is safe to say that such a revolutionary increase in speed has rarely been achieved in other fields.

(iv) The building of electronic computers has begun to throw some light on the working of the human brain. One striking contrast between the two is that the computer works much more rapidly than the human brain but as yet lacks the ability to correct itself or to discriminate if the information given it is not exactly accurate. Thus the human brain can insert the missing symbols in the following word, *NFORM*TION. According to J. Z. Young (1964, p. 6): 'As machines that assist our brains develop we shall learn from them a language much more useful than the present one for speaking about brains and their "products" '.

(v) The great speed of a computer makes it possible for the geographer to contemplate and make calculations that previously would have been impossible. For example the factor analysis calculations discussed in chapters 6 and 7 (Northern England) took the relatively small computer at Nottingham University approximately 2 minutes, and cost about £15. They would theoretically have taken a person many years to carry out on a mechanical calculating machine. By the mid 1960s, STRETCH, one of the largest computers in the world, could perform 1,500,000 calculations a *second*. Perhaps the sorting speed of even a modest sized computer can be better appreciated if it is realized that it might, if appropriately programmed, run through a suitably prepared text the length of the Bible in a matter of tens of seconds, count the frequency of many different words and word combinations during the process, and subsequently print them out.

(vi) The application of computers in the modern world has received enormous publicity, so it is superfluous to elaborate it here. The geographer should, however, consider them from two points of view: firstly, to what extent and in what ways can they help with calculations and sorting, and secondly, how do they actually affect the things the geographer is studying such as the regulation of traffic flows or the ability to plan an economy the size of that of the U.S.S.R.

On the first score, the words of D. F. Merriam (1965, p. 513) could be applied to geography as to geology: 'The advent of the computer is revolutionizing geology, just as it has other scientific disciplines. Application of the computer to geology, however, has been slow because of the reluctance of geologists — mostly due to their traditionally qualitative training — to accept more rigorous quantitative methods. Thus the complete potential of the computer has yet to be realized in the earth sciences, and it may be many years before a full evaluation of its impact can be made'.

Geographers in the U.S.A. have been much quicker than their British colleagues to grasp the significance of computers to their subject. Several, presumably with the help of colleagues from other disciplines, were already using them in the late 1950s. Swedish geographers, including T. Hagerstrand and S. Nordbeck, have also recommended them for the handling of data for points on the earth's surface. R. C. Kao (1963) discussed some of the possibilities of using computers in geography. In particular he stressed the need for giving the precise location by co-ordinates of each place for which information, such as census data, is gathered. He suggested that unlike some disciplines, which would require computers only for certain clearly defined purposes, geography would find a considerable range of uses. Probably one of the first large scale data processing operations of a geographical nature (though not done by geographers) was the processing of a considerable amount of 1951 census data for 157 British towns by C. A. Moser and W. Scott (1961). In his *Agricultural Atlas of England and Wales*, J. T. Coppock (1964, Appendix II) draws attention to the possibilities of the computer in geographical research. The use of computers will be illustrated by various examples in the present book, but no attempt is made to introduce methods of programming or to consider the merits of different languages or computers.

2.8 Tables and matrices

The word matrix has various meanings, one of which is a precise mathematical meaning, another a more general term for an array of information.

(1) 'A square array of *real* numbers' (J. T. Schwartz, 1961)

(2) 'An orderly array of symbols by rows and columns' (Marks, R. W., 1964). In contrast to (1), there does not have to be the same number of rows as of columns

(3) Any table with numerical data, or even non-numerical data. In this broad sense, matrix and table become virtually interchangeable. This extended use of the term matrix is not however permissible in a mathematical sense.

Matrices provide a very concise and convenient way of displaying data in the form of an array. In the broader sense of the term, they have been used by geographers to sort out relationships (as between branches of a discipline) and to store data for reference (as road distances between towns). They may also prove useful in helping to arrange a number of steps in some operation into the best or a good order (as in programmed learning, see C. A. Thomas (1963)). Computer mapping treats the map surface as a matrix with compartments, rather than as a continuous surface. In a more precise mathematical sense, various operations, including matrix addition and multiplication, may be performed on data arrayed in a matrix, and it is in this form that much data can most efficiently be processed by electronic computers

J. T. Schwartz (1961, p. 4) stresses the following aspect of a matrix: 'by regarding a [square] array of numbers as constituting a single object, a matrix, we will be able to handle large sets of numbers as single units, thereby simplifying the statement of complicated relationship'. Most mathematicians do not require a matrix to be *square*, the meaning of which will be shown below. The remainder of this section describes some of the essential features of matrices and of operations on them, using examples as far as feasible from spatial situations.

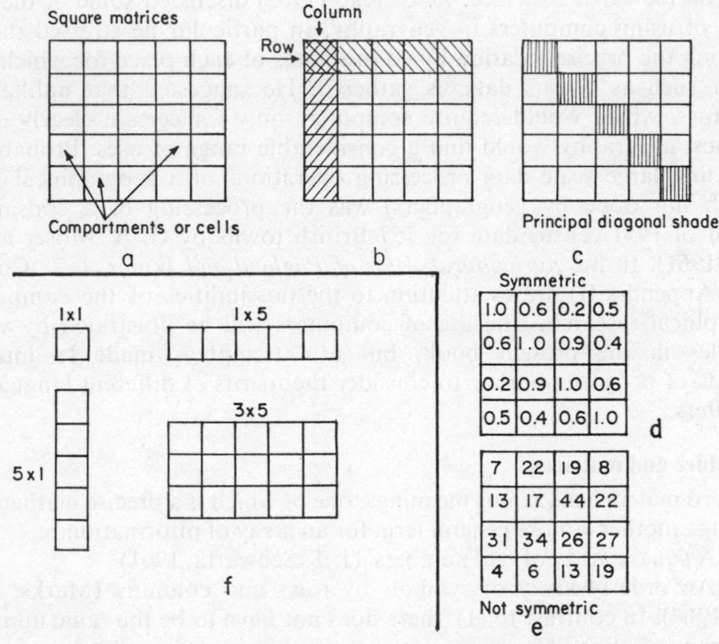

Figure 2.12 Introductory diagrams to show features of matrices. (*a*) A square matrix. (*b*) Rows and columns. (*c*) Principal diagonal. (*d*) Symmetric matrix. (*e*) Non-symmetric matrix. (*f*) Other cases.

Figure 2.12*a* shows a square divided into 25 compartments. This is a 5×5 matrix. Each space must contain one and only one *entry* (also called element or constituent). Matrices conveniently 'start' at the top left hand corner; graphs in contrast, conventionally 'start' at the bottom left. The horizontal sets of entries are called rows, the vertical ones columns (figure 2.12*b*). A matrix is *square* if it has the same number of rows as columns (e.g. 3×3, 5×5). In a square matrix, the entries running diagonally from top left to bottom right form the *principal diagonal* (figure 2.12*c*). If a square matrix has the same information in the compartments below and to the left of the principal diagonal as above and to the right of it, the matrix is called

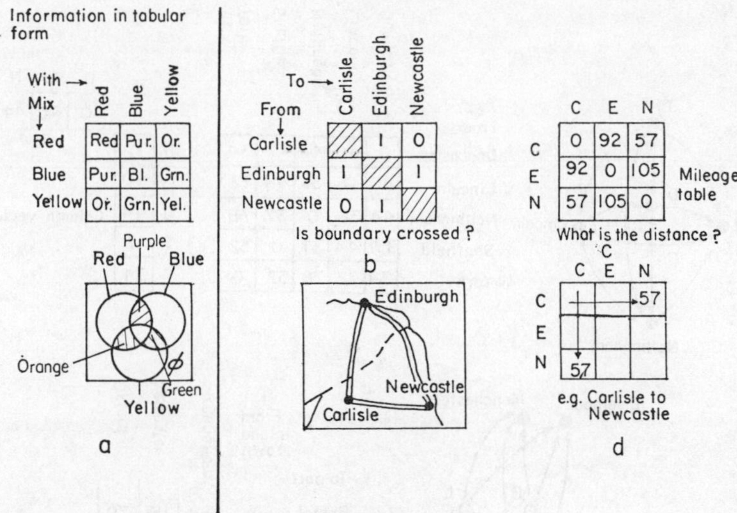

Figure 2.13 Information in tabular form. (*a*) Colour mixes. (*b*) Road travel examples.

symmetric (figure 2.12*d*). Only square matrices can be symmetric, but not all square matrices are symmetric (figure 2.12*e*). A matrix may have only one row or one column. Such a matrix is usually called a row vector or a column vector (figure 2.12*f*).

Some ways in which information may be stored are illustrated in figure 2.13. For simplicity, 3×3 matrices are used. Figure 2.13*a* shows the secondary colour expected when two primary colours are mixed. This is *nominal* information. Below, the same situation is illustrated with a Venn diagram. Figure 2.13*b* shows a *yes/no* situation which is nominal yet 'sub-quantitative' as 1 and 0 are used. It merely states whether yes (1) or no (0) the England–Scotland border is crossed on a road journey between 3 given places, Carlisle, Edinburgh and Newcastle (see map). Figure 2.13*c* shows *interval* information which records the actual (best) road distances between the same three places. Figure 2.13*d* shows how to read the information from the matrix. For example, the distance from Carlisle to Newcastle is found by the intersection of the row for Carlisle and the column for Newcastle, or vice versa.

In figure 2.14*a* quantitative data are used to illustrate shortest recommended road distances between all the members of a set of inland English towns. As in the example in 1.13*b–d*, the matrix is square and symmetric because the distance is given *both* from for example Lincoln to York and from York to Lincoln. Since the distance is identical, half of the matrix is superfluous and is usually omitted if being printed as a distance reference matrix. Contrast the matrix in figure 2.14*b*, showing shortest recommended road distances between a set of three inland English towns and a set of five major ports. Here interest focuses only on the distance between inland centre and port, not between pairs of inland centres, or pairs of ports. The matrix could have

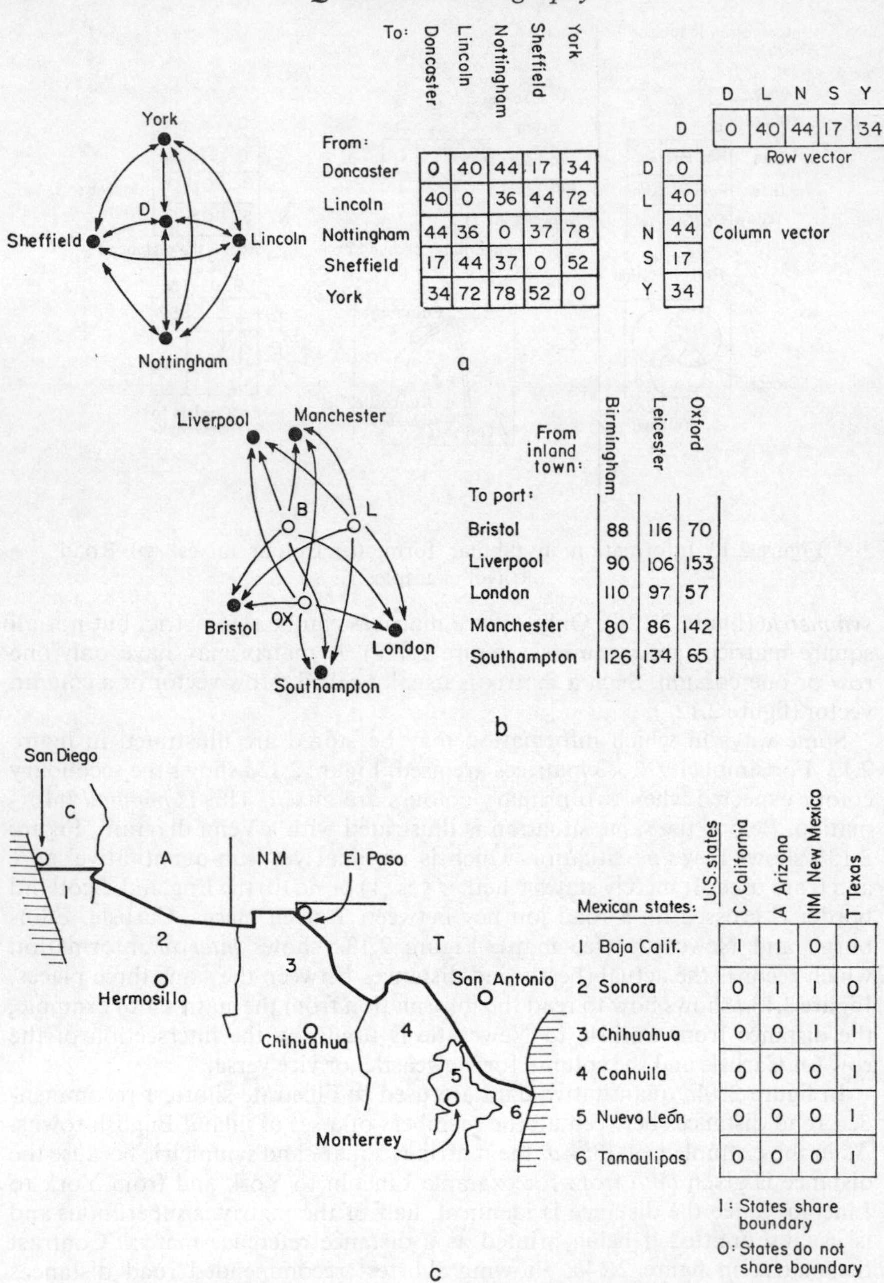

From: \ To:	Doncaster	Lincoln	Nottingham	Sheffield	York
Doncaster	0	40	44	17	34
Lincoln	40	0	36	44	72
Nottingham	44	36	0	37	78
Sheffield	17	44	37	0	52
York	34	72	78	52	0

Row vector

	D	L	N	S	Y
D	0	40	44	17	34

Column vector

	D
D	0
L	40
N	44
S	17
Y	34

To port: \ From inland town:	Birmingham	Leicester	Oxford
Bristol	88	116	70
Liverpool	90	106	153
London	110	97	57
Manchester	80	86	142
Southampton	126	134	65

Mexican states: \ U.S. states	C California	A Arizona	NM New Mexico	T Texas
1 Baja Calif.	1	1	0	0
2 Sonora	0	1	1	0
3 Chihuahua	0	0	1	1
4 Coahuila	0	0	0	1
5 Nuevo León	0	0	0	1
6 Tamaulipas	0	0	0	1

Figure 2.14 Information in tabular form. (*a*) Symmetric or repeated information. (*b*) Non-symmetric or non-repeated information. (*c*) Yes-no information.

been square if it had been concerned with the same number of inland towns as of ports, but it still would not have been symmetric. A further example is shown in figure 2.14*c*. Here the matrix shows concisely which U.S. states share a common international boundary with which Mexican states.

One of the ways in which a matrix can help the geographer is as a store of information that cannot easily be recorded in any other way. For example, information could be recorded about some relationship among the 100 principal states in the world. The matrix would have 10,000 compartments. If one wanted to know, for example, which countries had an embassy in which others at a given time, then it would be possible to store 10,000 pieces of information far more compactly in a matrix than in any other way so that it could be quickly referred to.

From a mathematical point of view, the possibility of performing operations on matrices is of more importance than the possibility of using them simply for reference. In the long run, it seems essential that geographers should become acquainted with matrix notation and should learn how the operations work. First, it is essential to note that each entry space in a matrix has a unique identification. A number is given to each row: 1, 2, 3, . . . or 01, 02, 03 . . . if two digits are needed and likewise to each column 1, 2, 3. . . . The row number precedes the column number. In matrix notation it is customary to refer to any particular row as the ith and to any particular column as the jth. The lower case a refers to the entries in some matrix A. The total number of rows in a matrix is referred to by m and the total number of columns by n. Figure 2.15*a* illustrates these features. Figure 2.15*b* shows the way matrices are usually printed.

Figure 2.15*c* gives a simple example of how the data in two matrices, PIGS and SHEEP, can be added, For addition, the matrices need not be square but they must have the same number of rows and of columns. The adding operation shows how many PIGS *and* SHEEP the given set of farms delivered to the given set of markets in a certain period (say a year).

Figure 2.15*d* shows *scalar* multiplication. The matrix shows how many tons of gravel a set of three lorries carried from a set of two gravel pits to a building site during a day. The problem is to find how many tons can be carried by the same lorries in a five-day week. It is merely necessary to multiply in turn each number in the matrix by 5.

Proper matrix multiplications is a more complicated operation, and can be lengthy if there are many columns and/or rows. In ordinary multiplication, whichever the order in which two numbers are multiplied, the result is the same. By the commutative law of multiplication, $a \times b = b \times a$ $(3 \times 2 = 2 \times 3)$. This is not so in matrix multiplication where order must be right. Moreover, one matrix can be multiplied by another only if the first matrix has the same number of columns as the second has rows. Thus a 4×2 matrix can be multiplied by a 2×2 matrix, but not by a 4×4 matrix. In figure 2.15*e*, the multiplication of matrices is shown by a simple example, and a matrix (like 1 in ordinary multiplication) is shown which leaves the first matrix unchanged when multiplied by it.

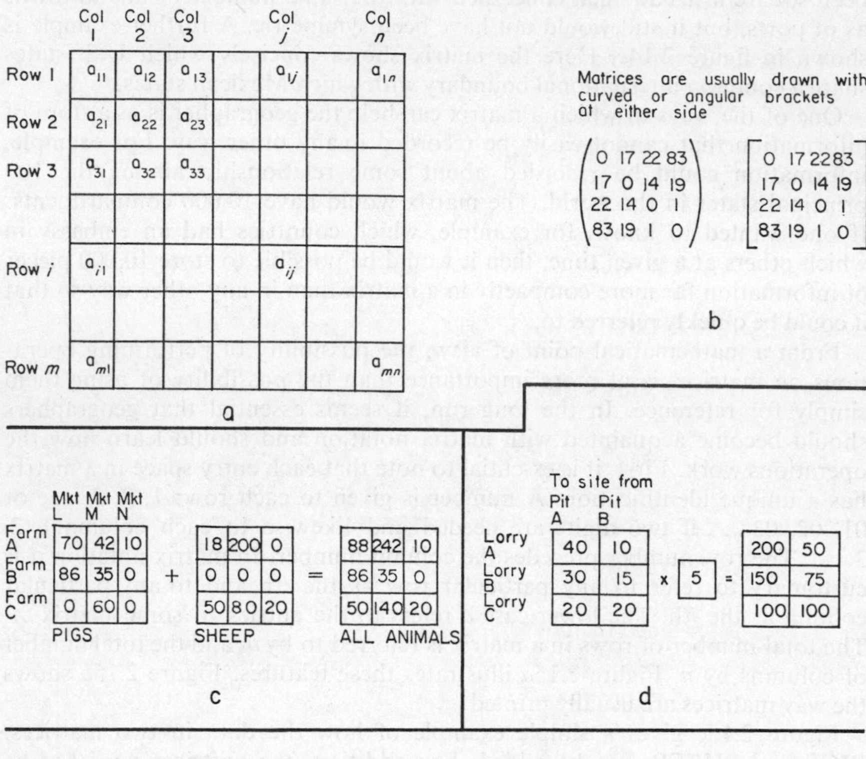

Matrix multiplication $\begin{pmatrix} 1 & 2 \\ 4 & 3 \end{pmatrix} \bullet \begin{pmatrix} a & b \\ c & d \end{pmatrix} = \begin{pmatrix} 1a+2c & 1b+2d \\ 4a+3c & 4b+3d \end{pmatrix}$

Thus $\begin{pmatrix} 1 & 2 \\ 4 & 3 \end{pmatrix} \bullet \begin{pmatrix} 1 & 0 \\ 0 & 1 \end{pmatrix} = \begin{pmatrix} 1\times1+2\times0 & 1\times0+2\times1 \\ 4\times1+3\times0 & 4\times0+3\times1 \end{pmatrix} = \begin{pmatrix} 1 & 2 \\ 4 & 3 \end{pmatrix}$

Procedure $\begin{pmatrix} 1 & 2 \\ 4 & 3 \end{pmatrix} \bullet \begin{pmatrix} a & b \\ c & d \end{pmatrix}$

Result

e

Figure 2.15 Matrix notation and simple operations. (*a*) Row and column reference numbers. (*b*) Less formal representation of matrices. (*c*) Matrix addition. (*d*) Scalar multiplication. (*e*) Ordinary matrix multiplication.

The application of matrix multiplication to a very simple situation of a spatial nature is now given. Sand and gravel are carried from two quarries by rail by two trains to a construction site. The pits are referred to as A and B, the trains as Q and R and the material as S and G. The problem is to calculate how much sand and gravel altogether are carried to the site in a day. The first matrix shows the number of return journeys made between the site and each pit by each train:

	Train Q	Train R
Pit A	10	4
Pit B	8	6

The second matrix shows the carrying capacity in tons of sand and of gravel by each train:

	Sand	Gravel
Train Q	100	20
Train R	60	40

The answer is obtained by multiplying the two matrices as follows:

$$\begin{matrix} & \begin{matrix} Q & R \end{matrix} \\ \begin{matrix} A \\ B \end{matrix} & \begin{bmatrix} 10 & 4 \\ 8 & 6 \end{bmatrix} \end{matrix} \times \begin{matrix} & \begin{matrix} S & G \end{matrix} \\ \begin{matrix} Q \\ R \end{matrix} & \begin{bmatrix} 100 & 20 \\ 60 & 40 \end{bmatrix} \end{matrix} = \begin{matrix} \begin{matrix} QAS \quad RAS & QAG \quad RAG \end{matrix} \\ \begin{bmatrix} 1000+240 & 200+160 \\ 800+360 & 160+240 \end{bmatrix} \\ \begin{matrix} QBS \quad RBS & QBG \quad RBG \end{matrix} \end{matrix}$$

$$= \begin{matrix} & \begin{matrix} S & G \end{matrix} \\ \begin{matrix} A \\ B \end{matrix} & \begin{bmatrix} 1240 & 360 \\ 1160 & 600 \end{bmatrix} \end{matrix} = 3360 \text{ tons}$$

The reader is invited to calculate the total number of ton miles carried in the course of the day by the two trains if pit A is three miles from the site and pit B is five miles.

Familiarity with what matrix multiplication is actually doing may be gained by following step by step what calculations are actually being performed on the trains, the pits and the materials carried. In particular, it should be noted that the trains Q and R are the columns in the first matrix, the rows in the second matrix.

2.9 Geometries

(a) *Introduction*

The mathematician Carl Friedrich Gauss (1777–1855) suggested that the uncurved plane surface of Euclid, the flat space on which geometry had hitherto been operating, 'was only a special case of a more general geometry that could just as well be applied to curved surfaces' (Life *Mathematics* (1965,

p. 160). Gauss and other pioneers of new geometries worked more than a century ago. To the layman, non-Euclidean geometries may lack respectability, even if they are known but, as with algebra and arithmetic, it is customary now to refer to geometries, rather than geometry. Different mathematical sources give different versions of how many kinds of geometry there are. This seems to be because they take different criteria to classify geometries. R. Courant (1964) discusses five geometries. They are illustrated in figure 2.16. In the 19th century the mathematician Felix Klein classified

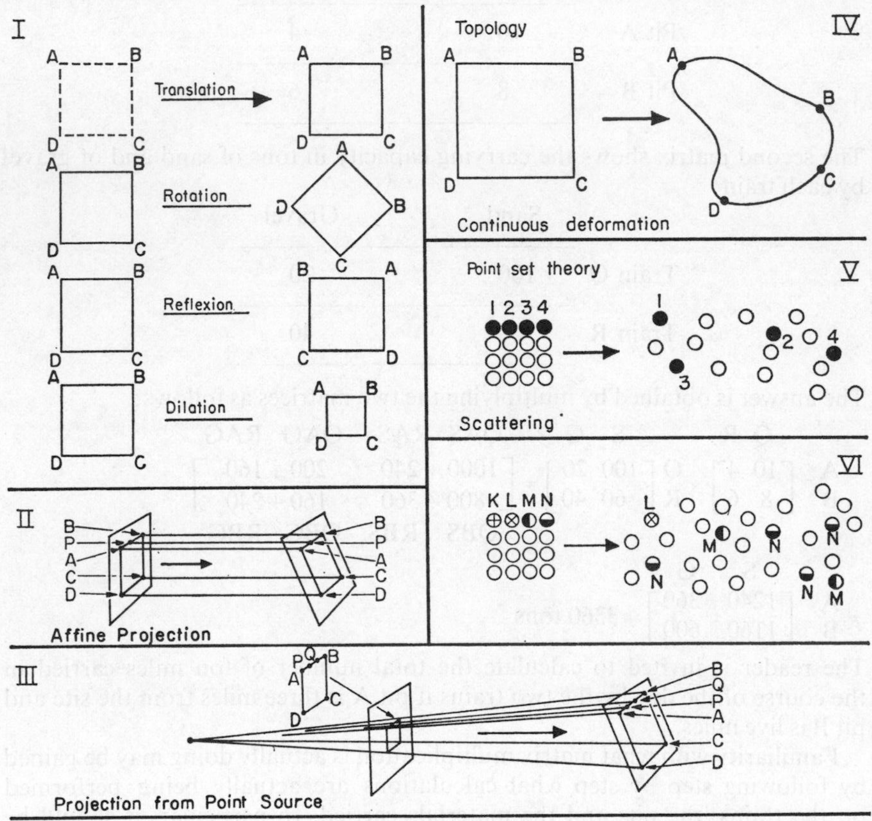

Figure 2.16 Types of geometry: I Euclidean, II Affine, III Projective, IV Topological, V Point set theory, VI Geographer's problem (I-V are based on figures in *Sci. Am*, Sept. 1964, pp. 42-43).

geometries according to the invariant groups of figures when they undergo various groups of transformations. Five are distinguished on this basis: Euclidean, affine, projective, topology and point set theory. The preservation of key features of the square (e.g. its right angles, and the equality of length of its sides), are shown in Euclidean geometry (I) as it undergoes certain transformations. Affine geometry (II) permits the same transformations, but allows also projection by parallel rays to a tilted plane; a feature shown in the illustration is the preservation, after transformation, of the ratio of AP to PB (collinear points). Projective geometry (III) permits point-source projection to a randomly tilted screen. 'An invariant property of figures projected in this way is the cross ratio of collinear points'. The ratio of AP/PQ: AD/PD is unchanged by the transformation.

Topology (IV) is altogether more flexible than the foregoing geometries, or perhaps 'elastic', for it is referred to as the *rubber sheet geometry*. Imagine a rubber sheet pinned on a surface to form the square ABCD. When this is deformed, ABC and D are still in the same order. Topology is discussed in section 2.10. Point set theory referred to already in section 2.2 is less obviously a geometry to the layman than the others. Nevertheless it seems to be extremely useful for geographers, at least as a background concept. Points 1, 2, 3 and 4 remain in a one–one correspondence after 'scattering'. A sixth situation (VI) is added to the five geometries outlined by Courant. It typifies many situations facing the geographer. Points K L M and N are not only scattered but are no longer in a one–one correspondence. For example N is in a one–four correspondence, K in a 1–0 correspondence. This situation will be reconsidered in chapter 5.

Affine and projective geometry concern the geographer in several ways, if only marginally. Projective geometry is relevant in map projections, and in aerial photography. Usually however aerial photographs are taken from such a great height that the height of the objects themselves is far less than the height of the viewer above the ground, and their vanishing point can therefore be ignored. A similar effect is achieved in a horizontal view by taking a photograph with a telescopic lens. The invariant property of the cross ratio of collinear points in projective geometry could also be useful in surveying.

Euclidean geometry and topology are of particular importance in geography. The rest of this section deals principally with Euclidean geometry. Topology is introduced in the following section. Both geometries will reappear frequently later in the book.

(b) The location of points and objects in space

Euclidean geometry is characterized in the eyes of the layman by its preoccupation with angle and distance. It is fundamental in geography for locating points in space. There is no special starting point in the universe against which other things may be located in terms of absolute distance and direction from it. Position must therefore be considered in relation to the sun, or, more commonly in geography, to some arbitrarily chosen place on the earth's surface. For example the poles and Greenwich serve for the whole of the earth's surface, a point somewhere to the southwest of Britain for the Ordnance Survey grid of Britain, and so on. Other points may be located in relation to the point of origin. Small objects may be considered single points, but larger objects, forming for example large surfaces, like lakes, need several or many points on the boundary to describe their location.

Points may be located in relation to a point of origin in 1-dimensional

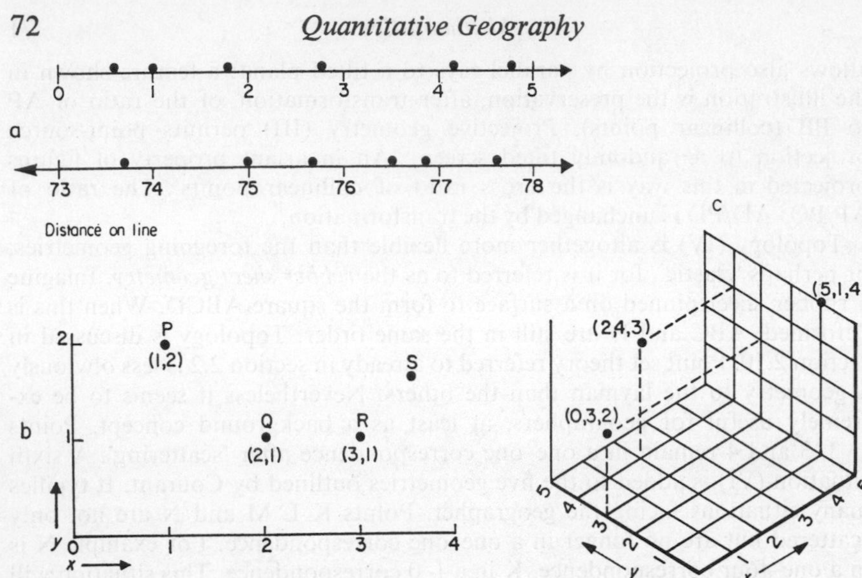

Figure 2.17 Locating points in space. (*a*) On a line. (*b*) With co-ordinates on a surface. (*c*) In 3-dimensional space.

space (a line), 2-dimensional space (a surface or patch) and 3-dimensional space (volume).

(i) 1-dimensional case. In figure 2.17*a*, dots are shown distributed on a straight line. Their positions are described in terms of units of distance from a point of origin, 0. In the lower line, this lies far to the left of the part of the line under consideration. The point of origin could be within the distribution, as on the integer number line in figure 2.8.

(ii) 2-dimensional case. Two axes, orthogonal to one another (crossing at right angles) are required to locate points on a surface. An example is given in figure 2.17*b*. Three points, P. Q, and R are located on the surface. They all have integer co-ordinates. The position along the 'horizontal' (x) axis is given before the position on the 'vertical' (y) axis. The co-ordinate position is called an *ordered pair*. The integers 1 and 2 appear for both P and Q, but differ in order. The point S cannot be given a precise location in terms of integers, but its position can be given in terms of real numbers to as many decimal places as desired. As shown in figure 2.19*e* and *f*, the point of origin, O, may lie outside the distribution of points to be located in relation to it, or it may lie within. In the latter case, one or both of the numbers in the ordered pair may be negative.

(iii) 3–dimensional case. A third axis may be drawn orthogonal to the first two to allow the location of points in three dimensions. An *ordered triad* or *triple* is then required to fix a point in space. This is illustrated in figure 2.17*c*. Three space dimensions will be considered again in chapter 9. It is useful to try to visualize further dimensions. Although it is not possible to add further axes to represent dimensions in space as popularly conceived, it is possible to do so mathematically. This fact forms the basis of multivariate analysis, which will be discussed in chapters 3 and 6.

There are clearly many applications of the concept of location of things in space. Some 2-dimensional examples will be briefly discussed. Figure 2.18

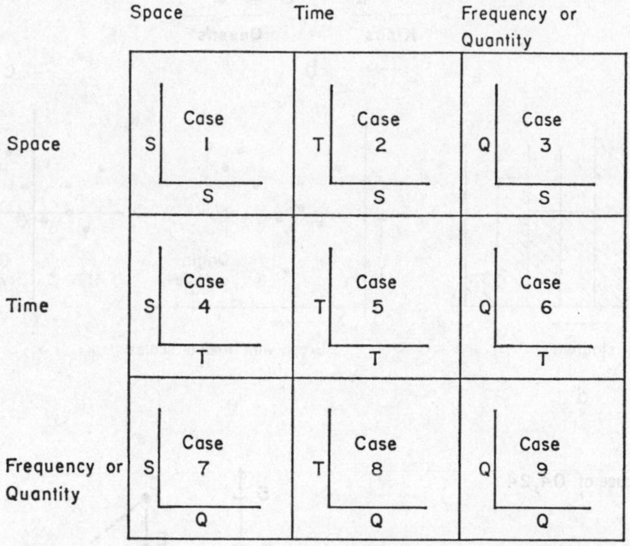

Figure 2.18 Some ways of using graphs to plot information.

shows that there are apparently nine ways in which spatial, temporal and other variations can be plotted on a graph in relation to themselves or to one another. Some examples follow:

Case 1 is a cartographical or topographical situation, the map, elaborated in 2.19*g*. Both axes, *x* and *y*, represent spatial distance from 0. Conventionally, the scale is linear, but this need not be so. For example a logarithmic scale may be useful for some purposes.

Case 2 might be used to show the time taken to travel given distances.

Case 3 can be exemplified by varying quantities (Q) of traffic at different places along a road (S) or railway.

Case 6 illustrates the change of some quantity during time.

Most of the nine cases will be used at some stage in the book.

In figure 2.19 it is shown that a nominal scale can be used on a graph and that discrete compartments as well as points can be given unique locations. Figure 2.19*a* shows the notation of matrices. Note that conventionally, the point of origin is at the top left, not the bottom left, of the diagram, as in most graphs (see figure 19*e* and *f*). However, the SYMAP computer mapping system

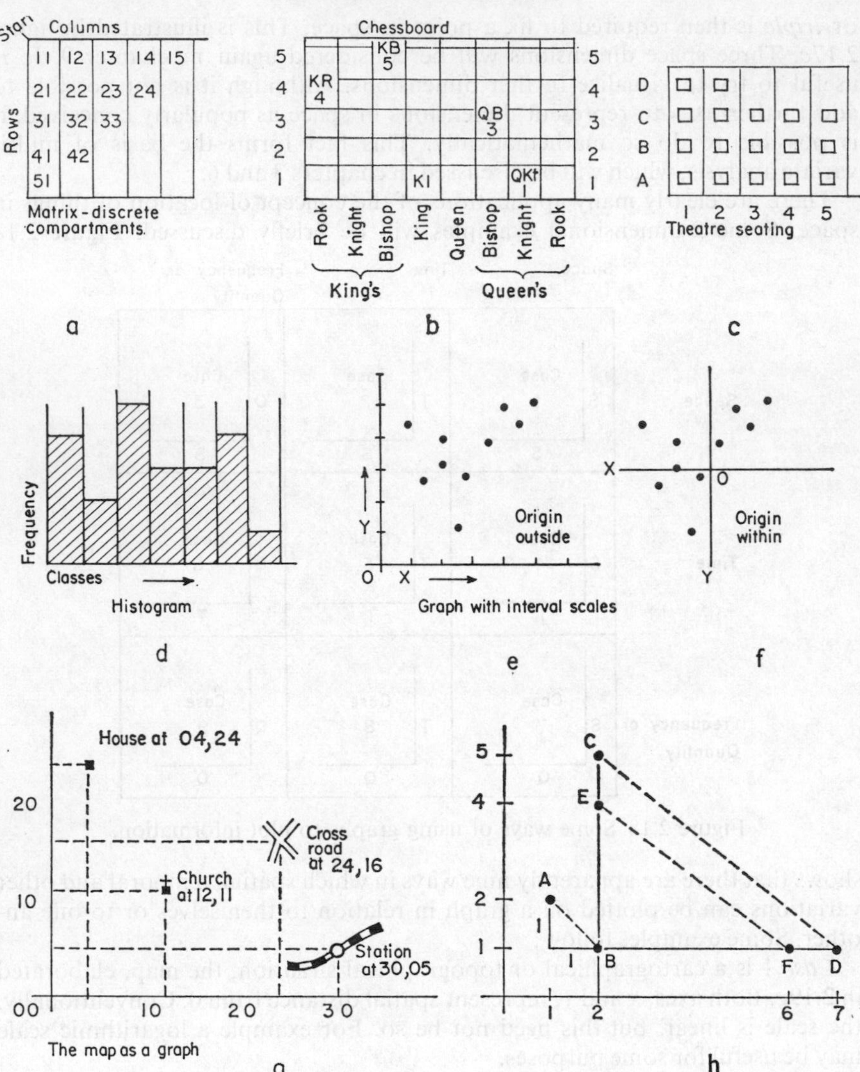

Figure 2.19 A comparison of various ways of recording information on a 2-dimensional surface. (*a*) Matrix notation. (*b*) Chessboard. (*c*) Theatre seating. (*d*) Histogram. (*e*) and (*f*) Graphs. (*g*) Cartograph. (*h*) The use of Pythagoras' theorem.

starts at the top left. The chessboard in figure 2.19*b* shows that a name can be given to the columns in a diagram. However a number would be satisfactory. As in figure 2.19*c*, to avoid confusion, a letter is often given to one of the directions in a diagram. Note that the histogram in figure 2.19*d* is a kind of graph, but the *x*-axis is divided into discrete classes, while the

y-axis may have either discrete steps or be continuous and therefore use real number values.

The remaining diagrams in figure 2.19 show the graph as a map (sometimes referred to as a cartographical or topographical diagram). In figure 2.19*g*, points are given Cartesian co-ordinate location. In figure 2.19*h*, a very important aspect of graphs, the basis of much recent work designed to computerize maps, is illustrated. Each of the points A to F in figure 2.19 can be given a unique co-ordinate position on the graph. Each point then has a particular relationship to every other one. This is of great geographical importance. When however there are very many points, it becomes laborious to calculate the distance from each to every other. For example when there are 100 points, there are $n(n-1)/2$ or almost 5000. The reader is invited to consider the following calculation with the help of Pythagoras' theorem

$$EF = \sqrt{(BF)^2 + (BE)^2} = \sqrt{16 + 9} = 5$$

All the information needed to calculate the distance between each pair of points is contained in the co-ordinate positions of the points.

(*c*) *Some simple transformations*

Basically, only one geometrical transformation is used in the construction of a map covering a small part of the earth's surface, that of dilation (expansion *or* reduction). Some part of the earth's surface is greatly reduced, but the relative distance between all places is preserved.

Two further Euclidean transformations may also be applied to maps, but their use meets opposition from various quarters: rotation and reflexion. These are illustrated in figure 2.16 in a generalized form.

(i) *Rotation.* A figure such as the square (ABCD) in figure 2.16 may be rotated through any number of degrees. On rotation through 360° it has returned to its original position. The rotation of various figures such as squares and equilateral triangles plays an important part now in the teaching of mathematics to children. For the geographer it implies turning a map round from its familiar orientation with north at the 'top' to some other position with another compass point at the top. There is no reason to do this unless the advantage obtained by showing some new feature or relationship more clearly, outweighs the disadvantage of losing the familiar north–south orientation. Two examples of rotation are given in figure 2.20*a*. The Soviet Far East is shown with north at the top and again after being rotated through 180°. It is a matter of opinion whether the remoteness of Vladivostok, and its vulnerability in relation to China, are clearer after rotation, but such a rotation is worth trying. The threat to Germany of interwar Czechoslovakia was a subject of Nazi propaganda. A rotated map perhaps brings out better the threatening appearance of Czechoslovakia. Rotations for purposes of comparison will be discussed in chapter 5.

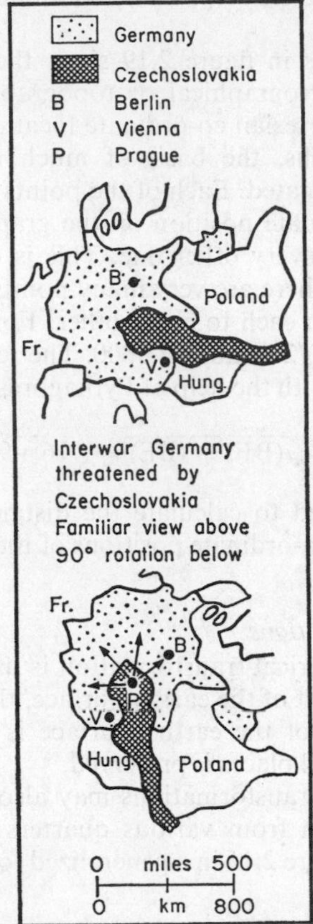

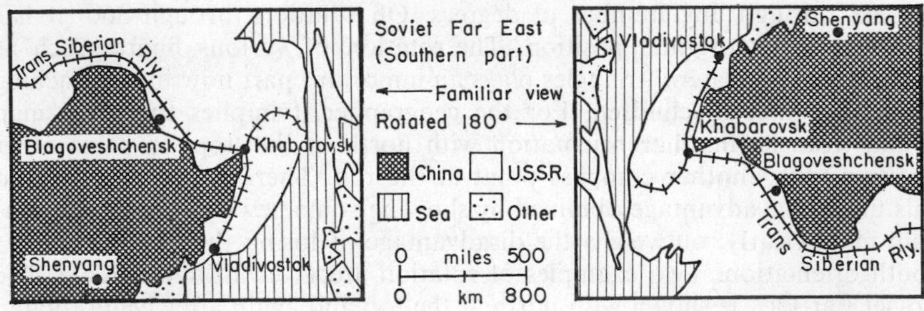

Figure 2.20*a* Two examples of rotation. Czechoslovakia is shown in the familiar view and again after rotation through 90°. The effect of a 180° rotation on the Soviet Far East is also shown.

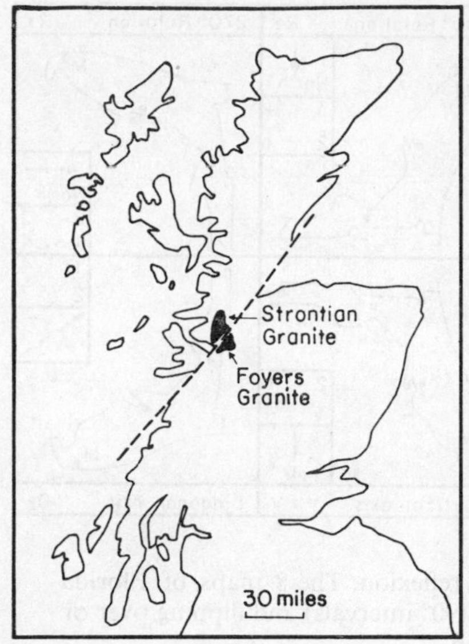

(b) Translation

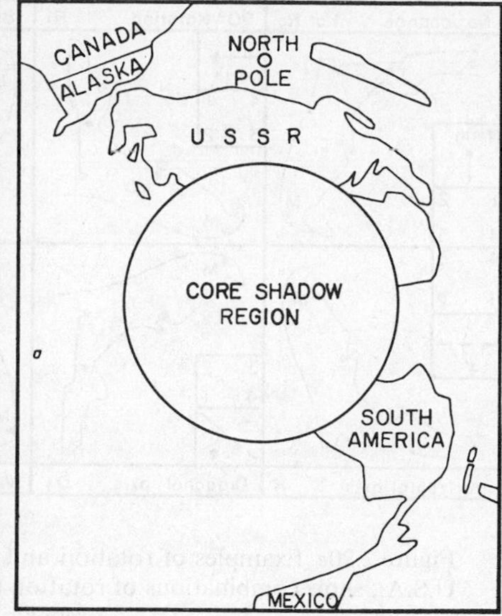

(c) Reflexion

Figure 2.20*b* An example of translation. The movement of N.W. Scotland relative to Central Scotland along the Great Glen Fault. The map shows the original position of the two areas in which the Foycrs and Strontian Granites are in contact with each other. They are now 56 miles apart. *After Kennedy*.

Figure 2.20c An example of reflexion. A mirror version of part of the world as seen from Montana, U.S.A., 'through' the core of the earth.

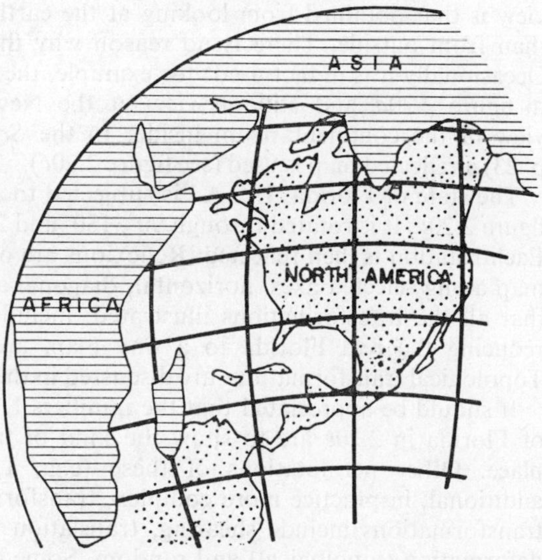

(d) Reflected view of North America

Figure 2.20*d* A reflected view of North America. North America is seen through a hollow globe (New York World Fair).

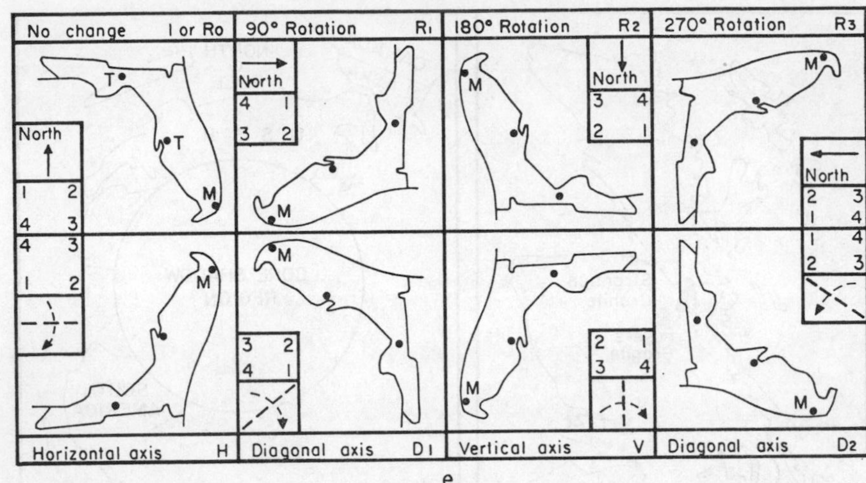

Figure 2.20*e* Examples of rotation and reflexion. The 8 maps of Florida U.S.A., show combinations of rotation (90° intervals) and flipping over or reflexion.

(ii) *Reflexion.* A reflexion of a familiar map may at times be valuable, either for its own sake, or for purposes of comparison (see chapter 5). The view is that obtained from looking at the earth's surface from inside rather than from outside. There is no reason why this view should not be taken. Occasionally it is in fact used: for example, the large hollow globe illustrated in figure 2.20*d*, and still on view at the New York World Fair, gives a view of the continents from inside. In the *Scientific American* (July 1966, p. 23) a reflected map is used (see figure 2.20*c*).

The state of Florida, U.S.A., is subjected to a series of transformations in figure 2.20*e*. It is rotated through 90°, 180° and 270° from its original position. Each rotation is then reflected. Reflexions are obtained by 'flipping' over the map along selected axes, horizontal, diagonal and vertical in this case. Note that all the transformations illustrated, including the original operation of reducing the real Florida to a tiny map, are Euclidean transformations. Topological transformations are discussed in the next section 2.10.

It should be appreciated that the numbers 1, 2, 3, 4 on each of the 8 maps of Florida in 2.20*e* are to show the kind of transformation that has taken place. Other permutations of these (e.g., 1,3,2,4, 1,4,2,3) indicate that additional, in practice more complex, transformations are possible. Further transformations include shearing, translation (see figure 2.20*b*) continuous deformation (topological) and random. Some of these are well described by Bunge (1966).

New methods of visual representation deserve more attention in geography. M. Scheerer (1963) stresses the point that a novel view of a familiar situation or problem may give an answer quickly. He quotes a well known test example.

A picture of two riderless horses is shown and on a separate piece of paper, the two riders are given. The riders do not fit the horses if placed in the apparently obvious way, but they do if rotated through 90°. Another problem that requires a drastic shift in thinking is to make four equilateral triangles out of six equal length matches. No amount of manipulating these on a (two-dimensional) surface will give the answer. It is only necessary to build them into the edges of a tetrahedron in three dimensions. Scheerer remarks: 'If insight is the essential element in intelligent problem solving, fixation is its arch-enemy. Fixation is overcome and insight attained by a sudden shift in the way the problem or the objects involved in it are viewed'.

(*d*) *Surveying*

A useful application of plane Euclidean geometry in geography is found in the methods adopted to survey a small area. If the area to be surveyed is small enough the curvature of the earth may be ignored and the principles of plane geometry applied to surveying. For many purposes geographers can make use of the official surveys but in some instances, particularly in geomorphology, it may be necessary to make special surveys. These surveys may have a variety of aims and the examples selected illustrate the 1-, 2- and 3-dimensional nature of these situations.

In the 1-dimensional example the height of a raised beach may be required at a special point. This measurement may be made with a surveying aneroid which only gives a difference of height. A more accurate value could be

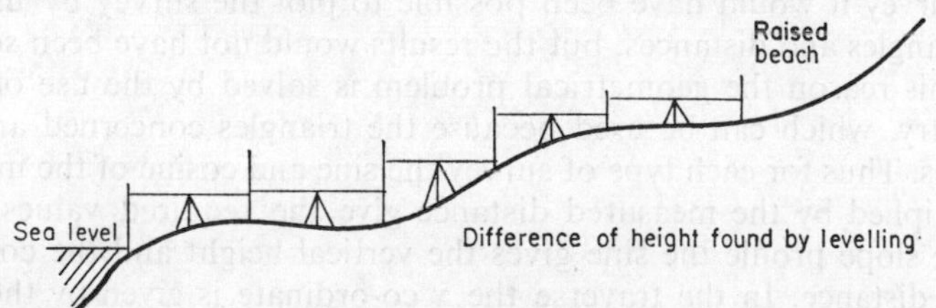

Figure 2.21 A diagram to show how differences of height are found by levelling.

obtained by levelling. This method (figure 2.21) also measures the difference in height as recorded on a staff and only this vertical difference of height enters into the calculations as the level line of sight is set horizontal before each measurement. In this survey only the vertical dimension is of any relevance.

The 2-dimensional problem can be of two types. A profile of a beach slope may be required, and this involves both horizontal and vertical measurement. Alternatively, the outline of a lake may be required, and this involves two horizontal dimensions. The same geometrical principles may be used for each type of survey although different instruments are required for the two surveys. In each type the survey could be based on a system of distance and angular measurements. In the first case of the profile the angles would be vertical ones measured with an Abney level, and the distance would be the slope distance between survey points. In the second case the angles would be horizontal ones and would be measured by a prismatic compass or theodolite and the distances would be measured between the angular measurement points by pacing or by tellurometer (see glossary) respectively.

In solving the problem of plotting the results in each case precisely the same method could be adopted, based on identical geometrical principles. In the slope profile it is necessary to work out the vertical change of height and horizontal equivalent of the slope distance that was measured, so that the results can be calculated cumulatively and plotted accurately. This calculation adopts the same procedure as the use of traverse tables to find the co-ordinates of the points in the traverse. The survey (figure 2.22) can

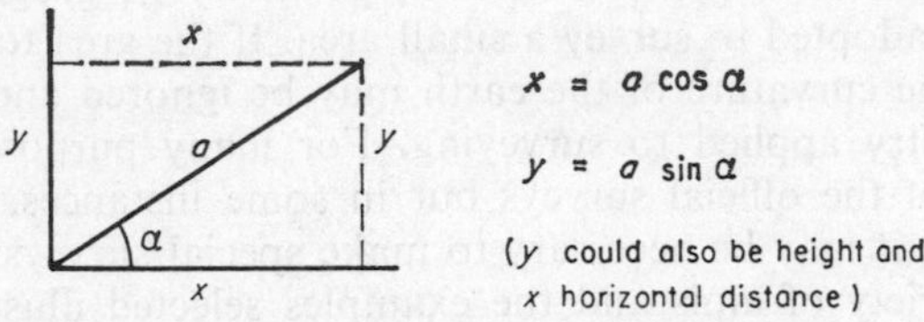

Figure 2.22 Diagram to show how the distances x and y can be found by simple trigonometry from the angle α and the distance a.

most accurately be plotted if the x and y co-ordinates are calculated. In both types of survey it would have been possible to plot the survey by using the measured angles and distances, but the results would not have been so accurate. For this reason the geometrical problem is solved by the use of simple trigonometry, which can be used because the triangles concerned are right-angled ones. Thus for each type of survey the sine and cosine of the measured angle multiplied by the measured distance give the required values. In the case of the slope profile the sine gives the vertical height and the cosine the horizontal distance. In the traverse the x co-ordinate is given by the cosine and the y co-ordinate by the sine if the angle is measured from the x-axis.

The most complex situation is the 3-dimensional one when both direction and height must be recorded on a map that covers a whole area, such a group of moraines, kettle holes or a spit, in which height as well as relative position is important. Plane-tabling is a survey method that can well be adapted for this type of problem and this depends on geometrical principles. It is a system whereby triangles are built up graphically and, if the map is an independent one, the scale is based on the measurement of one base line. The geometrical principle used is that if the side of a triangle is known as well as its angles then the other side must be fixed and can be calculated. The whole basis of plane-tabling is to set up the table in the correct orientation so that all the lines on the map are parallel to the equivalent lines on the ground. This can readily be done from known fixed points when others are also visible, but in the problem of resection the point from which the resection is carried out is not known so that the table cannot at once be correctly orientated.

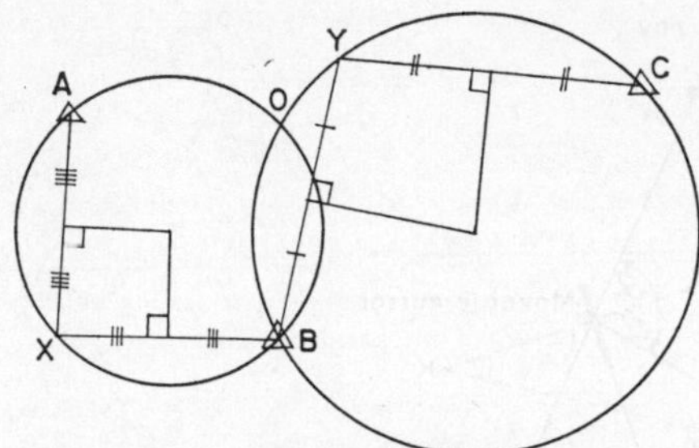
A
Y
O
C
X
B
Back rays from A & B meet at X
Back rays from B & C meet at Y
Circle AXB cuts circle BYC at O,
which is required position

Figure 2.23 Diagram to illustrate resection by geometrical methods. The centres of the two circles are found by bisecting the chords and drawing perpendiculars to them to meet at the circle centres.

Resection has a purely geometrical solution (figure 2.23), based on finding the point where two circles intersect by drawing back rays from two pairs of points with one in common. In order to fix the point it is necessary to obtain the correct angles on the paper as those that exist between the points on the ground. This means that another solution can be adopted that is easier to apply. Any point on a piece of tracing paper is marked and rays are drawn to the three known points. These angles are then correct and the tracing paper is moved about the map until the lines go through the three fixed points and the required position can then be pricked through to the plane-table map. This solution to the resection problem is also in fact used when navigating along a coast. A station pointer is used in navigation to set out the required angles and is then moved over the chart to establish the required position. The necessary angles can be found by using a sextant held horizontally.

The principle on which the sextant operates is a very good example of the application of geometrical principles to the development of survey instruments. A brief description of the sextant will, therefore, provide a useful application of geometry to surveying (figure 2.24). The sextant is based on the optical principle that a ray of light is reflected from two mirrors in succession, so that the angle between first and last ray is twice the angle between the mirrors. The geometrical reason for this relationship is shown in figure 2.24. It is based on the theorem that the exterior angle of a triangle is equal to the sum of the internal and opposite angles. A pointer is attached to the moveable mirror which records the angle between the first and last ray. The scale of the sextant must, therefore, be marked as if the angles were twice their true size. The scale covers $\frac{1}{6}$ of the circumference of the circle, hence the name sextant, although angles up to 120 degrees can be measured. The sextant can be used to measure both vertical elevations or horizontal angles. In the first case the sun or other body is brought down to appear to rest on the horizon, in the second two markers are brought into apparent contact, when the sextant is held horizontally.

Another simpler example of the application of geometrical principles to

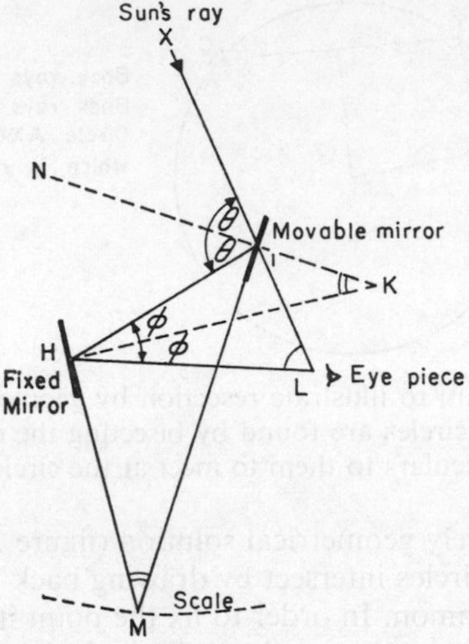
Sun's ray
X
N
Movable mirror
I
K
H
Fixed
Mirror
L
Eye piece
Scale
M
Ray path X I H L
∠HLX is angle between first ray and last ray
∠HMI is angle between mirrors = ∠HKI
∠HKI = θ−φ in Δ HKI
∠HLI = 2θ−2φ in Δ HLI
∴ angle between first and last ray is twice angle between mirrors

Figure 2.24 The principle of the sextant is illustrated diagrammatically. The angles on the scale are drawn half their true value because the angle at *M* is half the angle at *L*.

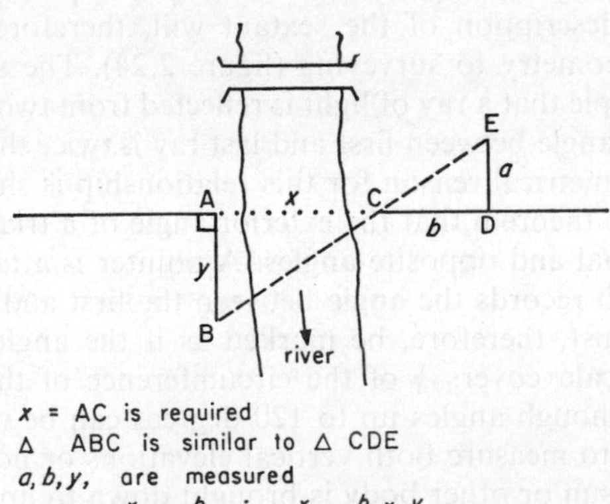
E
a
A
x
C
b
D
y
B
river
x = AC is required
Δ ABC is similar to Δ CDE
a, b, y, are measured
x/y = b/a ∴ x = b/a × y

Figure 2.25 A diagram to show how indirect measurements can be made to obtain the width of a river.

surveying is the method adopted to circumvent obstacles on a traverse. Figure 2.25 illustrates how the principle of similar triangles can be used to measure across a stream that lies in the line of a traverse. This necessitates the setting off of a line at right angles to the line of traverse and getting three points in line to make two similar triangles. The required distance x is then equal to $b/a \,.\, y$ or by/a. In setting out the right-angle in the field the principle of Pythagoras' theorem can be used (figure 2.26). The tape can be arranged in the proportions 3, 4 and 5, the latter being the hypotenuse of the right angled triangle.

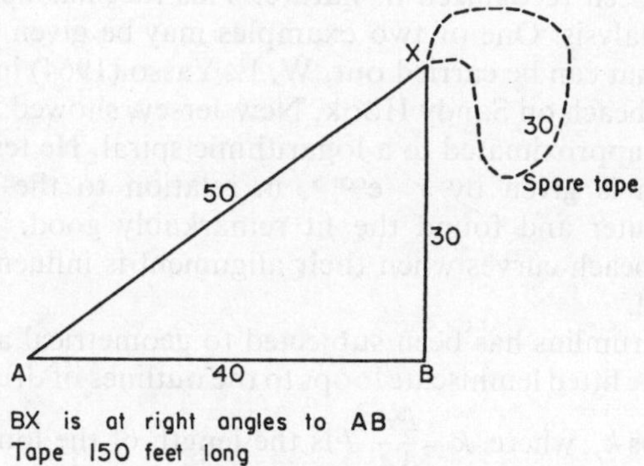

Figure 2.26 Setting out a right angle with a surveying chain by the use of Pythagoras' theorem.

Surveying involves the use of angular measurements so that it is worth considering the units of angular measurement. The familiar degrees minutes and seconds are arbitrary units derived by dividing one revolution into 360 degrees. In some instruments one revolution or a complete circle is divided into 400 grades, so that there are 100 grades in a right angle. The units of circular measure are not arbitrary in this way. They are known as radians and a radian is defined as the angle subtended by an arc equal in length to the radius of the circle. One radian is equal to 57° 17′ 45″. The total length of the circumference of a circle is $2\pi R$, so for the angle equal to 1 radian, the value is $360/2\pi$. The number of radians in a right angle is $\pi/2$. The length of any arc can be found by multiplying the radius by the angle in radians. Thus radians are connected with the important yet elusive geometrical constant π, which is a never ending figure in decimal notation, 3.14 . . . From Gregory's series π can be shown to be a convergent series by integral calculus in the formula $\frac{\pi}{4}=1-\frac{1}{3}+\frac{1}{5}-\frac{1}{7}+\frac{1}{9}-\frac{1}{11}+\ldots$. The value of π has been calculated to 10^5 places of decimals by computer and each digit shows an approximately equal number of occurrences, but no pattern in their arrangement emerges. Thus one of the constants in most common use is one of the most difficult to give an exact value to, except in terms of radians.

Simple plane surveying thus uses many of the principles of plane Euclidean geometry, both in the survey methods and in the principles on which some of the instruments operate. Geometry is also necessary to the calculations of the results although geometry may be taken a stage further to include the trigonometrical ratios that are a development of plane Euclidean geometry (see chapter 8 under Spherical surveying).

(*e*) *Geometrical forms in nature*

It is also of interest to note the many examples of Euclidean geometrical forms that have been recognized in nature. This fact has been used in geomorphological analysis. One or two examples may be given to illustrate the type of analysis that can be carried out. W. E. Yasso (1964) in a very detailed survey of a small beach on Sandy Hook, New Jersey, showed that the form of the beach closely approximated to a logarithmic spiral. He tested the form of this curve, which is given by $r = e^{\cot \alpha}$, in relation to the surveyed beach outline by computer and found the fit remarkably good. The same form also fitted other beach curves when their alignment is influenced by a neighbouring headland.

The shape of drumlins has been subjected to geometrical analysis by R. J. Chorley (1959). He fitted lemniscate loops to the outlines of drumlins, using the equation $\rho = l \cos k$, where $k = \frac{l^2 \pi}{4A}$, l is the length of the long axis and A is the area of the loop. This provides a more accurate shape than the symmetrical ellipse and suggests a connexion between the shape of the drumlin and the force that formed it. H. Svensson (1959) has related the form of a glaciated valley to a parabola and found that the formula $y = 0.000402\ x^{2 \cdot 046}$ fitted the valley that he had surveyed very closely. The exponent is close to 2 which indicates that the best fit curve is very close in form to a parabola (see glossary). This form of curve can be shown to fit very closely to many beach profiles as well.

Other geometrical patterns that probably have physical significance are some of the types of frost crack patterns. There are two fundamental ones, the tetragonal or right-angled pattern and the polygonal pattern, where the angles often approach 120°. A. M. Lachenbruch (1962) has suggested that the first type forms in inhomogeneous material and the non-orthogonal type in more homogeneous material subjected to uniform cooling, the size depending on the magnitude of the thermal stress.

Logarithmic curves have been extensively fitted to the longitudinal profiles of rivers in attempts to extrapolate them downstream and establish former base levels. Circles have been shown to fit the back wall of corries and this similarity with slump slip planes first led W. V. Lewis to put forward the idea of rotation in the formation of corries (Lewis, 1960). Straight lines form an important and diagnostic element of slope profiles. Other geometrical shapes that have significance are the sinusoidal and trochoidal form that sea waves assume according to the two main theoretical analyses of these features. The same forms may also be seen in the ripples that develop under some types of flow, those formed by oscillatory wave action being symmetrical while those formed by unidirectional flow are usually asymmetrical.

Harmonic analysis is important in tidal prediction and analysis. This is based on the separation of a large number of sine curves, which describe the nature of the tidal forces, and provides a link between geometry and trigonometry.

2.10 Topology

(*a*) *Introduction*

Very broadly, topology differs from Euclidean geometry in that it is less, or even not at all, concerned with such features as distance, orientation and straightness. It is vitally concerned however with order and contiguity.

Most subway or underground railway systems (e.g. New York, London) and many bus systems in the world display their routes and stopping places to the public with topological maps. Sketch maps, so familiar in geography, are also topological, while in a sense any map that includes more than a few million square miles of the earth's surface is bound to have either scale (distance) or shape (orientation) or both drastically distorted, by Euclidean standards. The use of topological maps has been discussed both by geographers and by non-geographers. Even a brief introduction to topology helps to show that Euclidean geometry has some serious limitations. Basically Euclidean geometry and topology may be thought of as rigid and elastic (or 'stretchable') respectively. Many things in the real world are elastic, not rigid. Topology not only includes an elastic (rubber-sheet) approach, but also has a vocabulary, lacking in Euclidean geometry, for describing relationships.

On the other hand, it must be appreciated there are situations in which correct scale and orientation are vital and topology is quite unsuitable: for example in property maps, in navigation charts. There are other times, however, when a topological layout communicates the message better than a scale layout. For example, the New York Subway map or the London Underground map is much clearer after the elimination of interfering symbols or 'noise' (see glossary) created by bends in lines, by the close spacing of stations in some places, and so on.

Topology is considered to have originated in the 18th century. The Swiss mathematician L. Euler (1707–1783) was able to make a generalization from a problem already under discussion about crossing seven bridges (see figure 2.27*a*) in the town of Königsberg in East Prussia (now Kaliningrad, U.S.S.R.). Was it possible to cross each bridge once before crossing any bridge twice? Topology was developed in the 19th century. It has subsequently been widely applied in many fields, including for example engineering and biology, but it also continues to develop in pure mathematics. Like other geometries, topology has its own requirements and invariants. Many introductory books are now available on the subject. It is interesting to note that topology is introduced in the School Mathematics Project in Book 2 for children aged 12–13*. It has been found, moreover, that young children can form concepts

* See *School Mathematics Project, Book 2,* Cambridge University Press, (1966, pp. 1–20).

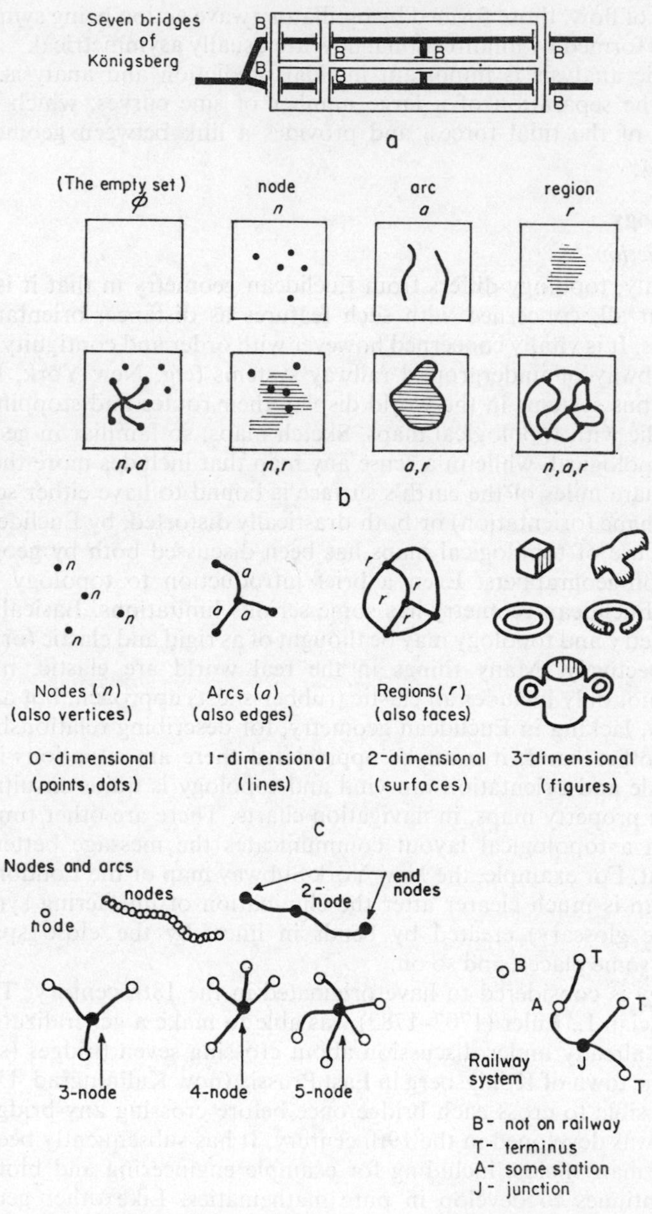

Figure 2.27 Some basic ingredients of topology. (*a*) The Königsberg bridge problem. (*b*) Nodes, arcs and regions. (*c*) Dimensions in topology. (*d*) The development of nodes.

of a topological nature before they appreciate Euclidean or projective geometry*.

Topology is not confined to two dimensions but it will be considered mainly in two dimensions in this book.

(b) Terms and transformations

Figure 2.27*b* shows some features of topology. Three ingredients may occur in 2-dimensional topology. *Nodes*, *arcs* and *regions* are roughly equivalent to points, lines and surfaces in Euclidean geometry. They are zero, 1- and 2-dimensional figures in 2-dimensional space. They may be considered individually, in pairs, or all three together. They will be referred to in lower case letters, or in lower case initials, as *n*, *a*, *r*. Further basic aspects of topology are illustrated in figure 2.27*c* and 2.27*d*.

While nodes, arcs and regions form the basis of topology on a 2-dimensional surface, 'solid' topology, in three space dimensions, has its own terms. In figure 2.27*c* the 3-dimensional figures may be imagined as made of plastic (plasticine) material. They can be shaped. The cube and the figure to the right of it are equivalent in topology because they have no holes through them. The beach ring (American-style doughnut) is of a higher *genus* than the cube because it has a hole through it. The test is that it can be sliced once without being cut into two parts. The mug is a higher genus still because it has two holes.

The idea of the node is extended in figure 2.27*d*. In topology the term *node* appears to be used in a general sense to refer to any 'point' that is under consideration. In a more restricted sense, a node is the intersection of three or more arcs. For the immediate purposes of the geographer, nodes may usefully be classified as follows:

(0) A node *alone* (no arcs).
(1) A node at the *end* of an arc (end or terminator node).
(2) A node *on* an arc (2-node).
(3) A node with *three* arcs leaving it (3-node).
(4), (5), etc., as 3, but with 4, 5, etc. arcs leaving it.

The above nodes are illustrated on a simple railway system in figure 2.27*d*.

Mathematicians are not consistent in their use of the terms described. Vertex (*v*), edge (*e*) and face (*f*) may be used instead of node, arc and region. Sometimes special connotations are given to the terms. For example, O. Ore, (1963) uses *edge* for a 'line' joining two nodes, but *arc* for a series of edges running through several nodes. The reader should make sure of the terminology adopted by the particular mathematician whose material he is reading. Unfortunately not all mathematicians state clearly how they are using the terms, and some do not appear to be consistent even within a single work.

* J. Piaget, 'How Children Form Mathematical Concepts', *Sci Am*, (Nov. 1953). 'A child begins with the last [topology]: his first geometrical discoveries are topological. At the age of three he readily distinguishes between open and closed figures: if you ask him to copy a square or a triangle, he draws a closed circle.' (p. 3 of Reprint.)

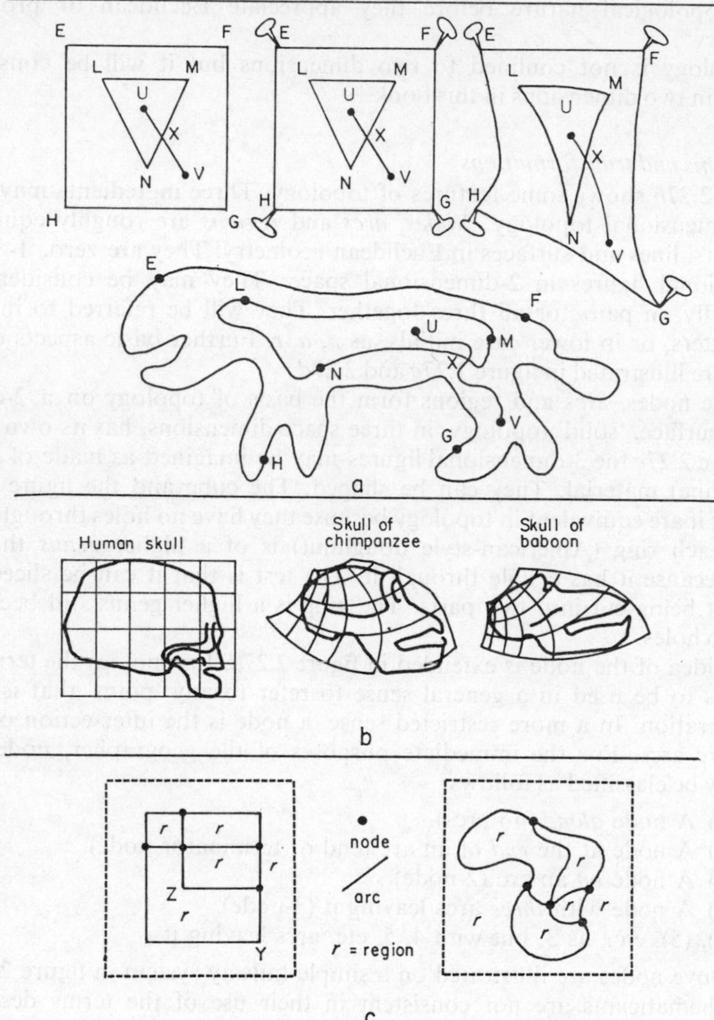

Figure 2.28 Transformation of regions. (*a*) Elastic sheet transformation. (*b*) Grid transformation. (*c*) Network transformation.

Figure 2.28*a* illustrates some of the properties of 2-dimensional topology. EFGH is a square sheet of elastic material. Its surface consists of an infinite set of points. (The geographer may do better to imagine a *very large number* of small dots instead of an infinite set of points). Within EFGH is the smaller *region* enclosed by the *arc* LMN. The elastic sheet is nailed at its four corners to a board. The lower right hand nail is then removed, and the corner (G) tugged some distance to the lower right. The original corner point G is then nailed to the board again. The operation performed on the elastic sheet is called a transformation. It is quite feasible that every point on the sheet except E, F and H, which remain nailed, has moved some distance from its original location during the transformation. But on the whole, the points originally nearest to G moved further from their original positions than those nearest to E. With an even more flexible material than elastic, it would be possible to transform the square EFGH into the animal below it.

Whereas in Euclidean geometry the tugged square and the animal are no longer equivalent to the original square or to one another, in topology they are. Why they are, may be illustrated by reference to LMN, within EFGH. The region LMN contains within it a node U, while external to it is a node V. The arc UV crosses MN at X. Throughout the transformations, U remains within the region LMN, V outside it. Another feature that remains without varying is the order of the points LMXN on the arc. Topology is not concerned with the distance or direction between points, but it *is* concerned with preserving *order* during a transformation. Correct scale and shape disappear on the Underground map of London, but the order of the stations is preserved correctly, and interchange stations are also correct, which means that the individual lines cross correctly.

It seems largely but perhaps not entirely true to say that Euclidean geometry has more requirements than topology, or that topology has only some of the requirements of Euclidean geometry. Hence its greater flexibility. On the other hand, topology does not appear to have any requirements of its own that would not be needed also in Euclidean geometry.

Figure 2.28*b* shows the use of a topological transformation of a Cartesian co-ordinate grid to compare the skulls of a human, a chimpanzee and a baboon. The grid brings out clearly the relative contraction of the jaw and expansion of the skull in the human compared with the baboon. The diagrams are based on diagrams in *On Growth and Form* by D'Arcy Thompson. The book was first published in 1917. It contains many other transformations and also other applications of topology. Such transformations might be of use in comparing different places in geography. For example, in figure 2.28*c* the two networks are equivalent topologically. A system of nodes, arcs and regions is transformed, but the two systems remain equivalent topologically.

(*c*) *A basic relationship in topology*

One of the most striking relationships in 2-dimensional topology is summarized as follows: $a+2=n+r$. There are always two more nodes and regions combined than arcs. (Note that an outside region must always be counted). This relationship is illustrated by examples in figure 2.29*a*. It should be noted, however, that the nodes to be counted are always 3-nodes or higher than 3-nodes (4-nodes, etc.). The diagrams *a–c* show how when node *L* is moved to *N*, one node and one region disappear but two arcs disappear. When a new node is added at *P* (see diagrams *A* and *D*), two new regions and three arcs are also created.

If the faces of a cube (figure 2.29*b*) are considered to be regions, the corners

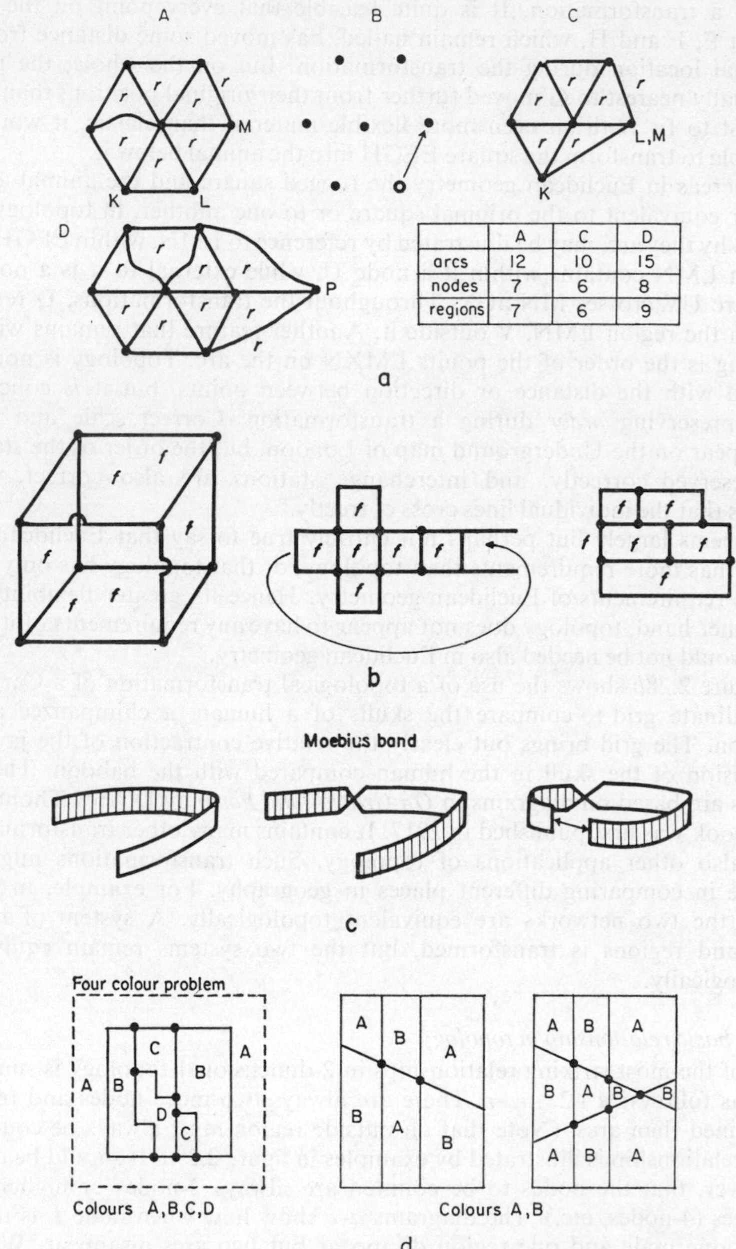

	A	C	D
arcs	12	10	15
nodes	7	6	8
regions	7	6	9

Figure 2.29 Some basic relationships in topology. (*a*) Ratio of arcs to nodes and regions. (*b*) Topological views of a cube. (*c*) Moebius band. (*d*) Four colour problem and a modification.

nodes, and the edges joining the corners, arcs, then $a=12$, $r=6$ and $n=8$, and the formula $a+2=r+n$ is satisfied. This exercise, unfortunately, confuses 3-dimensional solid Euclidean geometry with topology. Note that the upper attempt to undo the cube or box is not satisfactory whereas the lower attempt is.

(d) Contiguity of regions

A Moebius band is illustrated in figure 2.29*c*. A strip of paper is dark on one side, light on the other. It is possible to join the ends so that light joins light, dark dark. The band then has two faces (sides). If, however, the band is given one twist (see figure) so that a light end is joined to a dark and vice versa, then the band has only one face. It has only one face because it is possible to travel from any point to any other on the face without crossing an edge (the joined ends do not count as an edge). The band may also be twisted two, three and more times.

The Moebius band introduces the interest of topology in the contiguity of regions. The so-called four colour problem in topology illustrates one aspect of the subject of interest to the geographer,* firstly from the point of view of cartography, and secondly, with regard to the contiguity of the regions in a network.

It has been found from experience, but has not been proved, that only four colours are needed to colour a map with regions such as counties, for no two contiguous regions to be of the same colour. An example is given in figure 2.29*d*. There are four colours, *A*, *B*, *C* and *D*. In a special case when arcs cross making only 4-nodes, then only two colours are needed (*A* and *B*), though a third colour would be necessary for the outside region.

2.11 Combinations and permutations

An appreciation of the meaning of combinations and permutations is essential for the geographer who uses quantitative methods. Ordering, classifications and relationships are closely connected and probability and statistics cannot be appreciated without them.

Combinations are subsets of objects from a given set. Order is not relevant in combinations. Permutations are arrangements of objects in which order is relevant. For example, the letters A, B and C may be *combined* in eight ways: ABC; AB; AC; BC; A; B; C; ϕ (the empty set). In everyday experience, it is possible to meet eight combinations of a family of three people: all three, 3 pairs, 3 singly or none.

Permutations of A, B and C are as follows: ABC; ACB; BAC; BCA; CAB; CBA. Thus there are 6 possible permutations of three things. In everyday experience, the planning of a visit to three shops involves a choice between six alternative routes.

* According to O. Ore, (1963 p. 109): 'The British mathematician Cayley, himself one of the pioneers of graph theory, in 1879 published an article on the four color problem, appropriately enough in the first volume of the Proceedings of the Royal Geographical Society.'

A B C may be *first* combined and *then* permuted. For example, they can be taken two at a time and then each pair may be permuted twice: AB; AC; BC; BA; CA; CB.

The total number of combinations of n things is 2^n. Thus, for example, 5 things can be arranged in 2^5 or 32 ways, 5 at a time + 4 at a time + 3 at a time + 2 at a time + 1 at a time + 0 at a time.

It is often useful to know how many combinations of n things are possible r at a time. This is calculated by the formula.

$$^nC_r = \frac{n!}{r!\,(n-r)!}$$

The sign like an exclamation mark is called factorial (e.g. factorial 5 is 5! which means $5\times4\times3\times2\times1$). For example, the number of combinations of 5 things (A B C D E) 2 at a time is calculated as follows.

$$\frac{n!}{r!\,(n-r)!} = \frac{5!}{2!\,(5-2)!} = \frac{5\times4\times3\times2\times1}{2\times1\,(3\times2\times1)} = \frac{120}{12} = 10$$

These can clearly be seen in a 5×5 table.

	A	B	C	D	E
A		AB	AC	AD	AE
B			BC	BD	BE
C				CD	CE
D					DE
E					

The total number of permutations of n things is n (!) or n factorial. Thus 5 things can be arranged in $5\times4\times3\times2\times1 = 120$ ways. Tables of values for permutations are available in some books on probability and also in books with mathematical tables.

The number of things taken r at a time and permuted (order being relevant) is

$$^nP_r = \frac{n!}{(n-r)!}$$

For example, 5 things taken 2 at a time and permuted is

$$\frac{5!}{(5-2)!} = \frac{5!}{3!} = \frac{120}{6} = 20$$

The other half of the table above can be filled to give a re-ordering of each of the combinations.

An appreciation of combinations and permutations in geography is useful in many circumstances. Some brief examples are given. Suppose you have decided that there are 10 points that you want to make about some question under discussion. Since these cannot be dealt with simultaneously they must be put in order. But there are more than 3 million ways (10!) that the points could be arranged. There are 50! or about 3×10^{64} different orders in which the 50 state capitals of the U.S.A. could be visited in turn.

Combinations accumulate less rapidly than permutations, but soon become impressively large. For example, a party of 10 people can be grouped into 2^{10} subsets, 1024 altogether. The 90 departments of France can be grouped into approximately 10^{26} subsets. Grouped just thirty at a time there are about 10^{24} subsets. Remember that 10^{26} is 10^{2} (or one hundred) times bigger than 10^{24}. Allowing groupings only of contiguous departments thirty at a time there are about 10^{22} ways. In a city with 1 million people, there are $2^{1,000,000}$ or about $10^{320,000}$ possible subsets. Since it is generally agreed that the quantity of atomic particles in the universe does not exceed about 10^{100} the potential number of groups of people that a sociologist could encounter in a city with 1 million people, or merely with 1000 (10^{320}) must be called ultra-astronomical.

References

Ackerman, E. A. (1958). *Geography as a Fundamental Research Discipline*, University of Chicago, Dept. of Geography, Research Paper No. 53.

Adler, I. (1962). *Thinking Machines*. Signet Books, New York, Chapt. 8.

Alker, H. R. (1965). *Mathematics and Politics*, Macmillan, New York.

Bailey, C. A. R. (1964). Sets and Logic 1, Sets and Logic 2 in *Contemporary School Mathematics*, Arnold, London.

Barr, S. (1965). *Experiments in Topology*, John Murray, London.

Berge, C. (1962). *Theory of Graphs and its Applications* (translation Alison Doig), Methuen, London.

Biggs, E. E. (1965). *Mathematics in Primary Schools*, Curriculum Bulletin No. 1, The Schools Council, H.M.S.O., London.

Blumenthal, L. M. (1961). *A Modern View of Geometry*, Freeman, San Francisco and London.

Bochner, S. (1966). *The Role of Mathematics in the Rise of Science*, Princeton University Press, N.J.

Boehm, G. A. W. (1959). *The New World of Mathematics*, Faber, London, p. 124.

Bowran, A. P. (1965). *A Boolean Algebra*, Macmillan, London.

Breuer, J. (1958). *Introduction to the Theory of Sets*, Prentice Hall, Englewood Cliffs,

Bullard, Sir Edward (1966). The detection of underground explosions. *Sci. Am.*, July 1966, **215,** No. 1, pp. 19–29.

Carter, H. C. (1965). *Modern Basic Mathematics*, Methuen, London, p. 1.

Chinn, W. G. and Steenrod, N. E. (1966). *First Concepts of Topology*, Random House (New Mathematical Library), New York.

Chorley, R. J. (1959). The shape of drumlins, *J. Glaciol.* 3, 339–344.

Collins, D. A. and Booth, H. E. (1965). *Binary Arithmetic*, Blackwell, Oxford.

Coppock, J. T. (1964). *An Agricultural Atlas of England and Wales*, Faber and Faber, London.

Courant, R. (1964). Mathematics in the modern world, *Sci. Am.*, **211**, No. 3 pp. 40–49.

Dubisch, R. (1964). *Lattices to Logic*, Blaisdell, New York.

Edwards, E. (1964). *Information Transmission*, Chapman and Hall, London, Ch. 2, Some basic mathematics.

Eisenhart, C. (1952). The reliability of measured values — Part I Fundamental Concepts, *Photogramm. Eng.*, **18,** 542–554.

Evans, D. C. (1966). Computer logic and memory, *Sci. Am.*, **215,** No. 3, p. 75.

George, F. H. (1965). *Basic Principles of Digital Computing*, Teaching Programmes Ltd., Bristol.

Golledge, R. G. and Amadeo, D. M. (1966). Some introductory notes on regional division and set theory, *Profess. Geog.*, **XVIII,** No. 1, 14–19.

Gray, J. F. (1962). *Sets, Relations and Functions*, Holt, Rinehart and Winston, New York.

Hamill, L. (1966). A note on tree diagrams, set theory and symbolic logic, *Profess. Geog.*, **XVIII,** No. 4.

Hoernes, G. E. and Heilweil, M. F. (1964). *An Introduction to Boolean Algebra and Logic Design* (Programmed), McGraw-Hill, New York.

Information, A Scientific American Book, 1966. Freeman, San Francisco and London. See also *Sci. Am.*, September 1966.

Kao, R. C. (1963). The use of computers in the processing and analysis of geographic data, *Geog. Rev.*, **LIII,** 1963, No. 4, 530–547.

Kline, M. (1964). Geometry, *Sci. Am.*, **211,** No. 3, p. 63.

Lachenbruch, A. M. (1962). Mechanics of thermal contraction cracks and ice-wedge polygons in permafrost, *Geol. Soc. Am. Spec. Papers*, **70,** 65.

Lewis, W. V. (1960). Investigations on Norwegian cirque glaciers, *Roy. Geog. Soc. Res. Ser.* **4,** 98–99.

Lietzmann, W. (1965). *Visual Topology*, Chatto and Windus, London.

Life Science Library (1965). *Mathematics*, Time-Life International, Nederland,

Lipschutz, S. (1964). *Set Theory and Related Topics*, Schaum Publishing Company, New York.

Lytel, A. (1964). *ABC's of Boolean Algebra*, Foulsham-Sams, Slough, Bucks.

Macfarlane Smith, I. (1964). *Spatial Ability, Its Educational and Social Significance*, University of London Press Ltd., London.

Marks, R. W. (1964). *The New Mathematics Dictionary and Handbook*, Bantam Books, New York, p. 93.

Merriam, D. F. (1965). Geology and the computer, *New Scientist*, **26,** 444 pp. 513–516, 20 May 1965.

Moser, C. A. and Scott, W. (1961). *British Towns. A Statistical study of their social and economic differences*, Oliver and Boyd, Edinburgh and London.

Murphy, B. (1966). *The Computer Society*, The Great Society Series, Blond, London.

Neumann, J. von (1958). *The Computer and the Brain*, Yale University Press, New Haven.

Niven, I. (1965). *Mathematics of Choice*, Random House, New York.

Nystuen, J. D. (1966). Effects of boundary shape and the concept of local convexity, Discussion Paper No. 10, referring to J. Perkal, On the length of empirical curves and an attempt at objective generalisation, *Michigan Inter-University Community of Mathematical Geographers.*

Ore, O. (1963). *Graphs and Their Uses*, Random House (New Mathematical Library), New York.

Piaget, J. (1953). How children form mathematical concepts, *Sci. Am.*, **189,** November 1953.

Piaget, J. (1960). *The Child's Conception of Geometry,* Routledge and Kegan Paul, London.

Rosenbaum, E. P. (1958). The teaching of elementary mathematics, *Sci. Am.*, **198,** May 1958.

Rosenthal, E. B. (1966). *Understanding the New Maths,* Souvenir Press, London, p. 15.

Sawyer, W. W. (1964). Algebra, *Sci. Am.*, September 1964, **211,** No. 3, 70–78.

Scheerer, M. (1963). Problem-Solving, *Sci. Am.*, April 1963, **208,** No. 4, 118–128.

Schwartz, J. T. (1961). *Introduction to Matrices and Vectors,* McGraw-Hill, New York.

Shear, J. A. (1966). A set-theoretic view of the köppen dry climates, *Ann. Assoc. Am. Geog.*, **56,** No. 3.

Stephenson, G. (1965). *An Introduction to Matrices, Set and Groups,* Longmans, London.

Stoll, R. R. (1961). *Sets, Logic and Axiomatic Theories,* Freeman, San Francisco and London.

Svensson, H. (1959). Is the cross-section of a glacial valley a parabola?, *J. Glaciol.*, **3,** 362-3.

Teitz, M. B. (1962). Regional theory and regional models (section set-theoretic foundations for a regional concept), *Reg. Sci. Assoc. Papers,* **9,** 35–50.

Thomas, C. A. (1963). *Programmed Learning in Perspective,* Lamson Technical Products Ltd.,

Tobler, W. R. (1963). Geographic area and map projections, *Geog. Rev.*, **LIII,** No. 1, 59–78.

Yasso, W. E. (1965). Plan geometry of headland-bay beaches, *J. Geol.*, **73** (5), 702–714.

Young, J. Z. (1964). *A Model of the Brain,* Clarendon Press, Oxford.

Bibliography

This is a special list of mathematical texts selected from those consulted by the authors. Many more possible works of reference have not been included. The titles are broadly arranged from easy to difficult, but also by content. Many of the topics covered in chapter 2 of the present book are discussed in *several* of the publications that follow.

Historical

Wolff, P. (1963). *Breakthroughs in Mathematics,* Signet Science Library, New York.

General Reader

Makins, V. (1967). The New Maths Crusade, *The Observer* (Supplement).

Exploring Mathematics on Your Own (Series), Murray, London.

D. A. Johnson and W. H. Glenn (1964) (Reprint 1965), No. 1 *Invitation to Mathematics.*

D. A. Johnson and W. H. Glenn (1964) (Reprint 1967), No. 6 *Sets, Sentences and Operations.*

D. A. Johnson and W. H. Glenn (1960), No. 12 *Topology — The Rubber Sheet Geometry.*

D. A. Johnson and D. F. Taylor (1966), No. 13 *Logic and Reasoning in Mathematics.*

M. Scott Norton (1966), No. 16 *Basic Concepts of Vectors.*

Life Science Library (1965). *Mathematics*, Time-Life International, Nederland.
Boehm, G. A. W. (1959). *The New World of Mathematics*, Faber, London.
Adler, I. (1958, 1964). *The New Mathematics*, Signet Science Library, New York.
Adler, I. (1966). *A New Look at Geometry*, Signet Science Library, New York.
Sawyer, W. W. (1963). *Mathematician's Delight*, Penguin Books, Harmondsworth.
Rosenthal, E. B. (1966). *Understanding the New Maths*, Souvenir Press, London.
Mathematics in the Modern World, *Sci. Am.*, **211**, No. 3, September 1964.
Montague, H.F and Montgomery, M. D. (1961, 1963). *The Significance of Mathematics*, Merrill, Columbus, Ohio.
Reid, C. (1965). *A Long Way from Euclid*, Routledge and Kegan Paul, London.
Bochner, S. (1966). *The Role of Mathematics in the Rise of Science*, Princeton University Press, Princeton, New Jersey.

For Parents

Heimer, R. T. and Newman, M. S. (1965). *The New Mathematics for Parents*, Holt, Rinehart and Winston, New York.
Allendoerfer, C. B. (1965). *Mathematics for Parents*, Macmillan, New York.

For Teachers

Mathematics in Primary Schools, Curriculum Bulletin No. **1**, The Schools Council, H.M.S.O., London.
Association of Teachers of Mathematics (1967). *Notes on Mathematics in Primary Schools*, Cambridge University Press, Cambridge.
Felix, L. (1966). *Modern Mathematics and the Teacher*, Cambridge University Press, Cambridge.
Fletcher, T. J. (1964). *Some Lessons in Mathematics*, A Handbook on the Teaching of 'Modern Mathematics', Cambridge University Press, Cambridge.

Junior Children

Sealey, L. G. W. *et al.*, *Beginning Mathematics, Let's work it Out!*, and others, Blackwell, Oxford.

Secondary School Level

Heritage, R. S., *Learning Mathematics* (The Shropshire Mathematics Experiment), Penguin Books, Harmondsworth, Book 1, 1966, Book 2, 1965, Book 3, 1966.
Contemporary School Mathematics, Arnold, London.
G. Matthews (1964), *Matrices* 1 and 2.
C. A. R. Bailey (1964), *Sets and Logic* 1 and 2.
F. B. Lovis (1964), *Computers and Logic* 1 and 2.
J. A. C. Reynolds (1964), *Shape, size and place.*
A. J. Sherlock (1964), *An Introduction to Probability and Statistics.*
School Mathematics Project, University Press, Cambridge.
Book 1, 1965.
Book 2, 1966.
Book 3, 1964.
Book 4 (Experimental Text), 1964.

More Advanced Texts

Carter, H. C. (1965). *Modern Basic Mathematics*, Methuen, London.
Ward, M. and Hardgrove, C. E. (1964). *Modern Elementary Mathematics*, Addison-Wesley, Reading, Mass.

Good, R. A. (1966). *Introduction to Mathematics*, Harcourt, Brace and World, New York.

Waisman, F. (1959). *Introduction to Mathematical Thinking*, The Science Library, Harper Torch, Harper and Row, New York.

Reference Works

Marks, R. W. (1964). *The New Mathematics Dictionary and Handbook*, Bantam Books, New York.

Sippl, C. J. (1966). *Computer Dictionary and Handbook*, Sams, Indianapolis.

3

Statistics

'The mark of an educated man is the ability to make a reasoned guess on the basis of insufficient information.'

(*Unattributed*)

3.1 Introduction

In this chapter the aims of using statistical methods in dealing with geographical data are considered. Statistics have been divided into the following main types: firstly descriptive statistics, secondly sampling, and thirdly inferential statistics. Each of these types has a different purpose but each is applicable to many different kinds of data. Some methods are suitable only for data on the nominal or ordinal scale of measurement, while others can be used with interval and ratio data.

Statistical techniques can be used for data consisting of one, two or more than two variables and it is on this basis that the subsections of the chapter are divided. The data may be expressed in one, two or three dimensions. Measurements in area are particularly important in geographical data. Some special adaptations of conventional techniques are applicable to this type of data. Other methods are particularly relevant to data expressed on a time scale and these will be considered in chapter 10.

The first section deals with descriptive statistics, the purposes of which are widely known. The second deals with sampling, a more controversial aspect of statistics. In some situations it is possible to measure or record all the individuals that make one statistical population, in which case it is not necessary to use sampling methods. However, a population to be studied is often so large that it is either impossible or impracticable to measure it all. Sampling is then essential. Sometimes there is a reason why the whole population cannot feasibly be tested. A frequently quoted example is the testing of flash bulbs before they leave the factory. Economically it is not possible to test them all and some, therefore, must be selected. In some circumstances the whole population cannot be observed. Caves for example may be inaccessible or undiscovered. The minerals throughout a body of rock cannot all be examined.

The third section of this chapter describes some of the tests that have been developed to infer the relationship between the sample and the whole population from which it has been drawn and to compare it with one or more similar samples of one variable. Another important aspect of statistical inference is to examine relationships between two or more variables by making use of the various correlation tests that have been developed both for two variables and for multivariate situations. When samples are being considered there must be some uncertainty concerning the character of the total population. Statistics of the inferential type, therefore, are based not on certainty, as is true of mathematics, but on probability. Most statistical results are expressed at a specific confidence level, or level of significance, which implies that the result is not certain because the relationship shown could sometimes occur by chance. For this reason it is necessary to consider briefly some elementary aspects of probability theory as this forms the basis of inferential statistics and is also important in sampling.

The development of the quantum mechanics theory in atomic physics has shown that, at least on the microscale, science rests not on the deterministic relationship of cause and effect, but on an essentially statistical basis. There is no ultimate certainty in quantum mechanics; when one aspect, such as position is known, then speed cannot be accurately known, and vice versa. The elementary particles obey the laws of chance and their movement can only be expressed in terms of statistical probability. It seems therefore that the behaviour of real objects on the earth is probabilistic and should be studied by statistical laws. If this principle applies on the microscale there is no apparent reason why it should not apply on the mesoscale of geographical analysis. The macroscale of immense interstellar distances also illustrates the falsity of the deterministic approach to scientific analysis. Einstein's relativity theory has shown that on the scale of the universe and dealing with speeds approaching that of light the position of the observer plays a vital part. It is quite possible from different positions for the time sequence of events to be reversed. Effect can then precede cause, showing that a deterministic law is impossible. On the mesoscale of this earth the law of gravitation holds and the complexities of quantum mechanics are not apparent. Nevertheless the principles of indeterminacy, or uncertainty, and the need to use statistical thinking is implicit in the workings of nature and man. When numbers are very large, statistical probability can become virtual certainty, but the principle of indeterminacy still holds good (see chapter 12).

3.2 Probability and statistical distributions

(*a*) *Probability*

It has been said that 'statistics is a mathematical study of probability'. It is, therefore, necessary to discuss some of the basic principles of probability before considering the application of statistics in dealing with various problems. Probabilities can range from absolute impossibility (a probability of 0) to absolute certainty (a probability of 1). For example it is impossible for a river to flow uphill along its whole course and it is certain that everyone alive now will die. In between, however, lie most of the events that take place from day to day. Some of them are extremely unlikely, others are virtually certain, but in the centre are many events in which the probability is about 0.5.

Probability theory is basic to statistics. It is essential in interpreting the results of inferential statistics, all of which are expressed in terms of probability, not certainty. The number of events under consideration plays an important part in establishing the accuracy of probability. Probability can be defined as follows. If there are s successes in n trials and if the ratio s/n approaches a limit as the number n grows, then this limit is defined as the probability of success in one single trial.

It is important to differentiate between the pure conceptual idea of probability and the actual results of probability in action. Mathematically, if there is an equal chance of two events happening then each has a probability of $\frac{1}{2}$ according to the definition of probability. If however a coin is tossed many times, although the number of heads should theoretically equal the number of tails, in practice in most trials the number of heads will only approximate to the number of tails and they will not therefore be exactly equal.

There are a number of laws that are important in probability theory. If the probability of an event occurring is p and the only other alternative has a probability q, then $p+q=1$. This means that the sum of all the possible occurrences must be equal to unity. If $p_1, p_2, p_3 \ldots$ are the separate probabilities of mutually exclusive events occurring, then the probability that one of these events will occur is given by $p=p_1+p_2+p_3 \ldots$. Events are mutually exclusive, for example, if the probability of getting a 1 or a 2 on one throw of one dice is considered, as only one number can come up on each throw. The probability of getting a 1 or a 2 on the first throw is $\frac{1}{6}+\frac{1}{6}=\frac{1}{3}$.

If $p_1, p_2, p_3 \ldots$ are the separate probabilities of the occurrence of a succession of independent events, the probability p that all these events will occur is given by $p=p_1 \times p_2 \times p_3 \ldots$. For example if a coin is tossed 3 times the probability of getting 3 tails is $\frac{1}{2}\times\frac{1}{2}\times\frac{1}{2}=\frac{1}{8}$, but if 6 tails are obtained together the probability of this occurring is only $\frac{1}{2}\times\frac{1}{2}\times\frac{1}{2}\times\frac{1}{2}\times\frac{1}{2}\times\frac{1}{2}=\frac{1}{64}$ and 9 tails in succession should only occur $\frac{1}{512}$. This probability also applies to any other specific arrangement of 9 tosses, e.g. HTHHTTHHH.

Another rule concerns replacement. The probability of drawing an ace from a pack of cards is $\frac{4}{52}$, but if the card drawn is not an ace and it is *not* replaced in the pack then the probability of drawing an ace on the second draw is $\frac{4}{51}$. If the first card had been an ace then the probability of drawing a second ace would be $\frac{3}{51}$. The larger the number of possibilities the smaller will be the probability that a single event will occur. For example in a lottery with 1000 tickets, a person with one ticket has a chance of winning of $\frac{1}{1000}$. The hypothesis that he will win it might be rejected with a 99.9 per cent confidence level. If on the other hand, another person has 100 tickets then his chance of winning is $\frac{100}{1000}$ or $\frac{1}{10}$.

One of the difficulties of probability theory is to know when to add and when to multiply the probabilities according to the laws given already. This difficulty can be considered in terms of sets and by the use of dice. Figure 3.1*a* illustrates all possible permutations of throws of two dice. The rows indicate the result for the red dice and the columns for the blue dice. The probability of getting a 1 on both dice can be indicated by the intersection of the two sets that indicate a 1 on the dice. This is also shown by the

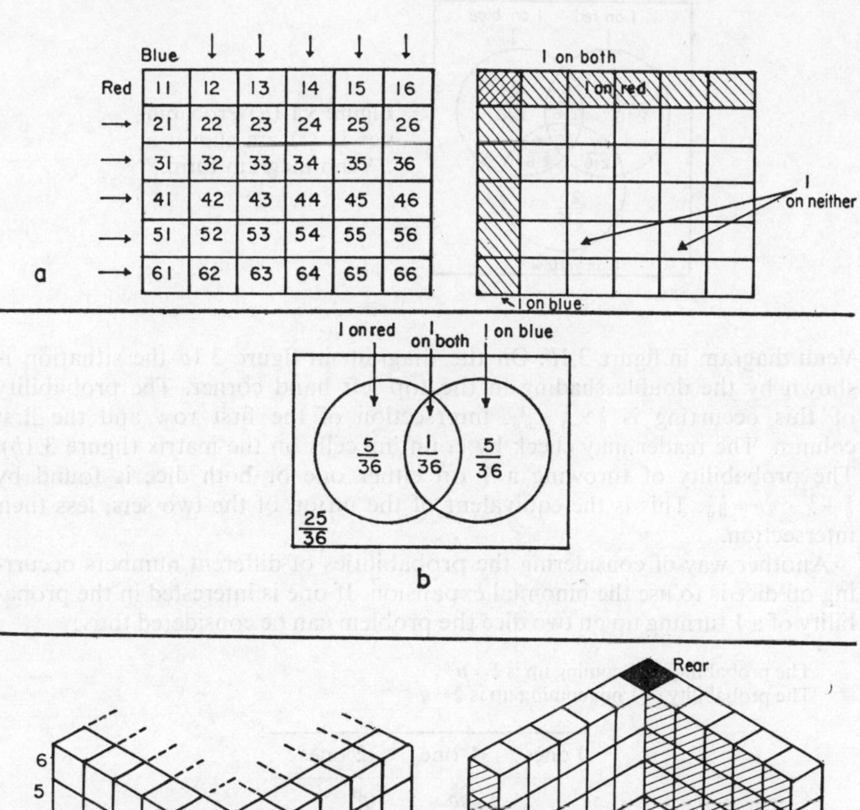

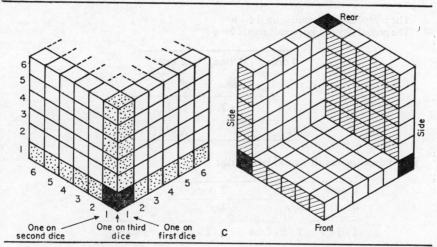

Figure 3.1 (*a*) Matrix to show all possible outcomes of throws with two dice; right hand matrix shows the cases where a 1 is obtained on the first dice, the second dice, and both dices. (*b*) Probabilities in (*a*) are shown in Venn diagram form. (*c*) Block diagram of part of a 3-dimensional matrix with 216 entries, to show cases where a 1 is obtained on one dice, two dice and three dice.

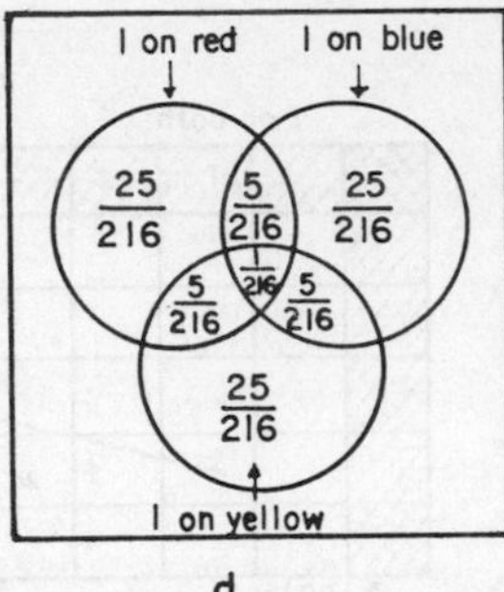

Figure 3.1 (*d*) Probabilities in (*c*) are shown in Venn diagram form.

Venn diagram in figure 3.1*b*. On the diagram in figure 3.1*a* the situation is shown by the double shading in the top left hand corner. The probability of this occurring is $\frac{1}{6}\times\frac{1}{6}=\frac{1}{36}$, intersection of the first row and the first column. The reader may check by counting cells on the matrix (figure 3.1*b*). The probability of throwing a 1 on either one or both dice is found by $\frac{1}{6}+\frac{1}{6}-\frac{1}{36}=\frac{11}{36}$. This is the equivalent of the union of the two sets, less their intersection.

Another way of considering the probabilities of different numbers occurring on dice is to use the binomial expansion. If one is interested in the probability of a 1 turning up on two dice the problem can be considered thus:

The probability of 1 coming up is $\frac{1}{6}=p$
The probability of 1 not coming up is $\frac{5}{6}=q$

0 ones	1 one	2 ones
q^2	$2qp$	p^2
$\frac{5}{6}\times\frac{5}{6}$	$2\times\frac{5}{6}\times\frac{1}{6}$	$\frac{1}{6}\times\frac{1}{6}$
$\frac{25}{36}$	$\frac{10}{36}$	$\frac{1}{36}=\frac{36}{36}=1$

For 3 dice the result would be similar

0 ones	1 one	2 ones	3 ones
q^3	$3q^2p$	$3qp^2$	p^3
$\frac{5}{6}\times\frac{5}{6}\times\frac{5}{6}$	$3\times\frac{5}{6}\times\frac{5}{6}\times\frac{1}{6}$	$3\times\frac{5}{6}\times\frac{1}{6}\times\frac{1}{6}$	$\frac{1}{6}\times\frac{1}{6}\times\frac{1}{6}$
$\frac{125}{216}$ +	$\frac{75}{216}$ +	$\frac{15}{216}$ +	$\frac{1}{216}=\frac{216}{216}=1$

Any required combination of values can thus be selected. Thus 1 one or 2 ones would give $\frac{90}{216}$, while no ones would be $\frac{125}{216}$. The situation is already too complex to be easily illustrated by a diagram, but an attempt is made in 3.1*c*. Imagine a structure of $6\times6\times6$ (or 216) small square blocks. These may be coloured. Each of 3 dice is represented by one of the three dimensions. Each small cube is a triad of numbers. One corner cube for example is (1, 1, 1). This is marked black on the left hand diagram. The three dotted rows represent a 1 on 2 out of 3 dice, and the three faces a 1 on 1 out of 3 dice (see right hand diagram). The cut out residue on the right hand diagram represents all the permutations that have no 1 at all. The eight outcomes regarding the presence or absence if a 1 on three dice are catered for in the Venn diagram in figure 3.1*d*.

(b) Application of probability to real situations

Probability theory is concerned with the likelihood of occurrence of chance events. The theory can be used to give confidence limits to a set of observations. The aim is to determine how likely it is that the set of observations could have occurred by chance or randomly. If the likelihood or probability of the events occurring randomly is low, then the result can be considered significant at the appropriate level of confidence.

The degree of confidence is closely related to the numbers of values in the data analysed. For example four out of five farms on soil A may grow a specific crop, but none of five farms on soil B grows this crop. It may be suggested that there is a relationship between the soil type and the crop. The small number of occurrences would not, however, result in a high probability. If the numbers of farms were 40 and 50 respectively in the first group of farms, then the probability would be much higher. The relationship between the soil and crop could then be stated with greater confidence.

In a random sample each member of the total population has an equal chance of being picked. Drawing a card from a pack or tossing a coin or rolling a dice are all random or chance operations. Hence their frequent use in explaining probability theory, which depends on the chance of the event occurring being unbiased. In any situation where choice is possible, a number of possibilities must exist. Tossing a coin provides 2 choices, rolling a dice provides 6, drawing a card from a normal pack 52, for a single operation. The operations are rarely single and when two or more are combined in different ways the choice becomes very great. The number of possibilities or choices can be calculated by permutations and combinations. For example 6 objects or numbers can be arranged in 6! ways; this is $6 \times 5 \times 4 \times 3 \times 2 \times 1 = 720$. The probability of drawing 6 numbers in the same order or any other specific order, such as 1, 2, 3, 4, 5, 6 is $\frac{1}{6!}$ or $\frac{1}{720}$. This is assuming the numbers are replaced before each draw. For this reason numbers chosen randomly can only very rarely correlate exactly with each other or with a non-random set of values. This principle is important in hypothesis testing and inferential statistics as the hypotheses are usually stated in definite terms, but the results are stated in probability terms, at various different levels of significance. These levels determine the proportions of times that the results arrived at could have occurred purely by chance. Only when the numbers of trials or objects involved are very large is it usual for the probabilities to approach 1 or a state of certainty. A large number of repeated events or measurements develop characteristic patterns of distribution and these various patterns are recognized and named as statistical distributions. Some of the more common ones will be considered.

(*c*) *Statistical distributions*

(i) *The binomial distribution* A way of considering the build up of the binomial triangle pattern is by examining the pin-ball machine and the probabilities that a ball will reach each of the 7 bottom cells shown in figure 3.2. To reach the extreme end cells, a ball must take the correct choice on a $\frac{1}{2}$

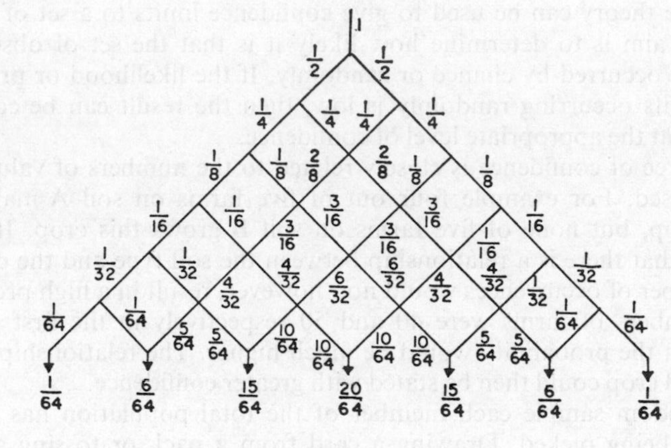

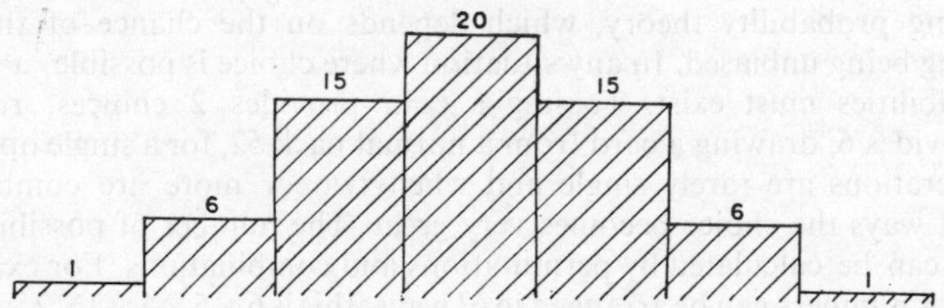

Figure 3.2 The symmetrical binomial probability curve built up by means of the addition of successive probabilities. This tends towards a normal curve.

basis at 6 points. Its chances of doing this are only $\frac{1}{2}\times\frac{1}{2}\times\frac{1}{2}\times\frac{1}{2}\times\frac{1}{2}\times\frac{1}{2}=\frac{1}{64}$. Its chance of reaching the middle cell is much greater because balls can feed in from both sides so that the total probability as shown on the diagram is $\frac{20}{64}$. The adjacent cells can each hope to receive $\frac{15}{64}$ balls. In time the balls will build up and tend towards a symmetrical pattern that is in fact a binomial distribution. It can be obtained mathematically by using the binomial expansion of $(p+q)^n$. p is the probability of one event occurring and q of another event, the two events being the only possible ones so $p+q=1$. If p and q are equally likely to occur, as in the example above, $p=q=\frac{1}{2}$.

If there is only one object then either p or q must happen and the probability of p happening is $\frac{1}{2}$ and q also has a probability of $\frac{1}{2}$. If for example two stones are picked out of a large collection of stones in a stream, half of which are moving and half of which are stationary, the probability that both stones are moving is $p^2=\frac{1}{2}\times\frac{1}{2}$, the probability that both stones are stationary is $q^2=\frac{1}{2}\times\frac{1}{2}$. When these two possibilities are eliminated there remains the possibility of one stone moving and one remaining stationary, which can occur in two different ways according to which stone is moving. The joint probability of either of the last two events must be $1-(p^2+q^2)$ or $1-\frac{1}{2}$, and as the last two events have equal probability their individual probability must be $\frac{1}{4}$. This can be shown in table 3.1*a* thus:

Table 3.1*a* Binomial probability

Result	2 stones moving		1 moving 1 stationary		2 stationary
Probability	p^2	+	$2pq$	+	q^2

Table 3.1*b*

Result	Ways of arising	Binary check	Probability	Probability of type of result
3 moving	m m m	1 1 1	p^3	p^3
2 moving 1 stationary	s m m	0 1 1	qp^2	$3p^2q$
	m s m	1 0 1	pqp	
	m m s	1 1 0	p^2q	
1 moving 2 stationary	s s m	0 0 1	q^2p	$3pq^2$
	s m s	0 1 0	pqp	
	m s s	1 0 0	pq^2	
3 stationary	s s s	0 0 0	q^3	q^3

Table 3.1*b* gives $p^3+3p^2q+3pq^2+q^3$. The process can be carried on in this way indefinitely but it becomes increasingly complex. The resulting probability pattern is based on the expansion of $(p+q)^n$. This is seen from the fact that $(p+q)^2=p^2+2pq+q^2$ and $(p+q)^3=p^3+3p^2q+3pq^2+q^3$, and these are the results obtained. This expansion is called the binomial expansion and the triangle of values that can be built up is called Pascal's triangle. The first few values of the coefficients which give the whole number before the p and q in the individual terms in the equation, the powers of p decreasing to the right and q increasing in this direction by 1 each term.

This pattern is the same as that derived from the pin ball probability pattern shown in figure 3.2, and is called the binomial distribution. As long as p and q

Table 3.2 Pascal's triangle

Row			
1	1 1	/2	/2
2	1 2 1	/4	$/2^2$
3	1 3 3 1	/8	$/2^3$
4	1 4 6 4 1	/16	$/2^4$
5	1 5 10 10 5 1	/32	$/2^5$
6	1 6 15 20 15 6 1	/64	$/2^6$
7	1 7 21 35 35 21 7 1	/128	$/2^7$
8	1 8 28 56 70 56 28 8 1	/256	$/2^8$

are equally probable the values given by working out the terms will be symmetrical. Supposing a sample of 8 stones is examined in the river, then there is only a $\frac{1}{256}$ probability that they are all moving and the same probability that they are all stationary, but there is a $\frac{70}{256}$ probability that 4 will be moving and 4 stationary and so on. Each term is found by adding the two terms in the line above that lie on either side of it. It is possible to deduce by this reasoning the number of objects which may be expected to fall into each group in terms of the probability p that an event will occur and q that it will not occur where $p+q$ is equal to 1.

These probabilities may be illustrated in the form of a histogram, which will be symmetrical if $p=q$. The number of steps in the histogram (see figure 3.2) can be made of any required magnitude. If there are 16 objects, then a histogram with 5 steps should contain 1, 4, 6, 4, 1 of the objects, if there are 512 objects, then to get convenient whole numbers the histogram should have 10 steps containing 1, 9, 36, 84, 126, 126, 84, 36, 9, 1 objects. This process can go on indefinitely until the steps become so small that the line outlining them approximates more and more closely to a particular smooth curve. If the number of objects increases to infinity then the curve becomes quite smooth.

(ii) *Normal curve* The symmetrical smooth version of the binomial curve is called the *normal curve*. This curve can deal with continuous events better than the either/or type of event considered in relation to the binomial distribution. The distribution of height of people is a continuous one which follows the normal curve. There are more people of roughly average height and only a few very tall, on the one hand, or very short, on the other. A large sample will give the mean or average height and many people will approximate to this height, but there may be very few of precisely the mathematical mean, if this is expressed in extremely precise terms. Similarly to get exactly 50 heads in 100 tosses has a probability of only 0.0796.

The normal curve was first put forward by De Moivre in 1733 but was also independently discovered by Gauss and Laplace. It is sometimes called the Law of Errors, as it was found that errors in astronomical observations followed this distribution. There are many sources of errors, but only a few very large ones which give either much too large or much too small a value. The addition of all errors should approximate a normal distribution.

The total area under the normal curve is made equal to 1. This is the total probability. The form of the curve is such that it never finally reaches the horizontal axis, but for practical purposes it can end at 3 standard deviations on either side of the mean. At +3 and -3σ the area under the curve is 99.73 per cent of the possible 100 per cent. At +4 and -4σ the area rises to 99.994 per cent. Tables are provided that give the area under the normal curve in terms of z, the standard deviation. Thus when z is 1, the area is 0.6826 from the axis, and where z is 2 the area is 0.9546 including both sides of the curve. These tables for z are referred to in some of the tests discussed later. The proportions of the area for given values of z indicate the probability proportions, because the whole area has the probability 1 and the probability is uniformly spread throughout the area under the curve. Using the values of z it is possible to calculate the probability for any point on the curve in terms of the area to the right or left of that point. In some tests only one end of the normal curve is used in rejecting a hypothesis (figure 3.3(*b*)). In figure 3.3(*a*) a two-tailed test is used as both ends of the curve are considered. In this type of test both extremes of the distribution are included. Normally a two-tailed test is safer.

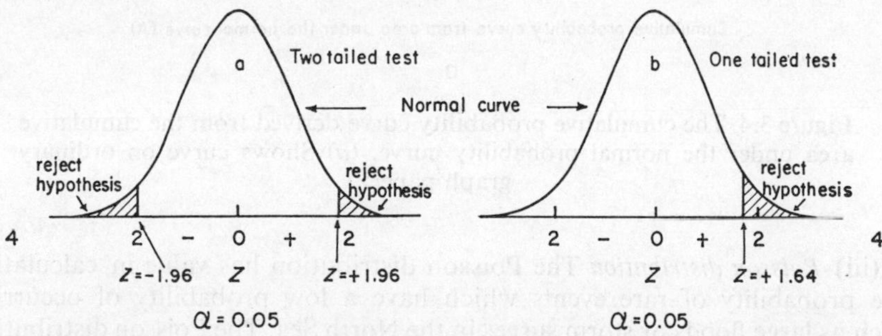

Figure 3.3 Diagrams to illustrate areas under the normal curve in which a two-tailed test and a one-tailed test imply rejection of a hypothesis.

The normal curve shown in figure 3 is a bell shaped curve centred on the mean and spread out according to the pattern of probability expectation. This curve is shown in a cumulative form in figure 3.4(*a*). The curve can be made into a straight line by suitable adjustment of the scale, which must be expanded at the lower and upper part where the values change slowly and compressed in the middle where they change fast. This modification of the scale is made on probability paper (see figure 3.4(*b*)) which therefore has the property that a normal distribution is plotted as a straight line. Probability paper is used for plotting the size distribution of sediment and this enables the normality of the distribution to be tested at a glance by the straightness of the line.

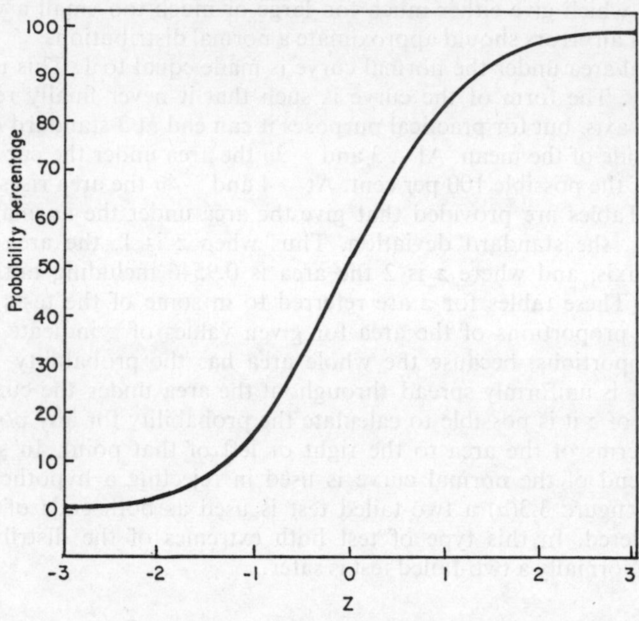

Figure 3.4 The cumulative probability curve derived from the cumulative area under the normal probability curve. (*a*) Shows curve on ordinary graph paper.

(iii) *Poisson distribution* The Poisson distribution has value in calculating the probability of rare events which have a low probability of occurring such as large floods or storm surges in the North Sea. The Poisson distribution deals with isolated events in a continuum of time or length, area or volume, so that it is not possible to say how often the event did not occur. It is based on the natural law of growth called the exponential law. In compound interest a small sum is added each year; the following year the addition on the rather larger capital increases by a smaller amount and so on. If 100 is the capital (P) and $\frac{1}{100}$ the growth rate (r) the interest is 1 per cent, then after 1 year the value is given by $P\left(1+\frac{r}{100}\right)$, which equals $100+\frac{100\times 1}{100}=101$, by the second year the value is $P\left(1+\frac{r}{100}\right)^2$ and so on. However the interest can be added continuously and not annually, and then the amount will be $P\left[\left(1+\frac{1}{n}\right)^n\right]^{rt/100}$ in time $t.n$ is the number of times interest is added, so for

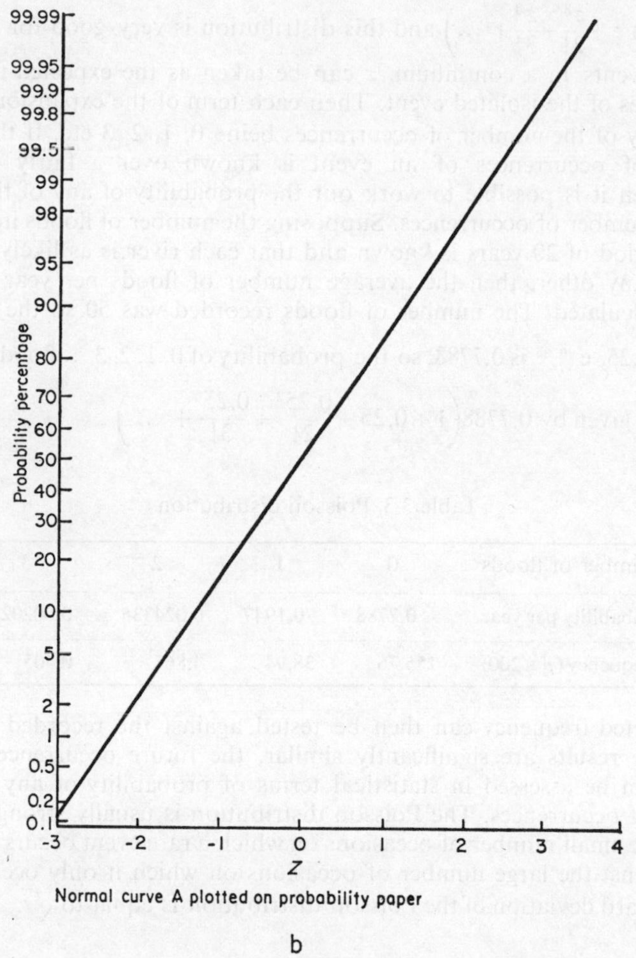

Figure 3.4(*b*) Shows curve on probability paper on which it becomes a straight line.

continuous growth $n = \infty$ and the value $\left(1+\frac{1}{n}\right)^n$ must be found by calculus. This can be done via the binomial expansion and gives $1+\frac{1}{1}+\frac{1}{2!}+\frac{1}{3!}+\frac{1}{4!}+\dots$. This is a continuous sequence which gives the value e, which for the first seven terms is e = 2.718282. This value e can be raised to any power, z, to give $e^z = \frac{z^0}{0!}+\frac{z^1}{1!}+\frac{z^2}{2!}\dots$ which simplifies to $e^z = 1+z+\frac{z^2}{2!}+\frac{z^3}{3!}+\frac{z^4}{4!}+\dots$. To make a useful statistical distribution the terms must sum to unity. This can be done by multiplying by e^{-z} as $e^{-z}.e^z = e^0 = 1$. Thus

$1 = e^{-z}\left(1 + z + \frac{z^2}{2!} + \frac{z^3}{3!} + \ldots\right)$ and this distribution is very good for describing isolated events in a continuum. z can be taken as the expected number of occurrences of the isolated event. Then each term of the expansion gives the probability of the number of occurrences being 0, 1, 2, 3 etc. If the average number of occurrences of an event is known over a fairly long time period then it is possible to work out the probability of any of the various possible number of occurrences. Supposing the number of floods in ten rivers over a period of 20 years is known and that each river is as likely to have a flood as any other, then the average number of floods per year per river can be calculated. The number of floods recorded was 50 so the average is $\frac{50}{200}$. $z = 0.25$, $e^{-0.25}$ is 0.7788, so the probability of 0, 1, 2, 3 ... floods per river per year is given by $0.7788\left(1 + 0.25 + \frac{0.25^2}{2!} + \frac{0.25^3}{3!} + \ldots\right)$.

Table 3.3 Poisson distribution

Number of floods	0	1	2	3
Probability per year	0.7788	0.1947	0.024335	0.002025
Frequency ($p \times 200$)	155.76	38.94	4.867	0.405

The expected frequency can then be tested against the recorded frequency and if the results are significantly similar, the future occurrence of flood danger can be assessed in statistical terms of probability of any particular number of occurrences. The Poisson distribution is usually strongly skewed so that the small number of occasions on which a rare event occurs frequently is set against the large number of occasions on which it only occurs rarely. The standard deviation of the Poisson distribution is equal to $\sqrt{z}$.

3.3 Descriptive statistics

Geographical data frequently occur in long lists of values, and these values provide the raw material of statistics. The aim of descriptive statistics is to describe these data numerically so that their magnitude and spread can be expressed concisely and meaningfully. The most important aspects of the data are their mean or average size and the extent and nature of the spread of the data. These aspects are described under the headings (*a*) Central tendency and (*b*) Dispersion. Similar methods can be adapted for use with data expressed as individual values or as grouped values. The data can also refer to no, one, two or three space dimensions, as well as time data. Examples of the statistical description of 2-dimensional areal data will be discussed in chapter 4 and the analysis of 3-dimensional data is mentioned in chapter 9 and time data in chapter 10.

(*a*) *Central tendency*

There are three commonly used measures of central tendency, the mean, the median and the mode. The former is only applicable to data that can be expressed on the interval or ratio scales, but it is the most versatile statistically. The *mean* for individual values can be found simply by summing all the values and dividing by the number of values. Two methods of calculating the mean of grouped data are shown on the worksheet and described in the instructions given on pp. 113–114.

Another method of expressing the central tendency is the *median*. This value can be found by arranging all the values in order. Then the middle value is the median if the set contains an odd number of values. If the number of values is even then the mean of the two central values is the median. The median is not affected by the extremes of the observations and it does not take individual values into account. The median can be found in grouped data by plotting a frequency curve. This is illustrated in figure 3.5 from which it can be seen that the 50th percentile on the cumulative frequency curve lies at 271 ft. This value is then the median value, as 50 per cent of the observations lie above this point and 50 per cent below it. In this particular example the diagram in figure 3.5 shows that the mean and median coincide. The significance of this will be discussed in the section dealing with dispersion. Until fairly recently the median value was frequently used to express the size

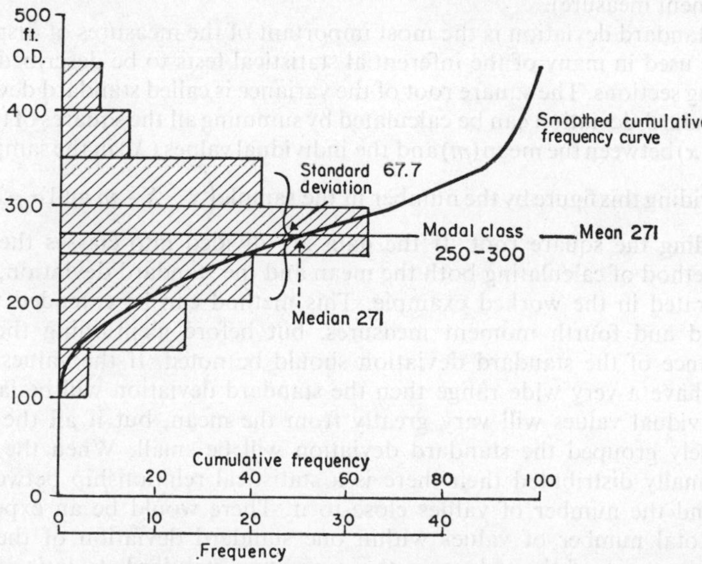

Figure 3.5 Altimetric curve for part of Yorkshire Wolds shown (*a*) in 50 ft. groups. (*b*) in a smoothed cumulative frequency curve. The mean, modal class and median are shown and the standard deviation is also indicated. The distance of one standard deviation on either side of the mean includes 66.5 per cent of the total curve.

of sediment samples. Now, however, the central tendency is often expressed as the mean $M_z = 16\text{th} + 50\text{th} + 84\text{th}$ percentile values divided by 3. This value gives the mean in a way similar to that used for single values.

The other central tendency is the *mode*. This form of expressing the central tendency applies mainly to grouped data and is defined as the class containing the greatest number of values. In the example given this is the class 250–300 ft.

The mean and median central tendencies can also be applied to areal data and this aspect is discussed in chapter 4. The mean is generally the most useful of these measures of central tendency. It makes use of all the values and can be used in many statistical formulae. The median does not take the extreme values into account. It can however be used when precise values are not known. The mean cannot be calculated in this instance.

(*b*) *Dispersion and further moment measures*

It is often necessary to describe the spread of the values in a sample. This is an essential element of the methods used to assess whether two samples come from the same population and it also determines the width of confidence bands within which a certain percentage of values may be expected to fall. The spread of the data and the pattern of the spread can be expressed in several descriptive statistical terms. Firstly there is the standard deviation (also called the 2nd moment measure, the mean being the first). Secondly there is the skewness (or 3rd moment measure) and thirdly the kurtosis (or 4th moment measure).

The standard deviation is the most important of the measures of dispersion and it is used in many of the inferential statistical tests to be described in the following sections. The square root of the variance is called standard deviation. The standard deviation can be calculated by summing all the squares of the differences (x) between the mean (m) and the individual values (X) in the sample and then dividing this figure by the number in the sample $\left(x = X - m \text{ and } \sigma = \sqrt{\frac{\sum x^2}{N}}\right)$ and finding the square root. If the data are divided into classes there is a short method of calculating both the mean and the standard deviation, which is illustrated in the worked example. This method can be extended to give the third and fourth moment measures, but before mentioning these the significance of the standard deviation should be noted. If the values in the sample have a very wide range then the standard deviation will be large, as the individual values will vary greatly from the mean, but if all the values are closely grouped the standard deviation will be small. When the values are normally distributed then there is a statistical relationship between the mean and the number of values close to it. There would be an expected $\frac{2}{3}$ of the total number of values within one standard deviation of the mean and 95 per cent of the values in the range two standard deviations either side of the mean. This is shown in the example in figure 3.5 in which 66.5 per cent fall within 1 standard deviation of the mean and 94.6 per cent within 2 standard deviations. This can be stated as 271 ± 67.7 ft where 271 is the mean and 67.7 is the standard deviation.

Skewness The skewness of a frequency distribution curve is a measure of its asymmetry. It can be expressed in terms of the relationship of the mean to the median. One formula used to express the skewness of a sediment size distribution illustrates the point. This can be given as (mean value – median value), divided by standard deviation $\frac{mean - median}{\sigma}$. When the median is greater than the mean the skewness is negative. In a negatively skewed distribution the tail extends further to the small values. The reverse also applies, as a positively skewed distribution has a higher mean than median and a tail at the large value end. The skewness can be calculated relatively easily for a frequency distribution divided into classes and a worked example is given on the worksheet. The example given refers to the data illustrated in figure 3.5. It can be seen that the very low value of the skewness of –0.097 agrees with the close correspondence of the position of the median and mean on the cumulative frequency curve. The median is very slightly higher than the mean.

Kurtosis The kurtosis indicates the spread of the frequency curve. The kurtosis of a normally distributed curve should be 3 if calculated by the method illustrated on the worked example. If the value exceeds 3 the distribution is less peaked than the normal curve and if it is less than 3 the distribution is more peaked. The sample illustrated has a kurtosis of 2.8 which shows that the curve is nearly normal.

The skewness and kurtosis are valuable in the analysis of sediments. They have been used to differentiate different sedimentary environments. These methods of describing the character of a frequency distribution are sometimes called moment measures. The first moment is the mean, the second is the standard deviation, the third is the skewness and the fourth is the kurtosis. The first two moment measures play an important part in many of the statistical tests that are discussed in this chapter.

Table 3.4 Mean, standard deviation, skewness and kurtosis of grouped data
Work Sheet

1	2	3	4	5	6	7	8	9	10	11
ft		cumulative								
X	f	f	Xf	d	df	d^2f	d^3	d^3f	d^4	d^4f
425	4	100	1700	−3	−12	36	−27	−108	81	324
375	7	96	2625	−2	−14	28	−8	−56	16	112
325	21	89	6825	−1	−21	21	−1	−21	1	21
275	32	68	8800	0	0	0	0	0	0	0
225	20	36	4500	+1	+20	20	1	20	1	20
175	13	16	2275	+2	+26	52	8	104	16	208
125	3	3	375	+3	+9	27	27	81	81	243
	100	100	27100		−47	184		−185		Σd^4f 928
			ΣXf		+55	Σd^2f		+205		
					Σdf +8			Σd^3f +20		

Mean $\frac{Xf}{N}=271$ ft $N=100$ $C=50$ $A=275$

Mean $=m=A-\frac{C\,df}{N}=275-\frac{50(8)}{100}=271$ ft

Standard deviation $=s=\frac{C}{N}\sqrt{N\,\Sigma\,d^2f-(\Sigma\,df)^2}=\frac{50}{100}\sqrt{100\times184-64}=\frac{1}{2}\sqrt{18336}$

$=\frac{1}{2}\times135.4=67.7$

$v_1=\frac{8}{100}=\frac{\Sigma\,df}{f}$ $v_2=\frac{184}{100}=\frac{\Sigma\,d^2f}{f}$ $v_3=\frac{20}{100}=\frac{\Sigma\,d^3f}{f}$ $v_4=\frac{928}{100}=\frac{\Sigma d^4f}{f}$

Moments about X_0 v_1 0.08 v_2 1.84 v_3 0.20 v_4 9.28
$X_0=275$

Moments about M $m_2=C^2(v_2-v_1^2)=2500\ (1.84-0.08\times.08)=2500\ (1.8336)=4584$

$m_3=C^3(v_3-3v_1v_2+2v_1^3)=0.2-3\times0.08\times1.84+0.001$

$=.2406\times125000$

$m_4=C^4(v_4-4v_1v_3+6v_1^2v_2-3v_1^4)=2500\times2500\ (9.28-(4\times.08\times0.2)$

$+(6\times.0064\times1.82)-(3\times0.0064\times0.0064)$

$=9.286\times6250000=58{,}000{,}000$

Standard deviation $s=\sqrt{m_2}=\sqrt{4584}=67.7$

Skewness $\alpha_3=m_3/s^3=-30075\div4584\times67.7=\frac{-30075}{310336.8}=-.0969$

Kurtosis $\beta_2=m_4/s^4=\frac{58{,}000{,}000}{21{,}003{,}889}=2.761$

Mean and Standard Deviation of Grouped Data There is a relatively quick and easy method of finding the mean and standard deviation of grouped data.

Step 1. List classes as on work sheet

Step 2. Enter number of items in each class. In this example the numbers refer to number of heights in each 100 ft. class recorded in each 1 km. square of the National Grid over an area of 100 sq. km.

Step 3. Find $\Sigma\,xf$ by multiplying each value in column 1 by the adjacent value in column 2 and summing column 4 (This step can be omitted, but it provides a check)

Step 4. Assume a mean value and enter 0 opposite this value in column 5; enter −1, −2, −3 upward and 1, 2, 3, downward from the 0 row.

Step 5. Multiply columns 2 and 5 to find *df*, noting sign and enter in column 6.

Step 6. Find the algebraic sum of *df*. This is $\Sigma\,df$.

Step 7. Multiply columns 5 and 6 to find d^2f and sum the values to give $\Sigma\,d^2f$ in column 7.

The mean is now found from the formula $m=A-C\frac{\Sigma df}{N}$ or by $\frac{\Sigma Xf}{N}$. These results should be the same. (Step 3).

The standard deviation is found from the formula

$$s=N\sqrt{N\Sigma\,d^2f-(\Sigma\,df)^2}$$

A = selected mid-point (Step 4)

C = class difference (50 ft. in this example)
N = total number of values in sample (100 in this example)
Note that if the values in column 1 are arranged in ascending order

$$m = A + \frac{Cdf}{N}$$

3.4 Sampling

It is often desirable to base hypotheses on data that are not a complete set of the total population. This may be due to the inaccessibility of part of the population or perhaps to the very large size of the whole population. One of the objects of sampling is to obtain a representative group of the total population often with the smallest group that will give the desired reliability in the results. The method of sampling adopted in geographical data, which are very diverse in character, must vary with the nature of the data and where the data are variable a larger sample will be required. The aims of the study must always be considered when the sampling plan is drawn up. It is important to be clear whether the whole population has the attribute that is of interest or whether this attribute only belongs to some of the population. This point will be elaborated later with an example. If one is comparing different attributes possessed by different members of the total population then a larger sample will be required than if the whole population possesses all the attributes, provided the attribute is a variable.

The population may consist of discrete objects, such as people, drumlins or villages, or it may be a continuous feature, such as a land surface consisting of continuously varying slopes. The population may consist of dimensionless values not associated with position, it may be distributed along a line (a river or railway for example) that has one space dimension, it may be scattered over a 2-dimensional surface, or within a 3-dimensional one, or it may be spread over time. All these different populations can be sampled by use of appropriate sampling plans. The samples must first be collected and then they must be related to the parent population. The first of these objectives necessitates the use of a sampling plan and the second the estimation of the size of the sample required to give information concerning the parent population with an appropriate degree of confidence.

(*a*) *Sampling plan*

It is assumed that the samples are required to provide information about the population. For this reason it is necessary that every member of the population should have an equal chance of being selected. A sample based on this principle is called a random sample and it can be used safely to infer qualities of the total population. Different sampling plans can be adapted to the various dimensions in which data exist but certain basic patterns are frequently used. The main principle is that bias should not be allowed to enter the sampling procedure. One method of avoiding bias is to take a regular sample, by taking for example every tenth (or some other interval) individual in the population, if it consists of discrete objects with no fixed position, as in an opinion poll. If sampling is done along a line the samples can be equally spaced or if in area, a regular grid can be used and the same method of systematic sampling can be applied to volumes. Where the area has a regular pattern itself, such as the grid road system in parts of U.S.A., then a regular systematic sampling plan can give a biased sample. This is an instance when it is better to choose a random method of sampling.

A random sample can be obtained by selecting the sample with the aid of a table of random numbers. A single number will locate a point sample on a line or from a list of values. A pair is needed to select a point sample on an area and three numbers are needed to select a point in a volume. If the environment being sampled is variable it is necessary to represent the variable areas or strips in proportion to their extent. This can be done by taking the numbers in the sample in proportion to the extent of the different environments. For example in a coastal survey the sedimentary environments may include the dunes, the intertidal beach and the shallow water offshore in varying proportions and systematic or random samples should be taken in proportion to their relative areas along the coast.

Nested or multistage sampling can also be used in some situations. For example if a sample of drainage basins is to be selected from a large area, a number of 3rd order basins can be selected and these can then be used themselves to select a number of 1st order basins, the selecting being made randomly at both stages. This then is a nested sample as the smaller units are included within the larger. The same method of nested sampling can be carried out over an area divided by a grid into regular units and sub-units. See figure 3.6.

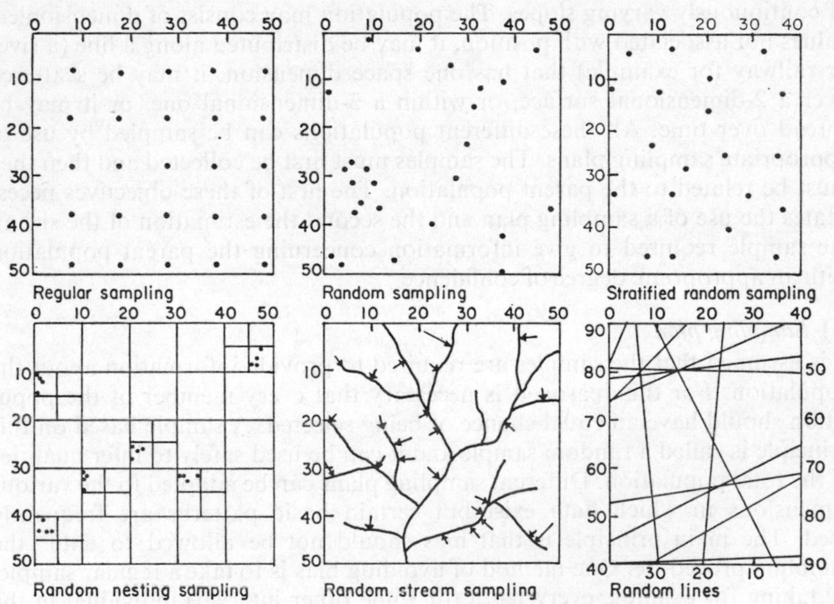

Figure 3.6 Patterns of sampling in area.

It is important that the aim of the investigation is borne in mind when the sampling plan is devised or it may be found to provide insufficient evidence. For example if the change of stone shape vertically and horizontally in a deposit is to be differentiated, it is essential that the samples include separately samples from the different horizons as well as over the areal extent of the deposit. This pattern is called stratified sampling. The population to be sampled must be clearly defined and the variability assessed, if necessary by taking a pilot sample, to ensure that the number in the sample is adequate.

(*b*) *Number in sample*

Some statistical models allow several variables to be considered and the samples should be collected in sufficient numbers to allow valid results to be given. The number in the sample that is collected depends on the degree of certainty that is required. The sample size should be such that it is possible to infer from it sufficiently accurately the character of the whole population. The mean and standard deviation of the sample can be used to estimate the mean of the whole population. If the sample is about 30 or more it is possible to give confidence limits of the mean of the population from the mean of the sample. This is because a large number of sample means form a normal distribution centred on the true mean. The values of the individual sample means will have a standard deviation equal to $\sigma/\sqrt{n}$. σ = standard deviation of population. n = number in sample. The value of $\sigma/\sqrt{n}$ is called the standard error. It provides an estimate of the values within which the true mean must lie. However the population standard deviation is not known so that the formula can only be used in a modified form. σ^2 is estimated by $\frac{n}{n-1}s^2$ where s is the sample standard deviation.

It is, however, possible to calculate the number of values required in a sample to give a specific confidence level from the results obtained by taking an experimental sample. The formula that provides this result is $n = (st_{\alpha/2}(m))^2/d$, n = required number, s is standard deviation of sample, $t_{\alpha/2}$ is the upper Student's t value for probability, m is number in sample less one and d is the required degree of accuracy in units from the true population mean.

An example will clarify the use of the formula. In the 92 provinces of Italy it is required to obtain a mean within 50 of the population mean for the area of the provinces. An experimental sample of 25 provinces is taken and the mean and standard deviation calculated. The values were mean 332, s = 161.5. $t_{\alpha/2}(m)$ from tables is given as 2.06 at 95 per cent confidence level. The formula then gives

$$\frac{(161.5\times 2.06)^2}{50} = 44.2 = n.$$

This value must be a whole number so a sample of 45 will give the required accuracy. The large sample needed in this example is caused by the large standard deviation relative to the mean. If an accuracy within 100 had been required then a sample of 12 would have sufficed.

If the sample is large and the standard deviation low then the true mean of the population can be specified accurately with a high degree of confidence. The accuracy increases less rapidly than the size of sample. The law of diminishing returns sets a limit beyond which it is not economical to go. The standard deviation also clearly influences the size of sample required and where the spread is small, a small sample of 20 to 30 will give good results.

3.5 Inferential statistics

(*a*) *Introduction*

When data have been collected by sampling it is necessary to formulate hypotheses concerning the data. The validity of these hypotheses can then be tested by various statistical methods. This section deals with data about only one variable, values of which are normally obtained by sampling, although in some circumstances the whole population can be included in the hypothesis. Many of the tests to be considered, however, are concerned with inferring from the sample the characteristics of the whole population. The size of sample required for this purpose has already been mentioned.

In testing statistical hypotheses certainty cannot be achieved so the results are given in terms of probability. The terms significance level and confidence limit are used to describe the degree of probability of the result. The idea of confidence limits is based on the probability of the results occurring by chance. The confidence limit gives a measure of the accuracy with which a population mean is known in relation to the sample mean. The limits are often expressed at a 95 per cent probability level, although this is quite arbitrary. This means the 95 per cent of the outcomes will be right on average. The significance level indicates the percentage of times that the relationship tested could be expected to occur by chance. The 95 per cent or 99 per cent levels are usually used and these indicate that there is a 1 in 20 or 1 in 100 probability respectively of the given result occurring by chance. Again any level can be selected and the nearer 100 per cent the less likely is the result to occur by chance. At 99 per cent a result can be considered to be highly significant, and 95 per cent is usually considered as significant, but below this the result is only suggestive.

When a relationship is to be tested a null hypothesis is set up. This hypothesis assumes that there is no relationship between the values being tested. If the statistical result shows that the value is significantly greater at the required significance level then the null hypothesis is rejected and the alternative is accepted. Two types of error can occur in this procedure. Either the sample indicates that the hypothesis should be rejected when in fact it is true; this is known as a type 1 error. Alternatively the sample indicates that the hypothesis should be accepted when in fact it is false; this is a type 2 error. The higher the significance level the smaller the probability of committing a type 1 error. The probability of committing a type 2 error depends on the size of the sample among other things. If the type 1 error probability or α has been fixed and the number in sample (n) is also fixed then β, the probability of a type 2 error can be calculated as there is an inverse relationship between α and β, an increase in n will reduce the likelihood of both types of error. The power of a test is defined as the probability of rejecting the null hypothesis when it is in fact false (Power $= 1 - \beta$).

Different tests have different powers and are suited to various types of data for which special assumptions may apply. Tests may be subdivided broadly into non-parametric and parametric tests. The former can be applied to data which are only available on the nominal or ordinal measuring scale and on the whole these tend to be less powerful than the parametric tests. The non-parametric tests do not, however, make any assumptions concerning the statistical distribution of the data. The parametric tests on the other hand, can be used for data on the interval and ratio scales but they assume that the items in the population are distributed normally.

A wide variety of tests is available for a large number of different types of problem. A few examples will be selected from both the non-parametric and the parametric tests. One important inference that must often be made is to infer the character of the total population from that of the sample in terms of the sample mean and sample standard deviation. The t test (p. 124) gives an accurate value of the population mean in the form

$$\overline{X} = \bar{x} \pm t \frac{\hat{\sigma}}{\sqrt{n}}$$

$\overline{X}$ is the population mean, $\bar{x}$ is the sample mean
t value obtained from t tables
$\hat{\sigma}$ population standard deviation
n number in sample.

The standard deviation of the sample will generally be less than that of the population, which can be given fairly closely by the relationship $s\sqrt{\frac{n}{n-1}} = \hat{\sigma}$, where s is the standard deviation of the sample and $\hat{\sigma}$ the estimate of that of the population and n is the number in the sample. This method provides values for the population mean in terms of confidence limits.

Example of finding confidence limit for a large sample A sample of 50 stones selected randomly from an esker had a mean length of $\overline{X} = 34.48$ cm and a standard deviation, $s = 9.77$. The confidence limits of the mean of the population at 95 per cent and 99 per cent significance levels are required. It is assumed that the population standard deviation σ is the same as s, the sample standard deviation, $\sigma_{\bar{x}}$ is the standard deviation of the sample means. Then

$$\sigma_{\bar{x}} = \frac{s}{\sqrt{n-1}} = \frac{9.77}{\sqrt{49}} = \frac{9.77}{7} = 1.396.$$

The population mean, m, can be calculated from the formula $(\overline{X} - m)/1.4 = \pm 1.96$ for the 95 per cent significance level or $(\overline{X} - m)/1.4 = \pm 2.58$ for the 99 per cent significance level. The values 1.96 and 2.58 are found from tables which relate the significance levels to the appropriate area under the normal curve. 1.96 is the value of z which corresponds to an area $\frac{1}{2}(0.95) = 0.475$, and 2.58 is the area corresponding to $\frac{1}{2}(0.99) = 0.495$.

The mean of the population is then found to be

$m = 34.48 \pm 2.74$ at the 95 per cent significance level and

$m = 34.48 \pm 3.58$ at the 99 per cent significance level.

The results depend on the size of the standard deviation of the sample and the number in the sample.

Example of finding confidence limit of a population mean from a small sample Twelve values for the roundness of pebbles from deltas have a mean of 334 and standard deviation, s, of 11. The estimated standard deviation of the population $\hat{\sigma}$ is given by $\hat{\sigma} = \sqrt{\frac{n}{n-1}}\, s = \sqrt{\frac{12}{11}}\, 11$

$\hat{\sigma}_{\bar{x}}$ (best estimate of standard deviation of the sample means) is given by

$\frac{\hat{\sigma}}{\sqrt{n}}$. This is then $\frac{1}{\sqrt{12}} \times \sqrt{\frac{12}{11}} \times 11 = 3.32$

t for 95 per cent significance level 11 df is 2.201

99 per cent 11 df is 3.106

$$\frac{\bar{X} - m}{3.32} = \pm 2.201 \text{ for 95 per cent level of confidence}$$

$m = \bar{X} \pm 7.3$ or $m = 334 \pm 7.3$

$$\frac{\bar{X} - m}{3.32} = \pm 3.106 \text{ for 99 per cent significance level}$$

or $m = 334 \pm 10.3$

(*b*) *Tests for differences between means and standard deviations of two or more samples of one variable*

A wide variety of tests is available to test the significance of the difference between two or more samples of one variable. Some non-parametric examples of these will first be given and then some parametric tests will be described. The null hypothesis states that two samples do belong to the same population. If the hypothesis is rejected then it may be assumed with the appropriate degree of probability that the differences between the two samples indicate that they come from different populations.

(i) *Non-parametric tests* Non-parametric tests may be used to assess the probability that two samples are drawn from the same population. The advantage of non-parametric tests is that no assumptions are made concerning the normal nature of the population distribution and their applicability to nominal and ordinal data has already been mentioned.

The sign test One of the simplest non-parametric tests that can be used to test the significance of the difference between *paired* variables is the sign test. A worked example of the test sets out the method of working and the results for this example. It is a comparison of the slope angles taken at various points down a valley on both sides of the valley wall. It is assumed in this instance that the accuracy of the observations does not warrant the use of a more powerful test. The sign test is such that there is a rather large possibility of committing a type 2 error. This would be reduced if the t test were used.

Sign test The data consist of pairs of variates, and the null hypothesis states that there is no significant difference between the pairs. The data consist of pairs of slope angles taken at the same point on a stream up both sides of the valley and at various positions along the valley. The test assesses whether there is a significant difference in slope on either side of the valley.

Table 3.5 Data for sign test

Pair	X NE Slope	*Y* SW Slope	Sign of *D* *X* – *Y*
1	36	34	+
2	36	35	+
3	35	36	–
4	37	35	+
5	34	32	+
6	33	34	–
7	35	34	+
8	32	31	+
9	30	29	+
10	31	30	+
11	29	30	–
12	27	26	+
13	28	25	+
14	25	24	+
15	26	25	+
16	24	22	+
17	20	18	+
18	15	16	–
19	16	14	+
20	14	12	+

20 Differences 16 + 4 –

$n=20$ (n must exceed 15) $p=q=\frac{1}{2}$

$m=np=10 \qquad s=\sqrt{npq}=\sqrt{\dfrac{20}{2\times 2}}=\sqrt{5}=2.24$

$$z=\frac{(16-10)-0.5}{2.24}=\frac{5.5}{2.24}=2.46$$

$$P=2\times(+0.5000-0.4931)=2\times 0.0069=0.0138$$

Procedure

1. List data in pairs
2. Enter sign of $X-Y$ opposite each pair.

3. Sum + signs, − signs
4. Calculate m by multiplying n x 0.5 (n = no. of pairs − any 0s i.e. when $X = Y$)
5. Calculate s by square root of ($n \times 0.25$)
6. Work out equation for z by deducting m from value equal to the number of + signs − 0.5 and dividing by s
7. Enter z tables for this value, these give area under the probability curve between the value $z = 0$ at the axis and the value entered. Apply 2 tailed test by deducting value of z table from 0.500 and multiplying result by 2.
8. If this value is less than 0.05 then the null hypothesis is rejected, and the samples can be considered significantly different.

$$z = \frac{(\text{no. of + signs} - m) - 0.5}{s}$$

This value is less than 0.05 and therefore the NE slopes are significantly steeper and the null hypothesis is rejected, at the 95 per cent confidence level.

The Mann–Whitney U test This test can be used with data on ordinal or higher scales and it is also applicable to samples of unequal size, i.e. the values cannot be paired as in the sign test since the number of values in the two samples need not be equal. The worked example illustrates the application of the test to a fairly small sample. If larger samples are available the distribution of U approaches the normal distribution, and the tables for z, the area under the normal curve, can be used to test the significance of the value of U. This test can be quite powerful and often provides a useful alternative to the t test and is much quicker to compute.

Mann–Whitney U Test Example
Procedure for sample size 9–20 for n_2.

Table 3.6 Data for Mann–Whitney U test

Data	Beach roundness	Rank	Delta roundness	Rank
	325	6	349	12
	485	23	258	1
	456	22	335	7
	438	20	317	4
	455	21	310	3
	338	8	304	2
	369	16	371	17
	389	18	340	5
	365	15	344	10
	351	14	345	11
	350	13	405	19
			324	5
	$n_1=11$	$R_1=176$	$n_2=12$	$R_2=100$

Step 1. Rank data using all figures so that lowest rank $= n_1 + n_2$

Step 2. Sum rank orders for R_1 and R_2

Step 3. Calculate U from formula $U_1 = n_1\,n_2 + \dfrac{n_1(n_1+1)}{2} - R_1$

$$U = 11 \times 12 + \frac{11(11+1)}{2} - 176$$

$$= 132 + 66 - 176 = 22$$

U' can also be found by $U_2 = n_1 n_2 + \dfrac{n_2(n_2+1)}{2} - R_2$

$$= 132 + 78 - 100 = 110$$

The smaller value is required i.e.

$U = n_1 n_2 - U'$, where U' is the larger value. This is $22 = 132 - 110$.

The value of U is less than that given in tables for a significance level of 95 per cent in a one-tailed test, as 22 is less than 28, and the null hypothesis that the two samples are drawn from the same populations is rejected at this level of significance.

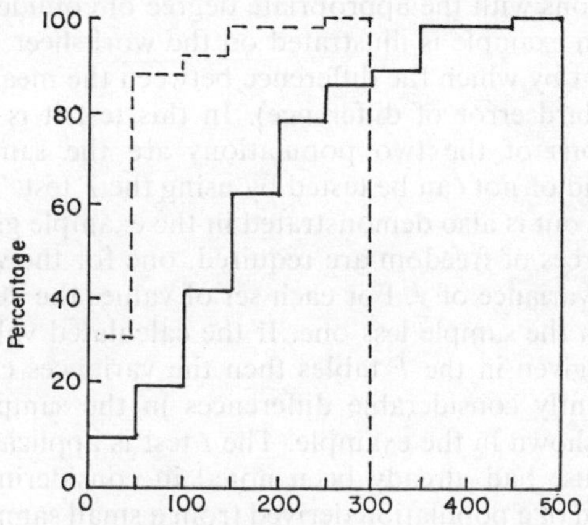

Figure 3.7 Cumulative frequency graphs for roundness of limestone pebbles on Foley Island off Baffin Island to illustrate the use of Kolmogorov–Smirnov non-parametric test.

Kolmogorov–Smirnov test This is a very simple non-parametric graphical test but it is not very powerful. However it is so easy and quick to apply that it often provides a useful check. It can best be applied to data divided into classes and plotted as a cumulating graph between 0 and 100 per cent as shown in the example in figure 3.7. R. L. Miller and J. S. Kahn (1962) provide a graph to test the significance of the maximum difference between the two graphs. A value N must first be calculated by the formula $n_1n_2/(n_1+n_2)$ where n_1 and n_2 are the numbers in each set of data. In the example illustrated $n_1=n_2=50$, so $N=25$, the maximum difference in relation to 1 is 0.68. The table shows that for $N=25$ the difference must exceed 0.32 to indicate that the two populations from which the samples were drawn are different at the 99 per cent level. It can thus be concluded that the roundness of the stones at the margin of the polygon is significantly less than that of the stones in the centre. This supports the suggestion that the central stones have moved up from below and have suffered less surface weathering, a process which reduced their roundness.

(ii) *Parametric tests. Student's t test* The student's t test was worked out by a brewer who used the name Student. Its purpose is to establish whether two samples could have been drawn from the same population. The null hypothesis states that the two samples come from the same population and if the null hypothesis is rejected it is assumed that the two samples come from different populations with the appropriate degree of confidence. The method of working out an example is illustrated on the worksheet. (The value given for t is the amount by which the difference between the means of the samples exceeds the standard error of difference). In this test it is assumed that the standard deviations of the two populations are the same. Whether this assumption is valid or not can be tested by using the F test. The way in which this test is carried out is also demonstrated in the example given. In the F test two different degrees of freedom are required, one for the variance of x and the other for the variance of y. For each set of values the degrees of freedom are the number in the sample less one. If the calculated value of F does not exceed the value given in the F tables then the variances can be considered identical. Apparently considerable differences in the sample variances still pass this test, as shown in the example. The t test is applicable to samples of all sizes and its use had already been noted in considering the confidence limits for the mean of a population derived from a small sample.

Student's t test Null hypothesis states that the two samples come from the same population.

Data table 3.7 consists of 12 values for roundness of pebbles from raised beaches and 12 from deltas.

Table 3.7 Data for t test

	Beaches x				Deltas y		
	x	$x-\bar{x}$	$(x-\bar{x})^2$		y	$y-\bar{y}$	$(y-\bar{y})^2$
	325	−73	5329		349	+15	225
	485	+87	7569		258	−76	5776
	456	+58	3364		335	+1	1
	456	+58	3364		317	−17	289
	438	+40	1600		310	−24	576
	455	+57	3249		304	−30	900
	338	−60	3600		371	+37	1369
	369	−29	841		340	+6	36
	389	−9	81		344	+10	100
	365	−33	1089		344	+10	100
	351	−47	2209		405	+71	5041
$n_x=12$	350	−48	2304	$n_y=12$	324	−10	100
	4777	−299 +303			4001	−157 +150	

$\bar{x}$ 398 $\quad \Sigma(x-\bar{x})^2$ 34599 $\quad \bar{y}$ 334 $\quad \Sigma(y-\bar{y})^2$ 14513

$$S=\sqrt{\frac{34599+14513}{22}}=\sqrt{\frac{49112}{22}}=\sqrt{2232}$$

$$S_{\bar{x}}=\frac{\sqrt{2232}}{\sqrt{12}}=\sqrt{186} \qquad S_{\bar{y}}=\frac{\sqrt{2232}}{\sqrt{12}}=\sqrt{186}$$

$$S_{\bar{x}-\bar{y}}=\sqrt{372}=19.2$$

$t=\dfrac{64}{19.2}=3.33$ degrees of freedom 22.

$t_{0\cdot 01}=2.818 \quad t>t_{0\cdot 01}$

null hypothesis rejected with 99 per cent confidence

F test for homogeneity of variances

$$S_x^2=\frac{34599}{11}=3145 \quad S_y^2=\frac{14513}{11}=1319$$

$F\ \dfrac{3145}{1319}=2.38$

$F_{0\cdot 99}\ \nu_1\ 11,\ \nu_2\ 11=4.5 \qquad F<F_{0\cdot 99}$

Therefore variances can be considered homogeneous and the t test result is valid from this point of view.

Procedure

Step 1. List data in 2 columns x and y

Step 2. Sum values and find means of the 2 columns

Step 3. Find difference from mean of each value and sum to check = 0

Step 4. Square differences from mean for the 2 columns

Step 5. Sum squared differences for x and for y and then sum these two totals

Step 6. Calculate S by dividing total sum of squared differences (step 5) by no. in sample $(x+y)-2$ (22 in this example) and find square root

Step. 7 Calculate $S_{\bar{x}} = \frac{S}{\sqrt{n_x}}$ where n_x is number in column x (12 in this example.)

Step 8. Calculate $S_{\bar{y}} = \frac{s}{\sqrt{n_y}}$ where n_y is number in column y (12 in this example.)

Step 9. Calculate $S_{\bar{x}-\bar{y}} = S_{\bar{x}} + S_{\bar{y}}$ i.e. square root of step 7 + Step 8. (Note — there is no need to work out square roots until step 9)

Step 10. Calculate t by dividing difference between means of column x and column y by $S_{\bar{x}-\bar{y}}$ (step 9)

Step 11. Consult tables for t at required level of significance with $n-2$ degrees of freedom. If t is greater than tabled value then the null hypothesis can be rejected with the appropriate degree of confidence.

The test is based on the homogeneity of the variances. This can be tested using the F test.

Calculate variance for x, which is S_x^2 (sum of squared differences for x divided by $n_x - 1$) and

variance for y, S_y^2 (sum of squared differences for y divided by $n_y - 1$)

divide $\frac{S_y^2}{S_x^2} = F$ putting larger value on top. Check result in tables for F

F should be entered with appropriate degrees of freedom for x and y i.e. $n_x - 1$ and $n_y - 1$. If the tabled value exceeds the calculated value then the variances can be considered homogenous.

Analysis of variance This is a parametric test, which assumes normality in the population distribution. This test makes it possible to consider several samples of one variable simultaneously. There are a number of different models of the test to suit different problems. The test can be used to examine the differences between the means of several samples and to assess whether they belong to the same or different populations. Analysis of variance is a method by which the variation *between* the different sample means is compared with the variation of items *within* each sample. The ratio of these values gives the value of F which can then be checked against tabled values of F for the appropriate degrees of freedom and at the required significance level. The details of working an example and the results are shown on the instructions and work sheet. The example considers whether five sets of readings on the elongation of drumlins in different parts of the drumlin field of north-west England could all have come from the same population. The value of F

was found to be considerably higher than the tabled value at the 99 per cent significance level. It is thus concluded that there is a significant difference in elongation of the drumlins in different parts of the drumlin field. The analysis of variance test is adaptable to a wide range of situations and is a powerful test when used on suitable data, (see Moroney for further details).

Analysis of variance The null hypothesis states that all the samples could have been drawn from the same population. In this example there are 5 samples. Each consists of 20 values for the elongation of drumlins in 5 separate areas, (1) Around Appleby, (2) In the Eden Valley, (3) Around Carlisle, (4) In Tyne Gap and (5) Solway Lowlands West of Carlisle. The total population consists of all the drumlin elongation values in this part of Northwest England.

Table 3.8 Data for analysis of variance

Five samples of 20 variates each

	A	A^2	B	B^2	C	C^2	D	D^2	E	E^2
1	2.5	6.25	2.0	4.00	2.1	4.41	2.7	7.29	3.4	11.56
2	2.4	5.76	4.9	24.01	1.5	2.25	3.0	9.00	4.4	19.24
3	2.6	6.76	3.3	10.89	1.3	1.69	3.2	10.24	4.0	16.00
4	1.7	2.89	3.8	14.44	1.2	1.44	2.4	5.76	4.9	24.01
5	2.0	4.00	4.0	16.00	1.6	2.56	3.4	11.56	4.9	24.01
6	2.6	6.76	3.5	12.25	2.2	4.84	2.8	7.84	4.0	16.00
7	2.2	4.84	2.6	6.76	1.8	3.24	3.7	13.69	4.7	22.09
8	5.1	26.00	3.5	12.25	1.7	2.89	3.5	12.25	4.0	16.00
9	3.2	10.24	3.3	10.89	2.7	7.29	4.7	22.09	3.4	11.56
10	2.2	4.84	3.8	14.44	3.9	15.21	1.9	3.61	2.5	6.25
11	3.4	11.56	3.4	11.56	2.6	6.76	2.0	4.00	7.2	51.80
12	2.0	4.00	1.5	2.25	2.5	6.25	2.0	4.00	4.5	20.25
13	2.0	4.00	2.8	7.84	2.1	4.41	2.0	4.00	3.7	13.69
14	2.3	5.29	2.0	4.00	2.2	4.84	3.6	12.96	5.0	25.00
15	3.0	9.00	2.4	5.76	2.3	5.29	2.5	6.25	6.4	41.00
16	2.5	6.25	2.7	7.29	1.4	1.96	1.7	2.89	4.9	24.01
17	1.8	3.24	2.3	5.29	1.3	1.69	2.8	7.84	3.5	12.25
18	2.4	5.76	2.9	8.41	1.4	1.96	2.6	6.76	9.7	94.00
19	2.9	8.41	2.0	4.00	1.1	1.21	2.6	6.76	5.0	25.00
20	2.2	4.84	4.0	16.00	1.5	2.25	3.0	9.00	3.2	10.24
$N=$ 100	51.0	140.69	60.7	198.33	38.4	82.44	56.1	167.79	93.3	484.17

$T=51.0+60.7+38.4+56.1+93.3=299.5$ $T^2/N=897.04$ correction factor

Total sum of squares $140.69+198.33+82.44+167.79+484.17-897.04$
$=1073.23-897.04=176.19=SS$

Between sample sum of squares $=\frac{1}{20}(51.0^2+60.7^2+38.4^2+56.1^2+93.3^2)-897.04=82.9=SST$

Within sample sum of squares $=176.19-82.9=93.3=SSE$

	Source of variation		Sum of squares	*df*	*V*	
SST	Between sum	Step 8 result	82.9	1 4 (No of column − 1)	$\frac{82.9}{4}$	1
SSE	Within sum	Step 9 result	93.3	2 95 (3 − 1)	$\frac{93.3}{95}$	2
SS	Total	Step 7 result	176.2	3 99 (N − 1)		

Check step 8 + step 9 = step 7 $V_1 = 20.7$ for between sum
$V_2 = 0.984$ for within sum

$$\therefore F = \frac{V_1}{V_2} = \frac{20.7}{0.932} = 21.03$$

F_{99} (for $V_1 = 4$, $V_2 = 95$) = 3.52 $\therefore F = 21.03$ is significant because $F > F_{99}$

Procedure

Step 1. List samples in 5 columns
Step 2. Add up each column
Step 3. Add column sums to give T = Total sum
Step 4. Calculate T^2/N N = Total number of variates. This is the correction factor.
Step 5. Square all values in the 5 columns and sum each column
Step 6. Sum 5 column squares
Step 7. Calculate *total sum of squares* (*SS*) by subtracting T^2/N (step 4) from total given in step 6. This gives total sum of squares
Step 8. Calculate *between sample sum of squares* (*SST*) by squaring sums of the value found in step 2 for each of the 5 columns. Divide this by the number in each column (20 in this case), and deduct the value T^2/N (step 4)
Step 9. Calculate *within sample sum of squares* (*SSE*). This is total sum of squares (step 7) less between sample sum of squares (step 8).
Step 10. Set up table as above

(*c*) *Testing distributions*

Non-parametric tests The tests that have been described briefly already in the last sub-section were mainly designed to test whether two or more samples were significantly different or not. In this section a few of the tests that are available to establish the characteristics of a distribution are mentioned. Most of the examples given in this section are 1-dimensional, but the application of similar methods to 2- and 3-dimensional and time situations are considered in the later chapters. Many of the methods can be used on data in different dimensions with suitable adaptations.

Random turning point It is sometimes useful to establish whether a set of distances such as the thickness of varves, or the length of successive inter-moraine distances, are distributed randomly or whether they show some trend. The distance between the objects is plotted as an ordered sequence, as in the example shown (see figure 3.8). The points are joined and the turning

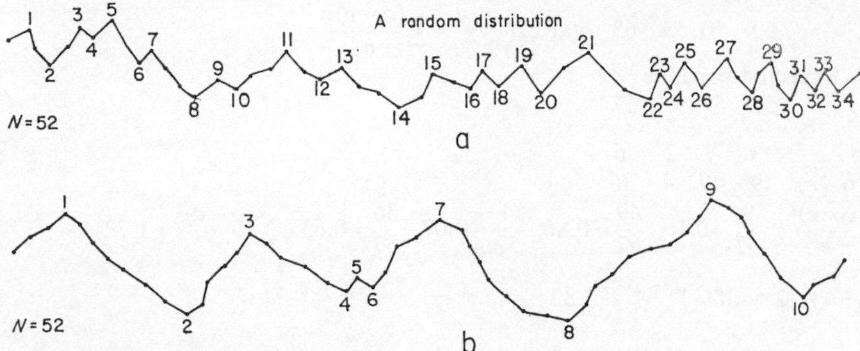

Figure 3.8 A series of observations to illustrate random turning point test. The upper series is a random one with 95 per cent confidence; the lower one is a non-random series. (*a*) Number of turning points expected is $\frac{2}{3}(n-2)=33.3$. in this example. Variance $(16n-29)/90=9.2$. 95 per cent confidence limit is $33.3\pm\sqrt{9.2}\times1.96=33.3\pm3.2\times1.96=33.3\pm5.96$. No. of turning points $=34$. (*b*) Turning points 10. A non-random distribution.

points counted. A turning point is defined as a point which is a peak or trough on the graph. The number of turning points that would be expected in a random distribution is $\frac{2}{3}(n-2)$ or 33.3 in this example. Its variance is $(16n-29)/90$. The 95 per cent confidence limit is found by multiplying the square-root of this value by 1.96. If the actual value lies outside the 95 per cent confidence band, then it can be assumed that the pattern is not random. There should be at least 50 observations for this test to be applied. In the example given the distribution is definitely random as the value of 34 lies close to the calculated value of 33.3. The value would have to be greater than 40 or less than 27 to be significantly non random.

Chi (χ) *square test* There are many uses for the chi square test, which can be applied to any data that can be expressed as frequencies. The test measures the discrepancies between the expected and the observed frequencies and can be used for one, two or many variables. Only the one-variable cases will be considered in this section. The chi square test only gives reliable results if the minimum expected frequency in any one group is no less than 5.

A simple application of the chi square test would be to compare the actual spacing of objects with a regular spacing. For example the number of moraines per mile can be counted in areas where many occur in parallel sequence and where they may be regarded as annual features. The total number can be averaged to give the expected frequency per mile assuming that they are evenly spaced. The observed value can then be compared with the expected at the required significance level. The degrees of freedom are equal to the total number of groups less one.

Observed per mile

60 50 55 63 total 228 $\chi^2 = \sum \frac{(o-e)^2}{e}$

Expected per mile

57 57 57 57

$o - e$ 3 7 2 6

$(o-e)^2$ 9 49 4 36

$\frac{(o-e)^2}{e}$ $\frac{9}{57} = 0.16$, $\frac{49}{57} = 0.86$, $\frac{4}{57} = 0.07$, $\frac{36}{57} = 0.63 = \sum \frac{(o-e)^2}{e} = 1.72$

$\chi^2 = 1.72$ df $= 3$ $\chi^2_{95} = 7.82$

The calculated result is less, therefore the hypothesis that the moraines are regularly spaced is not rejected.

It is sometimes necessary to determine whether a particular distribution is normal. This will help to decide what type of test to use, whether it is parametric or non-parametric. This problem can be solved by using the chi square test to compare the observed and expected frequencies. A fairly large sample is required and the degrees of freedom are the number of classes less three. Frequency distributions can also be tested against other types of statistical distribution such as the binomial or Poisson. An example of testing a frequency for the normal distribution is given. The use of probability paper is a quicker way of testing a distribution for normality usually, but the significance of the result cannot be ascertained.

Testing distribution for normality

Data Measurement of stone length in a raised beach crest in Åland Island, Finland.

Test Null hypothesis is that the sample frequency is normally distributed.

Procedure First the expected frequencies in the different classes must be calculated using the area under the normal curve given in the z tables. The class boundaries must be known for this purpose and they are less than 6.5 cm, 6.5–8.5, 8.5–10.5, 10.5–14.5, 14.5–18.5, 18.5–22.5, and over 22.5. The mean and standard deviation of the sample must also be known and these values are $m = 12.24$ cm and

$s = 6.37$ cm, $n = 50$, z is obtained by the formula

$$z_1 = \frac{\text{class boundary} - m}{s}.$$

Thus $z_1 = \frac{6.5 - 12.24}{6.37} = \frac{-5.74}{6.37} = 0.903$ giving A in tables as 0.3165 (area under normal curve) area outside this is $0.5000 - 0.3165 = 0.1835$ and this value is multiplied by n to give expected 9.175)

$$z_2 = \frac{8.5 - 12.24}{6.37} = \frac{-3.74}{6.37} = 0.586 \quad A = 0.2208 \quad e = (0.3165 - 0.2208) \times 50 = 4.785$$

$$z_3 = \frac{10.5 - 12.24}{6.37} = \frac{-1.74}{6.37} = 0.273 \quad A = 0.1074 \quad e = (0.2208 - 0.1074) \times 50 = 5.670$$

$$z_4 = \frac{14.5 - 12.24}{6.37} = \frac{+2.26}{6.37} = 0.354 \quad A = 0.1383 \quad e = (0.1074 + 0.1383) \times 50 = 12.285$$

$$z_5 = \frac{18.5 - 12.24}{6.37} = \frac{+6.26}{6.37} = 0.985 \quad A = 0.3377 \quad e = (0.3377 - 0.1383) \times 50 = 9.970$$

$$z_6 = \frac{22.5 - 12.24}{6.37} = \frac{+10.26}{6.37} = 1.61 \quad A = 0.4463 \quad e = (0.4463 - 0.3377) \times 50 = 5.430$$

$$z_7 = >22.5 \qquad e = (0.5000 - 0.4463) \times 50 = 2.685$$

$$\overline{50.000}$$

Chi square test

	o	e	$o - e$	$(o - e)^2$	$\frac{(o - e)^2}{e}$
<6.5	8	9.2	−1.2	1.44	0.16
6.5 − 8.5	7	4.8[a]	+2.2	4.84	1.01
8.5 − 10.5	12	5.7	+6.3	39.69	6.97
10.5 − 14.5	7	12.3	−5.3	28.09	2.29
14.5 − 18.5	8	10.0	−2.0	4.00	0.40
18.5 − 22.5 } >22.5 }	8	8.1	−0.1	0.01	0.001

$$\sum \frac{(o - e)^2}{e} = 10.831 = \chi^2$$

[a]This value is near enough 5 to be allowed

degrees of freedom $6 - 3 = 3$

$\chi^2_{0.01} = 11.34 \quad \chi^2_{0.05} = 7.82$

The null hypothesis is rejected at the 95 per cent level but accepted at the 99 per cent level.

One other example of the use of the chi square test in a one variable situation will be mentioned. This is the slightly modified test that can be applied to orientational data in which the distribution is continuous. This method is called the Tukey chi square and is useful in analysing the significance of stone orientation data on a 2-dimensional distribution. The test provides a measure of the mean orientation as well as a method of determining whether there is a preferred orientation in the data. If this is strong, so that the calculated χ^2 has a high value, then it can be concluded that there is a strongly preferred orientation that could not readily have developed by chance. An example of the application of this test and the appropriate procedure is given.

Tukey chi square test for orientational data The value of χ^2 for C^2+S^2 for $2df$ is required. The mean orientation is given by $\tan\theta=\frac{S}{C}$ for 360° pattern and $\tan 2\theta=\frac{S}{C}$ for 180° ∴ add mid -point of class interval to θ. Class interval is 20° ∴ add 10°. C is given by $\frac{\Sigma x\cos 2\theta}{\sqrt{\Sigma\cos^2 2\theta}}$. S is given by $\frac{\Sigma x\sin 2\theta}{\sqrt{\Sigma\sin^2 2\theta}}$ $x=\frac{o-e}{\sqrt{e}}$ (e is 5.6 as it is expected that the distribution be regular for $n=50$ i.e. 50 ÷ 9. $\sqrt{e}=2.36$). For 20° interval $\sqrt{\Sigma\cos^2 2\theta}=2.117$ $\sqrt{\Sigma\sin^2 2\theta}=2.117$. The data are derived from the till fabric measured in a drumlin.

Work-sheet

θ	obs	exp	$o-e$	x	$x\cos 2\theta$	$x\sin 2\theta$	$\cos 2\theta$	$\sin 2\theta$
0	9	5.6	+3.4	1.44	+1.44	+0	+1.000	0.000
20	15	5.6	+9.4	4.00	+3.08	+2.57	+0.766	+0.643
40	6	5.6	+0.4	0.17	+0.03	+0.17	+0.174	+0.985
60	6	5.6	+0.4	0.17	−0.03	+0.15	−0.500	+0.866
80	5	5.6	−0.6	0.25	+0.23	−0.08	−0.940	+0.342
100	0	5.6	−5.6	2.37	+2.22	+0.81	−0.940	−0.342
120	3	5.6	−2.6	1.10	+0.55	+0.95	−0.500	−0.866
140	2	5.6	−3.6	1.52	−0.26	+1.50	+0.174	−0.985
160	4	5.6	−1.6	0.68	−0.52	+0.44	+0.766	−0.643
					+7.55	+6.59		
					−0.81	−0.08		
					+6.74	+6.51		

$C=+\frac{6.74}{2.117}$ $S=+\frac{6.51}{2.117}$ $C=3.18$ $S=3.08$

$\tan 2\theta=\frac{3.08}{3.18}=0.972$ $2\theta=44°$ $\theta=22°$

mean orientation = 22° + 10° = 32°

$\chi^2=10.1+9.5=19.6$

Procedure

Step 1. List observed values against appropriate class

Step 2. List expected values

Step 3. Calculate $o-e$

Step 4. Calculate x where x is $\frac{o-e}{\sqrt{e}}$ $\sqrt{e}=2.36$ for n = 50 and 20° div.

Step 5. Calculate $x\cos 2\theta$, (values of $\cos 2\theta$ can be obtained from tables and are constant, the same applies to $\sin 2\theta$)

Step 6. Calculate $x\sin 2\theta$

Step 7. Sum $x\cos 2\theta$, noting signs

Step 8. Sum $x\sin 2\theta$, noting signs

Step 9. Calculate C and S by dividing $\Sigma x\cos 2\theta$ and $\Sigma x\sin 2\theta$ by $\sqrt{\Sigma\cos^2 2\theta}$ and $\sqrt{\Sigma\sin^2 2\theta}$ respectively (both values = 2.117)

Step 10. Calculate tan 2θ by dividing S and C, look up 2θ in tan tables, divide by 2 and add 10 degrees. This gives mean orientation

Step 11. Calculate χ^2 by squaring C and S and adding the results

χ^2 can now be compared with tabled values for 2 *df*

The significance levels are 90 per cent $\chi^2 = 4.61$
95 per cent $\chi^2 = 5.99$
99 per cent $\chi^2 = 9.21$
99.5 per cent $\chi^2 = 10.60$

The calculated χ^2 exceeds the tabled value for a 99.5 per cent significance, therefore the null hypothesis is rejected with a very high degree of confidence. This hypothesis states that the distribution is uniform.

3.6 Inferential statistics: 2 variables — correlation

(a) *The working of three important tests*

Most of the tests that have been described so far are intended to assess the significance of the difference between two or more sets of data or assess the

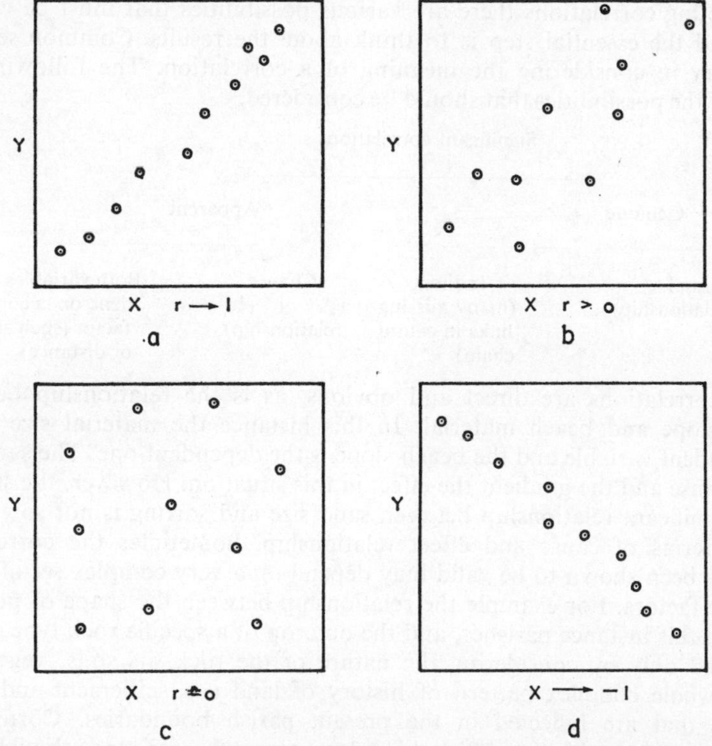

Figure 3.9*a* Four scatter graphs to show correlations ranging from *(a)* a strong positive correlation, *r* is nearly 1, through *(b)* a weak positive correlation, $r > 0$, and *(c)* no correlation, $r \simeq 0$, to *(d)* a strong negative correlation, *r* is nearly -1.

characteristics of the distribution of the samples. Correlation methods provide measures whereby the relationship between two or more variables can be calculated. The degree of relationship can be expressed in precise terms as a coefficient of correlation. In this section the correlation of two variables only will be considered and the more complex situation of a large number of variables will be considered in the next section. The simple correlation coefficient provides the raw material for the multivariate analysis processes.

The correlation coefficient is made to range from +1 to −1 except for the chi-square test. The former implies a perfect positive correlation and the latter a perfect negative one, in which one variable increases as the other decreases. In these situations the values of the two variables plotted on a graph against each other can be linked by a straight line (see figure 3.9*a*). When the correlation coefficient is zero there is no relationship between the two variables at all. However when the value of the correlation coefficient is significantly high then a relationship between the variables can be suggested. It does not mean that the two variables are necessarily causally related. In interpreting correlations there are various possibilities that must be considered and the essential step is to think about the results. Common sense is necessary in considering the meaning of a correlation. The following are some of the possibilities that should be considered.

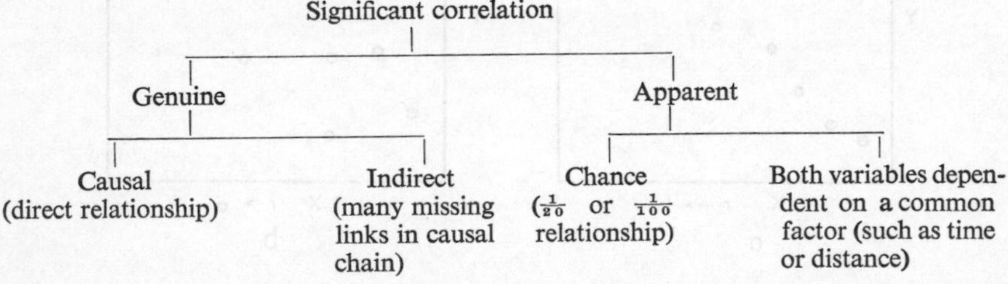

Some correlations are direct and obvious, as is the relationship between beach slope and beach material. In this instance the material size is the independent variable and the beach slope is the dependent one. The sand size is the cause and the gradient the effect in this situation. However, the statistically significant relationship between sand size and sorting is not so easy to see in terms of cause and effect relationship. Sometimes the correlation that has been shown to be valid may depend on a very complex set of intermediate factors. For example the relationship between the shape of political units, in this instance parishes, and the outcrop of a specific rock type can be explained only by considering the nature of the rock, its soils, vegetation and a whole complex pattern of history of land use, settlement and other matters that are reflected in the present parish boundaries. Correlation methods can establish relationships but commonsense and thought are needed to explain these relationships.

Most situations in reality are extremely complex and reflect the interaction of a large number of dependent and independent variables, so that high correlations cannot be expected to occur often. Errors in measurements and sample selection can also affect the level of correlation. This complexity can be reduced to some measure of order by the more complex methods of sorting out many of variables, such as factor analysis. Three of the possible methods of correlation will be considered first, including both non-parametric and parametric tests, then the relative merits of the three methods discussed will be assessed by comparing the results of the 3 tests on the same data.

(i) *Chi-square test for correlation* The chi-square test is a very flexible one that can be adapted to test if two variables are correlated if they are given in frequencies. The data can be on the nominal or any higher scale and the test is non-parametric. The variation in relative frequency in the various classes is listed. The null hypothesis states that there is no difference between the two variables. An example illustrates how the test can be applied. It is adaptable and a wide range of different patterns can be set up. It is possible to have only 2×2 cells as in the example given later or more than two frequencies in both rows and columns. To be valid the chi-square test demands that no cell has an expected frequency of less than 5 and the total numbers involved should be greater than 20. This test can indicate whether the whole range of values compared differ significantly and if they do the classes that contribute most to the difference can be identified by considering the chi-square value for each cell individually. Its significance can be tested by using only one degree of freedom which gives $\chi^2_{0.05}$ as 3.841. If this value is exceeded by the calculated value then the difference is significant in this particular class. In the example given none of the individual cells has a significant result but the consistency of the trend leads to the significance of the total result. The first cell, which has the smallest values, provides the most important contribution to the difference. This test has the advantage that it is easy to compute but it lacks in power.

Chi-square test — 2 variables Frequency of northeast facing corries correlated with frequency of southwest facing corries at different altitudes.

Table 3.9 Data for chi-square correlation

	SW		NE		Totals
	o	*e*	o	*e*	
below 1000′	10	16.0 e_1	25	19.0 e_3	35
1000–2000′	33	35.3 e_2	44	41.7	77
above 2000′	67	58.7	61	69.3	128
Totals	110		130		240

calculate $e_1 = \frac{35+110}{240}$ $\quad e_2 = \frac{77 \times 110}{240}$ $\quad e_3 = \frac{130 \times 35}{240}$ etc.

then

$$\sum\frac{(o-e)^2}{e}=\frac{6^2}{16}+\frac{6^2}{19}+\frac{2.3^2}{35.3}+\frac{2.3^2}{41.7}+\frac{8.3^2}{58.7}+\frac{8.3^2}{69.3}$$

$$\sum\frac{(o-e)^2}{e}=2.25+1.89+0.15+0.13+1.17+0.99=6.58$$

$$\therefore \chi^2=6.58$$

k = columns r = rows

There are 2 degrees of freedom because $(r-1)(k-1)=1\times2=2$. $\chi^2_{0.05}=5.99$ so as $\chi^2=6.58$ is greater than this value the null hypothesis is rejected and the two directions can be considered to be significantly different at the 95 per cent level, but they are not significantly different at the 99 per cent level as $\chi^2_{0.01}$ is 9.21.

(ii) *Spearman rank correlation coefficient** Spearman rank correlation is a test for correlation using ranked data on the ordinal scale of measurement and is therefore non-parametric. The data may be collected in ranked form (e.g. the placings of competitors by judges in a competition) or interval or ratio data may be converted to ranked form. The following example shows how the index of correlation, ρ (rho), is calculated. Two persons are asked to rank the ten countries listed in alphabetical order below (table 3.10) according to their importance (in the view of each judge) as world powers. A significant positive correlation in this example suggests that there is close agreement between the judges.

Table 3.10 Data for rank correlation

	Judge A	Judge B	d	d^2
1. Austria	7	8	−1	1
2. Burma	9	9	0	0
3. Costa Rica	10	10	0	0
4. Ghana	6	7	−1	1
5. India	4	5	−1	1
6. Japan	5	3	2	4
7. New Zealand	8	6	2	4
8. U.S.S.R.	2	2	0	0
9. U.K.	3	4	−1	1
10. U.S.A.	1	1	0	0
			$\Sigma d=0$	$\Sigma d^2=12$

The index of correlation is derived from the following formula

$$\text{rho}(\rho)=1-\frac{6\Sigma d^2}{n^3-n}$$

* The reader concerned with teaching statistical tests to geographers is advised to teach Spearman first. Chi-square test initially causes difficulties when it is being set up. Product-moment involves long manual calculations. Spearman is relatively simple *and* appeals immediately to commonsense.

In the formula

ρ is the index of correlation. It falls between -1 and $+1$
d is the difference in ranking between the two judges
Σd^2 means sum (add) all values of d^2
n is the number of cases (objects) being ranked
1 and 6 remain unchanged in any application of the formula

Steps:

Step 1. If the data are not already ranked, rank them, placing the pair of values side by side for each case. In the above example, the data are already ranked, but in an example later in this section they are derived from ratio data. Note that when equal rankings are given by the judges, it is necessary to calculate the positions the equally ranked items would have on the list, sum them and find their means. For example, judge A might have put New Zealand, Burma and Costa Rica all last equal. They should occupy places 8, 9, 10. As $8+9+10=27$, they should all be given a rank of 9 in this case. Had he put just Burma and Costa Rica last equal, then their correct ranking should be calculated by summing $9+10$ and finding the mean, namely 9.5. When an even number of items receive equal ranking the answer should end in .5

Step 2. Calculate the difference in ranking. The sum of the d values should be 0. This is an optional initial check of the calculation

Step 3. Calculate the values of d^2 and sum these

Step 4. Calculate as shown below

$$\rho = 1 - \frac{6\Sigma d^2}{n^3 - n}$$

$$\rho = 1 - \frac{6 \times 12}{1000 - 10}$$

$$\rho = 1 - \frac{72}{990}$$

$$\rho = +0.927$$

In this case there is a high positive correlation since the judges are in close agreement.

When the rankings are identical a correlation of $+1.0$ is obtained. It may be useful for the reader to refer back to section 3.2 on Probability. Since there are 10! ways of arranging the positive integers 1–10 and only one of these is identical with a given ordering, then the probability of obtaining $+1.0$ by chance is $\frac{1}{10!}$ or approximately $\frac{1}{3{,}629{,}000}$. This is evident from a consideration of the formula, since each d is 0 so $6\Sigma d^2$ is also 0. When the rankings of one judge are in reverse order to those of the second, a correlation of -1.0 is obtained.*

* It is interesting to note that if one of the judges gives *all* the cases equal rank (which would be 5.5 with 10 cases), the index of correlation is always $+0.5$, whatever the ranking of the second judge. Therefore this suggests that the test is inappropriate when more than a few cases are equal (tied).

The purpose of the test is to establish whether a hypothesis can be accepted or rejected with given confidence. If the numbers 1–10 were twice chosen randomly (without replacement) and put side by side, a correlation near 0 would usually occur. The larger the number of cases (n), the more often the randomly derived numbers would give correlations near 0. Thus when n is 10, the 95 per cent confidence limit is about ± 0.600, whereas when n is 30 it falls closer to 0, at ± 0.31.

The following are selected significance levels for the Spearman rank correlation tests. S. Siegel (1956). (Use $n - 1$).

	95 per cent	99 per cent
$n = 10$	0.56	.75
15	0.44	.60
20	0.38	.53
25	0.34	.47
30	0.31	.43

(iii) *Product–moment correlation* If two sets of variables are measured on the interval or ratio scale it is possible to plot the pairs of values against each other on a graph. The scatter of points then gives a visual impression of the relationship between the two variables. A more exact value can be obtained by calculating the coefficient of correlation mathematically. The Pearson product–moment correlation provides a method for doing this. It is also possible to enter regression lines on the graph. These straight lines are drawn through the scatter of points in such a way that the squares of the distances between the points and the regression line are reduced to a minimum. The equations for the regression line, particularly the regression coefficient, provide a means of predicting the value of the dependent variable which would be expected for any given value of the independent variable. It is strictly dangerous to extrapolate far beyond the range of values covered in the original set of data. It is assumed in this test that there is a single linear regression between the two variables. This may not always be true, but it is sometimes possible to allow for this by using the log of the values or converting the data into other forms, such as the ϕ scale used in sediment studies.

An example of the computation necessary to calculate the correlation coefficient is given with instructions. When the correlation coefficient has been found it can be used to indicate what percentage of the variation of the dependent variable depends on the independent one by working out the value of $100r^2$. Thus if r is 0.8 then 64 per cent of the variation in y is 'explained' by variations in x, if x is the independent variable.

The linear regression formula is given in the form $Y = a + bX$ where a is the intercept on the Y axis. This is the value of Y when X is zero, and b gives the slope of the regression line in the form of the tangent of the angle of slope. If a and b have been calculated then Y can be predicted from X as shown in the example.

Example of product–moment correlation and regression analysis In this example, as is customary, X is the independent variable on which Y, the dependent variable, depends. X is the mean stone size in millimetres at different points on a series of deltas in Baffin Island. Y is the slope of the delta, expressed as a cotangent of the slope angle, which is equal to a gradient of 1 in Y, at the points where X was measured. The regression is expressed in the form $Y = a + bX$ so that Y can be calculated from known values of X. It is suggested, in this instance, that there is a causal relationship between the two variables, so that Y is determined, at least partly, by the value of X.

The pairs of values are first plotted against each other on a scatter graph. This gives an indication of the type and degree of association between the two variables. Four scatter graphs are shown in figure 3.9*a* to illustrate possible relationships ranging from a strong positive association (a), through a weak positive association (b), and no association (c) to a strong negative association (d). The degree of association or correlation is given by the value of r, the correlation coefficient. Figure 3.9*b* shows the plot of the observations used in this example and the scatter of points suggests a negative correlation between the two variables.

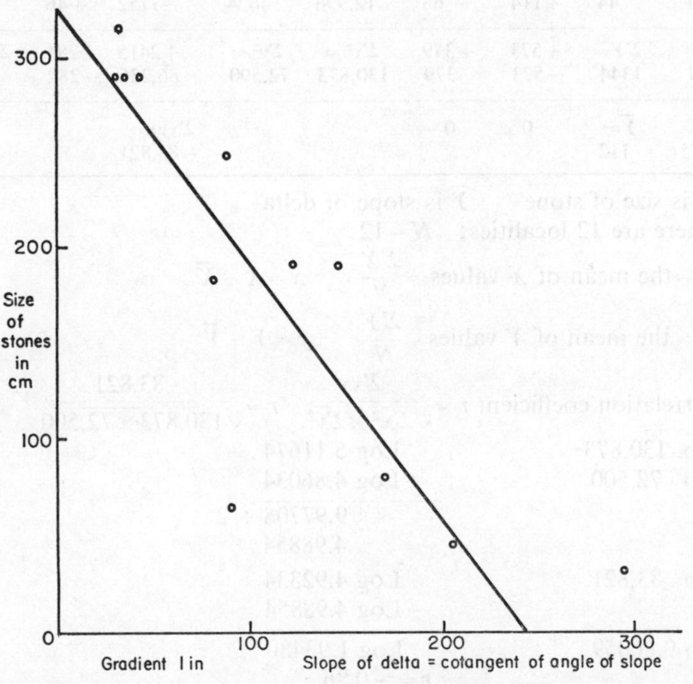

Figure 3.9*b* Graph to show relationship between slope of delta and size of stones from observations in West Baffin Island. The regression line equation is $Y = 232.8 - 0.654X$ where Y, the delta slope, is the dependent variable and X, the stone size, is the independent variable.

The calculation of the coefficient of correlation, r, can be set out as shown in table 3.11. Note that the upper case X and Y refer to the individual observations, and the lower case x and y refer to the differences between the individual observations and the means of each set of observations.

Table 3.11 Product moment data and worksheet

1 X	2 Y	3 x	4 y	5 x^2	6 y^2	7 xy	8 $x+y$	9 $(x+y)^2$
32	295	−145	+183	21,025	33,489	−26,535	+38	1444
291	35	+114	−77	12,996	5929	−8778	+37	1369
316	33	+139	−79	19,321	6241	−10,981	+60	3600
44	207	−132	+95	17,424	9025	−12,540	−37	1369
89	147	−88	+35	7744	1225	−3080	−53	2809
191	118	+14	+6	196	36	+84	+20	400
248	89	+71	−23	5041	529	−1633	+48	2304
81	172	−96	+60	9216	3600	−5760	−36	1296
65	91	−111	−21	12,321	441	+2331	−132	17,424
289	32	+112	−80	12,544	6400	−8960	+32	1024
184	81	+7	−31	49	961	−217	−24	576
291	44	+114	−68	12,996	4624	−7752	+46	2116
$\Sigma X=$ 2121	$\Sigma Y=$ 1344	+573 −573	+379 −379	$\Sigma x^2=$ 130,873	$\Sigma y^2=$ 72,500	+2415 −86,236	+281 −282	$\Sigma(x+y)^2=$ 35,731
$\overline{X}=$ 176.75	$\overline{Y}=$ 112	0	0			$\Sigma xy=$ −83,821		

$\overline{X}$ is size of stone $\quad Y$ is slope of delta

There are 12 localities; $N=12$

$\overline{X}=$ the mean of X values $=\dfrac{\Sigma X}{N} \qquad x=X-\overline{X}$

$\overline{Y}=$ the mean of Y values $=\dfrac{\Sigma Y}{N} \qquad y=Y-\overline{Y}$

The correlation coefficient $r=\dfrac{\Sigma xy}{\sqrt{\Sigma x^2\times\Sigma y^2}} \qquad r=\dfrac{-83,821}{\sqrt{130,873\times 72,500}}$

No. 130,873	Log 5.11674
No 72,500	+Log 4.86034
	9.97708 ÷ 2
	= 4.98854
No 83,821	Log 4.92334
	−Log 4.98854
No 0.86059	Log $\bar{1}$.93480
	$r=-0.86$

Instructions for calculation of r (Pearson product-moment correlation coefficient)

Step 1. Add up column 1. Divide it by N (12 in this example) to give $\overline{X}$ (Symbol for the mean of X's)

Step 2. Add up column 2. Divide it by N to give $\overline{Y}$

Step 3. Complete column 3 by taking $\overline{X}$ from X. Put in + or − as appropriate. Check that +'s and −'s are equal

Step 4. Complete column 4 by taking $\overline{Y}$ from Y and check as in step 3

Step 5. Square each number in column 3 and enter it in column 5

Step 6. Square each number in column 4 and enter it in column 6

Step 7. Multiply numbers in column 3 by numbers in column 4 and enter results in column 7.

Note columns 8 and 9 provide a check for the calculation but are not part of the procedure for finding the correlation coefficient. The check is made in steps 13 to 15.

Step 8. Add numbers in columns 3 and 4 and enter result in column 8 (Note signs correctly)

Step 9. Square numbers in column 8 and enter in column 9

Step 10. Add numbers in column 5 to give Σx^2

Step 11. Add numbers in column 6 to give Σy^2

Step 12. Add numbers in column 7 to give Σxy (find difference between + and − values and note sign correctly)

Step 13. (Check) Add +'s and −'s separately in column 8, the results should be equal

Step 14. Add numbers in column 9 to give $\Sigma(x+y)^2$

Step 15. Equate $\Sigma x^2 + \Sigma y^2 + 2\Sigma xy$ with $\Sigma(x+y)^2$. The results should be equal.

The figures required to calculate r are now available. The correlation coefficient, r, is given by $r = \dfrac{\Sigma xy}{\sqrt{\Sigma x^2 \ \Sigma y^2}}$. r always lies between + 1 and − 1 In this example it is −0.86. This is significant at the 99 per cent confidence level. (R must exceed 0.71 to be significant at the 99 per cent level for 12 observations (pairs of values).

Instructions for calculation of the regression coefficient and regression line formula The equation for the regression line is given by $Y = a + bX$. a gives the intercept on the Y axis and b is the regression coefficient, which gives the slope of the regression line as a tangent value. a is calculated by using the equation $\overline{Y} = \overline{X} \times \dfrac{\Sigma xy}{\Sigma x^2} - a$ (1) (note that $b = \dfrac{\Sigma xy}{\Sigma x^2}$)

Step 1. Substitute the known values in the formula (1) and calculate the value of a.

Step 2. Substitute a and b in the regression formula $Y = a + bX$ (2)

Step 3. Using equation (2) calculate values for Y corresponding to values for X of 0, 100, 200 and 300. Plot these co-ordinates for X and Y on the scatter graph. They should lie on a straight line, which is the regression line of Y in terms of X. This line is shown in figure 3.9(b).

Note that it is also possible to draw the regression line of X in terms of Y by using the equation $\overline{X} = \overline{Y} \times \frac{\Sigma xy}{\Sigma y^2} + a_1$. This line may be of value when there is no direct dependence of one variable on the other, for example when both are dependent on a third variable.

Instructions for calculation of the standard error of estimate It is useful to be able to assess the reliability of the regression lines and this can be done by calculating the standard error of estimate. Each regression line has a standard error of estimate which depends on the standard deviation of the appropriate variable and the correlation coefficient. The standard error of estimate gives a measure of the reliability of the sample to predict the relationships of the whole populations of the two variables.

In this example the regression of Y in terms of X has been calculated, this

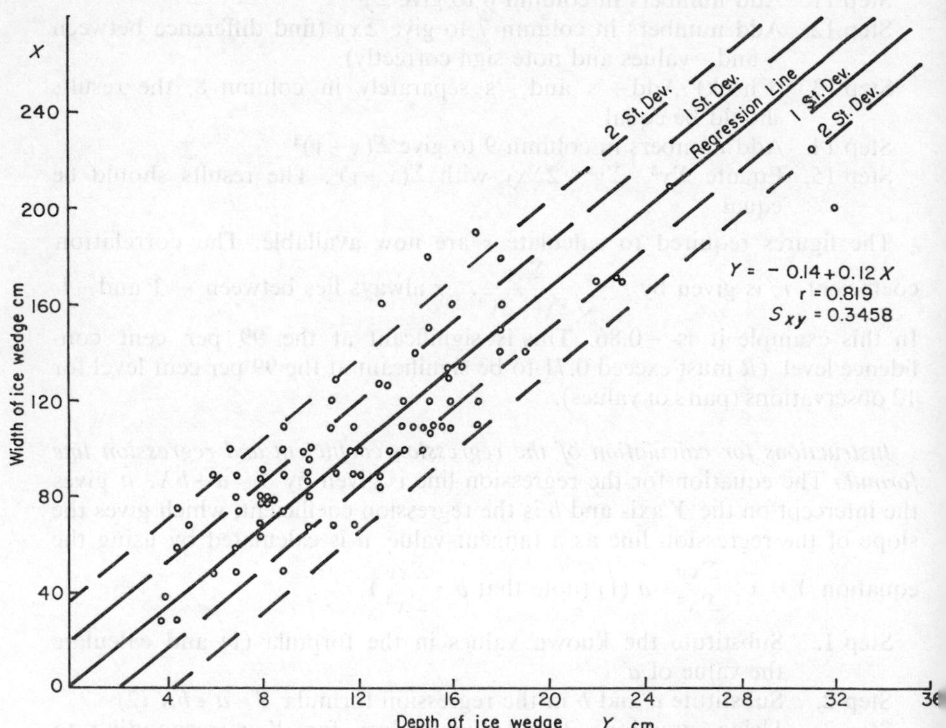

Figure 3.10 Graph to show relationship between width of ice wedges (X) and depth of ice wedges (Y). The regression line shown gives Y the dependent variable. The confidence limits given by the standard error of estimate S-- are shown at 1 and 2 standard deviations from the regression line.

can be called the regression of Y on X, and the regression formula has the form $Y=a+bX$. The standard error of estimate for this regression line is given by $S_{yx}=\sigma_y\sqrt{1-r^2}$. This can be stated as $\frac{\Sigma y^2}{N}\times\sqrt{1-r^2}$. If two lines are drawn parallel to the regression line, on either side of it, at distances equal to $2S_{yx}$ then 95 per cent of the total population would be expected to fall between these two lines. An example of the standard error of estimate for the 1 standard deviation and the 95 per cent levels (2 standard deviations), is shown in figure 3.10.

This subsection on product–moment correlation concludes with the description of a method for finding the coefficient with the help of matrices, discussed in 2.8. Although manual methods of working out the product–moment correlation coefficient are quicker for limited data, the matrix method is of interest as it is the way large amounts of data are handled in a computer program. The results of such an operation are given in chapter 6.4, for the pairwise intercorrelation of 25 variables for the 50 states of the U.S.A.

In order to show how this method is carried out a very simple example is worked by both the standard method and by matrix algebra.

In table 3.12 the normal method of calculation r is given first. The values have been chosen to provide a very simple result in order to illustrate the method. The two columns X and Y are two variables to be correlated, and it is from them that the calculation starts. The value of r, given by

$$r=\frac{\Sigma xy}{\sqrt{\Sigma x^2\,\Sigma y^2}}=0.95$$

Table 3.12 Simple product–moment correlation

X	Y	x $X-\bar{X}$	y $Y-\bar{Y}$	x^2	y^2	xy	$r=\frac{\Sigma xy}{\sqrt{\Sigma x^2\,\Sigma y^2}}$
3	2	0	0	0	0	0	$r=\frac{6}{\sqrt{10\times4}}$
4	3	+1	+1	1	1	+1	
2	1	−1	−1	1	1	+1	$r=\frac{6}{6.32}$
1	1	−2	−1	4	1	+2	
5	3	+2	+1	4	1	+2	
$\Sigma X=15$	$\Sigma Y=10$	$\Sigma x=0$	$\Sigma y=0$	$\Sigma x^2=10$	$\Sigma y^2=4$	$\Sigma xy=+6$	$r=0.95$

$\bar{X}=3$ $\bar{Y}=2$ =Means

Matrix method of calculating correlation coefficient

The data matrix D is first set out as a 5×2 matrix.

$$\text{Data matrix } D=\begin{matrix} X\;Y \\ \begin{bmatrix} 3 & 2 \\ 4 & 3 \\ 2 & 1 \\ 1 & 1 \\ 5 & 3 \end{bmatrix}\end{matrix}$$

The column matrix (or vector) $j = \begin{bmatrix}1\\1\\1\\1\\1\end{bmatrix}$ and its transpose $j^T = [1\ 1\ 1\ 1\ 1]$

are also required.

The 5×5 matrix J is found by multiplying j by j^T

$$j\,j^T = \begin{bmatrix}1&1&1&1&1\\1&1&1&1&1\\1&1&1&1&1\\1&1&1&1&1\\1&1&1&1&1\end{bmatrix}$$ This and the identity matrix I are required

for computing the correction factor.

Matrix I is a 5×5 matrix $I = \begin{bmatrix}1&0&0&0&0\\0&1&0&0&0\\0&0&1&0&0\\0&0&0&1&0\\0&0&0&0&1\end{bmatrix}$

The sums of X and Y are found by multiplying vector j^T by D

$$j^T D = [1\ \ 1\ \ 1\ \ 1\ \ 1] \cdot \begin{matrix} X & Y \\ \end{matrix} \begin{bmatrix}3&2\\4&3\\2&1\\1&1\\5&3\end{bmatrix} = [15\ \ 10] = [\Sigma x\ \ \Sigma Y]$$

The mean values of X and Y are then found by multiplying $j^T D$ by $1/N$, where N is the number of pairs of variates.

$$j^T D \times 1/N \text{ gives } [\bar{X}\ \ \bar{Y}] = \tfrac{1}{5}\,[15\ \ 10] = [3\ \ 2]$$

The transpose matrix D^T is then multiplied by the data matrix D to give the uncorrected sum of squares.

$$D^T D = \begin{bmatrix}3&4&2&1&5\\2&3&1&1&3\end{bmatrix} \cdot \begin{bmatrix}3&2\\4&3\\2&1\\1&1\\5&3\end{bmatrix} = \begin{bmatrix}55&36\\36&24\end{bmatrix} = \begin{bmatrix}\Sigma X^2 & \Sigma XY\\ \Sigma XY & \Sigma Y^2\end{bmatrix}$$

The correction factor is next required. This is given by $I - \dfrac{1}{N}J$.

The corrected sum of squares is then given by $D^T(I - \dfrac{1}{N}J)D$.

This formula can be adjusted as follows:

$$D^T D - 1/N D^T jj^T D = D^T D - \frac{1}{N}(j^T D)^T (j^T D)$$

The terms of the correction factor are now in the form $(j^T D)^T \cdot j^T D$
These values must then be divided by $1/N$.

$$(j^T D)^T \cdot j^T D = \begin{bmatrix} 15 \\ 10 \end{bmatrix} \cdot [15 \ 10] = \begin{bmatrix} 225 & 150 \\ 150 & 100 \end{bmatrix}$$

$$1/N(j^T \mathrm{D}) \cdot j^T D = 1/5 \begin{bmatrix} 225 & 150 \\ 150 & 100 \end{bmatrix} = \begin{bmatrix} 45 & 30 \\ 30 & 20 \end{bmatrix} = \begin{bmatrix} \bar{X}\Sigma X & \bar{X}\Sigma Y \\ \bar{Y}\Sigma X & \bar{Y}\Sigma Y \end{bmatrix}$$

The corrected sum of squares is now found by deducting the correction matrix from the uncorrected sum of squares matrix.

$$\begin{bmatrix} 55-45 & 36-30 \\ 36-30 & 24-20 \end{bmatrix} = \begin{bmatrix} 10 & 6 \\ 6 & 4 \end{bmatrix} = \text{corrected sum of squares}$$

These values give the following

$$\begin{bmatrix} \Sigma X^2 - \bar{X}\Sigma X & \Sigma XY - \bar{X}\Sigma Y \\ \Sigma YX - \bar{Y}\Sigma X & \Sigma Y^2 - \bar{Y}\Sigma Y \end{bmatrix} = \begin{bmatrix} SS \ X^2 & SS \ XY \\ SS \ YX & SS \ Y^2 \end{bmatrix}$$

The sample variance–covariance matrix is next obtained by multiplying the corrected sum of squares by $\frac{1}{N-1}$

This is given by $\frac{1}{N-1} \begin{bmatrix} SS \ X^2 & SS \ XY \\ SS \ YX & SS \ Y^2 \end{bmatrix} = \frac{1}{4} \begin{bmatrix} 10 & 6 \\ 6 & 4 \end{bmatrix} = \begin{bmatrix} 2.5 & 1.5 \\ 1.5 & 1.0 \end{bmatrix}$

$$= \begin{bmatrix} s_{X^2} & s_{XY} \\ s_{YX} & s_{Y^2} \end{bmatrix}$$

The correlation coefficient, r, is given by multiplying the resulting matrix twice by the matrix $\begin{bmatrix} 1/s_X & 0 \\ 0 & 1/s_Y \end{bmatrix}$ as shown below

$$r = \begin{bmatrix} 1/s_X & 0 \\ 0 & 1/s_Y \end{bmatrix} \cdot \begin{bmatrix} s_{X^2} & s_{XY} \\ s_{YX} & s_{Y^2} \end{bmatrix} \cdot \begin{bmatrix} 1/s_X & 0 \\ 0 & 1/s_Y \end{bmatrix}$$

Multiplying the first two matrices gives

$$r = \begin{bmatrix} s_{X^2}/s_X & s_{XY}/s_X \\ s_{YX}/s_Y & s_{Y^2}/s_Y \end{bmatrix} \cdot \begin{bmatrix} 1/s_X & 0 \\ 0 & 1/s_Y \end{bmatrix}$$

multiplying these two matrices gives

$$r = \begin{bmatrix} s_{X^2}/s_{X^2} & s_{XY}/s_X s_Y \\ s_{YX}/s_Y s_X & s_{Y^2}/s_{Y^2} \end{bmatrix}$$

substituting the values in the example gives.

$$r = \begin{bmatrix} 1/\sqrt{2.5} & 0 \\ 0 & 1/\sqrt{1} \end{bmatrix} \cdot \begin{bmatrix} 2.5 & 1.5 \\ 1.5 & 1.0 \end{bmatrix} \cdot \begin{bmatrix} 1/\sqrt{2.5} & 0 \\ 0 & 1/\sqrt{1} \end{bmatrix}$$

$$= \begin{bmatrix} 2.5/\sqrt{2.5} & 1.5/\sqrt{2.5} \\ 1.5/\sqrt{1.0} & 1/\sqrt{1.0} \end{bmatrix} \cdot \begin{bmatrix} 1/\sqrt{2.5} & 0 \\ 0 & 1/\sqrt{1} \end{bmatrix}$$

$$= \begin{bmatrix} 2.5/2.5 & 1.5/\sqrt{2.5}\times\sqrt{1} \\ 1.5/\sqrt{2.5}\times\sqrt{1} & 1/1 \end{bmatrix} = \begin{bmatrix} 1 & 1.5/1.58 \\ 1.5/1.58 & 1 \end{bmatrix}$$

$$r = \begin{bmatrix} 1 & 0.95 \\ 0.95 & 1 \end{bmatrix} = \begin{bmatrix} r_{XX} & r_{XY} \\ r_{YX} & r_{YY} \end{bmatrix}$$

The resulting correlation matrix has ones on the principal diagonal. These are the correlation of X with X and Y with Y. The correlations of X with Y and Y with X are shown in the off diagonal in this 2×2 matrix.

The same coefficient of correlation is found by both methods. Although in this simple example the matrix method is much longer, when a great many correlations are required the matrix method is much more suitable. This type of computation necessitates the use of a computer, which operates well by matrix methods. The procedure also supplies useful values as the calculations are made. These values are the sample means, the sample sum of squares, the sample variance and the sample standard deviation.

(*b*) *The Application of the three tests*

In this subsection the verdicts of a visual correlation and three statistical correlation techniques are compared. Four procedures are applied in turn to a set of data for 20 provinces of northwest Italy. The number of provinces

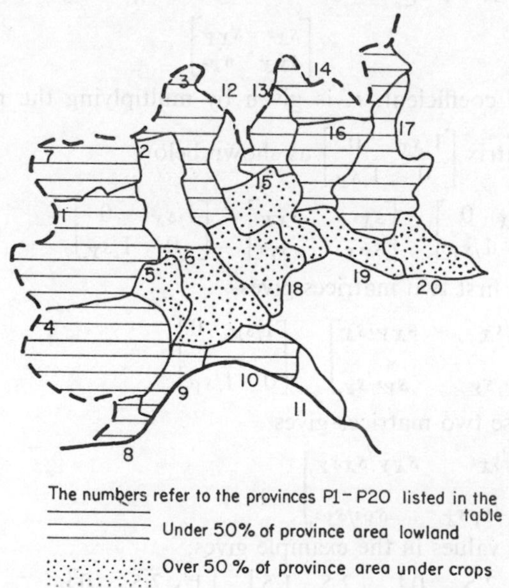

Figure 3.11 Key map of provinces of Northwest Italy.

(=cases) has been limited to 20 to keep the size of the example viable. An increase in the number of provinces would make the visual correlation more difficult compared with the statistical ones. Two variables are used, percentage of total area of province classed as lowland or low hill (as opposed to mountain or high hill) and percentage total area of province classed as cropland (field plus tree crops). This basic information is shown in table 3.13. It is supplemented by a map (figure 3.11) showing the distribution of the provinces. In practice (though not in the case of this example), the person applying tests to a given problem would also, it is hoped, already know something about the topic under consideration. Thus to appreciate the problem, some knowledge of both the relief features and agriculture of northwest Italy is desirable. The definition of lowland — low hill will not be discussed here, but will be assumed to be satisfactory for the needs it was devised to satisfy. Note that for all three statistical tests to make sense it is vital to express lowland and cropland as percentages of the total area of provinces, thereby holding area constant; variations in area of the provinces do not then intrude on the correlation. In other words, if actual area of lowland and cropland is used in the correlation, results will be affected by variation in size of province areas. In some circumstances, the actual area of the provinces might be of interest.

It must be pointed out that the purpose of this subsection is to compare the methods and results of three tests. Strictly speaking, if information were available, smaller units (communes, of which there are about 3,000) could be used instead of the provinces. The confidence levels used would be greatly modified even by taking a 10 per cent sample of smaller units from the 3,000. Strictly, therefore, the indices of correlation to be discussed are descriptive in nature, and are not the basis for statistical inference.

(i) *Non-statistical approach* Commonsense suggests that in northwest Italy, cultivation will become more difficult as altitude and ruggedness increase. A positive correlation will therefore be expected between lowland and much cropland. (It is however important to appreciate that this is not a law. In fact, quite the opposite occurs in both Mexico and Peru for example). A visual comparison of the 4th and 5th columns of data in the table confirms that where the percentage of lowland is high the percentage of cropland also tends to be high, and vice versa. There is a relationship between the two distributions or variables. If it is a causal relationship, however, it can only be in one direction (relief influencing cropland, not the reverse).

The map in figure 3.11 suggests a strong correlation between lowland and cropland. The eleven provinces with *less than* 50 per cent of their area classed as lowland are shaded horizontally. The six provinces with *more than* 50 per cent of their area under cropland are shaded with dots. The two sets have no members in common. They are disjoint. Visually, then, mountain appears to repel cropland. Remember, though, that the geographer is often dealing with much more complex and varied situations than the one illustrated. A visual appreciation may then be more difficult to achieve.

Table 3.13 Data for application of correlations

Province	Basic data			Basic data processed		Chi-square test		Spearman rank correlation test			
	Total area	Lowland area	Cropland area	As percentage of total area: Lowland	Cropland			Rank for Lowland	Rank for Cropland	d	d^2
P 1 Torino	683	326	188	48	28	0	0	11	11	0	0
P 2 Vercelli	300	187	135	62	45	1	1	8	7	1	1
P 3 Novara	359	125	78	35	22	0	0	17	15	2	4
P 4 Cuneo	690	339	241	49	35	1	1	10	9	1	1
P 5 Asti	151	151	114	100	75	1	1	2.5	3	0.5	0.25
P 6 Alessandria	356	312	233	88	65	1	1	6	6	0	0
P 7 Aosta	326	0	10	0	3	0	0	19.5	19.5	0	0
P 8 Imperia	116	48	28	41	24	0	0	13	14	1	1
P 9 Savona	154	62	32	40	21	0	0	14.5	16.5	2.5	6.25
P 10 Genova	183	73	39	40	21	0	0	14.5	16.5	2.5	6.25
P 11 La Spezia	88	54	22	61	25	1	0	9	13	4	16
P 12 Varese	120	82	31	68	26	1	0	7	12	5	25
P 13 Como	207	68	36	33	17	0	0	18	18	0	0
P 14 Sondrio	321	0	9	0	3	0	0	19.5	19.5	0	0
P 15 Milano	276	100	193	100	70	1	1	2.5	5	2.5	6.25
P 16 Bergamo	276	101	80	37	29	0	1	16	10	6	36
P 17 Brescia	475	209	169	44	36	0	1	12	8	4	16
P 18 Pavia	296	267	215	90	73	1	1	5	4	1	1
P 19 Cremona	176	100	144	100	82	1	1	2.5	2	0.5	0.25
P 20 Mantova	234	100	196	100	84	1	1	2.5	1	1.5	2.25
All 20 provinces of northwest Italy	5787	3094	2193	57 (mean)	39 (mean)						122.5 $\Sigma d^2 = 122.5$

Source: *Annuario statistico italiano 1961*, Rome 1962, pp. 2–5 (relief data modified).

(ii) *Chi-square test* A chi-square test may be applied in the example under consideration. For convenience the provinces may be divided into two subsets of ten each, first according to percentage lowlands, second according to percentage cropland. These are marked 1 or 0 in the 6th and 7th columns of table 3.13.

	Ten with highest percentage lowland		Ten with lowest percentage lowland	
Ten with highest percentage cropland	11 Vercelli Cuneo Asti Alessandria	Milano Pavia Cremona Mantova	01 Bergamo Brescia	
Ten with lowest percentage cropland	10 La Spezia Varese		00 Torino Novara Aosta Imperia Savona	Genova Como Sondrio

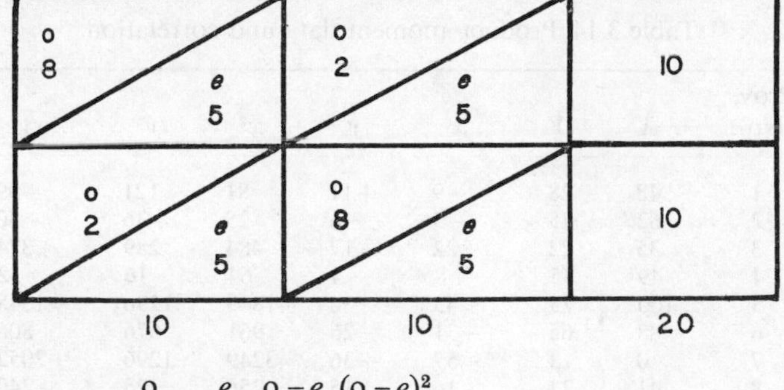

o	e	o − e	$(o-e)^2$
8	5	3	9
2	5	−3	9
2	5	−3	9
8	5	3	9

$\Sigma(o-e)^2 = 36$
$e = 5$

Chi-square = 36/5 = 7.2
With one degree of freedom this is significant.

(iii) *Spearman rank correlation index* In the table the province data for percentage lowland and percentage cropland have formed the basis for a ranking of the provinces 1–20 for each variable.

d^2 is 122.5, n is 20 (provinces).

$$\text{Thus } \rho = 1 - \frac{6 \times 122.5}{20^3 - 20} = 1 - \frac{735}{7980} = 1 - 0.09$$
$$= +0.91$$

This shows a strong positive correlation and suggests a relationship between the two sets of figures that could have occurred by chance only very rarely indeed. The 99 per cent confidence level for Spearman rank correlation coefficient for 20 objects ($n=20$) is ± 0.55, approximately.

(iv) *Pearson product-moment correlation* Columns X and Y show the percentage of the total area of each province occupied by hill/lowland and by cropland respectively. These are the data displayed in columns 4 and 5 of the main data table. Note that as with Spearman it is incorrect to use the actual area of hill/lowland and of cropland, since size of province affects the result. First the influence of the area of each province on these other variables would have to be eliminated. By taking the percentage of each province occupied by hill/lowland and by cropland on the other hand, the actual area of the province does not directly intrude in the calculation, but nevertheless may distort the result somewhat, since equal weight is now given to provinces differing considerably in area. Ideally the initial data would be available for divisions of equal area.

The mean of the values in X ($\bar{X}$) and Y ($\bar{Y}$) are calculated by summing and dividing by 20.

Table 3.14 Product–moment data and correlation

Prov. No.	X	Y	x	y	x^2	y^2	xy
1	48	28	+9	+11	81	121	+99
2	62	45	−5	−6	25	36	+30
3	35	22	+22	+17	484	289	+374
4	49	35	+8	+4	64	16	+32
5	100	75	−43	−36	1849	1296	+1548
6	88	65	−31	−26	961	676	+806
7	0	3	+57	+36	3249	1296	+2052
8	41	24	+16	+15	256	225	+240
9	40	21	+17	+18	289	324	+306
10	40	21	+17	+18	289	324	+306
11	61	25	−4	+14	8	196	−56
12	68	26	−11	+13	121	169	−143
13	33	17	+24	+22	576	484	+528
14	0	3	+57	+36	3249	1296	+2052
15	100	70	−43	−41	1849	1681	+1763
16	37	29	+20	+10	400	100	+400
17	44	36	+13	+3	169	9	+39
18	90	73	−33	−34	1089	1156	+1122
19	100	82	−43	−43	1849	1849	+1849
20	100	84	−43	−43	1849	2025	+1935
Sum:	1136	784			18,706	13,568	+15,481
Mean:	$\bar{X}=57$	$\bar{Y}=39$					−199
							15,282

In columns x and y, the difference from the mean ($\bar{X}$ and $\bar{Y}$) of each value in columns X and Y has been recorded. As a check these should add approximately to 0 at the foot of x and y columns; the rounding of numbers has prevented this from happening exactly.

The x and y values are squared and summed.

x is multiplied by y and the sum of the negative results subtracted from the sum of the positive ones.

Note that whereas x^2 and y^2 must always be positive, some values in xy may be negative.

Thus we now have:

$x^2 = 18{,}706 \quad y^2 = 13{,}568 \quad xy = 15{,}282 \quad X = 57 \quad Y = 39$

The correlation coefficient (r) is derived from the following formula:

$$r = \frac{\Sigma xy}{\sqrt{\Sigma x^2 \cdot \Sigma y^2}} = \frac{15{,}282}{\sqrt{18{,}706 \times 13{,}568}} = \frac{15{,}282}{\sqrt{253{,}803{,}008}}$$

$$r = \frac{15{,}282}{15{,}931} = +0.96$$

This is an even stronger positive index of correlation between lowland and cropland than the index of +0.91 obtained by Spearman in (iii).

Table 3.15 Correlation matrix

	L	C	PD	PG	I
Percentage Lowland	1.00				
Percentage Cropland	0.844 0.890	1.00			
Population Density	0.387 0.255	0.223 −0.010	1.00		
Population growth 1951–61	−0.537 −0.342	−0.509 −0.400	0.245 0.298	1.00	
Income per capita	−0.099 0.038	−0.180 −0.074	0.233 0.539	0.562 0.673	1.00

Comparison of Spearman (upper) and Pearson (lower) for 10 pairs of correlation indices for North Italy ($N = 40$).

The foregoing examples are presented to give the reader a clearer picture of the comparative procedures and outcomes of four methods of correlation, rather than to demonstrate that any particular method is superior. It is obvious, however, that given that the initial data are interval or ratio, the interval (product-moment) test makes fullest use of the information, the ranking test loses some, and the chi-square test a great deal. All three tests reject with 99 per cent confidence the null hypothesis that there is no correlation between the distribution of lowland and of cropland. If the correlation had not been so high (it is not, for example, in sets of provinces in peninsular Italy), then the finer tests might have rejected the null hypothesis while the coarser ones failed to do so. It is the general impression of the authors from work they have done in which the Spearman rank correlation test and product-moment test have been applied on the same data that the results do not usually differ appreciably. The following table compares the rank and product-moment correlation coefficients for each pair of 5 variables for 40 provinces of North Italy. The same provinces are used in chapter 7.

3.7 Partial and multiple correlation

Many situations that occur is geographical work are complex and involve more than two variables. The fact that correlations between pairs of variables are usually fairly small and one variable only 'explains' a relatively low percentage of the variation in the other one indicates that in geography there are usually other variables that should be taken into account. There are several statistical methods for dealing with many variables but this section will be concerned mainly with factor analysis as a method of sorting out a large number of variables, although partial correlation will be mentioned first. The methods of multivariate analysis require a computer to handle the large volume of data. No attempt will be made to explain the steps in the program that does the analysis. All that is intended is to give some indication of the meaning of the results. A brief mention will first be made of partial correlation. This draws attention to a fairly simple and useful technique that can be calculated easily without the use of a computer. A multivariate analysis using a multiple Spearman rank correlation to provide a correlation matrix is also considered in chapter 6.

(*a*) *Partial correlation*

A simple example using 3 variables illustrates the principle of partial correlation. A partial correlation coefficient between two variables is the correlation between these two when a third variable, on which both the others depend, is controlled. In this way the share of each variable in turn can be calculated.

Example of partial correlation The formula, which can be used for three variables only, used to calculate the partial correlation coefficient is

$$r_{ij,k} = \frac{r_{ij} - (r_{ik} r_{ik})}{\sqrt{(1 - r_{ik}^2)\ (1 - r_{jk}^2)}}$$

r represents the correlation coefficient, the subscripts i, j and k refer to three individual variables, thus r_{ij} is the correlation coefficient of i and j, k is the variable that is controlled so that the influence of i and j on each other can be measured. In the first example the value for i is L_3 (length of 3rd order stream), j is N_2 (number of 2nd order stream), k is the area of 23 drainage basins. The correlation coefficients are $r_{ij}=0.59$

$$r_{ik}=0.53$$
$$r_{jk}=0.64$$

then

$$r_{ij,k}=\frac{0.59-0.53\times 0.64}{\sqrt{(1-(0.53)^2)(1-(0.64)^2)}}$$

$$=\frac{0.59-0.34}{\sqrt{0.72\times 0.58}}=\frac{0.25}{\sqrt{0.42}}=\frac{0.25}{0.63}=0.385$$

When area is controlled, the relationship between L_3 and N_2 falls from 0.59 to 0.385, showing that their higher correlation is partly dependent on their individual high correlations with area, which was found to be the single most important variable. Another example shows the importance of the area of the basin.

$i=H-h$ (gradient of basin), $j=H$ (height of highest point in basin)
$k=$ area of basin (sample of 23)
$r_{ij}=0.91$ $r_{ik}=0.10$ $r_{jk}=0.02$ $r_{ij,k}=0.90$

In this instance the correlation of area with the other two variables is small so that their intercorrelation is little reduced by controlling the area. Area does not, in this instance, influence the correlation r_{ij} noticeably.

(b) *Introduction to factor analysis*

Factor analysis was developed mainly to analyse the results of tests in the behavioural sciences, but it can be and has been used for a wide variety of problems, in which many variables are obtained for a number of individual units, either people or areas. It is, therefore, relevant to geographical analysis of problems for which data can be assembled in matrix form. In geographical examples the variables are usually properties of the individual areas or objects that make up the total area or problem involved in the analysis. Factor analysis attempts to account quantitatively for the differences in characteristics among the areas or objects. Common sources of variance are sought amongst the properties that are the variables used in the test. Factor analysis thus attempts to identify the characteristics which the variables have in common and which result in their intercorrelation. It is helpful to consider the problem as the intersection of sets. Where there is no correlation the variables are disjoint sets, but where they intercorrelate they have something in common that can be expressed by the intersection of the sets. The degree of intersection depends on the closeness of the correlation. When the amount of intercorrelation is high then the area of intersection will be great compared with the rest of the sets, (see also chapter 6).

Before considering the stages of factor analysis some of the basic assumptions will be mentioned. It is assumed that in the matrix of intercorrelated variables there are some common factors running through the data. It is these common factors that are extracted and expressed by factor analysis. These indicate the most important common characteristics that are inherent in the properties that are used as variables. The variance (standard deviation squared) of the individual variables is a measure of the extent of the spread of values of the variables in the individual areas. The variance can be divided into three types. These are the *common variance*, which is that portion of the *total variance* that correlates with other variables, *specific variance*, which is that portion of the total variance that does not correlate with any other variable. Finally there is the *error variance*, which is the chance variance due to errors of sampling, measurement or other error sources. The common characteristics are expressed by common variances of each variable in relation to the others.

The total variance can be shown diagrammatically

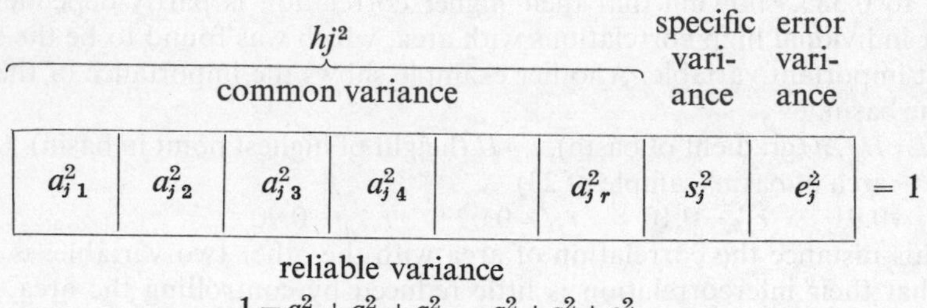

reliable variance

$$1 = a_{j1}^2 + a_{j2}^2 + a_{j3}^2 \ldots a_{jr}^2 + s_j^2 + e_j^2$$

the values of the $\sqrt{a_{j1}^2}$ ... are the factor loadings, they represent the correlation of that variability j, with each factor. It is the common variances which are important as they represent common elements running through the data and these result in high correlations. The specific variance is only of significance in the variable under consideration and is unrelated to the other variables. Example from B. Fruchter:

Table 3.16 Factors

Variable	Factor						
1	I	II	III	IV	h^2	s^2	r_{jj}
1	0.7	0.0	0.4	0.3	0.74	0.12	0.86
2	0.0	0.6	0.0	0.2	0.40	0.10	0.50
3	0.8	0.1	0.5	0.0	0.90	0.05	0.95
C	0.3	0.5	0.6	0.0	0.70	0.10	0.80

correlation $r_{12} = 0.7 \times 0.0 + 0.0 \times 0.6 + 0.4 \times 0.0 + 0.3 \times 0.2 = 0.06$ the common loading on factor IV is the only thing that produces the correlation.

Test 1

a_{11}^2 a_{13}^2 a_{14}^2 s_1^2 e_{12}^2 $\qquad$ $r_{jj} = h^2 + s^2 = 1 - e^2$

0.49 0.16 0.09 $\qquad$ $0.86 = 0.74 + 0.12 = 1 - 0.14$

$\underbrace{\hspace{3cm}}_{0.74 = h^2}$

r_{jj} is reliability coefficient, i.e. 1 – error variance. Given r_{jj} then s^2 and e^2 can be computed.

It is assumed in factor analysis that the total variance is the sum of the component variances, including the specific variance, the error variance, and the individual common variances the latter together making up the common variances or *communality*. The individual common variances are the variances of each variable with all the other variables in turn. The value of the square root of the sum of the common variances for each variable is called the factor loading. Each variable has a factor loading for each factor. They represent the amount of correlation of that particular variable with the corresponding factor. The common variance is the most important as this is of most use for prediction because it correlates with other variables. The specific variance is related only to itself and is therefore of little diagnostic value. The *uniqueness* of the particular variable is the sum of the specific variance and the error variance. Alternatively it is equal to 1 – the communality, since the communality plus the specific variance plus the error variance equals one, (see diagram in text).

Another assumption of factor analysis is that the correlation of the variables can be accounted for by their common factor loadings. Thus the correlation matrix provides the data from which the factor loadings are calculated and the reverse operation is also possible. The summation of the cross products of the common variances of two variables gives their correlation, (see example). Where variables are strongly correlated they will have similar loadings of high value on several factors. The communality of the variable can be found by summing the squares of the loadings on the factors of that area.

The stages of factor analysis can be enumerated as follows:

1. Areas or units must be chosen.
2. Variables must be chosen.
3. The variables for each area must be arrayed in a matrix.
4. The means and standard deviations of the variables are computed. From these values the variances can be obtained.
5. The variables are all correlated with each other by product-moment correlation, giving a correlation matrix.
6. The factors are calculated from the correlation matrix.
7. The loadings of the individual variables on each factor are given.
8. The weightings of the individual areas on each factor are given.

The information that is obtained from these individual steps and how they follow from each other will be briefly commented upon.

1. The *choice of areas* clearly depends on the nature of the problem being investigated. In some problems it is useful to obtain uniform areas, if it is desired to hold this property constant as in the example described in chapter 7. In some cases the areas must be irregular; for example if a particular type of unit is being studied, such as a drainage basin or parish. There

is no need for the individual units to consist of areas. They can be people or samples of some sort. There is a very large variety of possibilities.

2. *The choice of variables* is important if the results are to be valuable. The variables should be independent as far as possible, and should not be mathematically related. However the purpose of factor analysis is to establish common characteristics amongst the variables so that the results will not be satisfactory if there is no common variance, but only specific variance and error variance. For this reason it is worth thinking carefully before the variables are selected. They often, however, depend on the data available and it is not always possible to obtain satisfactory variables. They should be standardized as far as possible before they are used. They need not be in all the same units of measurement, but as product-moment correlation is used it is assumed that they are normally distributed.

3. The arranging of the *data* in *matrix* form is merely for convenience of working, especially if the data are to be put on a computer. It also facilitates the interpretation of the results if the data are arranged systematically. By arraying the data in a matrix it can be ascertained that the matrix is complete. The analysis requires a complete set of data, so that each area has all the variables, even if a zero value is entered in some instances. If several different areas are being compared, as in the example discussed in chapter 7, then it may happen that one of the areas has no values for some of the variables. If one variable consists only of a whole series of zeros in the matrix, the product-moment correlation cannot be established and the analysis is impossible. It is therefore essential to have a complete matrix with no sets of zeros.

4. The calculations of the *means and standard deviations* for the variables for each area provides a useful guide to the degree of variation of the individual measurements of each variable as well as a measure of the absolute value of the variable. The standard deviation will be shown to be closely related to the mean in many instances, the two values increasing together. The range of variability of the data does play a significant part in the way in which one variable correlates with another. When the range of variability is small the variable tends towards a constant and will not correlate with other variables so highly. Thus the standard deviation may be useful in considering the value of the correlations. In some instances it can be shown that there is a relationship between the relative standard deviation, found by dividing the mean by the standard deviation, and the factor loading.

5. The *correlation matrix* provides the data from which the factors and factor loadings are computed. The matrix provides a measure of the relationship between all the possible pairs of variables.

A further step from product-moment correlation provides the regression line that minimizes the square of the distance between the variables and the line. This is easy to plot when two variables are plotted against each other on a graph. It can also be seen that when the correlation is high all the points will fall near the regression line. It is possible to draw an ellipse round these points so that the relative length of the long axis of the ellipse, lying along the line of regression, is great (relative to the short axis) when the correlation is high and the scatter of points small. When the correlation is low the points are scattered widely and the ellipse approximates to a circle with many equal axes. It can be seen that if the variables cover a wide range of values the length of the long axis of the ellipse for a high correlation will be longer than if the range of variation among the variables is small. This helps to account for the low factor loadings on variables with small standard deviations.

If three variables are inter-correlated it is possible to visualize the scatter of points in three dimensions with a plane of best fit running through the points. This plane must now be defined with both x, y and z co-ordinates. The figure that outlines the scatter of points now becomes a solid figure, an ellipsoid, but the same considerations apply to its shape and size. It is not possible to visualize the situation with more than three variables, but it is possible to continue this sort of analysis mathematically to any number of variables by using n dimensions. However, the calculations become very long if the number of variables increases beyond about 20. The idea can be obtained from an axis of best fit through a many dimensional scatter of points. The length of this axis can be defined by an eigenvalue which leads on to the next stage in the analysis.

6. *Factors* are derived from the correlation matrix and they are based on the length of the axes necessary to define the scatter of points in many dimensions. Thus the factors are based on the eigenvalues. The first factor is determined by the degree of inter-correlation of the variables. If they are all closely associated the axis of the hyperellipsoid will be long and most of the points will fall near it. This will give a high factor value on the first factor. The percentage of the variability of the variables that this factor accounts for can be worked out if the factor value is multiplied by 100 and divided by the number of variables. The same applies to the second and subsequent factors. In this way a cumulative percentage can be worked out as each subsequent factor is added in turn.

The second factor is related to the length of an axis orthogonal to the first factor axis. The position of the second factor is fixed in such a way that the distance to those points which lie furthest from the first axis lie close to the second factor axis. For this reason the second factor shows different loadings from the first. The length of the second factor axis is normally shorter than the first factor axis, and this is increasingly true as the values of the correlation coefficients in the matrix increase. In situations where all the variables are completely random in their distribution in respect of the other variables, and all the correlation coefficients approach zero, the length of the axes of the factors are all about the same and the values of all the factors are approximately equal.

The third factor axis is orthogonal to the other two and this is true of all the subsequent rotations through right angles that are necessary to produce the second and subsequent factors. Each subsequent axis should get shorter as the variability in the data is increasingly accounted for and the percentage variation 'explained' approaches 100. The factors summarize the influence of all the variables on the variability of the data, but the analysis can be carried further to show how each variable is related to the factor axes at each stage of the rotation. This analysis gives the factor loadings on each variable.

7. The *factor loadings* on each variable are closely related to the correlation coefficients. When the coefficients are high then the factor loadings are high. The closer some particular variable under consideration lies to the factor axis the higher will be its factor loading. When several variables all have high loadings they all lie close to the factor axis and are themselves highly inter-correlated. This means that they have much in common and that their variation together accounts for the variability amongst the areas. This will give a high first factor value.

One of the merits of the factor loadings is that the most important variables can be selected as those having the highest factor loadings. These variables can be used to characterize the area. In the example to be considered in chapter 7 it will be seen on table 7.10, which gives all the first and second loadings for the variables used, that some of the variables occur with high factor loadings in all the areas. This applies to the first variable and almost to the seventh. These two variables may therefore be deemed to describe the areas and to differentiate them.

If it is desired to test the difference between the areas statistically, to ascertain whether they are significantly different, then it is to these two variables that the analysis of variance test should be applied. Some variables will be seen to have low factor loadings in all the areas. This means that they are not useful diagnostically owing to their small variance.

The loadings on the second factor are usually considerably different from those on the first factor for most of the variables. This means that the variables that are far from the first factor axis lie closer to the second factor axis as this is positioned so as to minimize the distance to the most outlying of the points with respect to the first factor axis. In the same way many of the points that lie very close to the first factor axis will lie far from the second factor axis in many instances, though points close to the intersection of the two axes will have a high loading on both factors. The same arguments can be applied to the third and subsequent axes but the results become less significant and are not usually considered much beyond the third or fourth factor.

A useful result of the factor loadings is to indicate what type of variable is causing the variation within the different areas and by analysing several areas separately it is possible to see how each responds to variations in the different variables selected. Within each individual area the subareas can be distinguished in the same way by examining the individual factor weightings for each of the subareas or other unit for which the variables were collected.

8. *The individual factor weightings* show how each area contributes to the variations of each factor. A useful method of considering the relationship between the individual weightings on the first and second factors is to plot these values against each other on a graph. An example of this type of graph is shown in chapter 7 for 120 subareas which are divided into six distinct areas of differing character, 20 areas belonging to each of the regions. The

figure shows that it is possible to distinguish the six different major areas by their factor weighting characteristics. If the individual subareas that make up each major area are considered, then these show the same tendency to be distributed according to the value of the most important of the variables on which the differentiation depends.

Factor analysis can show both how the variables combine to account for the variability of the data and also how the individual units respond in terms of the first few factors to be extracted. This provides a very useful means for establishing, on the basis of the most important variables, the significant differences between the areas or other units used in the analysis. Examples of the application of this method of analysis are discussed in chapters 6 and 7.

References

Adler, I. (1966). *Probability and Statistics for Everyman*, Signet Science Library, New York.

Alder, H. L. and Roessler, E. B. (1962). *Introduction to Probability and Statistics*, Freeman, San Francisco and London.

Baggaley, A. R. (1964). *Intermediate Correlational Methods*, Wiley, New York.

Blalock, H. M. (1960). *Social Statistics*, McGraw-Hill, New York.

Blalock, H. M. (1964). *Causal Inferences in Nonexperimental Research*, North Carolina University Press, Chapel Hill.

Chayes, F. (1946). Application of the coefficient of correlation to fabric diagrams, *Trans. Am. Geophys. Union*, **27**, 400–405.

Chorley, R. J. (1966). *The Application of Statistical Methods to Geomorphology*, in (Ed. G. H. Dury) *Essays in Geomorphology*, Heinemann, London.

Coleman, S. (1964). *Introduction to Mathematical Sociology*, Free Press of Glencoe, New York.

Duncan, O. D. and Others (1961). *Statistical Geography*, Free Press of Glencoe, London.

Flinn, D. (1958). On tests of significance of preferred orientation in 3-dimensional fabric diagrams, *J. Geol.*, **66**, 526–539.

Fruchter, B. (1954). *Introduction to Factor Analysis*, Van Nostrand, Princeton, New Jersey.

Goldberg, S. (1960). *Probability, An Introduction*, Prentice-Hall, Englewood Cliffs, New Jersey.

Gregory, S. (1963). *Statistical Methods and the Geographer*, Longmans, London.

Huntsberger, D. V. (1964). *Elements of Statistical Inference*, Allyn and Bacon, Boston, Mass.

Krumbein, W. C. and Graybill, F. A. (1965). *An Introduction to Statistical Models in Geology*, McGraw-Hill, New York.

Mackay, J. Ross (1959). Comments on the use of chi-square. *Ann. Assoc. Am. Geog.*, **49**, No. 1, p. 89.

McCollough, C and Van Atta, L. (1963). *Statistical Concepts, A Program for Self-Instruction*, McGraw-Hill, New York.

Melton, Mark A. (1958). Use of punched card to speed statistical analyses of geomorphic data, *Bull. Geog. Soc. Am.*, Vol. 69, 355–358.

Middleton, S. V. (1965). The tukey chi-square test, *J. Geol.* **73** (3), 547–549.

Miller, R. L. and Kahn, J. S. (1962). *Statistical Analysis in the Geological Sciences*, Wiley, New York.

Moroney, M. J. (1962). *Facts from Figures*, Penguin Books, Harmondsworth. Various editions.

Mosteller, F., Rourke, R. E. K. and Thomas, G. B. (1961). *Probability: A First Course*, Addison-Wesley, Reading, Mass.

Reichmann, W. J. (1964). *Use and Abuse of Statistics*, Penguin Books, Harmondsworth.

Robinson, A. H., Lindberg, J. B. and Brinkman, L. W. (1961). A correlation and regression analysis applied to rural farm population densities in the great plains, *Ann. Assoc. Am. Geog.*, **51**, No. 2, 211–220.

Siegel, S. (1956). *Non-parametric Statistics for the Behavioral Sciences*, McGraw-Hill, New York.

Spiegel, M. R. (1961). *Theory and Problems of Statistics*, Schaum's Outline Series, New York.

Steinmetz, R. (1962). Analysis of vectorial data, *J. Sed. Pet.*, **32**, (4), 801–812.

Watson, G. S. and Irving, E. (1957). Statistical methods in rock magnetism, *Monthly Notices Roy. Astron. Soc. Geophys. Suppl.*, **7**, No. 6.

Zobler, L. (1958). The distinction between relative and absolute frequencies in using chi-square for regional analysis, *Ann. Assoc. Am. Geog.*, **48**, 456–457.

Part 2

Spatial distributions and relationships

4

Methods of describing single distributions

'He had bought a large map representing the sea
  Without the least vestige of land:
And the crew were much pleased when they found it to be
  A map they could all understand.'

The Hunting of the Snark Lewis Carroll (1876)

The mapping and study of single distributions in area forms a major part of geography. A distribution in geography seems to suggest to many a distribution of dots on a surface, rather than of lines or patches on a surface. The distribution of dots on a surface is however a special case. The study of distributions has been considered in geography particularly in what have variously been called economic mapping, statistical mapping, and point patterns on a uniform plane. Some, such as J. Dickinson (1963) and F. J. Monkhouse and H. R. Wilkinson (1963) have recently gone at length into various methods of obtaining the best or at least a good visual representation of distributions. The ability of people to appreciate what is being shown visually has also been investigated.

In a somewhat different direction, J. F. Hart (1954), M. F. Dacey (1964), and R. Bachi (1963) have considered aspects of distributions of points on surfaces and of information about points on surfaces more from the point of view of descriptive statistics. Just as a distribution on a line has a mean, median, variance and other features that can be measured (see chapter 3), so a 2-dimensional distribution has a mean centre (often referred to as centre of gravity), a standard distance, roughly equivalent to standard deviation, and other measureable features.

Both approaches go back a considerable way. Some thematic maps of Ireland were produced in the 1830s. They may be seen in the Library of the British Museum. Proportional symbols, shading and other techniques were used on them in a very sophisticated way. Similarly, already in the 19th century such features as centre of gravity of distribution of population were discussed, if not widely used. The so-called centrographical method was developed particularly in the interwar period (e.g. E. E. Sviatlovsky and W. C. Eells 1937).

The purpose of this chapter is to introduce at a fairly elementary level some of the more statistical aspects of distributions, mainly of points in areas. The theme is that a visual interpretation of a distribution, however well the distribution is portrayed on a map, is often not adequate. Numerical indices may provide more precise and penetrating information.

4.1 Dots on a patch

The table below summarizes the various combinations of dimensions that may be encountered in distribution maps. To avoid confusion with the precise terms *point* and *surface* in Euclidean geometry, dot and patch have been used instead. Together with line, for which a satisfactory substitute does not come to mind, they suggest real 3-dimensional objects, collapsed for convenience to 2, 1 or 0 dimensions, as already explained in chapter 1.

	0	1	2	3
This \ On this	Dot (point)	Line	Patch (surface)	3D space
0 Dot (point)	0.0	0.1	0.2	0.3
1 Line	not possible	1.1	1.2	1.3
2 Patch (surface)	not possible	not possible	2.2	2.3
3 3D space	not possible	not possible	not possible	3.3

Dots on a patch are *both* a subset of the general situations of dots on some base (shown in row 0 in the table), *and* a subset of the general situations of patches as the base for dots, lines or smaller patches (column 2 in the table).

While attention is devoted mainly to dots on patches in this chapter, reference will be made to some of the other situations in the table, both for useful comparison and for general interest. Chapter 9 will however deal more fully with the 3-dimensional situations. Cases 0.1, 0.3, 1.2 and 2.2 of the table are all illustrated in figure 4.1. The 3-dimensional situation, Case 0.3, may be exemplified by the distribution of water-fleas in a glass tank at the zoo.

The dot distribution may be considered in set vocabulary as follows: in some universal patch set (U), there is a patch set (S), on which a number of dots are located. Generally, U is a map, and the limit of U, the frame of the map. The objects that the dots represent are usually both small (in relation to the total area of S), and compact (otherwise they would be considered as line sets). In other words, their own dimensions are small enough to be irrelevant, and only their location need be considered. A dot therefore stands in for each object (or given number of objects when there are many). The dot itself may be considered to be a point occupying a location that can be

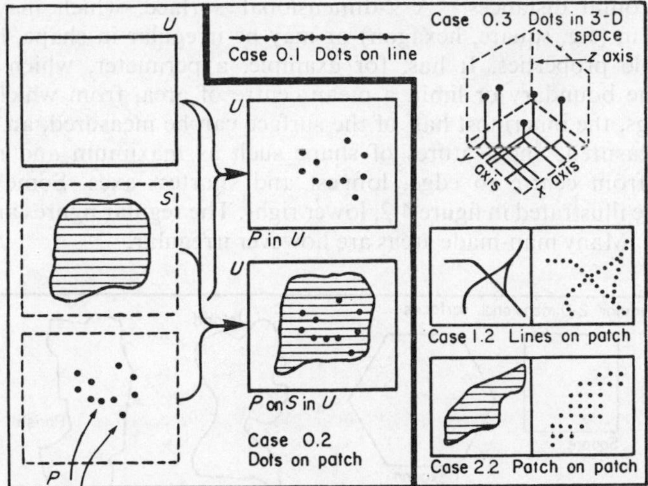

Figure 4.1 Types of distribution. Refer to the table in section 4.1 for key to the numbering of the cases.

identified with required precision by co-ordinates on the patch set *S*, a surface consisting of an infinite number of points. The set of dots will be called *P*. The situation is illustrated in figure 4.1. If the dots are considered to be a finite set of points, then as subset *P*, they form a subset of the total set of points on the surface *S*.

In the real world, the situation might be the set of sheep in a field. The field itself is a subset of a bigger patch, the land around it. The situation under consideration may be a set of points on a surface that covers the whole of the universal set chosen (the frame): some trawlers, for example, in part of the North Atlantic Ocean. While not essential, it may be useful to think of a completely empty 2-dimensional space (as in Lewis Carroll's verse — see chapter heading) which forms the setting for the distribution in which interest is concentrated.

In the above situation, the geographer may be interested *either* in the relationship of the dots to the surface on which they are located, or in the relationship of the dots to one another, or in both. Ultimately, of course, the geographer will also be interested in the relationship between the first set of objects (dots) and some other set or sets of objects. This situation will be discussed in chapter 6.

It must be emphasized again that the base patch on which the dots are located itself has inherent spatial features even before the dots are located on it. The surfaces in which geographers are interested are usually considered by convention to be flat.* Vertical variations are usually small compared

* Spherical surfaces involving large parts of the earth's surface will be discussed in chapter 8.

with horizontal distances.* A 2-dimensional surface, which may have a regular form (e.g. square, hexagon) or may be irregular in shape, has many measureable properties. It has, for example, a perimeter, which is a line forming the boundary or limit, a mean centre of area, from which, among other things, the innermost half of the surface can be measured, an area that can be measured, and features of shape such as maximum and minimum distances from centre to edge, longest and shortest axes. Some of these features are illustrated in figure 4.2, lower right. The regular figures are usually man-made. Many man-made areas are however irregular.

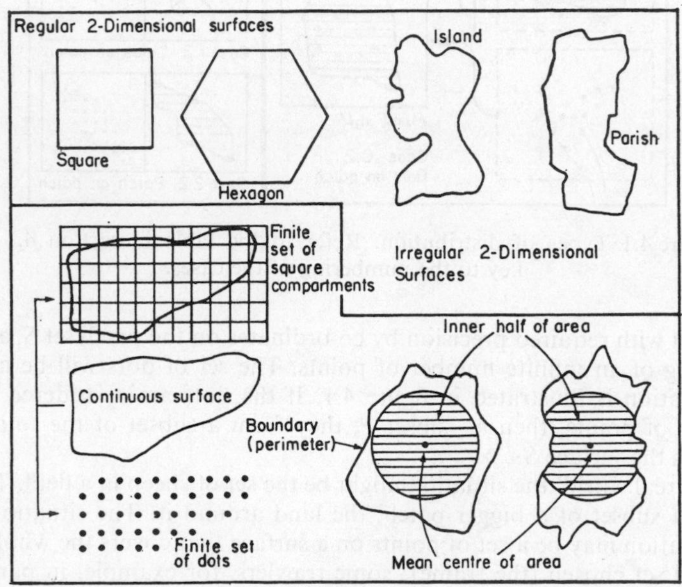

Figure 4.2 Selected features of surfaces or patches.

The base patch on which the distribution of dots is deployed (see figure 4.2) may be represented as continuous, by a grid with compartments of appropriate shape and fineness, or by dots representing the grid by standing in for it at the centre of each compartment. The stand-in dots for the grid could also be the actual nodes of the grid itself. A method for making a dot distribution on a surface is described in the latter part of this chapter.

In the case of a set of dots on a bounded 2-dimensional patch it is possible to have, in very general terms, both a regular and irregular distribution of dots and a regular or an irregular shaped patch on which the dots are located. The four situations are illustrated in figure 4.3*a*. Situation I is easy to describe and once described is of little further interest in itself. Situations II and III may occur sometimes in the real world. However, situation IV, an irregular

* The situations in which this is not so are considered in chapter 9.

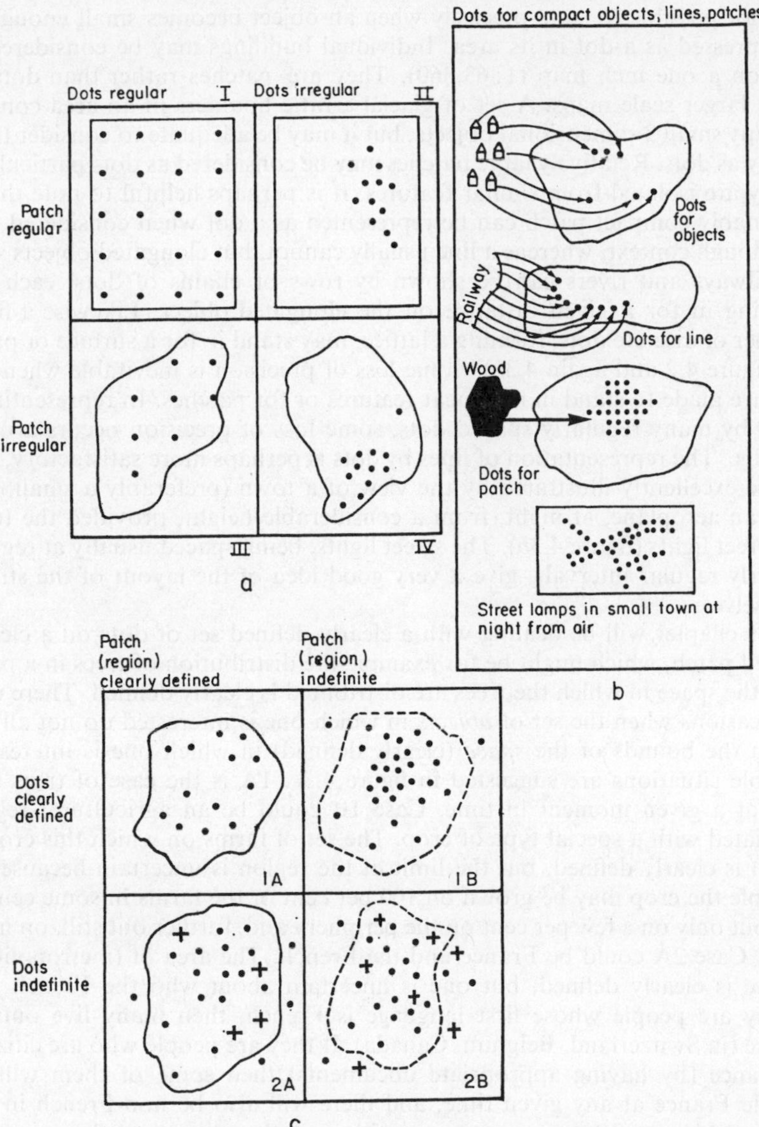

Figure 4.3 Dot distributions on patches. (*a*) Regular and irregular. (*b*) Representation of objects by dots. (*c*) Clearly defined and indefinite dots and patches.

arrangement of dots on an irregular shaped patch, is the arrangement that most frequently confronts the geographer.

It was suggested earlier in this section that a single object or a set of like objects in an area could be considered as a dot or as dots if small enough.

It is not possible to state precisely when an object becomes small enough to be expressed as a dot in its area. Individual buildings may be considered as dots on a one inch map (1: 63,360). They are patches rather than dots on much larger scale maps. A set of glacial erratic boulders in an area consists of many small 3-dimensional objects, but it may be adequate to consider them simply as dots. Relatively large patches may be considered as dots particularly if they are isolated from similar features. It is perhaps helpful to note that a reasonably compact patch can be represented as a dot when considered in a big enough context, whereas a line usually cannot, but elongated objects such as railways and rivers can be shown by rows or chains of dots, each dot standing in for a given distance on the elongated object. Likewise a finite number of discrete dots, forming a lattice, may stand in for a surface or patch (see figure 4.2 and again 4.3*b*). Some loss of precision is inevitable whenever dots are made to stand in for linear features or for patches. In representing a patch by many regularly spaced dots, some loss of precision occurs around the edge. The representation of lines by dots is perhaps more satisfactory, and can be excellently illustrated by the view of a town (preferably a small one) from an aeroplane, at night, from a considerable height, provided the town has street lights (figure 4.3*b*). The street lights, being spaced usually at regular or fairly regular intervals, give a very good idea of the layout of the streets themselves.

This chapter will be dealing with a clearly defined set of dots on a clearly defined patch, which might be for example the distribution of trees in a park. Here the space in which the trees are distributed is clearly defined. There may be occasions when the set of *objects* in which one is interested do not all fall within the bounds of the *space* (clearly defined) in which one is interested. Possible situations are suggested in figure 4.3*c*. 1A is the case of trees in a park at a given moment in time. Case 1B could be an agricultural region associated with a special type of crop. The set of farms on which this crop is grown is clearly defined, but the limit of the region is uncertain because for example the crop may be grown on 100 per cent of the farms in some central area but only on a few per cent on the periphery and, further out still, on none at all. Case 2A could be France and the French. The area of (metropolitan) France is clearly defined, but one is uncertain about who the French are. If they are people whose first language is French, then many live outside France (in Switzerland, Belgium, Canada). If they are people who are citizens of France (by having appropriate documents) then some of them will be outside France at any given time, and there will also be non-French in the country. In case 2B, one is not sure either of the objects or the area: for example neither cultural Wales (as opposed to politico-administrative Wales), nor the Welsh people, are easy to define. Much effort is spent by geographers defining regions and/or criteria in cases 1B, 2A and 2B. Perhaps some of this effort could be saved if greater attention were given to sorting out 1A first.

4.2 Characteristics of sets of dots

(a) Static (fixed) and mobile (in time) distributions

There is obviously a fundamental difference between sets of objects that move during a given period of time and sets of objects that are fixed at particular points in space. A distinction should also be made between permanently fixed objects, and mobile objects that remain fixed in one place over a particular given period of time (e.g. cars left in a car park by their drivers for the night), but may move at other times. Figure 4.4a shows an example of static and mobile objects, trees standing at the sides of a clearway road, and vehicles moving along the clearway. A physical example might be boulders at the bottom of a stream bed not moving at the time in question and particles of material being carried along in the water. However, apparently fixed objects may in reality be changing their location slowly over time, or after a period of being static may move some distance over a short period. Examples of these situations will come to mind. Time, which is of course essential to detect whether objects are mobile or fixed, will be discussed more fully in chapter 10.

(b) The number of items in the distribution

As suggested in chapter 1, more can be inferred about a distribution with a considerable number of dots (say at least 50 to 100) than about a distribution with only a few (say 1 to 10). It is illuminating, however, to start with a distribution of no dots. In mathematics this is quite a reasonable concept, and in set terms would be the empty set. For example, in figure 4.4*b* the first map shows the distribution of towns in Iceland with more than 250,000 people; there are none. There are many empty sets in the world or in parts of it: for example, land over 5,000 ft above sea-level in the British Isles, steel works in Liberia, active volcanoes in the Netherlands, and so on.

Next is the unit set, about which geographers have much to say. For example, the second map in figure 4.4*b* shows towns in Iceland with more than 50,000 people; there is one. It is possible to speculate interminably as to whether the coastal location of the capital has been important in its growth there. Often one encounters sets of two (dyads), three (triads) or four items. Oxford and Cambridge were the only universities in England in the Middle Ages, while Canterbury and York were the two leading ecclesiastical centres (figure 4.4*d*). The three capitals of Scandinavia, the three counties of Jamaica (figure 4.4*c*), the four basic regions of Italy, and so on, are examples of sets with small numbers of members. Geographers should be wary of making too much of the spatial arrangements of small sets of objects in their areas, since, as should have become clear in chapter 3, arrangements that look 'significant' can come about randomly.

Figure 4.4*e* shows the distribution of clubs in major league baseball in the U.S.A. These are divided into two subsets (which are disjoint although they 'share' certain cities), the clubs belonging to the American League and the clubs belonging to the National League. Both these distributions (each

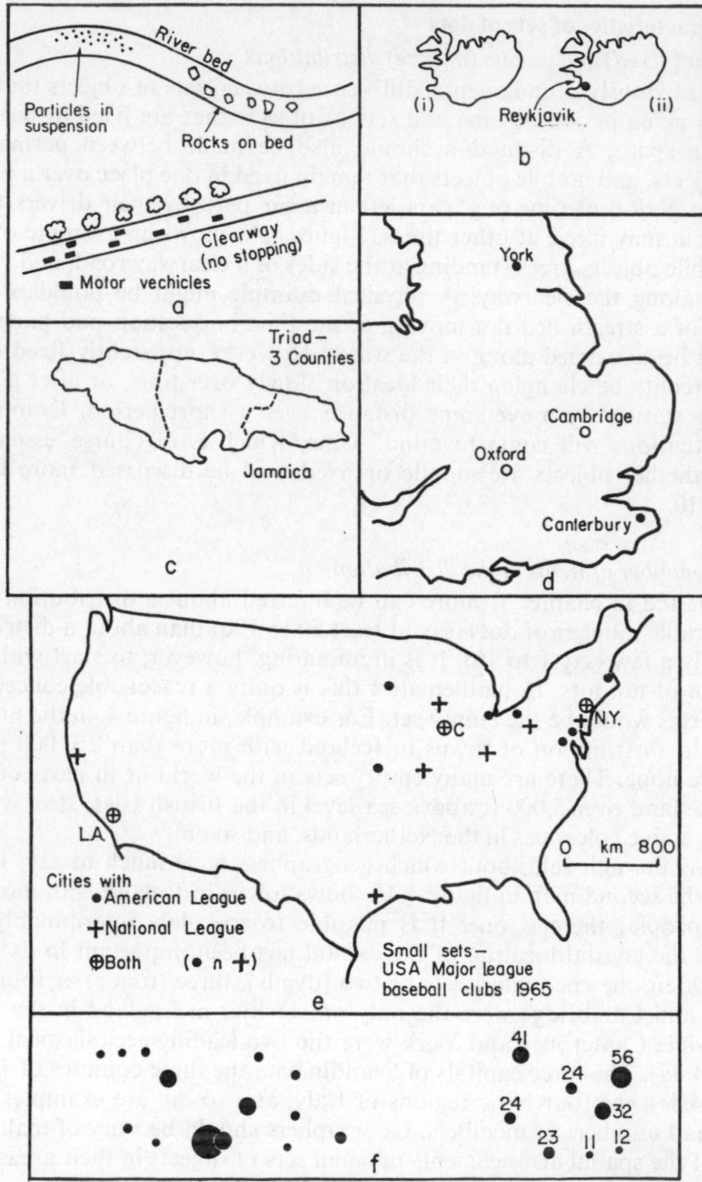

Figure 4.4 (*a*) Mobile objects. Examples of sets of dots with a small number of members. (*b*) (i) and (ii) Empty and unit sets. (*c*) Triad. (*d*) Two dyads. (*e*) Two sets each with ten members. (*f*) Weighting of dots according to their importance.

subset contains ten elements) are scattered over the surface of the U.S.A. They are sets with just about a large enough number of elements for something to be said about their distribution. In this case, they reflect the distribution of large urban agglomerations.

A time comes in the construction of a dot map where for one reason or another the actual set of objects in the distribution to be represented is too large for each object to be represented by a dot. It would be feasible to show the location, by place of residence, of every individual in a village (if adequate data were available). In contrast, it would not be possible to put dots on a map in a one-one correspondence with the whole population of several million people of a country. It becomes necessary to have a one to several or a one to many correspondence. Then a dot would be standing in for many people and might not be located where any person actually lived.

(*c*) *Weighting the objects in a distribution*

Often the geographer is confronted in an area with a set of objects that is clearly defined but has members of differing 'weight'. In a given area there might, for example, be a set of power stations, steelworks, or villages. In each case, there would probably be large differences in the capacity or size of the items. In geography the distribution of such objects is customarily mapped with symbols (circles, 'spheres', squares and so on) proportional in some appropriate way to the importance of the objects they represent (see figure 4.4*f*). This method has an advantage visually except where symbols crowd together excessively. Unfortunately, however, such a distribution is difficult to manipulate beyond the visual stage. For example, if the distribution is represented on the map by dots each equal to one object or to a standard number of objects, the dots can be counted easily. If, however, they are weighted, then unless the appropriate quantity involved is indicated on or alongside each symbol, the exact importance of each symbol cannot be assessed without reference to the original data from which the map was made. However the inclusion of quantitative information with the symbols (see figure 4.4*f*) introduces 'noise' and interferes with the visual interpretation of the distribution.

4.3 Subset distributions

Many maps are unsatisfactory visually because their compilers have failed to distinguish between sets of objects in a given space that are free to be anywhere in that space, and sets of objects that are not free to be anywhere in it *because they are subsets* of larger sets of objects. The problem will be considered with the help of set vocabulary. Suppose a general area in which one is interested is called the universal set. Within this universal set there may be a particular subset (patch(es), line(s) or dot(s)) with which one is concerned. The treatment of this subset is reasonably straightforward. Often, however, one is concerned with a further subset, contained within the larger subset. The argument that follows is that in the latter case the geographer should look at the second (smaller) subset in relation to its proper base, the first subset, not in relation to the universal set. It is of course possible, though not necessarily convenient, to convert the base subset into a new universal set. The new universal set will not however often be rectangular in shape, and may in fact be quite unviable as the base for a map.

In figure 4.5*a* and *b*, two distributions are shown, trees in a park, and traffic lights in a fictitious town (figure 4.5*g* and *h* shows a comparable situation with mobile objects in space). There are several basic differences in the two dis-

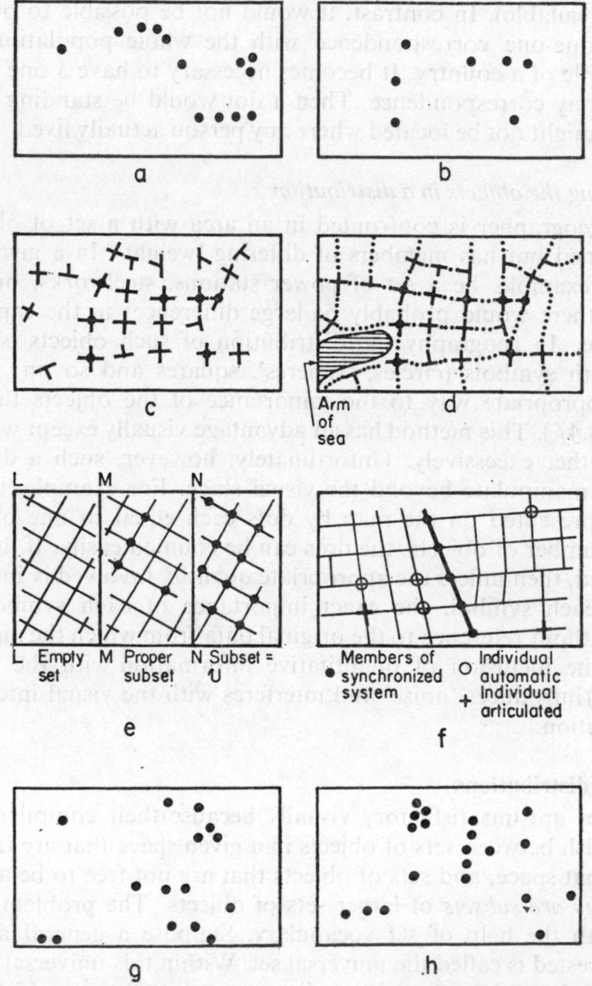

Figure 4.5 Subset distributions. (*a*) Distribution of trees. (*b*) Distribution of traffic lights. (*c*) Traffic lights as a subset of road intersections. (*d*) Road intersections as a subset of roads. (*e*) Road intersections all, some and none without traffic lights. (*f*) Subsets within the set of traffic lights. (*g* and *h*) Mobile examples of above e.g., children in playground at different times.

tributions of dots in 4.5*a* and *b*. The number of dots (objects) differs; the set of trees is more numerous than the set of traffic lights. Trees are physical objects, traffic lights are man-made. The actual spatial arrangement differs (though neither arrangement is easy to describe verbally and a comparison of the two is also difficult). Even then one cannot help feeling uneasy about the two. In this example the trees could have been planted *anywhere* in the park when the piece of land (a field) was converted into a park; they were put, however, somewhere specific. The traffic lights could not have been put anywhere in the Charlesville area because by definition here, a traffic light (a set of striped poles with lights on the top) occurs *only* at a road intersection. One could erect the poles and put lights at the top in the middle of a field but (by definition) they would not be traffic lights. Thus unless the whole space in the rectangle in figure 4.5*b* is road intersection, the traffic lights must be confined only to a certain part of it. In short, there is a set of road intersections in the Charlesville area. A subset of these has traffic lights. The relationship of a subset to a set can be very flexible. Thus in figure 4.5*c* there could be three different basic situations in a given area with road intersections: case *L*, where the subset of traffic lights is *empty* because there are *none*, case *N*, where every road intersection has a traffic light (the subset then equals the set). or case *M*, where some road intersections have a traffic light and some do not.

At all events, the trees in figure 4.5*a* are not a subset of anything but the total space of the park. The distribution of trees can therefore be considered in its own right. In contrast, the dots in the Charlesville area (figure 4.5*b*) can only be considered as a distribution within the limits their *base*, the set of road intersections (figure 4.5*c*), not the whole space. Indeed, visually it is pointless to map a distribution of traffic lights without showing roads as well. Carrying the argument further, the road intersections themselves are a subset of all roads (figure 4.5*d*). Roads are a subset of land, land of the earth's surface (figure 4.6). Furthermore, the traffic lights themselves may be divided into subsets (figure 4.5*e*): for example, those that change automatically, following a set time schedule, and those that are articulated by traffic as it approaches them. Some could even be synchronized along a busy road in a *system*.

It will be noted that as one follows the traffic light subset relationships (a kind of nesting of sets) back to the earth's surface, the dots at one stage become a subset of lines (road intersections of all roads), and the lines at a later stage become a subset of a patch, the land.

If one disregards the complication of 3-dimensional space, then it appears possible to have six relationships of subsets to sets : dot(s) on dots, dot(s) on lines, line(s) on lines, dot(s) on patches, line(s) on patches and patch(es) on patches. It is *not* possible to 'subset' back on a lower number of dimensions. One cannot, for example, have a patch subset on a line set or a line subset on a finite set of dots.

The distribution of *large collieries* in the Yorkshire coalfield is a further illustration. The large collieries may be a subset of another set of dots, the set

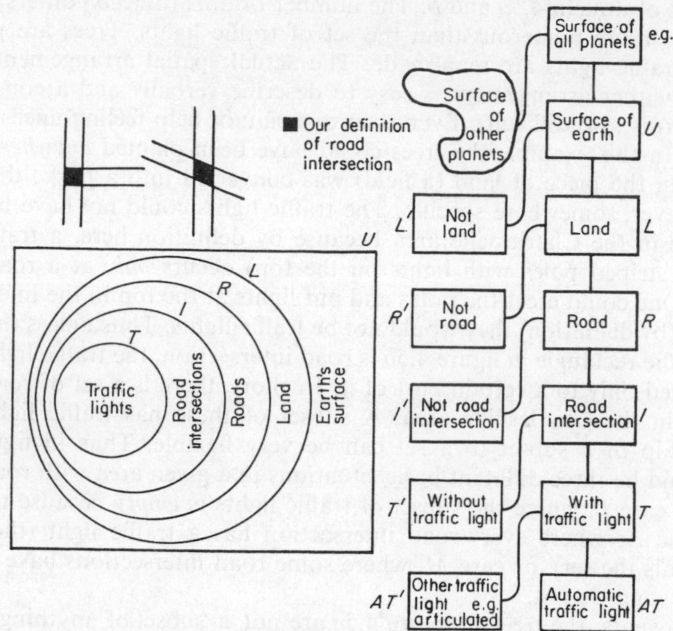

Figure 4.6 Diagrams to illustrate the relationship of traffic lights to bigger distributions.

of all the collieries. All the collieries may be regarded as a subset of the coalfield. The first subset is a dot nesting in another dot set, but the second subset (all collieries) is a subset of dots nesting in a patch set (the exposed coalfield). The patch set, the outcrop of coal measures in this example, could be considered a subset of all Carboniferous rocks, themselves a subset of Palaezoic rocks. These in turn could be a subset of the total land surface of Great Britain. The important thing is that the dot subset of all collieries has nested into a subset which is a patch (thereby by-passing a line set in this case) and this patch subset can only nest into another patch subset or set, not into a dot set or subset (unless for cartographical convenience the continuous surface is represented as a lattice of dots). In short, the nesting can only be from dot to dot or to a larger number of space dimensions, from line to line or a larger number of dimensions, from patch to patch or a larger number of dimensions, and from 3-dimensional space only to 3-dimensional space. Figure 4.7*a*, *b* illustrates the situation. Sometimes the geographer may be considering several disjoint subsets in a given area. For example (figure 4.7*c*) the patches in the U.S.S.R. representing areas in which over 50 per cent of the total surface is under forest, pasture and arable land respectively, are by definition disjoint sets.

It is suggested that geographers should pay more attention to the relationships of a particular distribution (spatial variable) that interests them to

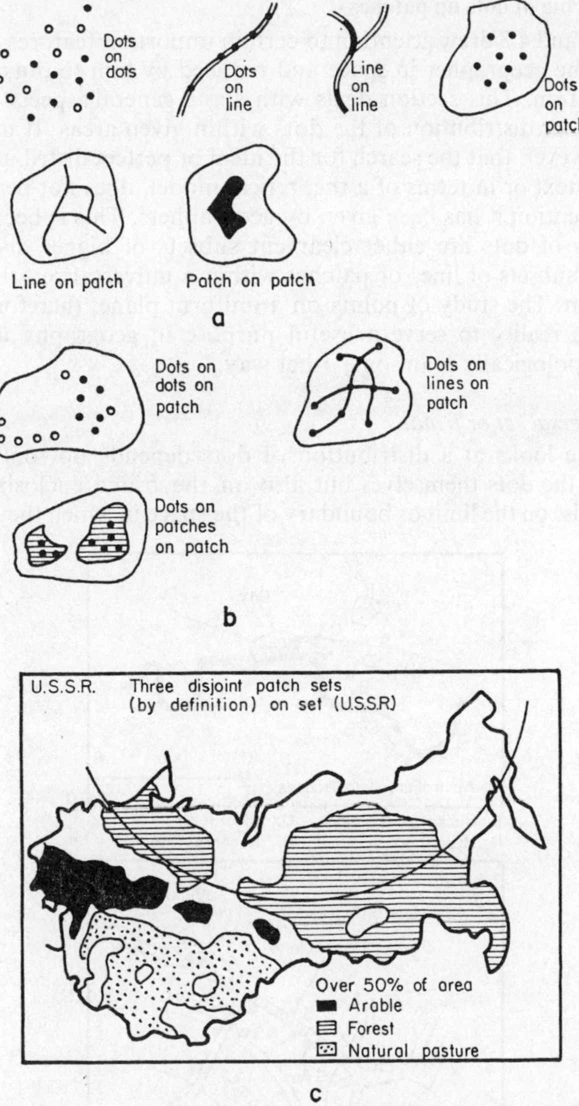

Figure 4.7 Examples of subset distributions. (*a*) Subsets on sets. (*b*) Subsets on subsets on sets. (*c*) Disjoint patch sets.

other variables, *not only* in causal terms, but also in 'subset' terms. When for example, some feature such as per capita income is being mapped, it is absurd to shade (say) high income *areas*, since people with (or preponderance of people with) high or low income are a subset of all people, not of area. The matter is taken up again in section 9 of this chapter.

4.4 The spacing of dots on patches

Sections 4.2 and 4.3 drew attention to certain important features of the objects studied by the geographer in space and reduced by him to dots for purposes of manipulation. This section deals with some general aspects of the actual arrangement or distribution of the dots within given areas. It must be borne in mind, however, that the search for the ideal or perfect distribution, whether in a real context or in terms of a theoretical model, does not perhaps deserve the great attention it has been given by geographers. This is because so many distributions of dots are either clear cut subsets of bigger distributions of dots, or are subsets of lines or patches within a universal set, the area under consideration. The study of points on a uniform plane, therefore, seems too remote from reality to serve a useful purpose in geography unless greatly modified, topologically or in some other way.

(a) The universal set or frame

The way one looks at a distribution of dots depends not only on the distribution of the dots themselves but also on the frame enclosing the distribution, that is, on the limit or boundary of the space in which they are situated.

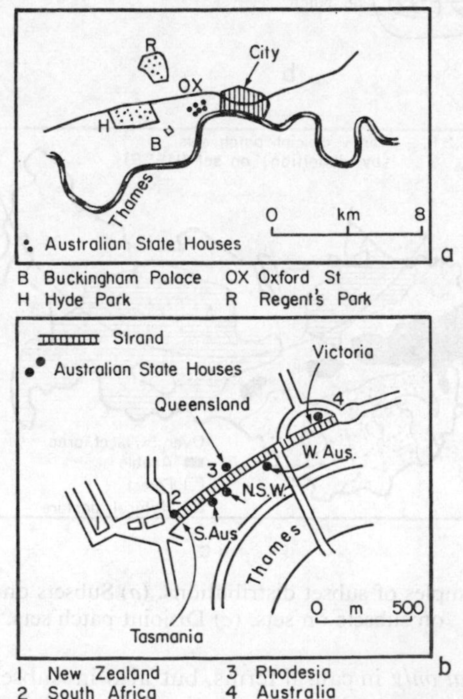

Figure 4.8 The distribution of Australian State houses in London in (*a*) a broad context. (*b*) Within the Strand. Note the presence here also of other Commonwealth or ex-Commonwealth representation.

Consider, as a very simple example, the distribution in London of State houses representing the six States of Australia (figure 4.8). These are all located on the Strand, a street in London. When considered in the context of Greater London (figure 4.8*a*) they are obviously highly clustered. When considered in the context of the Strand itself (figure 4.8*b*), they appear to be rather spaced out. This extreme example will suffice for the moment for the question of the frame around a distribution.

(*b*) *The grid or mesh superimposed on an area*

When a histogram is constructed in order to put data spaced along a line into a selected number of classes, some precision is lost. Moreover, careful

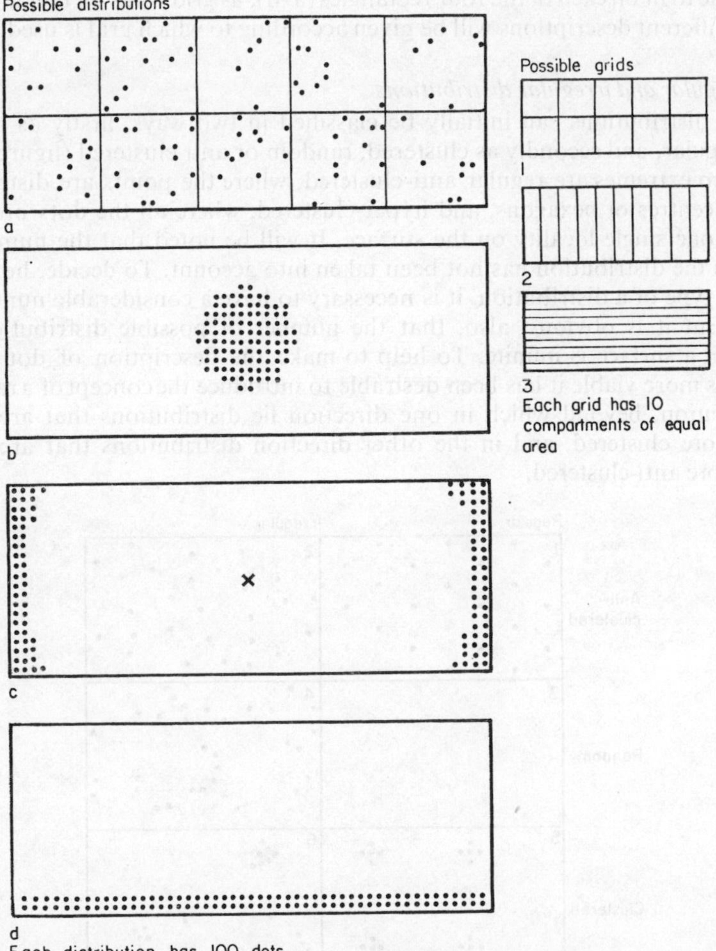

Figure 4.9 The effects of different grids (e.g. 1–3) on the assessment of different distributions of 100 dots (e.g. a–d).

consideration is needed both of the number of compartments (classes) and of their spacing along the line. The superimposition of a grid on a surface to facilitate the study of objects distributed over the surface is similar, but it involves two dimensions, not one. The choice of number of compartments in relation to the number of observations (dots) in the area is important, as is the shape of the compartments (e.g. square, hexagonal).

Figure 4.9 shows four distributions of 100 dots (a–d). A fifth, a regular distribution of 100 dots in the same rectangular space, can also be imagined. Grids could be superimposed on the distribution of dots to facilitate a description of them. Three possible regular grids (with compartments of uniform area) are shown (1–3). Appropriately enlarged, each of the three grids could be made to fit on each of the four rectangles (a–d), as grid 1 has on rectangle (a). Very different descriptions will be given according to which grid is used.

(*c*) *Regular and irregular distributions*

Dot distributions can initially be classified in two ways, firstly as regular or irregular, and secondly as clustered, random or anti-clustered (figure 4.10). The two extremes are regular anti-clustered, where the points are distributed as the centres of hexagons, and hyper-clustered, where all the dots are piled up on one single locality on the surface. It will be noted that the number of dots in the distribution has not been taken into account. To decide, however, on the type of a distribution, it is necessary to have a considerable number of dots. But it is obvious, also, that the number of possible distributions of dots on a surface is infinite. To help to make the description of dot distributions more viable it has been desirable to introduce the concept of a random distribution, beyond which in one direction lie distributions that are more and more clustered, and in the other direction distributions that are more and more anti-clustered.

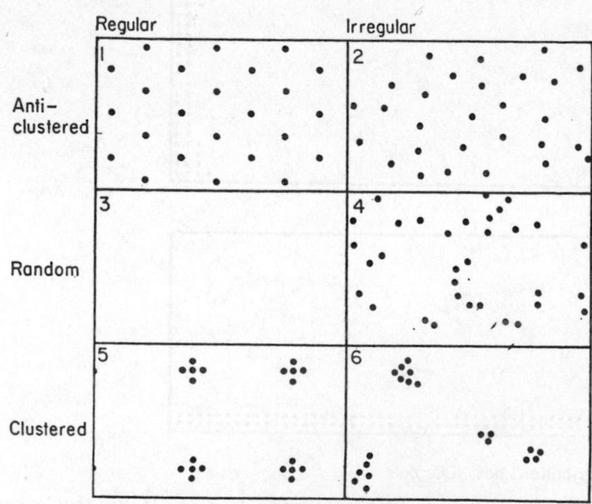

Figure 4.10 Types of dot distribution.

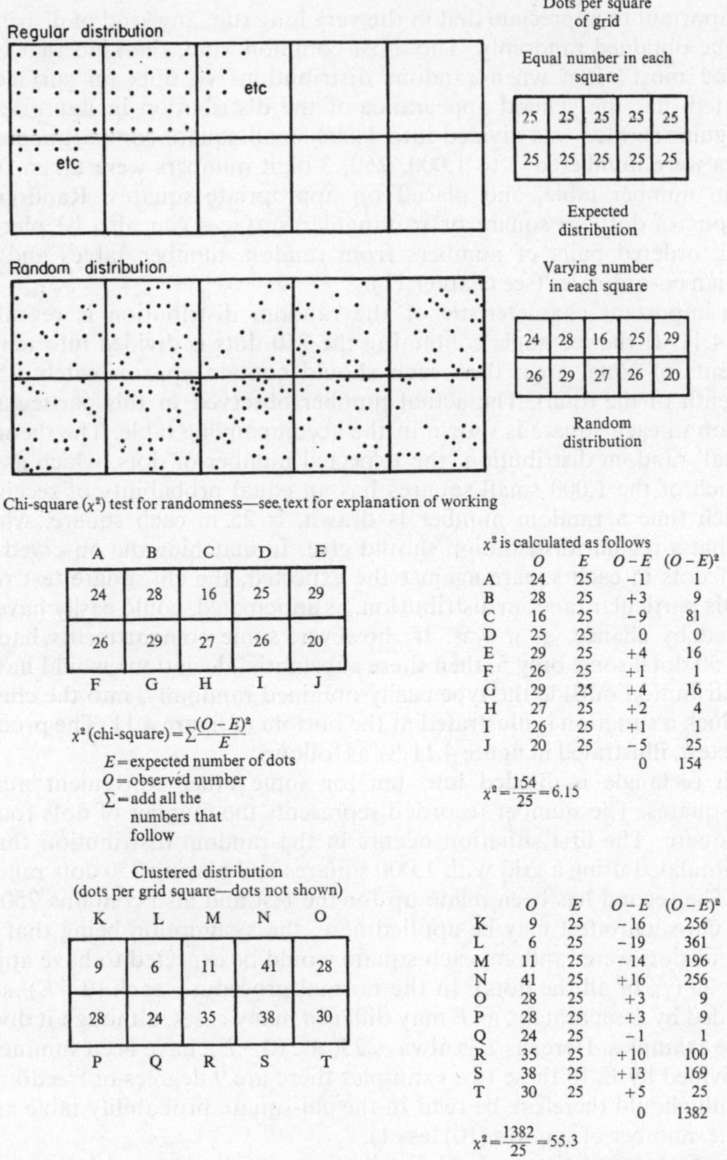

Figure 4.11 The use of the chi-square test to test for randomness in a distribution.

(*d*) *Random, clustered and anti-clustered*

It is important to appreciate that in the very long run, any kind of distribution could be obtained randomly. The most common kind, the kind that can be expected most often when random distributions of dots on surfaces are simulated, has the general appearance of the distribution in figure 4.11. A rectangular surface was divided into 1,000 small square compartments. The squares were numbered 1 to 1,000. 250, 3-digit numbers were taken from a random number table, and placed on appropriate squares. Random distributions of dots on square or rectangular surfaces can also be placed by reading ordered pairs of numbers from random number tables and using Cartesian co-ordinates (see chapter 11).

One important characteristic of the random distribution is revealed in figure 4.11. If the rectangle containing the 250 dots is divided into ten compartments of equal area, then each should contain approximately 25 dots (one tenth of the total). The actual number observed in this particular distribution in each square is shown in the accompanying table. The theoretical or 'ideal' random distribution, the expected number of dots, which assumes that each of the 1,000 small squares has an equal probability of receiving a dot each time a random number is drawn, is 25 in each square, which is also what a regular distribution should give. In matching the observed numbers of dots in each square against the expected, the chi-square test reveals that this particular random distribution, as anticipated, could easily have been obtained by chance, as it was. If, however, some compartments had contained 50 dots, some only 5, then these substantial deviations would have put the distribution outside the type easily obtained randomly, into the clustered type. Such a situation is illustrated at the bottom of figure 4.11. The procedure for the test, illustrated in figure 4.11, is as follows:

Each rectangle is divided into ten (or some other convenient number) equal squares. The number recorded represents the number of dots found in each square. The first situation occurs in the random distribution that has been simulated using a grid with 1,000 squares and placing 250 dots randomly on it. The second has been made up for the test and also contains 250 dots.

The chi-square test may be applied here, the assumption being that if the spread of dots were random, each square would be expected to have approximately 25 ($\frac{1}{10}$ of all the dots). In the normal procedure, each $(0-E)^2$ should be divided by E separately, as E may differ in many cases, although it does not in these examples. Here, as E is always 25, the $(0-E)^2$ have been summed and then divided by E. In these two examples there are 9 degrees of freedom, and the result should therefore be read in the chi-square probability table against $n=9$ (i.e. number of squares (10) less 1).

Other features of the random distribution may be assessed by comparing the distance between each dot and its nearest neighbour (or, in turn, two or more nearest neighbours). Random distributions are often characterized by having some dots very close. For example in this case, when the distribution was made one small square received 4 of the 250 dots whereas more than 750 of the thousand squares obtained none. Anti-clustered distributions have fewer dots close together, while clustered ones have many. The two essential features of a random distribution are (1) that the dots should have an equal chance of falling anywhere on the surface in question, or in the squares of a grid, if this is used in place of a continuous surface, and (2) that the placing of a dot should not be affected by the placing of any others (see chapter 3). More precise locations could bc obtained by using real number co-ordinates to several decimal places.

(*e*) *Real world examples of distributions of dots on patches*

Do random distributions of objects (represented by dots on maps) occur in reality? In the Second World War the Germans are reputed to have claimed on one occasion that they bombed Random heavily. This apparently was

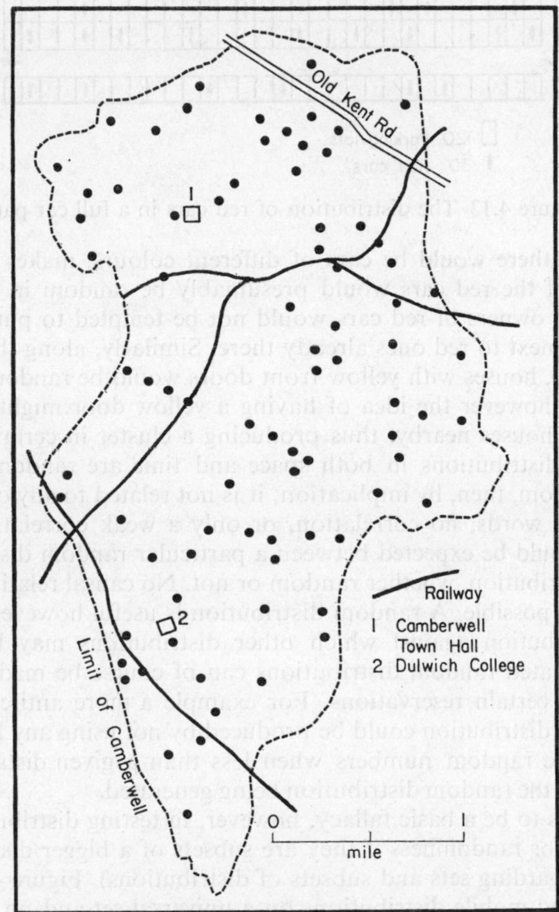

Figure 4.12 The distribution of V1 (flying) bombs in the borough of Camberwell, South London, 1944.

because the British Broadcasting Corporation often announced the dropping of bombs at random. Ironically, an investigation of the distribution of V1 and V2 bombs dropped in south London by the Germans in 1944–45 showed that there was a random distribution of these. Figure 4.12 shows the distribution of V1 bombs in part of South London (source: *The Evening News*, Monday, October 2nd, 1944). The area covered is the former London borough of Camberwell.

Many random spatial distributions that one might expect to find in reality tend to be trivial. For example (figure 4.13) in a car park with lots spaced in a

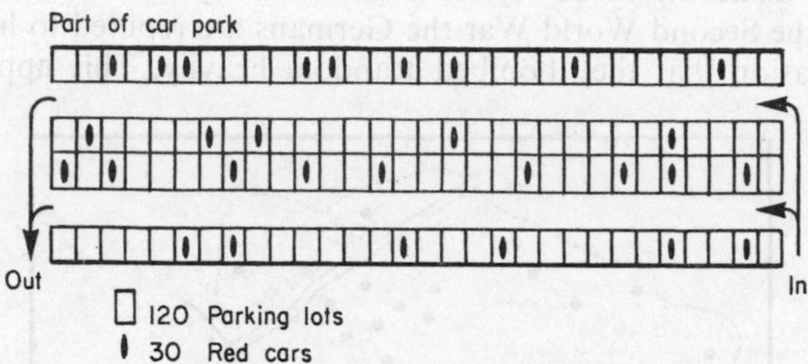

Figure 4.13 The distribution of red cars in a full car park.

regular array there would be cars of different colours, makes or sizes. The distribution of the red cars would presumably be random in the car park, assuming that owners of red cars would not be tempted to put their cars in empty spaces next to red ones already there. Similarly, along the streets of a housing estate, houses with yellow front doors would be randomly arranged. Possibly here however the idea of having a yellow door might spread from one house to houses nearby, thus producing a cluster in certain places. No doubt many distributions in both space and time are random. If a distribution is random, then, by implication, it is not related to any other distribution. In other words, no correlation, or only a weak correlation, obtained by chance, would be expected between a particular random distribution and any other distribution, whether random or not. No causal relationship would apparently be possible. A random distribution is useful however as a general type of distribution against which other distributions may be measured. More complicated random distributions can of course be made (simulated) by putting in certain reservations. For example a more anti-clustered than usual random distribution could be produced by not using any locations that came from the random numbers when less than a given distance from an existing dot in the random distribution being generated.

There seems to be a basic fallacy, however, in testing distributions of dots on a surface for randomness if they are subsets of a bigger distribution (see section 4.3 regarding sets and subsets of distributions). Figure 4.5*a* and 4.5*g* show static and mobile distributions on a universal set and on subset bases. It is feasible to test the distribution of the trees in the park or the children in the playground for randomness of distribution on the surface U. But it is meaningless to test the traffic lights or the motor vehicles for randomness of distribution against their total surface U. The traffic lights must be tested for randomness on the base of road intersections, the cars against all roads. To determine how to do this is not straightforward, but it is not impossible. The situation has not been seriously considered by geographers, many of whom have preferred to work from the theoretical viewpoint on a uniform plane, or who have disregarded the concept of randomness altogether. An example will illustrate how the test might be arranged.

Figure 4.14*a* shows the distribution of garages in the vicinity of the town of Grangerville. Their distribution, if tested, would undoubtedly come out as clustered. But in reality these garages are closely tied to the road system (fig. 4.14*b*). It is quite possible that their distribution in relation to their base, the road system, is anti-clustered, or simply random. How could this be checked? By dividing the roads into a number of equal length sections (say

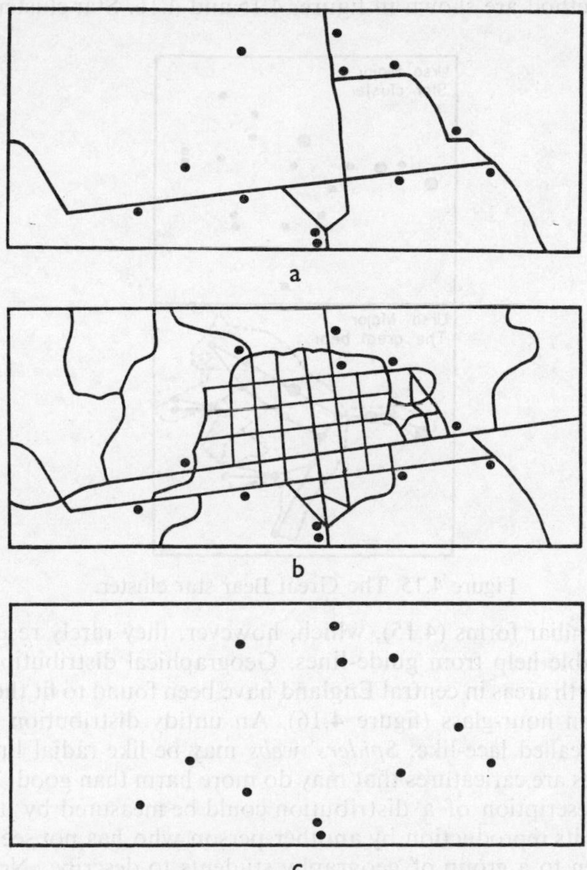

Figure 4.14 Distribution of garages in Grangerville. (*a*) In relation to main roads. (*b*) In relation to roads. (*c*) Alone.

50 or 100 yards), numbering these, and locating randomly an appropriate number of 'garages', the distribution could be simulated. The nearest neighbour distances of the random garages could then be compared with the nearest neighbour distances of the real ones. Or the model might be further refined to omit places impossible for the construction of a garage such as buildings unlikely to be pulled down, to give emphasis to main roads (figure 4.14*c*), and so on.

It is hoped that this section will have made clear the great complexity of distributions of dots on patches. Moreover, what has been said about dots on patches mostly refers in a general way to any distribution in any space. Indeed a distribution of dots in 3-dimensional space is far more complex to handle. It is not surprising, then, that geographers have developed a makeshift way of describing distributions verbally. This is usually done either by quick analogy with some familiar object, or with rather laborious words. Examples of the first method are shown in figures 4.15 and 4.16. Star clusters are often

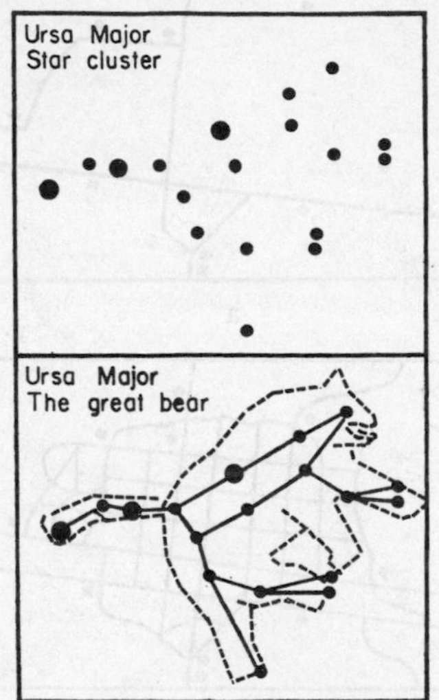

Figure 4.15 The Great Bear star cluster.

likened to familiar forms (4.15), which, however, they rarely resemble without considerable help from guide-lines. Geographical distributions, such as towns or growth areas in central England have been found to fit the shape of a coffin or of an hour-glass (figure 4.16). An untidy distribution of built-up area is often called lace-like. *Spiders' webs* may be like radial line patterns. Such analogies are caricatures that may do more harm than good. The success of a verbal description of a distribution could be measured by its adequacy as a basis for its reproduction by another person who has not seen it. Figure 4.17 was given to a group of geography students to describe. No indication was given of what the dots stand for, what the area is, nor of scale or orientation. As expected, the result was disappointing from various points of view.

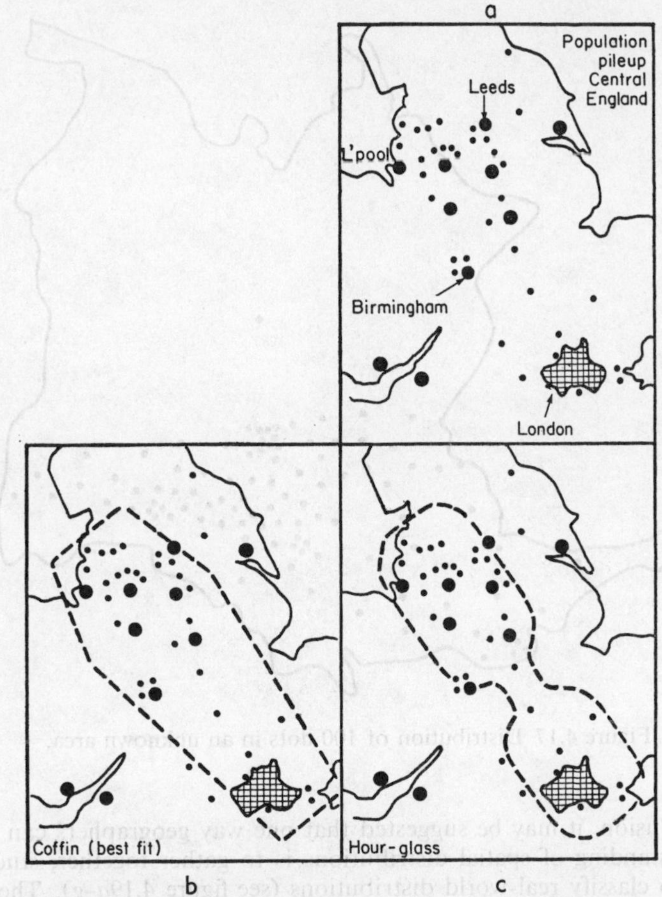

Figure 4.16 Population in Central England. (*a*) As illustrated by the larger towns. (*b*) Described as a coffin. (*c*) As an hour glass.

There was a lack of agreement on the use of terms, preoccupation with one dot far outside the main cluster, and general lack of ability to cope. The actual area in figure 4.17 is Colombia, South America. Each dot = 1 per cent of its total population.

Even, however, if one had an adequate vocabulary to give a general description of a distribution of dots on a patch, one could still not do justice in terms of required precision to the situation. Figure 4.18 shows the distribution of woodland (part of Sherwood Forest) north of Nottingham. There are 200 squares. Each square that has more than 50 per cent woodland is given a 1. The other squares have 0. The number of possible arrangements of woodland on this area of 200 squares is 2^{200}, or about 10^{64}. It would indeed be difficult to find enough different verbal descriptions to cater for all these contingencies.

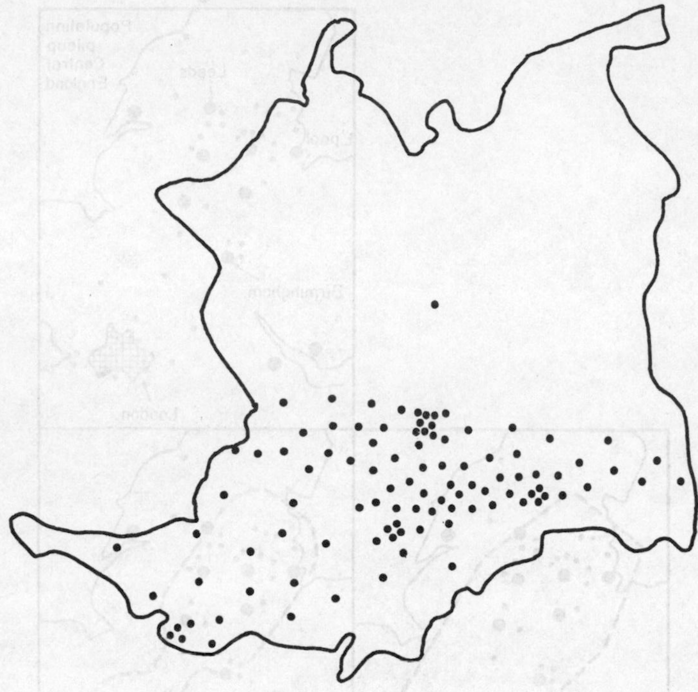

Figure 4.17 Distribution of 100 dots in an unknown area.

In conclusion, it may be suggested that one way geographers can further the understanding of spatial distributions is to gather together, study, and attempt to classify real world distributions (see figure 4.19*a–g*). These may be non-geographical (e.g. figure 4.19*b* and *c*) as well as geographical. They may be man-made, some of which have been illustrated already in this chapter, or they may be physical.

(f) The nearest neighbour concept

One way of describing a distribution with a sufficiently large number of dots is to study nearest neighbour features. In fact each dot may have not only a first nearest neighbour but a second, third and so on. Sometimes the nearest neighbour of a particular dot is located outside the patch on which the dots are located, or beyond the frame of the map (e.g. in figure 4.21*d*).

Very broadly, it is possible to have extreme situations in which at one extreme all dots form nearest neighbour pairs, of which there will thus be N/2 (e.g. twin settlements along stream, figure 4.20*a*), and at the other, all dots form one cluster. This happens in figure 4.20*b* — stations on the railway, on which, because the distance between stations increases successively to the

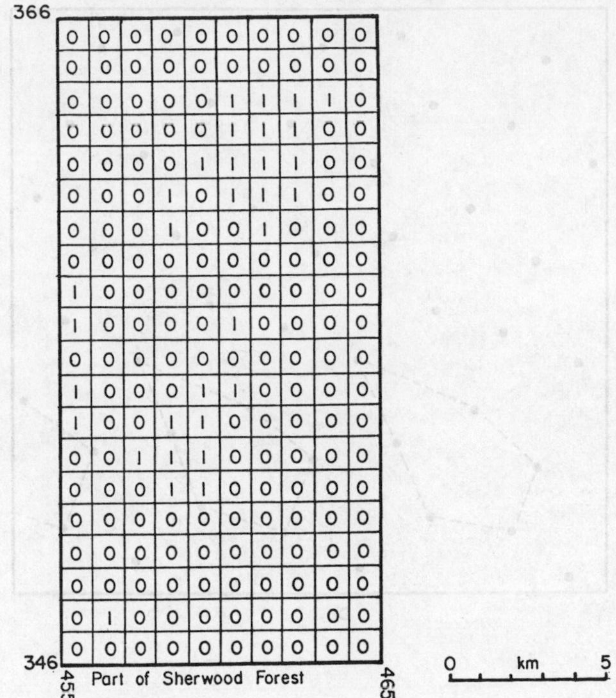

Figure 4.18 Distribution of woodland in part of Nottinghamshire.

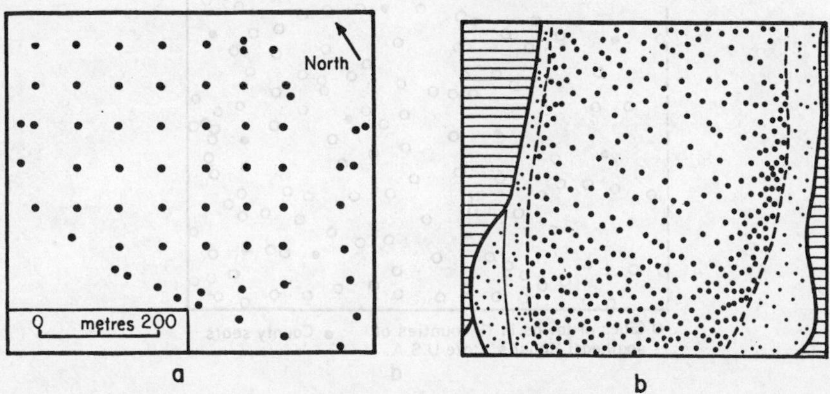

Figure 4.19 Examples of real world dot distributions. (*a*) Distribution of road intersections in the central part of the town of La Paz, Bolivia. (*b*) Distribution of spots on part of the trunk of Sahib, outsize elephant in Swiss Krie circus (*Times*, June 20th, 1964).

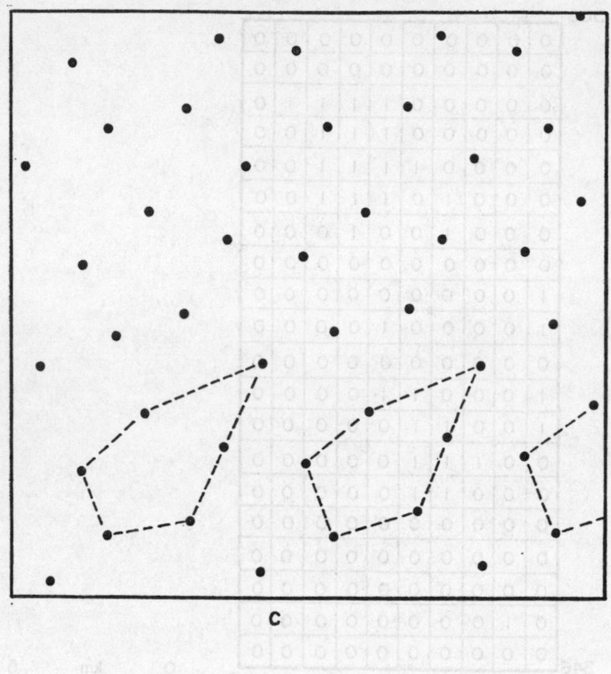

c

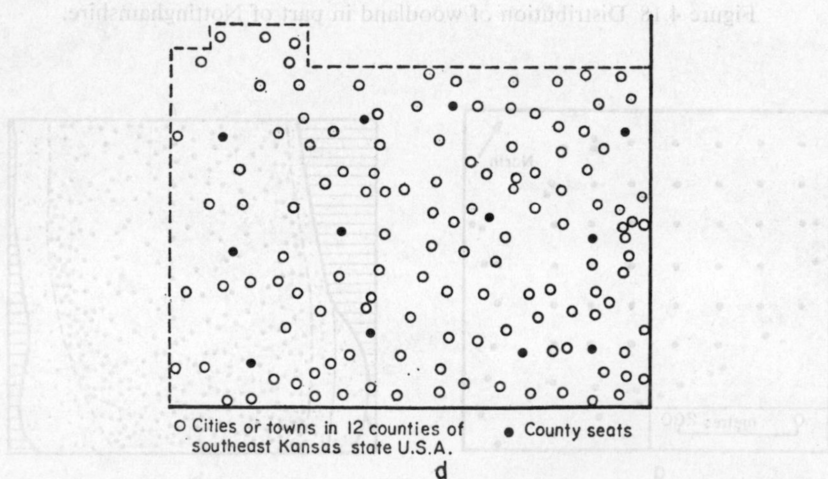

d

Figure 4.19 Examples of real world dot distributions. (*c*) An anti-clustered, apparently irregular, distribution. Broken lines, however, pick out the regularity of a 6-dot pattern. Spotlite frosted glass. (*d*) Cities or towns in 12 counties of South-east Kansas State, U.S.A.

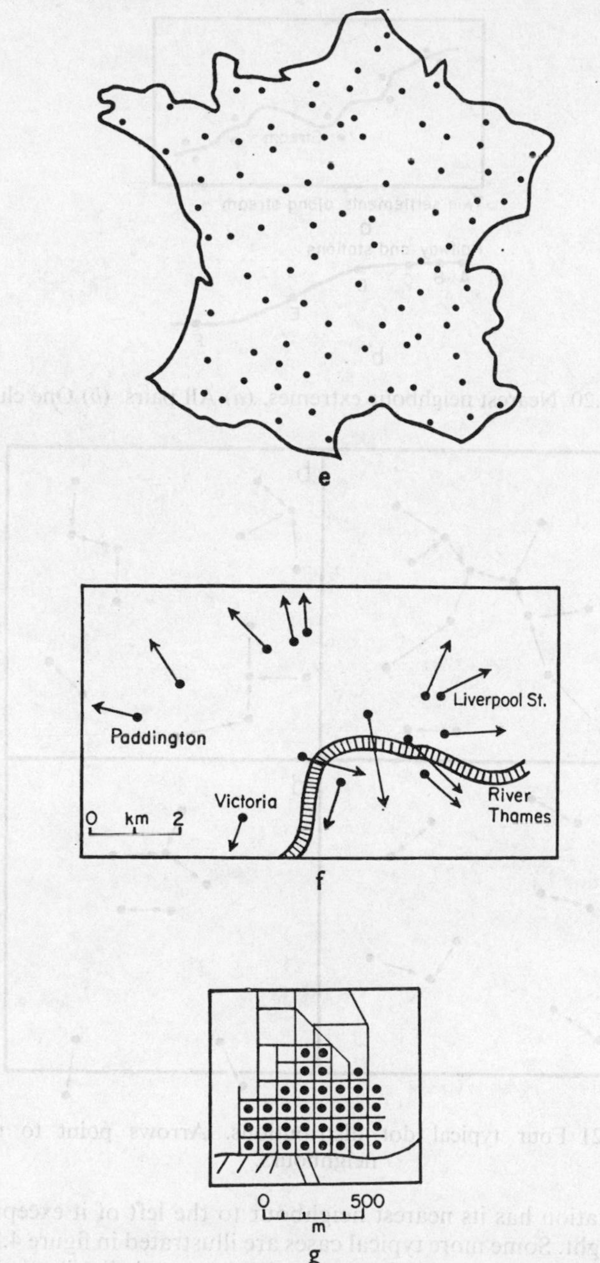

Figure 4.19 Examples of real world dot distributions. (*e*) 89 Department Capitals of mainland France. (*f*) 14 rail terminus stations in London. (*g*) Oil tanks by a refinery, Talara, Peru.

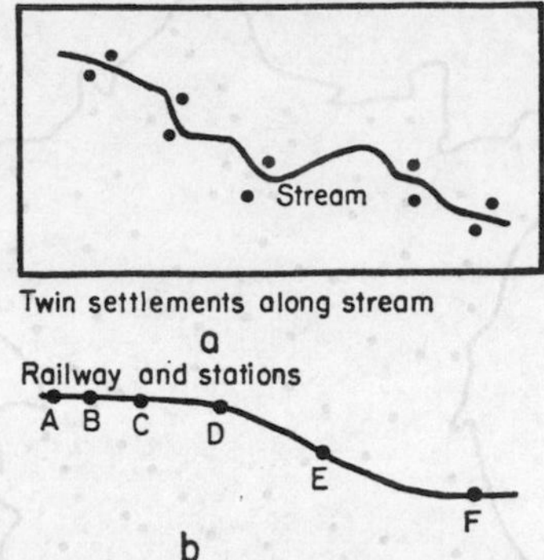

Figure 4.20 Nearest neighbour extremes. (*a*) All pairs. (*b*) One cluster.

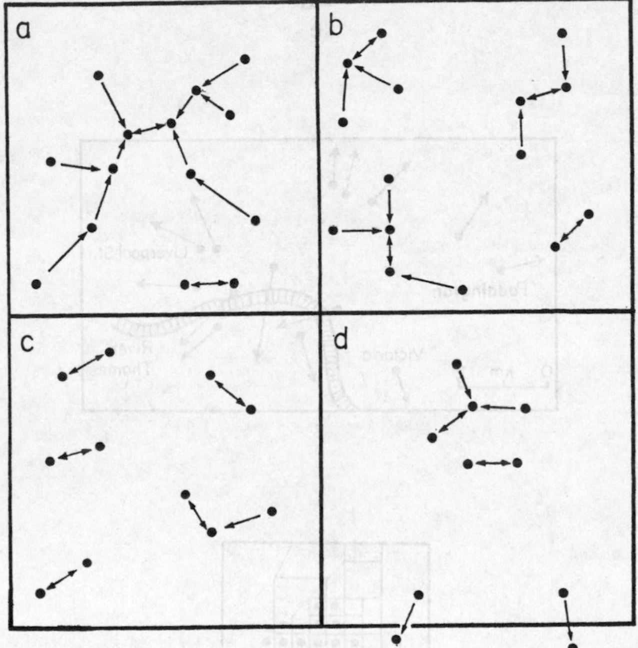

Figure 4.21 Four typical dot distributions. Arrows point to nearest neighbours.

right, each station has its nearest neighbour to the left of it except A, which has B to its right. Some more typical cases are illustrated in figure 4.21

In both the real world and in randomly generated distributions of dots, there tend to be both isolated pairs and clusters. The distributions of towns with over 40,000 inhabitants (commune population) in North Italy (figure 4.22) serves as an example.

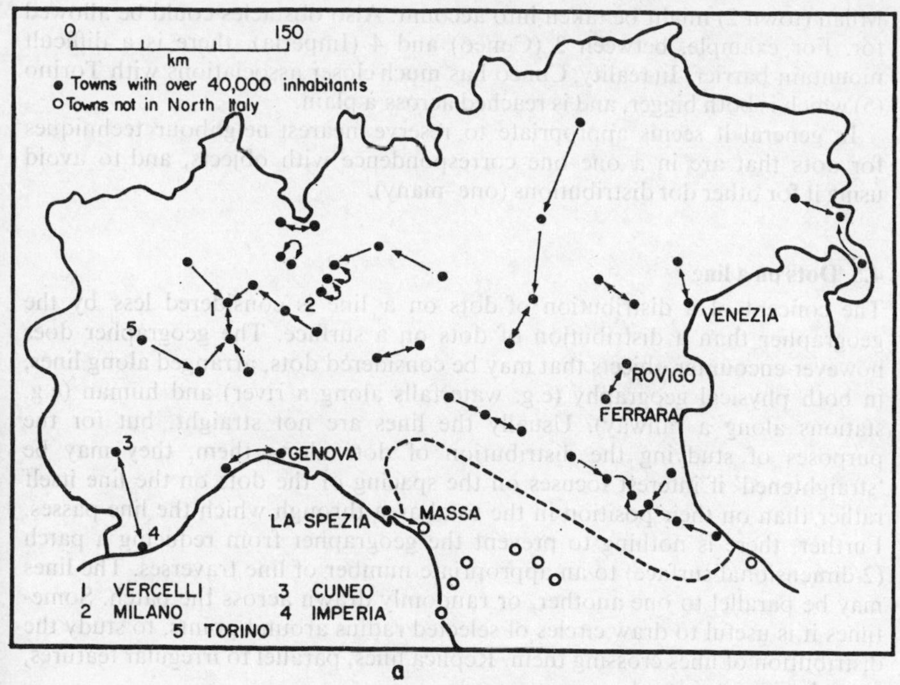

Figure 4.22 Nearest neighbours among the larger towns of North Italy.

Clusters

Number	Number of members
1	9
1	7
1	5
1	4
2	3
8	2
1	1*

* Nearest neighbour lies outside North Italy, Massa.

The nearest neighbour idea is a useful technique for giving a rough appraisal of spacing features of dots. In particular, it serves as a test for randomness. There is an expected proportion of reciprocal nearest neighbour pairs; this proportion should not be affected by the number of dots in the distribution. It serves also as a descriptive method, with appropriate numerical information to allow a comparison of two or more distributions. Thus the number of clusters with different numbers of members (figure 4.22) for North Italy could be compared with those of Peninsular Italy.

Refinements of the crude dot distribution could be introduced in nearest neighbour procedure. Thus, in the case of North Italy the great size of

Milan (town 2) might be taken into account. Also obstacles could be allowed for. For example, between 3 (Cuneo) and 4 (Imperia), there is a difficult mountain barrier. In reality, Cuneo has much closer associations with Torino (5) which is both bigger, and is reached across a plain.

In general it seems appropriate to reserve nearest neighbour techniques for dots that are in a one–one correspondence with objects, and to avoid using it for other dot distributions (one–many).

4.5 Dots on a line

The concept of a distribution of dots on a line is considered less by the geographer than a distribution of dots on a surface. The geographer does however encounter objects that may be considered dots, arranged along lines, in both physical geography (e.g. waterfalls along a river) and human (e.g. stations along a railway). Usually the lines are not straight, but for the purposes of studying the distribution of dots along them, they may be 'straightened' if interest focuses on the spacing of the dots on the line itself rather than on their position in the total area through which the line passes. Further, there is nothing to prevent the geographer from reducing a patch (2-dimensional surface) to an appropriate number of line traverses. The lines may be parallel to one another, or randomly drawn across the patch. Sometimes it is useful to draw circles of selected radius around points, to study the distribution of lines crossing them. Replica lines, parallel to irregular features, may also be considered.

In some ways, dots on lines are simpler to study than dots on patches, but much of what has been said so far in this chapter applies just as well to dots on lines as to dots on patches, at least when appropriately modified. Thus, for example, it is possible to have regular and irregular distributions. These may also be random, clustered or anti-clustered. Some examples of dots on lines are given in figure 4.23.

4.6 The construction of a dot map

The distributions considered so far have largely been of limited numbers of objects represented by dots in a one–one correspondence. Once the number of objects to be mapped by dots exceeds several hundred, it usually becomes desirable to represent them by a reduced number of dots. Often therefore the geographer makes a dot map in which *each* dot represents *several* or *many* objects. Such maps appear frequently in geographical publications (e.g. cattle in Argentina, rice cultivation in China). Rarely, however, is there an explanation and justification of the actual placing of dots achieved. Many methods varying in precision and in criteria as to the placing of dots have been used. One procedure is described here because it is felt that all geographers should be aware of the drawbacks in constructing a dot map where there are many objects to each dot. At the same time he should be proficient in the construction at least of a simple dot map.

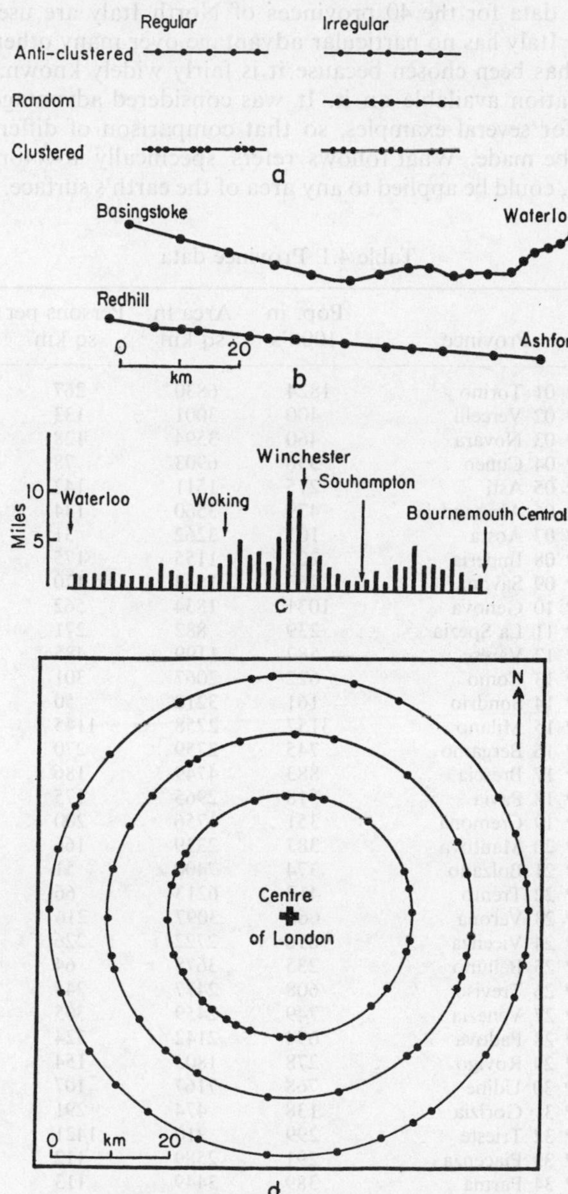

Figure 4.23 Examples of distributions of dots on lines. (*a*) Types. (*b*) Examples of stations on two stretches of railway. (*c*) Distances between stations on the railway between Waterloo and Bournemouth. (*d*) Crossing points of railways and circles at 20, 30 and 40 km from the centre of London.

Population data for the 40 provinces of North Italy are used in this example. North Italy has no particular advantage over many other parts of the world but it has been chosen because it is fairly widely known, and there is much information available on it. It was considered advantageous to keep to this area for several examples, so that comparison of different methods could easily be made. What follows refers specifically to North Italy but, within reason, could be applied to any area of the earth's surface.

Table 4.1 Province data

Province	Pop. in 1000's	Area in sq km	Persons per sq km
P 01 Torino	1824	6830	267
P 02 Vercelli	400	3001	133
P 03 Novara	460	3594	128
P 04 Cuneo	536	6903	78
P 05 Asti	215	1511	142
P 06 Alessandria	479	3560	134
P 07 Aosta	101	3262	31
P 08 Imperia	202	1155	175
P 09 Savona	263	1544	170
P 10 Genova	1031	1834	562
P 11 La Spezia	239	882	271
P 12 Varese	582	1199	485
P 13 Como	622	2067	301
P 14 Sondrio	161	3212	50
P 15 Milano	3157	2758	1145
P 16 Bergamo	745	2759	270
P 17 Brescia	883	4749	186
P 18 Pavia	518	2965	175
P 19 Cremona	351	1756	200
P 20 Mantova	387	2339	166
P 21 Bolzano	374	7400	51
P 22 Trento	412	6213	66
P 23 Verona	668	3097	216
P 24 Vicenza	616	2722	226
P 25 Belluno	235	3678	64
P 26 Treviso	608	2477	245
P 27 Venezia	749	2459	305
P 28 Padova	694	2142	324
P 29 Rovigo	278	1803	154
P 30 Udine	768	7167	107
P 31 Gorizia	138	474	291
P 32 Trieste	299	210	1421
P 33 Piacenza	291	2589	112
P 34 Parma	389	3449	113
P 35 Reggio Emilia	380	2291	166
P 36 Modena	511	2690	190
P 37 Bologna	841	3702	227
P 38 Ferrara	403	2632	153
P 39 Ravenna	330	1860	177
P 40 Forli	521	2910	179

Table 4.1 (source: *Annuario statistico italiano* 1963, p. 13) contains two sets of data for the 40 provinces of North Italy, number of inhabitants (pop. in 1000's) and territorial extent (area in sq km). From these data, density of population (persons per sq km) has been derived. The map in figure 4.24*a* shows the location of the 40 provinces and province capitals.

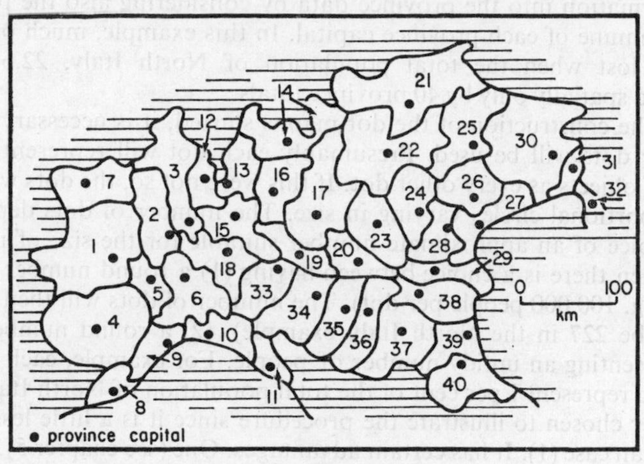

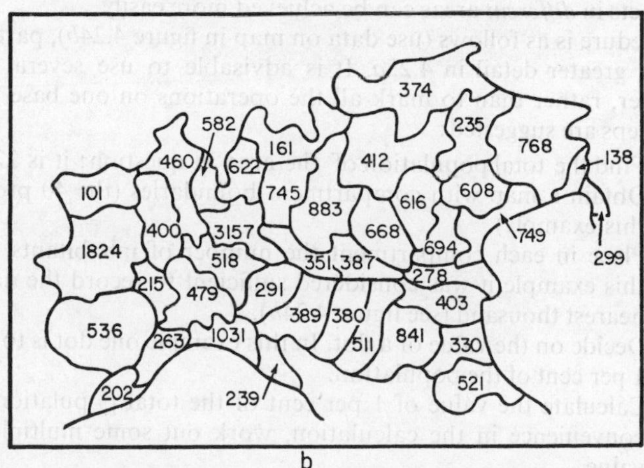

Figure 4.24 North Italy. (*a*) Location map of 40 provinces. Numbers tally with list of provinces P 01 to P 40 in table 4.1. (*b*) Population in 1963 of the 40 provinces to nearest thousand inhabitants.

This serves as an example of a one–one correspondence situation, since the province capitals can for some purposes be considered comparable elements in a clearly defined set. The dots could of course have been inflated as appropriate to represent the differing number of inhabitants of each capital.

The map in figure 4.24*b* shows the population data from the table transferred to the province map. The dot distribution map will be constructed from these data. It would have been possible to construct a much more detailed and accurate dot map using information for the several thousand communes into which the 40 provinces are divided.* It would also be possible to introduce extra information into the province data by considering also the population of the commune of each province capital. In this example, much precision is obviously lost when the total population of North Italy, 22,660,000, is represented spatially only by 40 province totals.

Before the construction of the dot map is started, it is necessary to decide how many dots will be used. Presumably each dot will represent the same number of objects as every other dot. If this were not so, the dots would have to be proportional circles varying in size. The number of dots depends first on the choice of an approximate number suitable for the size of map to be drawn. Then there is a choice between having (1) a round number of people per dot (e.g. 100,000 people per dot). The number of dots will then be untidy (it would be 227 in the North Italy example). (2) a round number of dots each representing an untidy number of people. For example, each dot could be made to represent 1 per cent of the total population of North Italy. This is the number chosen to illustrate the procedure since it is a little less straightforward than case (1). It has certain advantages. One (see chapter 5) is that if a standard number of dots is used, visual comparison of the distribution of *similar* objects in *different* areas can be achieved more easily.

The procedure is as follows (use data on map in figure 4.24*b*), part of which is shown in greater detail in 4.25*a*. It is advisable to use several layers of tracing paper, rather than to mark all the operations on one base map. The following steps are suggested:

Step 1. Find the total population of the area in question: it is 22,660,000.

Step 2. Obtain a map with compartment boundaries (the 40 provinces in this example).

Step 3. Place in each compartment the number of inhabitants it has. In this example it was considered sufficient to record the data to the nearest thousand (see figure 4.24*b*).

Step 4. Decide on the value of a dot. In this example one dot is to represent 1 per cent of the population.

Step 5. Calculate the value of 1 per cent of the total population, and for convenience in the calculation, work out some multiples of this value.

1 dot	226.6	(approx. 227)
2 dots	453.2	(,, 453)
3 dots	679.8	(,, 680)
4 dots	906.4	(,, 906)
5 dots	1133.0	(,, 1133)

* An example of a compromise between province and commune is in U. Giusti's work. Giusti uses sub-regions within the provinces. *Caratteristiche ambientali italiane . . .*, (1943.)

6 dots	1359.6	(approx.	1360)
8 dots (Torino)	1812.8	(approx.	1813)
14 dots (Milano)	3172.4	(approx.	3172)

Step 6. Calculate the number of whole dots each province is entitled to receive by comparing its population with the number of people per one dot or per appropriate multiple thereof.

Very rarely will each province have precisely or even approximately enough people to be 1 per cent or a multiple of 1 per cent of the total population. There will have to be some 'exporting' and 'importing' of population among provinces, and compensation as a result. In other words there is some loss of precision in the operation of deciding where the dots can go. It is therefore desirable to proceed as consistently and honestly as possible, starting at one side and trying to compensate in all directions, not just one, as the area is worked over.

The North Italy example is worked for the first few western provinces in the western part (see figure 4.25*a* and accompanying notes).

Step 7. When the above procedure has been carried out across the area, the number of dots should be checked. If the required number (e.g. 100) is not obtained, the procedure should be started again.

Step 8. When 100 dots have finally been obtained, some common sense must be used in placing them. It is of course possible to spread them evenly over the surface of each province. Step 6 has then been largely superfluous. It is advisable to take into account known features of the area and arrange dots accordingly. In the North Italy example, the Alpine portions of several northern and western provinces may deliberately be starved of dots because they have few people. Milano commune, in contrast, can claim some eight dots, since it is known to have about 1,700,000 people. It would however be spuriously precise to pretend to find a perfect location for each dot. In fact, a somewhat different arrangement of dots will be obtained each time the operation is carried out for North Italy, particularly if step 6 starts, for example, from the east rather than the west. This understandably disturbs many geographers. However the dot map obtained has several advantages, which will emerge in later sections of the book. The resulting map done in this exercise is shown in figure 4.25*b*.

Those who object to the arbitrary character of the above procedure should bear in mind that it is both useless and virtually impossible to record over 20 million people as dots on one map. Therefore some loss of precision is inevitable. It is with some satisfaction therefore that one reads the view of Lewis F. Richardson (1961), who did a great deal of work on the distribution of population in countries of the world.

> 'I find that there is not one uniquely best map. Geographers, being accustomed to uniquely determined maps, may feel that any arbitrariness is unsatisfactory. Physicists however are accustomed to equipotentials and

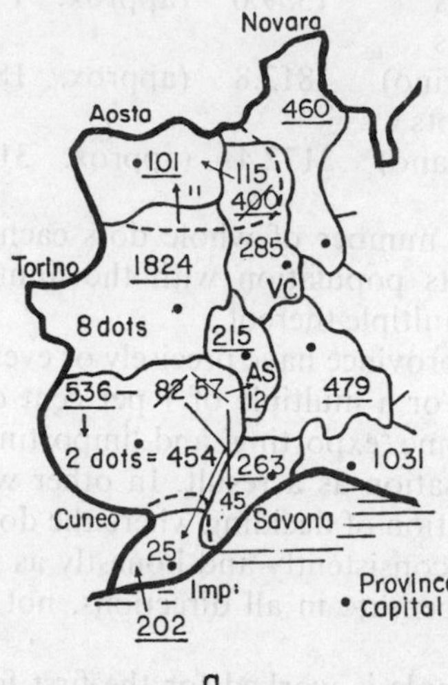

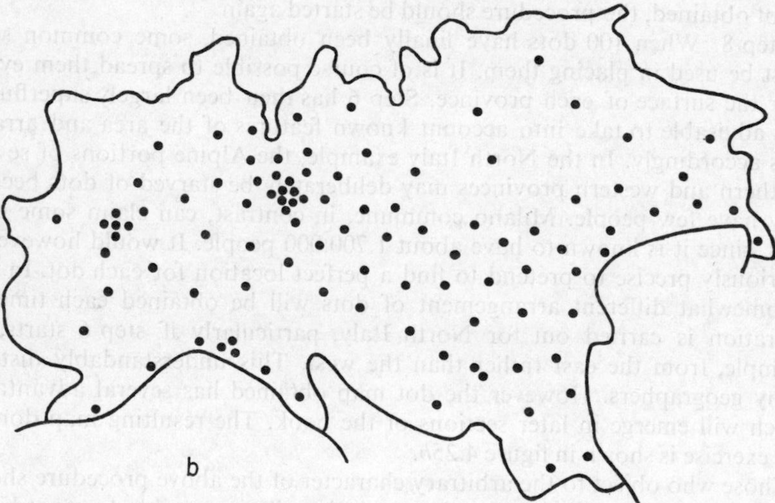

Figure 4.25 The construction of a dot map for population in North Italy. (*a*) Procedure: start at Turin which has 1824, but needs 1813 to have 8 dots. Therefore combine Turin residue (11) with Aosta (101) and (115) from Vercelli (VC) to make 227. Vercelli will then have 285. This will make another dot plus 48 residue, etc. (See also text.) (*b*) Result: distribution of population in North Italy. Each dot = 1 per cent of the total or approximately 227,000 persons.

stream-lines which represent a field in a manner partly arbitrary, being selections from two infinite families of lines. The important relation, I suggest, is not that the census should uniquely determine the map, but almost conversely that the map should quickly give a good idea of the census.'

This section concludes with brief reference to some associated techniques:

(1) It is interesting to compare with the dot map in figure 4.25*b* the visual impression of the distribution of population in North Italy as achieved by maps on the basis of the same data showing density (see figure 4.26). Note that in this case the shading scale may be related to the mean density for all North Italy, or may be arranged about the median density of the provinces. As when 1 dot = 1 per cent, so when the mean (or median) density is used as a base for shading, comparison with other areas is facilitated.

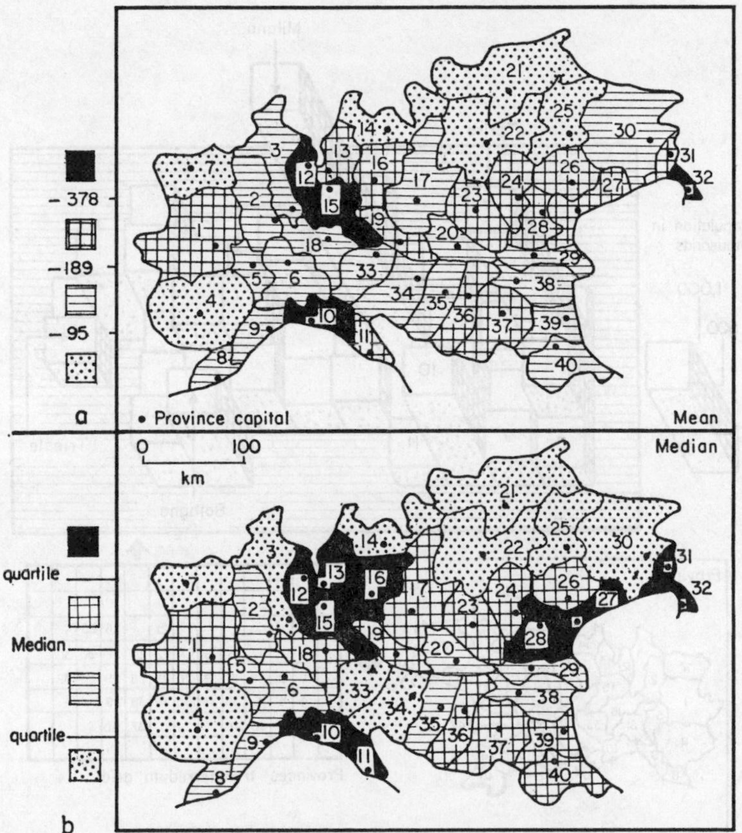

Figure 4.26 Density of population in North Italy using data displayed in the table in section 4.6. (*a*) A scale of shading arranged about the *mean* density, 189 persons per sq km. (*b*) A scale of shading arranged in quartiles about the median, which lies between the 20th and 21st provinces in order of density of population.

(2) It is possible to imagine from a topological viewpoint that provinces are equal in area. A striking visual impression of the real variations in density of population can then be achieved by raising a column on a third dimension axis at an appropriate angle from the plane surface to give a 'factory chimney' picture (see figure 4.27). It is possible, though untidy, actually to do this above the true shape of each province. It will be shown in the next section that the dot map with an appropriate grid superimposed on it may also be used.

(3) A by-product of the dot map is a map of cells (figure 4.28) each (in this example) containing 1 per cent of the total population. While this represents a further modification and abstraction of the original distribution of population, and depends on subjective construction of the cells, it may give a valuable visual picture for general comparative purposes.

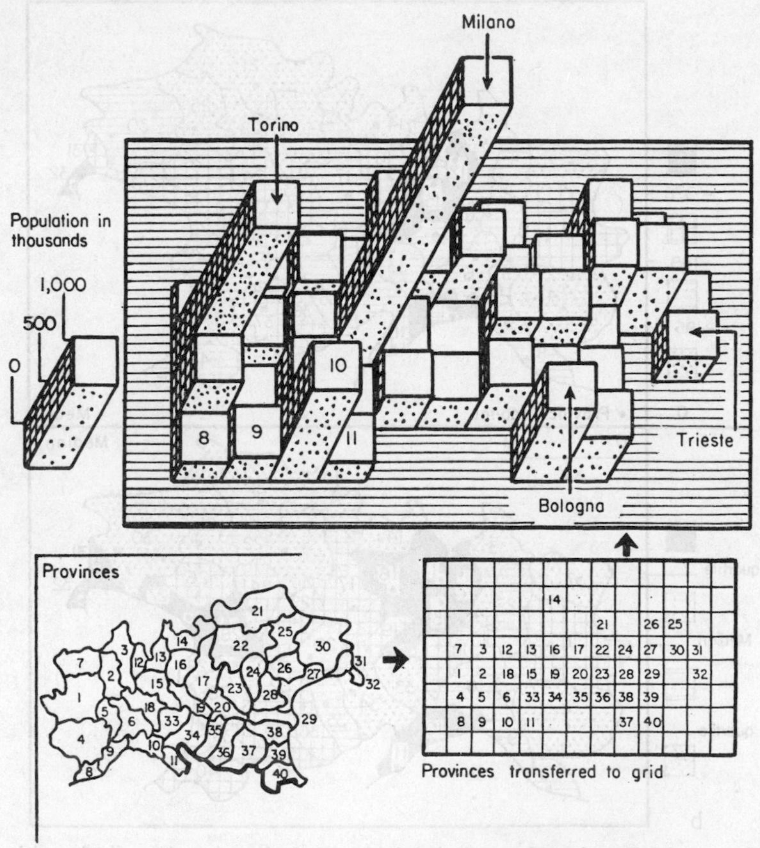

Figure 4.27 Variations in density of population of North Italian provinces as illustrated by columns raised above the plane surface and proportional in height to population. The diagram may be considered a histogram in three dimensions.

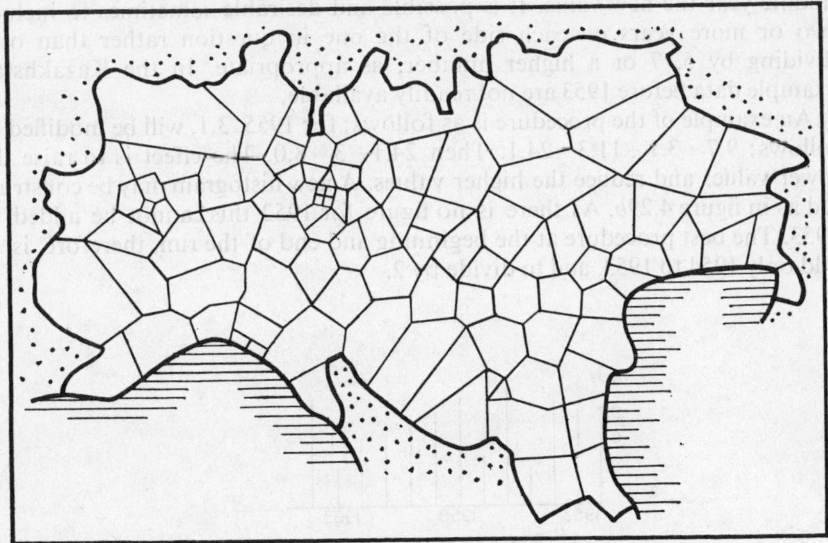

Figure 4.28 Distribution of population in North Italy represented by 100 equal population cells.

4.7 Generalizing a dot distribution

Two techniques that modify dot distribution maps are now described. The first is an extension of the idea of a running mean and can be applied on any suitable dot map. The second calculates the potential of a distribution and can be used with uniform dots or with a distribution of dots of different weight. The first technique has not been developed widely in geography (J. P. Cole 1965). The second has been used mainly to show population potential (C. D. Harris 1954).

(*a*) *The running mean idea*

The method of calculating the mean of a set of values was shown in chapter 3. The smoothing effect of a so-called 'running' mean (sometimes moving average) has often been applied to distributions on time scales to bring out general or long term trends by removing 'noise' caused by short term fluctuations (W. G. Hoskins 1964). A simple example (see figure 4.29) will illustrate the technique as used with time: it is thought that wheat yields in Kazakhstan (U.S.S.R.), as well as showing violent year to year fluctuations, have shown a gradual downward trend as a result of excessive use of the soil since the new lands campaign started in 1953. The information, yields in tsentners per hectare*, is shown in table 4.2. The running mean procedure takes each year, adds the values for the years on either side, and divides by three, giving the

* Source: *Narodnoye khozyaystvo SSSR*, Moscow, various years.

middle year the new value. It is possible and desirable sometimes to include two or more years on each side of the one in question rather than one, dividing by 5, 7 or a higher number, as appropriate. In the Kazakhstan example data before 1953 are not readily available.

An example of the procedure is as follows: for 1955, 3.1, will be modified as follows: 9.7 + 3.1 + 11.3 = 24.1. Then 24.1 ÷ 3 = 8.0. The effect is to raise the lower values and reduce the higher values. A new histogram may be constructed as in figure 4.29*b*. As there is no figure for 1952 this cannot be added to 1953. The best procedure at the beginning and end of the run, therefore, is to add only 1954 to 1953, and to divide by 2.

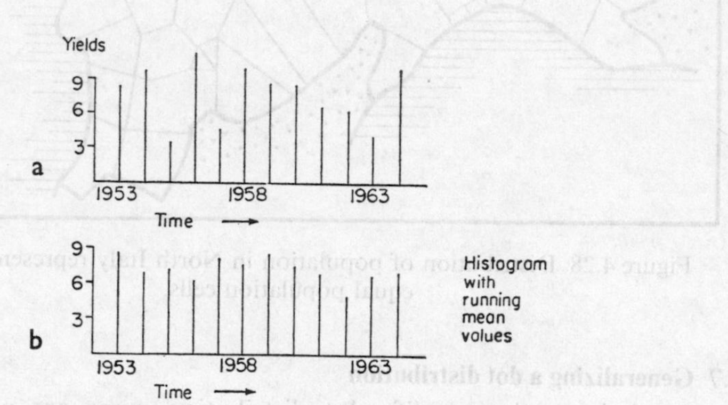

Figure 4.29 Wheat yields in Kazakhstan, 1953–1964, in tsenters per hectare. (*a*) Year by year. (*b*) Using three year running mean.

Table 4.2 Kazakhstan wheat yields

Year	Value: One year	Value: Three year running mean
1953	8.2	8.9
1954	9.7	7.0
1955	3.1	8.0
1956	11.3	6.3
1957	4.5	8.5
1958	9.6	7.7
1959	8.9	8.9
1960	8.4	8.0
1961	6.8	7.2
1962	6.5	5.7
1963	3.9	6.8
1964	9.9	6.9

The same technique can be applied to data on a 1-dimensional space (distance). Table 4.3 shows data for traffic frequency collected at half mile intervals along a main road out of a large town. For the example, a 5-value running mean is applied. The reader is invited to continue the calculations.

Table 4.3 Traffic frequency data

Distance out			Distance out		
Centre	450	—	5.5	500	—
0.5	600	—	6.0	300	—
1.0	750	630	6.5	250	—
1.5	550	720	7.0	300	—
2.0	800	780	7.5	350	—
2.5	900	—	8.0	250	—
3.0	650	—	8.5	200	—
3.5	500	—	9.0	150	—
4.0	500	—	9.5	150	—
4.5	350	—	10.0	100	—
5.0	450		10.5	150	—

To transfer the technique in a satisfactory way to two-space dimensions, and to see the connexion, takes some thought. Basically, the 1-dimensional case takes a class on a distribution scale and adds to it the values of the classes on either side. In a 2-dimensional case, the surface in question might be divided into grid squares. Each square would contain a certain numbers of dots. Instead of attributing to each square the number of dots in it, one could add the number of dots in each of the four contiguous squares and divide by 5. It might be better, however, to use the 8 nearest squares around a given square, thus covering a more compact area. 1- and 2-dimensional cases are compared in figure 4.30.

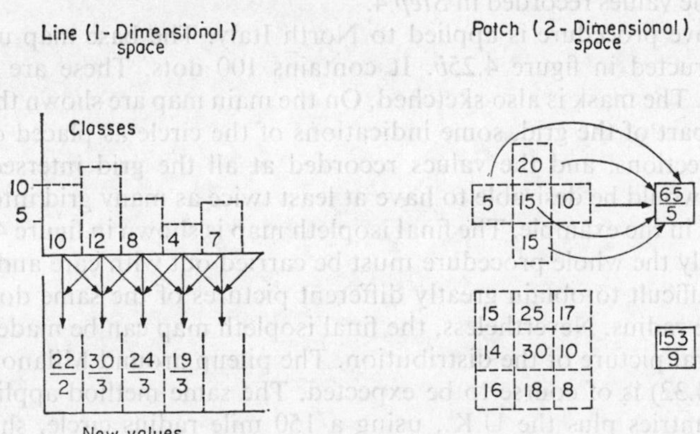

Figure 4.30 A comparison of the running mean idea applied in 1- and 2-dimensional space.

Since a circle is even more compact than a square, it has a slight advantage over a square for adapting the running mean technique to a 2-dimensional surface. A procedure using a circle will now be outlined step by step. Some comments will follow. The reader may then modify and refine the method as required. It is advisable to use several layers of tracing paper.

Step 1. Take a dot map. It is useful, though certainly not essential, to have a round number of dots: for example 100 (see section 4.6). If there are too many dots (say over 250) the task of carrying out the procedure becomes very laborious.

Step 2. Place on this dot map base a tracing-paper grid with squares of appropriate size. The corners of the squares, the intersections of the lines on the grid, will be used in the procedure. The accuracy of the result will depend on the fineness of the grid. A fairly coarse grid, giving about 200 intersections may be used to start with, and a finer one nested into this if required.

Step 3. Make a 'mask', preferably of thin cardboard. The mask will have a circle cut out. The position of the centre of the circle will be marked by sticking down two pieces of thread as two diameters crossing at right angles. See instructions on figure 4.31. The radius of the circle is important. In practice it may be illuminating to apply circles with different radii to the same distribution in turn (J. P. Cole, 1963). If the radius is very small, the original dot distribution map will be approximately reproduced. If it is very large, then the picture will be too generalized. The radius of the circle used in figure 4.31 is 80 km.

Step 4. Put a further piece of tracing paper on the original dot map base and grid covering it. Take the mask circle and place the centre of it (shown by the crossing of the two threads) successively on each grid intersection. Count the number of dots falling within the circle on each occasion and record the number at the centre of the circle. In effect, the operation involves counting how many dots fall within a given radius of each grid intersection.

Step 5. Isopleths may then be inserted on the basis of information provided by the values recorded in *Step* 4.

The above procedure is applied to North Italy. The base map used is the one constructed in figure 4.25*b*. It contains 100 dots. These are shown in figure 4.31. The mask is also sketched. On the main map are shown the original 100 dots, part of the grid, some indications of the circle as placed on certain grid intersections, and the values recorded at all the grid intersections. In practice it would be desirable to have at least twice as many grid intersections as are used in the example. The final isopleth map is shown in figure 4.32.

Obviously the whole procedure must be carried out with care and thought. It is not difficult to obtain greatly different pictures of the same dot map by varying the radius. Nevertheless, the final isopleth map can be made to give a good general picture of the distribution. The pileup around Milano in North Italy (see 4.32) is of course to be expected. The same method applied to the E.E.C. countries plus the U.K., using a 150 mile radius circle, showed the greatest pileup in the Cologne area (over 25 per cent of the total area population within 150 miles) and very low values in peripheral places (under 2 per

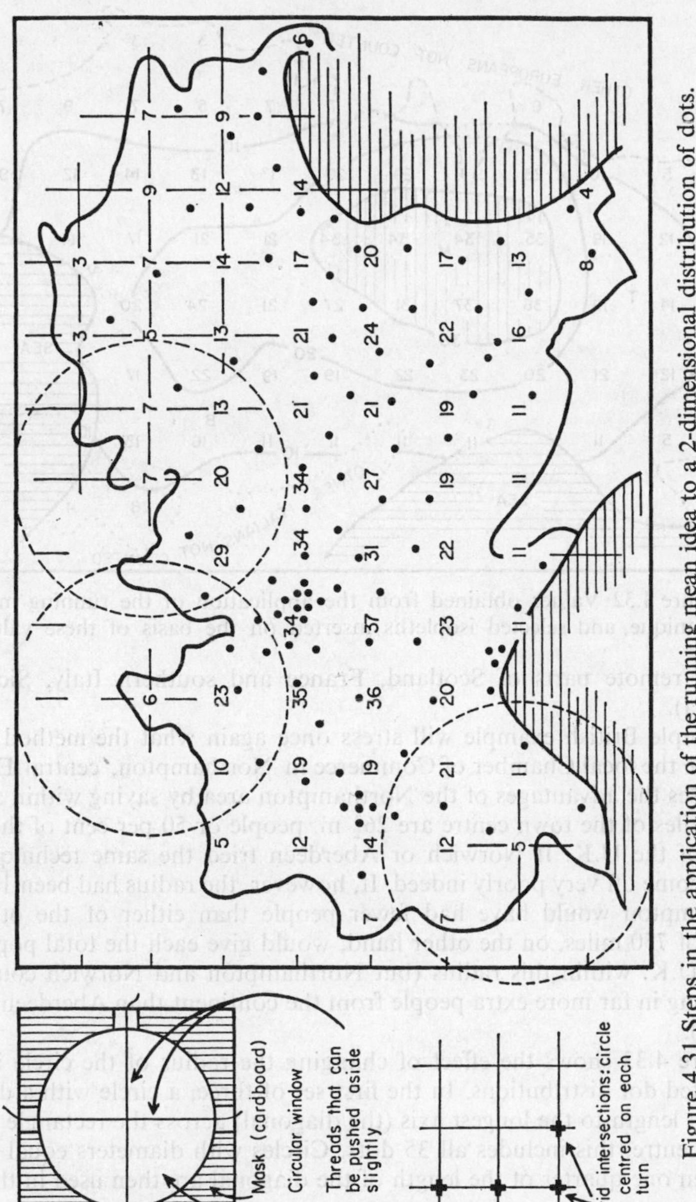

Figure 4.31 Steps in the application of the running mean idea to a 2-dimensional distribution of dots.

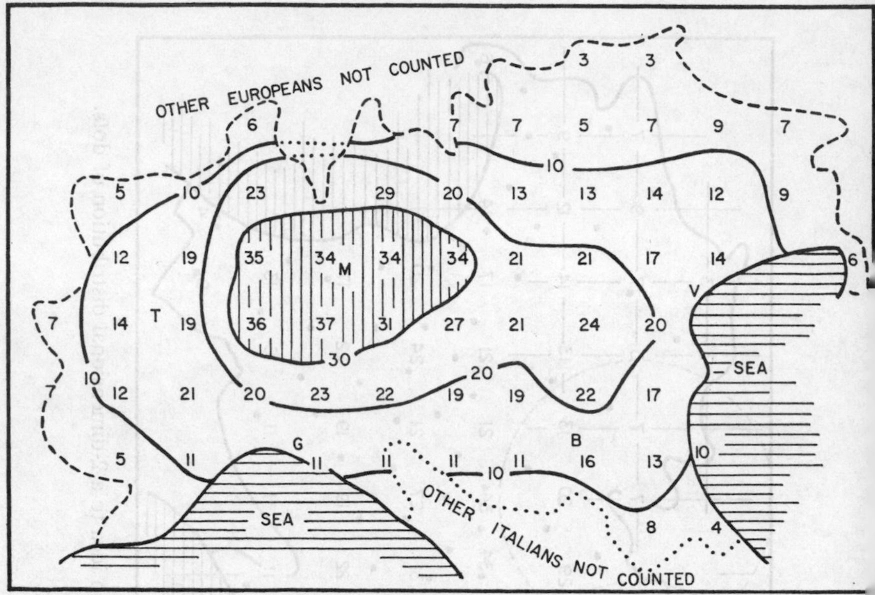

Figure 4.32 Values obtained from the application of the running mean technique, and selected isopleths inserted on the basis of these values.

cent in remote parts of Scotland, France and southern Italy, Sicily an Sardinia).

A simple British example will stress once again what the method shows Suppose the local Chamber of Commerce in Northampton, central England advertises the advantages of the Northampton area by saying within a radiu of 70 miles of the town centre are 26½ m. people or 50 per cent of the popu lation of the U.K. If Norwich or Aberdeen tried the same technique the would come off very poorly indeed. If, however, the radius had been 10 mile Northampton would have had fewer people than either of the others. A radius of 750 miles, on the other hand, would give each the total populatio of the U.K. within this radius (but Northampton and Norwich could the also bring in far more extra people from the continent than Aberdeen, if the desired).

Figure 4.33 shows the effect of changing the radius of the circle in thre concocted dot distributions. In the first set of three, a circle with a diamete equal in length to the longest axis (the diagonal) across the rectangle is use In the centre, this includes all 35 dots. Circles with diameters equal to hal and then one quarter of the length of the diagonal are then used in the oth two sets.

(*b*) *Population potential*

Another method of giving a generalized visual picture of a distribution on patch is described briefly below with a simple example. In practice, t

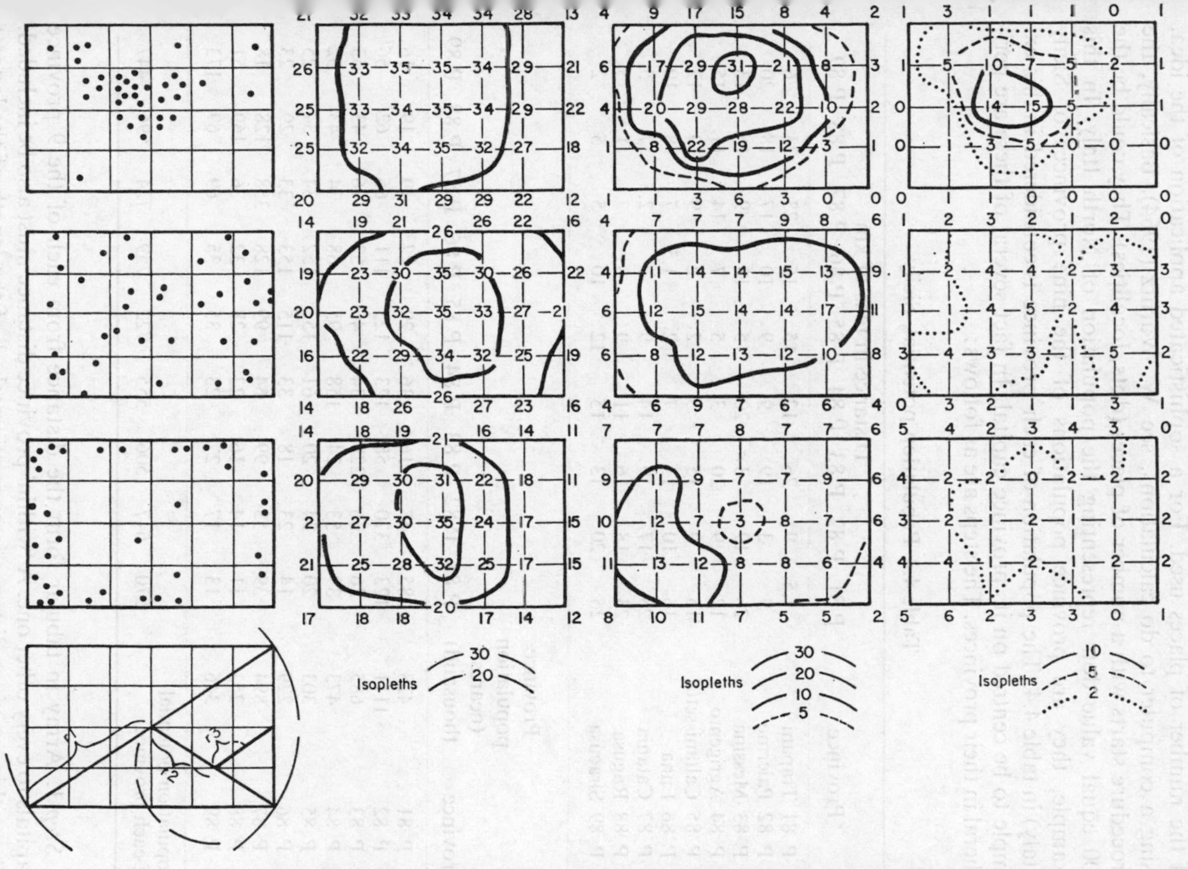

Figure 4.33 Examples of the effect of change of radius of the circle in the application of the running mean idea to dot distributions.

procedure requires information for considerably more than 9 places before it becomes useful. However, the amount of calculating increases with the square of the number of places used. For a sophisticated application of the idea, using a computer to do calculations, see W. Warntz (1964). Basically, the procedure starts with a number of places (dots, localities). They could be the 100 equal value dots representing the population of North Italy. In this example, they are province populations of the nine provinces of Sicily (Italy) in table 4.4. The population of each province is considered in this example to be centred on its province capital. In fact several of these are peripheral in their provinces. The steps are as follows:

Table 4.4 Population potential, Sicily

Province	Distance in tens of km								
	P 81	P 82	P 83	P 84	P 85	P 86	P 87	P 88	P 89
P 81 Trapani	1.5	9	26	12	15	16	23	23	26
P 82 Palermo	9	3	19	9	9	10	17	18	20
P 83 Messina	26	19	3	20	15	13	9	16	13
P 84 Agrigento	12	9	20	3	5	7	14	11	15
P 85 Caltanissetta	15	9	15	5	2	2	9	9	12
P 86 Enna	16	10	13	7	2	1.5	7	8	10
P 87 Catania	23	17	9	14	9	7	2.5	7	5
P 88 Ragusa	23	18	16	11	9	8	7	1.5	5
P 89 Siracusa	26	20	13	15	12	10	5	5	2

Province	Province population (nearest thousand)	P 81	P 82	P 83	P 84	P 85	P 86	P 87	P 88	P 89
P 81	428	285	47	16	36	29	27	19	19	16
P 82	1111	123	370	58	123	123	111	65	62	56
P 83	685	26	36	228	34	46	53	76	43	53
P 84	473	39	53	24	158	95	68	34	43	32
P 85	303	20	34	20	61	152	152	34	34	25
P 86	229	14	23	18	33	115	153	33	29	23
P 87	894	39	53	99	64	99	128	358	128	18
P 88	253	11	14	16	23	28	32	36	169	51
P 89	346	13	17	27	23	35	35	69	69	173
Population potential of each province		570	647	506	555	722	759	724	596	447

Step 1. Array in tabular form the distance from each of the 9 province capitals to every other one. A 'within' province distance must also be included. The within province distance is approximately $\frac{1}{4}$ of the length of the longest axis across the area of the province. Some thought should be given to calculating this distance since it affects the results considerably. The upper table contains distances to the nearest ten km.

Step 2. The population of each province is then taken, and can conveniently be placed in column form. It is given in thousands in the lower part of the table, first column.

Step 3. A second 9 × 9 table is then produced. This contains the results of a calculation whereby in turn the population of each province is divided by the distance of that province from every other province as well as from itself (principal diagonal). The effect is to reduce the value according to distance. This need not be done linearly, however. It is possible to make the division, for example, by the square of the distance.

Step 4. The potential for each province is the sum of its relationship, derived as described above, to all provinces. It is the sum of each column in the second table. The values are shown in figure 4.34. When sufficient values are obtained, an isopleth map may be constructed.

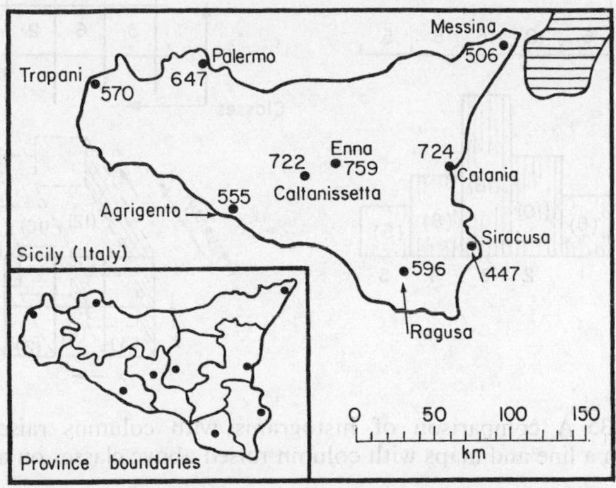

Figure 4.34 Population potential values for nine provinces of Sicily.

The map gives a very general picture of the original distribution on which it is based. It shows quickly which areas have a high potential and which have a low potential. Like the running mean method already described, when used with population data, it gives an idea of the number of people within reach of different localities. But instead of giving an idea of the number of people within a chosen radius of different places, it gives a view of the relationship of each place to the total population of the area considered. The actual values of the potential must be considered in a relative, not an absolute sense. In the example given, the potential value could have been inflated a thousand times merely by using total population figures instead of population to the nearest thousand. The value of the procedure lies in its ability to put a relative value to different localities.

4.8 Central tendency and dispersion in a 2-dimensional distribution

Among other things, a map has been referred to as the basic tool of the geographer, and as a model of the real world. The theme of this section is that whatever else it is, a map is often the 2-dimensional counterpart of the familiar histogram of the statistician. Histograms were discussed in chapter 3. Data that can be handled statistically also often have measures of central tendency and measures of dispersion. Similarly, distributions in 2-dimensional

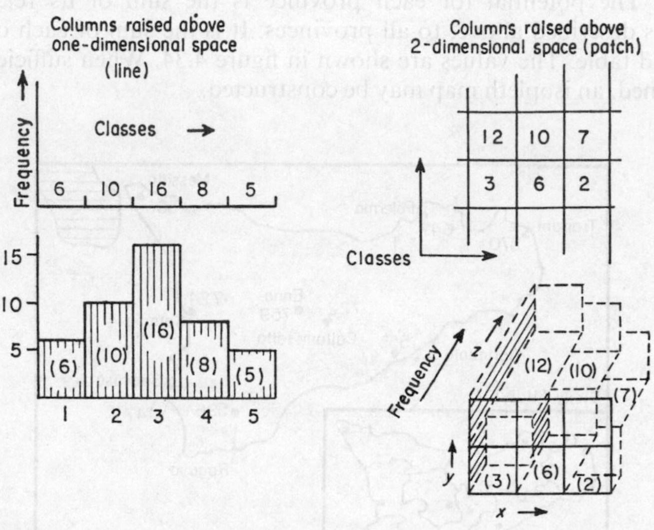

Figure 4.35 A comparison of histograms with columns raised above classes on a line and maps with column raised above classes on a surface.

space have measures of central tendency and dispersion. There is a considerable literature on this subject, some of which was noted in section 4.1. This section describes simple procedures for finding measures of central tendency and dispersion in 2-dimensional distributions.

Figure 4.35 compares the 1-dimensional and 2-dimensional cases, and shows how in the 1-dimensional case the x-axis carries the distribution and the y-axis the frequency. In the 2-dimensional case both x- and y-axes are used to show the spatial distribution and the z-axis has to be used to show frequency. Figure 4.36 shows the 100 dots of North Italy (see figure 4.25*b*) converted to a map histogram. The representation of the columns, including the angle at which the z-axis leaves the base plane, the shading and so on are very important, since the only merit of the whole exercise is to give a good visual picture. The map was constructed on the basis of a grid of squares placed over the dot distribution map.

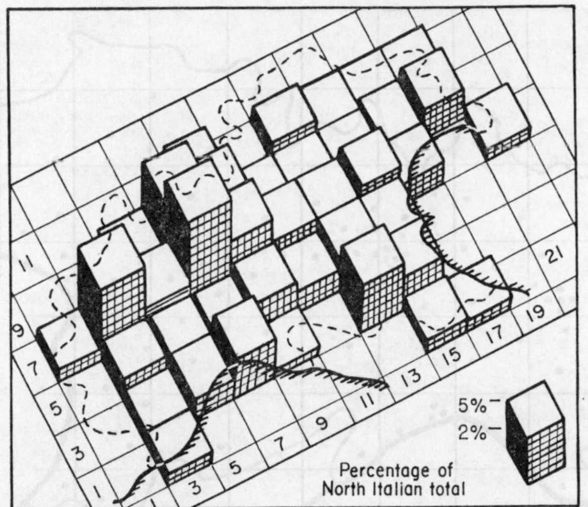

Figure 4.36 The distributon of population in North Italy represented by columns raised above the squares of a grid.

(*a*) *Measures of central tendency*

(i) *A quick method of finding the mean centre of a patch* (e.g. of a country) (see figure 4.37*d*). Cut out the shape as carefully as desired in cardboard. Make a small hole near one edge. Put a drawing pin through this and into a pinboard. Allow the cardboard to swing until it settles. Then drop a perpendicular line from the drawing pin across the card. Repeat from two or three other points near the periphery. The lines should intersect at the mean centre of area.

A more laborious way is to divide the patch into small equal area squares and to place a dot at the centre of each. This regular distribution of dots is then treated in the same way as the irregular distribution to be discussed below.

(ii) *A quick method of finding the mean centre of a distribution of dots.* Pair off dots that are close to each other, putting the value of two midway between them. Pair off the 2s, and so on. At some stage it will become desirable to collapse two points of different value: for example 16 and 32. The new value of 48 for these will be $\frac{16}{16+32}=\frac{1}{3}$ of the way from the 32 towards the 16. This method is subjective, but if care is taken to work across the distribution of dots, collapsing near neighbours as far as possible, then a good approximation of the mean centre is achieved.

(iii) *A more precise way of finding the mean centre of a dot distribution* is to place a grid of squares* over the distribution so that 0, the intersection

* Note the difference between a grid of squares and a square grid. The latter could have another mesh in it, such as hexagons.

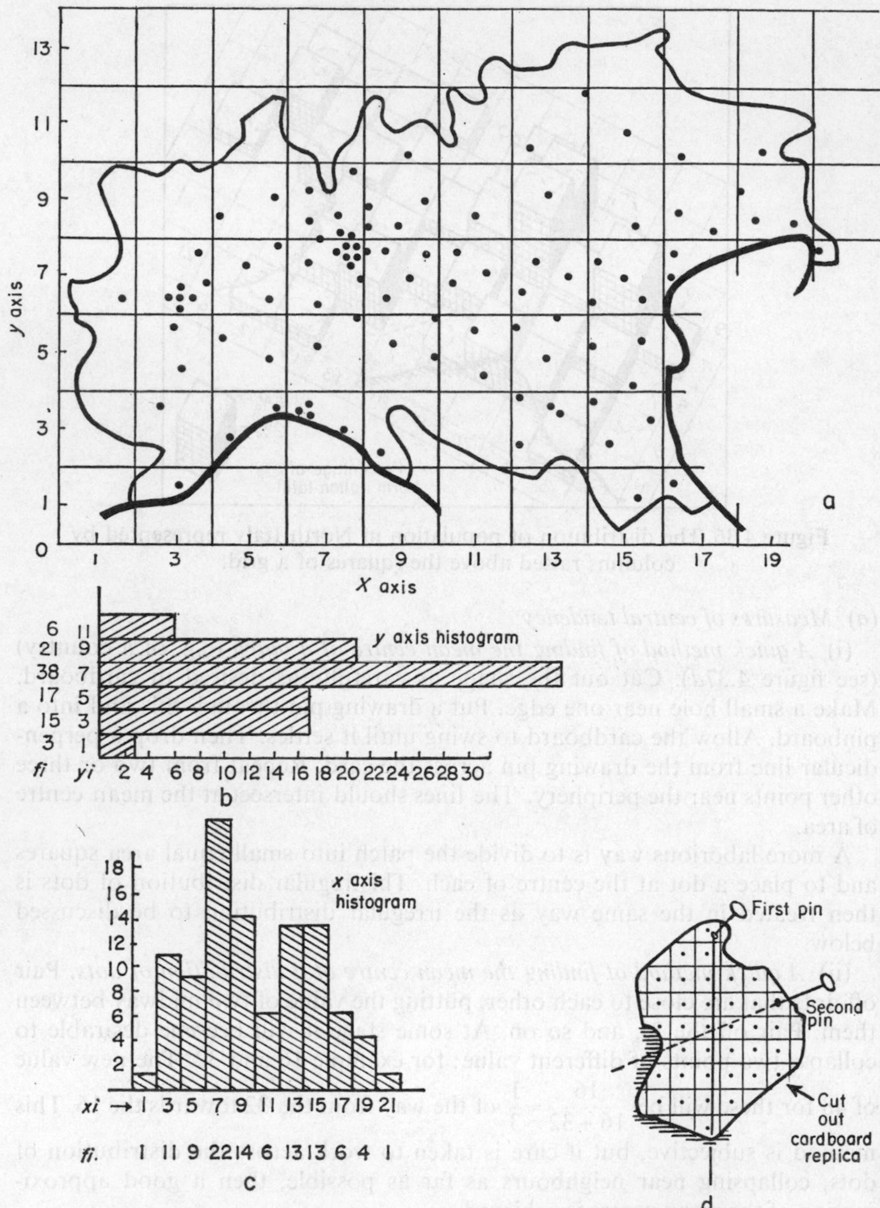

Figure 4.37 (*a–c*) A method of discovering the mean centre ('centre of gravity') of a dot distribution map, illustrated by North Italy. (*a*) The original distribution with superimposed grid. (*b*) Finding the mean on the *y*-axis. (*c*) Finding the mean on the *x*-axis. (*d*) A quick method of finding the mean centre of the area of a patch.

Table 4.5 Grouped data

X_i	f_i	$X_i f_i$	
1	1	1	
3	11	33	
5	9	45	
7	22	154	
9	14	126	$N=100$ (dots)
11	6	66	
13	13	169	$\frac{\Sigma X_i f_i}{N}=9.88$
15	13	195	
17	6	102	
19	4	76	
21	1	21	
		$\Sigma X_i f_i=988$	

of x- and y-axes, falls both to the 'south' and 'west' of the distribution. In fact, the orientation is irrelevant; the same result is achieved whatever the orientation of the axes. The x- and y-axes must of course intersect at right angles. Procedure (follow in fig. 4.37*a–c*)

Step 1. Treat in turn the x- and y-axes as 1-dimensional distributions. This requires an appreciation of the use of classes and grouped data. It is convenient, though not essential, to have whole number class intervals and class midpoints. Each class may therefore be 2 units of distance in 'width'. The mid-points of successive classes to the right of 0 along the x-axis and above 0 on the y-axis become 1, 3, 5 … . This facilitates calculation.

Step 2. Take the x-axis (the procedure for the y-axis is identical) and find the frequency of dots in each class. Thus there is 1 dot in class 1, 11 dots in class 2, and so on. Repeat for the y-axis. There will be a mean value on the x-axis and a mean value on the y-axis. $\bar{X}$ (mean on the x-axis) will be $\frac{\Sigma X_i f_i}{N}$

An example is given, table 4.5, of the method for calculating the mean of

Table 4.6 Data for areal distribution

x-axis				y-axis		
X_i	f_i	$X_i f_i$		Y_i	f_i	$Y_i f_i$
1	10	10		1	10	10
2	10	20		2	19	38
3	15	45	$N=100$	3	13	39
4	15	60		4	13	52
5	37	185		5	26	130
6	13	78		6	19	114
$\Sigma X_i f_i=398$				$\Sigma Y_i f_i=383$		
$\bar{X}=\frac{398}{N}=\frac{398}{100}=3.98$				$\bar{Y}=3.83$		

grouped data. X_i is the midpoint of each class, f_i the frequency of dots in it. In the example there are 100 dots.

The reader is invited to set out the calculations for finding the mean on the y-axis. The answer is 6.54.

A simple example, table 4.6, is now given to illustrate the method of finding the mean centre, median centre and mode of a dot distribution. This should

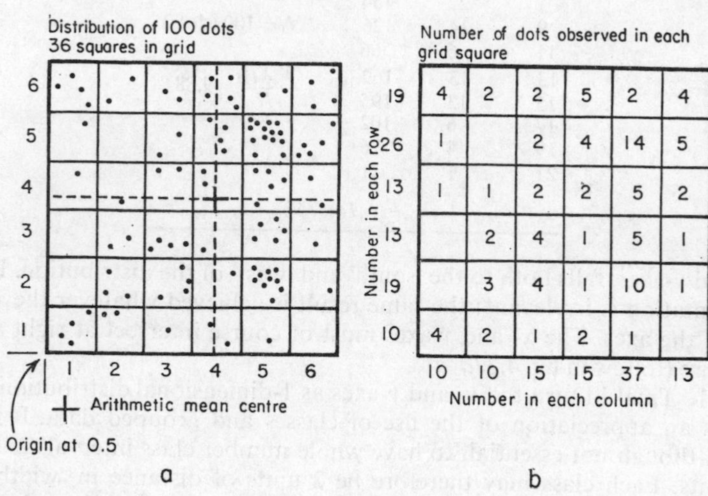

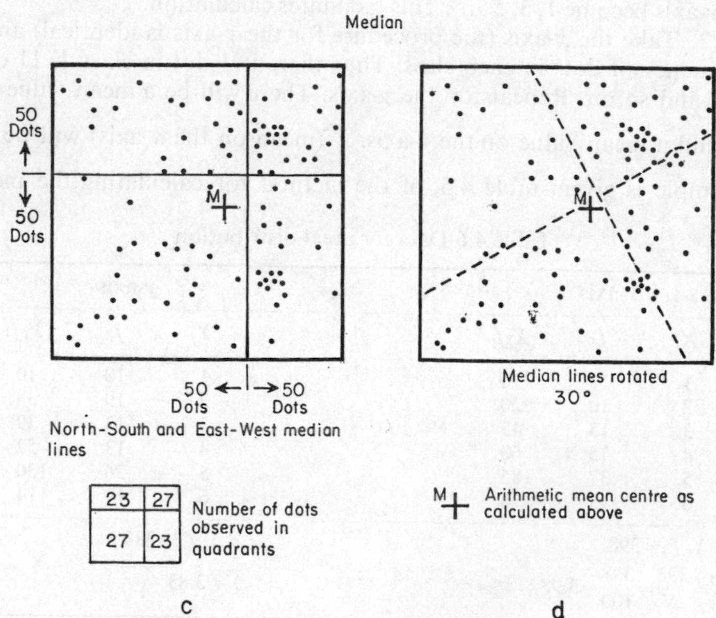

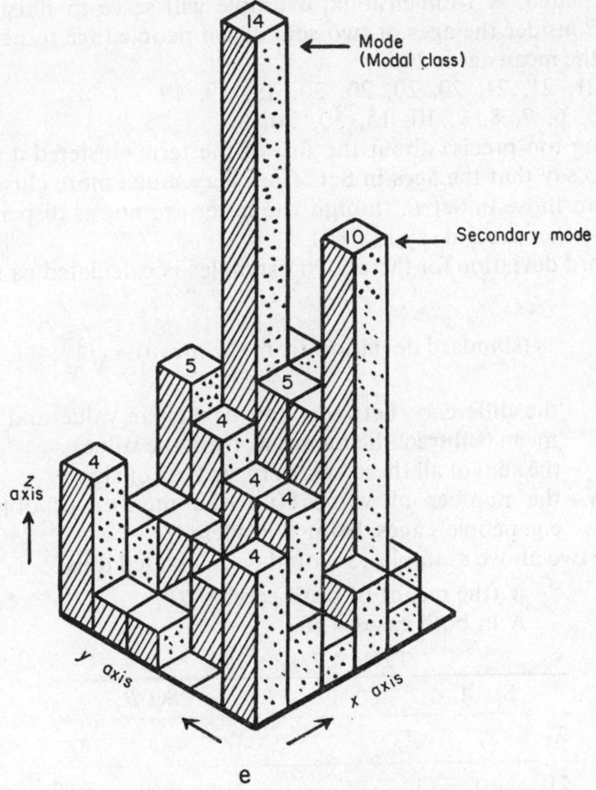

Figure 4.38 A comparison of the mean centre, median centre and modal class. The same distribution of 100 dots is used throughout. (*a*) The 100 dots and the grid superimposed on the distribution. (*b*) The number of dots observed in each grid square. (*c*) and (*d*) Median centres. (*e*) Columns raised above grid squares, proportional in height to the frequency of dots in each 'class'.

be followed on figure 4.38, using the data in figure 4.38*b*. The calculations are shown in table 4.6 for finding the mean centre.

The (arithmetic) mean centre of the distribution of 100 dots is shown in figure 4.38*a*. Its position was calculated above. The median centre is shown in figures 4.38*c* and *d*. Note that whereas the mean centre is the same whatever the orientation of the grid, the median centre varies in position somewhat according to the orientation of the median lines. The modal class is most apparent in figure 4.38*e*, the distribution shown in histogram form.

(*b*) *Introduction to the idea of dispersion*

Once the mean centre of a distribution has been found, it is possible to describe in various ways the dispersion about this of the values from which the mean is calculated. A 1-dimensional example will serve to illustrate simple dispersion. Consider the ages of two sets of ten people (age to nearest year); in each case the mean age is 20:

Set *A*: 21, 21, 21, 20, 20, 20, 20, 19, 19, 19
Set *B*: 5, 6, 7, 8, 9, 10, 15, 30, 30, 80

Without being too precise about the use of the term clustered it would seem reasonable to say that the ages in Set *A* are very much more clustered about the mean than those in Set *B*, though the latter are not as dispersed as they might be (all either around say 1 or 80).

The standard deviation for the above examples is calculated as follows (see table 4.7):

$$s \text{ (standard deviation from the mean)} = \sqrt{\frac{x_i^2}{N}}$$

x_i = the difference between any particular value and the mean (subtract the mean, X, from the value)
Σx_i = the sum of all these values regardless of sign
N = the number of values (i.e. the number of objects, e.g. people's ages, being considered)

Taking the two above examples, s would be calculated thus:

$\bar{X}$ (the mean) in both cases = 20
N in both cases = 10

Table 4.7 Age data

Set *A*			Set *B*		
X_i	x_i	x_i^2	X_i	x_i	x_i^2
21	+1	1	80	+60	3600
21	+1	1	30	+10	100
21	+1	1	30	+10	100
20	0	0	15	−5	25
20	0	0	10	−10	100
20	0	0	9	−11	121
20	0	0	8	−12	144
19	−1	1	7	−13	169
19	−1	1	6	−14	196
19	−1	1	5	−15	225
		6			4780
$x_i^2 = 6$			$x_i^2 = 4780$		
$s = \sqrt{\frac{6}{10}} = \sqrt{0.6}$			$s = \sqrt{\frac{4780}{10}} = \sqrt{478}$		
$= 0.775$			$= 21.9$		

In practice this example involves too small a number of people for the finding of the standard deviation to be particularly meaningful.

The above example may be employed within reason to a dot distribution on

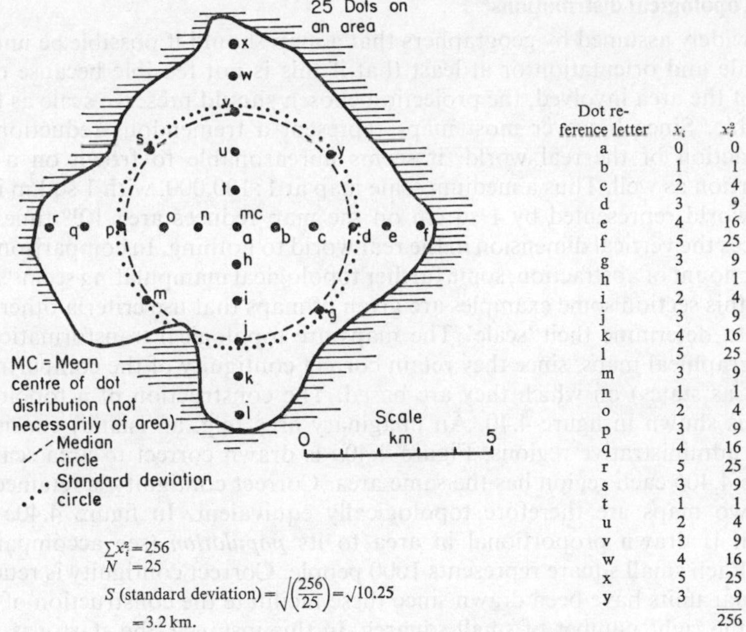

Dot reference letter	x_i	x_i^2
a	0	0
b	1	1
c	2	4
d	3	9
e	4	16
f	5	25
g	3	9
h	1	1
i	2	4
j	3	9
k	4	16
l	5	25
m	3	9
n	1	1
o	2	4
p	3	9
q	4	16
r	5	25
s	3	9
t	1	1
u	2	4
v	3	9
w	4	16
x	5	25
y	3	9
		256

Figure 4.39 A measure of dispersion of dots about the mean centre of a distribution in two dimensions.

a patch. A simple example is worked in figure 4.39. Here for convenience the dots have been arranged regularly. There are too few dots to give a sensible result. The dots have been arranged so that dot *a* is actually at the mean centre of the distribution. In a real case it would be necessary to calculate the mean centre by the method already shown earlier in this section (4.8*a*). The mean centre of the dot distribution is then the 2-dimensional equivalent of the mean of a distribution of values along a line, as already illustrated in this section. The standard deviation of the dot distribution gives the distance of the radius of a circle that can then be described, centred on the mean centre. Roughly one-third of the dots will be left outside the circle. It is also possible to draw other circles: for example, one that leaves half the dots outside itself. This is comparable to a median line. Many refinements may be added to the procedure for finding 'standard distance' in an area, but they will not be developed here.

One important implication of transferring a technique from a distribution on a 1-dimensional base to a 2-dimensional base must not be overlooked. Whereas two standard deviations on a line distribution cuts off twice as much space as one standard deviation, two standard distances radius gives a circle four times as large (2^2) as the circle given by one standard distance radius.

4.9 Topological distributions*

It is widely assumed by geographers that a map should if possible be uniform in scale and orientation or at least that if this is not feasible because of the size of the area involved, the projection chosen should preserve scale as far as possible. Since however most maps represent a tremendous reduction and abstraction of the real world, it seems unreasonable to frown on a little distortion as well. Thus a medium scale map at 1:100,000, with 1 sq km in the real world represented by 1 sq cm on the map reduces area 10^{10} times and reduces the vertical dimension in the real world to nothing. In comparison with this amount of abstraction, some further topological manipulating seems small.

In this section some examples are given of maps that use criteria other than area to determine their 'scale'. The maps are topological transformations of topographical maps, since they retain correct contiguity of the compartments (such as states) on which they are based. The construction of a topological map is shown in figure 4.40. An imaginary area (e.g. an island) consists of eight administrative regions. Figure 4.40*a* is drawn correct to area scale. In figure 4.40*b* each region has the same area. Correct contiguity is retained and the two maps are therefore topologically equivalent. In figure 4.40*c* each region is drawn proportional in area to its *population* (see accompanying list). Each small square represents 1000 people. Correct contiguity is retained. Angular units have been drawn since these facilitate the construction of units with the right number of small squares. In this instance, the start was made with region 4, which is central, and is the largest, having 100,000 inhabitants.

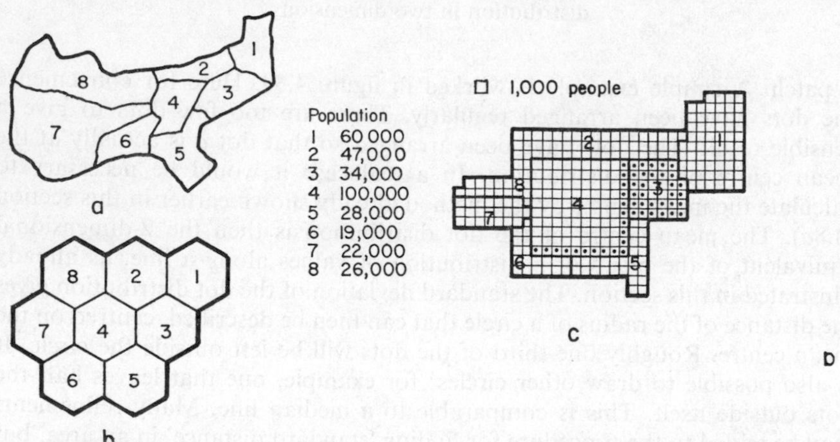

Figure 4.40 The construction of a topological map based on the number of inhabitants of compartments. (*a*) The original map, correct to scale. (*b*) The same compartments with each compartment given the same area and correct contiguity preserved. (*c*) The same compartments drawn proportional to their population.

* For introduction to topology, see chapter 2.10.

It could therefore conveniently be represented by 10×10 small squares. The small population of regions 7 and 8, which are large territorially and therefore prominent in map (*a*), is revealed in map (*c*), which shows their true importance in terms of people.

The United States is used to illustrate a topological transformation. Figure 4.41*a* shows the U.S.A. with its 48 contiguous mainland states, plus Hawaii and Alaska (not to scale). Figure 4.41*b* shows the same 50 states represented toplogically, with each state equal to one unit of area (actually 24 small squares on a piece of graph paper, on which the map was built up). While this map is unusual in concept it is not entirely irrelevant because each state in fact is represented by two senators. In this respect and in certain others, the states are all equal, regardless of territorial extent, number of inhabitants, income or any other variable. It is impossible to construct this map correctly on a network of hexagons (as was done in 4.40*b*) because not

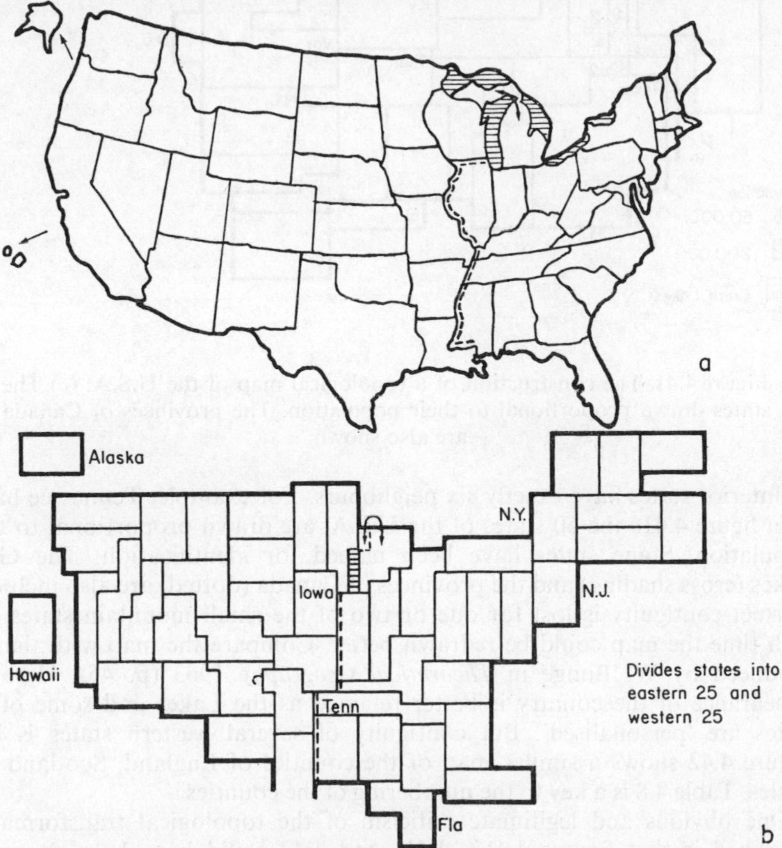

Figure 4.41 The construction of a topological map of the U.S.A. (*a*) The 50 States correct in scale (except Hawaii and Alaska). (*b*) The states equal in area.

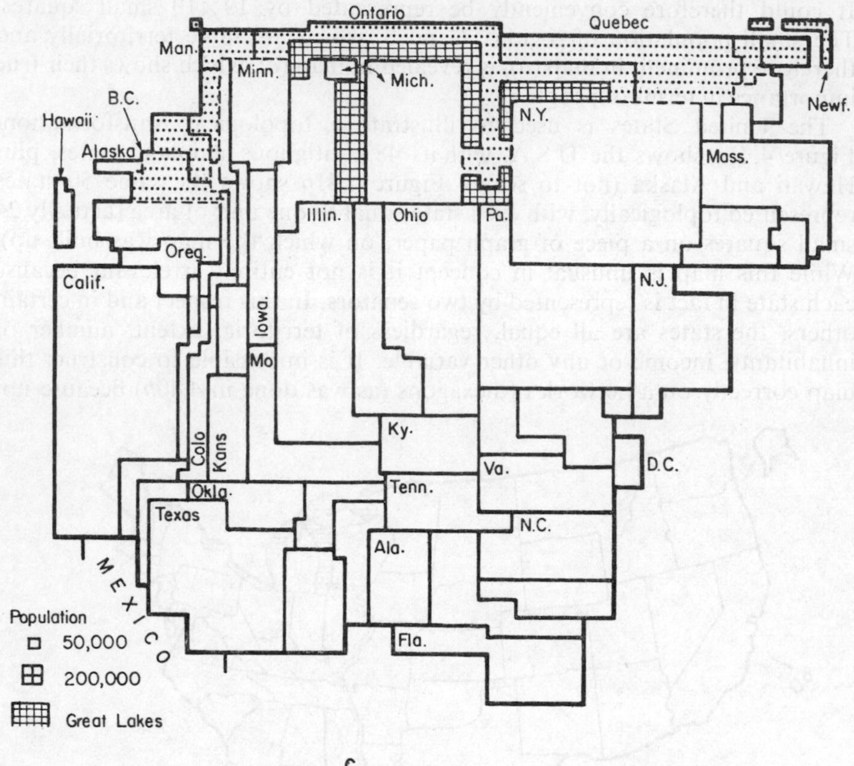

Figure 4.41 The construction of a topological map of the U.S.A. (*c*) The states drawn proportional to their population. The provinces of Canada are also shown.

all interior states have exactly six neighbours. For example, Tennessee has 8.

In figure 4.41*c* the 50 states of the U.S.A. are drawn proportional to their population. Some states have been named for identification. The Great Lakes (cross shading) and the provinces of Canada (dotted) are also included. Correct contiguity is lost for one or two of the small mountain states, but with time the map could be redrawn better. Compare the map with that reproduced by W. Bunge in *Theoretical Geography* 1963 (p. 45). Here the appearance of the country is better retained as the Lakes and some of the states are 'personalised'. But contiguity of several western states is lost. Figure 4.42 shows a similar map of the counties of England, Scotland and Wales. Table 4.8 is a key to the numbering of the counties.

One obvious and legitimate criticism of the topological transformation described is that figures 4.41*b*, 4.41*c* and 4.42 could have been drawn in many different ways. Interestingly, then, the conventional topographical map, with its territorial scale correct, has one special property and advantage.

Table 4.8 British counties

County	Pop. in 1961 in 1000's	County	Pop. in 1961 in 1000's
1 London (old county)	3195	49 Westmorland	67
2 Middlesex	2230	50 Monmouth	444
3 Essex	2287	51 Glamorgan	1228
4 Kent	1701	52 Pembroke	94
5 Surrey	1733	53 Carmarthen	168
6 Sussex (E)	665	54 Cardigan	54
7 Sussex (W)	411	55 Brecknock	56
8 Hampshire	1336	56 Radnor	18
9 Isle of Wight	96	57 Montgomery	44
10 Berkshire	503	58 Merioneth	39
11 Oxfordshire	310	59 Caernarvon	121
12 Buckinghamshire	486	60 Anglesey	52
13 Bedfordshire	381	61 Denbigh	174
14 Hertfordshire	832	62 Flint	150
15 Cornwall	342	63 Wigtown	29
16 Devon	832	64 Kirkcudbright	29
17 Dorset	309	65 Dumfries	88
18 Somerset	599	66 Roxburgh	43
19 Wiltshire	423	67 Selkirk	21
20 Gloucester	1001	68 Peebles	14
21 Herefordshire	131	69 Berwick	22
22 Worcestershire	569	70 East Lothian	53
23 Shropshire	297	71 Midlothian	580
24 Warwickshire	2023	72 West Lothian	93
25 Northamptonshire	398	73 Fife	321
26 Leicestershire	682	74 Kinross	7
27 Rutland	24	75 Clackmannan	41
28 Soke of Peterborough	74	76 Stirling	195
29 Huntingdonshire	80	77 Dunbarton	185
30 Isle of Ely	89	78 Renfrew	339
31 Cambridgeshire	190	79 Lanark	1626
32 Suffolk (W)	130	80 Ayr	343
33 Suffolk (E)	343	81 Bute	15
34 Norfolk	562	82 Perth	127
35 Holland	103	83 Angus	278
36 Kesteven	135	84 Kincardine	49
37 Lindsey	505	85 Aberdeen	299
38 Nottinghamshire	903	86 Banff	46
39 Derbyshire	878	87 Moray	49
40 Staffordshire	1734	88 Nairn	8
41 Cheshire	1368	89 Shetland	18
42 Lancashire	5132	90 Orkney	19
43 West Yorkshire	3641	91 Caithness	27
44 East Yorkshire	527	92 Sutherland	13
45 North Yorkshire	554	93 Ross and Cromarty	58
46 Durham	1517	94 Inverness	83
47 Northumberland	819	95 Argyll	59
48 Cumberland	294		

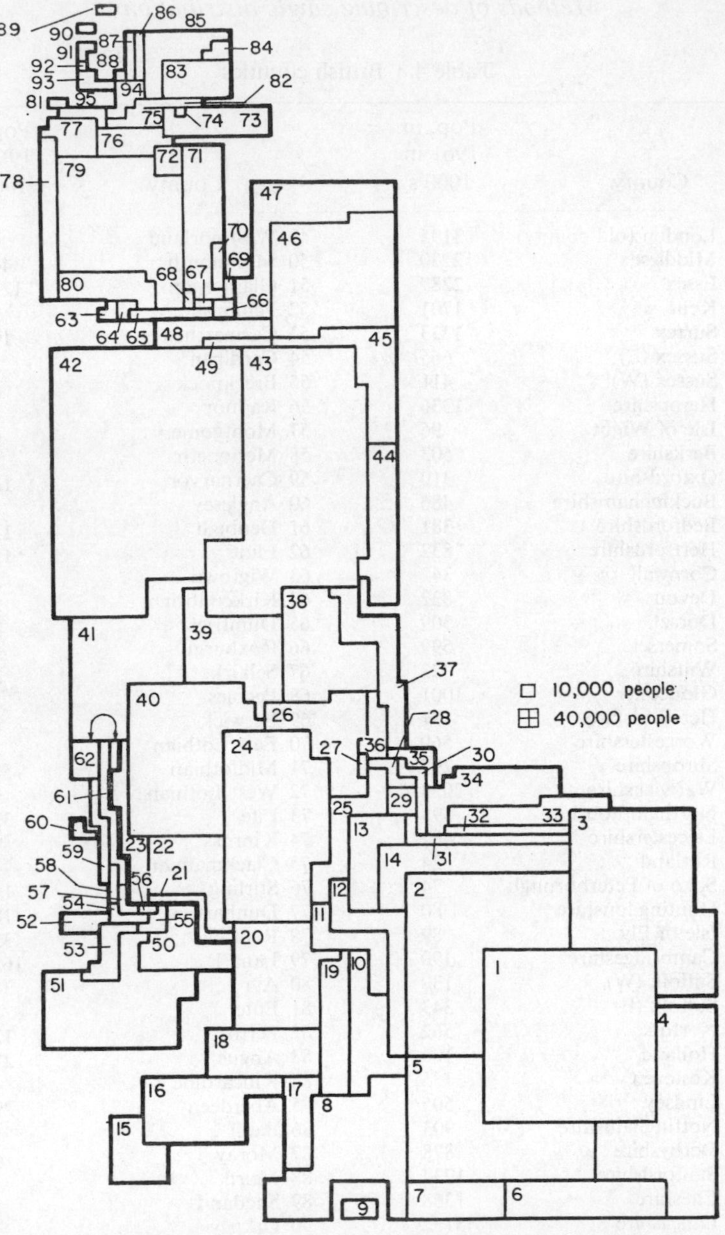

Figure 4.42. Topological map of the population of Great Britain, based on the population of counties. See text for explanation and list of county populations. The map shows 95 administrative counties of England, Scotland and Wales. The area of each county is proportional to the population of the county in 1961. The map is correct topologically since contiguity of counties has been preserved. Note: English counties are 1961 administrative counties.

Apart from variations possible with different projections, it remains fundamentally a fixed shape. Another disadvantage of the topological map is its unfamiliarity. It is the purpose of the maps in this section to show the kind of visual pictures that can be given topologically. Two advantages may briefly be noted:

(1) Striking new facts and relationships may be revealed visually with topological maps, not evident on topographical ones.

(2) For some purposes topological maps form a more consistent and indeed visually honest base for maps showing certain kinds of distribution by colouring or shading. For example, on a conventional map of the world showing per capita income, both Australia and Belgium would be heavily shaded or conspicuously coloured because both are among the most prosperous countries in the world. But visually Australia would be about 300 times as conspicuous because it is 300 times as large territorially. Yet the two countries have roughly the same number of people and roughly the same total national income. The income is distributed among people, not among pieces of land. It would of course be possible to draw a circle proportional to population on each of the two areas and then to shade or colour this. Or a topological map could be drawn. In short, for a topological map to have shading on units of area drawn proportional to population is a visually more honest base for all distributions based on people than is a topographical map with its areas shaded.

A topological transformation need not be confined to population based maps. Thus regions can be drawn proportional to the area they have under cropland, to their income, or even to their annual precipitation. In figure 4.43 the procedure has been used to show Mexico in the eyes of Villahermosa, Tabasco state. Tabasco is one of the major civil divisions of Mexico (there are 29 states, two territories and the Federal District (D.F.)). The state telephone directory (Directorio Telefónico, No. 29, Feb. 1966, Villahermosa), contains a list of the telephone charges to 11 places in Tabasco and to 23 places outside. Presumably this 'short' list contains the places most often called from Villahermosa. Thus the 'host' state has 11/34 of the places listed. As distance from Tabasco increases, so the importance (size) of the places listed also increases. Thus only 5 places are listed in the northwestern 3/5 of the national area. South Mexico is greatly inflated in relation to the rest when the map of Mexico is redrawn proportional to the 34 exchanges listed in Villahermosa. In map 4.43*b*, each exchange has around it an equal sized piece of area.

Finally 4.44 shows a small part of a most interesting, extremely elaborate map of the Parliamentary Constituencies of the United Kingdom. The map was compiled by Dr. T. H. Hollingsworth of Glasgow University and appeared in *The Times* at the time of the 1964 General Election (*The Times*, Monday, October 19th, 1964). Although it is called a demographic map, and does broadly reflect the distribution of population in the U.K., it is in fact similar in concept to the U.S.A. map in figure 4.41*b* in which the states are each equal to one unit of area, regardless of how many inhabitants they have

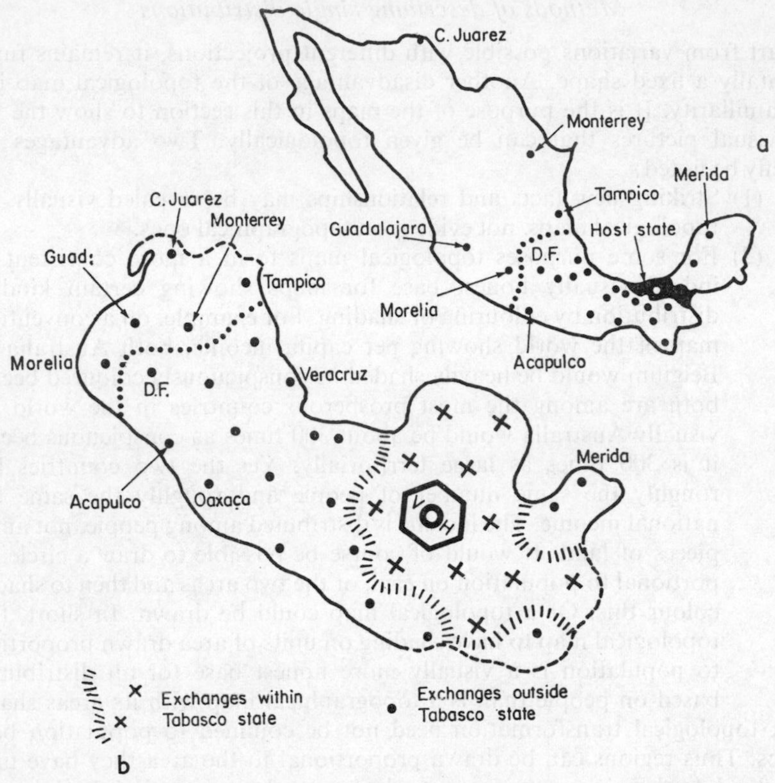

Figure 4.43 A view of Mexico from the state of Tabasco based on places given in the list of charges in the Tabasco Telephone guide. (*a*) The places listed on a map in which scale is correct. (*b*) The places listed, each within a piece of land of equal area. Mexico viewed from Villahermosa. D.F. includes Mexico City.

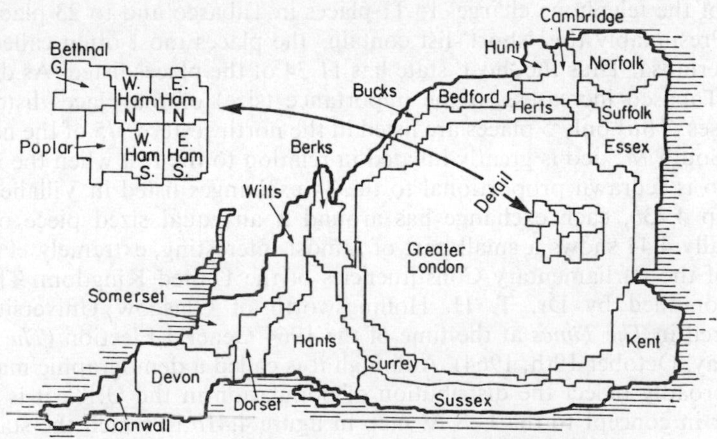

Figure 4.44 Part of a map of the constituencies of the United Kingdom, with only county boundaries shown. The new Greater London is shown.

and of their territorial extent. Although the U.K. parliamentary constituencies do not vary so much in population as the states of the U.S.A., there is a difference of several times between extremes. The map is not, therefore, truly a demographically based topological map but a constituency based map, each constituency returning one Member of Parliament. Correct contiguity has been preserved but shape has naturally been greatly distorted. In general with maps of this kind, the likelihood of having to distort shape grows both with the number of compartments to be handled and as the difference in density increases between extremes in the area under consideration. Thus in England the thinly peopled counties between the Greater London area and the Midlands tend to become greatly elongated, as do the mountain states in the U.S.A.

References

Bachi, R. (1963). Standard distance measures and related methods for spatial analysis, *Reg. Sci. Assoc. Papers*, **10**.

Cole, J. P. (1963). *Geography of World Affairs*, Penguin Books, Harmondsworth, p. 320 (3rd ed. 1964).

Cole, J. P. (1965). *Latin America: Economic and Social Geography*, Butterworths, London, p. 434.

Dacey, M. F. (1960). The spacing of river towns, Letter in *Ann. Assoc. Am. Geog.*, **50,** No. 1, 59–61.

Dacey, M. F. (1964). Two-dimensional random point patterns: a review and an interpretation, *Reg. Sci. Assoc. Papers*, **13,** 41–55.

Dacey, M. F. (1964). Imperfections in the uniform plane, Discussion Paper No. 4, *Michigan Inter-University Community of Mathematical Geographers*.

Dickinson, G. C. (1963). *Statistical Mapping and the Presentation of Statistics*, Arnold, London.

Getis, A. (1964). Temporal land-use pattern analysis with the use of nearest neighbor and quadrat methods, *Ann. Assoc. Am. Geog.*, **54,** No. 3, 391–399.

Giusti, U. (1943). *Caratteristiche ambientali italiane*, Istituto Nazionale di Economia Agraria, Studi e Monografie, N.27, Roma.

Harris, C. D. (1954). The market as a factor in the localization of industry in the United States, *Ann. Assoc. Am. Geog.*, **44,** 315–348.

Hart, J. F. (1954). Central tendency in areal distributions, *Econ. Geog.*, **30,** No. 1, pp. 48–59.

Hoskins, W. G. (1964). Harvest fluctuations and English economic history, 1480–1619, *Agricultural Hist. Rev.*, **12,** Part 1.

Jannello, C. (1963). Texture as a visual phenomenon, *Architectural Design*, p. 394.

Jenks, G. F. (1963). Generalization in statistical mapping, *Ann. Assoc. Am. Geog.*, **53,** No. 1, 15–26.

McCarty, H. H. and Salisbury, N. E. (1961). *Visual Comparison of Isopleth Maps as a Means of Determining Correlations between Spatially Distributed Phenomena*, Department of Geography, State University of Iowa, Iowa City, No. 3.

Monkhouse, F. J. and Wilkinson, H. R. (1963) (first published 1952). *Maps and Diagrams*, Methuen, London.

Porter, P. W. (1963). What is the point of minimum aggregate travel?, *Ann. Assoc. Am. Geog.*, **53,** No. 2, 224–232.

Richardson, L. F. (1961). The problem of contiguity, *Gen. Systems Yearbook*, **VI**, 139–187.

Sviatlovsky, E. E. and Eells, W. C. (1937). The centrographical method and regional analysis, *Geog. Rev.*, **27,** 240–254.

Tobler, W. R. Undated paper (duplicated). *Spectral Analysis of Spatial Series.* Supported by National Science Foundation.

Warntz, W. (1964). A new map of the surface of population potentials for the United States, 1960. *Geog. Rev.*, **54,** 170–184.

5
Comparisons

'Do, do: he'll but break a comparison or two on me; which, peradventure not marked or not laughed at, strikes him into melancholy; and then there's a partridge wing saved, for the fool will eat no supper that night.'

Much ado about nothing

5.1 Difficulties of comparison

Selected techniques for describing spatial distributions were described in chapter 4. The emphasis almost throughout the chapter was on a single distribution of objects in a given space. In chapter 5 some techniques for comparing two or more similar distributions in different spaces will be outlined.

Comparison of roughly similar things is a technique that has been used in many disciplines. In biology, for example, features of two or more plant or animal species may be compared. An example is illustrated in chapter 2, figure 2.28*b*. Comparative government is a well developed aspect of political science. Comparison has also been applied in the historical study of architecture (e.g. by Sir Banister Fletcher). In geography, the comparison of a few things has been employed less frequently, and perhaps less rigorously, than the exhaustive study of unique things or the quantitative (statistical) study of many things (see chapters 6 and 7).

The need for comparison in geography arises when the geographer is considering from two to about 6–10 fairly similar objects or situations, or arrangements of these. The Antarctic ice cap, the Pacific Ocean, and New York are perhaps too unique in their own ways to be usefully compared with any other things. On the other hand, glaciers, lakes and large towns in general are numerous enough to allow the application of statistical methods when they are studied in quantity. But in geography, as certainly in history, there are some things that occur in small quantities yet are of considerable importance: for example island arcs, large sovereign states, large towns in very cold or very high localities. These may often be usefully compared.

The purpose of comparing two things or a few things is to throw light from one to the other or others, or sometimes to throw light both ways. Thus in history, a comparison of Napoleon and Hitler might be rewarding. Similarly, London and Paris have enough features in common to merit comparison. This chapter considers methods of detecting and measuring similarities and differences between objects or situations that are broadly comparable. It does not make analogies between situations that may be fundamentally different in most respects and similar in a few or even in only one respect (see chapter 11). In most cases, a comparison of *two* things is used to illustrate techniques of comparison. The techniques could however be extended to three or more things.

Very broadly, comparisons in geography may be worked out and expressed verbally (in language), visually in map form (topographical or topological) or numerically with the help of appropriate mathematical operations and indices. These methods may be combined.

When a comparison is to be made, at least the following questions must be asked:

(1) Are the same things being compared (e.g. people in Iceland and people in Newfoundland)?
(2) Are the things being compared contemporaneous or not (e.g. Iceland in 1965 and Newfoundland in 1965)?
(3) Is the comparison being carried out between different places or in the same place?

The above yes/no questions give eight possible combinations. These are tabulated below: 0 = same, 1 = different.

Table 5.1 Comparability table

Thing	Time	Place	Usefulness
0	0	0	Superfluous
0	0	1	Good
0	1	0	Good
0	1	1	Difficult
1	0	0	Impossible (virtually)
1	0	1	Dubious
1	1	0	Dubious
1	1	1	Remote

Of the eight combinations, cases (001) and (010) seem to be the most common and useful. They may be illustrated by two examples:

Case 001: a comparison of the distribution of population (same thing) in 1960 (same time) in North Italy and Peninsular Italy (different places).

Case 010: a comparison of the distribution of population (same thing) in 1931 and 1961 (different times) in North Italy (same place).

Other cases may also be of interest. For example case (011) could be a comparison of the distribution of bandit activity in Colombia in 1960 A.D.

with that in England in the 12th century (things the same, time and place different).

The above combinations may occur in various forms. The geographer may indeed be contemplating any of the ten following situations for his comparison: dots on dots, lines, patches or in 3-dimensional spaces; lines on lines, patches, or in 3-dimensional spaces, patches on patches or in 3-dimensional spaces; and 3-dimensional objects in 3-dimensional spaces. Of these ten possible situations, comparisons of four kinds will be considered in this chapter: dots on lines, lines on patches, dots on patches and patches on patches.

Unfortunately, comparison is not straightforward, even when the kind of comparison to be made has been sorted out. Firstly, the spaces (bases) being compared may differ in form. Two (or more) areas to be compared may differ in territorial extent, shape (in particular compactness), elongation, and in other ways. Contrast, for example, North Italy and Peninsular Italy. Secondly, the objects on them may differ in number, as well as in the way they are arranged on their respective bases. Thus the number of people in North Italy and in Peninsular Italy is not identical, though in this instance it is fairly similar.

Differences in the bases may to some extent be reduced by the following operations, already outlined in chapter 2: *dilation* (change of scale of one base), *rotation* (e.g. of one or both bases, to bring longest axes parallel), *reflexion* (this tends to cause visual confusion by giving a new image of the base area, but on occasions could be useful) and *topological transformations* (e.g. straightening of lines). When the same base is being used in the comparison at different times, then the base shape usually remains relatively unchanged, though if a political unit is being used for the base it may undergo boundary changes over decades if not years.

Differences in the original number of objects on the bases may to some extent be overcome by reducing them to an equal number of symbols (e.g. 100 dots) for each area. It was suggested in chapter 4 that in two areas with different total numbers of people, the distribution of population could be expressed in each area by 100 dots, each representing one per cent of the total population. It is not however always convenient to do this.

Even when the above problems have been satisfactorily overcome, the general difficulty of comparison remains. The problem is illustrated in figures 5.1 and 5.2. Figure 5.1 shows a single object that has changed its location in space after a period of time. An animal at rest in relation to its space is suddenly induced to move by the sound of an explosion at time 0. In the 1-dimensional case (5.1*a*), there is only one straight line path for it to follow to run away from the sound. In the 2-dimensional case (5.1*b*) it may choose any direction to the 'east' of the explosion, on a flat surface. In the 3-dimensional case (5.1*c*) it can fly in a vertical direction as well as over two horizontal dimensions. In each of the three examples its new position at time 1 is predicted. This is then compared with its observed position at time 1. The predicted and observed positions can easily be expressed by Cartesian

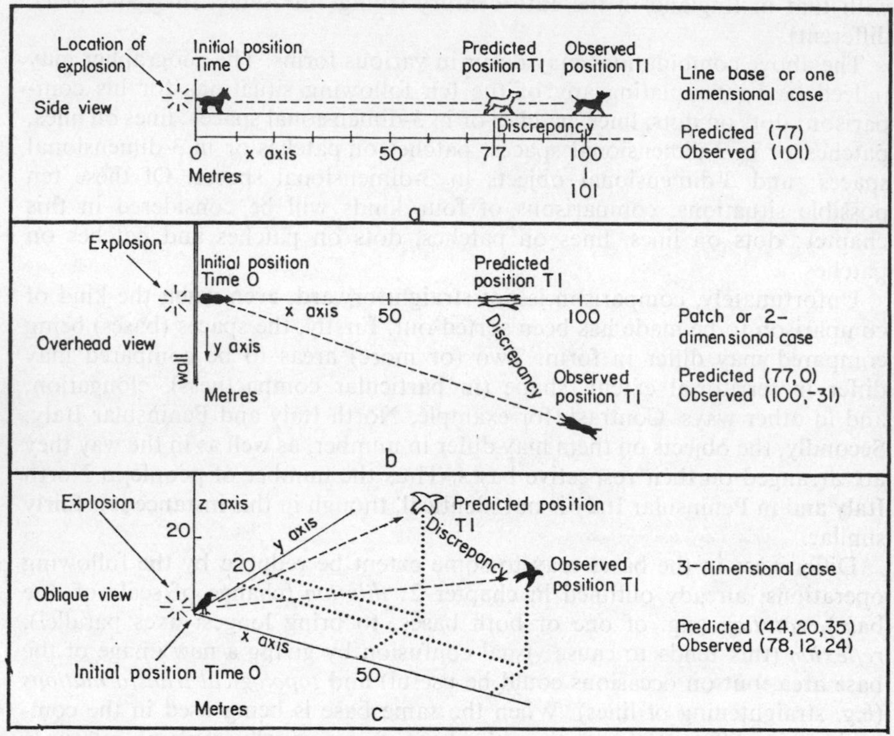

Figure 5.1 Problems of comparing predicted with observed movements. (*a*) Along a line. (*b*) In 2 dimensions. (*c*) In 3 dimensions.

co-ordinates, expressing the initial position of the animal as 0. In the line example, it has moved 101 metres east instead of 77 as predicted. There is a discrepancy of 24 metres. The discrepancy might be expressed as an error of nearly 24 per cent. The discrepancy is less easy to describe in the more complex 2- and 3-dimensional examples.

In practice the geographer is usually concerned with a change over time in the position of several or many dots, rather than one (the animal in this example). He is often concerned also with new positions at *several* instants in time rather than with just one new position after one time interval. Is each new position to be taken afresh as zero in the co-ordinate system, or can several intervals be accommodated in one diagram?

Figure 5.2 shows how a cloud has changed position between time 0 and time 1 (for example over 1 hour). Its position, characterized for convenience by the centre of gravity of the mass, has shifted a given distance, altering its position on *x*, *y* and/or *z* co-ordinates. In addition its own size and shape will probably have changed. Suppose its position and form has been predicted for 1 hour after time 0. Then its recorded position at time 1 can be compared

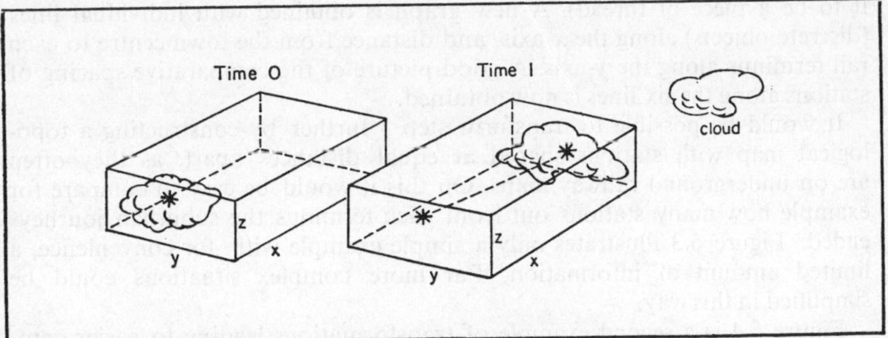

Figure 5.2 The movement of a cloud between times 0 and 1.

with the predicted position at time 1. How can the predicted and observed positions be compared when three co-ordinates, together with size (volume) and shape, five variables, at least, are involved? This situation was suggested in conversation by T. Hagerstrand. All this assumes that in fact its boundary surfaces can be satisfactorily defined.

The rest of this chapter will do no more than offer tentatively some simple ways in which comparisons may be attempted in geography. In view of the lack of literature on this aspect of geography, the reader will do well to develop his own ideas and techniques. Attention is also drawn to an important contribution on the subject by Karl G. Grytzell (1963) who deals mainly with Stockholm, Copenhagen, Paris, London and New York. He illustrates his use of population density to demarcate comparable city areas.

5.2 Comparing dots on lines

This situation is concerned with line bases and with devices for comparing these. Often lines have dots on them that are of interest; for example waterfalls on a stream network, international boundaries across a river, stations on a railway network. It is feasible, however, that the line base may have line subsets on it. Thus it might be of interest to compare the distribution of roads with restricted speed (e.g. 30 m.p.h.) in relation to the total road network in two or more towns.

Figure 5.3 shows stages of modification of a map that lead to a visual picture designed to facilitate comparison. Six railways leave the central area of a town in generally radial directions. A visual comparison of the spacing of stations is desired. A mathematical one could of course be obtained directly by tabulating data for intervals between stations. It will be seen that information is lost as the modifications are made:

Step 1: Arrange the six lines again (figure 5.3*b*), ignoring *orientation.* Detailed changes of *direction* are retained and scale is correct.

Step 2: In the second step (figure 5.3*c*) each line is 'pulled' straight (imagine it to be a piece of thread). A new graph is obtained with individual lines (discrete objects) along the *x*-axis, and distance from the town centre to each rail terminus along the *y*-axis. A good picture of the comparative spacing of stations along the six lines is now obtained.

It would be possible to transform step 3 further by constructing a topological map with stations spaced at equal distances apart, as they often are on underground railway maps. On this it would be easy to compare for example how many stations out from each terminus the suburban journeys ended. Figure 5.3 illustrates only a simple example with, for convenience, a limited amount of information. Far more complex situations could be simplified in this way.

Figure 5.4 is a second example of transformations leading to easier comparison. It shows two inlets in the coast of England, the Thames Estuary and Southampton Water. More could be added. The geographer is interested in

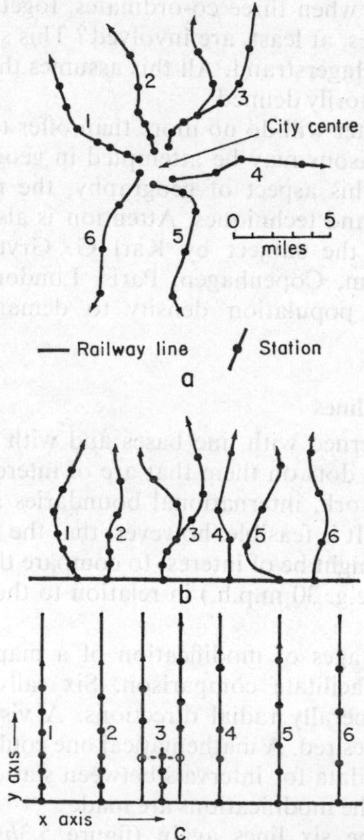

Figure 5.3 A comparison of six radiating suburban railways. (*a*) Map of the railways studied. (*b*) Their arrangement side by side. (*c*) Straightened.

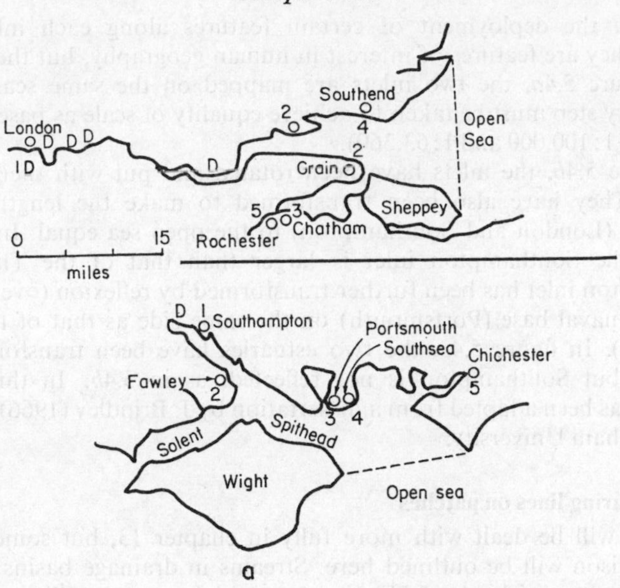

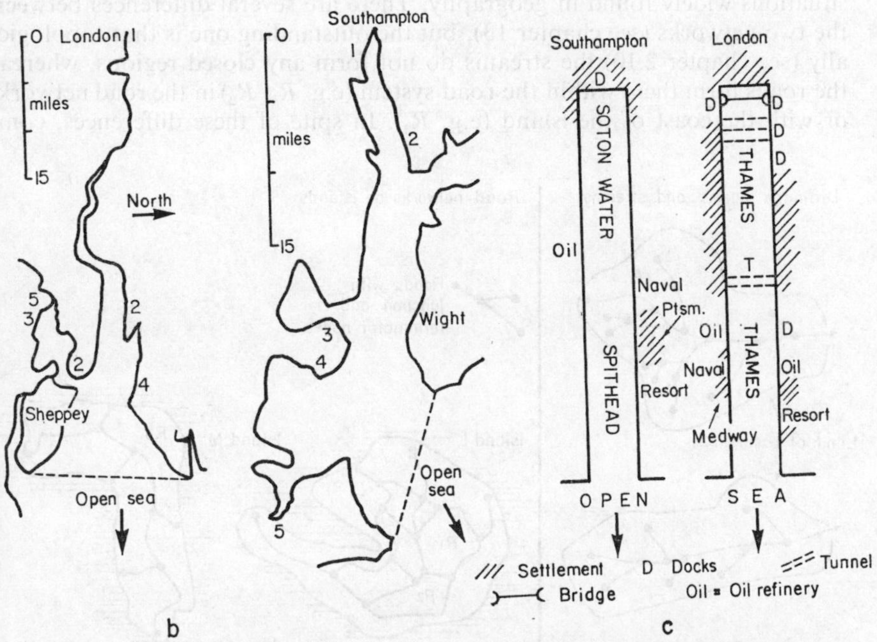

Figure 5.4 A comparison of two inlets in southeast England. (*a*) Topographical maps on same scale. (*b*) After rotation, plus dilating and reflexion of Southampton Water. (*c*) Topological transformation (without reflexion).

comparing the deployment of certain features along each inlet. In this example they are features of interest in human geography, but they need not be. In figure 5.4*a*, the two inlets are mapped on the same scale. Often a preliminary step must be taken to achieve equality of scale as base maps may differ (e.g. 1:100,000 and 1:63,360).

In figure 5.4*b*, the inlets have been rotated and put with their long axes parallel. They have also been transformed to make the length from the inlet head (London and Southampton) to the open sea equal. In effect, the scale of the Southampton inlet is larger than that of the Thames. The Southampton inlet has been further transformed by reflexion (over long axis) to put its naval base (Portsmouth) on the same side as that of the Thames (Chatham). In figure 5.4*c*, the two estuaries have been transformed topologically, but Southampton is not reflected, as in 5.4*b*. In this example, material has been adapted from a dissertation by J. Brindley (1966), submitted at Nottingham University.

5.3 Comparing lines on patches

Networks will be dealt with more fully in chapter 13, but some problems of comparison will be outlined here. Streams in drainage basins (5.5*a*) and road networks on islands (5.5*b*) are two special cases that illustrate general situations widely found in geography. There are several differences between the two networks (see chapter 13), but the outstanding one is that topologically (see chapter 2.10) the streams do not form any closed regions, whereas the roads form them within the road system (e.g. R_1, R_2) in the road network, or with the coast of the island (e.g. R_3). In spite of these differences, com-

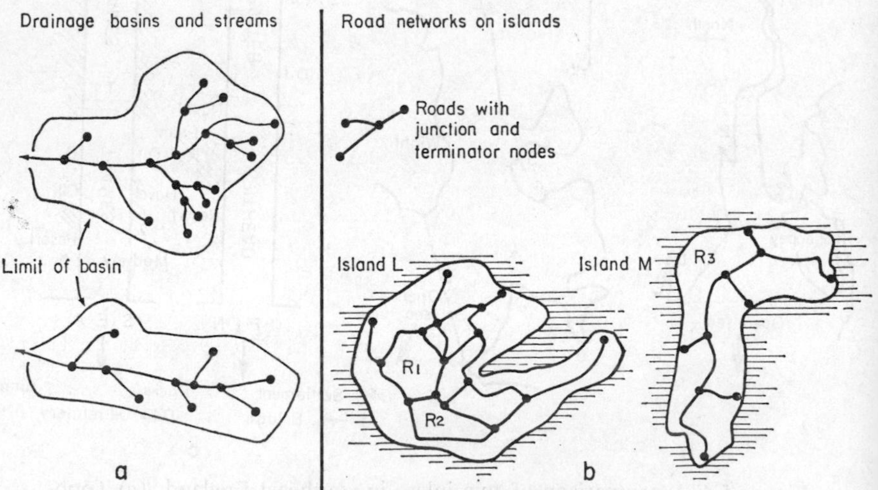

Figure 5.5 Comparing networks. (*a*) Drainage basins and streams. (*b*) Road networks on islands.

parison of the drainage basins and of the islands presents several similar difficulties:

1. The bases (drainage basins or islands) differ in *size* and in *shape*.

2. From a Euclidean point of view, the length of the networks (streams or roads) differs from area to area. Comparison may be achieved for example by dividing the length of the network by the area of the base (e.g. 15 miles of stream or road, base area 6 sq miles, length of network per sq mile 2.5). This, in other words, is the mean density of 'channel' per suitable unit of area.

3. From a topological point of view, the network can be conceived in terms of nodes, arcs and, where relevant, regions. Various ratios may be calculated. These can be compared without regard to the size of the base or the length of the network though some ratios are of dubious value. The two road networks may be compared on the basis of the proportion of terminator nodes to total nodes on the road network. Contrast the two islands, L and M:

	Total nodes	Terminator nodes	Other nodes	Percentage terminator nodes
L	16	4	12	25%
M	10	6	4	60%

The percentages facilitate comparison of efficiency of the networks, if 3-nodes and higher order nodes are considered a sign of efficiency and terminator nodes a sign of inefficiency of the networks. In the above example for simplicity there are too few nodes to make a strikingly significant difference.

The stream networks may be compared according to the different orders of stream and the proportions of the total network in each basin accounted for by each order. Other measures are also possible (see chapters 11, stream simulation model, and 13, on stream networks).

Topological maps may also be useful for assisting in the visual comparison of lines on patches. An example is provided by figure 5.6 which shows the 'railway' network, as in the early 1960s, in southwest Wales, and in Cornwall. The coast is shown in a broken line. Although all stations might be considered as nodes, only junction stations (3- or more nodes) and terminal nodes (1-nodes) are shown. The topological maps make visual comparison of important features easier than it is on the topographical maps because of the removal of 'noise'.

5.4 Comparing dots on patches

The geographer often wishes to compare distributions of dots on patches. The dots may each represent a discrete object (one–one correspondence), or they may each represent several or many objects (one–many correspondence).

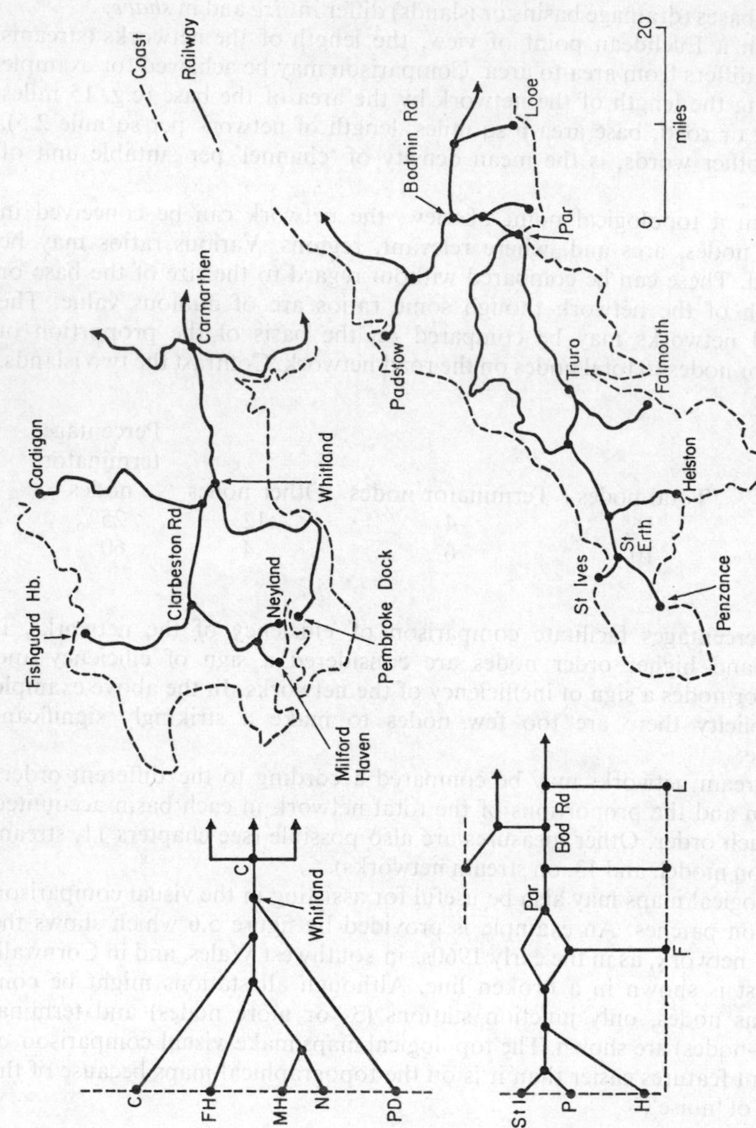

Figure 5.6 Topographical and topological representation of rail networks in southwest Wales and Cornwall.

Thus it may be of interest to compare the distribution of province capitals in North Italy with the distribution of province capitals in Peninsular Italy. There are only 40 capitals in each case. Each can therefore easily be represented on a map (figure 5.7). The comparison might, however, be between the distribution of total population in the two areas (figure 5.8). In this case it would be convenient to have either one dot per so many people for each area (the number of dots would then differ, because total population differs) or one dot per so much per cent (say 1 per cent) of the total population, in which case the number of dots would be the same but a North Italian dot would not represent exactly the same number of people as a dot for Peninsular Italy.

In the comparison of the dot maps for the two parts of Italy in figures 5.7

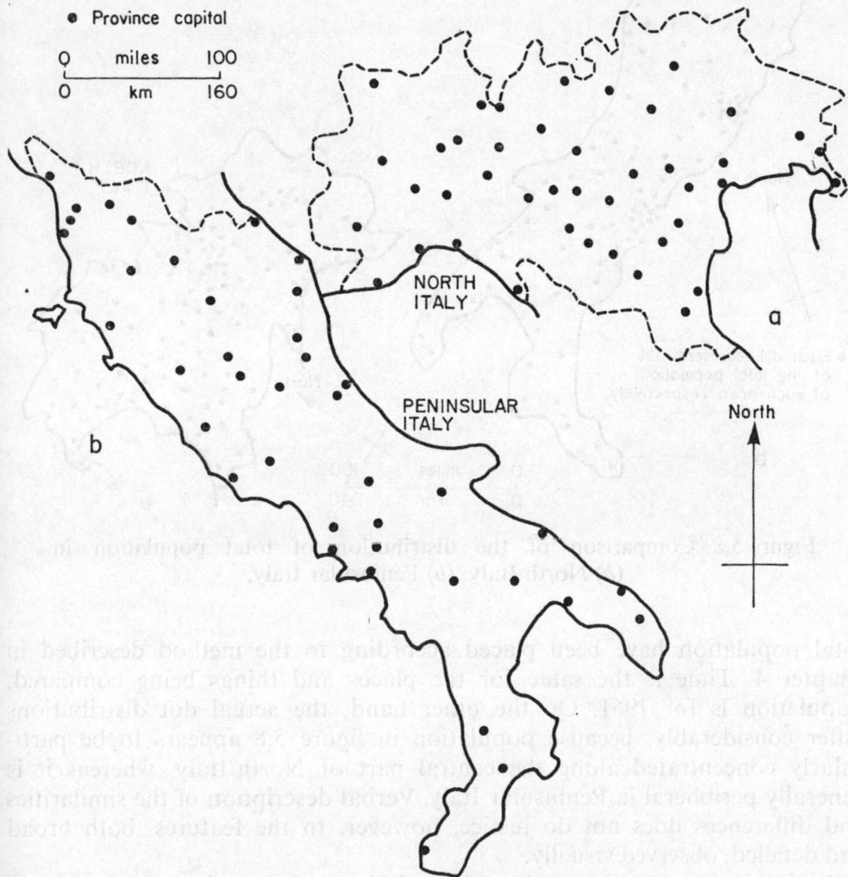

Figure 5.7 Comparison of the distribution of province capitals in (*a*) North Italy (*b*) with Peninsular Italy.

and 5.8, the following features may be noted: the base areas differ considerably both in territorial extent, and in shape, particularly degree of elongation. The orientation of the longest axis also differs but the map in figure 5.8 has been rotated. The distributions on the base consist of similar items (province capitals of the same country, Italian citizens) and in each of the two comparisons, the number of dots is identical (40 province capitals by chance, 100 dots for total population by design). Note that the dots for

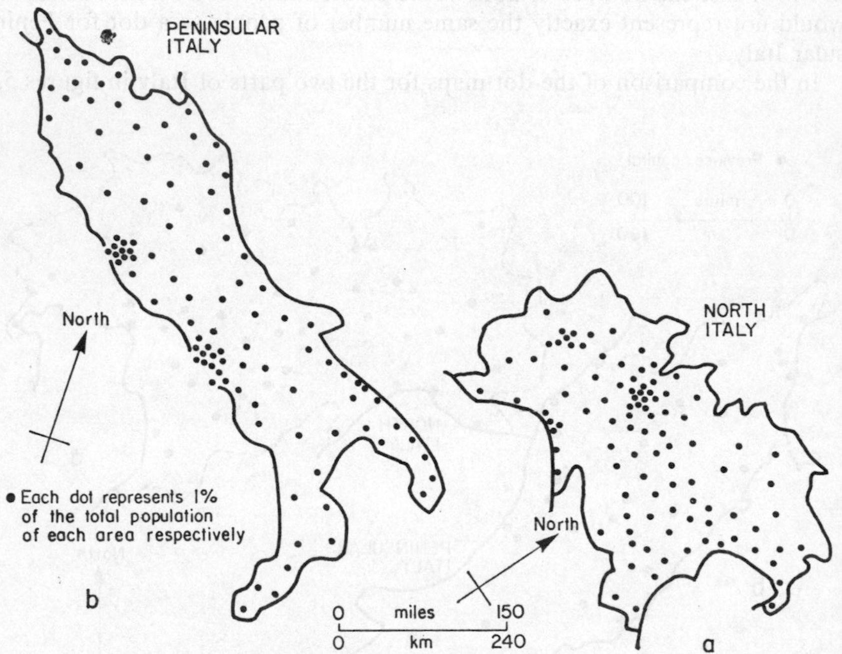

Figure 5.8 Comparison of the distribution of total population in (*a*) North Italy. (*b*) Peninsular Italy.

total population have been placed according to the method described in chapter 4. Time is the same for the places and things being compared. Population is for 1961. On the other hand, the actual dot distributions differ considerably, because population in figure 5.8 appears to be particularly concentrated along the central part of North Italy, whereas it is generally peripheral in Peninsular Italy. Verbal description of the similarities and differences does not do justice, however, to the features, both broad and detailed, observed visually.

A visual comparison may be achieved simply by putting two (or more) dot maps side by side. Further processing and re-representation of the basic date, whether in the form of new maps, or in the form of diagrams, may however improve the scope of the general visual comparison or facilitate the comparison of special features, such as, for example, the relationship of population to the mean centre of area. Some ideas will be put forward in the remainder of this section and in the following section.

A way in which a comparison of two dot distributions may be developed along particular lines is illustrated by the following procedure. Suppose it is desired to show clearly the relationship of distribution of population to mean centre of area. The method is illustrated by its application in Colombia. The same method could be applied in North Italy, Peninsular Italy or elsewhere. Procedure:

Step 1. Find the mean centre of the area under consideration by the method described in chapter 4.8.

Step 2. Draw concentric circles (see figure 5.9*a*) around this, including (for example) the innermost 10 per cent of the area, the next 10 per cent, the third 10 per cent and so on. These may be termed innermost or 1st tenth of the area, 2nd, and so on. Quite soon, part of the circle may pass outside an irregularly shaped unit of area. It is therefore necessary to adjust the radii of this and each subsequent circle to obtain the next 10 per cent of the area within them. The task may become laborious, but by no means impossible. In practice, a fairly rough calculation will suffice.

Step 3. Count the number of dots that fall in each 'ring'.

Step 4. Draw a histogram (5.9*b*). The vertical scale of the histogram represents the number of dots counted in each ring. The classes along the horizontal axis (ten in this case, because territory was divided into tenths) represent concentric rings of increasing distance from the centre.

This method needs considerable preparation and mapping of data. It gives a result in a highly standardized form, however, and this facilitates quick comparison between several areas to each of which the above steps may be applied. The mid points of the ten (or other number chosen) columns in the histogram may be joined to give a single curve. Comparable curves for several areas may then be put together on a single diagram and easily compared visually. In this particular example of the modification of data for comparison, information is retained about the *distance* of population from the mean centre of area, but is lost for direction.

Concentric circles could be drawn around other localities instead of the mean centre of area: for example, mean centre of population, national capital, chief port. Zones of increasing distance from some line, such as the coast, or a major river or railway could also be drawn with the help of *replica* lines.

5.5 Comparing patches on patches

In this section, a visual and a numerical comparison will be developed along lines somewhat similar to those discussed in section 5.4, to show how features of varying density of population in different areas may be compared. Use is made of population density data for major civil divisions. The major civil divisions therefore provide ready made patches. The techniques discussed could however be modified to deal with other patches on patches.

(a) Visual comparison

Figure 5.10 shows the first method to be described. The aim is to compare the distribution of population in Colombia, Chile and Brazil about the respective mean centre of area of each country. The following steps should be applied to each country in turn:

Step 1. Find the mean centre of area of the country.

Step 2. Arrange (rank) the major civil divisions of the country in order of increasing distance from the mean centre of area. The distance

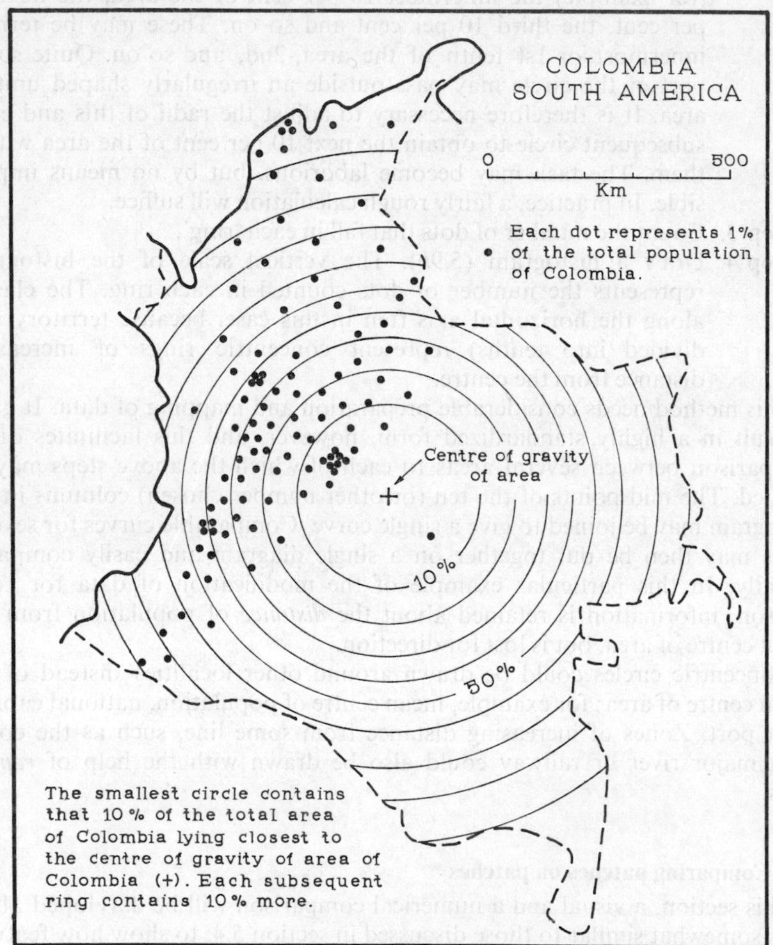
COLOMBIA
SOUTH AMERICA
0
500
Km
Each dot represents 1% of the total population of Colombia.
Centre of gravity of area
10%
50%
The smallest circle contains that 10% of the total area of Colombia lying closest to the centre of gravity of area of Colombia (+). Each subsequent ring contains 10% more.

Figure 5.9(a).

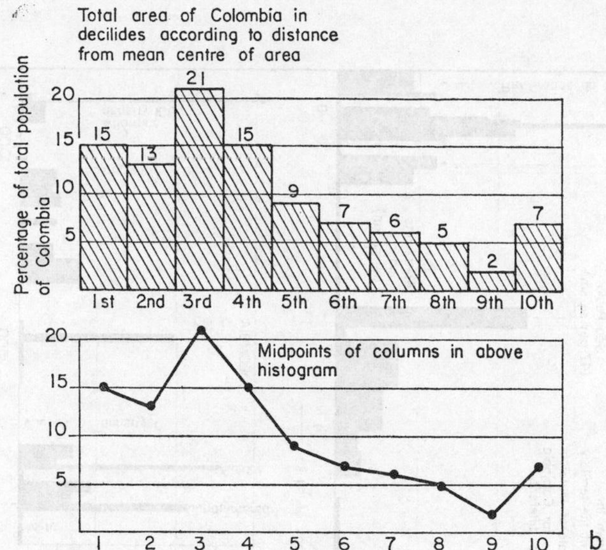

Figure 5.9 The distribution of population in Colombia and its relation to the mean centre of area of Colombia. (*a*) In map form. (*b*) Transferred to a histogram.

of the centre of area of each individual division may be measured from the centre of the country.

Step 3. On a histogram of appropriate size, mark off on the horizontal axis distances proportional in *length* to the area of each division, starting with the one nearest to the centre of *area* of the country on the left, and each in turn thereafter, in order of distance from centre.

Step 4. On the vertical axis make an appropriate scale proportional to the density of population of the divisions.

Step 5. On the histogram, raise a column above its appropriate portion of the horizontal axis, proportional in height to the density of population of the division in question.

The result of the above procedure applied to Colombia, Chile and Brazil is shown in figure 5.10.

As Brazil differs greatly in population and in area from the other two countries, the horizontal scale is different. Similarly, the scales for Chile and Colombia differ. There is room for further refinements. The horizontal and vertical scales could be made equal for all three areas, thus inflating Brazil to its proper size. Alternatively, the horizontal distance could be made uniform in length for all three countries. Whatever modifications are made, however, it is not the aim to remove the irregular columns, which look rather like factory chimneys. They do in fact reflect the irregularities of size and density of population of major civil divisions. What the comparison does allow is a

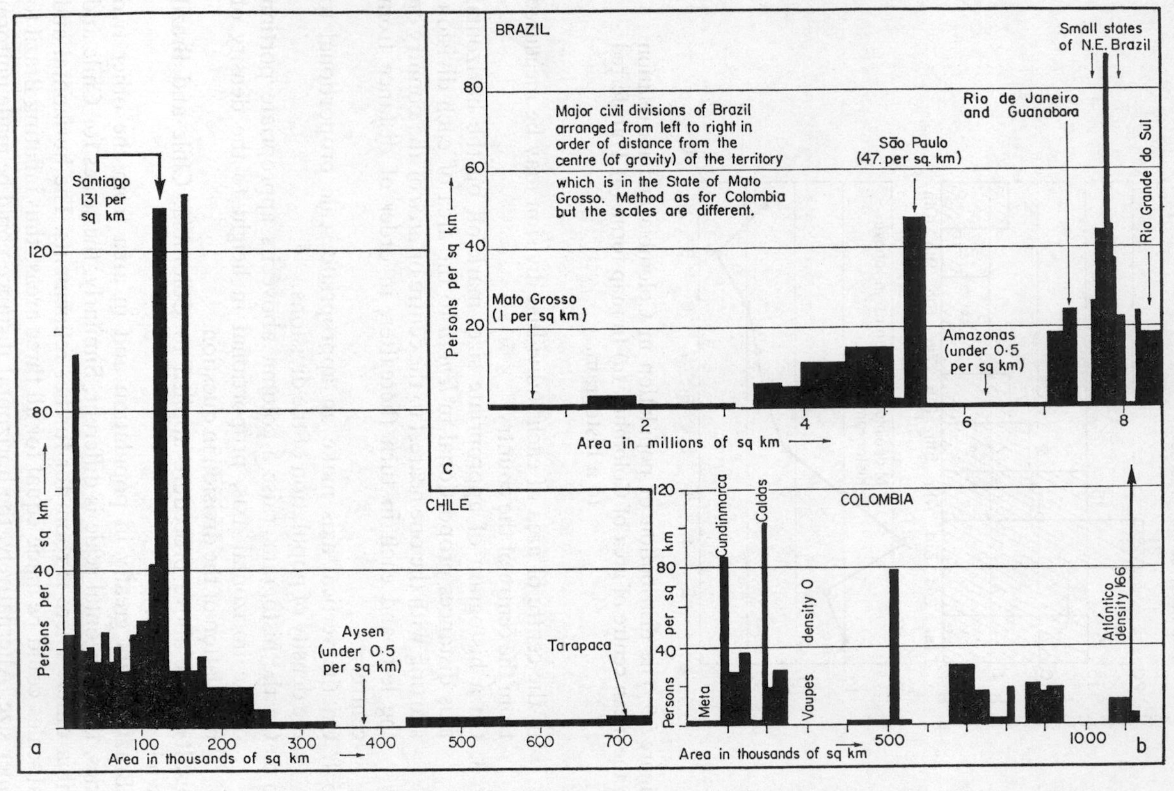

Figure 5.10 The distribution of population in (*a*) Chile, (*b*) Colombia, (*c*) Brazil in relation to their respective mean centres of area and compared by histograms.

quick visual appreciation of the 'pile up' of population in Chile near the centre, of Brazil towards the periphery, and of Colombia roughly midway.

(*b*) *Numerical comparison, supported by graph*

A fairly widely used method of comparing distributions on 2-dimensional surfaces in geography is the Lorenz curve. Basically this uses a square graph with *x*- and *y*-axes having comparable scale such as per cent units. Appropriate data are collected for subdivisions of the total area being considered. If the density of the objects distributed over the area is uniform, a straight line curve will result, running from 0, at an angle of 45° across the

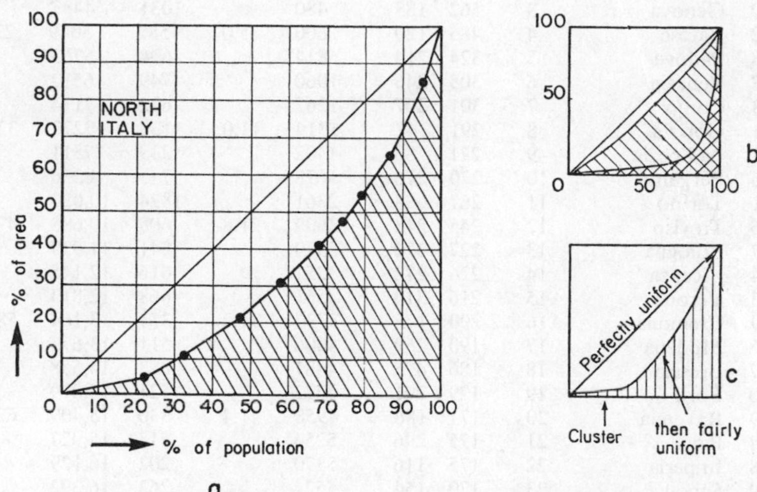

Figure 5.11 Lorenz curves (*a*) The distribution of population of North Italy expressed graphically to describe the degree of concentration. (*b*) and (*c*) Types of distribution. (*b*) Contrasts — clustered (cross-hatched) — with near uniform (diagonal shading).

graph. Variations of lesser or greater degree in the density will be reflected in the degree of concavity of the curve.

An example is worked out for the 40 provinces of North Italy. The data used are set out in table 5.2. The procedure will be outlined step by step. The reader may then adapt this to his own material and situations.

Step 1. Rank the 40 provinces according to density of population, from high to low.

Step 2. Record and sum the area cumulatively of each province.

Step 3. Record and sum the population cumulatively of each province.

Step 4. Calculate at appropriate intervals (in this example, every fourth province), the percentage of area and of population contained by the cumulative totals of area and population respectively.

Table 5.2 Data for provinces of North Italy*

Province Ref. No.	Name	Rank	Density of population	Area	Area cumulative	% of area	Population	Population cumulative	% of population
P 32	Trieste	1	1421	21	21		299	299	
P 15	Milano	2	1145	276	297		3157	3456	
P 10	Genova	3	562	183	480		1031	4487	
P 12	Varese	4	485	120	600	5.0	582	5069	22.4
P 28	Padova	5	324	214	814		694	5763	
P 27	Venezia	6	305	246	1060		749	6512	
P 13	Como	7	301	207	1267		622	7134	
P 31	Gorizia	8	291	47	1314	11.0	138	7272	32.1
P 11	La Spezia	9	271	88	1402		239	7511	
P 16	Bergamo	10	270	276	1678		745	8256	
P 01	Torino	11	267	683	2361		1824	10,080	
P 26	Treviso	12	245	248	2609	21.8	608	10,688	47.2
P 37	Bologna	13	227	370	2979		841	11,529	
P 24	Vicenza	14	226	272	3251		616	12,145	
P 23	Verona	15	216	310	3561		668	12,813	
P 19	Cremona	16	200	176	3737	31.2	351	13,164	58.1
P 36	Modena	17	190	269	4006		511	13,675	
P 17	Brescia	18	186	475	4481		883	14,558	
P 40	Forli	19	179	291	4772		521	15,079	
P 39	Ravenna	20	177	186	4958	41.4	330	15,409	68.0
P 18	Pavia	21	175	296	5254		518	15,927	
P 08	Imperia	22	175	116	5370		202	16,129	
P 09	Savona	23	170	154	5524		263	16,392	
P 35	Reggio E.	24	166	229	5753	48.0	380	16,772	74.0
P 20	Mantova	25	166	234	5987		387	17,159	
P 29	Rovigo	26	154	180	6167		278	17,437	
P 38	Ferrara	27	153	263	6430		403	17,840	
P 05	Asti	28	142	151	6581	54.9	215	18,055	79.7
P 06	Alessandria	29	134	356	6937		479	18,534	
P 02	Vercelli	30	133	300	7237		400	18,934	
P 03	Novara	31	128	359	7596		460	19,394	
P 34	Parma	32	113	345	7941	66.3	389	19,783	87.3
P 33	Piacenza	33	112	259	8200		291	20,074	
P 30	Udine	34	107	717	8917		767	20,841	
P 04	Cuneo	35	78	690	9607		536	21,377	
P 22	Trento	36	66	621	10,228	85.4	412	21,789	96.2
P 25	Belluno	37	64	368	10,596		235	22,024	
P 21	Bolzano	38	51	740	11,336		374	22,398	
P 14	Sondrio	39	50	321	11,657		161	22,559	
P 07	Aosta	40	31	326	11,983	100.0	101	22,660	100.0

* Provinces ranked in descending order of density of population

Source: *Annuario statistico italiano* (1964, pp. 2–3).

For example, the first four provinces in density (Trieste, Milano, Genova, Varese), together contain 5.0 per cent of the area, but have 22.4 per cent of the population. Using co-ordinates 5.0 on the *y*-axis and 22.4 on the *x*-axis, plot this on the graph. Carry on for the values at the 8th, 12th and subsequent multiples of 4. Join the points on the graph, as shown in figure 5.11*a*. The same calculation has been made for Peninsular Italy but as an almost identical curve is produced, this has not been plotted. Thus although the higher density of population tends to be central in North Italy and peripheral in Peninsular Italy, the degree of concentration and dispersal is almost identical.

Table 5.3 Comparison of North and Peninsular Italy

Province order	North		Peninsular	
4th	5.0	22.4	8.8	29.8
8th	11.0	32.1	17.5	42.8
12th	21.8	47.2	23.1	49.6
16th	31.2	58.1	31.6	59.0
20th	41.4	68.0	38.7	65.8
24th	48.0	74.0	48.8	74.4
28th	54.9	79.7	56.2	79.4
32nd	66.3	87.3	74.0	89.7
36th	85.4	96.2	88.0	95.8
40th	100.0	100.0	100.0	100.0

An index of degree of general uniformity of clustering of the distribution may be calculated by calculus or simply by counting small squares on the graph. The area under the curve is found and then expressed as a percentage of the right angled triangle into which it fits in the graph. An index of 100 per cent would represent a completely uniform distribution. One of only 20–30 per cent a very considerable degree of clustering. Figures 5.11*b* and 5.11*c* show the kind of curves that would be encountered. This method, like the others illustrated in this chapter, shows how with appropriate visual methods (maps and graphs) or numerical indices, steps may be taken to compare essential features of distributions that differ greatly in their size, layout and in other ways. A great deal of both thought and common sense are necessary to obtain results that are still only unfortunately approximate. The Lorenz curve has however been widely used in various disciplines and its merits and drawbacks will not be discussed here.

5.6 Further ideas on comparison

The techniques for comparison so far described allow either a broad and general but very approximate comparison between all (or most) of the features of two or more things, or a more limited, more precise comparison of one or a few features of two or more things such as distance of population from the mean centre of the patch on which it is distributed. Serious and to some extent successful attempts have been made in other disciplines, particularly in biology, to develop rigorous methods of comparison between distributions of recognizable features on similar objects. In 1917 D'Arcy W. Thompson developed a method for facilitating the visual comparison of similar features, such as skulls, jaw-bones and limbs of different animal species. His diagrams are called co-ordinate diagrams. One example, a comparison of the skulls of three hominoids, human, chimpanzee and baboon, is illustrated in chapter 2.10. The method is to put recognizable morphological points (e.g. eye socket) on each of the skulls in a one–one correspondence. A rectangular grid is then drawn over one of the skulls (e.g. the human skull). Grids are then drawn over the other skulls (two in this case) with the recognized points retained at correct positions in relation to the new grids. The grids become distorted as the other skulls are distorted. Thus the fore part of the baboon skull becomes relatively greatly inflated compared with the rear part, when compared with the human skull. In fact, a case of *continuous transformation* compatible with topology, and using the set theoretical concept of one–one correspondences or matchings, is being used. The success of the operation depends on there being identical fixed objects on each surface, and only the surface itself being transformed in shape.

Others have discussed, criticized and further developed various concepts of D'Arcy Thompson including his transformations. P. H. A. Sneath (1967) has commented on the feasibility of transformations for comparing shapes, and some of his introductory ideas are of relevance to those geographers contemplating making comparisons.

According to Sneath, there are at least two drawbacks in the method described, the difficulty of drawing the grids, and the difficulty of expressing the deformations in mathematical terms that would allow general conclusions to be drawn. In other words, more rigorous mathematical techniques and numerical statements are needed.

Sneath lists five questions that must be considered. The first involves methods of finding corresponding points in each of the objects to be compared. The remainder involve the manipulation of grids and the derivation of information about similarities of shape and so on. The first question: 'Can automatic methods identify the corresponding points in such diagrams?' Sneath himself considers the most difficult to answer. In biology, however, there is no question that in different but related forms, many points *do* exist that may be put in one–one correspondences. In geography, this is by no means so obvious. In general, this may be because the things to be compared in geography are less similar than those in biology. Sneath remarks (p. 67): A basic step in comparing two diagrams is the recognition of pairs of corresponding points (e.g. the akanthion on each skull). If this were impossible one could not get very far. Similar considerations apply both to points and to lines (though the latter alone are in general not sufficient, and are not treated in this paper). Thompson (1942, p. 1035) calls such diagrams *isotropic* and when isotropy is poor, comparison becomes difficult'.

The remainder of this section discusses tentatively and with the help of diagrams some of the difficulties of finding correspondences in geographical situations. Sneath suggests that at least 20 corresponding points are desirable, but 100 or more are a satisfactory quantity to handle.

Perhaps the most basic difference to note is that in the biological examples, points bear the same general spatial relationship to one another, whereas in geography they need not. This is pointed out by Sneath, quoting D'Arcy Thompson (1942, pp. 1085–1086): 'We cannot fit both beetle and cuttle-fish into the same framework, however we distort it; not by any co-ordinate transformation, can we turn either of them into one another or into a vertebrate type. . . . Eyes they all have, and mouth and jaws; but what we call by these names are no longer in the same order or relative position; they are no longer the same thing, there is no *invariant* basis for transformation'. In fact, the transformation may be random, and therefore not topological, though perhaps related to point set geometry.

Consider now the following geographical cases. They illustrate first comparison of the same things at different times and secondly of different (but comparable) things at the same time.

Case 1. An imaginary town. The town has an agreed central point (1), say the central square. Six main roads leave the central part of the town and lead towards other places outside the area under consideration. Between time 0 and time 1 the built-up area (and population) of the town has grown. Can points in the town at time 0 be put in a one–one correspondence with points in the town at time 1? Consider figure 5.12*a*. The centre, 1, remains fixed. 2, the town hall, has been moved to 2′, 3, the original market, has been rebuilt at 3′. The original school has however become (say) six schools, so one–one correspondence cannot be established. The old cotton mill 16 has been demolished, and replaced by a new mill at 16′. From the examples given and other examples the reader will easily concoct, it is clear that while objects may often be represented by points that can be put in a one–one correspondence, these may vary in location so drastically that continuous transformation is impossible. To make matters worse, one–one correspondences may be replaced by one–many.

The comparison of the limit of the built-up area at the two times is a little more promising. This is only because the same six main roads still exist at the second time. The points where the limit of the built-up area crosses each outgoing main road are 4–9 at time 0, 4′–9′ at time 1. Similarly, arbitrarily, six points (10–15 and 10′–15′) may be recorded for places half way along each section of the limit of the built-up area between each of the pairs of neighbouring outgoing roads.

All the points, both those representing objects in the town and those representing places on the limit of the built-up area, may be given co-ordinate positions on the grid. If the existing grid were used for the town at time 0 and a new one for the town at time 1, the grid for the second would be relatively inflated on the east side because the town has grown more on this side than on the west.

Case 2. Two villages in figure 5.12*b* are roughly similar in number of inhabitants. Conventionally they would be represented in map form with

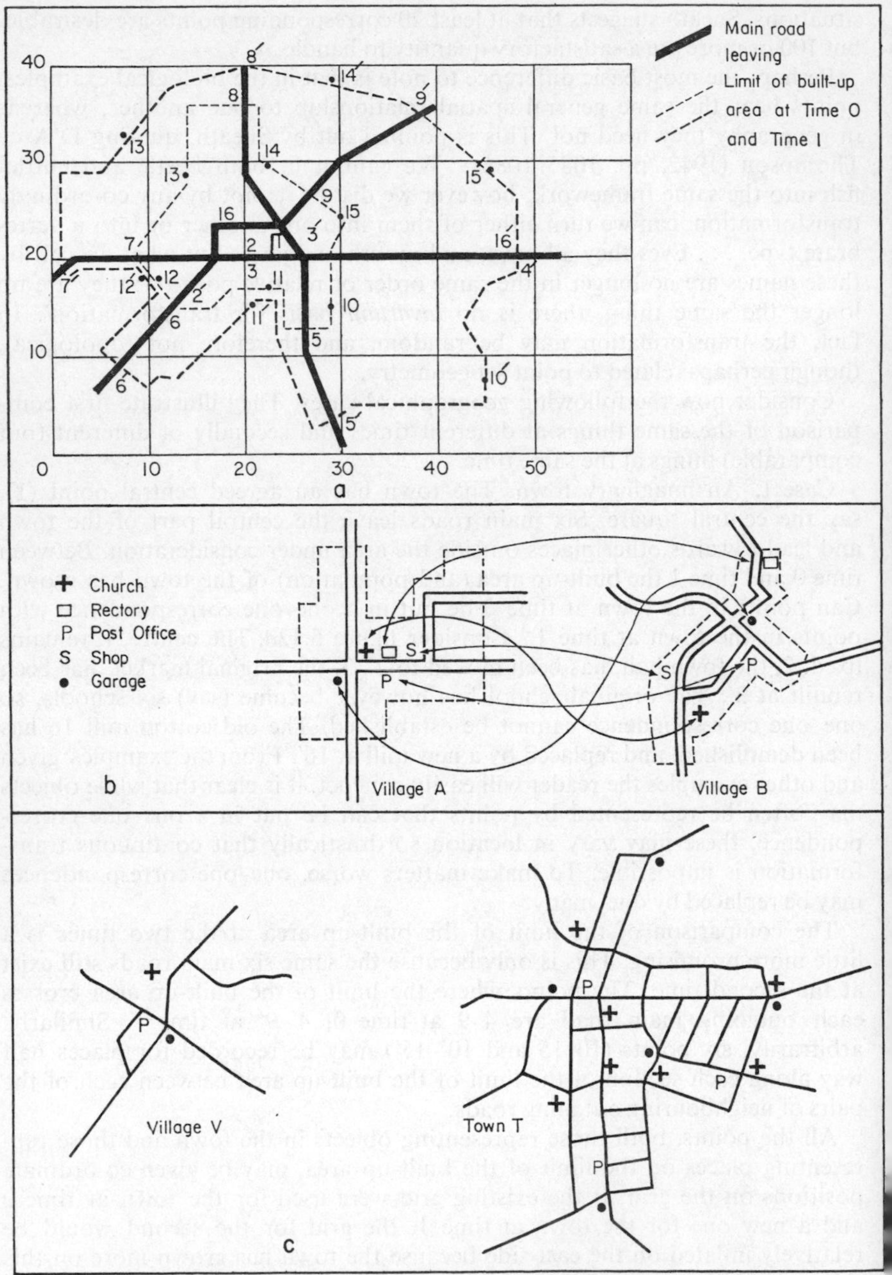

Figure 5.12 Problems of comparison. (*a*) Comparing the distribution of features of a town with features of the same town at a later date. (*b*) Comparing the distribution of features in two villages. (*c*) Comparing the distribution of features in a village and a town (key as for (*b*)).

north at the top. One or both could be rotated. They are also drawn on the same scale, but could be altered in scale. Reflexion is also possible. Even after some or all of these operations were performed, however, the street layout of the two villages would still be different both topographically *and* topologically. The streets, however, form the base on which many of the main objects in the village are distributed. In this example, five items have been put in a one–one correspondence (church, rectory, post office, shop and garage). This is impossible for some other items (e.g. school, farms) since these may exist in one village but not in the other, or may be more numerous in one than the other. Even the roads out differ in number, and the advantage of having six out in the town already discussed in Case 1 is lost. Even, however, if there were say 4 roads leaving each village, there would be no special reason for putting these in one–one correspondences, because there is no reason why any road in one village should be matched by one particular counterpart in the other. The situation here is not of four one–one correspondences, but of a many–many situation. This situation applies in general to the case of the 40 province capitals and the 100 dots in North Italy and Peninsular Italy discussed in section 5.4. Which province capital in the North is to be matched with which in the Peninsula? Milano and Roma as the two largest, Genova and Napoli as the chief ports? Aosta and Potenza as the highest above sea level? One–one correspondences here are debatable, to say the least. But as for the 100 dots, there seems no reason at all for matching these in particular pairs.

The geographer is faced in his comparisons, therefore, with serious problems. While bases (e.g. provinces, drainage basins) may provide points that may be put in one–one correspondences, the objects on the bases may be difficult to handle either because they cannot be put in one–one correspondences, or, if they can, because they are indiscriminately distributed over the base in such a way that no continuous transformation is possible. The geographer must at present be content either with the very general but approximate or with specialized but precise comparisons illustrated earlier in the chapter. Perhaps, however, it will be possible to develop means of usefully comparing areas and objects which cannot be transformed continuously, and which must be thought of in terms of one–many and many–many correspondences rather than one–one. The means may be built on those outlined above with reference to D'Arcy Thompson and Sneath, or completely afresh. The scattering of points allowed in point set theory (see chapter 2.9) may offer a starting point for studying random transformations. Perhaps, too, topology, and particularly networks, may be referred to. N. Rashevsky (1956) shows how simple organisms and complex ones may be compared with the help of topological concepts.

Figure 5.13 shows two countries, Mexico and Peru, that have some features in common, including broadly similar environmental regions, and a large (relative to their populations) national capital. They have been greatly simplified to show the essential problem, the fact that the features that may be matched between the two countries are arranged very differently spatially.

Figure 5.14*a* shows a very simple situation with a limited number of dots and patches in two areas, K and L. Most can be put in a one–one correspondence but the mining areas are 1–2 and the arid area 1–0. Is it possible to develop some method of comparing the two areas by drawing lines, for example straight lines, between matched places and then counting the number of crossings made by the lines? Unfortunately, by rotating one of the areas the number of crossings of lines would increase or decrease. It might be possible, however, to rotate one of the areas to several different positions, and then consider the results. The model in figure 5.14*b* illustrates how this might be done, but is too cumbersome to use as more than a device for seeing quickly the operations involved.

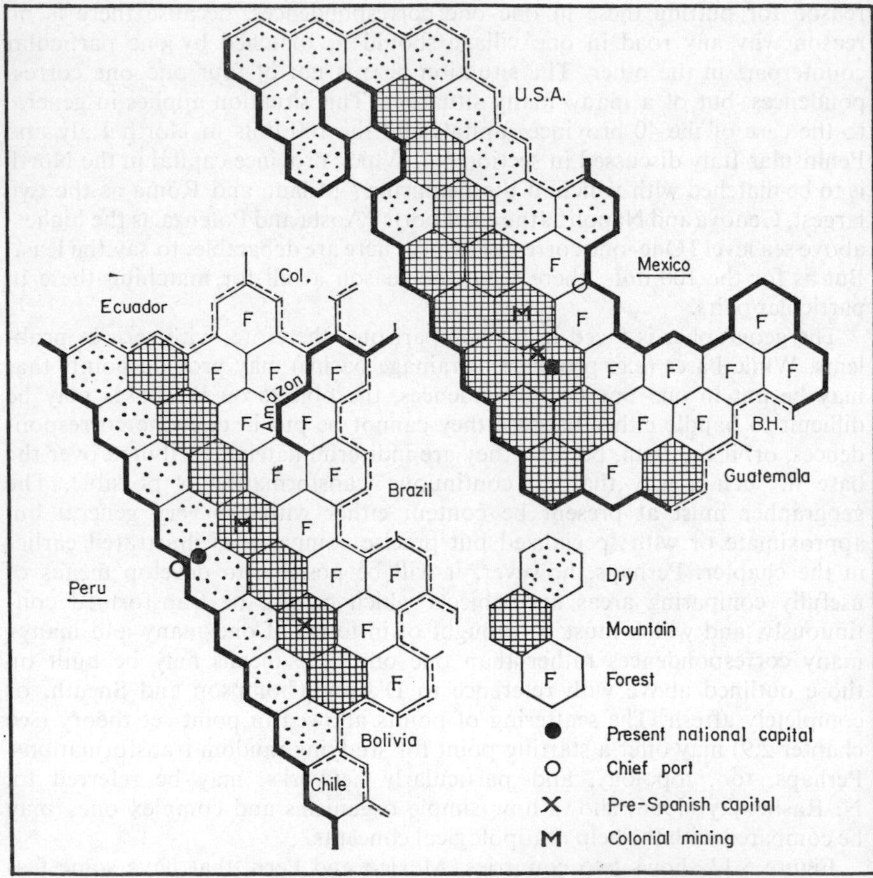

Figure 5.13 Comparing the distribution of similar features in Mexico and Peru. Note that the maps are not true to scale either between the two countries or within them. Hexagons are used to suggest one way in which data on a continuous surface could be more easily manipulated.

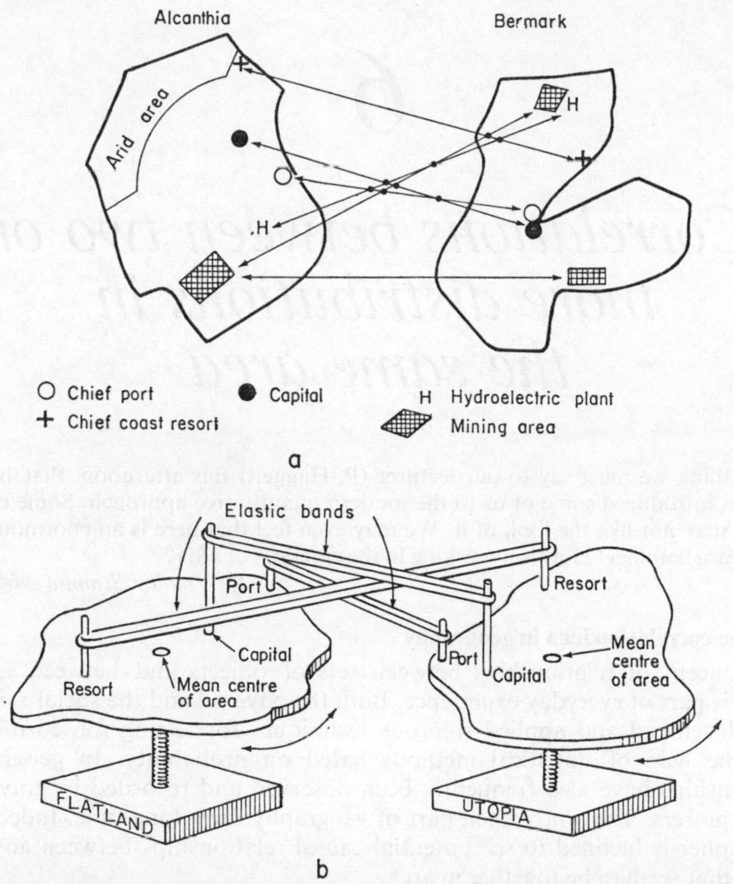

Figure 5.14 (*a*) Comparison of Alcanthia and Bermark by finding the crossing points of straight lines joining things matched. (*b*) Model for allowing the rotation of Flatland and Utopia to permit study of the changing number of crossing points of straight lines joining places matched.

References

Bunge, W. (1966). *Theoretical Geography*, Lund Studies in Geography, Lund, Sweden, ch. 8.

Fletcher, Sir Banister (1948). *A History of Architecture on the Comparative Method*, Batsford, London (14th ed.).

Grytzell, K. G. (1963). *The Demarcation of Comparable City Areas by Means of Population Density*, Ser. B. Human Geography No. 25. Lund Studies in Geography, Lund, Sweden.

Rashevsky, N. (1956). Topology and life: in search of general mathematical principles in biology and sociology, *Gen. Systems Yearbook*, **1**, 123–138.

Sneath, P. H. A. (1967). Trend surface analysis of transformation grids, *J. Zool., London*, **151**, 65–122.

Thompson, D'Arcy W. (1961). In *On Growth and Form* (Ed. J. T. Bonner), abridged edition, Cambridge University Press.

6

Correlations between two or more distributions in the same area

> 'I think we must say to our lecturer (P. Haggett) this afternoon, that he has introduced some of us to the modern quantitative approach. Some of us may not like the look of it. We may even feel that here is an enormous steam hammer! How is it working in the cracking of nuts?'
>
> *Sir Dudley Stamp* (1964)

6.1 The correlation idea in geography

The concept of relationships between sets of objects and between sets of events is part of everyday experience. Both the physical and the social sciences have developed and applied rigorous techniques for testing for correlation with the help of statistical methods based on probability. In geography, relationships have also frequently been observed and recorded by travellers and explorers. They have been part of geography for a long time. Indeed the geographer is inclined to see potential causal relationships between any two things that seem to be together in area.

The possibility of using more rigorous mathematical and statistical methods in geography to study relationships has only recently received widespread attention among geographers. Several techniques, widely used already in other disciplines, and now being used in geography, were described in chapter 3. It is assumed that the reader will have followed the procedures required to use the techniques. The purpose of chapters 6 and 7 is to show the application and findings of certain techniques on generally familiar geographical situations. The reader will in particular be considering Spearmen rank correlation coefficients, Pearson product-moment correlation coefficients, and certain steps in factor analysis. The above techniques, it must be stressed, are ones used in many disciplines. They may not be the best for dealing with spatial data. It is to be hoped that more appropriate geographical statistical methods will eventually be developed for handling spatial situations.

Very broadly, the correlation coefficients to be discussed are numerical statements about the co-occurrence of objects and the co-variance of values attached to them. These may supplement or replace verbal statements about correlations. A correlation coefficient expresses numerically relationship between two sets of numerical data which for convenience will be referred to as variables. Many geographical situations are so complex, however, that several or many variables rather than two are necessary to do justice to a situation under consideration. It is possible to extend a study to include more than two variables, either by considering various pairs in turn (partial correlation, ramifying linkage analysis), or by using factor analysis, which involves many dimensions, one for each variable.

A correlation coefficient is a *descriptive* numerical statement. If used appropriately, however, it can also become *inferential* in scope. Thus the correlation coefficient between two variables in a set of sample data may be used as a basis for inference, with a given level of confidence, about a relationship in the whole population from which the sample is taken. As such it is probabilistic in nature and, if high, is considered to reflect a closeness of relationships (co-variance) that could only very rarely have occurred by chance.

Although the correlation coefficients to be considered in chapters 6 and 7 are being used only as descriptive numerical statements, they may still be considered high and low relative to one another, and the confidence limits may at least be consulted, if not used as cut off limits for accepting or rejecting specific hypotheses.

In general, therefore, if an index of correlation is found to be reasonably high, either positively or negatively, it may be interpreted as follows:

(1) It reflects some degree of causal relationship between two variables, either direct or through other variables, which may be present in the study or not yet considered in it. H. M. Blalock (1964) is one of several social scientists who has gone into the meaning of causal relationships.

(2) It could be the rare occasion when a high index of correlation comes by chance.

(3) Particularly in temporal and spatial data, two general trends may produce a correlation without there being an obvious causal relationship.

(2) and (3) must be considered as possible dangers in correlation work. But they defeat the purpose of finding correlations at all if they are used indiscriminately to reject any high correlation that does not fit a hypothesis based on a preconceived idea. An apparently spurious correlation may in fact be genuine, but difficult to appreciate because intermediate, linking variables have not been taken into consideration.

Unfortunately, both in temporal and in spatial data, general trends over time or area may generate correlations that are not causally related though are technically correct and feasible. An example often quoted is the high degree of correlation over time between changes in teachers' salaries and in beer consumption. Similarly (see section 6.4), *in general*, precipitation diminishes in a westward direction in the U.S.A. So also does employment in manufacturing as a percentage of total employment. On the basis of data for the 50 states of the U.S.A., the two are highly correlated. But it would be naive to assume on this basis a causal relationship in either direction.

Much has already been written on the above dangers and drawbacks of correlation coefficients. It must be appreciated, however, that everything said applies equally to intuitive — visual — verbal correlations. The justification for pursuing mathematical methods of correlation may be summarized as follows:

(1) More consistent, precise *and concise* statements of correlation can be given mathematically than verbally. But the precision should not be taken to excessive limits. Often two, or no more than one decimal place, can be warranted in a coefficient index, given the approximate nature of the data used by geographers.
(2) Often relationships that have been worked out non-mathematically are confirmed.
(3) On the other hand, preconceived ideas may distort visual perception. These may be corrected. Two examples may be quoted: in correlation work on Mexico and France, illegitimate births and left vote respectively were variables that were expected to correlate highly with certain other variables. The indices of correlation clearly showed that they do not.

The geographer is faced with many different situations in which a correlation may be expected. Spatial correlation is summarized by a series of simple diagrams in figure 6.1. Imagine a draughts board with 64 squares but with the 32 black squares all in the 'western' half. Take 32 black draughts. Each draught will be placed on one square, and no square will contain more than one draught.

(6.1*a*) shows every black square with a draught on it. (6.1*b*) shows such a high proportion of the draughts on black squares ($\frac{27}{32}$) that there seems to be a high positive correlation. In (6.1*c*) only $\frac{18}{32}$ of the draughts are on the black squares. In (6.1*d*) the draughts are mostly on the white squares. (6.1*e*) the draughts are all on white squares.

Figure 6.2 (see its caption) shows a situation similar to that lying somewhere between 6.1*c* and 6.1*d*.

The opportunity may be taken here to recall from chapter 3 that the correlation index in both Spearman rank correlation and Pearson product-moment correlation increases away from 0 in both a positive *and* a negative direction. The negative situation can easily be switched to a positive counterpart. Imagine a complementary set of 32 white draughts that occupy the squares *not* held by black draughts. They represent 'not black', and occupy *all* the black squares in 6.1*e*. In general it is sufficient to 'reverse' one of a pair of variables to alter the sign in a correlation between them. At this stage, also, it is important to stress that a low index of correlation, suggesting no relationship between two variables, is as *important*, if not as satisfying, as a high correlation, since it suggests that a given line of investigation may be abandoned or reconsidered.

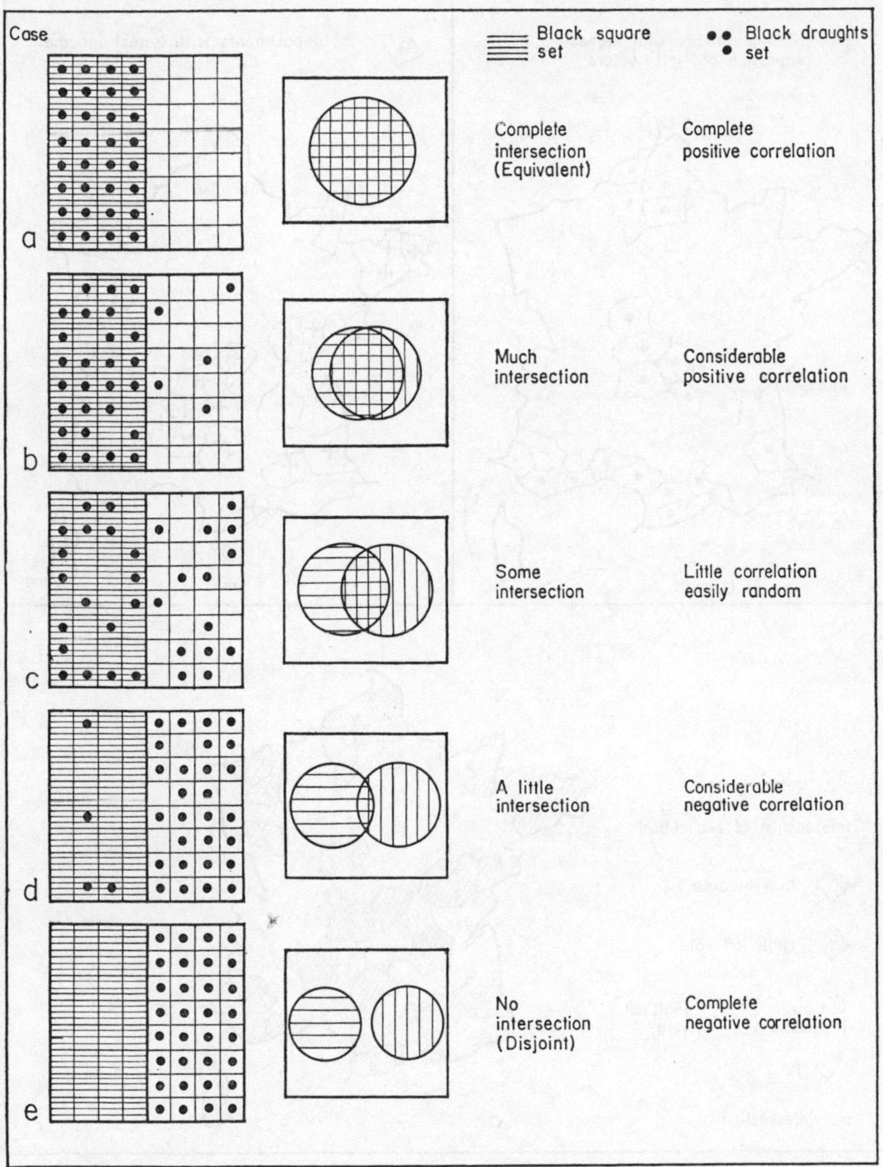

Figure 6.1 The correlation idea illustrated by dots on patches and by Venn diagrams; (*a*)–(*e*) show varying degrees of correlation.

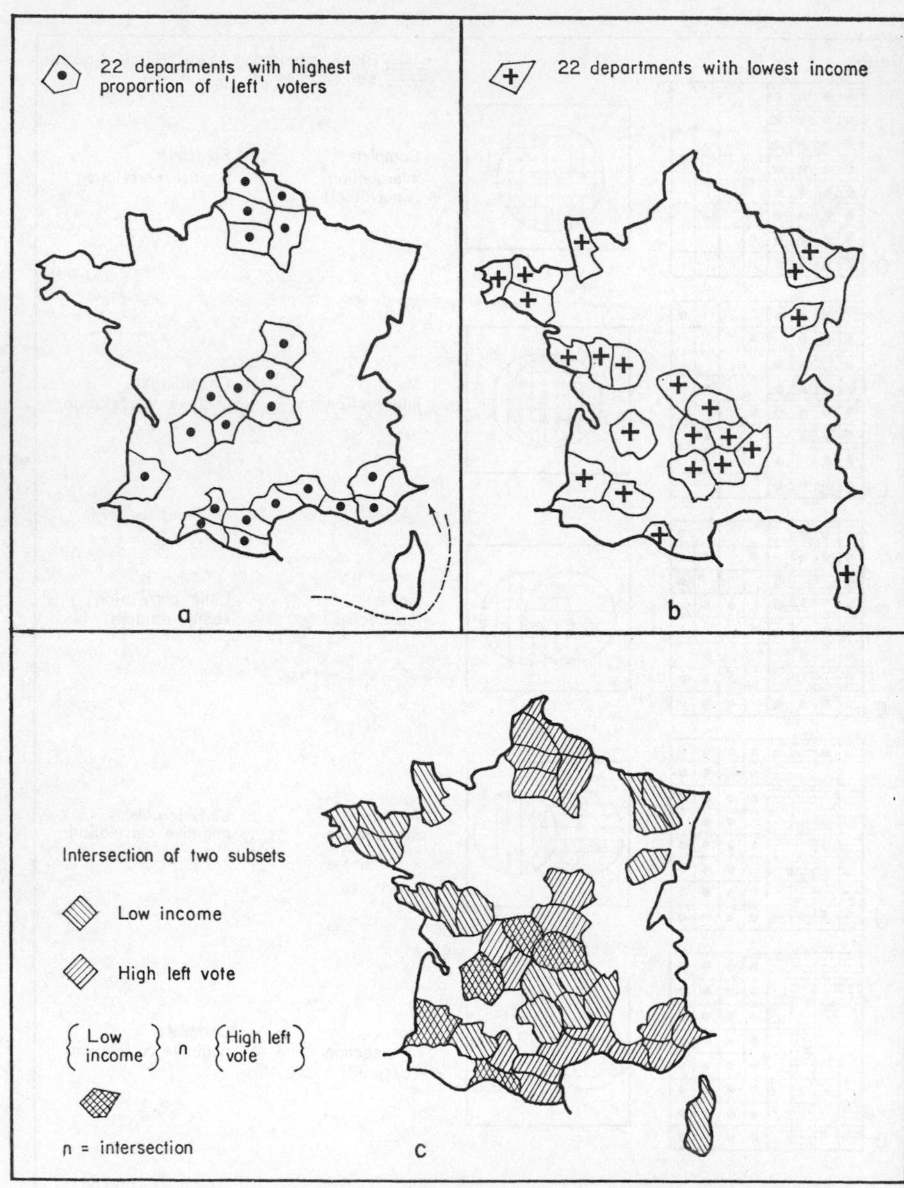

Figure 6.2 The correlation idea illustrated by two distributions in France. (*a*) The subset of 22 departments with highest proportion of left voters (1958). (*b*) The subset of 22 departments with lowest income (1961). (*c*) Showing the two distributions superimposed on one map. Those departments that belong to both sets form the intersection of the two sets.

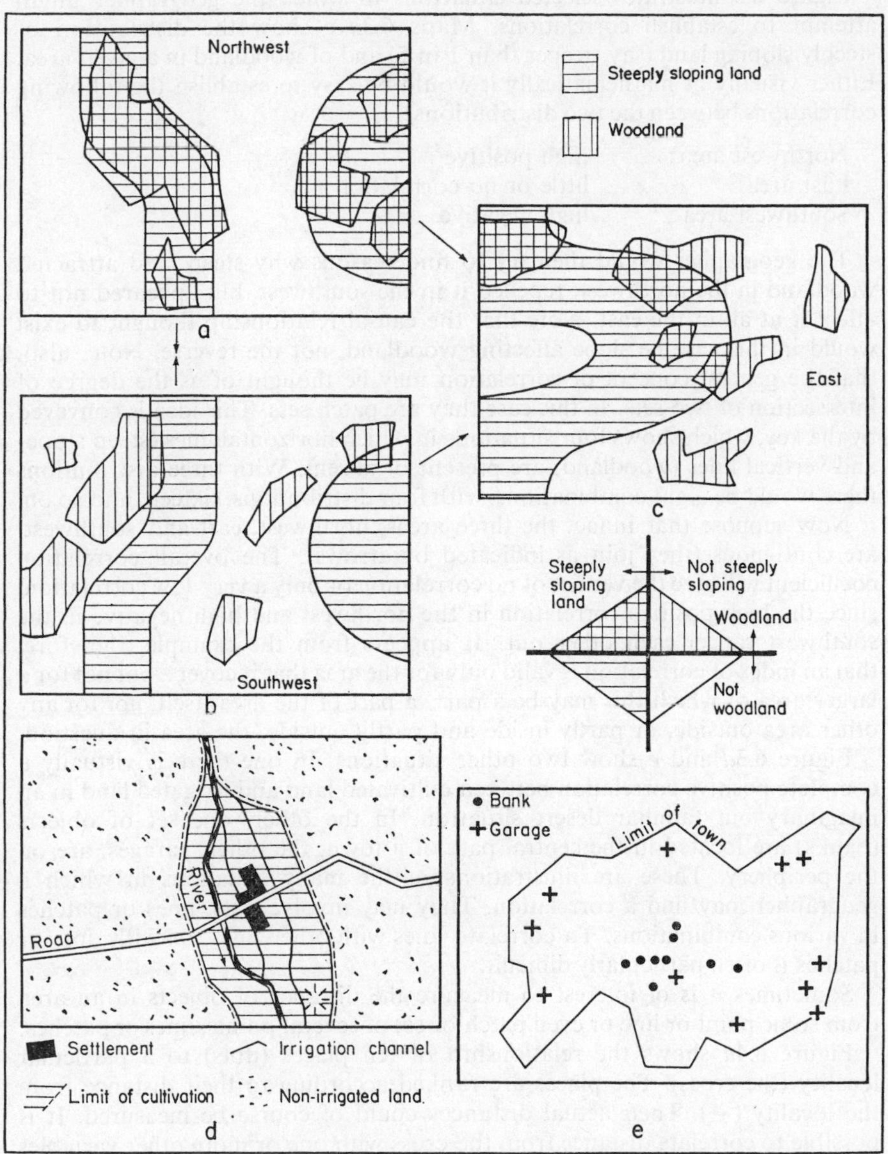

Figure 6.3 Examples of simple situations in which positive, negative and no correlations appear to exist between pairs of distributions. (*a*)–(*c*) Woodland and steeply sloping land in three areas. (*d*) Cultivated and irrigated land. (*e*) Banks and garages in a town.

Figure 6.3 illustrates selected situations in which the geographer might attempt to establish correlations. Maps 6.3*a–c* show the distribution of steeply sloping land (say steeper than 1 in 5) and of woodland in a given area. Either visually or mathematically it would be easy to establish the following correlations between the two distributions:

Northwest area:	high positive
East area:	little or no correlation
Southwest area:	high negative

The geographer would then try to find reasons why steep land attracted woodland in the northwest, repelled it in the southwest, but appeared not to affect it at all in the east. Note that the causal relationship thought to exist would in this case be slope affecting woodland, not the reverse, Note, also, that the general concept of correlation may be thought of as the degree of intersection of two sets. In this case they are patch sets. This idea is conveyed by the key, which shows four situations in which horizontal lines (steep slope) and vertical lines (woodland) are present or absent. With three distributions there would be eight combinations, with four distributions, sixteen, and so on.

Now suppose that in fact the three areas, northwest, east and southwest, are contiguous (they join as indicated by arrows). The overall correlation coefficient will give the verdict of no correlation or only a very low correlation, since the high positive correlation in the northwest and high negative in the southwest cancel each other out. It appears from the example, therefore, that an index of correlation is valid only for the area that it covers, but not for a larger area of which this may be a part, a part of the area itself, nor for any other area outside, or partly inside and partly outside, the area in question.

Figure 6.3*d* and *e* show two other situations. In one there is virtually a complete positive correlation between cultivated land and irrigated land in an imaginary but familiar desert situation. In the other, one set of objects (banks) are located in the central part of a town, the other, garages, are on the periphery. These are illustrations of the many situations in which a geographer may find a correlation. They may involve dots, lines or patches in various combinations. To correlate lines with other lines, or with dots or patches is often particularly difficult.

Sometimes it is of interest to measure the distance of objects in an area from some point or line or even patch or set of several points, lines or patches.

Figure 6.4*a* shows the relationship of ten places (dots) to a particular locality (the cross). The places are ranked according to their distance from the locality (+). Their actual distances could of course be measured. It is possible to correlate distance from the cross with one or more other variables for the ten dots. The cross could be a major industrial centre, and the objects around it other smaller settlements. The distance variable could be correlated with the variable for manufacturing population as a percentage of total population within each settlement. Although distance is usually measured from a point *within* a distribution, it is possible to measure distance from a point outside it.

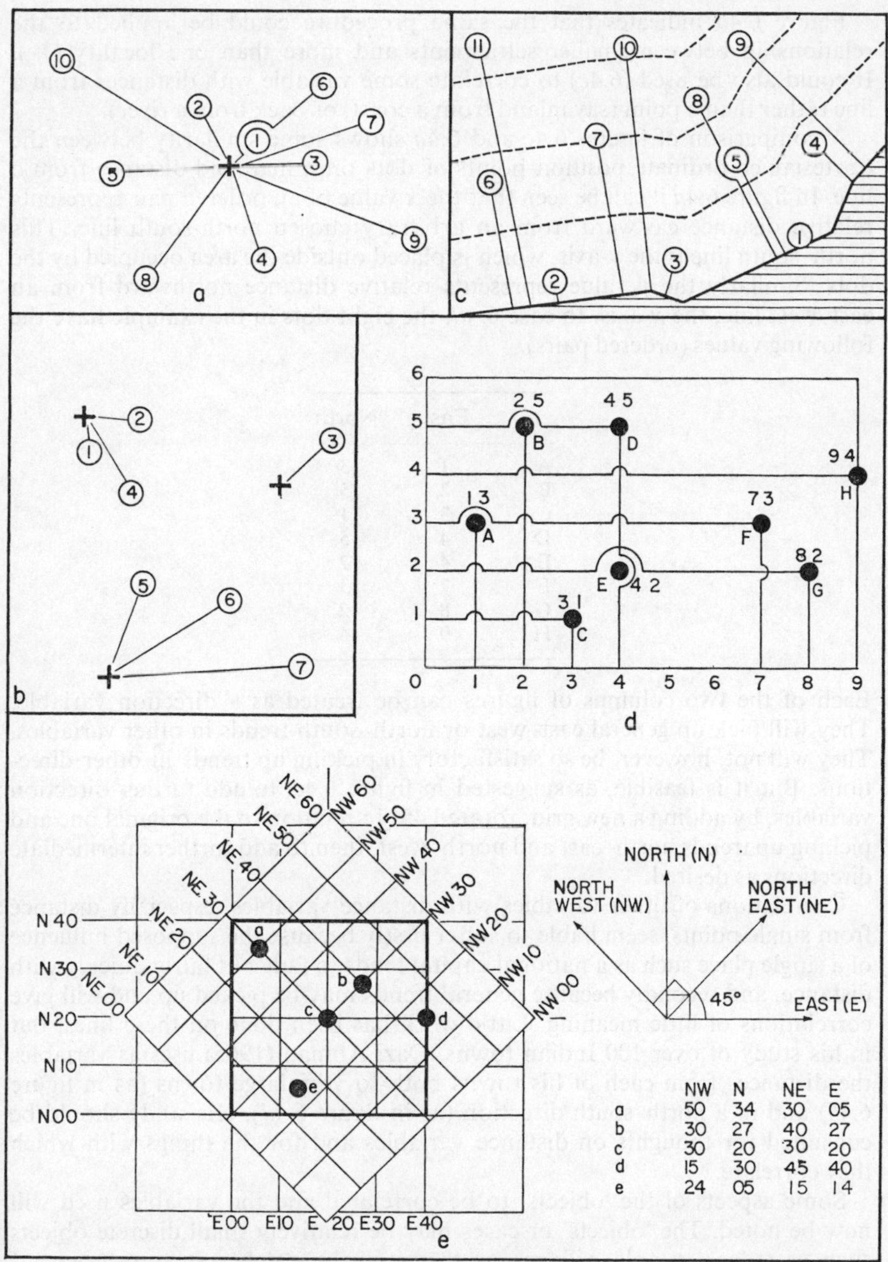

Figure 6.4 Examples of distance variables. (*a*) Distance from a point. (*b*) Distance from several points. (*c*) Distance from a line (e.g. the coast). (*d*) Distances from a parallel and a meridian or from *x* and *y* co-ordinates. (*e*) A grid to measure distance from four lines orientated as indicated in the diagram.

Figure 6.4*b* indicates that the same procedure could be applied to the relationship between smaller settlements and more than one locality (+). It could also be used (6.4*c*) to correlate some variable with distances from a line rather than a point (say inland from a coast, or back from a river).

A comparison of figures 6.4*c* and 6.4*d* shows some similarity between the Cartesian co-ordinate position points of dots on a map and distance from a line. In figure 6.4*d* it can be seen that the *x* value of an ordered pair represents relative distance eastward from an arbitrary chosen north-south line. This north–south line is the *y*-axis, which is placed outside the area occupied by the dots. Similarly the *y* value represents relative distance northward from an east–west line, the *x*-axis. In case 6.4*d*, the eight dots in the example have the following values (ordered pairs).

	East	North
A	1	3
B	2	5
C	3	1
D	4	5
E	4	2
F	7	3
G	8	2
H	9	4

Each of the two columns of figures can be treated as a direction variable. They will pick up general east–west or north–south trends in other variables. They will not, however, be so satisfactory in picking up trends in other directions. But it is feasible, as suggested in figure 6.4*e*, to add further direction variables, by adding a new grid, rotated 45° in relation to the original one and picking up trends north–east and north–west, then to add further intermediate directions as desired.

Correlations of other variables with distance variables, especially distance from single points, seem liable to suffer firstly because the supposed influence of a single place such as a national capital tends to fade out fairly quickly with distance, and secondly because general trends may be picked up and will give correlations of little meaning. Little work has been done on these lines, but in his study of over 100 Indian towns, Qazi Ahmad (1965) uses as variables the distances from each of his towns both to very large towns (as in figure 6.4*b*) and in a north–south direction (as in figure 6.4*d*). His study should be consulted for thoughts on distance variables and for the things with which they correlate.

Some aspects of the 'objects' to be correlated and the variables used will now be noted. The 'objects' or cases may be relatively small discrete objects such as stones, people, villages or they may be patches such as tracts of woodland, political units. The patches may cover only part of the total base area under consideration, or they may between them cover exactly the whole area. There have been many studies of correlations between discrete small objects. In geography, towns in particular have been studied (Moser and Scott (1961), Qazi Ahmad (1965)). Less has been done on the correlation of data for compartments. This will form the basis for several of the examples in chapters 6 and 7.

A grid of compartments that covers a whole area may either be regular (e.g. squares, rectangles, hexagons) or irregular (e.g. political units). A regular grid is usually one that has been superimposed for some purpose such as locating places. It is therefore unlikely that data will be published exactly for each compartment in the grid. This can however often be collected from existing maps, or gathered from field work or aerial photographs. The irregular grid is characteristic of political units. Many data are available for these in published form. A regular grid is however more representative of the area on which it is superimposed since area is constant for each compartment. However, many data used by geographers are based on people rather than area and this advantage is therefore to some extent reduced. Sometimes the irregular compartments in a system of political units are similar enough in both area and population to be regarded as satisfactory. Thus the departments of France, the provinces of Italy and Spain, and the counties of many states in the U.S.A. are satisfactory in this respect. Some countries, including the U.S.A. divided into states at national level, the United Kingdom divided into about 100 counties, and the U.S.S.R. into its 15 republics, are examples of very badly divided countries, since the units differ greatly in area and/or population. The countries of the world themselves also form a bad grid. In spite of this, several studies have been based on all or on some subset of them (e.g. by B. J. L. Berry, in N. Ginsberg (1961) and by B. M. Russett (1964)).

It is often desirable to modify data to eliminate the influence of differences in area or population among compartments. Thus arable land should not be expressed in absolute terms but as a proportion (percentage) of the total area of a compartment. Persons over 65 years of age should not be expressed absolutely but as a percentage of total population in the compartment. Moreover it is important to give constant thought to the influence the grid itself may be having on the data collected for its compartment.

Another question that inevitably arises in a piece of correlation work is: How many cases should be considered? The significance of the correlation coefficient is related to the number of cases taken. A significant correlation coefficient moves at first rapidly towards zero from about 8 cases to about 20. After that the decline is less sharp. After about 50, the advantages of increasing the number of cases is gradual. However, the number of cases that may be used varies very widely, depending on the scope of a given study. It may be less than 10. On the other hand, in the 1930's E. L. Thorndike (1939) studied 300 U.S. cities (and also about 300 variables). Recently there has been talk of correlating variables for approximately 3000 U.S. counties. A modern computer could handle thousands, even tens of thousands of cases. In chapters 6 and 7 of this book, the following numbers of cases are used in illustrations:

Departments of France	90
States of U.S.A.	50
Provinces of North Italy	40
Grid Squares in North England	20 and 120

No less important than the number of cases is the number of variables. This depends on the scope of the exercise to be undertaken and the calculating equipment available. About the study of some 1400 administrative units in greater New York, R. C. Wood (1964) writes:

> 'Choosing a sample of Community characteristics was a more formidable task than choosing the communities themselves. . . . The instrument by which we began to disentangle the snare is the mathematical technique called factor analysis.'

To calculate correlations between several variables is laborious even with a mechanical calculating machine. On the other hand, the correlation coefficient between each pair of 100 (or considerably more) variables can be calculated rapidly on an electronic computer. The exercise may involve only the testing of a few hypotheses between selected pairs of variables. On the other hand, it may be of a blanket nature, aimed simply to see what is related to what else, with a view, as often in factor analysis, to selecting a reduced number of key variables.

With the advent of the electronic computer it has become less necessary to economize in the number of variables considered. Nevertheless it is important to ask the question: which variables are meaningful in a given context? The following broad types of variables may be suggested:

(1) Variables about the compartments themselves (e.g. some index of their shape or compactness, their size). Note that these features would be constant in a uniform grid (e.g. of squares).

(2) Distance variables, discussed above: distance to given point or points, line or lines.

(3) About things on the ground. These may be subdivided into physical features, man himself, works of man, or in other ways desired. They form the main material for correlation. Very broadly, relationships may be thought of as follows:

1	Physical	–	physical	One way
2	Physical	–	human	
3	Human	–	physical	
4	Human	–	human	
5	Physical	–	physical	Two way
6	Physical	–	human	
	Human	–	physical	
	Human	–	human	

Possible examples are:

1	Altitude	–	precipitation
2	Soil	–	cultivated land
3	Mechanized ploughing	–	soil erosion

4 Urbanization – air pollution
5 Soil – vegetation
6 Farming – soil quality
7 Urbanization – industrialization

It is important to appreciate that 'human' activities can be the cause of changes in physical features, just as physical features can (as so often stressed in geography) affect man. The complex causal relationships will be reconsidered in chapter 15 under a discussion of regional geography.

Some further dangers and drawbacks of correlation may now be added briefly. Once it has been decided which variables are desirable in a given project, and suitable figures for these have been found, the following situations may arise:

(1) The variables may vary only over part of the area being studied. Some examples for France will illustrate this danger. All 90 departments of France have some arable land. 90 different values are therefore available for arable land as a percentage of total land. Only about 30, however, have more than a negligible proportion of land under the vine. In many of the remaining 60, the proportions are very near or exactly 0. This variable cannot correlate highly with other variables. Similarly, every department has some persons employed in transport, but not many have more than a negligible proportion engaged in, for example, the iron and steel industry. Again, all have some precipitation, but only some have outcrops of granite rocks. This difficulty may to some extent be overcome if only a selected part of the total area is considered: thus only the 30 vine growing departments would be chosen for a study of the vine and its relationship to physical conditions and to other activities.

(2) Many variables express information about subsets of things already used as variables. For example, to use percentage of population aged over 65 years of age as one variable, and percentage over 75 years of age as another, invites at least some degree of positive correlation by the very nature of the data, as does the inclusion of variables giving the proportion of total area over 1000 metres above sea level, then the proportion over 1500 metres.

Many other dangers could be listed, including those concerned with the independence of variables. The reader can only appreciate the various drawbacks and also the advantages by doing correlation work himself.

6.2 Ranking and rank correlation

In this section some applications of ranking and rank correlation are illustrated with examples from France. Space prevents the inclusion of all the data used. Most tables have therefore been cut off after a few of the 90 departments have been shown. The reader is referred back to chapter 3 for an outline of the Spearman rank correlation test. Other rank correlation tests are also available including Kendall's tau (W. H. Kruskal).

(a) Which variables?

Before the data and the way they have been processed are considered, it is important to give some attention to the choice of variables used. Since the exercise to be described was built up in an experimental way it does not claim to contain all the important spatial variables in France; for example, transportation networks are ignored entirely. It was necessary to ask more than once, however, what variables make a complete and balanced regional geographical study of an area such as France. All the time in looking for data and in processing it one should be thinking what one is trying to show. In this exercise little effort was made to collect data not readily available in published sources of one sort or another.

Very broadly, the 31 variables used here fall into four classes. They are referred to by letters or Roman numerals in the text and in tables in this section (6.2). There is no significance in the use of the letters A–S for the first nineteen variables and I–XII for the remainder. This was done for organizational convenience.

(1) Variables derived directly from the existence of the politico-administrative system of departments: A (area of department), B (population of department), C, D (compactness of department) I, II.
(2) Distance variables, notably V (distance of departments from Paris), VIII, IX.
(3) Variables representing objects or recorded values varying in their distribution over the surface of France.
Physical geographical features: III (altitude), X (mean annual precipitation) and XI (mean annual temperature)
Human geographical features: e.g. E (arable land), IV (new cars registered)
(4) Composite variables, derived already from a consideration of a combination of existing variables: notably XII, the meaning of which will be explained below.

The distance variables are particularly difficult to interpret. Some of the human geographical variables, notably G (forest), have a physical basis.

A consideration of the actual data for the variables raises many questions and doubts. One asks: Why were these variables chosen? Are they all sensible? In one important respect, almost all are satisfactory, for they vary all over France.

(b) The organization of the data

The Spearman rank correlation coefficient requires data in an ordered or ranked form, as already shown in chapter 3. For the 90 departments of France, it requires these to be ranked, using the numbers 1–90, for each variable. Lack of space makes it possible only to reproduce here some specimen data. The reader will find the original data in the sources given with table 6.1. Table 6.2 includes the raw data for selected variables. Table 6.3 includes the data for selected variables processed where necessary ready for ranking.

Table 6.4 shows extreme values for six selected variables. Table 6.5 shows the ranking of the first ten and last ten departments alphabetically on variables A–S.

The work of ranking was carried out in the following way. A paramount punched card was allocated to each department. Data of the kind displayed in table 6.3 were recorded on each card, a particular place being reserved (and noted on a master card) for each variable. It was easy, then, to sort the cards into the order 1–90 for each variable in turn. This can be done by ranking the values either low to high or high to low, as desired, *but it is most important to bear in mind which has been done* since the sign plus or minus before the eventual correlation coefficients depends on this. In chapter 3 it was explained what must be done when two or more departments tie and are eligible for the same rank.

Additional advantages of using the cards may be noted. By appropriate punching into subsets of four quartiles it was possible later to draw these out quickly. Further, it was useful for getting the 90 cards back into alphabetical order after ranking, to be able to sort them into four preliminary piles for an alphabetical ranking. The sorting of cards can of course be done mechanically by a card sorter, or by computer, but there is much to be learned by using them manually at first, as outlined in this section. Moreover some readers will not have ready access to a computer.

(*c*) *Mapping by quartiles*

To avoid confusion with other numbers (e.g. ranks, correlation coefficients) the alphabetical list placing numbers for each department (e.g. Corse is 20) is preceded by a D in the text.

In this exercise, 31 variables have been used, but attention has been focused mainly on 19 of these. It would be possible to map each of the 19 variables (or any other number desired) and to attempt to discover visually the degree of correlation between each pair. When 19 variables are used, there are altogether 19^2 or 361 possible pairs of correlations, e.g. variable 1 with 1,1 with 2, 1 with 3 and so on, 2 with 1, 2 with 2, and so on. 1 with 2 and 2 with 1 obviously give the same result. Making a correlation between each variable and itself (e.g. 2 with 2) is superfluous since there is of course a perfect correlation. The actual number of different pairwise correlations is therefore 171 (derived from $n(n-1)/2$). In other words, even if the possibility of comparing variables three at a time, four at a time, and so on, is not considered, 171 different combinations of two maps would have to be made, given 19 maps each with one variable, to see visually all pairwise correlations.

In view of the complexities noted, visual correlation would be very difficult. This does not mean, however, that some sort of visual impression of the distribution of values for each of the variables is completely unhelpful. There is no reason why a series of pairwise visual comparisons among, say, only four or five maps, should not allow some approximate statement about correlation. In itself such an assessment could only be imprecise since no careful measurement is involved. The method adopted in this exercise was to

Table 6.1 The 31 variables for France

Variable	Source	Date	Basic data	Form ready to rank
A	S1 p. 19		Area in square km	OK
B	S1 p. 19	62	Population in thousands	OK
C	S1 p. 19	62	Persons per square km	(B/A) OK
D			Index of compactness[a]	OK
E	S1 p. 108	62	Arable land (*terres labourables*)	As % of total area ((E÷A)×100)
F	S1 p. 108	62	Permanent pasture (*surfaces toujours couvertes d'herbe*)	As % of total area ((F÷A)×100)
G	S1 p. 108	62	Forest (*bois et forêts*)	As % of total area ((G÷A)×100)
H	S1 p. XVIII	62	Employment in agriculture (*agriculteurs+salaries agricoles*)	As % of total employed population[b]
I	S1 p. XVIII	62	Workers (*ouvriers*)	As % of total employed population[b]
J	S3 pp. 51–2	61	Purchasing power	In dollars per capita[c]
K	S1 p. 19		Population change 1954–62	1962 total as percentage of 1954
L	S4 p. 11	58	Left votes	As a percentage of all votes[d]
M	S2 p. 276	60	Electricity consumption	Per inhabitant
N	S2 p. 282	62	Cars in circulation (*voitures particulières et commerciaux*)	Per 1000 inhabitants
O	S2 p. 278	61	Building permits	Issued per 1000 inhabitants
P	S2 p. 290	60	Department budgets (*recettes budgetaires*)	Francs per inhabitant
Q	S2 p. 246	60	Infant mortality	Per thousand live births
R	S2 p. 242	62	Urban population	As % of total population[e]
S	S2 p. 244	62	People over 65 years of age	As % of total population
I	S1[f] p. 20	62	Population of Department capital	
II	S1 p. 20	62	Population of Department capital	As % of total population of Department
III	S5		Lowland	As % of total area
IV	S1 p. 276	62	New cars registered	Per 10,000 inhabitants
V			Distance from Paris[f]	
VI	S1 p. 110	62	Milk production	Per 1000 hectares
VII	S1 p. 108	62	Vines	As % of total area of Department
VIII			Distance from mean centre of area of France[g]	
IX			Distance from mean centre of population of France[g]	
X	S5		Mean annual precipitation	Ranked high to low
XI	S5		Mean annual temperature	Ranked high to low
XII			Deviation from 'average' Department[h]	Ranked small to large deviation

S1 *Annuaire Statistique de La France, 1963, Résultats de 1962* (Ministère des Finances et des Affaires Economiques), Paris (Imprimerie Nationale), 1963.

S2 *Tableaux de l'Economie Française* (Institut National de la Statistique et des Etudes Economiques), Paris 1963.
S3 *The European Markets*, The Chase Manhattan Bank, January 1964.
S4 P. Gourou and L. Papy, Cours de Géographie, Géographie (Classe de première), *France métropolitaine*, Classiques Hachettes, Paris, 1962.
S5 *Atlas de France*, Comité National de Géographie, 1933 and various subsequent years. Map 13, Temperatures, 14 Precipitation.

[a] Area of department divided by area of smallest possible circle enclosing it, multiplied by 100.
[b] Total not given here. See S1 p. XVIII, Total. *Ouvriers* are taken to represent industrial workers.
[c] See S3 for account of derivation of data. Although the data may be considered to represent per capita income, it has been collected to represent market potential and is therefore referred to in the context of France as purchasing power.
[d] See S4 p. 11, *pourcentage de voix obtenues par les partis socialiste et communiste* (Elections législatives de 1958).
[e] The complement of rural population, from which derived.
[f] S1 and other sources. Data mainly for 1962.
[g] Distance of mean centre of area of each department from given point (Paris, mean centre of area of France, mean centre of population of France).
[h] See text for explanation.

map the top and bottom quartiles in each column of ranking for the nineteen selected variables (A–S). With ninety subdivisions the median value for any variable would in fact be found between those subdivisions ranking 45th and 46th. Since there is no actually observed value for this position the median lies between the two, splitting the list into two equal subsets, each with 45 elements. The data have been divided into four groups as follows:

Department ranking	1st –	22nd	Highest values	(22)
	23rd –	45th	Upper quartile	(23)
	46th –	67th	Lower quartile	(22)
	68th –	90th	Lowest values	(23)

The figures in parentheses indicate how many departments (or subdivisions) fall into each group.

Figure 6.5 illustrates the use of mapping by quartiles for two variables. It is these two maps that were superimposed in figure 6.2. The department names are given in table 6.6.

(*d*) *The 'Average' department* (This section concerns variable XII)
It is apparent from a consideration of the ranked data in table 6.5 that certain departments tend to have similar rankings for several variables. Thus Alpes-Maritimes ranks between 1st and 10th for some of the variables and between 81st and 90th for several others. For several of the nineteen variables (A–S), Alpes-Maritimes has an extreme value. The same trend is discernible for example also for Seine, the department containing Paris (D75), Corse (Corsica) (D20) and Creuse (D23). In the same way other departments tend to have

Table 6.2 Selected absolute data for first 15 and last 5 departments alphabetically

	A	B	C	D	E
1 Ain	5826	327	56	66	162
2 Aisne	7428	512	69	51	375
3 Allier	7382	380	51	51	350
4 Alpes (Basses)	6988	91	13	44	81
5 Alpes (Hautes)	5643	87	15	40	54
6 Alpes-Maritimes	4298	618	144	40	14
7 Ardèche	5556	248	45	45	62
8 Ardennes	5253	300	57	57	153
9 Ariège	4903	137	28	52	83
10 Aube	6026	255	42	59	282
11 Aude	6342	269	43	58	138
12 Aveyron	8771	290	33	56	301
13 Bouches-du-Rhône	5248	1248	238	36	87
14 Calvados	5693	480	84	44	129
15 Cantal	5779	172	30	39	94
	.	.	.	.	.
16–85 follow	.	.	.	.	.
	.	.	.	.	.
86 Vienne	7044	331	47	53	419
87 Vienne (Haute)	5555	332	60	71	213
88 Vosges	5903	380	64	45	97
89 Yonne	7461	269	36	51	305
90 Belfort (Territoire de)	608	109	180	40	8

A = Area of department in square kilometres.
B = Population in thousands in 1962.
C = Persons per square kilometre (B/A).
D = Compactness index.
E = Arable land in thousands of hectares.
[a] These figures show values before conversions to amount per unit of area or per so many people to eliminate the influence of differences of area and number of people in the departments.

many rankings quite close to the mid-point, or median, value (say the value ranking 45th); among these are Aube (D10), Saône-et-Loire (D71) and Tarn (D81).

Observation of this tendency leads to the supposition that if some department were to be ranked 45th or 46th for all the variables considered, it would be the most 'average' department in France, in the sense that for every variable considered it represented the median value. To obtain an assessment of how near to such a hypothetical case the departments of France actually do come, some way was sought to assign to each department an index of its deviation from the median for each variable.

It was necessary first to ignore the fact that departments when ranked might tend towards the extreme in two directions, either towards the highest

Table 6.3 Selected data processed ready to rank

	E	H	J	K	IV
1 Ain	28.0	31.4	998	105.1	570
2 Aisne	52.5	20.4	936	104.8	480
3 Allier	48.0	31.8	1047	101.6	506
4 Alpes (Basses)	11.7	27.6	1069	109.2	665
5 Alpes (Hautes)	10.1	35.1	1184	104.1	563
6 Alpes-Maritimes	3.3	8.6	1869	119.0	879
7 Ardèche	11.8	35.5	706	98.5	456
8 Ardennes	30.4	18.6	911	105.9	414
9 Ariège	17.1	38.5	740	96.4	410
10 Aube	48.5	18.9	1008	104.8	526
11 Aude	22.9	38.7	934	99.8	550
12 Aveyron	34.1	48.0	762	98.0	452
13 Bouches-du-Rhône	17.8	4.2	1609	118.4	765
14 Calvados	23.4	28.5	1096	107.9	686
15 Cantal	17.0	53.0	775	96.4	399
	•	•	•	•	•
16–85 follow	•	•	•	•	•
	•	•	•	•	•
86 Vienne	62.0	37.2	795	103.1	523
87 Vienne (Haute)	40.0	31.8	1007	101.4	555
88 Vosges	17.1	17.2	875	102.0	462
89 Yonne	41.2	30.4	923	100.9	584
Belfort (Territoire de)	13.9	5.3	1024	110.0	604

E = Arable land as a percentage of total area of department.
H = Persons employed in agriculture as a percentage of total employed population.
J = Purchasing power in dollars per inhabitant.
K = Population change 1954–62 (1962 as percentage of 1954).
IV = New cars registered in 1962 per 10,000 inhabitants.

value (ranked 1st) or the lowest (ranked 90th). The important consideration was the amount of variation from the median ranking, not the direction of deviation. To measure this, some of the original thirty variables were discarded as unsuitable for this operation, and twenty were left. Given that if a department was in fact the median department in rank for a variable it would rank 45th (actually 45.5th, but 45 was taken for ease of calculating), the deviation from this median position could be measured by finding the difference between actual ranking of the department, and 45. For the 'average' department, the difference from 45 for each of the twenty variables would be 0. For another hypothetical case ranking 90th or 1st for every variable the summed difference from 45 would equal 900. It is unlikely that any one particular department would repeatedly rank 45th, 1st, or 90th.

The index of deviation from the median ranking was calculated on this

Table 6.4 Data ranked: abbreviated list (H = high, L = low)

A *Area of department* (sq km)

H–L

Rank		
1	Gironde	10,726
2	Landes	9364
3	Dordogne	9224
4	Côte d'Or	8787
5	Aveyron	8771
45	Orne	6144
	Median	6099
46	Sèvres (Deux)	6054
86	Vaucluse	3578
87	Rhin (Haut)	3531
88	Rhône	2859
89	Belfort (Terr.)	608
90	Seine	480

Range: 10,246

B *Population of department* ('000)

H–L

Rank		
1	Seine	5646.4
2	Seine et Oise	2298.9
3	Nord	2293.1
4	Pas de Calais	1366.3
5	Bouches du Rhône	1248.4
45	Dordogne	375.5
	Median	368.7
46	Eure	361.9
86	Ariège	137.2
87	Belfort (Terr.)	109.4
88	Alpes (Basses)	91.8
89	Alpes (Hautes)	87.4
90	Lozère	81.9

Range: 5564.5

I *Percentage ouvriers*

H–L

Rank		
1	Moselle	55.1
2	Nord	52.9
3	Belfort (Terr.)	52.0
4	Pas de Calais	51.4
5	Meurthe et Moselle	51.0
45	Var	32.9
	Median	32.9
46	Vienne (Haute)	32.9
86	Cantal	17.4
87	Lot	16.8
88	Lozère	15.8
89	Creuse	15.0
90	Gers	12.2

Range: 42.9

J *Per capita income, 1961* (in U.S. dollars)

H–L

Rank		
1	Seine	2019
2	Alpes Maritimes	1869
3	Rhône	1734
4	Bouches du Rhône	1609
5	Seine et Oise	1445
45	Aude	934
	Median	933
46	Cher	932
86	Gers	652
87	Côtes du Nord	636
88	Moselle	623
89	Vendée	614
90	Corse	532

Range: 1487

K *Population growth* (*1954–1962*) (1954 as percentage of 1962)

H–L

Rank		
1	Seine et Oise	134.4
2	Moselle	120.0
3	Alpes Maritimes	119.0
4	Bouches du Rhône	118.4
5	Var	117.5
45	Marne (Haute)	104.9
	Median	104.9
46	Sarthe	104.9
86	Loire (Haute)	97.3
87	Gers	97.0
88	Cantal	96.4
89	Ariège	96.4
90	Creuse	94.1

Range: 40.3

L *Percentage left, 1958*

L–H

Rank		
1	Orne	9
2	Ille et Vilaine	12
3	Vendée	12
4	Saône (Haute)	12
5	Vosges	12
45	Indre et Loire	29
	Median	29.5
46	Ardèche	30
86	Pas de Calais	57
87	Bouches du Rhône	57
88	Creuse	59
89	Pyrenées Orientales	65
90	Vienne (Haute)	83

Range: 74

Table 6.5 First and last ten departments alphabetically, with rankings on variables A–S

	A	B	C	D	E	F	G	H	I	J	K	L	M	N	O	P	Q	R	S
1 Ain	57	51	49	9	55	30	41	36	42	39	42	13	13	36	15	39	78	71	52
2 Aisne	13	25	36	38	21	51	55	67	15	44	48	75	32	74	58	40	10	53	57
3 Allier	14	44	54	39	31	31	72	34	52	31	66	80	49	8	65	60	85	40	14
4 Alpes (Basses)	24	88	90	63	80	26	18	49	62	27	20	79	2	7	5	76	76	47	27
5 Alpes (Hautes)	63	89	89	74	82	19	15	27	68	15	52	31	5	37	19	71	88	49	44
6 Alpes-Maritimes	83	16	13	71	89	28	16	82	44	2	3	32	77	3	1	14	84	4	4
7 Andèche	65	72	63	57	79	17	33	28	43	81	80	46	31	63	72	73	34	76	17
8 Ardennes	71	56	46	11	44	18	24	69	12	54	38	55	28	71	57	6	6	33	65
9 Ariège	78	86	87	37	73	23	28	22	57	76	89	85	3	80	62	59	47	69	1
10 Aube	48	67	68	16	28	78	35	68	16	37	47	57	61	29	34	25	80	36	35
11–80 follow																			
81 Tarn	58	53	51	31	34	62	54	31	40	52	59	62	37	39	68	51	65	29	24
82 Tarn-et-Garonne	85	82	60	48	22	74	68	9	82	66	68	38	51	43	75	82	32	57	36
83 Var	49	32	26	4	85	81	2	77	45	8	5	78	74	26	2	31	58	7	38
84 Vaucluse	86	55	22	85	69	89	21	48	48	6	12	60	58	1	11	45	39	16	54
85 Vendée	22	38	45	79	16	24	88	11	78	89	64	3	87	86	63	85	89	87	51
86 Vienne	20	48	58	34	6	76	24	24	71	71	58	17	70	19	53	77	67	67	28
87 Vienne (Haute)	66	47	44	2	37	21	58	35	46	38	67	90	34	44	18	49	87	44	9
88 Vosges	53	43	38	58	74	34	4	71	6	59	63	5	22	77	85	23	21	28	73
89 Yonne	12	63	75	40	36	82	27	39	59	48	73	50	42	16	41	66	62	73	3
90 Belfort	89	87	7	70	78	12	6	87	3	34	16	24	33	55	22	37	12	9	74

Table 6.6 Key to maps of France[a]

1 Ain	31 Garonne (Haute)	61 Orne
2 Aisne	32 Gers	62 Pas-de-Calais
3 Allier	33 Gironde	63 Puy-de-Dôme
4 Alpes (Basses)	34 Hérault	64 Pyrénées (Basses)
5 Alpes (Hautes)	35 Ille-et-Vilaine	65 Pyrénées (Hautes)
6 Alpes-Maritimes	36 Indre	66 Pyrénées-Orientales
7 Ardèche	37 Indre-et-Loire	67 Rhin (Bas)
8 Ardennes	38 Isère	68 Rhin (Haut)
9 Ariège	39 Jura	69 Rhône
10 Aube	40 Landes	70 Saône (Haute)
11 Aude	41 Loir-et-Cher	71 Saône-et-Loire
12 Aveyron	42 Loire	72 Sarthe
13 Bouches-du-Rhône	43 Loire (Haute)	73 Savoie
14 Calvados	44 Loire-Atlantique	74 Savoie (Haute)
15 Cantal	45 Loiret	75 Seine
16 Charente	46 Lot	76 Seine-Maritime
17 Charente-Maritime	47 Lot-et-Garonne	77 Seine-et-Marne
18 Cher	48 Lozère	78 Seine-et-Oise
19 Corrèze	49 Maine-et-Loire	79 Sèvres (Deux)
20 Corse	50 Manche	80 Somme
21 Côte-d'Or	51 Marne	81 Tarn
22 Côtes-du-Nord	52 Marne (Haute)	82 Tarn-et-Garonne
23 Creuse	53 Mayenne	83 Var
24 Dordogne	54 Meurthe-et-Moselle	84 Vaucluse
25 Doubs	55 Meuse	85 Vendée
26 Drôme	56 Morbihan	86 Vienne
27 Eure	57 Moselle	87 Vienne (Haute)
28 Eure-et-Loir	58 Nièvre	88 Vosges
29 Finistère	59 Nord	89 Yonne
30 Gard	60 Oise	90 Belfort (Territoire de)

[a] These numbers refer to the numbers on departments in figures 6.5*a* and *b* and 6.9*a* and *b* They refer to numbers 90 and below on the maps in figure 6.6*a* and *b*, but not to the higher numbers in larger print. They do not refer to numbers on maps in figures 6.7 or 6.8.

basis using twenty variables for each department. The results are shown on two maps (figures 6.6*a* and 6.6*b*) one of which shows values below 455, the other values above 455. As expected, Seine (D75), Corse (D20), Alpes-Maritimes (D6) and Creuse (D23) all have very high indices. Charente (D16), Saône-et-Loire (D11) and Tarn (D81) have very low indices and therefore come close to the hypothetical average department. The 'most average' department turned out to be Charente (D16) with 298. The 'least average' was Seine (D75) with 730. It appears from the maps that the more average or typical departments tend to be central, while the less average are peripheral. The mean centre of area (A) and the mean centre of population (P) are marked on the maps. It is of interest to note the marked clustering of values below 455 around these two points (A and P) on figure 6.6*a*.

(*e*) *Correlations*

In addition to providing a basis for quartile maps and for an examination of the variation of each department from a hypothetical median department,

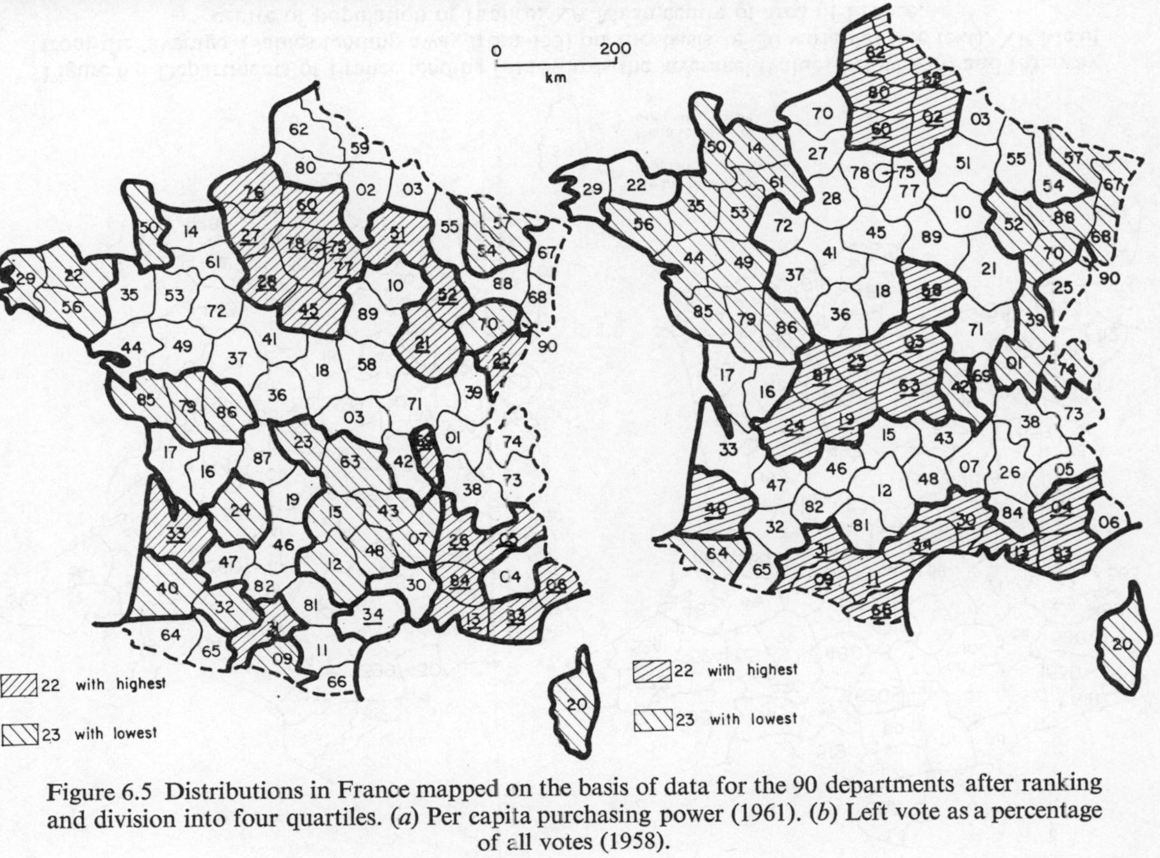

Figure 6.5 Distributions in France mapped on the basis of data for the 90 departments after ranking and division into four quartiles. (*a*) Per capita purchasing power (1961). (*b*) Left vote as a percentage of all votes (1958).

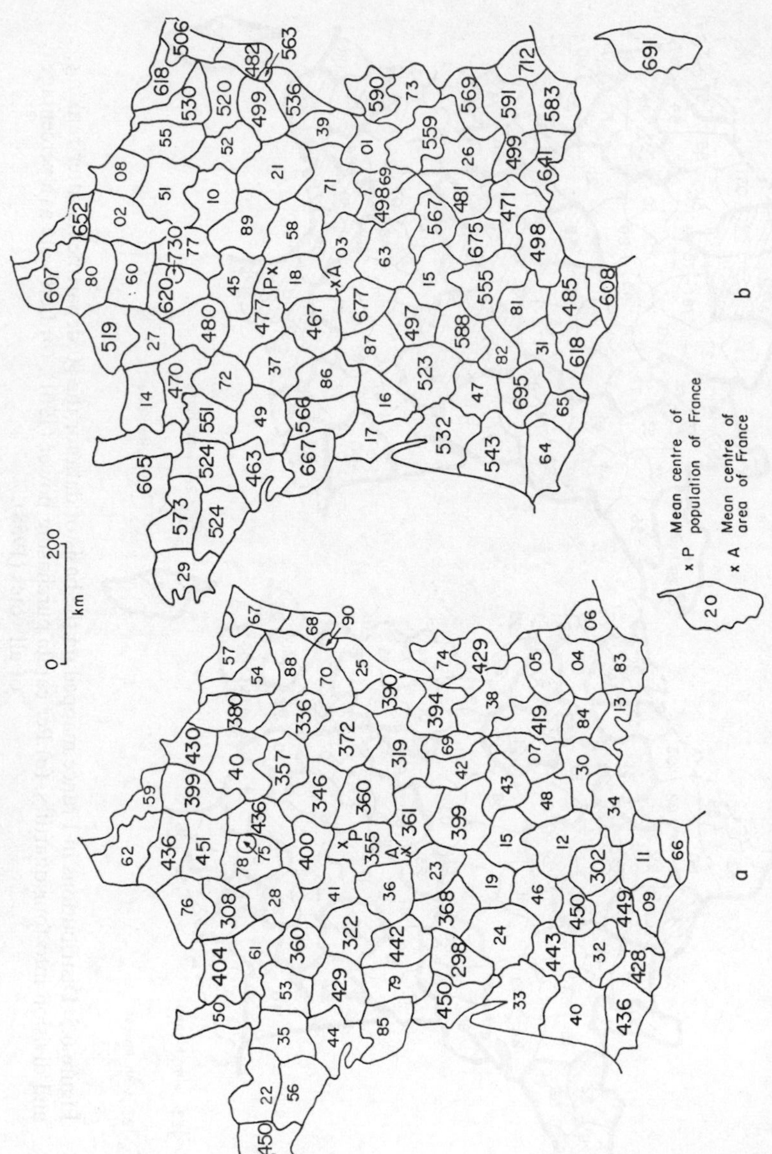

Figure 6.6 Departments of France tending (*a*) towards the 'average' (values below 455) and (*b*) away from the 'average' (values tending away from 455) on the basis of 20 variables (see text). XP Mean centre of population of France, XA Mean centre of area of France.

the ranked data in table 6.5 form the raw material from which the correlation matrix is produced. Table 6.7 is the matrix of correlation coefficients for the 19 variables A–S. There are 19^2 or 361 possible coefficients. All the coefficients on the principal diagonal are +1.00. The principal diagonal divides the matrix into two, the lower left hand side having the same coefficients as the upper right hand side (A with C, for example, gives the same as C with A). As well as being square, the matrix is therefore also symmetric (see chapter 2.8). The number of coefficients which can be different from each other is thus reduced to 171. Of these 171 coefficients only 73 appear in table 6.7 as being greater than +0.25 or less than −0.25. This value has been chosen because it is close to the 99 per cent confidence level for

Table 6.7*a* Correlation[a] matrix for 19 selected variables (A–S)

	Area	Population	Population density	Compactness	% area under crops	% area under pasture	% area under forest	% 'agriculteurs'	% 'ouvriers'	Per cap. purchasing power	'54–'62 popn. growth	% left vote, 1958	Elec. cons. per cap.	Motor vehicles per '000	Building permits per '000	Budget receipts per cap.	Infant mort. %	% Urban	% aged 65 +
	A	B	C	D	E	F	G	H	I	J	K	L	M	N	O	P	Q	R	S
A	100				33														
B		100	87				$\underline{30}$	$\underline{58}$	45	32	53				27	62		64	$\underline{53}$
C			100	$\underline{28}$				$\underline{66}$	54	39	63				37	69		76	$\underline{61}$
D				100							$\underline{31}$						$\underline{25}$	$\underline{25}$	25
E					100	$\underline{42}$	$\underline{52}$												
F						100				$\underline{29}$		$\underline{25}$		$\underline{39}$					
G							100	$\underline{31}$	32		25		35						
H								100	$\underline{90}$	$\underline{67}$	$\underline{81}$		$\underline{47}$		$\underline{49}$	$\underline{84}$		$\underline{86}$	49
I									100	49	61		56		28	74	26	69	$\underline{51}$
J										100	69		29	48	58	62		67	$\underline{25}$
K											100		34		73	72		80	$\underline{49}$
L												100							$\underline{43}$
M													100		27	39		39	
N														100	25		$\underline{52}$		
O															100	43		57	
P																100		80	$\underline{49}$
Q																	100		$\underline{34}$
R																		100	$\underline{41}$
S																			100

[a] Note that to save space, correlation indices have been rounded to two decimal places. The decimal point has then been moved two places to the right. Thus, e.g. 100 was +1.00 49 was +0.49; $\underline{81}$ was −0.81.

n = 90. But the indices are being used exclusively for descriptive purposes in this study. It has been chosen to mark the beginning of the 'strong' correlations. The fact that the majority of the coefficients, 98 in number, are not strong on this basis is an indication of the lack of conformity to be found among the French departments for the nineteen selected variables. Considerably fewer than 73 coefficients could be classed as very strong. Only 7 are further from zero than ± 0.80.

The fact that a correlation between two variables is weak may be just as interesting as the fact that it is very strong. Again, some of the high coefficients may be expected from the nature of the data and are of limited interest. Thus the highest coefficient of all in table 6.7 is −0.90, a negative one, between H (percentage '*agriculteurs*') and I (percentage '*ouvriers*'). Within the set of all people employed, if H (subset: people employed as '*agriculteurs*') is found to be large, I (the subset: people employed as '*ouvriers*') may be expected to be correspondingly low, since together these make up a considerable part, though not all, of the employed population (services are excluded), and presumably are mutually exclusive. Changes in the ranking of a department according to H and according to I are therefore not completely independent of each other. Moreover ranking according to H will tend to be the reverse of ranking according to I and a strong negative correlation coefficient must be expected. Some of the other high coefficients in table 6.7 fall into this category. The presence of such coefficients in the matrix may be

Table 6.7*b* Highest correlations

	Variables		Highest association with	
A	Area	E	% area under crops	.33
B	Population	C	Population density	.87
C	Population density	R	% urban	.76
D	Compactness index	K	'54–'62 population growth	−.31
E	% area under crops	G	% under forest	−.52
F	% area under pasture	E	% area under crops	−.42
G	% area under forest	E	% area under crops	−.52
H	%'*agriculteurs*'	I	%'*ouvriers*'	−.90
I	%'*ouvriers*'	H	%'*agriculteurs*'	−.90
J	Per capita income	K	'54–'62 population growth	.69
K	'54–'62 population growth	H	%'*agriculteurs*'	−.81
L	% left vote, 1958	S	% aged 65+	−.43
M	Electricity consumption per capita	I	%'*ouvriers*'	.56
N	Motor vehicles per '000 inhabs.	Q	Infant mortality (*per mil*)	−.52
O	Building permits per '000 inhabs.	K	'54–'62 population growth	.73
P	Budget receipts per capita	H	%'*agriculteurs*'	−.84
Q	Infant mortality %	N	Motor vehicles	−.52
R	%urban	H	%'*agriculteurs*'	−.86
S	% aged 65+	C	Population density	−.61

Note that some nearly 'reciprocal' variables have high negative correlations (e.g. *ouvriers*, *agriculteurs;* crops, forest).

explained by the 'blanket' nature of the computer program. Care and common-sense are needed to interpret the matrix.

The list in table 6.7*b* shows the coefficient of correlation between each variable in matrix 6.7 and its most closely associated variable. A bar under the number indicates negative correlation:

(*f*) *Exploiting the difference value, d*

When the 90 departments are ranked in the order 1 to 90 according to their position on the ordinal scale, the result is two columns of numbers for two variables as in table 6.8 where 20 out of 90 departments are shown, under column J and N. Column J is the ranking for purchasing power, N the column for motor vehicles in circulation. Column (JN) (*d*) is the difference between the ranking in columns J and N for each department. Thus Ain is ranked 39th for income per inhabitant, 36th for motor vehicles in circulation per 1000 inhabitants. *d* is only 3. The difference for Aisne, between 44 and 74, is much greater. The presence of a limited number of departments (e.g. D86 Vienne) with a large difference in ranking (*d*) is preventing the correlation between J

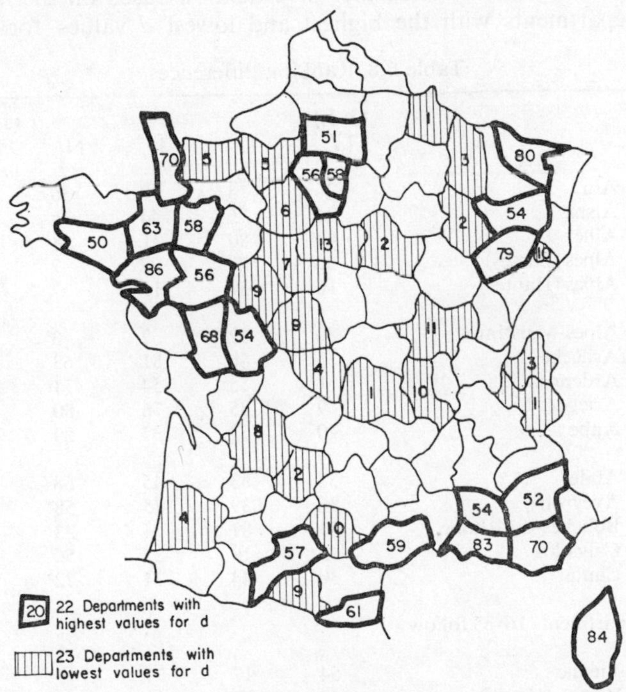

Figure 6.7 Departments of France showing similar ranking (lowest of values) and very different ranking (highest of values) in a comparison of two variables, income and left vote.

and N from being considerably higher than about +0.50. The value of *d*, therefore, is of interest in at least two ways.

(1) It draws attention to departments that are conforming least to a correlation that is generally strong positive (or negative). It is conceivable, of course, that all the departments might have a value of *d* roughly similar, in which case none would be particularly responsible for upsetting the close positive correlation.

(2) It is possible to map those departments that have a value of *d* that is exceptionally high (or exceptionally low). It is then possible to see if the departments conforming least or most are clustered in a particular part of the national area. These are similar to residuals derived from product-moment correlation and comparable with residuals in trend surface analysis (see chapter 9). The biggest anomalies have been mapped in figure 6.7 using the difference in ranking (*d*) between variable J, per inhabitant, and L, left vote. Thus Corse is 90th in income but 6th for left vote, the proportion of left vote being ranked low to high. In other words, although it is poor by French standards, the left vote is very small, and *d* is very large (84). Figure 6.7 points to the possibility of further investigation based on the fact that the departments with the highest and lowest *d* values for purchasing

Table 6.8 Ranking differences

	(J–L) (*d*)	L	J	N	(J–N) (*d*)
1 Ain	26	13	39	36	3
2 Aisne	31	75	44	74	30
3 Allier	49	80	31	8	23
4 Alpes (Basses)	52	79	27	7	20
5 Alpes (Hautes)	16	31	15	37	22
6 Alpes-Maritimes	30	32	2	3	1
7 Ardèche	35	46	81	63	18
8 Ardennes	1	55	54	71	17
9 Ariège	9	85	76	80	4
10 Aube	20	57	37	29	8
11 Aude	38	83	45	68	23
12 Aveyron	36	39	75	58	17
13 Bouches-du-Rhône	83	87	4	23	19
14 Calvados	5	19	24	56	32
15 Cantal	30	44	74	72	2
Departments 16–85 follow	. .	. .	. .	. .	. .
86 Vienne	54	17	71	19	52
87 Vienne (Haute)	52	90	38	44	6
88 Vosges	54	5	59	77	18
89 Yonne	2	50	48	16	32
90 Belfort (Territoire de)	10	24	34	55	21

power and left vote are clustered in certain parts of the country. In northwest France departments tend to be poor yet have a low left vote, while several by the Mediterranean tend to be relatively prosperous yet have a high left vote.

(*g*) *Comparisons of the 90 departments*

One of the most interesting steps in the rank correlation exercise on France will now be described in some detail. The table below shows five selected departments with their ranks on 19 variables (A–S). A careful look along the row will show that each department has a rank (number) in each column similar (not identical) to that of certain departments but different from that of certain other departments. Some thought is needed to see why this is so. A comparison of the rows for the five selected departments shows what is involved:

(D1 Ain, D4 Alpes Basses, D5 Alpes Hautes, D20 Corse, D75 Seine (Paris))

Variables as from table 6.5

	A	B	C	D	E	F	G	H	I	J	K	L	M	N	O	P	Q	R	S
D1	57	51	49	9	55	30	41	36	42	39	42	13	13	36	15	39	78	71	52
D4	24	88	90	63	80	26	18	49	62	27	20	79	2	7	5	76	76	47	27
D5	63	89	89	74	82	19	15	27	68	15	52	31	5	37	19	71	88	49	44
D20	6	61	81	87	90	6	43	37	73	90	15	6	89	90	6	90	7	56	45
D75	90	1	1	5	88	90	90	90	37	1	29	59	44	2	73	1	86	1	70

From this example it is clear that for most variables, Alpes Basses (D4) and Alpes Hautes (D5) are ranked in a fairly similar position. Thus for density of population (C) they are 90th and 89th respectively, for electricity consumption per inhabitant (M) they are 2nd and 5th respectively. In other words, because they are similar for many things, they tend to have roughly the same ranking among all the 90 departments. In contrast, Corse and Seine tend to have very dissimilar rankings. Where Seine (D75) is high, as for density of population (C), Corse (D20) is low; Seine comes 1st, Corse 81st. For purchasing power per inhabitant (J), Seine is 1st, Corse 90th.

It is possible to re-rank the values along the rows and to obtain a new coefficient of correlation. The variables are ranked according to the values for each department in turn. The departments take the place of the variables. It is possible in this way to obtain a new matrix of correlation coefficients with 8,100 indices, 90×90. In the new matrix it is possible to see how closely each pair of departments is 'correlated' on the verdict of the 19 variables (or any other set of variables chosen).

For this exercise, 22 variables were chosen out of the 31 noted at the beginning of this section. Distance variables were dropped because the indiscriminate inclusion of these could make departments appear more similar to nearby ones than to distant ones. The following 22 variables were used (refer back to table 6.1):

A, B, D, III, E–M, VI, N–S, X, XI

Density of population was excluded because it is a product of A and B, area and population. Three physical variables, altitude (proportion lowland), temperature and precipitation were added, and also VI (milk production). It is regrettable that socio-economic variables are more numerous than physical ones. This shortcoming could be overcome either by including more physical variables or by weighting the existing ones. In fact it is possible to debate indefinitely over which is the best set of variables to use in an exercise of this kind. At all events, the 22 chosen are sensible enough, and representative enough to differentiate the departments of France and give a fuller picture of what is going on in France than the usual differentiation based largely on single features such as geological outcrops or types of agriculture.

Table 6.9 shows how the re-ranking has been carried out. The old variables are down the left hand side, while the departments (five selected departments in this example) are along the top. The old variables have been ranked in turn for each department (from high numbers to low numbers). Thus Ain variable Q, with old ranking 78, is ranked 1st in the new ranking, variable R with old ranking 71 is ranked 2nd and so on. It can be seen in the new rankings that Alpes Basses and Alpes Hautes are fairly similar, whereas Corse and Seine differ considerably. Only one further step is required now to obtain the correlation coefficients for each pair of departments. The difference in ranking

Table 6.9 Old ranking and new ranking

	Old ranking					New ranking				
Old variables	Ain	Alpes Basses	Alpes Hautes	Corse	Seine	Ain	Alpes Basses	Alpes Hautes	Corse	Seine
A	57	24	63	6	90	5	16	10	19	4
B	51	88	89	61	1	8	1	1	10	21
D	9	63	74	87	5	21	8	7	7	17
III	68	86	89	81	13	3	2	2	8	16
E	55	80	82	90	88	6	4	5	3	5
F	30	26	19	6	90	16	15	19	19	3
G	41	18	15	43	90	11	18	20	13	2
H	36	49	27	37	90	14	11	16	14	1
I	42	62	68	73	37	9	9	9	9	14
J	39	27	15	90	1	12	13	21	3	22
K	42	20	52	15	29	10	17	11	16	15
L	13	79	31	6	59	18	5	15	21	11
M	13	2	5	89	44	19	22	22	6	12
VI	12	86	79	90	87	20	3	6	3	6
N	36	7	37	90	2	15	20	14	3	18
O	15	5	19	6	73	17	21	18	21	9
P	39	76	71	90	1	13	6	8	3	19
Q	78	76	88	7	86	1	7	3	17	7
R	71	47	49	56	1	2	12	12	11	20
S	52	27	44	45	70	7	14	13	12	10
X	8	13	21	34	82	22	19	17	15	8
XI	67	58	88	1	41	4	10	4	22	13

between Ain and Alpes Basses, Ain and Alpes Hautes, and between every other department, is found, squared, and the Spearman rank correlation formula applied. To find the correlation coefficient for each pair of departments requires about 4,000 repetitions of this procedure (rather less than half of 90 × 90). It is unlikely that a geographer would undertake all this work manually, because no epoch-making results are expected. On a manually operated calculating machine, with checking, the work might take one person a year, full time.

Space does not permit the reproduction here of any of the correlation matrix (90 × 90) worked out by the computer, but some of the results are discussed and mapped. Just as in the matrix in table 6.7 each variable correlates with itself 1.000, so here each department correlates with itself 1.000. This forms a base against which the similarity of each other department can be measured in relation to the one in question. For example, Corse correlates with the other four departments in table 6.9 as follows:

			Short form
With	Ain	−0.142	14
	Alpes Basses	+0.235	24
	Alpes Hautes	+0.104	10
	Seine	−0.430	43

Thus it is least like Seine. Alpes Basses correlates with Alpes Hautes +0.775, a strong positive correlation, indicating that the two departments are very similar.

Two technical points should be noted:

(1) The correlation coefficient is regarded here *strictly* as an indicator of similarity or lack of similarity, ranging over a scale extending from 1.000 to −1.000. The concepts of significance and of causal relationships do not enter. A correlation around 0.000 simply implies an average difference, as opposed to say +0.500 implying considerable similarity and −0.500 implying considerable dissimilarity. The similarity and dissimilarity are of course relative, being within the context of France. Presumably if some provinces of South Italy had been included as tracers they would have come out greatly different from all the French departments, but similar to one another.

(2) The relationship of each department to all the others occupies one dimension. There are therefore 90 dimensions involved in the new matrix. These are mathematical, not spatial, though there is an analogy with the space dimensions. The concept of more than three dimensions is difficult to grasp but is essential for an appreciation of factor analysis, to be discussed in the following sections of this chapter. The following will illustrate what is meant.

The department Alpes Basses (D4) correlates with Alpes Hautes (D5) +0.775, but with Alpes Maritimes (D6) only +0.435. The correlation between Alpes Hautes and Alpes Maritimes is +0.388. Thus it is more similar to D5 than to D6, differing from D5 only by 0.225, but from D6 by 0.565 which is 1.000–0.435. Now it might be thought that D5 and D6 would differ between themselves by the difference between +0.775 and +0.435, or by 0.340. But this is not so because the value for Hautes Alpes must be taken as 1.000 when comparing it with Alpes Maritimes. Thus when Hautes Alpes is compared with Alpes Maritimes the value is 1.000–0.388, which is a difference of 0.612 and not 0.340.

The fact that any given department has a numerical index of degree of similarity to all other 89, ranging between +1.00 and −1.00, means that it is possible to map the values and find the distribution of departments similar to and not similar to that department (called for convenience the *host* department). Three departments have been selected for mapping, Finistère (D29) in Brittany, Cher (D18) in the central part of France and Savoie (D73) in the French Alps. All 90 could be mapped in a similar way.

The coefficients have been abbreviated on the maps to two decimal places. The negative coefficients are those underlined. Cut-off points at +0.50 and −0.50 have been fixed as indicative of quite a strong correlation. Those departments correlating with each of the selected departments further from 0.00 than 0.50 have been heavily outlined in figures 6.8*a–c*.

In theory one might expect the conditions prevalent in one place to change in some relationship to distance as the observer shifted from that place to

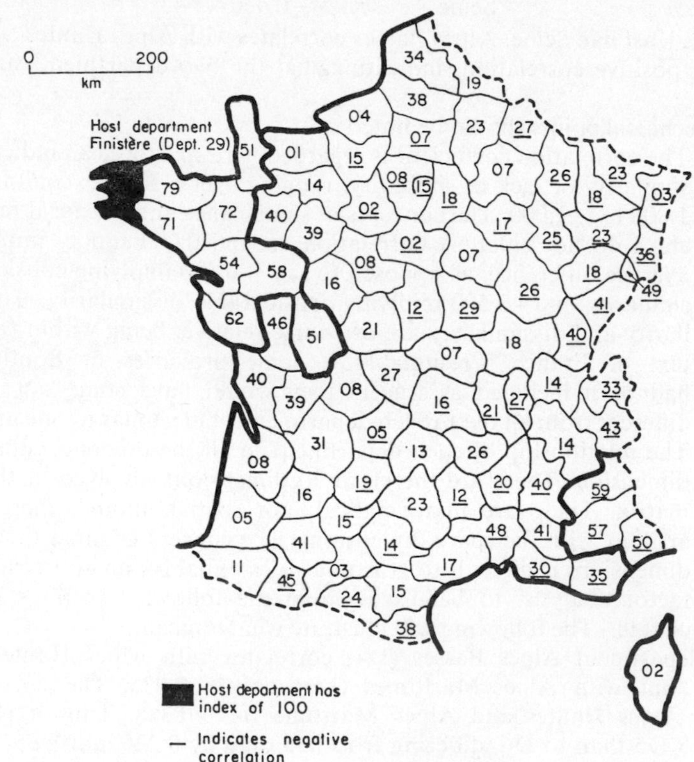

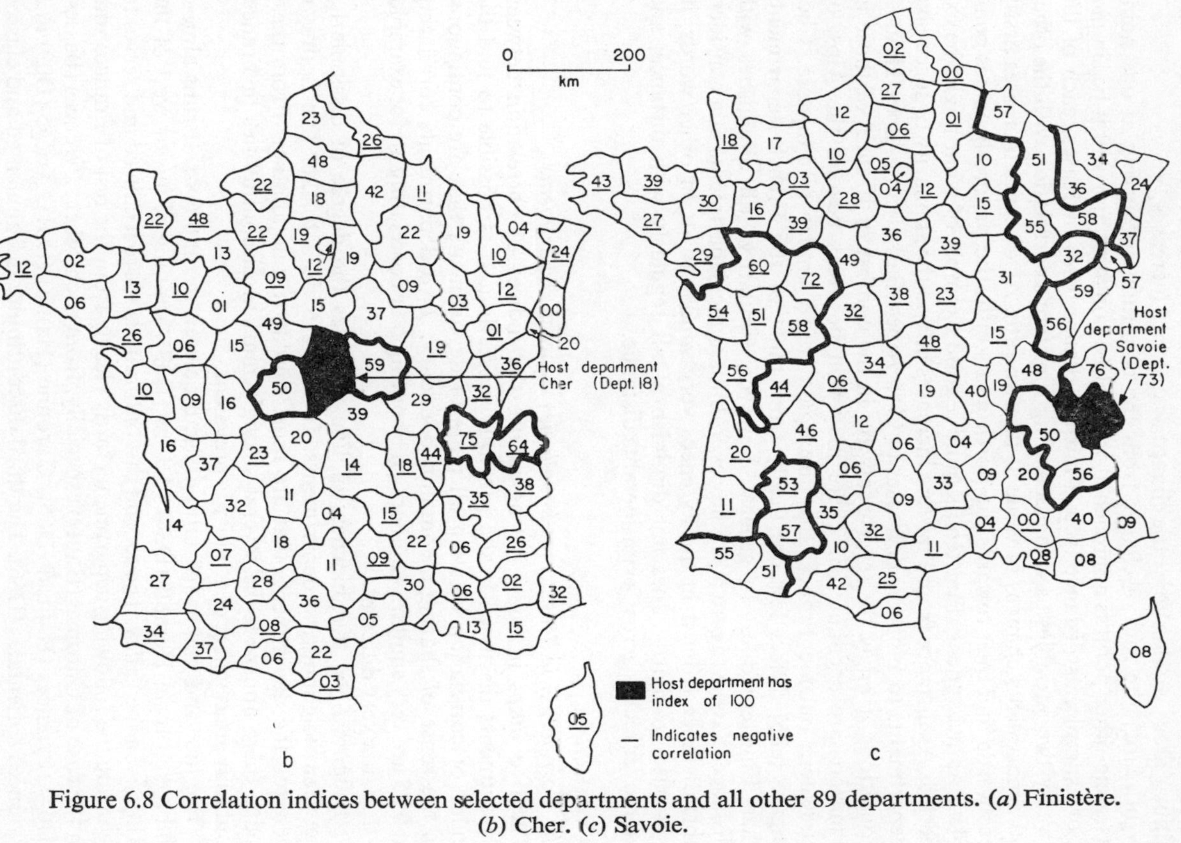

Figure 6.8 Correlation indices between selected departments and all other 89 departments. (*a*) Finistère. (*b*) Cher. (*c*) Savoie.

some other. Movement away from some strongly characterized place to another less highly distinctive place might be expected to show a rapid rate of change as distance increased. On the other hand differences between very similar areas would show little change as distance increases.

In reality the rate of change from the conditions occurring at one point as some other point is approached will not be constant, nor will it be simple. For example, as distance from Finistère (figure 6.8*a*) increases, each of the twenty-two variables used in calculating Finistère's correlation with the other 90 departments changes in value at its own rate. As distance increases firstly eastwards and then towards the north and south, the picture becomes complex. As it happens, Finistère is quite a good example of what might be expected. As distance from Finistère increases inland the similarity of successive departments to Finistère decreases until in Alpes (Basses) and Alpes (Hautes) conditions strongly the reverse (always in the context of France) of those in Finistère are encountered. There may well be areas beyond the Alps, or elsewhere outside France, that are again more similar to Finistère. Cher (figure 6.8*b*) differs less drastically from most of the French departments than Finistère does. In general, however, similarity still diminishes with distance. Savoie (figure 6.8*c*), however, finds some departments at a considerable distance from it in the Pyrenees very similar to it. In other words, in this case, similarity does not diminish at all regularly with distance, and moreover recurs across a considerable distance.

(*h*) *Correlation indices across boundaries between departments*

The two maps in figure 6.9 again use the indices of correlation between departments discussed in the previous subsection. It is possible to find the index of correlation between all pairs of departments that are contiguous, in the sense of sharing a common boundary. Two departments correlating at 0.85 are very similar. Another neighbouring pair correlating at, for example, – 0.25, are very different.

In figure 6.9*a*, the thickness of line indicates the degree of dissimilarity between neighbouring departments. In figure 6.9*b*, a thicker line is drawn between the departments that are most similar to one another. From these maps there appear to be several strongly marked discontinuities in France, and four areas where continuity is prevalent.

The first and major discontinuity can be seen in map 6.9*a*. It runs almost unbroken from Hérault (D34) to Jura (D39), running along or west of the Rhône–Saône valley. The second discontinuity is more localized and serves to separate the following departments fairly clearly from the rest of France, and, in the case of Garonne (Haute) especially, from each other: Pyrenées (Basses) (D64), Pyrenées (Hautes) (D65), Garonne (Haute) (D31), Ariège (D09) and Pyrenées–Orientales (D66). The third discontinuity is fragmented and classed as one only because it occurs broadly in western France. The major portion separates Orne (D61), Sarthe (D72) and Maine-et-Loire (D49) from Eure-et-Loir (D28) and the departments surrounding Paris. The second portion forms

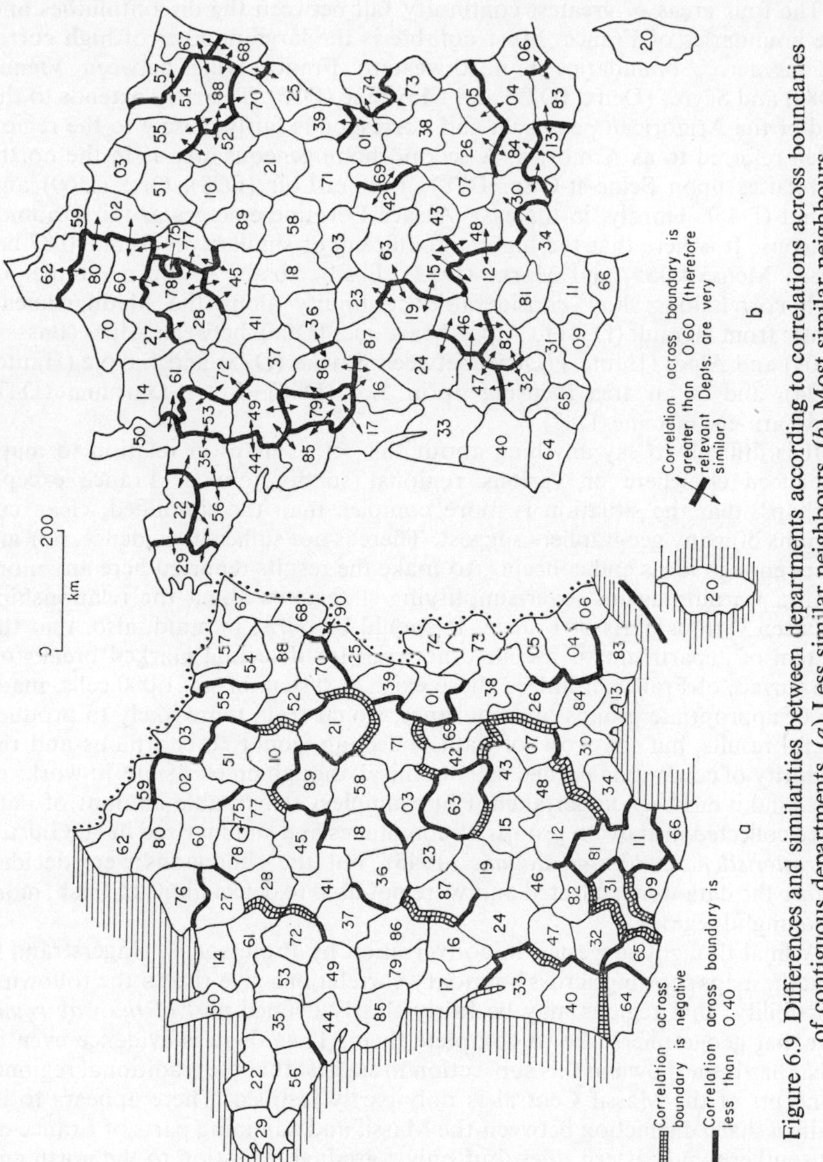

Figure 6.9 Differences and similarities between departments according to correlations across boundaries of contiguous departments. (*a*) Less similar neighbours. (*b*) Most similar neighbours.

the western boundary of Vienne (Haute) (D87) and the third portion differentiates Gironde (D33) from Lot-et-Garonne (D47).

The four areas of greatest continuity fall between the discontinuities and the boundaries of France. Most notable is the large number of high correlations across boundaries in northwestern France lying between Vienne (D86) and Sèvres (Deux) (D79) and Mayenne (D53). This area extends to the end of the Armorican peninsula and corresponds fairly closely to the region often referred to as Armorica. A second homogeneous area is in the north. It focuses upon Seine-et-Oise (D78), Eure-et-Loir (D28), Oise (D60) and Loiret (D45). Thirdly, in the east, Alsace-Lorraine emerges as fairly homogeneous. It is here that the highest coefficient of similarity is found, 0.87 between Meuse (D55) and Marne (D52). Lastly, to the south, a number of high correlations show considerable continuity along the Mediterranean coast: from Hérault (D34) to Alpes-Maritimes (D06), between Alpes (Basses) (D04) and Alpes (Hautes) (D05), between Savoie (D73) and Savoie (Haute) (D74), and in an area focusing upon Lot (D46), Lot-et-Garonne (D47) and Tarn-et-Garonne (D82).

It is difficult to say anything about this set of maps in relation to maps produced elsewhere or various regional subdivisions of France except, perhaps, that the situation is more complex than the simplified, clear cut regions of many geographers suggest. There is not sufficient evidence, nor are there enough ideas and concepts, to make the results reported here any more than a warning against over-simplifying statements about the relationships between various parts of France. It should be borne in mind, also, that the system of departments is a course network for detecting marked breaks on the surface of France, if indeed these exist. A system of say 1,000 cells, made up of appropriate groups of communes, would seem more likely to produce useful results, but the work of data collecting would be enormous and the quantity of computing needed astronomical, though on occasions in works of this kind it has been undertaken. For example a remarkable amount of data were collected in Italy on groups of communes and summarized in U. Giusti's *Caratteristiche ambientali italiane* (1943), but the subregions were decided *before* the data were collected and were not used to detect limits of fresh, more meaningful regions.

A final thought suggested in conversation by Professor T. Hagerstrand is that the maps showing across boundary correlations give rise to the following possibility: that regions may be fairly clearly defined *part of the way round* but that geographers like to complete them. Thus there is evidence even in what has been shown in this subsection to suggest that the traditional regional concepts of the Massif Central is only partly justified. There appears to be quite a sharp distinction between the Massif and adjoining parts of France on its southern and eastern sides, but only a gradual transition to the north and west. In the interests of simplicity or tidiness geographers seal off their regions into discrete compartments.

6.3 The product-moment correlation matrix

(*a*) *Introduction*

In section 6.2 it was shown how a convenient way of expressing indices of correlation between many pairs of variables was in the form of a square, symmetric matrix with the number of rows and columns equal to the number of variables. For the purposes of this section and the following section, material has been drawn from the *Statistical Abstract of the United States*, 1965 and 1966, to provide an example of a matrix of indices of correlation (Table 6.12*a*) produced from interval/ratio data, rather than ranked data as in the case of France. The exercise deliberately contains material that will be familiar to many geographers. Many of the findings are obvious, but for this very reason it should be easier for the reader to appreciate the various procedures in the application of factor analysis. Some care was given to the selection of 25 sets of information (variables) about the 50 states of the U.S.A. The district of Columbia was excluded because the inconvenience of including it seemed greater than any loss from its exclusion.

The 50 states of the U.S.A. do not make a good set of compartments for data collecting since they vary greatly both in area and in population. Virtually all the variables considered relate *either* to area *or* to population (density of population relates to both). In order to eliminate the direct influence of size it has been necessary to express data in terms of per unit area or per hundred (per cent) or per thousand (per mil) of total population. If this is not done, a very large amount of the total variation is likely to come from variations in the size of the states. Some indirect influence of size still seems likely to enter. Thus, for example, only states with over (say) 1 million inhabitants can include *large* urban centres. Large urban centres have certain features that smaller urban centres and rural areas cannot possess.

Since so many data are available for each of the states, they form a convenient set of compartments. They are also of interest in their own right as compartments, since they are considered as entities in the U.S.A. Under the federal system of the country they do in fact operate as entities to a greater extent than do the major civil divisions of countries with unitary governments, including France, discussed in the last section. A more thorough and searching investigation of the U.S.A. could however be carried out on the basis of the 3,000 counties (3,043 in 1962), on the basis of satisfactory groupings of these to form more and better shaped units for data collecting than the 50 states or on a suitable sample of the counties. The reader who feels uneasy about the material in this and the next section on account of the variations in size of the states should not condemn the techniques used *in general* for this.

(*b*) *The variables and their sources*

The first three variables (A–C) represent familiar physical features. Sharp variations in these features may often be so local, however, that the mean values given for the states are only very broadly representative. The physical variables do however correlate highly with some other variables. The next eight variables (D–K) are related to population. They reflect familiar features again. Variables L–V may broadly be thought of as economic and/or social. The last three are electoral, and represent regional variations in the expression of political opinion. Probably at one time or another all the variables used in this section have been mapped in some form or other for the U.S.A. One drawback of the data is that they are not all for the same year. However in most cases indices have not changed substantially over the last few years.

Imagine the 25 variables each mapped appropriately in turn as a separate map of the U.S.A. on transparent paper. They should be mapped in a standard form so that not only could each map be looked at individually, but maps could also be compared two at a time, three at a time and so on, as desired. Distributions that showed similar trends over the surface of the maps would be seen to correlate. An imaginary 'cabinet' for doing this is illustrated in chapter 15. There are several reasons why such a visual exercise runs into difficulties. Firstly, to view all possible combinations of the 25 maps, it would be necessary to consider 2^{25} or about 10^8 or 100,000,000 combinations. Allowing 10 seconds to view each combination, this would take a person working normal hours about 200 years. Secondly, any combination of more than three or four variables viewed together becomes very complex to handle visually. Thirdly, there seems to be no assurance that individual viewers will not see correlations they expect to see rather than correlations that exist.

Factor analysis already introduced in chapter 3 is a method by which the 25 maps described above may be combined. It gives numerical indices to complex relationships. But the numerical indices themselves must be interpreted by the geographer. The 25 variables are described in turn in table 6.10. Some are expressed in per thousand (per mil) or in other less familiar ways than per cent, in order to preserve a reasonable amount of precision without needing decimal places in the data. This has been done merely for convenience, and does not affect the raw data of the variables as far as the product-moment correlation is concerned.

The 25 variables are given in table 6.11 against an alphabetical list of the 50 states. For convenience, each state has been referred to by a three letter abbreviation.

(c) *The correlation matrix*

Table 6.12 is a matrix with the indices of correlation of each variable with every other (including itself). A corner of the original matrix is shown in figure 6.12*b*.

Altogether in the matrix, therefore, there are 300 unique indices of correlation. An index further from 0 than about ± 0.30 may be considered fairly strong, one further from 0 than ± 0.50 may be considered very strong.

It must be appreciated that some of the correlations could be expected to come out fairly high, For example, it could be expected that states with a high density of population (variable D) would be highly urbanized (variable F)

Table 6.10 Description of the 25 variables[a]

Reference letter	Page in source	Date if relevant	Topic	Expressed as
A	171		Approximate mean altitude of state above sea level	Feet above sea level to nearest 10
B	179	1931–60	Mean annual temperature	In degrees Fahrenheit
C	184	1931–60	Mean annual precipitation	In inches
D	13	1960	Population	Persons per square mile
E	26	1960	Negro	As per mil of total population
F	16	1960	Urban	Urban population as percentage of population total
G	48	1954	Birthrate	Births per 10,000 population
H	22	1960	Age of population	Median age in years × 10 to avoid decimal
I	11	1960–64	Population change	Average annual during 1960-64 per unit
J	34	1950–60	Migration index (white only)	Per mil of total white population
K	11, 16	1960	Rural farm population	As per mil of total population
L	225[b]	1965	Employment in manufacturing	As percentage of total population
M	571[b]	1965	Automobiles in circulation	Per mil inhabitants
N	517[b]	1965	Telephones	In use per mil inhabitants
O	334[b]	1964	Personal income	In dollars per capita
P	427[b]	1964	Federal revenue	As per mil of all state and local government revenue
Q	155	1963	Lawyers	Persons per lawyer
R	113	1960	School years completed	Median (for male only) years × 10
S	124	1964	Expenditure on education	In dollars per pupil per year
T	69	1963	Physicians	Per 100,000 people
U	61	1963	Infant mortality	(White only) per 10,000 people
V	760	1960	Housing quality	Percentage of houses with all plumbing facilities sound
W	375	1960	Republican Party Votes	Per mil of all votes cast in 1960 Presidential election (Nixon-Kennedy)
X	375	1964	Republican Party Votes	Per mil of all votes cast in 1964 Presidential election (Goldwater-Johnson)
Y	382	1962 or 1964	Republican Party Votes	Per mil of all votes cast in 1962/4 state governor elections

[a] Source of the 25 variables: the *United States Statistical Abstract*, *1965* or *1966* Washington D.C., 1965. (Obtainable from the U.S. Bureau of the Census.)

[b] *1966 Statistical Abstract*. per mil = per thousand.

Table 6.11 Raw data: the 25 variables for 50 states (see Table 6.10 for description)

	Altitude	Temperature	Precipitation	Population	Negro	Urban	Birthrate	Age	Population change	Migration	Rural farm	Manufacturing	Automobiles	Telephones	Income	Federal revenue	Lawyers	School years	Expenditure on education	Physicians	Infant mortality	Housing	1960 Republican party	1964 Republican party	1962 or 1964 Republican party
	A	B	C	D	E	F	G	H	I	J	K	L	M	N	O	P	Q	R	S	T	U	V	W	X	Y
01ALA	50	68	68	64	300	55	224	260	10	−69	116	312	390	328	1910	222	1181	89	277	79	245	54	418	695	37
02ALS	190	40	55	0	30	38	281	233	24	455	8	86	285	234	3375	355	973	121	683	69	248	58	509	341	477
03ARZ	410	69	7	12	33	75	229	257	46	519	34	160	395	374	2310	166	816	110	478	131	233	76	555	504	468
04ARK	65	62	49	34	218	43	216	290	19	−191	169	293	328	297	1781	239	1208	87	317	91	234	48	431	434	430
05CAL	290	60	18	100	56	86	207	300	33	282	19	244	450	543	3196	144	694	120	565	178	215	87	501	408	483
06COL	680	50	15	17	23	74	217	279	27	115	65	152	446	489	2706	169	526	119	470	169	256	76	546	382	567
07CON	50	50	43	518	42	78	202	318	21	100	9	424	441	549	3390	103	607	108	600	180	210	85	463	321	468
08DEL	6	54	45	226	136	66	233	287	22	210	44	370	402	533	3335	147	847	108	539	135	203	80	490	388	486
09FLA	10	72	56	91	178	74	198	312	33	700	19	154	454	450	2420	113	664	106	412	142	226	78	515	489	439
10GEO	60	62	47	68	285	55	237	259	20	−4	93	320	370	363	2156	190	805	88	330	102	230	58	374	541	0
11HAW	200	76	22	99	8	77	247	243	24	480	15	115	381	428	2906	226	1166	110	418	132	207	70	500	212	417
12IDA	500	51	11	8	2	48	203	260	9	−70	192	189	429	375	2338	208	1007	112	341	92	224	73	538	491	546
13ILL	60	51	34	180	103	81	209	312	9	−8	53	337	365	525	3245	119	529	104	551	140	209	78	498	405	481
14IND	70	52	39	129	58	62	221	289	8	5	99	412	403	441	2827	109	921	105	490	106	219	73	550	436	438
15IOW	110	50	30	49	9	53	204	303	0	−91	240	254	447	465	2595	117	769	104	464	117	204	70	567	379	320
16KAN	200	57	28	27	42	61	190	299	5	−27	143	202	445	453	2692	113	748	112	487	116	213	72	604	451	509
17KTY	75	56	41	76	71	45	222	276	9	−137	172	270	375	319	2043	212	918	85	324	95	263	53	536	357	493
18LSA	10	70	63	72	319	63	247	253	15	24	66	174	323	364	2061	206	778	86	418	114	206	61	286	568	393
19MNE	60	45	43	31	3	51	218	295	5	−75	48	368	342	375	2245	167	993	105	371	125	241	61	570	312	501
20MLD	35	58	44	314	167	73	204	287	24	145	31	250	368	478	3014	114	573	102	503	158	216	81	464	345	443

21MAS	50	51	43	655	22	84	242	321	8	−26	7	328	350	512	3023	110	482	113	528	196	202	81	396	234	503
22MIC	90	46	31	137	92	73	215	283	8	5	54	411	424	471	3009	117	845	104	510	141	215	79	488	331	441
23MIN	120	41	26	43	7	62	218	286	7	−34	165	241	433	467	2625	121	710	99	534	145	206	70	492	360	503
24MIP	30	65	49	46	420	38	243	242	14	−93	234	314	290	247	1566	213	989	86	273	76	229	45	247	871	381
25MIO	80	56	35	63	90	67	211	316	5	−43	120	283	366	467	2628	171	584	93	449	149	214	66	497	360	379
26MON	340	45	14	5	2	50	210	276	10	−40	150	122	409	397	2409	229	736	107	570	102	223	68	511	406	513
27NEB	260	52	28	18	21	54	206	302	11	−93	209	166	435	464	2573	162	599	109	407	115	206	72	621	474	400
28NEV	550	48	7	3	47	70	248	295	84	532	23	45	450	494	3289	181	783	120	543	107	283	81	488	414	332
29NHT	100	46	39	67	3	58	201	310	17	21	28	410	411	421	2570	154	949	105	440	143	226	73	534	361	332
30NJS	25	54	42	807	85	89	191	324	23	103	8	369	390	527	3242	94	669	106	607	143	219	85	492	339	496
31NMX	570	57	8	8	18	66	268	228	14	85	57	65	377	364	2227	248	951	110	475	94	282	70	494	404	398
32NYK	100	50	38	350	84	85	196	331	15	−5	18	282	292	560	3242	77	358	106	790	207	212	81	473	313	531
33NCA	70	60	44	93	245	40	219	255	15	−40	164	415	346	310	2028	146	1230	85	322	100	222	57	479	438	434
34NDK	190	42	15	9	1	35	222	262	5	−169	313	55	399	388	2304	184	842	89	424	90	238	61	554	419	443
35OHO	85	53	37	237	81	73	210	295	9	37	51	392	429	467	2816	122	660	106	465	136	216	78	533	371	589
36OKL	130	60	31	34	66	63	188	300	13	−95	104	159	420	433	2236	226	604	101	366	113	223	66	590	443	553
37ORG	330	55	42	18	10	62	179	308	13	7	73	259	477	458	2794	194	682	112	569	150	214	77	526	360	542
38PNA	110	52	39	252	75	72	192	320	3	−56	31	379	375	504	2728	125	941	100	479	153	214	79	487	347	554
39RIS	20	50	42	812	21	86	202	319	15	−36	4	384	405	447	2817	161	802	100	514	152	225	79	364	191	611
40SCA	35	64	47	79	348	41	223	234	16	−3	138	428	352	284	1838	164	1286	84	284	80	224	54	488	589	0
41SDK	220	46	25	9	2	39	219	277	12	−143	293	86	397	373	2055	192	938	90	444	73	227	62	582	444	517
42TEN	90	61	47	85	165	52	216	280	15	−78	153	349	348	353	1992	216	844	86	300	113	235	57	529	445	492
43TEX	170	67	29	37	124	75	219	270	19	21	66	196	416	408	2346	149	693	101	396	111	255	69	485	365	262
44UTA	610	51	14	11	5	75	241	229	25	14	43	163	409	440	2340	238	865	122	407	127	182	82	548	453	430
45VRT	100	43	33	42	1	39	201	293	11	−101	123	328	343	384	2340	237	836	100	522	172	247	68	586	337	351
46VGA	95	59	44	100	206	56	210	271	23	33	89	266	338	377	2392	164	864	92	380	108	241	66	524	462	362
47WAS	170	50	25	43	17	68	191	296	11	30	55	254	438	475	2864	149	804	118	534	143	209	78	507	374	442
48WVA	150	56	44	77	48	38	204	285	−8	−215	67	273	306	335	2007	178	1031	87	327	103	256	57	473	321	451
49WIS	105	45	30	72	19	64	216	294	9	−24	134	369	369	437	2682	107	675	98	543	117	218	75	518	377	506
50WYO	670	46	15	3	7	57	211	273	9	−65	126	72	435	456	2479	349	725	116	554	93	277	72	550	434	545

Table 6.12*a* Matrix of correlation coefficients (Pearson product-moment *r*). Indices are rounded to two decimal places then multiplied by 100 to save space in matrix. Negative indices are underlined

	A	B	C	D	E	F	G	H	I	J	K	L	M	N	O	P	Q	R	S	T	U	V	W	X	Y
A	100	$\underline{19}$	$\underline{78}$	$\underline{40}$	$\underline{45}$	05	19	$\underline{33}$	33	17	01	$\underline{66}$	40	07	02	44	$\underline{07}$	59	13	$\underline{11}$	38	23	35	01	20
B	$\underline{19}$	100	31	$\underline{06}$	62	14	11	$\underline{28}$	21	30	$\underline{14}$	$\underline{01}$	$\underline{12}$	$\underline{24}$	$\underline{37}$	$\underline{00}$	25	$\underline{31}$	$\underline{51}$	$\underline{14}$	$\underline{03}$	$\underline{25}$	$\underline{36}$	41	$\underline{38}$
C	$\underline{78}$	31	100	28	64	$\underline{15}$	$\underline{02}$	12	$\underline{24}$	$\underline{11}$	$\underline{13}$	54	$\underline{48}$	$\underline{28}$	$\underline{16}$	$\underline{10}$	19	$\underline{53}$	$\underline{23}$	$\underline{04}$	$\underline{17}$	$\underline{38}$	$\underline{49}$	20	$\underline{29}$
D	$\underline{40}$	$\underline{06}$	28	100	$\underline{03}$	57	$\underline{20}$	52	$\underline{02}$	01	$\underline{45}$	47	$\underline{07}$	45	46	$\underline{46}$	$\underline{32}$	05	36	53	$\underline{29}$	45	$\underline{34}$	$\underline{43}$	24
E	$\underline{45}$	62	64	$\underline{03}$	100	$\underline{21}$	19	$\underline{32}$	07	$\underline{05}$	10	31	$\underline{44}$	$\underline{44}$	$\underline{46}$	$\underline{00}$	33	$\underline{63}$	$\underline{52}$	$\underline{32}$	$\underline{01}$	$\underline{53}$	$\underline{66}$	70	$\underline{58}$
F	05	14	$\underline{15}$	57	$\underline{21}$	100	$\underline{17}$	46	31	42	$\underline{72}$	11	35	79	66	$\underline{50}$	$\underline{60}$	52	51	72	$\underline{32}$	82	$\underline{10}$	$\underline{41}$	26
G	19	11	$\underline{02}$	$\underline{20}$	19	$\underline{17}$	100	$\underline{70}$	31	32	$\underline{05}$	$\underline{33}$	$\underline{39}$	$\underline{44}$	$\underline{10}$	46	32	02	$\underline{10}$	$\underline{40}$	27	$\underline{31}$	$\underline{34}$	19	$\underline{27}$
H	$\underline{33}$	28	$\underline{12}$	52.	$\underline{32}$	46	$\underline{70}$	100	$\underline{10}$	$\underline{09}$	$\underline{25}$	37	27	71	50	$\underline{61}$	$\underline{60}$	16	44	67	$\underline{27}$	52	14	$\underline{47}$	39
I	33	21	$\underline{24}$	$\underline{02}$	07	31	31	$\underline{10}$	100	75	$\underline{38}$	$\underline{32}$	19	12	27	05	$\underline{06}$	36	16	06	26	28	$\underline{06}$	08	$\underline{12}$
J	17	30	$\underline{11}$	01	$\underline{05}$	42	32	$\underline{09}$	75	100	$\underline{56}$	$\underline{31}$	20	18	44	$\underline{02}$	$\underline{09}$	50	31	17	06	39	$\underline{02}$	$\underline{08}$	$\underline{00}$
K	01	$\underline{14}$	$\underline{13}$	$\underline{45}$	10	$\underline{72}$	$\underline{05}$	$\underline{25}$	$\underline{38}$	$\underline{56}$	100	$\underline{19}$	$\underline{01}$	$\underline{42}$	$\underline{59}$	18	26	$\underline{50}$	$\underline{44}$	$\underline{57}$	06	$\underline{58}$	18	44	$\underline{11}$
L	$\underline{66}$	$\underline{01}$	54	47	31	11	$\underline{33}$	37	$\underline{32}$	$\underline{31}$	$\underline{19}$	100	$\underline{20}$	11	07	$\underline{52}$	11	$\underline{34}$	$\underline{09}$	31	$\underline{36}$	03	$\underline{24}$	$\underline{07}$	$\underline{12}$
M	40	$\underline{12}$	$\underline{48}$	$\underline{07}$	$\underline{44}$	35	$\underline{39}$	27	19	20	$\underline{01}$	$\underline{20}$	100	53	30	$\underline{21}$	$\underline{29}$	51	15	19	$\underline{11}$	55	44	$\underline{14}$	16
N	07	$\underline{24}$	$\underline{28}$	45	$\underline{44}$	79	$\underline{44}$	71	12	18	$\underline{42}$	11	53	100	76	$\underline{60}$	$\underline{73}$	54	63	78	$\underline{40}$	89	20	$\underline{49}$	38
O	02	$\underline{37}$	$\underline{16}$	46	$\underline{46}$	66	$\underline{10}$	50	27	44	$\underline{59}$	07	30	76	100	$\underline{37}$	$\underline{55}$	68	83	61	$\underline{27}$	80	14	$\underline{59}$	36
P	44	$\underline{00}$	$\underline{10}$	$\underline{46}$	$\underline{00}$	$\underline{50}$	46	$\underline{61}$	05	$\underline{02}$	18	$\underline{52}$	$\underline{21}$	$\underline{60}$	$\underline{37}$	100	40	03	$\underline{21}$	$\underline{56}$	49	$\underline{49}$	01	$\underline{22}$	$\underline{08}$
Q	$\underline{07}$	25	19	$\underline{32}$	33	$\underline{60}$	32	$\underline{60}$	$\underline{06}$	$\underline{09}$	26	11	$\underline{29}$	$\underline{73}$	$\underline{55}$	40	100	$\underline{40}$	$\underline{60}$	$\underline{65}$	22	$\underline{63}$	$\underline{11}$	30	$\underline{39}$
R	59	$\underline{31}$	$\underline{53}$	05	$\underline{63}$	52	02	16	36	50	$\underline{50}$	$\underline{34}$	51	54	68	03	$\underline{40}$	100	62	40	08	73	36	$\underline{36}$	38
S	13	$\underline{51}$	$\underline{23}$	36	$\underline{52}$	51	$\underline{10}$	44	16	31	$\underline{44}$	$\underline{09}$	15	63	83	$\underline{21}$	$\underline{60}$	62	100	56	$\underline{16}$	69	14	$\underline{53}$	45
T	$\underline{11}$	$\underline{14}$	$\underline{04}$	53	$\underline{32}$	72	$\underline{40}$	67	06	17	$\underline{57}$	31	19	78	61	$\underline{56}$	$\underline{65}$	40	56	100	$\underline{41}$	72	01	$\underline{55}$	34
U	38	$\underline{03}$	$\underline{17}$	$\underline{29}$	$\underline{01}$	$\underline{32}$	27	$\underline{27}$	26	06	06	$\underline{36}$	$\underline{11}$	$\underline{40}$	$\underline{27}$	49	22	$\underline{08}$	$\underline{16}$	$\underline{41}$	100	$\underline{34}$	06	09	$\underline{14}$
V	23	$\underline{25}$	$\underline{38}$	45	$\underline{53}$	82	$\underline{31}$	52	28	39	$\underline{58}$	03	55	89	80	$\underline{49}$	$\underline{63}$	73	69	72	$\underline{34}$	100	24	51	$\underline{41}$
W	35	$\underline{36}$	$\underline{49}$	34	$\underline{66}$	$\underline{10}$	$\underline{34}$	14	$\underline{06}$	02	18	$\underline{24}$	44	20	14	01	$\underline{11}$	36	14	01	06	24	100	$\underline{31}$	26
X	01	41	20	$\underline{43}$	70	41	19	$\underline{47}$	08	$\underline{08}$	44	$\underline{07}$	$\underline{14}$	$\underline{49}$	$\underline{59}$	22	30	$\underline{36}$	$\underline{53}$	$\underline{55}$	09	$\underline{51}$	$\underline{31}$	100	$\underline{48}$
Y	20	$\underline{38}$	$\underline{29}$	24	$\underline{58}$	26	$\underline{27}$	39	$\underline{12}$	$\underline{00}$	$\underline{11}$	$\underline{12}$	16	38	36	$\underline{08}$	$\underline{39}$	38	45	34	$\underline{14}$	41	26	$\underline{48}$	100

on the whole. The index is +0.57. Similarly highly urbanized states (variable D) would be expected to have a small proportion of their population defined as rural farm population (variable K). The index is −0.72. If the findings of the correlation matrix are to be taken seriously, then a parallel matrix should be prepared to express the nature of the relationship that occurs between each pair of variables. This is not often done in published papers and books that use correlation matrices of this kind.

(d) Using the matrix further

The following matrix is the top left hand part of the main correlation matrix (table 6.12*b*). It shows the form of the correlation indices before they were condensed.

Table 6.12*b* Corner of previous matrix

	A	B	C	D	E	F	
A	1.00	−0.19	−0.78	−0.40	−0.45	0.05	1.87
B	−0.19	1.00	0.31	−0.06	0.62	0.14	1.32
C	−0.78	0.31	1.00	0.28	0.64	−0.15	2.16
D	−0.40	−0.06	0.28	1.00	−0.26	0.57	1.34
E	−0.45	0.62	0.64	−0.03	1.00	−0.21	2.18
F	0.05	0.14	−0.15	0.57	−0.21	1.00	1.12

The above matrix will be used to illustrate three quick techniques for detecting linkages among more than two variables at a time.

(i) Sum each row (or column) in turn, ignoring the minus sign and the 1.00 correlation each time. The results (see last column of table 6.12*b*) give an indication of the general amount of relatedness between each variable and every other. The higher the value, the more generally related is the variable. In the above example, the variables may be arranged as follows from most to least generally related: E, C, A, D, B, F. As further variables are introduced, this order may be changed.

(ii) When there are not too many variables (say up to 20), it is feasible to draw a circle and mark off on it, preferably at regular intervals, as many points as there are variables. A correlation index value considered high enough to be a strong correlation may then be decided upon. On the 'clock' pairs of variables may be linked by lines if they are correlated with a value above that chosen for the cut off. In this way it will become apparent visually which variables are linked to which others. Those variables that have few, several and many associated variables can quickly be seen.

(iii) Linkage analysis is a possible method of extracting related variables from a correlation matrix. B. J. L. Berry (1958) deserves credit for first using this technique (and probably several others) in geography. He himself refers to L. L. McQuitty (1957). In his Ph.D. thesis (p. 66), Berry claims:

'Since linkage analysis of this type has never previously been applied to research in geography, the methods, assumptions and procedures must be outlined with care. . . . A group of variables is defined as a category of such a nature that every variable in the category is more like *some* other variable in the category than it is like any other variable outside the category.' Linkage analysis involves looking along rows and down columns in the correlation matrix to locate the highest indices of correlation and to group variables appropriately on this basis. Berry himself was one of the first to use factor analysis in geography. This seems to have advantages over linkage analysis, but depends on the use of an electronic computer to do the very large amount of calculations it requires. The following section, 6.4, shows the steps in factor analysis that follow on from the correlation matrix discussed in this section.

6.4 Factor analysis

(*a*) *Introduction*

In the mesoscopic world of the geographer, where the microscope and the telescope are rarely needed, the complexity of the subject arises from the frequent occurrence of situations in which many variables must be taken into account simultaneously. Factor analysis is a general term for a sequence of several processes designed to detect 'families' of correlated variables. It is one kind of multivariate analysis that has already been fairly widely used in studies that would qualify either for a central position or at least for a peripheral position within the subject of geography.

The results of factor analysis do not seem any more difficult to appreciate than the results of many other techniques that have already been used in geography. But at least one big mental jump is required, an appreciation of the concept of *many* dimensions (rather than two, three or four). Like (one suspects) many other things, factor analysis is easy for those who know what it is doing. It must be pointed out, however, that the cost of applying the technique even to quite a small amount of data used to be exorbitant in terms of man hours on a calculating machine before the computer age. Factor analysis may be thought of in very broad terms in a geographical context as a method of finding mathematically the relationships between different combinations of maps that might have been discovered visually in the 'cabinet' suggested in the previous section. In a relatively early work on this kind of technique in geography, B. J. L. Berry (1958) (p. 64) in his study of the town of Spokane, U.S.A., writes:

'Consider that each of the remaining 49 types of business have been mapped, and that there are 49 maps of Spokane each containing information relating to numbers of stores of a particular type at 285 locations. The problem for empirical study can be thought of in these terms: What is the composite pattern of correspondence of the 49 geographical distributions? What geographical associations of business types occur?'

Some aspects of factor analysis have already been described in chapter 3, but without examples. This section to some extent repeats points made in chapter 3 as it applies the procedures described here to the example of the 50 states of the U.S.A. It must be pointed out that there are differences between principal component analysis and factor analysis. Since both use basically the same procedures to study relationships between many variables for many cases, no attempt will be made to distinguish between them. The difference is one of approach or scope of the study being made. According to D. N. Lawley (1953): 'In a sense then it can be said that whereas a principal component analysis is variance-orientated, a factor analysis is *covariance-orientated*'. In what follows, the term factor could with reservations replace the term principal component. M. G. Kendall (1957) and A. R. Baggeley (1964) may be consulted for discussions of factor analysis.

(*b*) *Factors and factor loadings*

The factors in factor analysis are sets of indices that summarize the intercorrelation or covariance of various sets of variables. One variable may be plotted on a single (x) axis and then characterized by a mean value, a point. Thus one variable occupies one dimension and is characterized by one zero dimension index. Two variables (the values appropriately normalized for comparability) may be plotted on a graph with x- and y-axes (see chapter 3). Their distribution may then be characterized by two lines of best fit, that is by two figures each of one dimension, orthogonal (at right angles) to one another. If the correlation is strong (either negatively or positively) the observations on the graph will show a marked ellipsoid arrangement, and one of the lines will be markedly longer than the other. Three variables may similarly be plotted rather clumsily in 3-dimensional space, with x-, y- and z-axes. Their distribution may then be summarized by three orthogonal planes of best fit, that is by three 2-dimensional figures. Though it is not possible to proceed further than 3-dimensional space with a graphical (or geometrical) representation of the correlation between three variables, the process may be taken algebraically into more dimensions with correspondingly more variables. Thus the present study of the relationship of 25 variables may be considered to involve fitting 25 figures, each of 24 dimensions, to a hyperellipsoid formed in 25 dimensions by 50 observations (the 50 states) each located on 25 co-ordinates.

When there are 25 variables, there is considered to be variation totalling 25 units. If there is zero correlation between all of the 25 variables, then each of the 25 figures of 24 dimensions fitting the cluster of 50 observations will be roughly the same size. Each eigenvector, as the figures are called, will have an eigenvalue of 1. If on the other hand all are entirely correlated, then the first eigenvector would absorb all the variation and would have an eigenvalue of 25. The remaining 24 would each be 0.

As would be expected there is neither complete intercorrelation, nor zero correlation in the U.S.A. example. Intercorrelation is quite strong, however, because 8.74 out of 25 units of variation accumulate on the first and highest eigenvector, another 4.58 units on the next, and 3.14 on the third. Expressed as percentages of the total variation, these reach 35, 18 and 13 per cent

Table 6.13 First five factors (total possible 25)

Factor (principal component) Eigenvalues	I	II	III	IV	V
	8.74	4.58	3.14	1.78	1.00
Cumulative proportion of total variance explained	35%	53%	66%	73%	78%
Factor matrix variable					
A Altitude	0.13	0.88	0.03	0.11	0.16
B Temperature	−0.36	−0.24	0.57	0.45	−0.06
C Precipitation	−0.32	−0.76	0.18	−0.19	−0.12
D Density of population	0.50	−0.58	0.14	−0.27	0.04
E Negro	−0.62	−0.51	0.44	0.20	0.18
F Urban	0.80	−0.17	0.41	0.15	0.10
G Birthrate	−0.38	0.33	0.54	−0.43	0.16
H Median age	0.69	−0.44	−0.30	0.08	−0.08
I Population change	0.18	0.36	0.71	0.16	−0.12
J Migration	0.31	0.29	0.76	0.07	−0.24
K Farm population	−0.56	0.15	−0.61	0.23	0.26
L Manufacturing	0.06	−0.81	−0.11	−0.04	−0.28
M Motor vehicles	0.48	0.34	−0.14	0.61	−0.13
N Telephones	0.92	−0.11	−0.03	0.22	0.12
O Income	0.86	0.00	0.19	−0.22	−0.04
P Federal expenditure	−0.51	0.58	0.07	−0.35	0.05
Q Lawyers	−0.72	0.06	0.06	−0.10	−0.46
R School years completed	0.70	0.53	0.19	−0.03	−0.01
S School expenditure	0.78	0.11	0.05	−0.36	0.18
T Physicians	0.81	−0.33	0.06	0.00	0.05
U Infant mortality	−0.33	0.47	0.10	−0.22	−0.21
V Housing	0.94	0.06	0.13	0.14	0.01
W President 1960	0.27	0.49	−0.51	0.20	−0.43
X President 1964	−0.67	−0.01	0.20	0.44	0.35
Y State governor 62–64	0.53	0.15	−0.33	−0.26	0.16

Table 6.14 First five rotated factors

Factor	I	II	III	IV	V
Rotated factor matrix variable					
A Altitude	−0.06	0.86	0.16	0.15	0.16
B Temperature	−0.06	−0.14	0.29	−0.03	−0.79
C Precipitation	−0.08	−0.75	−0.06	−0.33	−0.28
D Density of population	0.56	−0.51	0.17	−0.25	0.20
E Negro	−0.16	−0.33	−0.04	−0.38	−0.79
F Urban	0.74	−0.03	0.54	0.01	0.06
G Birthrate	−0.53	0.26	0.35	−0.60	−0.02
H Median age	0.75	−0.37	−0.08	0.29	0.26
I Population change	0.06	0.31	0.75	−0.00	−0.22
J Migration	0.02	0.17	0.89	0.01	−0.07
K Farm population	−0.33	0.26	−0.78	0.14	−0.19
L Manufacturing	0.12	−0.84	−0.10	0.06	−0.03
M Motor vehicles	0.32	0.37	0.16	0.69	−0.00
N Telephones	0.81	0.36	0.23	0.26	0.27
O Income	0.49	−0.36	0.51	0.00	0.54
P Federal expenditure	−0.58	0.44	−0.04	−0.29	0 08
Q Lawyers	−0.85	−0.22	−0.01	0.05	−0.18
R School years completed	0.26	0.47	0.53	0.16	0.46
S School expenditure	0.50	0.13	0.29	−0.17	0.62
T Physicians	0.75	−0.23	0.25	0.07	0.27
U Infant mortality	−0.29	0.28	0.09	−0.04	0.00
V Housing	0.67	0.12	0.46	0.25	0.35
W President 1960	−0.12	0.26	−0.06	0.74	0.43
X President 1964	−0.27	0.22	−0.22	−0.16	−0.77
Y State governor 62–64	0.41	0.17	−0.12	0.01	0.56

respectively, and together, in three out of 25 'directions', find about 66 per cent or nearly $\frac{2}{3}$ of all the variation.

Table 6.13 shows how each of the 25 variables (A–Y) 'loads' on the first five factors. The values are weighted in the factor matrix (6.13) to allow for the relative importance of each factor. Thus the values tend to spread less far from zero as the factors diminish in importance. The 'loading' of each variable on the factor is measured by the distance from zero either in a positive or a negative direction. Thus on factor I, Altitude (A) and Manufacturing (L) are loading only very lightly at 0.06 and 0.13. In contrast, Housing (V) and Farm population (K) are loading heavily, the former positively (0.94), the latter negatively (−0.56). In contrast, the direction of factor II picks out both Altitude (A) at 0.88 and Manufacturing (L) at −0.81, as the two variables loading most heavily.

Before any attempt is made to interpret the factor loadings in the factor matrix, it is essential to find out the way the data in the variable are arranged. Thus for example for the electoral data (W–Y), the variables express the percentage that voted Republican. Had percentage voting Democrat been used instead, then all the indices of correlation and the factor loadings of the three electoral variables would have been similar, but all the signs would have been changed. The reason is that in this case there are only two choices. Before any conclusions are drawn, therefore, a careful consideration of the nature and direction of the variables is vital.

In the long run, it is perhaps more important still to think what the relationships mean. Common sense suggested that the high correlation between income (O) on the one hand and housing (V) and the telephones (N) on the other is genuine. The higher the income of people, the more likely they are to have a house in good repair and a telephone (or two). On the other hand, the high negative correlation (−0.66) between high manufacturing and high altitude (conversely it would be positive between lowland and manufacturing) is not conceivably a causal relationship, but reflects two very general trends coinciding. The high correlation index does exist mathematically however and is itself quite real. There are other problems of interpretation as well. Some have already been discussed earlier in this chapter.

Often refinements are applied to factor analysis at the factor matrix stage to derive 'better' factors. Table 6.14 shows a rotated factor matrix. In this particular example it does not greatly modify the first three factors. These carry 66 per cent of all the variation. IV and V are however greatly modified. Interestingly the Kennedy–Nixon and Johnson–Goldwater presidential elections are highlighted on rotated factors IV and V respectively. It is not customary to consider factors with an eigenvalue less than 1.00. This is a convenient but in fact an arbitrary cut off point.

(c) *Individual weightings*

So far the 50 states have been lost from sight. It was suggested earlier that 25 distribution maps might be drawn, one for each of the 25 variables. One of the most interesting outcomes of factor analysis for the geographer is the

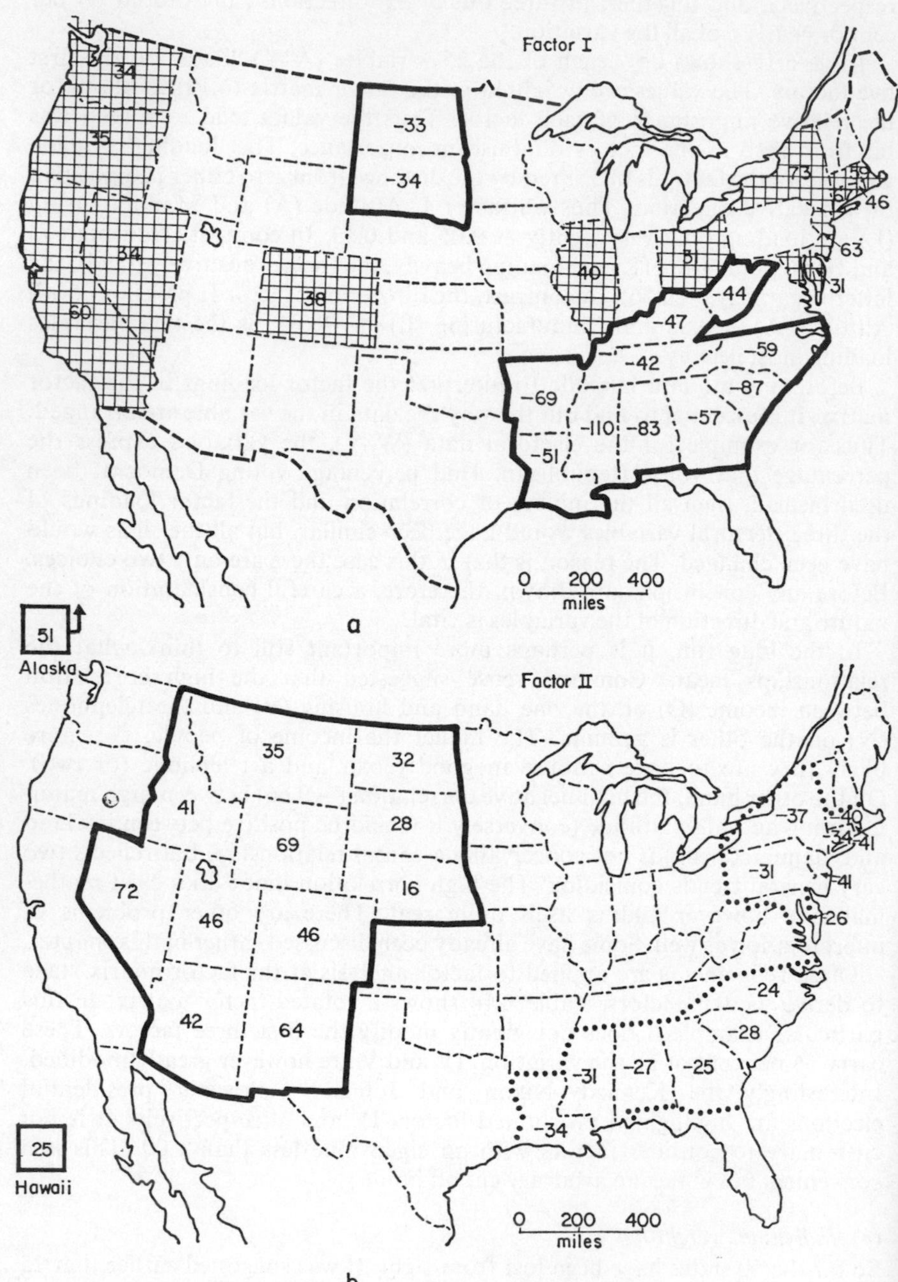
Factor I
34
35
34
60
38
-33
-34
40
51
73
59
46
63
31
-44
-47
-59
-42
-69
-87
-110
-83
-57
-51
0 200 400
miles
a
51
Alaska
Factor II
35
32
41
28
69
72
16
46
46
42
64
-37
-40
-34
41
-31
-41
26
-24
-28
-27
-25
-31
-34
25
Hawaii
0 200 400
miles
b

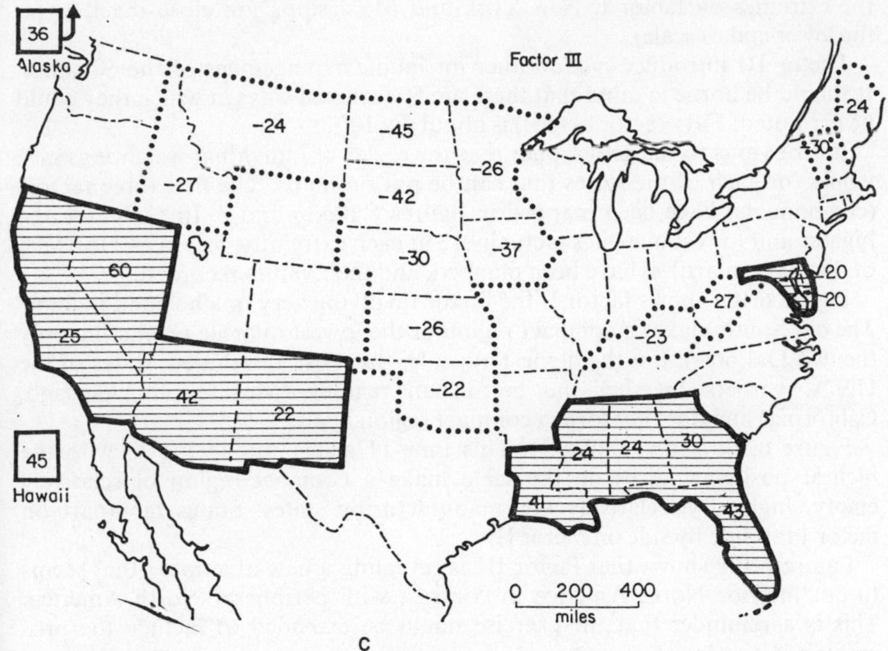

Figure 6.10 Individual weightings (factor scores) for the top and bottom quartiles of the 50 states of the U.S.A. on (*a*) Factor I, (*b*) Factor II, (*c*) Factor III. See text for explanation. Note that individual weightings (factor scores) are meaningless in an absolute sense and should only be considered relative to one another. The decimal point has been moved two places to the right from the original data in table 6.15.

reappearance of the cases (the 50 states in this example), assessed on the verdict not of individual variables, but of the factors, clusters of variables with variables loading on them as indicated in the factor matrix, table 6.13.

Table 6.15 contains the so-called individual weightings of each of the 50 states on each of the first three factors (I–III). For example factor I has the following variables 'loading' heavily on it.

V	Housing	0.94	T	Physicians	0.81
N	Telephones	0.92	F	Urban	0.80
O	Income	0.86	S	School expenditure	0.78

Factor I in table 6.15 shows that New York, Connecticut, New Jersey, California and so on are very similar to one another with respect to this factor, but are very different from Mississippi, South Carolina, Alabama, Arkansas and so on, at the other end of the scale However on factor II

A	Altitude	0.88
L	Manufacturing	−0.81 (lack of)
C	Precipitation	−0.76 (lack of)

the extremes on factor I, New York and Mississippi, are close together (at the lower end of scale).

Factor III introduces yet another intriguing arrangement of the 50 states. It should be borne in mind that there are 50 factorial ways in which they could be permuted. Fifty factorial (50!) is about 3×10^{64}.

It is obvious to the geographer that the scales of individual weightings give values for each of the states that can be put on maps. The first three factors (components) have been mapped in figures 6.10*a*, *b* and *c*. In each case the highest and lowest quartiles (actually 13 at each extremity, leaving 12 for each of the inner quartiles) have been mapped, and their values recorded.

Figure 6.10*a* shows factor I, the factor revolving very much round income. The old South makes a compact region of the lowest quartile (excluding only the two Dakotas). But the highest quartile, the most 'developed' states of the U.S.A. gathers together the big urban regions (Megalopolis, Chicago, California) and does *not* form a compact region at all.

Figure 6.10*b* shows factor II. This time 11 states among the 13 with the highest positive indices on the scale make a compact region of relatively empty, high, dry, relatively non-manufacturing states. States far apart on factor I fall side by side on factor II.

Figure 6.10*c* shows that factor III is revealing a new alignment that seems to put interior North America in contrast with peripheral North America. This is a reminder that the exercise might be extended to include the provinces of Canada.

The data on the scales in table 6.15 can also be plotted on a graph. Figure 6.12*a* shows factors I and II on the *y* and *x* co-ordinates respectively. Together factors I and II contain some 53 per cent of all the variation. In other words they are powerful together, but still provide only half the picture of all variation among the 25 variables. The triangular arrangement of the states in figure 6.12*a* suggests three extreme types, (1) New York–Massachusetts, (2) Mississippi and the Deep South, (3) Nevada–Wyoming–New Mexico. Not surprisingly, Iowa is one of the candidates for the position of most typical state of the U.S.A.

Although it may seem at first sight somewhat unethical to reverse the signs on the *x*-axis (factor II), the effect of this operation of flipping the graph over on its *y*-axis has a startling effect in 6.12*b*. In fact the operation does not affect the relative position of the states on the graph, but it produces a drastically transformed map of the U.S.A., one that is still however recognizable. It shows for example that whereas California is further in space distance from New York than Nevada is, it is only half as far away in terms of all the variables loading strongly on factors I and II.

Figure 6.11 shows the *census* divisions of the U.S.A. There are nine in mainland U.S.A. These are divisions of contiguous states forming fairly compact blocks. Figure 6.12*b* reveals that whereas some of the blocks contain member states that are similar in general to one another than to other states (e.g. the mountain states, the Pacific states), others contain states that differ greatly from one another. Thus Maryland and Delaware belong to the South

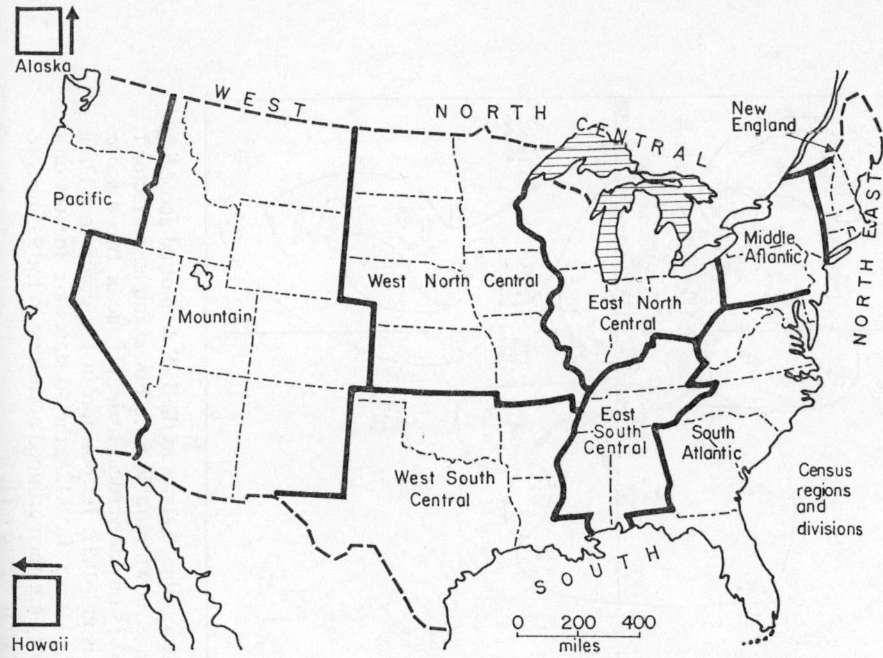

Figure 6.11 The nine census divisions of mainland U.S.A. This map is included for comparison with figures 6.10*a–c* and with the graphs in figure 6.12.

Atlantic group but are not at all like other members, including the Carolinas or Georgia. On the other hand they closely resemble Ohio and Pennsylvania.

An extension of factor analysis is to amalgamate the cases into the best groups. Thus a grouping of the 50 states into say 10 groups might be desired. This can be achieved by collapsing the most similar pair into one new unit (perhaps Ohio and Michigan), to produce 49, then the next most similar pair, and so on. If it is desired, the process can be made to collapse pairs of states only if they are contiguous. Thus although Delaware and Ohio are very similar, they could not be collapsed at least until they had become associated with intervening Pennsylvania. Cluster analysis (also called linkage analysis) will not be illustrated here, but in chapter 13, where the 50 states and first five factors from this section are used to illustrate it as a method of classification.

The reader may already have become suspicious of the question of the variables and factors. Is it not possible to inject the variables one wants into the mixture? The inclusion, for example, of several income variables would certainly produce a first factor that reflected income. It would even be possible to cultivate or cook artificial factors by putting some particular variable in

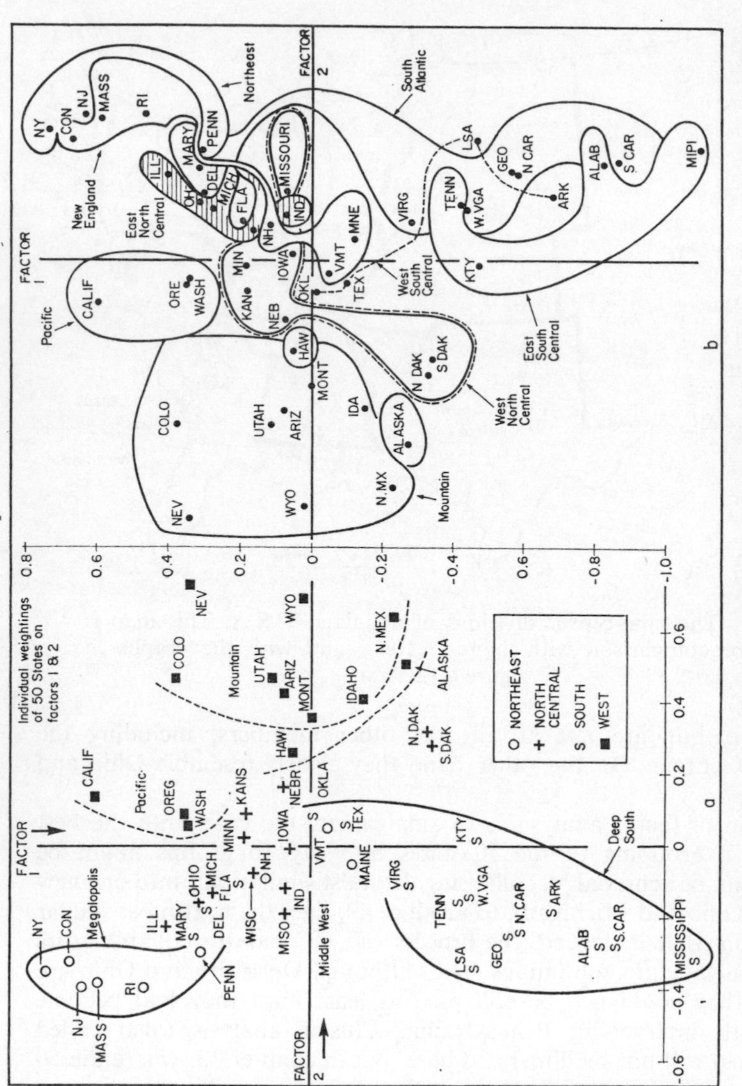

Figure 6.12 The individual weightings of each of the 50 states in the U.S.A. plotted according to their values on factors I and II. The data are shown in table 6.15. For example, on factor I, New York scores 0.73. It is therefore at +0.73 on the vertical scale. On factor two it scores −0.37. On the horizontal axis, therefore, it is at −0.37. Its position in the graph in 6.12*a* is therefore given by the ordered pair (−0.37, +0.73). In 6.12*b*, the 50 states are located on a transformed graph, the reflexion of 6.12*a*. They are then grouped according to the 9 census regions shown in figure 6.11.

Factor I case		Factor II case		Factor III case	
0.73	New York	0.72	Nevada	0.60	Nevada
0.67	Connecticut	0.69	Wyoming	0.45	Hawaii
0.63	New Jersey	0.64	New Mexico	0.43	Florida
0.60	California	0.51	Alaska	0.42	Arizona
0.59	Massachusetts	0.46	Utah	0.41	Louisiana
0.46	Rhode Island	0.46	Colorado	0.36	Alaska
0.40	Illinois	0.42	Arizona	0.30	Georgia
0.38	Colorado	0.41	Idaho	0.25	California
0.35	Oregon	0.35	Montana	0.24	Alabama
0.34	Washington	0.32	North Dakota	0.24	Mississippi
0.34	Nevada	0.28	South Dakota	0.22	New Mexico
0.31	Maryland	0.25	Hawaii	0.20	Maryland
0.31	Ohio	0.16	Nebraska	0.20	Delaware
0.30	Delaware	0.12	California	0.18	South Carolina
0.30	Pennsylvania	0.08	Kansas	0.17	Texas
0.27	Michigan	0.08	Oklahoma	0.13	Utah
0.19	Florida	0.07	Oregon	0.12	New Jersey
0.19	Minnesota	0.06	Texas	0.12	Massachusetts
0.18	Kansas	0.05	Washington	0.09	Virginia
0.16	Wisconsin	0.04	Vermont	0.08	Connecticut
0.12	New Hampshire	0.01	Kentucky	0.03	Colorado
0.11	Utah	0.01	Minnesota	0.03	Rhode Island
0.08	Arizona	−0.02	Iowa	0.02	New York
0.08	Nebraska	−0.06	Maine	−0.02	Illinois
0.07	Missouri	−0.09	Wisconsin	−0.04	North Carolina
0.07	Indiana	−0.10	New Hampshire	−0.05	Michigan
0.06	Iowa	−0.11	Virginia	−0.06	Arkansas
0.05	Hawaii	−0.11	Florida	−0.07	Ohio
0.02	Wyoming	−0.13	Indiana	−0.10	Indiana
−0.00	Montana	−0.14	West Virginia	−0.10	Tennessee
−0.01	Oklahoma	−0.15	Michigan	−0.10	Washington
−0.05	Vermont	−0.16	Tennessee	−0.11	Nebraska
−0.10	Texas	−0.17	Ohio	−0.13	Missouri
−0.12	Maine	−0.18	Arkansas	−0.14	Pennsylvania
−0.15	Idaho	−0.18	Delaware	−0.17	Oregon
−0.23	Mexico	−0.20	New Missouri	−0.18	Wyoming
−0.27	Alaska	−0.24	Illinois	−0.21	Wisconsin
−0.28	Virginia	−0.24	North Carolina	−0.22	Oklahoma
−0.33	North Dakota	−0.25	Georgia	−0.23	Kentucky
−0.34	South Dakota	−0.26	Maryland	−0.24	Maine
−0.42	Tennessee	−0.27	Alabama	−0.24	Montana
−0.44	West Virginia	−0.28	South Carolina	−0.26	Kansas
−0.47	Kentucky	−0.31	Pennsylvania	−0.26	Minnesota
−0.51	Louisiana	−0.31	Mississippi	−0.27	West Virginia
−0.57	Georgia	−0.34	Louisiana	−0.27	Idaho
−0.59	North Carolina	−0.34	Connecticut	−0.30	Vermont
−0.69	Arkansas	−0.37	New York	−0.30	Nebraska
−0.83	Alabama	−0.40	Massachusetts	−0.37	Iowa
−0.87	South Carolina	−0.41	New Jersey	−0.42	South Dakota
−1.10	Mississippi	−0.41	Rhode Island	−0.45	North Dakota

[a] Rank order of each standardized case ordered by size of factors (principal components) I, II and III.

several times. Factor analysis is rather like the making of a cake; the recipe and ingredients may be chosen at will; but is not much of geography rather like this at the moment anyway?

Factor analysis has been used already by geographers in a considerable number of papers and books. The following are noted as examples for the reader who wishes to consider at greater length the applications of factor analysis. C. A. Moser and W. Scott's *British Towns* (1961) has already been discussed. J. C. Fisher (1966) in *Yugoslavia—A Multinational State*, handles 26 variables for 55 towns and then 63 variables for 611 communes. Although many of his variables are similar, some interesting results emerge and are usefully brought into the general discussion of Yugoslavia. Q. Ahmad's *Indian Cities* (1965) was mentioned earlier in the chapter. W. L. Garrison and D. F. Marble (1964) have used factor analysis with the study of connectivity of transportation networks. Probably the first appearance of factor analysis in a British geographical journal was J. D. Henshall's paper *The Demographic Factor in the Structure of Agriculture in Barbados*. Thirty-two variables were collected for a sample of 116 farms. These examples suggest that geographers are likely to find factor analysis a useful technique (or series of techniques). They often spend a large amount of time gathering material, only to map each variable on a separate map and then make no more than visual and verbal assessments. Perhaps the most valuable function of factor analysis for the geographer is to provide an efficient way of assembling and sorting data and describing relationships in it, both more concisely and more precisely than is generally done on a visual and verbal basis. Indeed on some occasions, factor analysis might be the starting point for a geographical study, rather than the final stage.

References

Ahmad, Q. (1965). *Indian Cities: Characteristics and Correlates*, The University of Chicago, Department of Geography, Research Paper No. 102.

Berry, B. J. L. (1958). *Shopping Centres and the Geography of Urban Areas*, A Theoretical and Empirical Study of the Spatial Structure of Intra-urban Retail and Service Business, University of Washington Ph.D. thesis.

Blalock, H. M. (1964). *Causal Inferences in Non Experimental Research*, University of North Carolina Press, Chapel Hill.

Cole, J. P. and Smith, G. A. (1965). France: a rank correlation project. *Bull. Quantitative Mater. Geog.*, No. 1. Geography Department, University of Nottingham.

Fisher, J. C. (1966). *Yugoslavia — A Multinational State*, Chandler, San Francisco.

Garrison, W. L. and Marble, D. F. (1963). Factor-analytic study of the connectivity of a transportation network, *Reg. Sci. Assoc. Papers*, **XII**.

Ginsburg, N. and Berry, B. J. L. (1961). *Atlas of Economic Development*, University of Chicago Press, Chicago, Part VIII.

Giusti, U. (1943). *Caratteristiche ambientali italiane*, Istituto Nazionale di Economia Agraria, Studi e Monografie, N.27, Roma.

Gould, P. R. (1966). *On Mental Maps*, Michigan Inter University Community of Mathematical Geographers, No. 9.

Gould, P. R. (1967). On the geographical interpretation of eigenvalues, *Trans. Inst. Brit. Geog.*, **42,** 53–86.

Henshall, J. D. and King, L. J. (1966). Some structural characteristics of peasant agriculture in Barbados. *Econ. Geog.*, **42,** No. 1, 74–84.

Henshall, J. D. (1966). The demographic factor in the structure of agriculture in Barbados, *Trans. Inst. Brit. Geog.*, **38,** 183–195.

Kendall, M. G. (1957). *A Course in Multivariate Analysis*, Griffin, London.

King, L. J. (1961). A multivariate analysis of the spacing of urban settlements in the United States. *Ann. Assoc. Am. Geog.*, **51,** No. 2, 222–233 (June 1961).

Kruskal, W. H. (1958). Ordinal measures of association, *J. Am. Stat. Assoc.*, **53,** 814–861.

Lawley, D. N. and Maxwell, A. E. (1963). *Factor Analysis as a Statistical Method*, Butterworths, London, pp. 2–3.

McQuitty, L. L. (1957). Elementary linkage analysis for isolating orthogonal and oblique types and typal relevancies, *Educational and Psychological Measurement*, **17.**

Moser, C. A. and Scott, W. (1961). *British Towns. A Statistical Study of their Social and Economic Differences*, Oliver and Boyd, Edinburgh and London.

Robinson, A. H. The necessity of weighting values in correlation analysis of areal data, *Ann. Assoc. Am. Geog.*, **46,** 233–236.

Rummel, R. J. (1963). Dimensions of conflict behavior within and between nations, *Yearbook Soc. Gen. Systems Res.*, **VIII,** pp. 1–50.

Russett, B. M. and others (1964). *World Handbook of Political and Social Indicators*, Yale University Press, New Haven.

Stamp, Sir D. (1964). Discussion on P. Haggett's paper: Regional and local components in the distribution of forested areas in south-east Brazil: A multivariate approach, *Geog. J.*, **130,** Part 3, p. 380.

Thomas, E. N. and Anderson, D. L. (1965). Additional comments on weighting values in correlation analysis of areal data, *Ann. Assoc. Am. Geog.*, **55,** No. 3, pp. 492–505.

Thorndike, E. L. (1939). *Your City*, Harcourt Brace, New York.

Wood, R. C. (1964). *1400 Governments*, Anchor Books, New York, p. 37.

7

Comparing correlations between different areas

'Who suppes with the Devylle sholde have a long spoone!'
'A Lay of St Nicholas', *The Ingoldsby Legends*

Chapter 5 dealt with the *comparisons* of similar single distributions (variables) in different areas or in the same area at different times. Chapter 6 considered the *correlation* of different distributions (variables) in the same area. The purpose of chapter 7 is to look briefly at the possibility of *comparing* the correlation between two or more distributions (variables) in one area with the correlation between the *same* two or more distributions (variables) in one or more *other* areas. For example, chapter 5 compared the distribution of population in North Italy with the distribution of population in Peninsular Italy. Chapter 6 correlated income and motor vehicles in France. Chapter 7 considers comparing the correlation between income and motor vehicles in Italy with that in France. The chapter consists of the following examples, with comments. It is surprising that very little has so far been done on this aspect of geography using quantitative methods.

Pairwise correlations.

(1) Within Italy between selected subsets of provinces
(2) Between all Italy, all France and all Spain.

Multiple correlations.

(3) Within Italy, between North and Peninsular Italy.
(4) Between areas of North England.
Note: within and between are relative, e.g. Italy, Spain and France are within Europe.

7.1 Comparing correlations between one pair of variables.

In this section correlations of pairs of variables for six selected subgroups (subsets) of 30 provinces of Italy are compared. Indices for the same pairs of variables are also given for North Italy (40 provinces), the rest of Italy (52 provinces) and all Italy (92 provinces). The correlation indices are Spearman rank correlation indices. They are (1) between percentage of province area defined as lowland/low hill with percentage of each province defined as cropland (field plus tree crops), and (2) between per capita income for each province and motor vehicles registered per 1000 inhabitants in each province. The original data are not reproduced here but may be obtained from Italian sources (see references at end of chapter: *Annuario statistico italiano*, and *G. Tagliacarne*).

The purpose of this example is to show the feasibility of producing indices of correlation between a pair of variables for many selected parts of a total area under consideration. Data are used for 92 provinces of Italy. There are 2^{92} or approximately 10^{28} possible combinations of the 92 provinces. In this example they have been combined 30 at a time. There are approximately 10^{25} combinations of 92 provinces 30 at a time. Most of these would however be fragmented. Possibly the number of provinces that might be usefully combined 30 at a time to form reasonably compact units is 10^{22}. A mere six of these are illustrated.

In spite of the very large number of combinations possible, it is feasible to take successive groups in such a way that the total country is satisfactorily covered. Possible ways are suggested diagrammatically in figure 7.1. For example, (*a*) it would be possible to cover a whole area with subareas that accounted for every part exactly once. Thus Italy could be divided into 4 subgroups of 26 contiguous provinces. (*b*) Subgroups of provinces could overlap. Many provinces are considered at least twice. (*c*) The subgroups might cover less than the total area, as in the Northern England example in figure 7.4. Figure 7.2 shows the actual subgroups of provinces used in this study.

Suppose now an index of correlation has been obtained for a chosen pair of variables for each subarea. It is then possible to plot each index at the centre of its subarea. A number of point values may be built up, and an isopleth map constructed that reflects variations not in one variable as most maps do but variations in the *relationship between two*. The collection of enough indices of correlation between a pair of variables is laborious, and the work needed to write a program for a computer would be considerable. But the possibility is worth considering, for it opens new possibilities in data mapping.

In the Italy example (see table 7.1) the first correlation reflects an obvious one way causal relationship between altitude and cropland. The highest (and generally the most rugged) parts of Italy are the Alpine provinces of the North, the Apennine provinces of the Peninsula and parts of Sicily and Sardinia. The indices of correlation show a stronger correlation between lowland and cropland in North Italy than elsewhere. This seems to be the result of a sharper distinction between mountain and plain in the North than elsewhere. Also the Alps are generally higher than the Apennines. It is interesting to note, too, that the positive correlation between lowland and cropland drops very low ($+0.247$) for selected subarea (*d*). In this part of Italy, cultivation extends into the higher parts, while some low-lying areas are not used for cultivation, but are forested or pasture. It would be conceivable to find limited parts of Italy in which there is no correlation between the two variables lowland and cropland and indeed other parts of the world (e.g. Mexico, Ecuador,

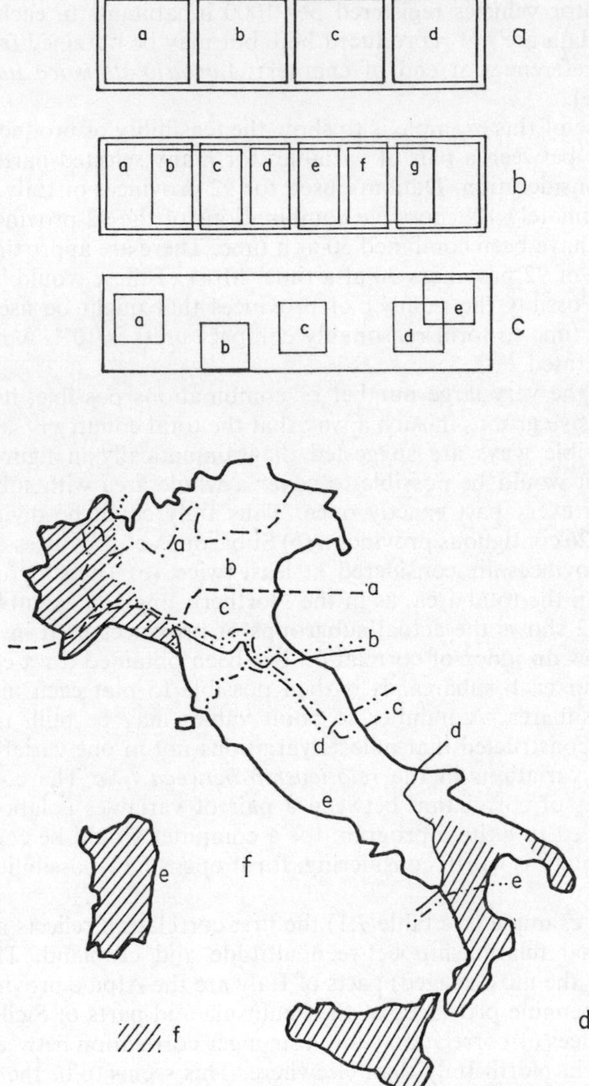

Figure 7.1 Possible ways of dividing an area into subsets. (*a*) Complete coverage no overlapping. (*b*) Complete coverage some overlapping. (*c*) Partial coverage with or without overlapping. (*d*) The actual six subsets, each of 30 provinces, used in this section. See maps in figure 7.2 for their actual distribution.

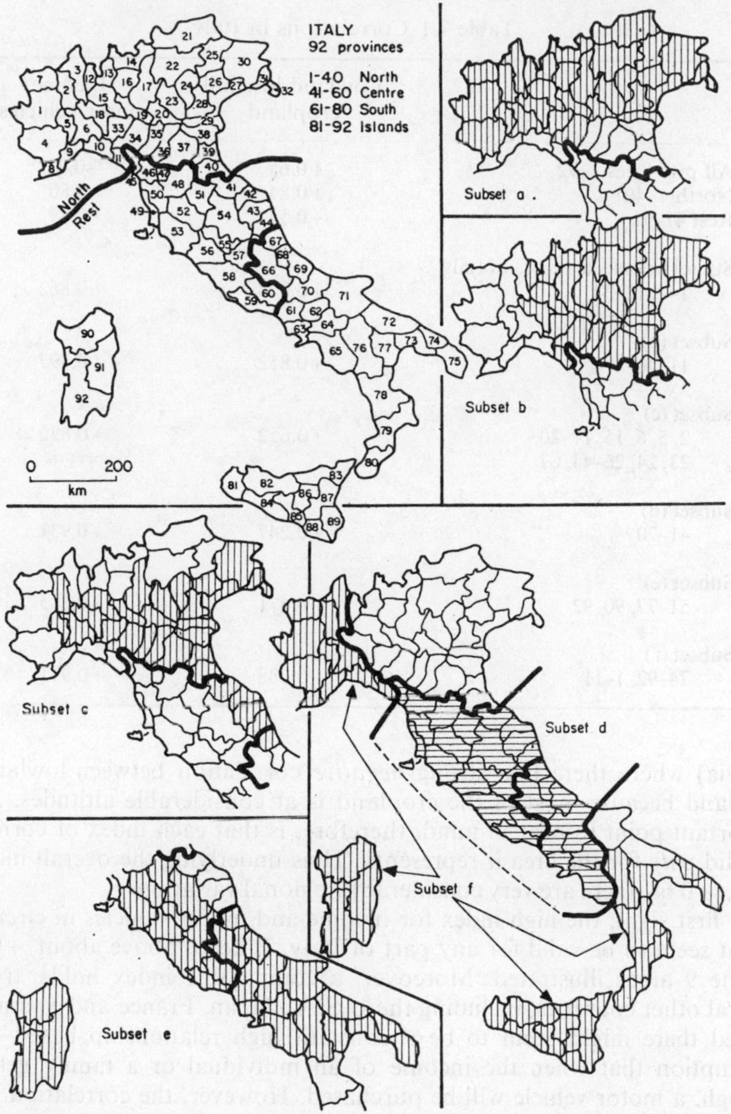

Figure 7.2 All 92 provinces of Italy are shown on the unshaded map. This may be referred to for the location of provinces shown in table 7.6. It also shows the division between provinces of North Italy and the rest of Italy. The remaining maps (*a–f*) show the distribution of the six subsets of 30 provinces selected for finding indices of correlation. Subsets *d* and *f* are shown together on one of the maps since they do not intersect.

Table 7.1 Correlations in Italy

	Lowland-low hill/ cropland	Income/ motor vehicles
All provinces 1–92	+0.68	+0.92
North 1–40	+0.84	+0.80
Rest 41–92	+0.54	+0.89
Subset (a) selected parts of Italy 1–7, 12–34	+0.873	+0.862
Subset (b) 11–40	+0.852	+0.797
Subset (c) 2, 5, 6, 15, 17–20, 23, 24, 26–44, 67	+0.652	+0.822
Subset (d) 41–70	+0.247	+0.931
Subset (e) 51–77, 90–92	+0.471	+0.885
Subset (f) 74–92, 1–11	+0.788	+0.921

Bolivia) where there is a strong *negative* correlation between lowland and cropland because most of the cropland is at considerable altitudes. A very important point to bear in mind, therefore, is that each index of correlation is valid *only* for the area it represents. Thus underlying the overall index for Italy, +0.68, there are very considerable regional variations.

At first sight, the high index for income and motor vehicles in circulation might seem to be valid for any part of Italy. It keeps above about +0.80 in all the 9 areas illustrated. Moreover, a comparable index holds true for several other countries, including the U.S.A., Japan, France and Spain. Here indeed there might seem to be a constant high relationship based on the assumption that when the income of an individual or a family gets high enough, a motor vehicle will be purchased. However, the correlation is near zero for the U.K. There appears to be no universal certainty, then, that high income areas have more motor vehicles per population than low income areas in the same country.

One further point is of interest. In the lowland/cropland relationship, the all Italy index lies roughly midway between the indices for the two parts North and the Rest. This does not necessarily always happen, as is shown in the income/motor vehicles correlation and as was found also to be the case with several other correlations for variables for Italy not discussed here.

7.2 Comparing correlations in Italy, Spain and France*

This section shows and discusses drawbacks in a comparison of correlation coefficient matrices for the same variables in Italy, Spain and France. Indices are for rank correlation. The variables are labelled according to the letters or the Roman numerals used for France in chapter 6.2.

The fact that statistical offices of national governments collect rather different sets of quantitative data means that variables drawn from government yearbooks and other published records for several countries must be compared with reservations. If however data are found to be reasonably comparable between countries then there is no reason why comparisons should not be made in an attempt to find differences and similarities in the relationships between some variables.

The network of compartments used for Italy was the subdivision of Italy into 92 provinces. This gives a similar number of units to those of France (90 departments). For Spain the 48 provinces were used.

Fifteen variables were included in the Italy and Spain matrices. Fourteen of these are similar enough to those used for France to make comparison possible. Lack of space prevents the inclusion of a list. It will be sufficient to note a few interesting relationships. Table 7.2, a correlation matrix, shows the comparable correlation coefficients for the three countries.

The index of compactness did not achieve any strong correlation in any of the three countries. There was a slight negative correlation in each case with population density (France −0.28, Italy −0.22, Spain −0.33), although there is no obvious reason why this should be so.

Published statistics for Italy include data showing the percentage area of each province occupied by mountain, hill and plain. These figures have been simplified to allow ranking according to percentage area occupied by mountain (or, in reverse, lowland). They are roughly comparable to rankings derived from the Statistical Yearbook for Spain and atlas maps for France.

In all three countries this lowland variable correlates positively with percentage of area under crops, quite strongly so for Italy and France but less for Spain (France +0.64, Italy +0.68, Spain +0.38). For France and Italy lowland also correlates quite strongly negatively with percentage area under pasture (France −0.42, Italy −0.56) and percentage area under forest (France −0.40, Italy −0.68), but this is not so for Spain, where in both instances the coefficients are near zero. For all three countries the unit with most lowland was ranked 1st. Apart from this relationship to lowland, the variables for cropland, pasture and forest were of little further interest for this comparison.

Per capita purchasing power was derived from the same source for France and Spain and from a second source for Italy. Again the figures are reasonably

* Chief Source: (i) for Italy: *Annuario statistico italiano*, 1962 (Istituto Centrale di Statistica), Roma 1963.
(ii) for Spain: *Anuario estadístico de España*, (Presidencia del Gobierno, Instituto Nacional de Estadística), Año XXXVIII, 1963.
(iii) for France: *Annuaire statistique de la France*.

Table 7.2 Correlation Matrix

A	B	C	I	II	D	III	E	F	G	X	I	J	K	N		
	0.41	−0.49	0.06	−0.25	0.28	−0.21	−0.22	0.37	0.10	−0.12	−0.30	−0.22	0.06	−0.21		
1.00	0.09	−0.59	−0.18	−0.37	0.42	0.16	0.32	0.10	−0.01	−0.45	−0.38	−0.45	−0.42	−0.51	A	Area of Province
	0.19	−0.18	−0.01	−0.17	0.16	0.32	0.33	−0.18	−0.19		−0.18	−0.16	−0.10	0.03		
		0.49	0.66	0.05	0.04	0.16	0.11	0.07	−0.16	0.32	0.12	0.01	0.30	0.03		
	1.00	0.67	0.74	0.24	−0.09	−0.11	0.08	−0.10	−0.07	0.34	0.13	0.00	0.47	0.23	B	Population of Province
		0.87	0.69	0.25	−0.17	0.39	0.21	−0.17	−0.30		0.45	0.32	0.53	−0.09		
			0.60	0.43	−0.22	0.30	0.17	−0.17	−0.15	0.53	0.41	0.30	0.33	0.32		
		1.00	0.73	0.46	−0.33	−0.24	−0.18	−0.13	−0.07	0.59	0.42	0.38	0.72	0.58	C	Density of population
			0.70	0.35	−0.28	0.30	0.10	−0.12	−0.21		0.54	0.39	0.63	−0.10		
				0.72	−0.18	0.26	0.06	−0.04	−0.05	0.60	0.35	0.36	0.36	0.35		
			1.00	0.80	−0.16	−0.07	0.08	−0.19	−0.11	0.64	0.39	0.34	0.66	0.55	I	Population of capital
				0.83	−0.15	0.20	0.09	−0.19	−0.13		0.42	0.43	0.58	0.15		
					−0.25	0.18	0.01	−0.16	0.01	0.56	0.35	0.51	0.28	0.49		
				1.00	−0.06	−0.02	0.07	−0.25	−0.11	0.66	0.46	0.50	0.53	0.60	II	Popn. of cap. as % of total population
					−0.12	−0.04	−0.12	−0.08	0.05		0.21	0.42	0.42	0.28		
						−0.22	−0.07	0.06	0.14	−0.20	−0.10	−0.21	−0.17	−0.13		
					1.00	0.17	0.24	−0.12	−0.12	−0.18	−0.07	−0.07	−0.17	−0.16	D	Compactness of Province
						0.13	0.25	−0.18	−0.09		−0.18	−0.05	−0.31	0.20		
							0.68	−0.56	−0.68	−0.12	0.04	0.08	0.08	0.05		
						1.00	0.38	−0.19	−0.04	0.04	0.05	0.05	0.13	−0.05	III	Lowland
							0.64	−0.42	−0.40		−0.09	−0.01	−0.03	0.06		
								−0.76	−0.84	−0.44	−0.32	−0.34	−0.18	−0.29		
							1.00	−0.62	−0.54	−0.24	−0.27	−0.20	−0.16	−0.04	E	Area under crops
								−0.42	−0.52		0.04	−0.15	−0.13	0.07		
									0.45	0.28	0.07	0.09	0.19	0.05		
								1.00	0.67	−0.05	0.08	0.06	−0.11	−0.14	F	% pasture
									−0.03		0.04	−0.29	−0.21	−0.39		
										0.35	0.30	0.28	0.10	0.32		
									1 00	−0.01	0.19	0.17	0.06	0.01	G	% forest
											0.32	0.24	0.25	0.07		
											0.64	0.58	0.42	0.57		
										1.00	0.79	0.61	0.76	0.60	X	Contrib. of non-agric.
												0.82	0.15	0.84		
											1.00	0.79	0.74	0.61	I	Employment in MF
												0.49	0.61	−0.04		
													0.32	0.92		
												1.00	0.66	0.83	J	Income per inhabitant
													0.69	0.48		
														0.29		
													1.00	0.68	K	Rate of popn. 51–61 growth
														0.20		

Comparable correlation coefficients for all Italy (upper rows) and all Spain (middle rows) and all France (bottom rows).

No suitable variable for France was found to make possible a comparison with variable X, contribution of non-agricultural activities to total income of comparable. In all three countries purchasing power correlates quite strongly with the variable for motor vehicles. In Italy and Spain the relationships of income to vehicles are +0.92 and +0.83 respectively, in France only +0.48. Again, in all three countries per capita purchasing power correlates very significantly with population growth (France +0.69, Italy +0.32, Spain +0.66), suggesting that either people have larger families in prosperous areas or that people are moving into more prosperous areas.

The French censuses are such that the calculation of percentage population growth had to be made for the period 1954–1962 whereas for Italy and Spain the period used was 1951–1961. The discrepancy is not serious enough to render comparisons impossible.

Other detailed points emerge from a consideration of table 7.2. The general idea also seems worth developing further.

7.3 Comparing relationships between many variables in North and in Peninsular Italy

This section gives examples of some of the difficulties of making comparisons between factors derived from factor analysis in two or more areas. North and Peninsular Italy are used to illustrate the areas. Sixteen variables are used, three with a physical basis, two on land use, two employment, two demographic and the rest economic, social and electoral. They have been prepared and processed on lines similar to those described for the U.S.A. in chapter 6.3 and 6.4. A complete definition of the variables and the raw data is therefore omitted. A brief definition is given in table 7.3. The 16 variables are fairly representative of the kind of distributions used by geographers studying Italy, though many more might be considered before justice could be done

Table 7.3 Data from correlation matrix

	Brief definition of variable	Sum of differences in correlation	Rank of sum of differences in correlation
N	Illiteracy	759	1
L	Stillbirths	723	2
M	Infant mortality	623	3
P	Communist vote	603	4
I	Population change (51–61)	569	5
D	Cropland	472	6
K	Radios	453	7
J	Income	448	8
F	Agricultural employment	438	9
O	Illegitimate births	408	10
G	Industrial employment	403	11
C	Lowland	373	12
E	Forest	347	13
A	Temperature (July)	298	14
B	Precipitation (Annual)	292	15
H	Density of population	244	16

Table 7.4 Correlation matrix of North and Peninsular Italy

NO—North
PE—Peninsula

Variable			A	B	C	D	E	F	G	H
Temperature	A		100							
Precipitation	B	NO	$\underline{52}$	100						
		PE	$\underline{63}$							
Lowland	C	NO	76	$\underline{56}$	100					
		PE	72	$\underline{63}$						
Cropland	D	NO	77	$\underline{73}$	89	100				
		PE	68	$\underline{53}$	62					
Forest	E	NO	$\underline{42}$	54	$\underline{81}$	$\underline{80}$	100			
		PE	$\underline{48}$	53	$\underline{45}$	$\underline{88}$				
Agric. emplt.	F	NO	24	$\underline{56}$	27	53	$\underline{42}$	100		
		PE	$\underline{01}$	$\underline{14}$	$\underline{04}$	46	$\underline{55}$			
Industrial emplt.	G	NO	$\underline{20}$	49	$\underline{19}$	$\underline{34}$	19	$\underline{79}$	100	
		PE	$\underline{19}$	35	$\underline{14}$	$\underline{53}$	67	$\underline{79}$		
Population density	H	NO	24	03	25	$\underline{01}$	$\underline{03}$	$\underline{57}$	24	100
		PE	15	00	27	17	$\underline{13}$	$\underline{39}$	19	
Population change	I	NO	$\underline{18}$	26	$\underline{34}$	$\underline{40}$	39	$\underline{55}$	49	30
		PE	52	$\underline{33}$	67	26	$\underline{18}$	$\underline{43}$	17	47
Income	J	NO	18	$\underline{12}$	04	07	10	$\underline{40}$	25	54
		PE	18	$\underline{14}$	22	$\underline{22}$	39	$\underline{78}$	47	17
Radios	K	NO	51	$\underline{16}$	28	13	04	$\underline{44}$	23	66
		PE	18	$\underline{07}$	22	$\underline{18}$	28	$\underline{83}$	47	38
Stillbirths	L	NO	05	$\underline{05}$	$\underline{27}$	$\underline{14}$	35	12	$\underline{12}$	$\underline{06}$
		PE	$\underline{03}$	$\underline{01}$	06	33	$\underline{46}$	64	$\underline{48}$	10
Infant mortality	M	NO	$\underline{25}$	03	$\underline{10}$	$\underline{08}$	$\underline{19}$	$\underline{07}$	27	$\underline{18}$
		PE	08	$\underline{10}$	19	34	$\underline{49}$	46	$\underline{44}$	26
Illiteracy	N	NO	57	$\underline{51}$	56	$\underline{61}$	$\underline{47}$	50	$\underline{49}$	$\underline{09}$
		PE	11	$\underline{03}$	10	40	$\underline{48}$	73	53	$\underline{07}$
Illegitimate births	O	NO	01	$\underline{10}$	07	$\underline{04}$	$\underline{03}$	16	$\underline{44}$	11
		PE	11	27	16	$\underline{03}$	08	$\underline{42}$	25	36
Communist vote	P	NO	59	$\underline{52}$	39	47	$\underline{23}$	26	$\underline{29}$	02
		PE	26	$\underline{30}$	20	$\underline{14}$	32	$\underline{37}$	30	$\underline{08}$
			A	B	C	D	E	F	G	H

[a] Indices of correlation have been rounded to 2 decimal places. The decimal point has then been moved two digits to the right. A bar under the number indicates a negative correlation.

Differences in correlations between North and Peninsular Italy

	A	B	C	D	E	F	G	H	I	J	K	L	M	N	O	P
A	00	11	04	09	06	25	01	09	70	00	33	08	33	46	10	33
B	11	00	07	20	01	42	14	03	59	02	09	04	13	48	37	22
C	04	07	00	27	36	31	05	02	101	18	06	33	29	46	09	19
D	09	20	27	00	08	07	19	18	66	15	31	47	42	101	01	61
E	06	01	36	08	00	13	48	10	57	29	24	81	30	01	11	55
F	25	42	31	07	13	00	00	18	12	38	39	52	53	23	58	63
G	01	14	05	19	48	00	00	05	32	22	24	30	71	04	69	59
H	09	03	02	18	10	18	05	00	17	37	28	16	44	02	25	10
I	70	59	101	66	57	12	32	17	00	21	12	09	16	22	68	17
J	00	02	18	15	29	38	22	37	21	00	06	100	48	73	16	23
K	33	09	06	31	24	39	24	28	12	06	00	89	26	89	27	10
L	08	04	33	47	81	52	30	16	09	100	89	00	88	93	10	63
M	33	13	29	42	30	53	71	44	16	48	26	88	00	81	09	40
N	46	48	46	101	01	23	04	02	22	73	89	93	81	00	30	100
O	10	37	09	01	11	58	69	25	68	16	27	10	09	30	00	28
P	33	22	19	61	55	63	59	10	17	23	10	63	40	100	28	00

100							
67							
	100						
46							
47	85						
		100					
59	91						
10	26	26					
			100				
01	74	63					
00	09	15	07				
				100			
16	57	41	81				
34	08	12	07	11			
					100		
12	81	77	86	70			
16	11	15	02	09	31		
						100	
52	27	42	08	18	03		
06	35	47	06	05	63	24	
							100
11	58	37	57	45	37	04	
I	J	K	L	M	N	O	P

to the complexity of the country. North Italy consists of 40 provinces, as does Peninsular Italy (Italia Centrale plus Italia Meridionale). Note that high negative indices of correlation are achieved between certain pairs of variables (e.g. D and E, F and G, because these are to some extent reciprocals).

The main correlation matrix (table 7.4) gives Pearson product-moment indices of correlation for North Italy and for Peninsular Italy, one above the other to facilitate comparison. The correlation index between certain pairs of variables (e.g. F and G) is the same or very similar in North Italy and in Peninsular Italy; but is very different for other pairs of variables. For example, N correlates with P (Illiteracy and Communist vote) +0.63 in North Italy, but −0.37 in Peninsular Italy.

It is possible to produce a new matrix in which the difference between North and Peninsular indices of correlation for each pair of variables is recorded. This has been done (smaller matrix in table 7.4). The sum of all the values in each column (or row) is an indication of the degree to which each variable correlates with all the others in a similar way or differently in North Italy and Peninsular Italy. Thus, for example, Illiteracy (N) is much less closely correlated with the other variables than is Density of population (H).

A little thought about the main correlation matrix in table 7.4 suggests that, given the lack of agreement between North and Peninsular Italy in the indices of correlation for many pairs of variables, it is not possible for identically or even similarly loaded factors to emerge. Given that the data are the same (the sources and information are identical for all Italy) it is

Table 7.5 Factors

	North factor 1	Peninsula factor 2	North factor 2	Peninsula factor 1
	34%	25%	22%	38%
July temperature	−0.74	+0.83	+0.45	−0.12
Annual precipitation	+0.78	−0.69	−0.10	+0.19
Lowland as %	−0.82	+0.87	+0.27	−0.11
Arable as %	−0.92	+0.67	+0.09	−0.57
Forest as %	+0.74	−0.54	+0.03	+0.71
Agricultural employment	−0.68	−0.18	−0.55	−0.90
Industrial employment	+0.60	−0.13	+0.36	+0.76
Population density	+0.07	+0.43	+0.74	+0.13
Population change 51–61	+0.53	+0.79	+0.59	+0.22
Income	+0.11	+0.38	+0.88	+0.86
Radios	−0.09	+0.45	+0.96	+0.81
Stillbirths	+0.10	−0.01	+0.15	−0.83
Infant mortality	+0.12	+0.18	−0.16	−0.71
Illiteracy	−0.79	−0.02	+0.03	−0.88
Illegitimate births	−0.22	+0.30	+0.04	+0.23
Communist vote	−0.61	+0.23	+0.39	+0.53

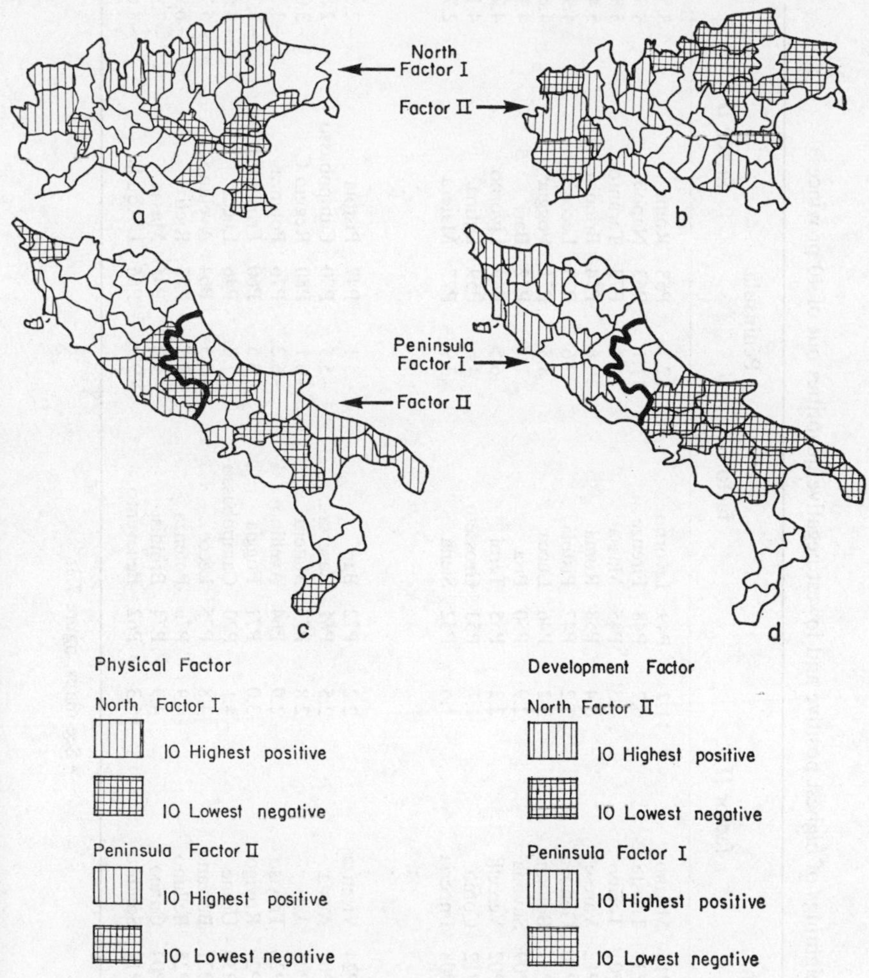

Figure 7.3 Individual weightings (factor scores) for provinces of North and of Peninsular Italy. The highest and lowest quartiles of the 40 provinces in each subarea are mapped. See table 7.6 for the data mapped. Note that factor I for peninsular Italy has been put with factor II for North Italy and vice versa.

feasible that if the relationships were the same in the two areas for the sixteen variables, similar factors would emerge. How different the factors actually are is shown by the data in table 7.5, the loadings of the 16 variables on the first two factors for North and Peninsular Italy respectively.

In North Italy, the first two (out of 16) factors (principal components), have 56 per cent of all the variation, while in Peninsular Italy, the first two

Table 7.6 Individual weightings of highest positive and lowest negative quantities out of 40 provinces[a]

North						Peninsula					
factor I			factor II			factor I			factor II		
P14	Sondrio	8.2	P15	Milano	11.7	P49	Livorno	12.7	P58	Roma	8.5
P13	Como	8.0	P31	Trieste	6.7	P48	Firenze	11.1	P63	Napoli	8.3
P22	Trento	8.0	P01	Torino	5.5	P45	Massa	11.1	P73	Taranto	5.8
P12	Varese	7.5	P12	Varese	5.4	P58	Roma	10.6	P74	Brindisi	5.4
P25	Belluno	6.9	P10	Genova	5.2	P47	Pistoia	9.6	P75	Lecce	4.9
P16	Bergamo	6.5	P37	Bologna	3.2	P46	Lucca	8.8	P71	Foggia	4.6
P21	Bolzano	6.1	P09	Savona	2.9	P50	Pisa	7.4	P72	Bari	4.3
P01	Torino	5.8	P02	Vercelli	2.2	P55	Terni	6.5	P49	Livorno	4.2
P07	Aosta	5.6	P12	Como	1.7	P53	Grosseto	4.6	P59	Latina	4.1
P10	Genova	5.6	P08	Imperia	1.6	P52	Siena	3.9	P77	Matera	2.7
	⋮			⋮			⋮			⋮	
P06	Alessandria	−3.9	P24	Vicenza	−2.2	P72	Bari	−5.1	P47	Pistoia	−2.6
P28	Padova	−4.9	P07	Aosta	−2.5	P61	Caserta	−5.5	P70	Campobasso	−2.8
P05	Asti	−5.0	P05	Asti	−2.8	P77	Matera	−6.1	P80	Reggio C.	−3.0
P19	Cremona	−5.0	P26	Treviso	−2.9	P64	Avellino	−6.2	P76	Potenza	−3.3
P27	Venezia	−5.5	P29	Rovigo	−3.0	P71	Foggia	−6.5	P60	Frosinone	−4.3
P40	Forli	−6.2	P30	Udine	−4.1	P70	Campobasso	−7.0	P46	Lucca	−4.6
P20	Mantova	−7.4	P21	Bolzano	−4.8	P75	Lecce	−7.5	P64	Avellino	−5.7
P39	Ravenna	−8.9	P25	Belluno	−4.9	P76	Potenza	−7.7	P57	Rieti	−6.9
P29	Rovigo	−10.5	P04	Cuneo	−4.9	P74	Brindisi	−9.0	P45	Massa	−7.8
P38	Ferrara	−10.8	P14	Sondrio	−5.3	P62	Benevento	−9.9	P66	L'Aquila	−8.6

[a] See maps, figure 7.3.

have 60 per cent. What is clear from the factor loadings, however, is that factors 1 and 2 for North Italy have mostly different variables loading highly on them from factors 1 and 2 for Peninsular Italy. However by interchanging 1 and 2 for Peninsular and putting factor 1 for North Italy against factor 2 for Peninsular Italy it is possible to see a closer resemblance. Similarly, 2 for North and 1 for Peninsular Italy are more similar. The reason is that physical features are exerting a stronger influence on other variables in North Italy than they are in Peninsular Italy. Economic and social considerations therefore take over factor 1 in Peninsular Italy, whereas factor 1 in North Italy is clearly dominated by physical and land use influences. The actual weightings of provinces are partially presented in table 7.6 and are mapped in figure 7.3.

There are certainly no formal methods of comparing the factors of North and Peninsular Italy. The reader is left, therefore, to consider for himself ways of exploiting the information in the very complex situation presented in this section. Reference should also be made to J. C. Fisher (1966), who compares factors in subareas in Yugoslavia. Another example is discussed at greater length in the next section (7.4).

7.4 An example of factor analysis applied to geomorphological data for six areas in northern England

The analysis is divided into the following sections. (*a*) The purpose of the analysis. (*b*) The choice of areas. (*c*) The choice of variables. (*d*) Analysis of the results. (i) Means and standard deviations; (ii) The correlation matrix; (iii) The factors — each area and the total area; (iv) The factor loadings on the variables for each area and the total area; (v) The factor weightings for subareas within each area; (vi) The factor weightings for each area within the total area. (*e*) Conclusions.

(*a*) *The purpose of factor analysis of areas in northern England*

Northern England contains areas of very different landform characteristics. One of the purposes of the analysis was to study quantitatively the differences and similarities between these different areas. The value of factor analysis, using a number of different variables to characterize the areas, is based on the analytical process which sorts out the combination of variables that best 'explain' the variation within the areas. The value of each factor indicates the amount of the variation that this factor accounts for, while the factor loadings provide a measure of the part the different variables play in accounting for this 'explanation'. From the factor loadings it is possible to obtain a measure of the most diagnostic and significant variable that can best be used to characterize the area concerned and to differentiate it from other areas. Variation within fairly homogeneous areas was studied as well as variation between very diverse areas.

(*b*) *The choice of areas*

Northern England can be divided into a number of different areas on the basis of the geology and geomorphological development. The major divisions include the Cheviot Hills, the Pennines north of the Tyne Gap, the coastal lowland area of Northumberland, the Alston and Askrigg Blocks of the Central Peninnes, the Lake District, the Eden Valley and Solway Lowlands. The areas chosen for this analysis covered as nearly as possible homogeneous areas within these major divisions. Six major regions covered a sufficient area with reasonable homogeneity to be used in the analysis. These areas are as follows:

A1 The Askrigg Block
A2 The Alston Block
A3 The Cheviots
A4 The Lake District
A5 The Solway Lowlands
A6 The Northumberland Coastal area.

The reader is advised to consult appropriate sheets of the 1 inch to 1 mile Ordnance Survey Map. Grid references of the centres are given for the 5 squares shown on figure 7.4.

Area 1 Askrigg Block Sheet 90
1, 95/95; 2, 85/95; 3, 85/85; 4, 95/85; 5, 85/75.
Area 2 Alston Block Sheet 84
1, 85/25; 2, 85/35; 3, 95/35; 4, 85/45; 5, 95/45.
Area 3 Cheviots Sheets 70 and 71
1, 85/25; 2, 95/25; 3, 85/15; 4, 95/15; 5, 75/05.
Area 4 Lake District Sheets 82, 88, 89
1, 25/25; 2, 15/15; 3, 25/15; 4, 35/15; 5, 25/05.
Area 5 Solway Lowlands Sheets 75, 82, 83.
1, 25/55; 2, 15/45; 3, 25/45; 4, 35/45; 5, 45/45.
Area 6 Northumberland Coast Sheets 71, 78.
1, 15/25; 2, 15/15, 3, 15/05; 4, 15/95; 5, 15/85.

There are thus four upland areas and two lowland areas. This distribution reflects the generally high level of Northern England, which is part of Highland Britain (figure 7.4).

Each of the areas chosen covers five 10 by 10 km squares on the 1 inch to 1 mile Ordnance Survey maps. Thus 500 square km of ground were covered in each major region. Each 10 by 10 km square was divided into four, a SW, SE, NE and NW quarter for the purposes of collecting the data. Each of the analysed subareas covered 25 sq km, 20 of these subareas being included in each major area. Equal sized units were chosen, rather than drainage basins, as it has been shown that the area of a drainage basin is one of the most important of the controlling variables of the basin. Other parameters associated with the drainage basin are usually very closely correlated with the area of the basin. By taking uniform areas, rather than drainage basins, which vary in area greatly, the variable of area is held constant and the true variation of the other parameters measured can be more easily established. In all instances the five 10 by 10 km squares were contiguous, at least at a corner, but the pattern varied so as to include as representative and homo-

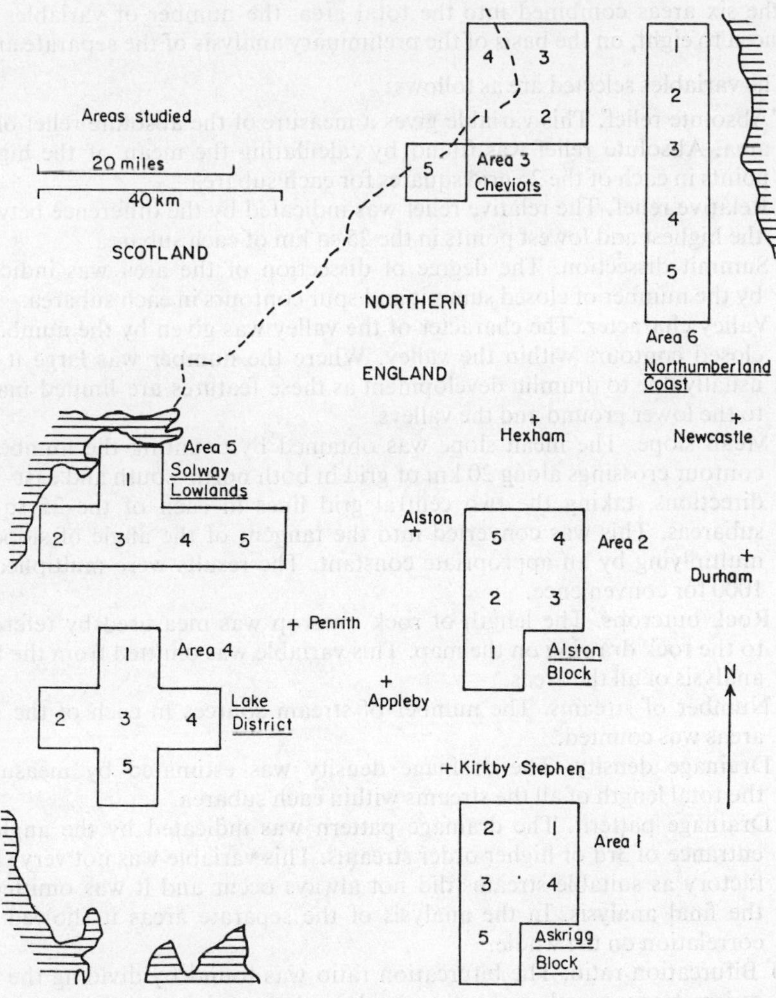

Figure 7.4. Location of the six subareas studied in the Northern England factor analysis example.

geneous an area as possible, so far as a visual inspection of the map allowed, within the limits set by the variables selected.

(*c*) *The choice of variables*

One of the criteria used in the selection of the variables was that they should be obtainable from the 1 inch to 1 mile Ordnance Survey maps. They are intended to measure the relief and drainage characteristics of the areas. Ten variables were collected for the six areas separately, while in the analysis of the six areas combined into the total area, the number of variables was reduced to eight, on the basis of the preliminary analysis of the separate areas.

The variables selected are as follows:

V1 Absolute relief. This variable gives a measure of the absolute relief of the area. Absolute relief was found by calculating the mean of the highest points in each of the 25 grid squares for each subarea.

V2 Relative relief. The relative relief was indicated by the difference between the highest and lowest points in the 25 sq km of each subarea.

V3 Summit dissection. The degree of dissection of the area was indicated by the number of closed summit and spur contours in each subarea.

V4 Valley character. The character of the valley was given by the number of closed contours within the valley. Where the number was large it was usually due to drumlin development as these features are limited mainly to the lower ground and the valleys.

V5 Mean slope. The mean slope was obtained by counting the number of contour crossings along 20 km of grid in both north–south and east–west directions, taking the two central grid lines in each of the 25 sq km subareas. This was converted into the tangent of the angle of slope by multiplying by an appropriate constant. The results were multiplied by 1000 for convenience.

V6 Rock outcrops. The length of rock outcrop was measured by reference to the rock drawing on the map. This variable was omitted from the final analysis of all the areas.

V7 Number of streams. The number of stream sources in each of the subareas was counted.

V8 Drainage density. The drainage density was estimated by measuring the total length of all the streams within each subarea.

V9 Drainage pattern. The drainage pattern was indicated by the angle of entrance of 3rd or higher order streams. This variable was not very satisfactory as suitable streams did not always occur and it was omitted in the final analysis. In the analysis of the separate areas it showed low correlation on the whole.

V10 Bifurcation ratio. The bifurcation ratio was found by dividing the first order stream numbers by the number of second order streams within each subarea. The range of values was too small to be significant.

(*d*) *Analysis of the results*

(i) *Means and standard deviations*. The means and standard deviations for the 10 variables for the six areas are shown in table 7.7. The means for the total area and the standard deviations are also shown. The units in which each is given are indicated. Each of the variables will be commented upon.

V1 *Absolute relief*. The four upland areas are of comparable mean height. The two Pennine areas are very similar, but the Cheviot area is considerably lower and the Lake District a little higher. The Solway area is the lowest. It is interesting to note that the two central Pennine areas have similar and

Table 7.7 Means and standard deviations

Areas	Variable 1 Absolute relief		Variable 2 Relative relief		Variable 3 Summit dissection		Variable 4 Valley character		Variable 5 Mean slope		Variable 6 Rock outcrop		Variable 7 Number of streams		Variable 8 Drainage density		Variable 9 Drainage pattern		Variable 10 Bifurcation ratio	
	Ft to nearest 10		Ft		Number		Number		Tangent ×1000		km		Number		Miles		Degrees			
	Mean	St. dev.	Mean	St. dev.	Mean	St. dev.	Mean	St. dev.	Mean	St. dev.	Mean	St. dev.	Mean	St. dev.	Mean	St. dev.	Mean	St. dev.	Mean	St. dev.
A1 Askrigg Block	1650	190	1210	140	3.2	1.5	3.9	6.7	136	30	14	1.5	39	15	20.2	6.2	75	15	3.7	0.9
A2 Alston Block	1690	220	1030	150	3.1	1.3	0.6	1.6	104	14	0.6	1.6	39	9	22.7	4.4	69	17	3.6	0.5
A3 Cheviots	1360	300	1190	300	11.2	3.9	0.5	1.2	170	38	0.4	0.8	35	10	21.0	2.6	62	12	3.5	0.8
A4 Lake District	1790	410	2250	370	11.8	4.4	0.9	1.9	277	71	8.9	57	26	11	18.0	3.5	82	24	3.7	0.8
A5 Solway Lowland	300	220	320	250	1.3	2.1	7.7	6.3	40	18	0	0	7	4	10.4	3.3	62	22	3.1	1.4
A6 Northumberland Coast	420	130	410	150	3.7	2.0	2.5	2.5	46	17	0	1	11	6	13.7	2.2	52	23	2.9	1.0
AA Total area	1200	660	1070	680	5.7	5.0	2.6	4.8	129	89	—		26	16	17.7	5.8	—		3.4	1.0

considerably lower standard deviations than the other two upland areas, particularly the Lake District. This reflects the plateau-like character of the Pennines compared with the Lake District.

V2 *Relative relief.* The same tendency is apparent in the relative relief figures for the mean and standard deviation for both the Pennines and the Cheviots as opposed to the Lake District. Again, however, the standard deviation is very considerably less in the two central Pennine areas than in the other uplands.

V3 *Summit dissection.* This variable mean value brings out very clearly the greater dissection in the Lake District and Cheviots than in the central Pennines.

V4 *Valley character.* The main areas of drumlins are revealed by this variable. They are very numerous in the Solway Lowland and occur sporadically in the Askrigg Block and along the Northumberland coast, but valleys are too steep and narrow on the whole in the Cheviots, Lake District and Alston Block.

V5 *Slope.* There is a close correlation between the mean values of variables 2 (Relative relief) and 5 (Mean slope) for the six areas. This is reasonable as relative relief should be related to mean slope under many conditions, although this need not necessarily be so. The two lowland areas have a large standard deviation relative to a small value for mean slope. This indicates considerable variation of slope within these areas.

V6 *Rock outcrops.* The outstanding rugged character of the Lake District is immediately apparent in this variable. The very large values of the standard deviation indicate a considerable range of values in the other upland areas. This variable is meaningless in the lowland areas and was omitted from the total area.

V7 *Number of streams.* The Pennines and Cheviots clearly have more streams than the other areas, particularly the lowland areas. The two central Pennine areas again show similar values.

V8 *Drainage density.* This variable shows very similar relationships to those shown by variable 7. The Pennines are the best watered areas and the lowlands the least. The Cheviots show rather less variability and the Askrigg Block and Solway areas rather more variability than the other three areas.

V9 *Drainage pattern.* There is little useful variation in this variable and it will not be analysed further. This is partly due to the difficulty of obtaining valid figures for all the areas.

V10 *Bifurcation ratio.* This variable also has little diagnostic value but for a different reason. There is little variation in the bifurcation ratio throughout the area and the standard deviations are on the whole low. The mean value is 3.4, which agrees closely with the value of 3.5 found in many U.S. drainage basins. It has been shown that this drainage characteristic tends to a uniform value in widely different areas. For this reason it is not of very great value in this analysis.

In most areas the relationship between the means and standard deviations appears to be approximately linear. However, the distance of any point from the straight line can be of importance in the analysis of the correlation and factors. The extra low standard deviations of areas 1 and 2 (Askrigg and Alston Blocks) in variables 1 and 2 (Absolute relief and Relative relief) are important. The analysis indicates that normally the standard deviation increases with an increasing value of the mean, which is quite reasonable in many instances as large values of the individuals will give rise to a large standard deviation, given a similar spread of data. In order to make the standard deviation comparable for different means, therefore, a relative standard deviation can be used. This is obtained by dividing the mean by the standard deviation. The significance of this value will be considered when the factor loadings are discussed in relation to the different areas. The relative standard deviation of some selected variables will be related to the relief characteristics of the area. The mean and standard deviations provide a useful summary of the value of the variables in the different areas and within the total area. The standard deviations increase greatly when all the areas are combined together in the total area.

(ii) *The correlation matrix.* The correlation matrix is given in table 7.8. The degree of correlation between the variables varies very considerably from one area to another. In the first area, the Askrigg Block, there are very few high correlations. Some variables, notably 1 (Absolute relief) and 2 (Relative relief) show no correlation in this area, whereas they show high correlations in most of the other areas. This is probably connected with the low standard deviations of the variables 1 (Absolute relief) and 2 (Relative relief) in the first two areas, (the central Pennines). This also results in a lack of correlation between variables 2 (Relative relief) and 5 (Mean slope), which is strong in nearly all the other areas. There is thus no systematic variation between any of the relief variables in the Askrigg Block (area 1). This seems to be a reflexion of the homogeneity of the relief features of this area. They vary very little and such variation as there is, is random in character. These comments also apply to the Alston Block (area 2), which is very similar in these respects to the Askrigg Block.

The only strong correlation within the Askrigg Block and the Alston Block (areas 1 and 2) is between variables 1 (Absolute relief) and 7 (Number of streams). This shows that, as would be expected, there are more streams rising on the higher ground. This relationship holds in all upland areas but not in the two lowland areas. Both areas 1 and 2 (the Askrigg and Alston Blocks) show a fairly strong correlation between number of stream sources (V7) and total stream length (V8). This again is to be expected and applies in all the upland areas and one of the lowland areas.

In these two Central Pennine areas the variation in some of the most important variables, as indicated by the analysis as a whole, is small. These variations tend to be random in relation to other variables. It is not, therefore, possible to characterize these areas by using only one or two variables, on which the others depend.

Area 3 (The Cheviots) has characteristics intermediate between areas 1 and 2 (Central Pennines) and area 4 (Lake District). The Cheviots show the same

Table 7.8 Correlation[a] matrices

Areas		V1		V2		V3		V4		V5		V6	V7		V8		V9	V10
V1	A1																	
Absolute	A2																	
relief	A3																	
	A4	100																
	A5																	
	A6																	
V2	A1	$\underline{00}$																
Relative	A2	02																
relief	A3	60																
	A4	77	81	100														
	A5	90																
	A6	$\underline{69}$																
V3	A1	31		$\underline{03}$														
Summit	A2	12		$\underline{17}$														
dissection	A3	$\underline{13}$		$\underline{17}$														
	A4	60	45	38	63	100												
	A5	66		55														
	A6	53		22														
V4	A1	$\underline{37}$		24		46												
Valley	A2	18		08		$\underline{28}$												
character	A3			$\underline{27}$		$\underline{02}$												
	A4	$\underline{77}$	$\underline{45}$	$\underline{59}$	$\underline{35}$	$\underline{43}$	$\underline{35}$	100										
	A5	$\underline{62}$		$\underline{50}$		$\underline{61}$												
	A6			$\underline{24}$		$\underline{21}$												
V5	A1	03		$\underline{08}$		$\underline{04}$		$\underline{38}$										
Mean	A2	08		$\underline{24}$		42		03										
slope	A3	46		44		57		$\underline{37}$										
	A4	81	75	68	91	42	74	$\underline{71}$	$\underline{38}$	100								
	A5	89		88		65		$\underline{42}$										
	A6	65		75		32		$\underline{27}$										
V6	A1	34		$\underline{08}$		08		$\underline{34}$		$\underline{02}$								
Rock	A2	34		23		00		$\underline{07}$		$\underline{25}$								
outcrops	A3	12		05		$\underline{48}$		$\underline{11}$		$\underline{24}$								
	A4	60		37		$\underline{37}$		42		57		100						
	A5	$\underline{08}$		$\underline{09}$		27		$\underline{22}$		$\underline{23}$								
	A6	$\underline{05}$		07		03		$\underline{14}$		13								
V7	A1	64		$\underline{31}$		25		$\underline{29}$		28		14						
Number of	A2	70		$\underline{02}$		32		03		09		59						
streams	A3	86		42		$\underline{06}$		$\underline{55}$		44		12						
	A4	62	80	31	44	$\underline{43}$	26	$\underline{64}$	$\underline{40}$	42	43	22	100					
	A5	28		35		$\underline{04}$		$\underline{29}$		15		$\underline{05}$						
	A6	62		12		34		$\underline{52}$		18		$\underline{15}$						
V8	A1	29		$\underline{11}$		$\underline{06}$		$\underline{21}$		18		09	78					
Drainage	A2	54		52		00		01		23		54	65					
density	A3	46		20		$\underline{36}$		$\underline{70}$		$\underline{11}$		26	60					
	A4	33	69	21	42	21	22	$\underline{52}$	$\underline{42}$	17	37	$\underline{03}$	81	85	100			
	A5	27		29		12		$\underline{17}$		28		$\underline{19}$	29					
	A6	33		27		$\underline{11}$		$\underline{38}$		31		$\underline{12}$	63					
V9	A1	$\underline{20}$		$\underline{19}$		19		31		$\underline{40}$		16	$\underline{40}$		$\underline{46}$			
Drainage	A2	$\underline{27}$		$\underline{09}$		$\underline{18}$		16		$\underline{20}$		$\underline{01}$	08		02			
pattern	A3	05		$\underline{04}$		$\underline{18}$		$\underline{25}$		$\underline{15}$		29	10		24			
	A4			$\underline{07}$		$\underline{53}$		17		$\underline{13}$		$\underline{35}$	$\underline{49}$		$\underline{21}$		100	
	A5	$\underline{36}$		$\underline{40}$		00		41		$\underline{36}$		29	$\underline{16}$		$\underline{34}$			
	A6	05		12		$\underline{14}$		$\underline{11}$		20		08	$\underline{19}$		$\underline{03}$			
V10	A1	30		$\underline{06}$		18		$\underline{01}$		$\underline{08}$		02	32		16		29	
Bifurcation	A2	$\underline{10}$		$\underline{10}$		37		03		27		$\underline{20}$	$\underline{01}$		$\underline{16}$		$\underline{08}$	
ratio	A3	07		18		27		40		09		21	16		$\underline{28}$		15	
	A4	00	32	$\underline{14}$	24	27	22	$\underline{26}$	$\underline{11}$	20	25	36	$\underline{07}$	33	$\underline{23}$	14	$\underline{27}$	100
	A5	$\underline{03}$		$\underline{10}$		01		$\underline{00}$		$\underline{07}$		$\underline{14}$	37		$\underline{35}$		00	
	A6	60		25		56		$\underline{28}$		13		$\underline{14}$	73		30		$\underline{06}$	

A1 Askrigg Block A3 Cheviots A5 Solway Lowland
A2 Alston Block A4 Lake District A6 Northumberland Coast

Values are multiplied by 100 and rounded, bar underneath indicating negative value

[a] The correlation coefficient for *all areas* is placed in the matrix to the right of the six correlation coefficients for each of the six districts. Variables V6 and V9 were not correlated for *all areas*.

strong correlations as areas 1 and 2, namely between variables 1 and 7 (Absolute relief and Number of streams), and between 7 and 8 (Drainage density), but there are in addition other correlations of importance. Variables 1 and 2 (Absolute relief and Relative relief) correlate positively, and variables 1 and 4 (Valley character) correlate negatively. The first of these shows that the higher land has greater relative relief, and the second that the higher land has fewer closed contours in the valleys. Variable 4 (Valley character) correlates negatively with variables 7 and 8 (Number of streams and Drainage density), which is explained by the positive correlation of 7 with 8, and 1 (Absolute relief) with 7 (Number of streams). In the third area, the Cheviots, therefore, there is some intercorrelation between relief and drainage variables.

In the fourth area (Lake District) the correlations are on the whole much higher. Variable 1 (Absolute relief) correlates highly with variables 2, 3, 4, 5, 6, 7 (Relative relief, Summit dissection, Valley character, Mean slope, Rock outcrops and Number of streams). There is an intercorrelation between all the relief variables through variable 1 (Absolute relief). Variables 2 (Relative relief) and 5 (Mean slope) correlate positively, showing that in this area slopes tend to be long as well as steep. Where this correlation is small, slopes are short and steep as in the youthfully dissected Alston Block (A2) and the generally higher and less deeply dissected Askrigg Block (A1). The Cheviots (A3) are intermediate in this respect. This correlation is, therefore, a measure of the depth of dissection. The number of low points (V4) correlates strongly negatively with the mean slope (V5) because where slopes are steep the ground is high and the valleys narrow. Hence few drumlins or other valley closed contours occur.

In the drainage variables there is again a strong correlation between variables 7 and 8 (Number of streams and drainage density). The negative correlation between 3 (Summit dissection) and 9 (Stream entrance angle) shows that as the area becomes more dissected into isolated summits the stream courses are more restricted in possible directions and smaller stream entrance angles occur. Variable 10 (Bifurcation ratio), and 11 and 12 (Number of waterfalls and lakes) do not correlate strongly with any other variable and are, therefore, omitted from the other areas, in which, in any case, they rarely occur.

In area 5 (The Solway lowlands), there are strong correlations amongst the relief variables, but no strong correlations amongst the drainage variables, probably reflecting the rather disturbed nature and limited development of the drainage of this lowland, drumlin-covered area. Variable 6 is also meaningless in this area as no cliffs occur. (A token amount of cliff had to be included in order to make the computer program work.) There is a strong correlation between variables 2 (Relative relief) and 5 (Mean slope), showing that the drumlins that provide most of the slope also cover the full relief range, forming the hill summits.

Area 6 (Northumberland coast) shows similar, but less strong, correlations among the relief features and some drainage correlations also occur in this area of more integrated drainage. Variable 10 correlates with the variables 7, 3 and 1. These correlations are probably not very significant, as in an area such as this where there are few streams the ratio is difficult to measure accurately. However it does suggest that first order streams tend to be longer in the lower parts of this lowland than in the complex relief of the uplands.

On the whole the individual areas show that there is more intercorrelation between the relief variables than the drainage ones with the exception of variables 7 and 8 (Number of streams and Drainage density) which nearly all show a strong positive correlation.

When all the areas are combined the same relationships hold. The correlations are fairly high, as would be expected when a greater range of variables are included from the highest areas to the lowest ones. This greater range among the values for each set of variables is seen in the much higher standard deviations for the combined data. Variables 1 and 2 correlate strongly, showing that absolute relief (V1) and relative relief (V2) are clearly associated. The relative relief (V2) also correlates with the number of high summits (V3) and there is a very strong correlation between variable 2 (Relative relief) and variable 5 (Mean slope). This is also dependent on the correlation of variables 1 and 2. Where the correlation of variables 1 and 2 (Absolute relief and Relative relief) is high, the correlation of variables 2 and 5 (Relative relief and Mean slope) is also high, a relationship reflected in the fairly high 1 and 5 (Absolute relief and Mean slope) correlation. The length of the slopes seems to be the determining factor in this relationship, as already mentioned. Variables 7 and 8 (Number of streams and Drainage density) correlate closely with each other and both correlate with variable 1 (Absolute relief). This reflects the greater drainage density and number of streams with increasing height. The only negative correlations, and these are consistently negative in all areas, are the number of valley closed contours (V4) with all other variables, but the correlations are not strong. This relationship is reasonable as most of the low closed contours occur in the lower areas (V1), with less relative relief (V2), fewer high summits (V3) and gentler slopes (V5), and also where the drainage is less well developed (V7 and V8).

The correlation matrices show that relationships vary between the individual areas. This variation is to be expected considering the great variety of relief forms in the different areas. However, in most areas some variables are consistently more strongly correlated with others. It is these variables that help to provide the loadings on the factors that are picked out by the process of factor analysis. The variables also account for the weights given to the factors. These will be considered next and then the part played by the variables in the factor loadings will be discussed.

(iii) *Factors.* Table 7.9 is based on the values of the first five factors. It shows the cumulative percentage of the variation that each factor in turn 'explains' in each area and in the whole area.

In considering the factors it must be remembered that areas 1 and 4 (Askrigg Block and Lake District) have 12 variables, areas 2, 3, 5 and 6 (Alston Block,

Table 7.9 Factors. The cumulative percentage of variation 'explained' areas

Factors	A1 Askrigg block	A2 Alston block	A3 Cheviots	A4 Lake District	A5 Solway	A6 Northumber-land coast	All areas
F1	25	28	34	42	41	38	55
F2	41	47	56	56	58	55	71
F3	56	61	71	70	72	68	83
F4	68	74	81	77	83	78	91
F5	78	82	87	85	91	87	96

Cheviots, Solway, and Northumberland coast lowlands) have 10 variables, and the total area has 8 variables. However it is the weakest variables that were omitted so that their omission should not make very much difference to the factors or factor loadings. The results for all the areas should, therefore, be approximately comparable.

The factor values are related to the degree of intercorrelation of the variables. Where the correlation coefficients are low, then the factors are smaller and the values more equal on each successive factor. This means that the axes are shorter and the hyperellipsoid tends more towards a hyperspherical form. The values are smaller in the first two areas, which show the least correlation amongst the variables. This is partly related to their homogeneity, as indicated by the low standard deviations in these areas in some of the most important variables. The Lake District and Solway areas (A4 and A5) show the greatest factor values. In these areas high correlations are most numerous. This means that one or two fundamental variables can be used to characterize these areas. This applies even more strongly to the total area. In the whole area the first factor accounts for about 55 per cent of the total variability, although the fact that there are only eight variables must be taken into account. However, for the reasons given this point is probably not very important.

The factors show the degree of intercorrelation of the variables and the amount of variation within the 20 different subareas that make each major area. The greater the homogeneity of the area, the less the factors can hope to expose by way of variation. Where intercorrelations are weak, no one variable can give a good characterization of the area concerned.

(iv) *Factor loadings.* The individual factors are related to the character of the variables. The loadings on the different variables gives useful information concerning the part the variables play in accounting for the variation within the area. These factor loadings for the first and second factors are shown in table 7.10. It should be noted that the values have been multiplied by 100 to avoid using the decimal point and they have been rounded to the nearest two places of decimals. If they are negative a bar is placed under the figure.

Table 7.10 Factor Loadings

Variables	Factor 1							Factor 2					
	A1 Askrigg block	A2 Alston block	A3 Cheviots	A4 Lake District	A5 Solway	A6 Northumberland coast	AA All areas	A1 Askrigg block	A2 Alston block	A3 Cheviots	A4 Lake District	A5 Solway	A6 Northumberland coast
V1 Absolute Relief	$\underline{71}$	$\underline{77}$	$\underline{90}$	$\underline{90}$	95	92	$\underline{93}$	42	$\underline{02}$	16	$\underline{00}$	$\underline{07}$	$\underline{13}$
V2 Relative Relief	20	$\underline{34}$	$\underline{61}$	$\underline{67}$	92	65	$\underline{86}$	$\underline{11}$	$\underline{07}$	24	07	03	$\underline{56}$
V3 Summit Dissection	$\underline{02}$	$\underline{14}$	18	$\underline{69}$	68	56	$\underline{66}$	67	83	80	$\underline{21}$	$\underline{64}$	09
V4 Valley Character	58	$\underline{07}$	80	86	$\underline{72}$	$\underline{63}$	57	34	$\underline{27}$	06	$\underline{08}$	25	$\underline{00}$
V5 Mean Slope	$\underline{38}$	$\underline{12}$	$\underline{45}$	$\underline{78}$	90	65	$\underline{86}$	$\underline{47}$	72	80	$\underline{17}$	01	$\underline{61}$
V6 Rock Outcrop	$\underline{27}$	$\underline{72}$	$\underline{23}$	$\underline{62}$	$\underline{08}$	$\underline{03}$	—	33	$\underline{21}$	$\underline{62}$	$\underline{54}$	$\underline{71}$	$\underline{45}$
V7 Number of Streams	$\underline{92}$	$\underline{87}$	$\underline{88}$	$\underline{76}$	35	77	$\underline{78}$	19	17	13	45	34	53
V8 Drainage Density	$\underline{78}$	$\underline{88}$	$\underline{72}$	$\underline{54}$	36	56	$\underline{72}$	$\underline{11}$	01	$\underline{45}$	70	61	20
V9 Drainage Pattern	53	$\underline{12}$	$\underline{21}$	45	$\underline{50}$	01	—	61	$\underline{40}$	$\underline{44}$	24	$\underline{50}$	$\underline{53}$
V10 Bifurcation Ratio	$\underline{26}$	18	05	$\underline{21}$	$\underline{05}$	72	$\underline{38}$	59	62	20	$\underline{75}$	$\underline{00}$	42

The values have been multiplied by 100 and rounded, the bar beneath denotes a negative value.

In variable 1 (Absolute relief) only the loadings on the first factor are of any significance. This variable is, therefore, important in accounting for the value of the first factor. The other variable that has high loadings in nearly all areas is variable 7 (Number of streams), with which variable 8 (Drainage density) is also closely associated. The inter-relationship of variables 1 (Absolute relief), 2 (Relative relief) and 5 (Mean slope) in four of the areas (A3, Cheviots, A4, Lake District, A5, Solway, and A6, Northumberland coast) leads to strong loadings on these variables in these areas. In order to assess the reason for the loadings on the different variables the relative standard deviations for some of them were plotted against the factor loadings for these variables. In the case of variables 1, 2, 3 and 5 there appears to be a negative correlation between the relative standard deviation, which is a measure of homogeneity (it is high in a homogeneous area) and the factor loading. Thus a high factor loading occurs in the least homogeneous areas. In these areas the relief variability within the area is large enough to ensure that it can be subdivided meaningfully. This does not apply to the most homogeneous areas, namely the Central Pennines (A1 and 2). For all the variables plotted in this way (1, 2, 3 and 5) the variability of the total area is large. This large variability in the whole area enables the different areas within it to be readily distinguished. They can be characterized according to the most significant variables.

If it is desired to test the difference between the areas by analysis of variance, then the factor loadings show that absolute relief (V1) is the best variable to use to characterize the relief. As far as the drainage is concerned the most significant variable is 7 (Number of streams). This variable is also related to variable 1 (Absolute relief) so that variable 1 alone could be used successfully to differentiate the areas and to assess the significance of the differentiation by analysis of variance. This applies only to the variables actually measured and used in this test. There might be some other variable even more diagnostic, and if it could be obtained readily from the ordnance survey map this would be an added advantage. It might well be necessary to undertake field work to arrive at a more reliable and diagnostic variable than the absolute relief. Variable 1 (Absolute relief) has the advantage that it is readily and accurately measureable on the map and does not involve any subjective judgement, as one or two of the other variables do, to a minor extent.

(v) *Factor weightings for each area.* In the relatively homogeneous areas it would not be expected that the individual factor weightings would exhibit a great spread or be very helpful in pointing to significant subdivisions within the area. This applies to the areas with low factor values, the Askrigg Block (A1) and the Alston Block (A2). The Lake District (A4) shows a greater variation in individual factor weightings. When these are plotted with factor 1 against factor 2, the variation on factor 2 is much smaller than that on factor 1. The variation in terms of factor 1 is closely associated with the absolute relief (V1). The variation in factor 2 is mainly dependent on the rock outcrops (V6), variable 8 (Drainage density) and variable 10 (Bifurcation ratio), which are of equal weight approximately. No clear cut grouping of the subareas emerge within the area.

(vi) *Individual factor weightings — Areas within the total area.* The individual factor weightings are of more value in the analysis that covers the whole area, combining 120 subareas, than those for the six areas separately, with only 20 subareas each. The purpose of the analysis of the individual factor weightings is to characterize the areas in terms of the value of the individual weightings on the first and second factors. A diagram (figure 7.5) can be produced by plotting the values of factor 1 weightings against those of factor 2 for all the 120 subareas. This diagram reveals the degree to which the different areas can be differentiated by this method. Then the previous analysis can point to the variables on which the characteristics of the areas depend. Each of the different areas is plotted by a different symbol. It can be seen that the individual areas on the whole form distinct and separate clusters of points.

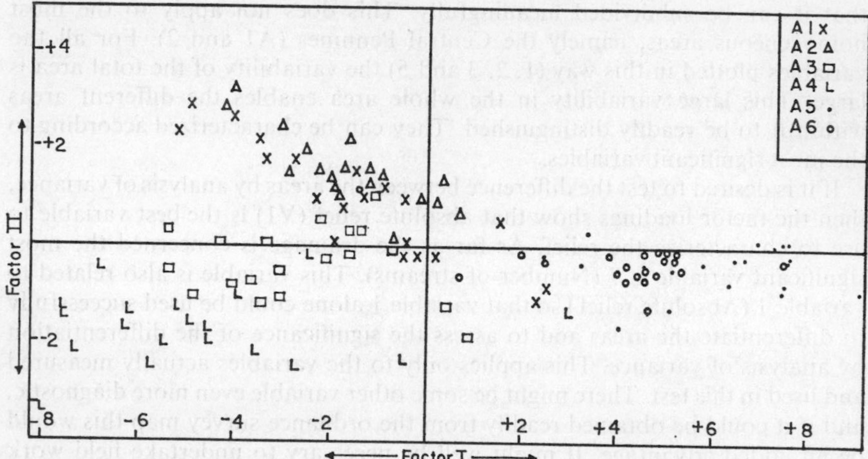

Figure 7.5 Individual weightings of 120 small areas on the first and second factors in the Northern England factor analysis example.

The only two that merge very closely are the Askrigg Block (A1) and the Alston Block (A2), which are geologically and structurally very similar. The Alston Block has slightly higher weighting on the second factor, in which variables 3 (Summit dissection), 7 (Number of streams) and 8 (Drainage density) play the most important part. There is also rather less spread along the factor 1 axis in the Alston Block compared with the Askrigg Block.

There is an interesting, apparently negative, relationship between the values of factor 1 and factor 2 individual weightings in both these areas (A1, A2), the minimum values of factor 2 corresponding with the maximum values of factor 1 and vice versa. This implies an apparent negative correlation between the variables that determine factor 1 and those that determine factor 2. Variables 1 (Absolute relief), 2 (Relative relief) and 5 (Mean slope) are dominant in factor 1, and variables 7 (Number of streams) and 8 (Drainage density) are also important. Variables 7 and 8 are common to both factors. It is interesting to note that these two variables, 7 and 8, are negative in factor 1 and positive in factor 2. They are almost equal owing to their close correlation with each other. The subareas at one end of the scatter of points in these two areas (A1 and A2), where factor 1 is negative and factor 2 is positive, are both higher and better drained than those at the other end. The reason for the negative value on factor 1 is that the areas are higher than average, and the positive value on factor 2 indicates greater drainage density (V8). Where the factor loading is negative the highest values of the variable are negative and vice versa. This is determined by the reflexion process in factor analysis and is not really significant. It does however account for the apparent inverse trend of the points for area 1 relative to factors 1 and 2. This is because the loadings on variable 1 (Absolute relief), 7 (Number of streams) and

8 (Drainage density) are all negative on factor 1, but 7 and 8 variables are positive on factor 2. Thus the linear trend emerges, with its apparent inverse character.

Area 3 (Cheviots) forms an intermediate group between areas 1 and 2 (Central Pennines) and area 4 (Lake District). All the upland areas lie on the negative side of the factor 1 axis with few exceptions (10 out of 80 points) and of these 5 are close to the axis. This is clearly related to the negative sign for the factor loading on variable 1 (Absolute relief) and those associated with it. In fact all the variables have negative signs except variable 4 (Valley character), which is a variable that reaches high values in the lowland areas. The scatter of points for the Cheviots (A3) has almost the same range of values on the factor 1 axis as areas 1 and 2 (Central Pennines), indicating a similar range on variable 1 (Absolute relief) because in all the areas this is the most important element of factor 1. For all the areas combined V1 has a loading on factor 1 of – 0.928. In area 3 (Cheviots) there is very little range on the factor 2 axis, indicating relative homogeneity with respect to the factor 2 variables, which are, variable 3 (Summit dissection), variable 7 (Number of streams) and variable 8 (Drainage density). Such variation as there is in area 3 is not so closely correlated with variable 1 (Absolute relief) as for areas 1 and 2 (Central Pennines).

Area 4 (Lake District) has points that lie entirely within the negative sector of factor 2 and with one exception within the negative sector of factor 1. This indicates that the area has the greatest values of variable 1 (Absolute relief), in other words it is the highest. However, as far as factor 2 is concerned it contrasts with areas 1 and 2 in having negative values. This must be mainly related to the variables 3 (Summit dissection), 7 (Number of streams) and 8 (Drainage density). This indicates that the drainage density is less in the Lake District than in the other upland areas and this can be seen to be correct if the values for the mean for variables 7 and 8 (Number of streams) and (Drainage density) are checked in table 7.7. Variable 3 (Summit dissection) is negative in factor 2. This means that areas with more peaks will tend to have a negative value on factor 2. It is true that area 4 (Lake District) has the maximum value of variable 3. This fact and the drainage values already mentioned account for the positioning of all the area 4 points on the negative side of factor 2. This balances the plateau-like character, with the low values of variable 3 (Summit dissection) in areas 1 and 2 (Central Pennines).

The two lowland areas, A5 and A6, provide the balance between the individual factor weightings on factor 1, as they all have positive values. The Solway Lowlands are lower in elevation and they form a group further along the positive factor 1 axis. Apart from 4 subareas, all the subareas of area 6 (Northumberland coast) lie very closely bunched together. This indicates little variation within this area. The standard deviations are small in absolute terms rather than relative terms, because the mean values are small for many variables. The negative values on factor 2 can be explained by the same reasoning already put forward for area 4 (Lake District) and this applies also to area 5 (Solway). The subarea with the lowest positive value for factor 1 is the highest and steepest. It also has the greatest drainage density (V8) and value of variable 7 (Number of streams).

The relationship between factor 1 and factor 2 provides a useful method of differentiating the six areas compared, in terms of their variability on the variables chosen. The most important of these variables are variable 1 (Absolute relief) and the associated variable 7 (Number of streams) and with variable 7, variable 8 (Drainage density) is closely linked on factor 1. The variables 3 (Summit dissection), 7 (Number of streams) and again variable 8 (Drainage density) occur on factor 2.

(*e*) *Conclusions*

The discussion has included an analysis of the various stages of factor analysis. The means and standard deviations provide useful preliminary descriptions of the areas concerned in terms of the variables used. The correlation matrix provides the data on which the values of the factors and factor loadings are based. The factors indicate the degree to which any area can be expressed in terms of the inter-relations between the variables. Where these inter-correlate strongly the factors are greater and the variation is 'explained' to a greater degree on the first factor. The factor loadings provide a valuable indication of the part each variable plays in accounting for the total variability. The individual factor weightings, particularly in this instance when all six areas were combined, provide the means to establish the nature of the difference between the areas in quantitative terms. In homogeneous areas the individual factor weightings for each area separately have less value, although they can indicate clearly those parts of an area that are atypical.

It has been shown that the six areas within northern England are very different in their characteristics, with the exceptions of the Askrigg and Alston Blocks, which together form the Central Pennines. These two areas show interesting similarities, which suggest similar geomorphological development. Their plateau-like character, with only short, steep slopes, is of great importance. They also tend to be better drained than the other highland areas. In these characteristics they contrast strongly with the Lake District and less strongly with the Cheviots, the latter forming an intermediate group. The two lowland areas are clearly very different. This is brought out and shown to be fundamentally due to their low elevation and lower drainage density. The absolute relief and number of stream sources are the most important variables and these are intercorrelated in the areas as a whole and in most of the individual areas. Thus the absolute relief could best be used to characterize these diverse areas since this variable could be tested by analysis of variance to assess its significance in the different areas and between the six areas within the total area.

References

Anuario estadístico de España (1963). Presidencia del Gobierno, Instituto Nactional de Estadística, Año XXXVIII.

Annuario statistico italiano, Annual publication, Istituto Centrale di Statistica, Roma, Italy.

Fisher, J. C. (1966). *Yugoslavia — A Multinational State*, Chandler, San Francisco.

Tagliacarne, G. (1965). Annual publication 'I Conti Provinciali'. *Moneta e Credito* (Review), Banca Nazionale del Lavoro, Roma, Italy N. **71**.

Part 3

Dimensions of space and time

8

The world as a spherical unit

'At the round earth's imag'd corners . . .'

John Donne, 17th century

'I'll go steadily south-west' says Tweedledum, 'while you go steadily north-east, and when we meet we'll have a battle.' 'Very well,' replies Tweedledee. 'It suits me, because I haven't got my armour today.'

F. Debenham, 1950

Often in geography it is necessary to consider and to map all or a large part of the earth's surface. It is then impossible to produce a map without distorting greatly the position or area of most places relative to one another. The purpose of this chapter is to suggest reasons why it is often important to take a global view of a problem. Situations when this is so will be illustrated. Problems of representation of phenomena on a flat surface may sometimes be avoided by methods which will then be discussed.

8.1 The earth's shell

The surface of the earth is not a plane 2-dimensional surface in the mathematical sense (Figure 8.1*a*). It has a finite area but as a spherical surface is unbounded in two dimensions (see figure 8.1*b*). The shell can be considered to extend through about 50 miles radially to include the bulk of the mass of the atmosphere on the outside and the crust and upper mantle within. The earth's shell is only about $\frac{1}{160}$ of the diameter of the earth. This continuously curved zone is influenced on both sides by forces external to itself (see figure 8.1*c*). The smaller volume is that enclosed within the earth's shell. From here the endogenetic forces emanate. These forces control the pattern of land and sea, including all diastrophic and volcanic manifestations in the earth's crust and other matters of great importance to the geographer. The larger volume is that outside. It extends outwards to include the solar system and space beyond it.

The earth's crust consists of relatively small, very thin spreads of light continental material set in an oceanic crust of denser material. Continental rocks are completely lacking in the oceanic sectors, where, according to some modern views, the mantle outcrops in a modified form. For this reason the Moho (see glossary) may be taken as the plane where geographical interest is taken over by geophysical interest and techniques. This plane, at the lower

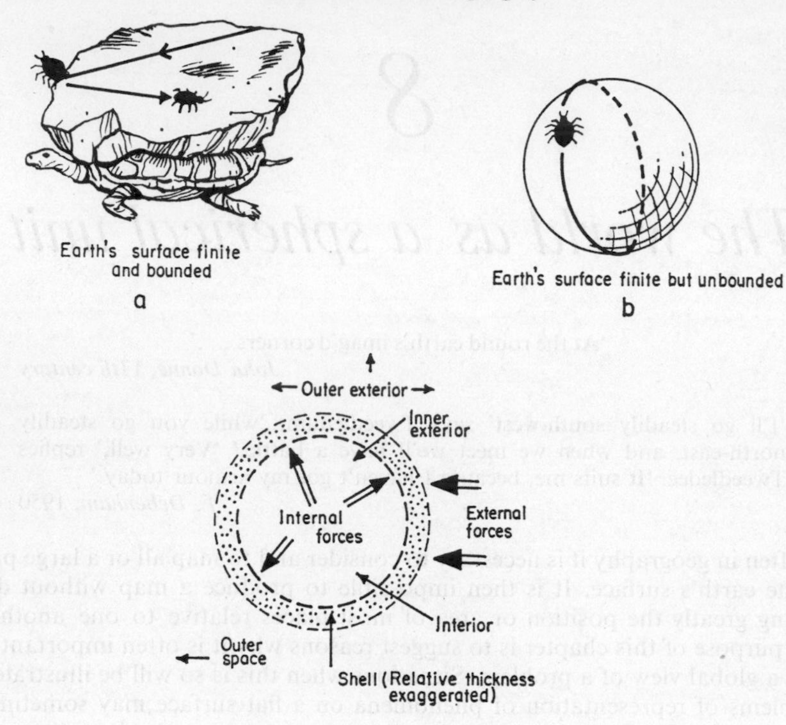

Figure 8.1 Views of the earth. (*a*) Flat earth. (*b*) Spherical earth. (*c*) The shell.

boundary of the earth's shell, divides the relatively homogeneous part of the mantle from the diverse continental crust and oceanic upper mantle. The Moho lies at about 12 km below the ocean surface but 35 to 50 km beneath the continental surface. Nevertheless, forces operating below the Moho play a vital part in determining the surface pattern of land and sea as well as major landform patterns. The effect of these forces is therefore of interest to geographers although the study of the forces is outside the geographical field.

At the other side of the earth's shell the atmosphere gradually thins and fades away. Fifty per cent of its mass lies within the lowest 3½ miles, 75 per cent within 7 miles, 90 per cent within 10 miles, 99 per cent within 20 miles, 99.9 per cent within 30 miles and 99.999 per cent within 50 miles (R. J. Ordway, 1966). Most weather phenomena are confined within the lowest layer, the troposphere, which extends to about 10 to 12 miles above the equator and 4 to 5 miles above the poles. Thus the tropopause which separates the troposphere from the stratosphere may be taken to mark the upper limit of the earth's shell. The most important of the external forces which operate outside this curved shell are those connected with the earth's membership of the solar system. Gravitational attraction between the members of the system keeps them rotating on their orbits around their common centres of gravity, and the earth rotating on its own axis.

Geography is concerned with the relatively thin shell, where atmosphere, hydrosphere and lithosphere meet and together allow the biosphere to exist. Sometimes it is necessary to consider the whole earth's shell as a spherical unit. In section 8.2, attention will be drawn to some of the forces operating externally to the earth's shell, both from outside and from within. Sections 8.3 to 8.5 mention some of those aspects of physical and human geography in which it is necessary to consider the earth as one spherical unit. Some patterns, both physical and human, that appear to have a world wide significance are mentioned. The mapping of the spherical surface and its representation on a 2-dimensional piece of paper in the form of a map are considered in section 8.6. Section 8.6 also includes a brief mention of geodesy and map projections. Section 8.7 is concerned with a variety of different methods of collecting and storing data covering the whole earth's surface. These methods are of varying degrees of abstraction, some being based on mathematical concepts introduced in chapter 2.

8.2 Forces external to the earth's shell

(a) *Outer external forces — the earth as a member of the solar system*

In some respects the earth is unique. It may be considered as the only member of the set of habitable planets known to man. Some astronomers have suggested that there may be about one million planets of the dimensions of the earth attached at suitable distances to other stars, but it is not known if these have developed life similar to that on the earth. For practical purposes, therefore, the earth may be considered unique in this respect. It is, however, a member of the set of planets attached to the solar system. It is useful to consider some of the ways in which this characteristic of the earth influences its geography. The planetary forces operating on the earth from within the solar system are usually taken so much for granted that geographers rarely stop to consider them. They do, however, exert a fundamental influence on the earth, on both animate and inanimate matter. In this section some of these influences will be mentioned. They include: (i) The earth's satellite the moon. (ii) Energy from the sun. (iii) Effects of the earth's rotation (1) round its own axis (2) round the sun in its orbit. (iv) Earth's magnetic field.

(i) *The earth's satellite the moon.* Until recently the earth's own satellite the moon rarely played an important part in the earth's affairs. The development of space travel has however now placed it within at least the indirect reach of man. No doubt it will soon be within the direct reach of man. The moon has a permanent and regular effect on the earth in its power to raise tides in the oceans. In its capacity as a tide producing force it is more than twice as powerful as the sun, on account of its proximity, which more than compensates for its much smaller mass. The reason is that the cube of the distance between the earth and moon or sun enters the formula for the tidal tractive force, while only one power of the mass enters. In other respects, however, the sun is by far the most influential element to the earth in the solar system.

(ii) *Energy from the sun.* It is from the sun that by far the greater part of the energy used on earth is derived. Some of this is natural energy, the energy of heat that keeps the circulation of the atmosphere and the oceans going and provides the water that falls as precipitation on the surface of the earth. Nearly all the energy used by man is derived ultimately from the energy of the sun, as pointed out by D. L. Linton (1965). The fossil fuels, coal and oil, come into this category, and also water power and wind power. The difference between them is that the former are an unrenewable source of energy, while the latter are self renewing as long as the sun shines. The sun's heat warms the water, causes evaporation and the transfer of the moisture upwards where it condenses. Lateral flow carries it over the land to fall as precipitation or to generate wind driven power. There are other, at present, minor sources of power, such as geothermal power and atomic power, although both of these depend on sources of supply that may not be inexhaustible in the long run. The moon provides a small amount of power where the tidal rise and fall can be harnessed. Thus apart from the relatively less important exceptions the earth is not self-sufficient as far as sources of energy are concerned and energy is vital to modern life. The earth depends on the sun and it is living rather heavily on capital, rather than current assets, in this respect.

(iii) *Effects of the earth's rotation*

(1) *Round its own axis.* Because the earth is a member of the solar system it rotates round the sun and it also rotates on its own axis, which is set at an angle of 23½ degrees to its plane of orbit round the sun. One of the effects of the earth's rotation about its own axis is to produce the equatorial bulge, which makes the equatorial diameter 27 miles longer than the polar diameter. The most obvious effect of the rotation of the earth about its axis is to produce day and night, except within the arctic and antarctic circles near the solstices. This provides the basic short unit of time, the day. Tidal friction is slowing the rotation of the earth and increasing the length of the day, but only by 1 second in about the last 100,000 years. In Silurian times there may have been about 400 days in a year, because it seems unlikely that the time the earth takes to complete its orbit has changed significantly. The effect of daylight and darkness is an important factor in regulating human life, at least in the middle and low latitudes, but it has less effect in the physical sphere. It can effect such phenomena as freezing and thawing, which in middle latitude areas have a diurnal rhythm, rather than the seasonal rhythm more typical of high latitudes. Certain meteorological phenomena are also closely associated with the change from day to night, such as land and sea breezes.

The rotation of the earth about its axis permeates the action of moving media on the surface, and causes many characteristic phenomena. The westward intensification of ocean currents, so well illustrated by the Gulf Stream system, is the result of the variation of the Coriolis force with latitude. The Coriolis force, which acts in the opposite direction to the pressure gradient force, causes deflexion to the right in the Northern hemisphere and to the left in the southern. This deflexion depends on the latitude, as shown by the formula $2v\omega \sin \phi$, where ϕ is the latitude and v is the velocity and ω is the angular velocity. The increase of the Coriolis force with distance from the equator causes the western intensification of ocean currents. The Cromwell current, which is a narrow ribbon of fast east-flowing water astride the equator in the Pacific beneath the west-flowing South Equatorial current, is also due to the change of direction of the Coriolis force across the equator. The zones of upwelling water along the equator and in eastern coastal areas in the oceans in middle latitudes are another result of the Coriolis force. This force is also responsible for the characteristics of the tidal amphidromic systems.

The effects of the Coriolis force in the atmosphere are well known. The geostrophic winds, which represent the balance between the pressure gradient force and the Coriolis force, illustrate the influence of the rotation of the earth on the moving atmosphere. The effect of cessation of rotation on the atmospheric and oceanic circulation would be enormous. It has been suggested that the Gulf Stream would cease to flow after 1600 days if the driving wind ceased (Stommel 1958 p. 148). This driving wind is influenced by the rotation of the earth and is dependent on the sun's energy in the form of heat. Without rotation and without the sun's energy the earth's major circulations would very rapidly cease.

(2) *Round the sun in its orbit*. The movement of the earth around the sun provides another major time unit on earth, the year. The angle between the earth's axis of rotation and the earth's orbital plane provides the seasons, as a result of the changing declination of the sun through the yearly cycle. The movement of the earth around its orbit is not strictly regular. It follows an elliptical path around the sun, which lies at one focus of the ellipse. This means that the intervals of time between successive meridional crossings of the sun are not even, so that days differ in length. For practical purposes an even mean time is kept and this must be connected with the true time by using the equation of time. This equation, which depends on the obliquity of the ecliptic and the ellipticity of the orbit of the earth, shows how far true time and mean time differ. The maximum amount is about 16 minutes.

(iv) *Earth's magnetic field*. The earth's membership of the solar system also probably provides its magnetic field. In the upper atmosphere this is made manifest in high latitudes by the aurora. It has a more practical use on the earth's surface by providing a useful means of direction finding in the magnetic compass. The fact that this provides a reasonably satisfactory point of reference is due to the earth's magnetic field, which resembles a dipole magnet set nearly along the axis of rotation. Thus the magnetic poles oscillate in position around the geographical poles.

A study of the palaeomagnetism of rocks has shown that the earth's field changes polarity fairly regularly at about $\frac{1}{2}$ to 1 million year intervals. The reason for this is obscure, but it has been suggested that the force which keeps the magnetic field in operation is related to movements of the solar system. If this movement were to cease then the magnetic field would run down and cease very quickly in about 10,000 years. Another suggestion is that the magnetic field external to the earth may protect it from cosmic rays, which could prove fatal to some organisms. Cosmic rays could reach the earth when the field dies down to change polarity, and may account for the extinction of some species. These reversals appear to have taken place 700,000 years ago, 2,400,000 and 3,500,000 years ago and if the present rate of decrease goes on the magnetic field would reach zero in 2000 years.

(*b*) *Inner 'external' forces*

The magnetic field provides a link between the outer external forces and the inner forces. Its original source of energy appears to be outside the earth's surface, but it invokes forces operating within the core. These forces in turn appear on the earth's surface as its magnetic field. It is interesting to note that it is especially through studies of palaeomagnetism that the theory of continental drift has come back into favour in a modified form (see chapter 12). The palaeomagnetic data provide some of the most convincing evidence of past continental displacement and also illustrate current crustal movement. The alternating bands of varying polarity on either side of the mid-Atlantic Ridge give strong support to the gradual widening of the ocean from the middle, since the rocks become younger as the ridge is approached from either side. The horizontal displacement by 160 km of magnetic anomaly belts that cross the Murray fracture zone in the northeast Pacific also illustrates the mobility of the crust in response to deep seated forces within the earth. The former evidence in particular provides valuable support for the convection current theory of ocean rifting and spreading, leading to continental displacement. This evidence, in conjunction with heat flow and seismic studies, and other geophysical work, suggests that the earth's crust responds on a large scale and actively to the forces operating external to it from within. The pattern of land and sea is related to the inner forces, while the shape of the earth is dependent on its rotation and its movement round the sun.

Most of the effects that depend on the forces external to the earth, both from outside and inside its shell, are either taken for granted or act so slowly that they are imperceptible within a human lifetime. As a result they are usually ignored by geographers. They are nevertheless vital and it is worth drawing attention to them, if only very briefly and incompletely.

8.3 The earth's shell as a single physical entity.

The whole of the earth's shell is the field of geographical study, and although many studies only consider a limited part of the shell, in some instances it is necessary to consider the earth as a whole. In these circumstances it is very important that the spherical character of the earth's surface is fully appreciated. The surface of the earth forms a unitary and closed surface of finite area. It is possible, however, to move continuously around the earth, in infinitely long paths, as for example along a great circle or parallel. Thus, the surface of a sphere is finite (in area) but unbounded (see figure 8.1*b*).

The atmosphere provides a good example of an enveloping surface that must be treated as a unit covering the whole sphere if its behaviour is to be properly understood. The movement of the air over the earth's surface is not restricted by any complete barriers. Partial restriction is caused by the mountain masses. The presence of land and water in varying proportions over the earth's surface also plays an important part in the nature of the atmospheric circulation and the resultant weather patterns and climatic types. The climatic zones, which emerge from a study of the action of the atmosphere over the whole of the earth, reflect distance from the equator and the distribution of land and sea. This latter factor is readily apparent in the difference between oceanic and continental climates. The whole system is in some respects a closed one although the heat energy to keep it going comes from the sun. Heat is distributed over the surface of the globe by the agency of the atmospheric circulation. It is impossible to appreciate the climate of Britain, for example, without taking into account what is taking place in other latitudes. In the tropics the warmer air masses form; in the arctic the cold air masses form. This is not enough, however, and the zonal aspects must also be taken into account. It is equally important to know what is happening over the Atlantic to the west and Eurasia to the east.

The oceans also illustrate particularly clearly the unity and spherical character of the earth's surface. There are only two deep water sources in the whole world ocean. Both are situated in the Atlantic Ocean. These two sources supply the whole of the world ocean with the deep water that fills the bulk of the oceanic basins. The circulation of the southern ocean, which is continuous around the Antarctic continent in high southern latitudes, is particularly important. Water can move unimpeded in this belt continuously from west to east under the influence of an equally continuous and unimpeded west wind circulation. This movement provides the link between the oceans. Water from the Atlantic is driven by the wind into the Indian Ocean and much also moves with the deep water circulation and from here water finds its way into the Pacific Ocean. The Pacific contains the oldest water, a fact shown by the very small amount of oxygen that it contains. Its long route is indicated by the homogeneity of the water, which has had a long period in which to become well mixed by the time it has travelled all the way from the surface in the north or south Atlantic to the depths of the Pacific Ocean. This oceanic circulation is strongly influenced by the distribution of land and sea, which accounts for the important difference between the Atlantic and Pacific Oceans.

8.4 Patterns on the earth's shell

The pattern of land and sea and other structural features of the earth's surface have intrigued geologists and geographers for a long time. One of the earliest attempts to envisage a global answer to this problem was that of Lowthian Green who in 1875 put forward the tetrahedral theory. He argued that as the earth contracted in cooling it would tend to change from spherical to tetrahedral form. The tetrahedron has a larger surface area in relation to volume than any other regular-sided 3-dimensional figure, while the sphere has the smallest surface area of all regular solids of equal volume. Evidence to support this view was found in the distribution of land and sea. The apices of the tetrahedron were placed at the south pole and symmetrically in the northern hemisphere, centred on the land masses of North America, Europe and Asia. The Arctic Ocean and Southern Ocean representing the low lying surfaces of the tetrahedron.

The tetrahedral idea was further developed by Vening Meinesz in his suggested pattern of primeval convection cells. The sinking cells, accumulating light material, became the continent nuclei and the rising cells the oceanic sectors. He argued that this pattern of convection currents would best fit a spherical shape of the dimensions of the earth when it was in a hot and convective state. Both these suggestions imply permanency of the positions of the continents and oceans.

More recently other patterns have been discerned in the physical feature of the earth by considering it as a spherical surface. The distribution of active volcanoes has been analysed by the 3-dimensional least square method by J. H. Tatsch (1964). He found that 93 per cent of all active volcanoes were situated where three mutually orthogonal planes passing through the centre of the earth cut the earth's surface, forming great circles on it. These great circles intersect at 5°N, 95°E; 30°S, 175°W; 55°N, 165°W; 5°S, 85°W; 30°N, 5°E; and 55°S, 15°E. In this way statistical analysis has brought out an interesting global pattern.

Other patterns on varying dimensions have been explored by B. B. Brock (1956), who used an empirical approach to the problem of the distribution of minerals. Brock considers that the patterns that he has discerned are related to mechanical laws at work in the earth's interior. He bases his analysis on the mosaic concept, suggesting that the larger patterns have within them smaller patterns. In this way patterns of varying dimensions can be found, from global ones to units with only 1 degree dimensions. He points out that arcuate structures are much more common than is usually appreciated and these frequently have radii of 7½, 10, 12½, 15 or 18 degrees. Larger small circles also have structural significance in some areas, for example the Alpine chains of Europe lie along a zone bounded by arcs of 60 and 75 degree radii. Great circles are also of major importance in structural patterns.

The external and visible patterns and symmetries on the earth's shell are the product of the inner external forces. When a plane within a sphere cuts its surface, the resulting line is arcuate. The radius of the arc depends on the angle of the plane. Thus many of the arcuate features, such as the island arcs and volcanic great circles, could be the result of activity acting along planes within the earth. These planes are manifest in the earth's shell as arcs of varying radii. A plane through the earth's centre is a great circle, but the angle of curvature becomes less as the chord distance becomes smaller.

In summing up his findings, Brock draws attention to the common occurrence of structural small circles with radii of 45, 60 and 72 degrees and also points out the importance of aliquot (see glossary) angles in a quadrant for small scale features. He shows that the vertices of the crossing great circle patterns have significance in delimiting continental areas. They now normally form structural basins, for example in the Baikal area and Bushfeld Complex. The vertices are normally spaced 45, 72 or 90 degrees apart. Individual structural features show preferences for definite sizes, for example continents for 72, 45 or 36 degrees, oceans for 120, 90 or 60 degrees, excluding the Pacific, and shields for 36, 22½, 18 and 15 degrees. The five radii symmetry of the continents and oceans relative to the centre of Africa is shown in figure 8.2.

Other patterns of symmetry on a world wide scale have been brought out in studying the mid-oceanic ridges. The ridges lie in a central position in the Arctic, Atlantic and Indian oceans. The pattern in the Pacific, however, is not so strongly developed and symmetrical as in the other oceans, as the east Pacific ridge runs into the Gulf of California. This pattern has been interpreted in terms of the movements of the continents away from the ridges as a result of the action of convection currents, which rise under the ridges and sink under the continental margin. This idea is supported by the high heat flow and vulcanicity under the ridges and the formation of island arcs and deep focus earthquakes at the Pacific continental margin. Thus the Atlantic and Indian Oceans are thought to be rift oceans, while lateral movements are common around the Pacific Ocean. One of the major difficulties is to reconcile this type of movement, which results in symmetry in the mid-oceanic ridges, with the worldwide symmetry as described by Brock and others. Further ideas on continental drift are discussed in chapter 12.

8.5 The world as a unit in human geography

The problem of representing distributions on the earth's surface in map form is considerable. It is customary to map distributions involving all or large portions of the earth on equal area projections, either cylindrical type projections with the equator across the middle from left to right or on zenithal or conic projections. All projections have their advantages and disadvantages. The equal area map projections fail to preserve the correct relative location of most places to one another. The globe could be used, but although it is more realistic, it cannot be viewed in total at one instant in time. The unusual expedient of making a hollow globe (New York World Fair), see figure 2.20*d*, is stimulating visually, since it gives the viewer an unusual mirror image views

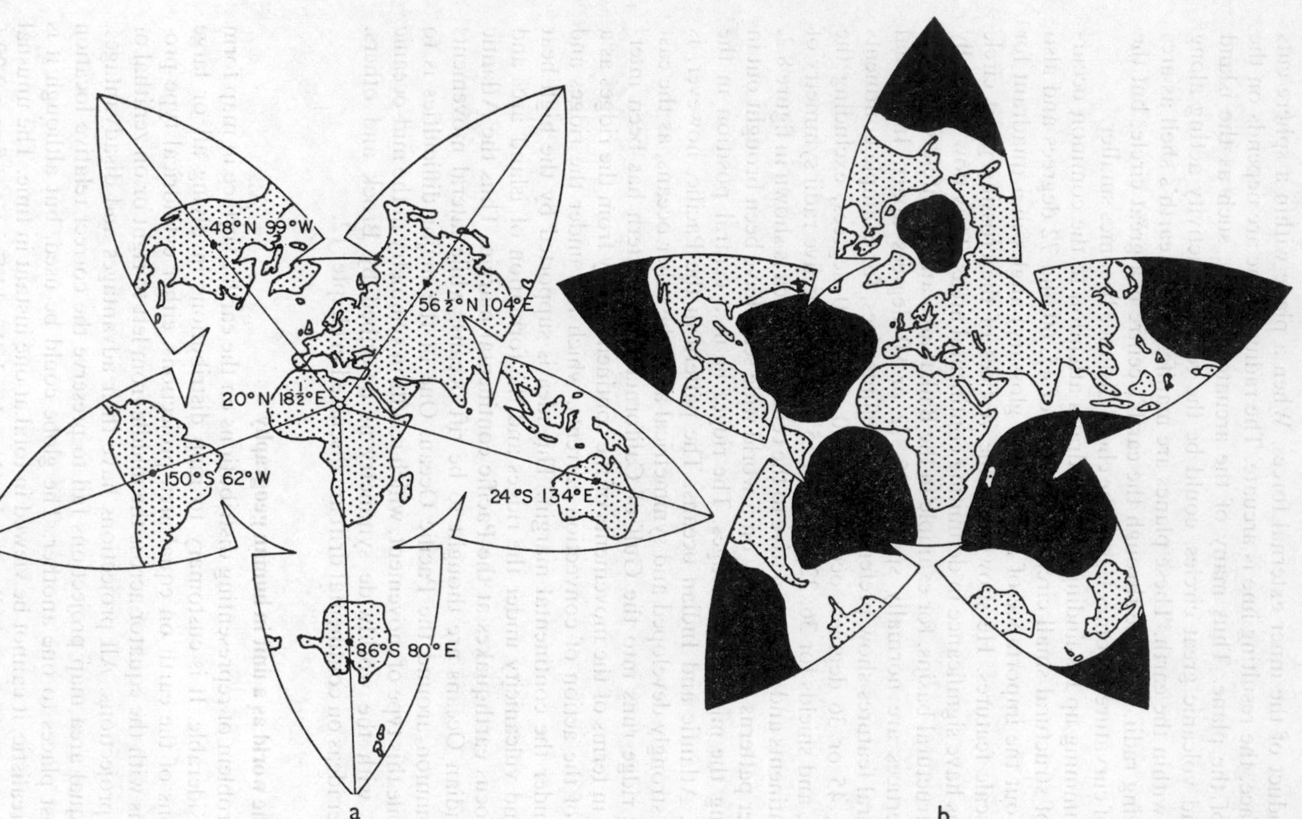

Figure 8.2 The geometrical arrangement of the continents and oceans relative to Africa (After B. B. Brock). (*a*) Geometrical arrangement of continents. (*b*) Geometrical arrangement of oceans.

of familiar areas, but it does not overcome the problem. The study of situations involving large parts of the earth's surface is further complicated by the size and complexity of the area involved (compared with small regions) and by the frequent lack of completeness and consistency of data. Most atlases manage to achieve a view that either has a land mass (such as a continent) central and the sea peripheral, or an ocean basin central and the land around the sides.

Many situations in human geography can, however, only be appreciated in their entirety from a world viewpoint. Perhaps the turning point was the Treaty of Tordesillas, 1493, by which the world, in effect the tropical world, was divided into two parts, the 'western' for Spain and the 'eastern' for Portugal. From the 15th century on, Portugal, Spain, England, France, Holland and Russia from their different bases in Europe, relentlessly spread their influence widely over the world, bringing it gradually into a single system for colonization and trade. This, however, has not always been so. Whether the human species has one or more than one place of origin, up to the 16th century it was still possible for sections of humanity to develop in isolation from other sections. Europe, India and China were virtually isolated from one another with a minimum of persons, goods and ideas filtering through the spaces between. The New World civilizations of Central America and of northwestern South America grew largely, if not entirely, independently of those of the Old World. Even now there is little direct contact between, for example, Argentina and Pakistan, but their Ambassadors may be only a mile or two apart in Washington or in Moscow, and their representatives share the U.N. building, so communication is feasible. Probably only a few million people now still remain basically outside the world system: in particular, many indigenous peoples in Amazonia and New Guinea.

Ideas can now spread quickly, even if they are not adopted quickly. Large scale trade takes place between countries many thousands of miles apart. A serious conflict in any place now may potentially have wide repercussions. A single long range continental ballistic missile somewhere in the U.S.S.R. or U.S.A. may be aimed with great accuracy at any place within a radius of several thousand miles, and may be sent there in a matter of minutes. Early this century, H. J. Mackinder (1904) noted that world affairs take place in a single arena. The establishment of the League of Nations and later the United Nations reflects the growing possibility of developing worldwide contacts, and the need to do so. It is hardly necessary, then, to point out that students of economic, social, political and other systematic branches of human geography, must think of what they are studying within a world framework or context, as must students of physical geography. A few brief examples will show the kind of global situations facing the human geographer, the kinds of questions he might ask and the kind of statements he might make:

(*a*) *Worldwide organizational systems*

The sea empires of Europe built and maintained their ocean channels of communication over vast distances. Russia alone worked within its own compact and continuous land area. The others had their lines of communication mingling in the Atlantic and Indian Oceans and elsewhere. Vast distances were covered: Spain had a regular link from the 16th century with Manila in the Philippines via Mexico. As the Russian empire spread eastwards in the 17th and 18th centuries, Russia was similarly concerned with organizational problems of hemisphere scale. What was the shortest distance between St. Petersburg, the Russian capital in the 18th–19th centuries, and Russian outposts in what is now California (see figure 8.3)?

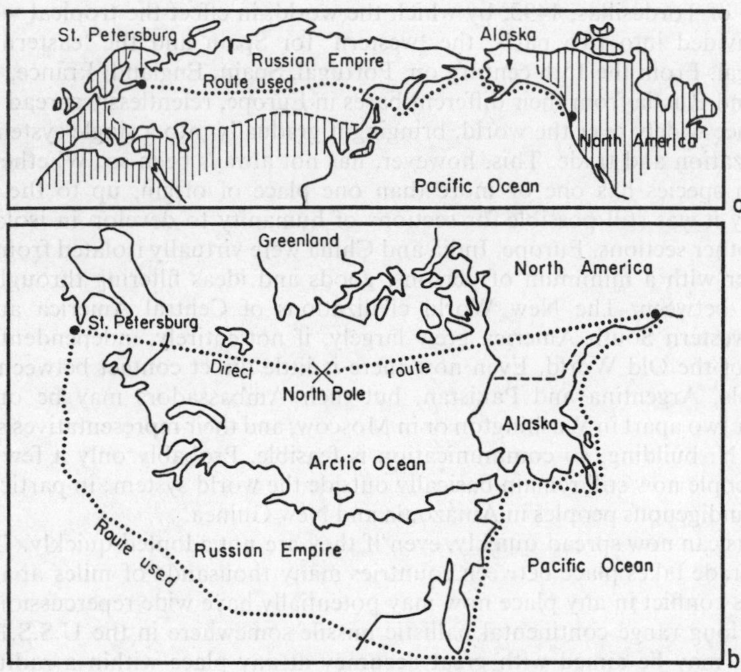

Figure 8.3 The journey from St. Petersburg to Northwest North America in the Russian Empire during the early 19th century. (*a*) On a cylindrical projection. (*b*) On a polar zenithal equidistant projection.

One immediately thinks of a line running left to right across the Russian Empire to the Pacific and across to California (figure 8.3*a*). In fact the shortest route is over the North Pole (figure 8.3*b*). A global view is necessary to appreciate the problem.

In the 20th century, the surviving European sea empires have faded away, but England and France maintain worldwide links. The U.S.A., however, is now the country with the most extensive and complex pattern of world links, both civil and military. Some idea of the complexity is illustrated in figure 8.4. This shows a representation of selected flights of Pan-American World Airlines, the U.S. Airline Company operating most extensively outside the

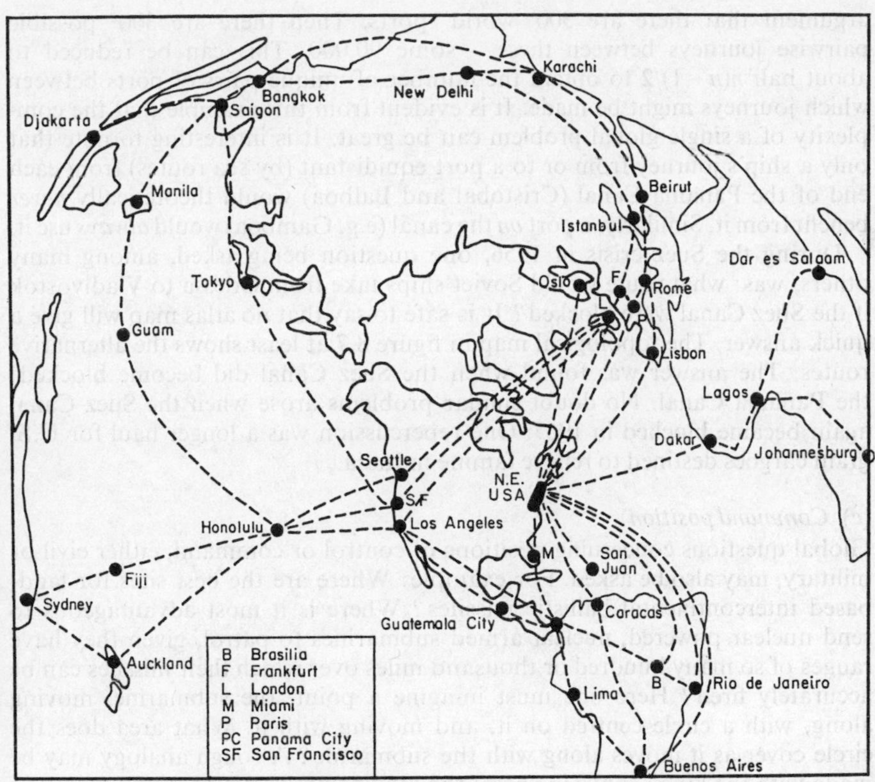

Figure 8.4 Selected flights of Pan-American World Airways.

U.S.A. The map is an interesting attempt to make the part of most interest to the majority of the passengers using Pan-American as recognizable in shape as possible. Distances on the periphery are greatly distorted (e.g. Djakarta–Sydney) but as Pan-American does not link these two places, such distortion is not too serious.

(*b*) *Routing problems*

Shipping and air services may be global in scope. The following example illustrates a specific problem:

When the Suez and Panama Canals were opened, each in turn immediately became of potential interest to almost every port in the world. One might ask: which journeys between pairs of world ports are more advantageously routed through the Panama Canal than not? Obviously a ship going from San Francisco to New Orleans would use the Canal because it would greatly reduce the length of its journey. Similarly, a ship sailing from say, Bombay to Singapore would certainly not use it. The magnitude of the task of answering the question, however, should not be overlooked. Suppose for the sake of argument that there are 300 'world' ports. Then there are 300^2 possible pairwise journeys between these — some 90,000. This can be reduced to about half $n(n-1)/2$ to obtain the number of unique pairs of ports between which journeys might be made. It is evident from this example that the complexity of a single global problem can be great. It is interesting to note that only a ship's journey from or to a port equidistant (by sea routes) from each end of the Panama Canal (Cristobal and Balboa) would theoretically *never* benefit from it. Similarly, a port *on* the canal (e.g. Gamboa) would *always* use it.

During the Suez crisis in 1956, one question being asked, among many others, was: what route would Soviet ships take from Odessa to Vladivostok if the Suez Canal were blocked? It is safe to say that no atlas map will give a quick answer. The topological map in figure 8.7 at least shows the alternative routes. The answer was found when the Suez Canal did become blocked: the Panama Canal. No doubt similar problems arose when the Suez Canal again became blocked in 1967. One repercussion was a longer haul for U.S. grain cargoes destined to relieve famine in India.

(*c*) *Command position*

Global questions concerning positions of control or command, either civil or military, may also be asked. For example: Where are the best sites for land-based intercontinental ballistic missiles? Where is it most advantageous to send nuclear powered, nuclear armed submarines to patrol, given they have ranges of so many hundred or thousand miles over which their missiles can be accurately fired? Here one must imagine a point, the submarine, moving along, with a circle centred on it, and moving with it. What area does the circle cover as it moves along with the submarine? A rough analogy may be made with the king's move in chess: the king commands a continuous *part* of the board from any point at which he is situated, since he can move to any of eight adjoining squares unless restricted by the edge of the board. The idea of the running mean extended to a 2-dimensional surface, discussed in chapter 4, will come to mind. Here a circle, centred at various points on a dot distribution, contains different numbers of dots at different places.

One interesting aspect of the earth as a closed finite surface is the fact that there is a limit to the expansion of the area enclosed by a circle of increasing radius from any point on it. This is relevant to the question of expansion as well as control of command from a point. As the radius increases, at first the area enclosed by a circle of this radius increases at a rate equal to the square of the radius. However, when the radius has increased to become equivalent to 90 degrees of latitude, a whole hemisphere will be included in the circle. When the radius increases beyond this point the amount of new area incorporated then begins to decrease at an increasing rate as the antipodal point is approached. By the mid 1960s, according to J. B. Wiesner and H. F. York (1964), the Minuteman II rocket had a range of 6,300 nautical miles, or roughly 10,000 km, the Atlas 9,000 nautical miles. 10,000 km is approximately 90 degrees from the launching site. The former rocket therefore commands a hemisphere.

8.6 Mapping on a spherical earth

The earth is almost a true sphere with a radius, in the equatorial plane, of approximately 6,378,300 m or nearly 4000 miles. It is worth remembering that the height of the mountains and the depth of the oceans are in fact very small compared with the size of the earth. The heights and depths appear great within the short space dimension that can be appreciated by a person, but they are only 65,000 feet or about 12.3 miles at their maximum combined value. This value is only $\frac{1}{325}$ of the radius of the globe; for a globe 2 ft in diameter this would be a little less than 1 mm. On a world scale the oceans are mere films of water and the greatest mountains only insignificant irregularities on the earth's surface, which does not depart very far from the true sphere. The common use of very great vertical exaggeration in profiles across large expanses of ocean floor or mountain ranges gives a very false impression of the true dimensions of the irregularities on the earth's surface in relation to its total size and shape.

The science of geodesy aims to measure the size and shape of the earth. To most people and for most purposes the earth can be considered to be a sphere. This however, does not satisfy the geodesist, who describes the figure of the earth rather evasively as a geoid, or earth shaped. The actual shape of the earth is fairly well known and has been since Erastothenes first measured it in the third century B.C. He used the method that has since been elaborated to give the present accepted value for the shape of the earth. The basis of the method is the comparison of measured lengths along lines of longitude in different latitudes and positions on the Earth. The length of a degree of latitude measured along the surface of the earth is compared with astronomical measurements of latitude. Very accurate angular determinations of latitude can be made by taking star sights with a good theodolite. These points can then be linked be geodetic ground survey. It is found that the average length of 1 degree of latitude is 111 km, but the value increases between the equator and the poles by about 1 in 100. The shorter polar radius is about 13 miles less than the equatorial one. The polar flattening is about $\frac{1}{296}$ $(a-b)/a$, where a is the equatorial diameter and b is the polar diameter.

The reason for this ellipticity in the shape of the earth is its rotation. It is necessary to provide a component of force that does not act directly towards the centre of the earth. If the earth were a true sphere the gravitational force would act directly towards the centre. As the earth is rotating, however, a force perpendicular to the axis of rotation is required to make the particles rotate around the axis. This is achieved if the spherical form is modified to become an oblate spheroid. The surface of the geoid is defined as that surface where the geopotential is everywhere the same. This is at sea level over the sea but it passes underground beneath the land.

Geodetic survey brought to light interesting characteristics of the earth's structure, by demonstrating the effects of isostasy. In the geodetic survey of India in the last century the attraction of the visible mass of the Himalayas was calculated in connexion with the deflection of the plumb bob giving the vertical. It was argued that the mountains would attract the plumb bob and thus change the vertical. However the actual observations showed that the attraction was only about one third of the expected amount. Thus it was suggested that the visible mass of the mountains was partly compensated by an invisible deficiency of density at lower levels in the crust. In this way the idea that mountains have roots was developed. Gravimetric surveys confirmed that there is an approximate balance between mass and density, by which approximate isostatic equilibrium over the whole earth is achieved. These observations of gravity have assisted in providing a better picture of the major structural patterns in the earth, which have been referred to already in this chapter. Thus the science of geodesy at one end links with geophysics to improve our understanding of the structure and shape of the earth. At the other end it links with problems of surveying on a curved surface and the representation of the surveys on maps through map projections. This aspect will be considered briefly next.

Some geometrical aspects of plane surveying in small areas have already been touched upon in chapter 2. The basis of the geometry must be modified slightly when the surveying covers large areas in which it is impossible to ignore the curvature of the earth. Because the surface of the earth is curved a distant object appears to be lower than it really is. There is a formula for correcting angles for curvature. The correction in feet is $\frac{2}{3}$ of the distance in miles squared. The value is 8 in. for a distance of 1 mile and 32 in. for 2 miles. Corrections have to be applied for curvature when the line of sight is long.

There are two main aspects in which the shape of the earth enters into geodetic surveying. There is firstly the surveying of the actual surface by means of a system of triangles. Secondly the survey demands the accurate measurement of latitude and longitude by reference to the sun or stars. In both aspects the sphericity of the earth must be taken into account.

Surveying over large areas aims to fix accurately the apices of triangles whose sides or angles are measured instrumentally. Geodetic survey is usually carried out by means of a theodolite with which angles are measured. This method was developed because of the difficulty of accurate linear measurement along the ground over great distances. The development of the tellurometer, however, has made linear measurement quick and accurate. This instrument is based on the travel time of micro-waves, which are sent out from one end of the line and received at the other. Whether the sides or angles or some of each are measured, the position of the apices of the triangles must be calculated by solving the triangles. The sides of the triangles lie along the earth's surface and are, therefore, curved. They are expressed in angular values so that spherical trigonometry must be used.

One of the aspects of the spherical triangle that must be allowed for in solving the triangles is the fact that angles of a spherical triangle add up to more than 180 degrees. This is called the spherical excess and increases as the sides of the triangles, which are arcs of great circles, increase in length. The spherical excess is equal to the area of the triangle divided by the radius of the earth squared and multiplied by the sine of 1 second. This amounts to only 0.13″ for a triangle of 10 sq miles, 1.32″ for a triangle of 100 sq miles and 6.59″ for a triangle of area 500 sq miles, which is about the size of a geodetic triangle, as these triangles have sides about 30 miles long. The spherical excess increases to 90 degrees for a triangle covering 90 degrees of latitude and 90 degrees of longitude, with sides along meridians and the equator. This shows one of the modifications that become necessary when a large portion of the earth's surface is being mapped.

In the measurement of latitude and longitude very large spherical triangles may be involved. The application of the spherical triangle to surveying is perhaps most clearly seen in field astronomy. The measurement of latitude and longitude, which defines position on the spherical earth, is based on the astronomical spherical triangle. This triangle, which is illustrated in figure 8.5,

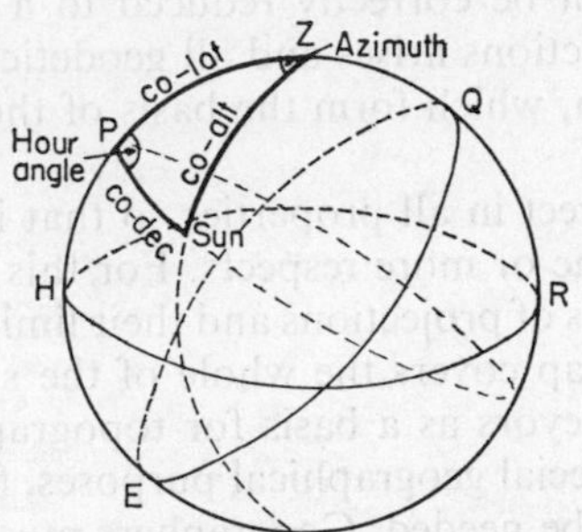

Figure 8.5 The astronomical spherical triangle. Co-dec. is co-declination, co-lat. is co-latitude and co-alt. is co-altitude.

has the pole at one apex, the zenith at another and the star or sun at the third. The three sides represent the co-altitude, the co-declination and the co-latitude. By measuring some of these variables it is possible to calculate the others. The declinations may be looked up in tables. To determine the longitude, it is necessary to calculate or measure the hour angle. The azimuth is needed to obtain the true direction of a line on the ground survey. The altitude can be measured. By solving the astronomical triangle the latitude and longitude can be established.

It is also of interest to note that the solution can often be found graphically by making use of the stereographic projection on which angles are preserved correctly. This projection is the basis of the astrolabe, which is an instrument used for working out latitude and longitude. It was employed frequently in earlier centuries before modern navigational aids were available. The astrolabe does not provide a very accurate result, although a stereographic projection may still be used for aerial navigation, where very rapid movement requires a very quick answer, and where accuracy is not so vital, owing to the speed of movement.

For accurate determination of latitude and longitude for geodetic survey work it is necessary to calculate the results by using spherical trigonometry. The accurate determination of latitude and longitude require observations of heavenly bodies beyond the earth, the sun or stars. The result of such a determination fixes latitude on the spherical surface of the earth in relation to its axis of rotation, the poles and equator. An arbitrary meridian must be used for longitude as there is no unique line from which to measure longitude in a continuously rotating sphere. Longitude is measured in terms of time. By geodetic survey the position of the points on the surface of the geoid can be established, in terms of their relative positions with respect to other points or in terms of their latitude and longitude. The aim of the survey, however, is partly to produce an accurate portrayal of the position of the points on a map. If the map is on a large scale, so that only a small area is shown, the relative position of the points can be shown accurately. However if the map covers a large area, accuracy in all respects is no longer possible as a curved surface cannot be correctly reduced to a plane surface. Thus the necessity for map projections arises and all geodetic surveys must be plotted on a suitable projection, which form the basis of the maps so much used by geographers.

No projection is correct in all properties so that inevitably all projections can be misleading in one or more respects. For this reason it is necessary to appreciate the properties of projections and their limitations. This is especially important where the map covers the whole of the spherical earth. The projections chosen by surveyors as a basis for topographical maps may not be the most suitable for special geographical purposes, for which special properties or viewpoints may be needed. Geographers must be aware of the properties of projection and be prepared to experiment with projections.

Projections can have a variety of useful properties. They can show relative area correctly, as in equal area projections, or shape correctly, as in orthomorphic projections. This property is elusive as strictly it applies only to points, because a map on which shape is true would correctly represent the globe and this is impossible. Orthomorphism implies that the meridians and parallels cross at right angles and the scale is equally exaggerated in all directions from each point. Maps can show correct bearing from one point or they may be constructed so that lines of constant bearing or rhumb lines are straight. Other maps can show the shortest distance or great circle distance between two points as a straight line. Again others may preserve the scale correct along certain meridians or parallels on the map.

Projections are based on a series of special geometrical shapes in relation to the generating globe. A plane may be placed tangentially to the earth at any point and the lines of latitude and longitude projected onto it. The plane may be converted into a cone if it is made to touch the earth along any small circle, usually a line of latitude. The cone in turn can change into a cylinder if it is made to touch the globe along a great circle, usually the equator. These three possibilities give the three main classes of projections, the zenithal, conical and cylindrical. All three may be drawn in their normal form, with the plane tangent at the pole, the cone on a parallel and the cylinder on the equator. However, they can also be drawn transversely and obliquely, although it is rare to find a transverse or oblique conical projection. This greatly increases the possible forms of the projection as they may be centred anywhere on the globe according to the nature of the problem they are intended to illuminate. Unfamiliar centering of a projection can often throw valuable new light on a problem where the relationship between different places is an important aspect of the problem. Figure 8.6 is an example of a zenithal equidistant projection centrcd on Wellington, New Zealand. Distances and directions are correct along straight lines from Wellington. These lines represent arcs of great circles. This property is achieved, however, at the sacrifice of both shape and area, which become increasingly distorted towards the edge of the map. Spain at the antipodes is stretched all around the bounding circle of the projection. The antipodal point is represented by the circumference of the circle containing the whole projection. This point can be reached equally easily from Wellington by setting out in any direction.

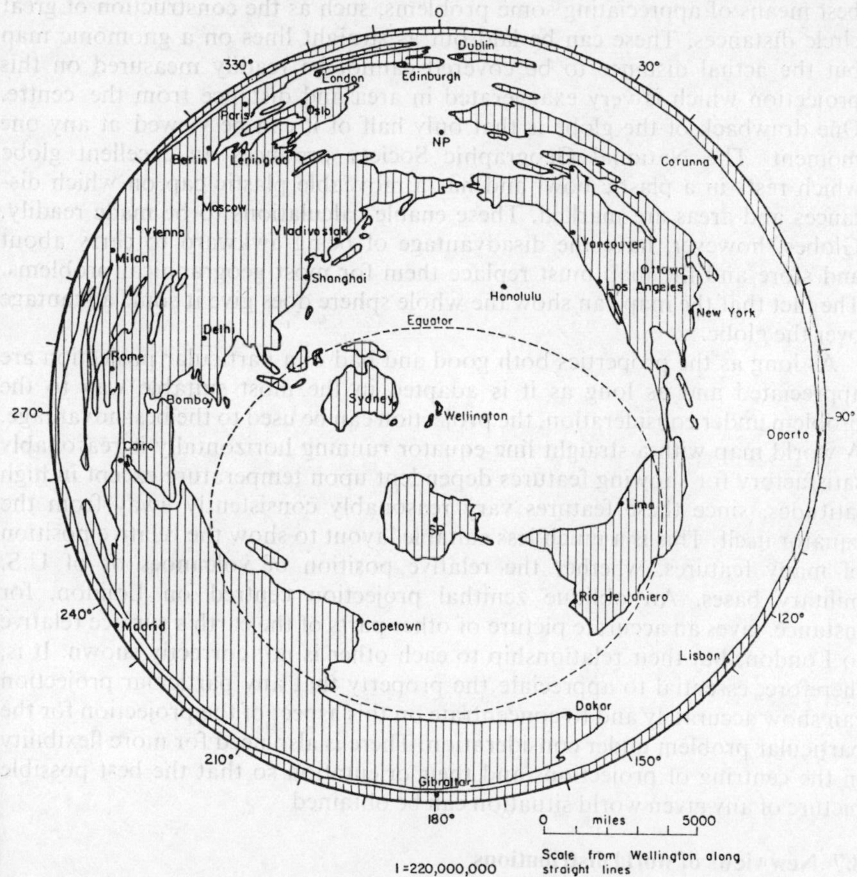

Figure 8.6 An oblique zenithal equidistant projection centred on Wellington, New Zealand. Distances and directions are shown correctly only from Wellington. Note the extreme distortion of Spain at the antipodes.

Apart from the three main groups of projections there is a fourth group of conventional projections that get away further from the strictly geometrical solution to the problem. Many of these projections are world networks, such as Mollweide, in which the earth is fitted into an ellipse to preserve the equal area property. In many of these conventional world nets it is possible to reduce the error of selected areas by interrupting the projection. This often, however, reduces the value of the overall picture that it gives of the earth as a whole.

From the point of view of appreciating the earth as a sphere on a map it is useful to get away from the familiar but grossly misleading Mercator map, which is only of value for navigation, or the Mollweide centred on the equator, and to construct maps with unusual centres and specific properties. Even the best map is always misleading and the globe itself may provide the best means of appreciating some problems, such as the construction of great circle distances. These can be laid out as straight lines on a gnomonic map but the actual distance to be covered cannot be readily measured on this projection which is very exaggerated in area and distance from the centre. One drawback of the globe is that only half of it can be viewed at any one moment. The National Geographic Society produces an excellent globe which rests in a plastic bowl and has a moveable plastic cap on which distances and areas are marked. These enable calculations to be made readily. Globes, however, have the disadvantage of being awkward to carry about and store and the map must replace them for most geographical problems. The fact that the map can show the whole sphere does give it some advantage over the globe.

As long as the properties both good and bad of a particular projection are appreciated and as long as it is adapted in the most suitable way to the problem under consideration, the projection can be used to the best advantage. A world map with a straight line equator running horizontally is reasonably satisfactory for showing features dependent upon temperature except in high latitudes, since these features vary reasonably consistently away from the equator itself. This is a much less suitable layout to show the relative position of many features, whether the relative position of volcanoes or of U.S. military bases. An oblique zenithal projection centred on London, for instance, gives an accurate picture of other parts of the earth's surface relative to London, but their relationship to each other is not correctly shown. It is, therefore, essential to appreciate the property that any particular projection can show accurately and to concentrate on this aspect of the projection for the particular problem under consideration. There is also need for more flexibility in the centring of projections and their orientation so that the best possible picture of any given world situation can be obtained.

8.7 New views of world distributions

(*a*) *A world view through topology*

Clearly the problem of representing the spherical surface of the globe or even just a sizeable part of it on a flat piece of paper has been a central problem facing cartographers and geographers for several centuries. Many projections distort or cut up the surface into strange forms. Almost invariably, however, attempts have been made to retain some Euclidean aspects in any world map. There seems scope for developing some world maps from a topological starting point (see chapter 2.10). One such map is illustrated in figure 8.7. The map shows the evolution of interoceanic navigation and the effect of the opening of the Suez and Panama Canals, and the Soviet Northern Sea Route. Each ocean is conceived as a node, and each interoceanic link, an arc. This is a very great abstraction since the oceans are popularly visualized as very extensive patches. The mainland masses occupy the *regions* between the arc. They are numbered I–VI. VII, Hawaii, stands in for the 'outside' region. There are three nodes, AT (Atlantic), IN (Indian) and P (Pacific). In figure 8.7, maps (*a*) to (*e*), and (*g*), the Pacific is represented by several nodes and arcs. These are 'dummy' ones used to facilitate the construction of the map. They stand in for the Pacific node, actually achieved in (*f*). The three ocean nodes are linked by seven proper arcs (listed 1 to 7). Map (*g*) shows the location of the items listed below.

REGIONS

I	Eurasia
II	Africa
III	North America
IV	South America
V	Australia
VI	Antarctica
VII	Hawaii

NODES

AT	Atlantic
IN	Indian
P	Pacific

(*a–e*) are 'dummy' nodes for Pacific

ARCS

1 Atlantic–Indian via Cape of Good Hope
2 Atlantic–Pacific via Cape Horn
3 Indian–Pacific via Tasman Sea
4 Indian–Pacific via East Indies
5 Atlantic–Indian via Suez Canal
6 Atlantic–Pacific via Panama Canal
7 Atlantic–Pacific via Northern Sea Route
(8) are 'dummy' arcs for Pacific node

This example illustrates the ability of topology to bring out clearly the essential features of a complex situation. It shows the implications for inter-oceanic navigation, of the opening of the Suez and Panama Canals and the Northern Sea Route (only part of the year). The pre-Suez situation is illustrated in maps (*a*) and (*b*).

(*b*) Pre-Suez situation. In its simplest form this is three arcs linking the Pacific and Indian Ocean nodes via East Indies, Tasman and Cape of Good Hope/Cape Horn. The Atlantic is not yet a node. Refer to (*f*) for arc names.

(*c*) The opening of the Suez Canal creates a new arc between Atlantic and Indian. The Atlantic becomes a 3-node.

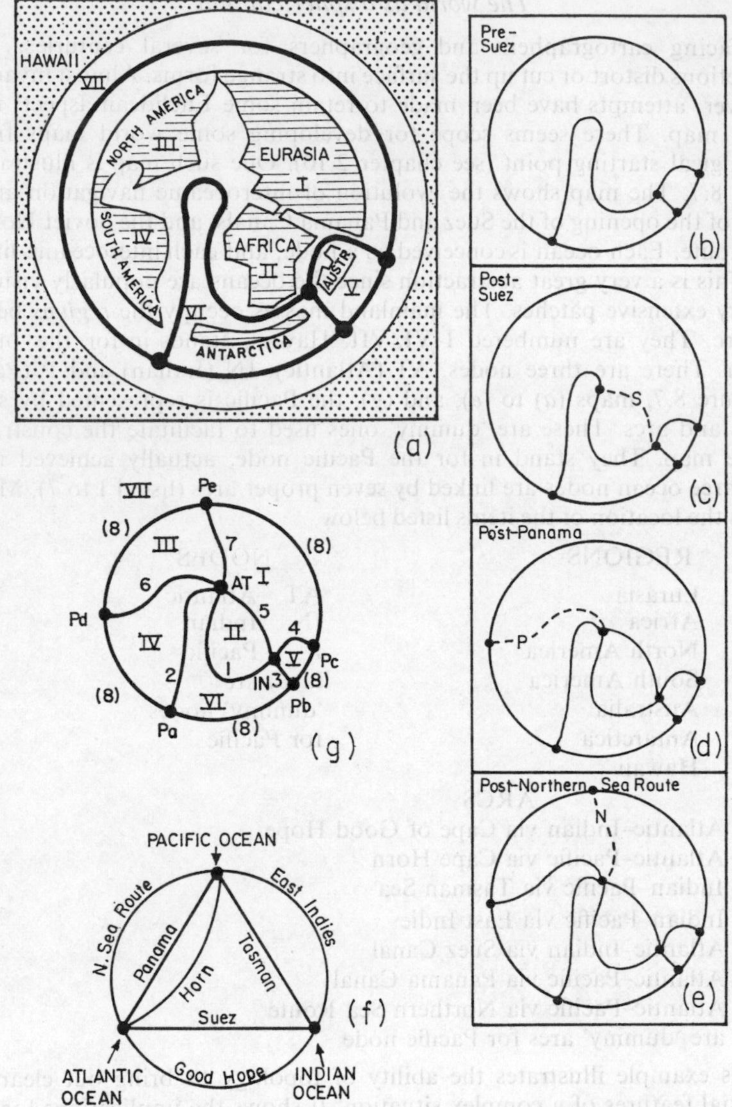

Figure 8.7 Topological view of inter oceanic navigation, see text for explanation

(*d*) The opening of the Panama Canal creates a new arc between Atlantic and Pacific.
(*e*) The 'opening' of the Northern Sea Route creates a new arc between Atlantic and Pacific.
(*f*) Here the situation is reduced to its most simple, revealing three nodes and seven arcs.

(b) A method of comparing access from European powers to their colonies
Topological maps discard information about direction and distance but retain correct linkage between nodes. It is possible with suitable diagrams to retain correct distance but to discard direction and correct linkages. Suppose it is desired to show visually the distance relationship of several of the European colonial powers to their respective colonies. Figure 8.8 shows a possible method of making a comparative diagram for this purpose.

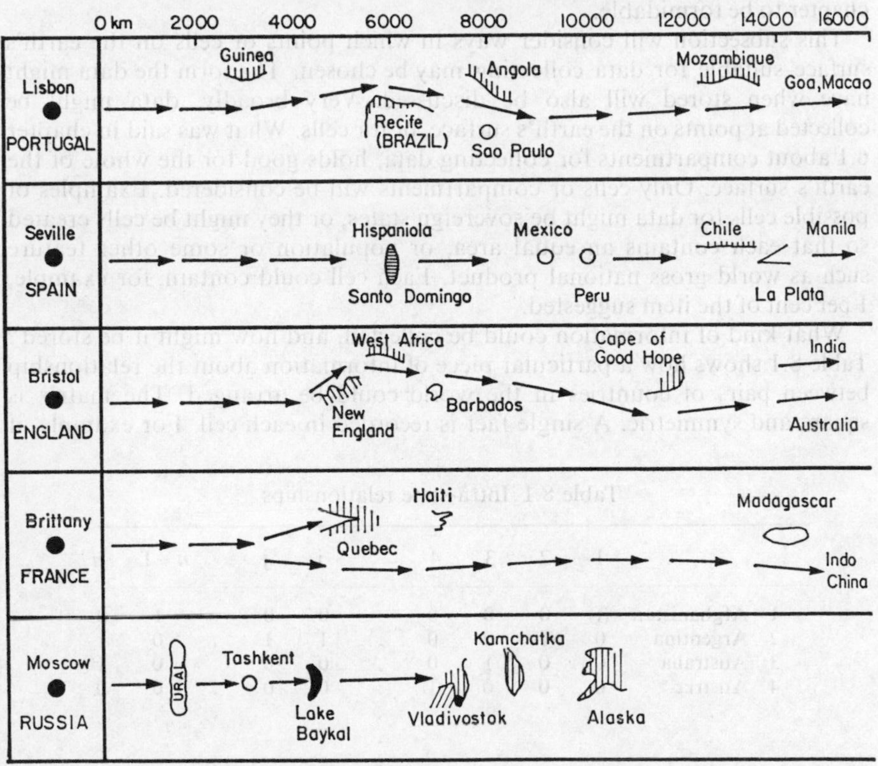

Figure 8.8 Distance relationships between five European powers and their former colonies.

Five European colonial powers are arranged one below the other. Selected colonies of each power are then strung out from left to right across the diagram, according to their distance from their respective mother countries. The eye can easily pick out which colonies of each European country are at similar distances. Distances are by sea for the four sea powers, by land for Russia. Not only may colonies of different powers be compared for distance (e.g. Santo Domingo of Spain with West Africa of England), but places belonging to the same colonial power and lying in different directions (e.g. Angola, São Paulo of Portugal) may also be compared. It should be possible to modify such a diagram to suit various purposes.

(*c*) *Non-visual representation of relationships on a world scale*

In the long run, rather than attempt to contrive maps and diagrams, it may prove more useful to collect data for appropriate places on the earth's surface, to store them in some compact form, such as a matrix, and then use the data as required. Such an idea is of course possible for relatively small parts of the earth's surface and is being developed, for example, in Sweden. The drawbacks of producing results in map form have already been shown in this chapter to be formidable.

This subsection will consider ways in which points or cells on the earth's surface suitable for data collecting may be chosen. The form the data might have when stored will also be discussed. Very broadly, data might be collected at points on the earth's surface, or for cells. What was said in chapter 6.1 about compartments for collecting data, holds good for the whole of the earth's surface. Only cells or compartments will be considered. Examples of possible cells for data might be sovereign states, or they might be cells created so that each contains an equal area, or population or some other feature such as world gross national product. Each cell could contain, for example, 1 per cent of the item suggested.

What kind of information could be collected, and how might it be stored? Table 8.1 shows how a particular piece of information about the relationship between pairs of countries in the world could be arranged. The matrix is square and symmetric. A single fact is recorded in each cell. For example, if

Table 8.1 Intra-state relationships

		1	2	3	4		i	j		$n-1$	n
1	Afghanistan	①	0	0	0	.	0	0	.	1	0
2	Argentina	0	①	0	0	.	1	1	.	0	1
3	Australia	0	0	①	0	.	0	0	.	0	1
4	Austria	0	0	0	①	.	0	0	.	0	0
	.					.			.		
	.					.			.		
	.					.			.		
i	Bolivia	0	1	0	0		①	1	.	0	0
j	Brazil	0	1	0	0	.	1	①	.	0	1
	.					.			.		
									.		
$n-1$	USSR	1	0	0	0	.	0	0	.	①	1
n	USA	0	1	1	0	.	0	1	.	1	①

there is a direct (not non-stop, but non-change) air flight between a pair of countries, then a 1 is recorded. If not, then 0 is recorded. The ①s along the principal diagonal represent within country flights. Suppose the information is recorded for flights involving $n = 100$ countries, then the matrix would have 10,000 compartments, of which $n(n-1)/2$ or 4450 would be different.

It would, however, be feasible to have, for example, ten spaces in each cell in the matrix, one for each of ten important Airlines, and to put a ten digit binary number in each cell, thus storing nearly 45,000 bits of information.

Table 8.2 shows another way in which information may be elaborated. There are four possible ways in which sovereign states (referred to as SS in the diagram) of the world may reach one another, physically speaking: (1) or 00 across a common land boundary (2) or 01 across the sea (void) (3) or 10 by crossing one other sovereign state (and perhaps also sea) (4) or 11 by crossing a minimum of 2 other sovereign states (and perhaps also sea).

Table 8.2 Intra-state accessibility

		SS_1	SS_2	SS_3	SS_4		SS_i	SS_j		SS_{n-1}	SS_n
SS_1	Afghanistan	—	10	10	11	.	11	10	.	10	10
SS_2	Argentina	10	—	01	10	.	00	00	.	01	01
SS_3	Australia	10	01	—	10	.	10	01	.	01	01
SS_4	Austria	11	10	10	—	.	11	10	.	10	00
	.					.			.		
	.					.			.		
	.					.			.		
SS_i	Bolivia	11	00	10	11	.	—	00	.	10	10
SS_j	Brazil	10	00	01	10	.	00	—	.	00	01
	.					.			.		
	.					.			.		
	.					.			.		
SS_{n-1}	Venezuela	10	01	01	10	.	10	00	.	—	01
SS_n	Yugoslavia	10	01	01	00	.	10	01	.	01	—

The need for a more satisfactory system of cells for collecting information than the sovereign states of the world has not passed unnoticed. For example, F. R. Pitts (1957) has discussed the possibilities of handling data on a world scale with the help of computers.

It would be particularly valuable for the collection of data to be able to cover the surface of the earth by a system of uniform sized and shaped cells. The simplest 'division' of the earth's surface into equal cells occurs in the baseball ball and tennis ball (8.9*a*) or in the cricket ball (8.9*b*). In the former case, the two parts 'interlock'. A larger number of cells would, however, be more useful. They should be further divisible into a number of equal shaped subdivisions. The pattern should be based on some form of regular polyhedron. There are only five regular polyhedrons. They have 4, 6, 8, 12 or 20 faces respectively. It is therefore reasonable to choose the figure with the largest number of faces. The figure has the added advantage that each of its faces can be readily subdivided into four equal parts. This figure, the icosahedron, consists of 20 equilateral triangles so arranged that two sets of five triangles are linked by a set of ten triangles.

It is also desirable that the surface area of the polyhedron should equal the surface area of the world. Given this assumption it is possible to calculate

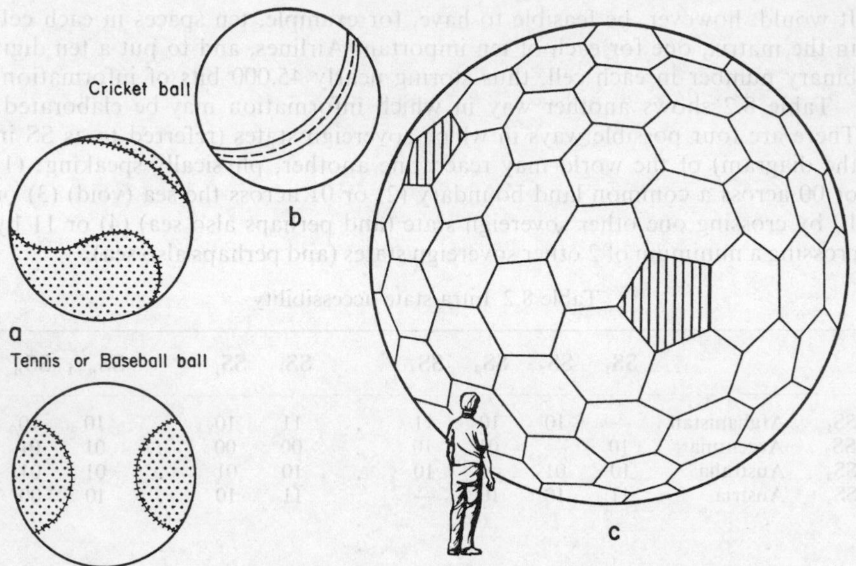

Figure 8.9 Examples of equal area cells on a spherical surface. (*a*) Baseball and tennis balls. (*b*) Cricket ball. (*c*) Hexagons on a giant balloon. The regularity is unfortunately occasionally marred by a pentagon.

the size of the triangles in relation to the earth's surface area. The area of a sphere is equal to that of the circumscribed cylinder or $4\pi R^2$. If R is taken as one unit then this value is 12.5664. From this it follows that the area of each triangle in the polyhedron must have a surface area of 12.5564/20, which is 0.6283. From this figure the length of the side of the triangle is 1.205 and the height of the triangle is this figure times the sine of 60 degrees giving 1.043. Thus the height of the triangles is very close in value to the radius of the earth to scale. It is convenient to fix the polyhedron so that the two points where five triangles meet are situated at the poles. Then the ten triangles are arranged symmetrically around the equator which lies at the mid-point of the vertical of the triangles. There must be some distortion of scale in some dimensions to produce the equal area property of the figure. This can be calculated for a line of longitude following one side and two heights of triangles. This value on the globe is πR or 3.1416 as R is 1, on the polyhedron it is $1.205 + 2.085$ which is 3.29, or a value 1.049 times the true length.

The latitude of the points *a*, *b*, *c*, *d*, *e* (see figure 8.10) must be determined. These points will not lie exactly on the sphere, but it is possible to calculate the latitude in whose plane they lie. This angle β is given by the sine of bz/R, where bz is $\frac{1}{2}$ the height of the triangle. Thus the angle β is 30° 25′ north. The points *f*, *g*, *h*, *i* and *j* similarly lie in the plane that cuts 30° 25′ south. The length of the sides of the triangles can be converted into miles by substituting the correct radius of the earth. If this value is taken to be 3960 miles, ignoring the polar flattening and taking the earth to be a true sphere, then the length

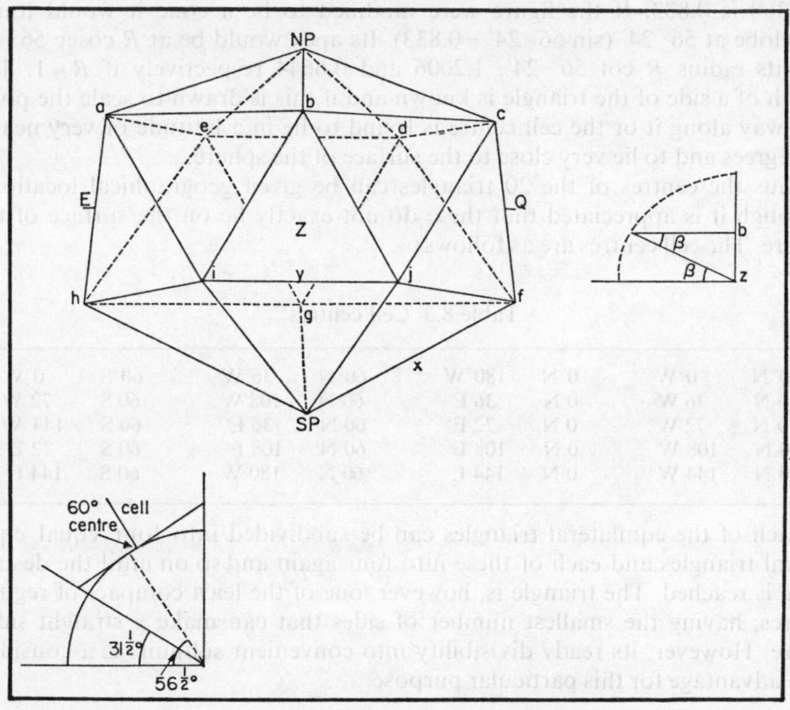

Figure 8.10 Diagram of icosahedron used to form uniform cells over the earth's surface. Standard parallel given by constant of the cone $5 \times 60/360 = 0.833$, $\sin^{-1} 0.833 = 56° \ 24'$, $R \cot 56° \ 24' = 1.33$, $R = 2$ ins; $R \operatorname{cosec} 56° \ 24' = 2.4$.

of the sides of the triangles can be calculated to be 4770 miles. The length round the five triangles at latitude 30° 25′ is 5 times this or 23,850 miles. This length can be compared with the distance round the globe in this latitude which is given by $2\pi R \cos 30° \ 25'$ or 21,200 miles. Thus the scale exaggeration at this parallel is 23,850/21,200 or 1.125.

It is also necessary to know the position of the centre of the 20 cells or triangles. The centres of the ten triangles that are fitted round the equator must lie on the equator, as this line lies half way along the height of the triangle, a property that has already been used. The centres must be spaced at 36 degrees apart. The other five centres must be spaced equally from the north and south pole respectively and five exactly must fit. At the latitude of 60 degrees the length of the parallel is exactly half that of the equator as the cosine of 60 degrees is 0.5. Thus the five cell centres can be fitted at this latitude at a distance of 72 degrees apart. This result was also arrived at by a semi-graphic method. It was argued that the angles at the apex of the figure at the pole add up to only 300 degrees so that there is some similarity with conical projections. The constant of the cone having the ratio

300/360 is 0.833. If the figure were modified to be a cone it would touch the globe at 56° 24′ (sin 56° 24′ = 0.833). Its apex would be at R cosec 56° 24 and its radius R cot 56° 24′, 1.2006 and 0.6644 respectively if $R=1$. The length of a side of the triangle is known and if this is drawn to scale the point half way along it or the cell centre is found to lie in a latitude of very nearly 60 degrees and to lie very close to the surface of the sphere.

Thus the centres of the 20 triangles can be given geographical locations, although it is appreciated that these do not exactly lie on the surface of the sphere. The cell centres are as follows:

Table 8.3 Cell centres

0°N	0°W	0°N	180°W	60°N	36°W	60°S	0°W
0 N	36 W	0 N	36 E	60 N	108 W	60 S	72 W
0 N	72 W	0 N	72 E	60 N	36 E	60 S	144 W
0 N	108 W	0 N	108 E	60 N	108 E	60 S	72 E
0 N	144 W	0 N	144 E	60 N	180 W	60 S	144 E

Each of the equilateral triangles can be subdivided into four equal equilateral triangles and each of these into four again and so on until the desired scale is reached. The triangle is, however, one of the least compact of regular figures, having the smallest number of sides that can make a straight sided figure. However, its ready divisibility into convenient subunits is a considerable advantage for this particular purpose.

References

Brock, B. B. (1956). Structural mosaics and related concepts, *Trans. Geol. Soc. S. Africa*, **LIX**, 149–197.

Leopold, L. C. and Langbein, W. B. (1962). The concept of entropy in landscape evolution, *U.S. Geol. Surv., Prof. Paper*, 500 A, U.S. Dept. Interior.

Linton, D. L. (1965). The geography of energy, *Geog.*, **50** (3), 197–241.

Mackinder, H. J. (1904). The geographical pivot of history, *Geog. J.*, **23**, No. 4.

Ordway, R. J. (1966). *Earth Science*, Van Nostrand, London, New York, Toronto.

Pitts, F. R. (1957). Regional Science Meeting, Tokyo.

Stommel, H. (1958). *The Gulf Stream*, University of California Press,

Tatchsh, J. H. (1964). Distribution of active volcanoes; summary of preliminary results of 3-dimensional least square analysis, *Bull. Geol. Soc. Am.*, **75** (8), 751–752.

Wiesner, J. B. and York, H. F. (1964). The test ban, *Sci. Am.*, **211**, No. 4, 27–35.

9

Geography and three space dimensions

'I call our world Flatland, not because we call it so, but to make its nature clearer to you, my happy readers, who are privileged to live in Space . . .
'In such a country, you will perceive at once that it is impossible that there should be anything of what you call a "solid" kind . . .'

Edwin A. Abbott (1884)

9.1 Three space dimensions

All objects in reality must be 3-dimensional themselves and must exist in 3-dimensional space. To consider a space of only two dimensions is therefore a simplification or symbolization of reality. For many purposes this simplification is adequate and necessary. In all situations where position on a 2-dimensional surface is the most important attribute of the objects concerned, the third dimension may be ignored. In 2-dimensional situations the distribution of the objects can be represented by rectangular co-ordinates. There are, however, many problems in both physical and human geography that necessitate a 3-dimensional analysis. Such problems include the following:

(1) Lines in a 3-dimensional space. In some situations the third dimension is important for its own sake as an independent variable. For example, the dip of a rock bed is as important as its strike.

(2) A surface in three dimensions. The third dimension is necessary in studying an undulating 2-dimensional surface. The term 'surface in three dimensions' is used here to denote an undulating 2-dimensional surface, on which points can be identified by using 3 space co-ordinates, *x*, *y* and *z* (see figure 9.1). The latter gives the elevation of the point above a fixed datum plane.

(3) 3-dimensional volume. The 3-dimensional volume of an object is sometimes of importance. This may include both the size and shape of the object under consideration.

Each of these problems will be dealt with in this chapter. The problems involve a consideration of methods of representing a 3-dimensional situation on a 2-dimensional piece of paper, and the interpretation of a 2-dimensional representation of a 3-dimensional situation. The importance of the third dimension in a few situations will also be briefly referred to.

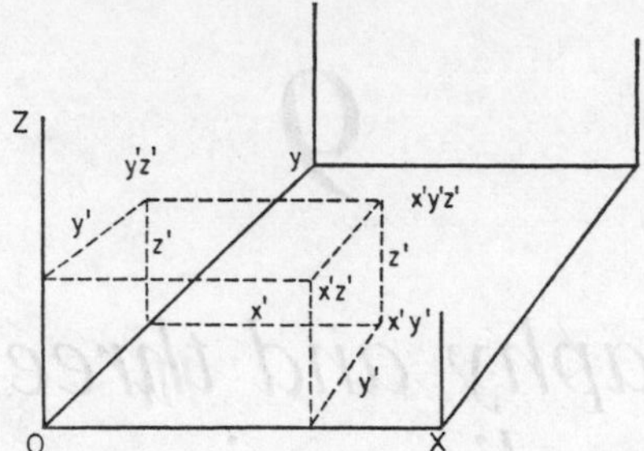

Figure 9.1 Diagram to illustrate co-ordinates in three space dimensions.

It has already been pointed out in chapter 8 that the shell of the earth, which is the field of geographical study, is very thin relative to its areal extent. Nevertheless there are certain processes in the operation of which the vertical dimension is essential. The necessity to consider the circulation of the atmosphere and ocean as a world-wide process has been stressed. It is no less essential to consider the vertical element in these circulations, because it is vital to their operation. The part played by the vertical dimension, small as it is relative to the horizontal ones, will be mentioned in section 9.4.

The third dimension is also clearly vital in all those processes in which the force of gravity plays a part and this, therefore, includes practically all natural processes. A consideration of the third space dimension is also necessary in human geography. Both the buildings and the communication channels of a large town exist and operate in 3-dimensional space.

9.2 Lines in three dimensions

In order to describe the position of a line in 3-dimensional space it is necessary to know both its orientation and its angle of dip. The orientation is normally measured on a horizontal plane relative to true or magnetic north. The dip is measured relative to the horizontal or vertical. There are a number of situations, particularly in physical geography, when both the orientation and the dip, or slope, of a linear feature must be taken into consideration e.g. the direction and slope of a stream. On a much smaller size scale the direction and dip of elongated stones in till or other deposits provide an example of this situation. An analysis of these stones provides information concerning the past flow of ice and processes of glacial deposition. Still smaller are the magnetized particles in an ancient rock. Their orientation and dip provide valuable data concerning the pattern of the earth's magnetic field and the position of the pole, relative to the area concerned, at the time when the particles received their magnetism. This study is the basis of palaeomagnetism, but on the whole it is less directly relevant to physical geography as a technique than till fabric analysis, although its world wide implications are very important in geography (see chapter 12, section 2).

The maximum amount of information can be obtained from a till fabric if the elongated stones within it are considered both in terms of their orientation and their dip. It is often possible to distinguish between a true till and a solifluction deposit by this means because the former shows an up-glacier preferred dip, while the latter shows a down-slope preferred dip. These criteria allow the two types of deposit to be differentiated in suitable situations.

Orientation can easily be measured in the field either with a compass or with a protractor. An aligned string, the orientation of which has been measured, is used as a datum from which the orientation is measured. The dip can be measured with a small hand level or protractor (see figure 9.2).

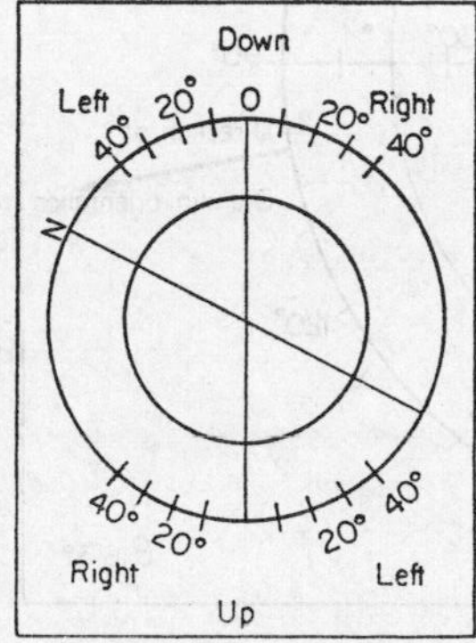

Figure 9.2 Diagram showing method of recording till fabric measurements, using lower hemisphere.

The values are best recorded on an equidistant zenithal projection as in figure 9.3. The correct orientation relative to north is found and each observation is plotted, according to its dip, as a point on the diagram. The point on the diagram is symbolic of both orientation and dip. Horizontal stones are marked on the circumference and a ten degree dip on the ten degree circle. The plotted pattern represents the lower hemisphere of a sphere. It gives a visual impression of the scatter of the points and some idea of the orientation and dip pattern can be obtained. It is more useful, however, to obtain an exact value for the distribution by statistical analysis. A method of analysis has been described by R. Steinmetz (1962) and a computer programme using this technique is given by J. T. Andrews and K. Shimizu (1966). This procedure will not be described in detail. It provides a value for the centre of gravity of the distribution of points in terms of orientation and dip and it also gives a measure of the strength of the pattern. This is given as an angular measure (see figure 9.3) denoting the radius of a circle of confidence.

Underground stream courses provide another example in which vertical relationships are important. Water disappearing down a sink hole will eventually reappear elsewhere. If a number of sink holes and springs exist it is not easy to appreciate the pattern of underground drainage unless this is considered 3-dimensionally. The lines of flow underground can be traced by putting fluorescein or lycopodium spores in the water at the sinks and watching the springs to establish where the coloured water emerges. A good

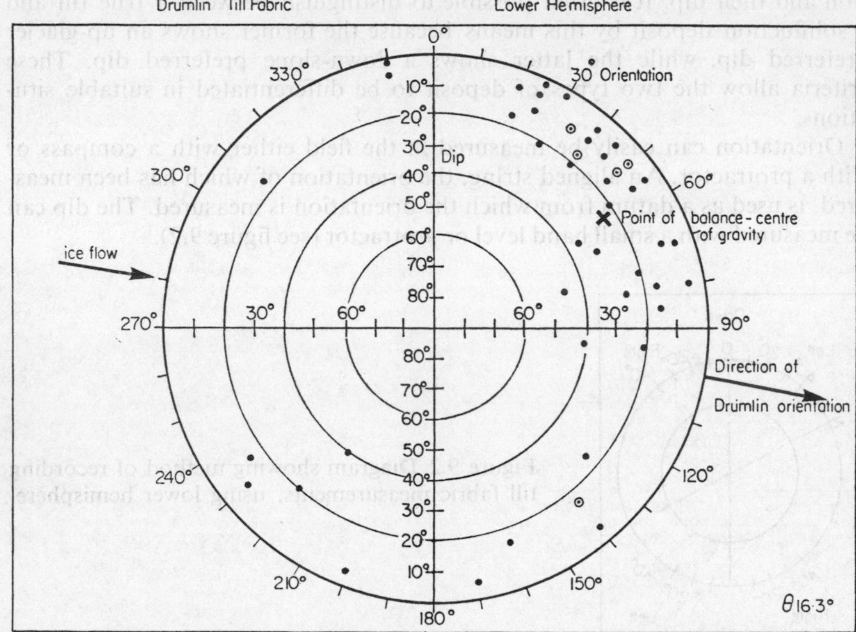

Figure 9.3 A till fabric analysis plotted as a 3-dimensional pattern in the lower hemisphere. The orientation is shown relative to north and the dip is given by the concentric circles. The outer circle indicates horizontal stones. The point of balance of the distribution has been computed and the direction of ice flow and direction of drumlin elongation is indicated. The fabric was recorded in a drumlin in Yorkshire.

instance of an underground pattern in which the vertical relationship is important is seen south of Malham Tarn, Yorkshire (see figure 9.4). The water that flows out of the tarn would be expected to reappear at the foot of Malham cove. However, experiments have shown that the water that comes out at the foot of the cove sank at the old smelt mills sink and that the water that disappears at the tarn sink re-emerges in Aire Head Springs near Malham village. The two streams must, therefore, cross underground. This means that their courses must be distinct and must lie one above the other at some point between the sinks and the springs. This situation of stream flow at two different levels can only occur with underground streams, although it may also happen that an underground stream crosses beneath a surface stream in some circumstances.

The superimposition of two moving media can also occur in glacier systems, where a tributary glacier may override or be overridden by the main glacier. Each ice mass will maintain its identity and differential horizontal movement may occur through the ice thickness. An example of this differential flow is

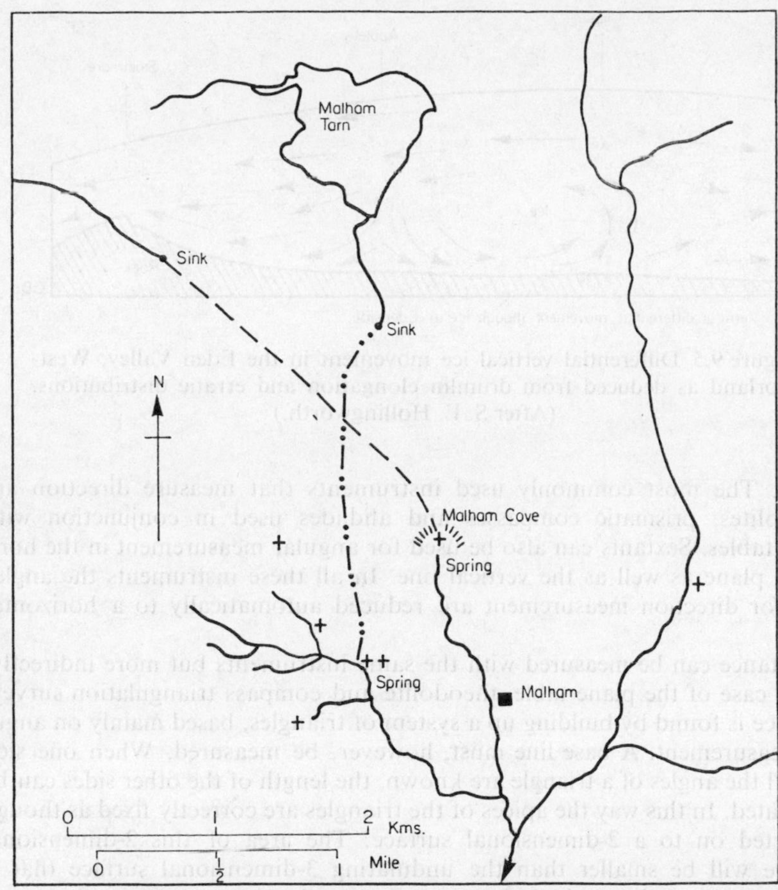

Figure 9.4 Map to show crossing of underground streams at different levels in the area near Malham Yorkshire. The underground parts of the stream are shown in dashed lines.

shown in figure 9.5, which is taken from the work done by S. E. Hollingworth in the Eden valley (Hollingworth, 1931).

9.3 Surfaces in three dimensions

(*a*) *Surveying*

A 3-dimensional undulating surface has already been defined. The land surface of the earth over small areas is the most familiar example. A common problem associated with a 3-dimensional surface is its representation on a flat piece of paper. The steps necessary to achieve this representation involve both surveying and cartography. In surveying it is necessary to measure direction, distance and height. Direction can be measured by several instru-

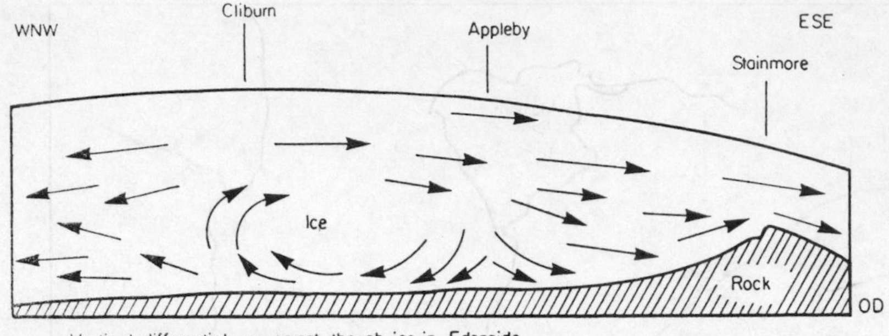

Figure 9.5 Differential vertical ice movement in the Eden Valley, Westmorland as deduced from drumlin elongation and erratic distributions. (After S. E. Hollingworth.)

ments. The most commonly used instruments that measure direction are theodolites, prismatic compasses and alidades used in conjunction with plane tables. Sextants can also be used for angular measurement in the horizontal plane as well as the vertical one. In all these instruments the angles used for direction measurement are reduced automatically to a horizontal plane.

Distance can be measured with the same instruments but more indirectly. In the case of the plane table, theodolite and compass triangulation survey, distance is found by building up a system of triangles, based mainly on angular measurement. A base line must, however, be measured. When one side and all the angles of a triangle are known, the length of the other sides can be calculated. In this way the apices of the triangles are correctly fixed as though projected on to a 2-dimensional surface. The area of this 2-dimensional surface will be smaller than the undulating 3-dimensional surface that it represents by an increasingly large amount as the slope angles increase and, with them, the area of the sloping surface.

The horizontal equivalent of a distance measured along sloping ground can be found by multiplying the measured distance by the cosine of the angle of slope. The resulting value will always be smaller than the measured distance, the difference increasing as the slope steepens. The difference in height between two points can be found by multiplying the horizontal equivalent of the distance between them by the tangent of the angle of slope. Alternatively, if the slope distance has been measured, it can be multiplied by the sine of the angle to give the height difference.

(*b*) *Cartography*

The problems of representing a small part of the 3-dimensional undulating surface of the earth on a flat piece of paper can be solved in a number of ways. The first reasonable attempt was by *hachuring*, using lines drawn in the direction of slope. An indication of degree of slope was given by the closeness of the hachures, but no accurate measure of slope was provided by this method. The most common and successful method of showing the 3-dimensional real surface in two dimensions is by the use of contours. A contour is a line joining points at the same altitude above sea-level, that is, with similar z co-ordinates above 0 ft O.D. Various other methods of mapping slopes and relative relief have been suggested. None of these is very satisfactory and few give as good a visual impression of the relief as a good contour map. However, for purposes of geomorphological research, a method of recording on a 2-dimensional map the essential elements of the relief has been developed.

(c) *Morphological mapping*

The essential features of relief that must be shown on a morphological map, which, as its name implies, aims to portray the shape of the ground, are the breaks and changes of slope. Symbols have been developed to record these features accurately. There are, however, difficulties when the features have a high gradient as their horizontal equivalent is then small compared to their morphological significance. For this reason special symbols have been developed to represent steep slopes. These are defined as slopes over 40°. The features that are shown on a morphological map differ from those shown on a normal contour map but they are the features that are morphologically significant. It is not possible to recognize these features from a contour map. For this reason a morphological map must be drawn in the field or from good stereo pairs of air photos, in both of which the 3-dimensional undulating surface can be properly appreciated. The most essential geomorphological features of the landscape can then be transferred to a 2-dimensional piece of paper by the use of suitable symbols (figure 9.6). These allow the 3-dimensional surface to be more satisfactorily reconstructed by those who are familiar with the symbols. A combination of a good contour map and a good morphological map, the former providing accurate elevations and the latter a better idea of slope, is the best means of appreciating a 3-dimensional undulating surface if this cannot be visited in the field. These two forms of map provide valuable data that can be analysed further as shown in the example of relief and drainage analysis discussed in chapter 7.

(d) *Underground surfaces in three dimensions*

It is sometimes possible to establish a 3-dimensional undulating surface beneath the ground level by making suitable measurements. An example is the establishment of the form of the water table in areas of suitable rocks. The chalk of southeast England illustrates such an area. In this area there are many wells dug to utilize the ground water. The positions and elevations of these wells are known, and can be plotted on a map. By measuring the depth to the water table in the wells, the height of the water at that point can be found by deducting the depth in the well from the ground level at the well site. If many such measurements are made simultaneously, then the level of the water table at all the points can be found at one time. The water table

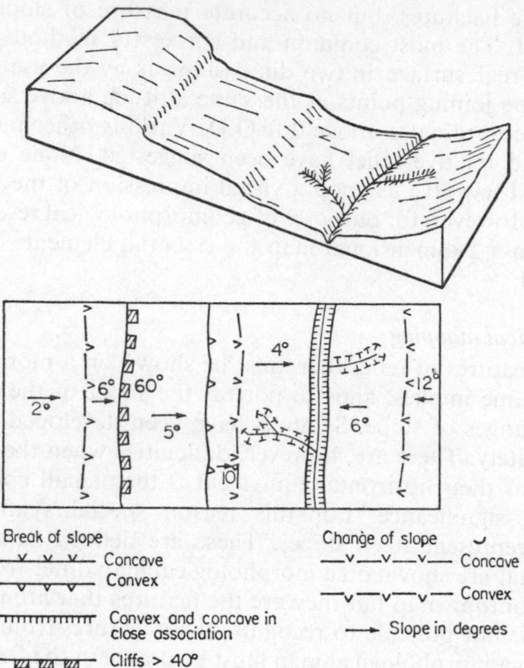

Figure 9.6 Block diagram and map to represent symbols used in morphological mapping.

surface can be contoured by interpolation. If observations are repeated at different time intervals, then the changes of the height and character of the water table with time can be found. Observations of this type have been made by W. V. Lewis on the chalk near Cambridge.

(e) *Isobases*

Many studies of raised shorelines treat the problem from a 2-dimensional point of view by projecting the observations of each shoreline on to a single tilted plane. This plane is intended to run normal to the direction of the isobases, which are lines joining points of equal isostatic uplift. A better method of dealing with this problem is to treat it 3-dimensionally. In this way it is possible to determine reasonably accurately the direction and degree of tilt of the plane. Figure 9.7 illustrates the basis of a method which consists of taking a number of triangles, based on 3 points on a single shoreline. The direction of the isobase relative to the elevation of the middle point B is determined and the tilt of the surface is at right angles to this line (2–2 on the diagram). The amount of tilt can be determined from the difference in level between A and B and B and C in a direction normal to the isobase. This method has been used successfully by J. T. Andrews (1966) to establish

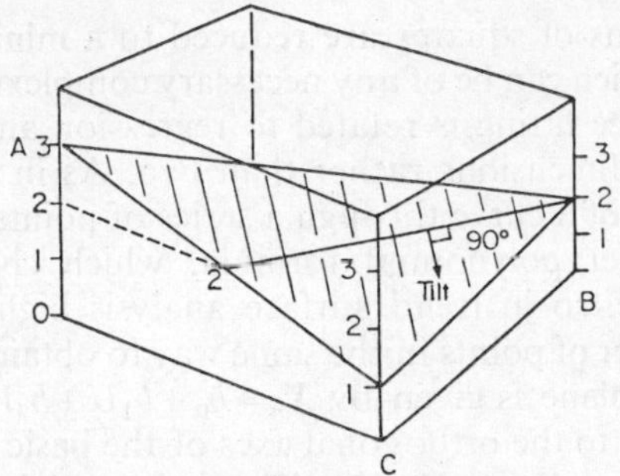

Figure 9.7 Block diagram to illustrate the location of isobases and the direction of tilt of unevenly uplifted marine features.

the tilt of the marine limit surface in part of west Baffin Island. The method is most successful where the points form well conditioned triangles and can only be applied to a synchronous shoreline.

The triangle method assumes that the shoreline is a plane tilted surface. It cannot produce anything but a plane surface. This may well provide a useful first approximation to the real form of the tilted surface. The technique of trend surface analysis provides a further stage, as this method can fit in turn plane, quadratic, cubic and higher surfaces to the points on the shoreline. S. B. McCann (1966) has applied this method of analysis to the raised shoreline on the west of Scotland around the island of Skye.

(*f*) *Trend surface analysis*

Geographical data often cover whole areas with continuously varying values. These values are usually only sampled at points, although it is the whole undulating surface which is of interest. Trend surface analysis provides a method whereby mathematical surfaces in three dimensions may be fitted to the data. The data may record height or depth at points on a surface, which occupies three space dimensions. Alternatively, the data may record other values which vary continuously over the area. The isopleths then illustrate variations other than the third space dimension. The data can, however, be treated in exactly the same way, and isopleths can be fitted to the data in the form of 3-dimensional surfaces.

Trend surface analysis is a mathematical technique by which surfaces of increasing complexity, starting with a plane, are fitted to point observations. The position of the plane is such that the sum of the squares of the vertical distances between the points and the plane is reduced to a minimum. The surfaces are, therefore, ones of best fit in the same way as a regression line is a line of best fit on a 2-dimensional surface. The positions of the points are defined relative to a rectangular co-ordinate grid. This technique is of particular geographical interest in that it treats data that are displayed on an areal basis. The data must be expressed numerically, on the ratio or interval scale, to allow the calculations of the sums of squares which forms the basis of the method. The sums of squares are reduced to a minimum by use of a polynomial function which can be of any necessary complexity to fit the data.

The method of surface fitting is related to regression analysis except that it takes place in three dimensions rather than two. As in 2-dimensional regression analysis the fit of a curve through a series of points can be improved by using a higher order polynomial formula, which changes the simple straight line to a curve, so in trend surface analysis higher order surfaces can be fitted to the scatter of points in the same way to obtain a better fit.

The formula for the plane is given by $X_n = b_0 + b_1U + b_2V$, where X is any point and U and V refer to the orthogonal axes of the basic grid, b_0, b_1 and b_2 are the coefficients. Thus $X = f\,(U_1\ V_1)$. The polynomial functions for the higher surfaces become increasingly more elaborate as the surface becomes more complex.

The linear plus quadratic surface has one change of direction of slope. It is defined by the formula

$$X_n = c_0 + c_1U + c_2V^2 + c_3U^2 + c_4UV + c_5V^2,$$

where all the c values are the coefficients for the various U and V values. As the surfaces become more complex so a greater proportion of the total sum of squares of the deviations between observed and computed values is accounted for.

The linear plus quadratic plus cubic surface has two changes of direction of slope and the surface is defined by an equation of the form

$$X_n = d_0 + d_1U + d_2V + d_3U^2 + d_4UV + d_5V^2 + d_6U^3 + d_7U^2V + d_8UV^2 + d_9V^3$$

where the d values refer to the coefficients. Higher surfaces of more complex form, the quartic, quintic and sextic can be fitted with increasingly long formulae and calculations. In practice it has been found that it is rarely necessary to go beyond the cubic surface as satisfactory fit can usually be obtained with either the quadratic or cubic surface and the increase in the goodness of fit is slow beyond the cubic surface.

The calculations required to reduce the sum of squares to a minimum for each surface are relatively simple but very long and tedious without the aid of a computer. The method basically involves matrix inversion to solve the equations, which give the coefficients. The basic formula for fitting the linear surface is given in the form

$$\begin{bmatrix} b_0 \\ b_1 \\ b_2 \end{bmatrix} = [U, V]^{-1} \cdot \begin{bmatrix} \Sigma X_{n\text{obs}} \\ \Sigma UX_{n\text{obs}} \\ \Sigma VX_{n\text{obs}} \end{bmatrix}$$ The U, V matrix is $$\begin{bmatrix} N & \Sigma U & \Sigma V \\ \Sigma U & \Sigma U^2 & \Sigma UV \\ \Sigma V & \Sigma UV & \Sigma V^2 \end{bmatrix}$$

There are 36 elements in the quadratic matrix and 100 in the cubic which illustrates the greatly increasing amount of calculation required for the higher order surfaces. Some computer programs for trend surface analysis plot the various surfaces, as well as giving the values of all the coefficients which define the equations for the surfaces and the deviations between observed and computed values. The total percentage of the variability accounted for by the surface is another useful value given by the computer output. This value indicates how much better the higher order surfaces fit than the linear surface. The deviations also supply useful information. These values may be plotted and the high deviations indicate those areas where the values are either considerably above or below those which best fit the surface. Local anomalies may be located in this way and their significance studied. Trend surface analysis is discussed in more detail in C. A. M. King (1967), Quantitative Bulletin for Geographers No 12.

Trend surface analysis has not been as widely used in geography as its potentialities suggest it should be. R. J. Chorley and P. Haggett (1965) enumerate some of the uses to which this method has been put. A simple example is included here to illustrate the application of the method to a geomorphological problem. Reference may also be made to Haggett's paper (1964) on vegetation and other features in South Brazil.

The purpose of the trend surface analysis in this example was to investigate the form of the summit surface of the Alston Block and Askrigg Block in the central Pennines. The first area covers the northern part of the Alston Block, bounded on the west by the Pennine Faults and the north by the Stublick Faults. The second area covers the southern part of the Askrigg Block, bounded on the west and south by the several Craven faults. It is believed that these uplands have been uplifted and warped in the relatively recent geological past. Early Pleistocene has been suggested as a possible date by F. M. Trotter (1929) for the Alston Block. If this supposition is true and the summit surface, as defined by summit heights, conforms to a warped surface, then the trend surfaces fitted through the points will indicate the form of warped surface. Monadnocks would show up as positive deviations and areas of excessive erosion as negative deviations. In this example the trend surface is a true 3-dimensional undulating surface.

The linear surface for the North Alston Block shows a plane sloping down to the northeast from 2800 ft to 400 ft over a north–south distance of 33 km. The quadratic surface slopes north at the western edge of the area and east along the southern edge, the change of slope indicating a possible warping of the surface (figure 9.8). The formula for the linear surface is

$$\text{Height} = 1253.19 - 18.17U + 42.86V$$

and for the quadratic surface

$$\text{Height} = 39.53 + 20.72U + 108.99V - 0.37U^2 - 0.63UV - 0.99V^2.$$

The cubic surface is defined by

$$\text{Height} = 142.62 + 8.56U + 116.73V - 0.53U^2 + 0.87UV - 2.34V^2 - 0.006 U^3 - 0.027U^2V + 0.01UV^2 + 0.013V^3.$$

The linear surface accounts for 83 per cent of the variability and the quadratic 90 per cent while the cubic surface only adds another 1 per cent bringing the total to 91 per cent. These values are higher than those for the Askrigg Block area.

The deviations for the quadratic surfaces are shown in figure 9.9. In all three surfaces (2 of which are not reproduced) the monadnocks of Cross Fell

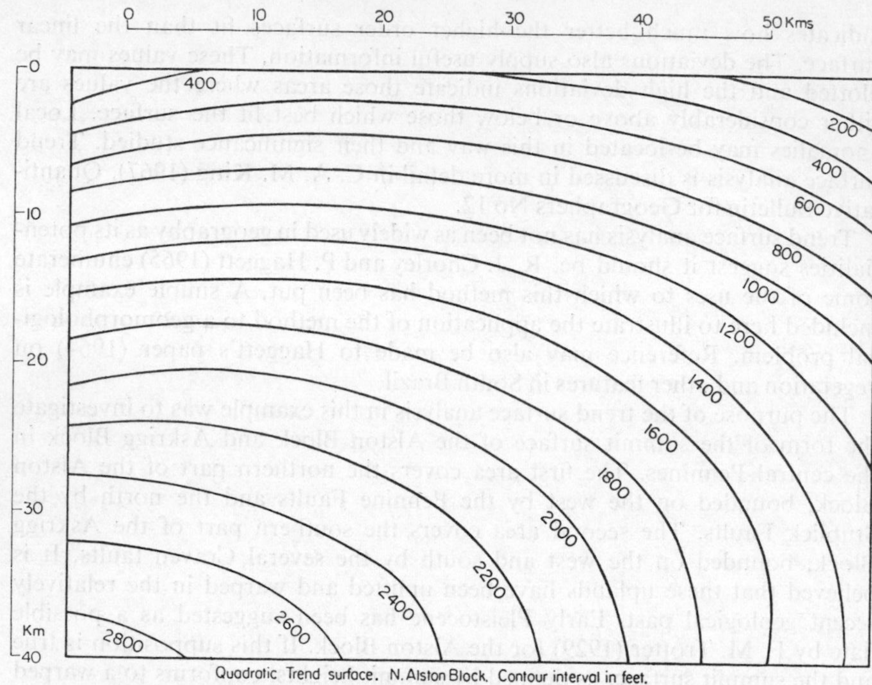

Figure 9.8 The quadratic trend surface fitted to the elevation of high points at the northern end of the Alston Block in the Pennines. The surface has been reproduced directly from the computer printout and has not been adjusted to rectify the scale. The lines indicate contours on the warped surface.

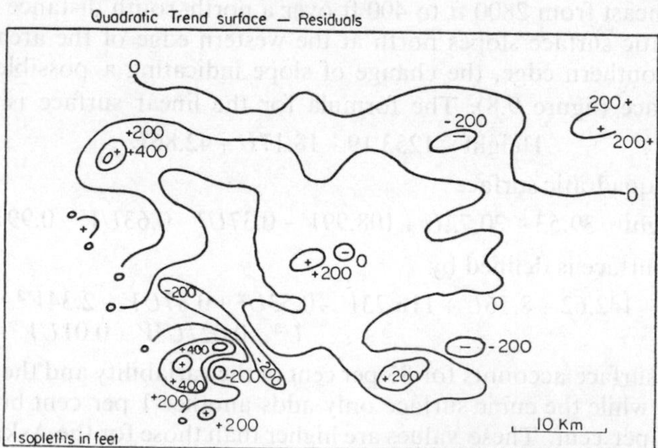

Figure 9.9 Isopleths of residuals from the quadratic trend surface in the northern Alston Block. The high points in the west represent monadnocks rising above the warped surface.

and Cold Fell stand up strongly as large positive deviations, which confirms the interpretation that they are monadnocks. Along the northern boundary of the area there is a strip of negative deviations, which is strongest in the linear deviations, but which appears in all three. It is probably due to the fact that the abrupt break of the Stublick fault cannot be accurately reproduced in the simpler surfaces. The deviations in the central part of the area are small on all the surfaces, showing that the summits of this area conform to a fairly simple sloping surface, trending down to the northeast. The deviations become large in the southeast of the area where the higher summits again form a group of monadnocks (393533) but other parts are lower than the trend surface giving negative deviations.

The results of the trend surface analysis confirm the suggestion that the summit surface of the northern Alston Block is warped. The major trend of warping is northwards in the west of the area, northeast in the northeastern and central part and more eastwards on the southern edge. The southern part of the Block, which is not discussed in detail, fits a surface trending down to the southeast. This is towards the downwarp of the Stainmore Syncline. Trend surface analysis provides a very useful method of fitting the best fit undulating 2-dimensional surface through the observed high points covering this particular area. It can be used in a wide variety of geographical situations, both physical and human.

9.4 Volume and 3-dimensional shape

The third dimension involves problems of volume as well as area. Shape is also involved in some cases. A number of situations in which these two aspects are important will be considered.

(*a*) Geological structure is essentially 3-dimensional. (*b*) The shape of pebbles is another problem which is considered. (*c*) On a very different scale is the measurement of volume in a landscape. Both have in common the necessity to measure in three dimensions. (*d*) Cave systems provide another 3-dimensional situation, in which height is as important as direction. (*e*) The landscape is modified by mass movement of material on slopes. This movement must be considered volumetrically to give useful data and results of this type of analysis are given. (*f*) The importance of the third dimension in two types of world wide circulatory systems is briefly commented upon. These are (i) atmospheric and (ii) oceanic circulation. In both the vertical element of the movement is essential to the functioning of the system as a whole. If there were no vertical movement of air there would be no precipitation. If water did not sink to deeper levels there could be no upwelling to renew the fertility of the oceans and the whole pattern of circulation in depth could not operate. The vertical and horizontal movements take place in masses of air and water which have volume, and for this reason, the discussion of these two circulations is included in this section.

(a) Geological structure

Geological structure is essentially 3-dimensional, but to portray this diagrammatically on a map necessitates reducing the data to two dimensions. Unless the 3-dimensional character of the structure can be appreciated from the map, however, it is not possible to obtain a clear picture of the geology. By providing information about the vertical dimension, vertical sections help in this respect.

The dip, strike and thickness of the strata are often of importance, as it is these factors that determine the pattern of outcrop in relation to the land surface. It is this pattern that is shown on the geological map. The dip is the maximum angle of slope along the bedding planes of stratified rocks and the strike is a horizontal line at right angles to the dip. The strike line is a contour line on a rock stratum, while the dip is analogous to a hachure line, which is drawn down the steepest slope normal to the contour. When the dip is very steep the outcrop will be narrow, as in a vertical dyke, which will appear often as a nearly straight strip on the map. Horizontal strata, on the other hand, will outcrop parallel to the contours. A knowledge of the dip and strike of rock strata is necessary for an appreciation of the relationship between structure and outcrop, which in turn determines the surface land forms in rocks of differing resistance. An ability to determine the dip and strike and to visualize the structure in three dimensions is very helpful in appreciating the relationship between geological structure and land form (see figure 9.10*a*). The strike can be determined in uniformly bedded strata by joining points where the same outcrop cuts the same contour (see figure 9.10*b*). The line joining these points will be straight if the dip is uniform in direction. In the figure 9.11*a* the strike line AB is drawn through A^1B^1 or A^2B^2 and the strike line CD through C^1D^1. The dip is at right angles to AB, CD, that is along the line XY. The outcrop between beds I and II falls 100 ft from 500 to 400 in passing from AB to CD, the dip therefore is to the north at 100 ft in 600 ft or 1 in 6. A section along the line xy shows the strata dipping north or upstream. The section shows that bed I is the youngest and III the oldest.

The thickness of bed II which lies between I and III can also be readily found as both its top and bottom outcrop are cut by the strike line AB. Along the line AB the top of the bed (I/II outcrop) is at 500 ft and the bottom of the bed (II/III) outcrop is at 400 ft. This is the difference in height between points A^1 and A^2 or B^1 and B^2. The thickness, as given in these terms, is the vertical thickness (a), such as would be encountered in a borehole. The thickness perpendicular to the dip (b) can also easily be found. The value of b is given by $b = a \cos x$ where x is the angle of dip, converted from a gradient to degrees.

The problem can also be worked the other way round. Figure 9.11*b* shows the positions of 3 coal mines. At A a coal seam is reached at 500 ft down the mine, at C the same seam is at 300 ft below the surface and at B it is also at 300 ft below the surface. A is at 600 ft above sea level, C at 700 ft and B at 500 ft. The levels of the coal seam above sea level at A, C and B are 100 ft, 400 ft and 200 ft respectively. This information enables the strike and dip to be

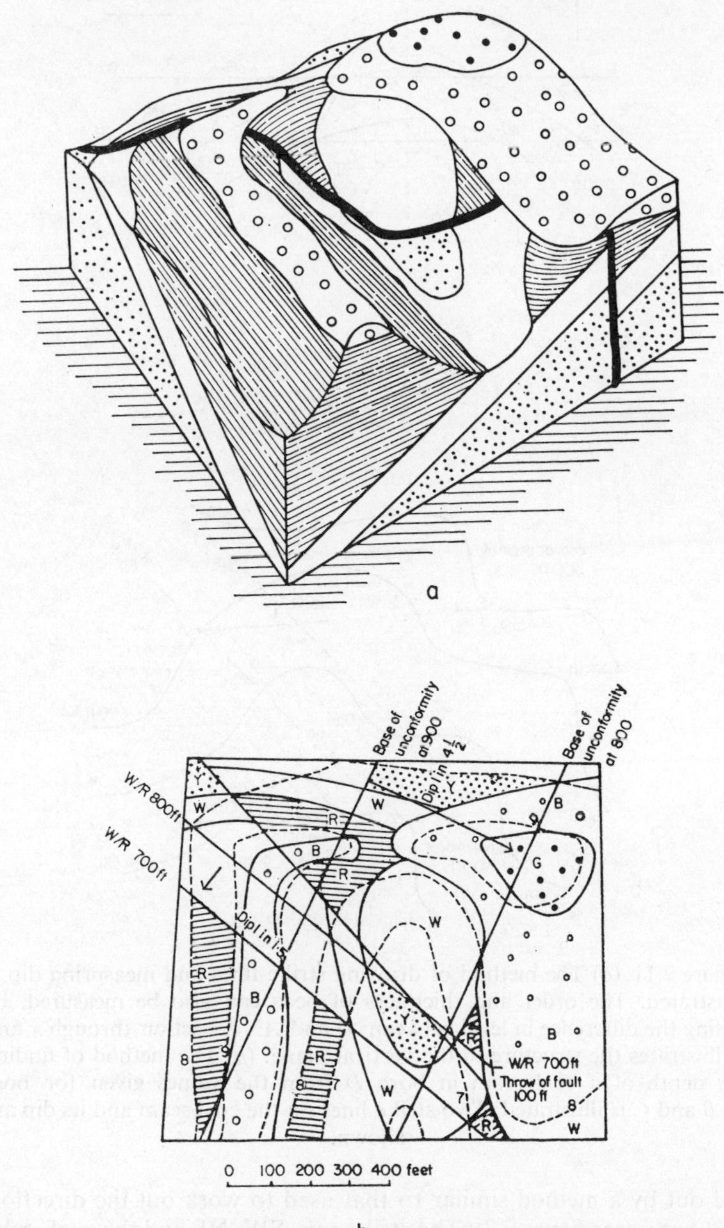

Figure 9.10 Map of geological structure, showing dip and strike and throw of fault. (*b*) The block diagram illustrates the structure in three dimensions and was drawn from a model (*a*).

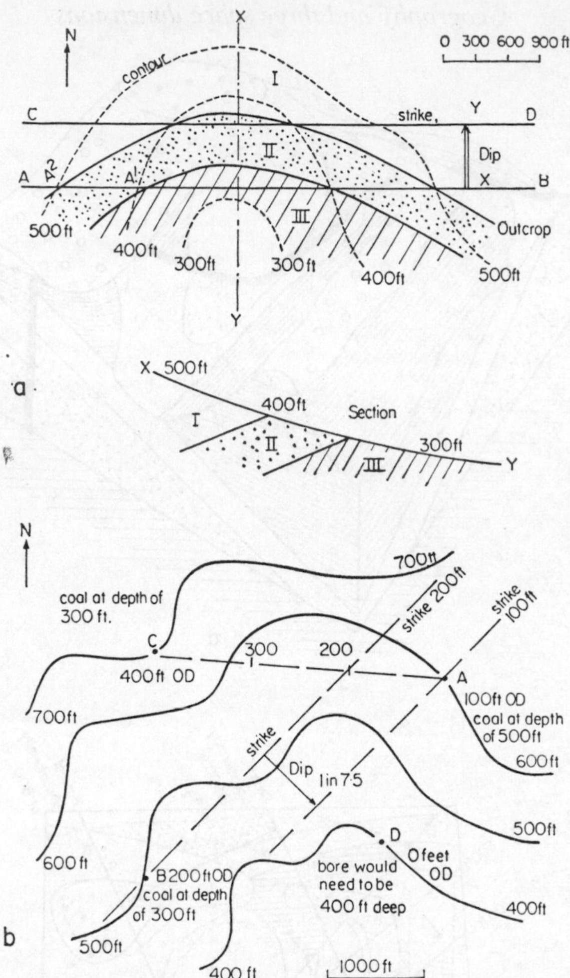

Figure 9.11 (*a*) The method of drawing strike lines and measuring dip is illustrated. The order and thickness of beds can also be measured, by noting the difference in level between A^1 and A^2. A section through *x* and *y* illustrates the structure in the vertical plane. (*b*) The method of finding the depth of a coal seam in bore *D* from the values given for bore *A*, *B* and *C* is illustrated. Two strike lines on the coal seam and its dip are shown.

worked out by a method similar to that used to work out the direction and tilt of isobases (see figure 9.7). The strike runs SW–NE and the coal seam dips at 1 in 7.5. It is now possible to work out how far it would be necessary to bore down to cut the coal seam at *D*, which is 400 ft above sea level. At this point the coal seam would be at sea level and therefore the bore hole would have to be 400 ft long to reach the coal.

The situation becomes more complex when the dip is not uniform in direction as the strike lines are no longer straight. The surface of any particular bed now becomes an undulating 2-dimensional surface rather than a sloping plane. Strike lines can, however, be drawn in a similar way by joining points where the same outcrop cuts the same contour. They will now be curved and not straight so that they cannot be extrapolated so accurately.

In dealing with all these situations it is a great advantage to be able to visualize the structure in three dimensions. This visualization may be aided by the use of block diagrams and the construction of sections. In some problems it is necessary to appreciate the structure in order to be able to reconstruct the original situation when the beds were laid down. This is necessary, for example, in the study of palaeomagnetic data. There are methods by which beds may be rotated back to their original positions by means of a stereographic projection. These methods are discussed by F. C. Phillips (1960).

(*b*) *Pebble shape*

Another aspect in which the 3-dimensional approach to a particular problem has yielded interesting results in the study of the shape of sedimentary particles. Techniques have been developed for dealing with particles from silt size up to boulders. Only the larger stones will be considered here as the small particles require special apparatus and lie beyond the scope of most geographical enquiry. The simpler measurements that have been applied to larger stones have produced interesting results.

The measures of sphericity and flatness that have been proposed make use of the long, intermediate and short axis lengths of the pebbles. These are designated the a, b and c axes respectively. W. C. Krumbein (1941) has developed a sphericity formula which is

$$s = \sqrt[3]{bc/a^2}$$

This value is unity for a completely spherical stone. It may be compared with A. Cailleux's (1945) flatness formula, which is given by

$$f = \frac{a+b}{2c} \times 100.$$

The results of these two formulae are strongly negatively correlated as shown in figure 9.12 on which are plotted the sphericity and flatness values for samples of 50 pebbles collected from a variety of depositional environments in west Baffin Island. The high flatness and low roundness were measured on platey Palaeozoic limestones but the bulk of the samples consisted of the much more spherical granite-gneiss stones. The range of values in the latter stones was small.

Zingg has developed a useful method of indicating the different forms of pebbles in terms of their a, b and c axis in graph form. Figure 6.13 illustrates the method of differentiating different shapes of pebbles. Spherical

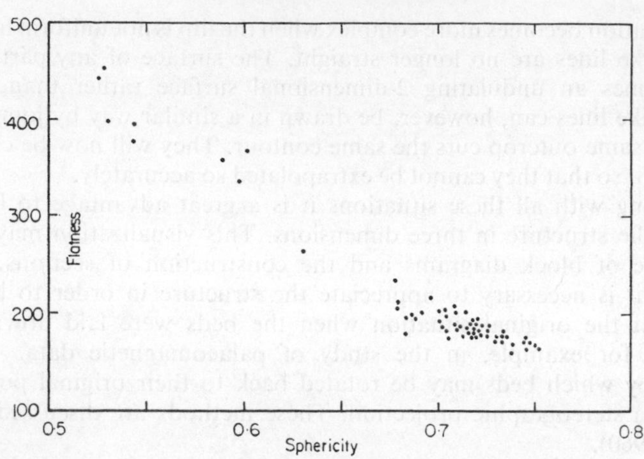

Figure 9.12 Correlation of sphericity and flatness of pebbles from west Baffin Island.

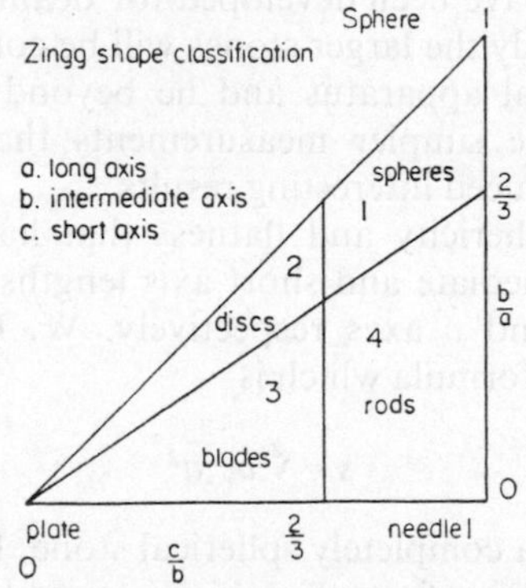

Figure 9.13 Zingg shape classification, showing four basic pebble shapes, *a*, *b* and *c* are long, intermediate and short axes respectively of the pebbles.

pebbles are defined as those in which the c/a ratio lies between $\frac{2}{3}$ and 1 and the c/b ratio lies between $\frac{2}{3}$ and 1. Rods have c/a between 0 and $\frac{2}{3}$ and c/b between $\frac{2}{3}$ and 1. Blades have the opposite characteristics to spheres, and discs to rods. By this means the main shape types can be differentiated and any pebble classified as belonging to one of the four shapes. A stone with values for *a*, *b* and *c* of 10, 5 and 2 would be a blade, if *a*, *b* and *c* were 10, 4, 3 it would be a rod, if *a*, *b* and *c* were 5, 4, 3 the stone would be spherical and if the values were 10, 8, 2 the stone would be disc shaped.

Various methods of geomorphological transport of different stone types have been distinguished by 3-dimensional measurements of this type. The roundness index of Cailleux has not been mentioned in this connexion as it only makes use of measurements in the principle plane of the pebble and is not, therefore, 3-dimensional. Nevertheless, it has been found superior in some analyses to the shape measures discussed in this section.

(c) *Available relief*

The measurement of pebble shape can give useful information concerning the operation of geomorphological processes. At the other end of the scale

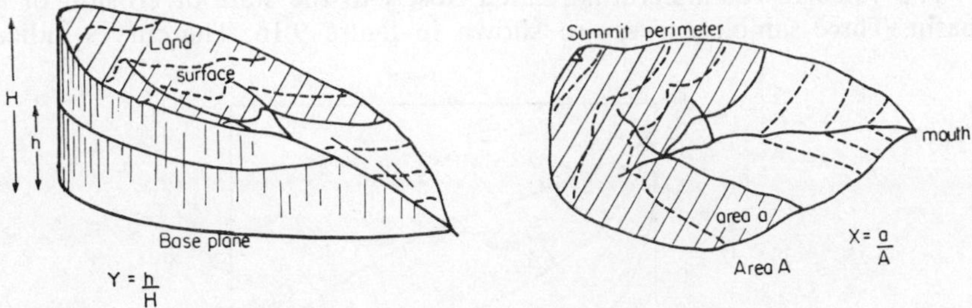

Figure 9.14 Plan and block diagram to illustrate dimensionless units used to plot hypsometric curves for a drainage basin. (After A. N. Strahler.)

the measurement of the shape of the whole landscape can also provide useful data. The relationship of area to height is a 3-dimensional problem that has received a great deal of attention by some geomorphologists. A. N. Strahler (1952) has developed a method of expressing the area–height relationship of a drainage basin dimensionlessly. The values he uses are shown in figure (9.14). x represents the ratio between area a and the whole area of the drainage basin. y represents the ratio of the height (h) between the mouth of the basin and the contour which defines the lower limit of area a, and the

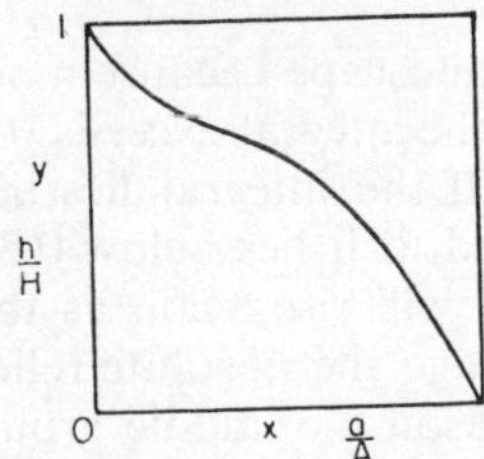

Figure 9.15 A sample hypsometric curve for a drainage basin on which a/A is plotted against h/H. a is area above a given contour. A is area of whole basin. h is height of given contour above lowest height in basin. H is total height range of basin. (After A. N. Strahler.)

total height range in the basin H. The values of x and y then vary between 0 and 1 and can be plotted on a graph as shown in figure 9.15. The hypsometric curve relating x and y must pass through $x=0; y=1$ and $x=1; y=0$ but its position within the graph is a measure of the state of erosion of the basin concerned.

The degree of erosion can be determined by the hypsometric integral $\int$, which is defined as the proportion of the area of the square graph lying below the curve relating x and y. The value of $\int$ can be between 0 and 1, where the whole area of the square is 1. The integral can be readily found in practice by measuring the area of the surface below the hypsometric curve. Mathematically the value of $\int$ can be found by integral calculus. The area can be considered to be made up of an infinite number of thin slabs of volume $a\,\Delta h$, where a is the area of the slab and h its thickness. These values can be integrated to give the total volume of the land above the base level. The integral can be found from the formula $V/HA = \int_0^{1.0} x \, dy$, where V is the volume, H the total height and range A the total area.

The value of the integral is related closely to the state of erosion of the basin. Three sample curves are shown in figure 9.16. The curves indicate

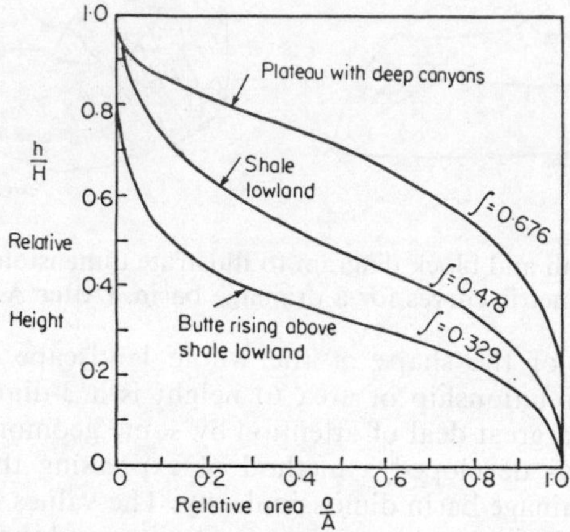

Figure 9.16 Hypsometric curves for three drainage basins in the Mesa Verde Region, New Mexico. (After A. N. Strahler.)

that a high integral value is associated with youthful landscape because most of the available volume is still present in the basin. If the integral exceeds 0.6 then the area is in a youthful stage of development. If the integral lies between 0.35 and 0.6, then the area is maturely dissected, if it lies below 0.35 then there are a few relict monadnocks. The integral will rise again as the remaining monadnocks are removed because at this stage the absolute relief of the basin will be considerably reduced. For this reason the middle group contains landscapes that include maturity as well as old age, in which the relief gradient is fairly constant throughout the basin. A. N. Strahler's (1954) paper should be consulted for details of other measures that can be used to characterize the hypsometric curves and the 3-dimensional landscape that they represent.

(*d*) *Speleology*

The study of caves and cave systems is essentially a 3-dimensional problem. The problems involved include both the survey of the cave in three dimensions and then its portrayal on a 2-dimensional plan. Both vertical and horizontal reference axes must be established. In many caves traversing provides the best means of measurement. To obtain a knowledge of the form of the cave passages it is necessary to take both horizontal and vertical offsets from the traverse lines. A map of the cave system can only be drawn by reducing the cave system to a horizontal plane. The vertical dimensions must be shown separately by sections along the cave passages. Separate diagrams are also needed to show the cross sectional plan of the passages. An example is shown in figure 9.17.

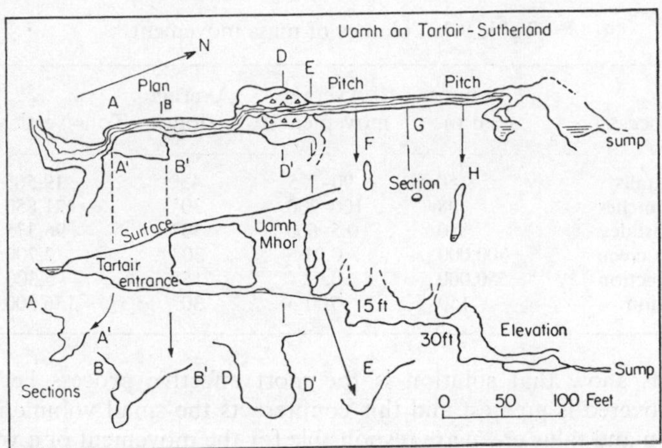

Figure 9.17 Plan, vertical profile and cross sections of a cave system. Uamh an Tartair — Sutherland, Scotland.

Some aspects of the geomorphological significance of cave systems has been pointed out by M. M. Sweeting (1950) in her work in Craven, Yorkshire. She has shown that there is an association between the level of the caves and the elevations of erosion surfaces in the landscape above. In this area where the limestone is lying nearly horizontally the height of the water table, which is related to the base level, determines the level at which caves form. The major passages are linked by near vertical pitches, which developed during the periods of rapidly falling water table accompanying the rejuvenation of the landscape.

(*e*) *Volume of mass movement*

Attempts are being made to calculate quantitatively the amount of material removed from an area by erosion. These measurements must be made by considering the volume of material moved. The volume implies a 3-dimensional approach to the problem. One method of assessing the volume of material evacuated from an area is to measure the load of the river draining the area. This load consists of bed load, suspension load and solution load.

Another method is to measure the actual rate of movement of material on a slope under various processes and to assess the volumes involved in the different types of movement. The effectiveness of a process depends both on the volume moved and the distance over which it is moved in a given time. Thus a large volume of material moving a short distance may not prove so effective in causing denudation as a smaller volume moved over a greater distance. The slope will also play a part in determining the amount of material moved down a given vertical distance. These relationships are shown in table 9.1 which gives results of the observations made by A. Rapp (1960) in Kärkevagge in North Scandinavia.

Table 9.1 Volumes of mass movement

Process	Volume cu m	Average movement m	Average slope	Tons/vertical m
Rockfalls	50	90–225	45°	19,565
Avalanches	88	100–200	30°	21,850
Earthslides	580	0.5–600	30°	96,375
Talus creep	300,000	0.01	30°	2,700
Solifluction	550,000	0.02	15°	5,300
Solution	150	700	30°	136,500

The results show that solution is the most effective process because the distance covered is greatest and this counteracts the small volume involved. Solifluction and talus creep are responsible for the movement of a very much greater volume of material, but it covers a much shorter distance. Hence these processes are relatively less effective as agents of denudation. Both volume and vertical movement must be considered in this type of study.

Most of the topics discussed so far in this chapter have been concerned with techniques of observing or analysing 3-dimensional phenomena. This section ends with a brief reference to certain situations in which it is important to appreciate the third space dimension. This dimension is vital to the functioning of the processes involved. The subsection is more theoretical, but it is put in this chapter on account of the emphasis that is placed on the third space dimension. Three situations will be mentioned. The third dimension in vegetation study, and the vertical component of the major circulations, already discussed in chapter 8, the atmospheric and oceanic circulations, will also be mentioned briefly.

(f) Vegetation

The 3-dimensional character of vegetation is best appreciated in forested areas. The larger trees penetrate furthest from ground level both upwards and downwards. Their roots reach deep into the ground and so do not compete with the shallow rooted plants. Below the tall trees stand the shorter bushes and small trees, forming the second layer of forest vegetation. Their roots penetrate to an intermediate level. The ground between the trees is covered by herbs and grasses, whose shallow roots do not disturb the deeper rooted plants. In this way the three layers of vegetation can grow together as an ecological whole which influences and modifies its own microclimate, through its relatively great vertical spread. The air near the ground is still and speed of air movement increases upwards. The amount of light increases upwards while the amount of moisture increases downwards. Thus many of the flowers of the tropical lianas and epiphytes are at the tops of the trees to take advantage of the light while their nourishment is obtained from the moisture layers below. The influence of the vegetation also extends downwards into the soil, in that forest soils can be distinguished from grassland and other soils.

(*g*) *Atmospheric and oceanic circulation*

The necessity to consider these major world circulations over the earth as a whole has already been stressed in chapter 8. It is no less necessary to take the vertical component of movement into account. The energy that generates the system is applied to the base of the atmosphere and the top of the ocean by the radiant heat from the sun. The circulation of the atmosphere is thus kept going by the rising of warmed air, while the oceanic circulation is maintained by the sinking of cooled water. The interface between the air and the water is thus very important to the circulation of both atmosphere and ocean.

(i) In the atmosphere air heated at the base rises, carrying with it moisture from the ocean. This moisture then condenses and may fall again as rain, although in the meantime the air may have been carried a long way horizontally so that the rain falls partly on the land.

In modern weather forecasting, the surface weather chart is supplemented by charts that show the pressure and temperature situation at various upper levels. These upper level charts are made by contouring a particular pressure surface by altitude. On these upper charts the fronts of the surface charts appear as sharp ridges and troughs. This demonstrates the 3-dimensional nature of the frontal lines that appear on the surface charts. These lines mark the position where gently sloping planes cut the earth's surface. The pattern of the frontal surfaces and the character of the air on either side of it can best be examined on T/ϕ graphs. These are 2-dimensional graphs on which are plotted the variation of the air temperature with altitude, representing a vertical line through the atmosphere.

The almost horizontal lines on the T/ϕ graph (figure 9.18) represent lines of equal pressure. The lines sloping from top right to bottom left are the dry bulb temperature lines. The lines nearly at right angles to these are saturated adiabatic lapse rate lines and the lines that slope from the top left to bottom right are dry adiabatic lapse rate lines. If the ground temperature and humidity are known then by using these lines it is possible to draw the path curve. This indicates the temperature and moisture content of the air at any height if it were to rise to that height from the surface. The relationship between the path curve and the actual observed air temperature indicates the stability or instability of the air. This determines whether it will rise spontaneously

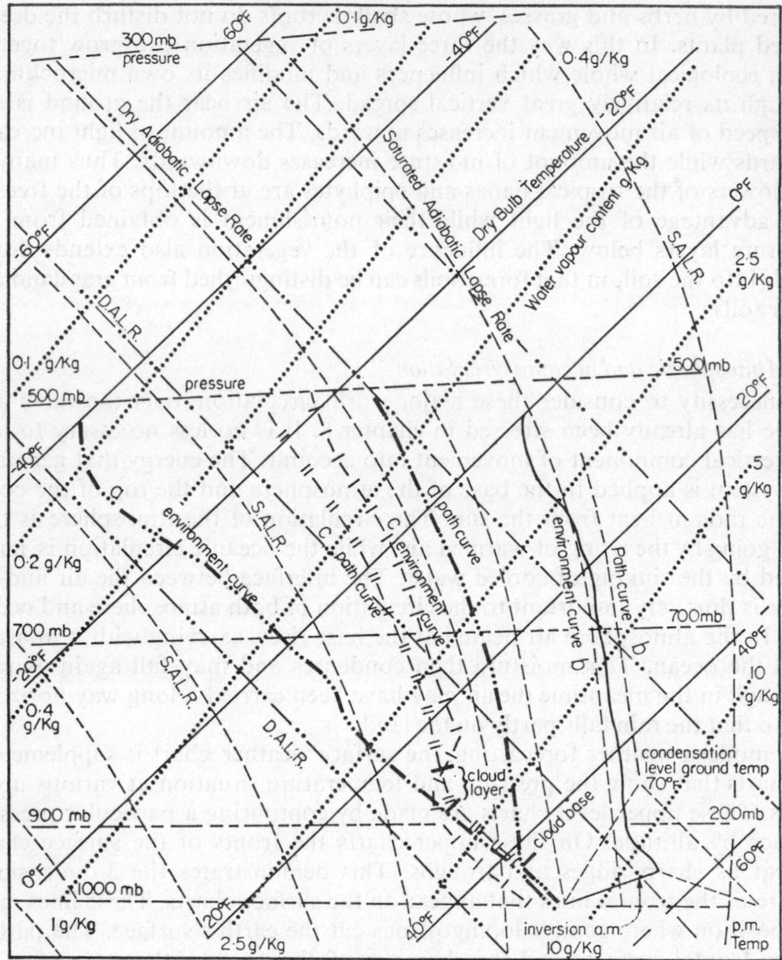

Figure 9.18 T/ϕ graph to illustrate sample upper air curves. Sample *a* is an unstable air mass. Sample *b* is stable when there is an inversion in the lower layers but could become unstable if the ground temperature rose to 70°F. The sample *c* is unstable below but stable above. A thin layer of cloud would form between the condensation level and the point where the path curve crosses to the left of the environment curve. The pressure lines, dry bulb temperature lines, dry and saturated adiabatic lines and the water vapour content lines are shown on the T/ϕ graph.

or not. In the examples on the diagram, the rising air of path curve (*a*) is everywhere warmer than the surrounding air, so it will go on rising and the air is unstable. The path curve lies to the left of the actual lapse rate curve. Curve (*b*) illustrates the opposite situation of a stable air mass. However, if this air were to be heated to 70°F at ground level instability and cumulus cloud activity would develop. The air is saturated at ground level in curve (*b*) so that its water vapour content is 7.8 g/kg as shown by the pecked lines. The condensation level can be found by following up the pecked line until it meets the dry adiabatic, assuming that the ground temperature rises above the value shown, which is 50°F. When the temperature reaches 70°F the path curve moves to the right of the lapse rate curve, and the air would become unstable. The third curve (*c*) is unstable below the point where the path curve cuts the actual curve, but cloud would only form between the condensation level and the crossing points of the two curves. The condensation level occurs where the path curve starts to follow the saturated adiabatic at 900 mb. The layer of cloud would thus be 50 mb thick and would be fair weather cumulus.

In order to understand the processes operating in the daily weather it is essential to consider the vertical dimension, as illustrated by reference to the T/ϕ graph. This applies both to frontal weather and to other activities such as convectional activity or anticyclonic activity, which may give rise to very stable conditions leading in winter to fog.

(ii) In the oceans the vertical element of the circulation is caused by the sinking of denser water to the bottom. This displaces water, and the sinking must be compensated by equivalent volumes of rising water elsewhere. The pattern of vertical circulation in the world ocean is dependent upon the fact that nearly all the water dense enough to sink to the ocean depths is formed in the high latitudes of the Atlantic Ocean and very little water sinks deeply in the Pacific or Indian Oceans. The sinking of water in the North Atlantic produces a well defined mass of cool, fairly saline water, called the North Atlantic Deep Water. In the South Atlantic, particularly in the Weddell Sea, is formed the densest of all water masses, the Antarctic Bottom Water. This water is a very cold but less saline water mass. Water also sinks to moderate depths at the Antarctic convergence in 52°S. This is called the Antarctic Intermediate Water. These water masses are displaced from their sources by further sinking of water but they maintain their identity for long distances. They form layers of water of differing temperature-salinity characteristics. The figure 9.19 illustrates the sandwich of water masses that occurs in a vertical section through the Atlantic Ocean. The Antarctic Bottom and Inter-

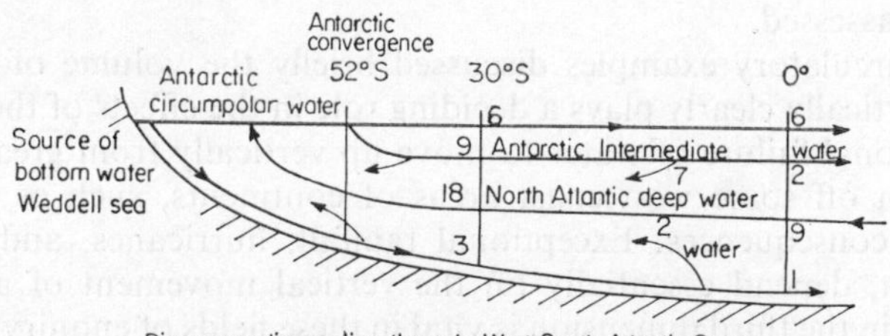

Figure 9.19 A diagrammatic cross section through the western South Atlantic to show the vertical layering of the major water masses in this ocean. The figures refer to water transport in millions of cubic metres per second.

mediate Waters move north below and above the south-moving North Atlantic Deep Water respectively. The southern water masses gradually lose water to the south-moving mass which increases in volume and becomes diluted as it moves south. It eventually rises to the surface south of the Antarctic convergence to form the Antarctic Circumpolar Water. Although the third dimension is small in actual size relative to the horizontal distances covered by the water masses, it is nevertheless vital to the circulation of the deep waters of the oceans. This circulation has been inferred from observations of temperature and salinity, it has been directly observed by acoustic techniques by J. C. Swallow and others (Swallow and Worthington, 1957) and it has been worked out theoretically by H. Stommel and others.

Only two other examples of the vertical dimension in the oceans will be mentioned. The upwelling of cooler water from the lower levels is very important off west coast mid-latitude areas in bringing nutrients to the surface. These nutrients then nourish very important fish stocks. The volume of fish caught off Peru is an illustration of the importance of this vertical water movement. The other example is the presence beneath the equator of a strong east-flowing ribbon of water that lies below the west-flowing equatorial current. This east-flowing current, called the Cromwell current, extends 2 degrees North and South of the equator but is only 200 m thick, reaching its maximum velocity at 100 m depth. Both these vertical patterns of flow are caused in part by the Coriolis force which affects moving media on the earth and which is brought about by the rotation of the earth.

These brief references must suffice to show the great importance of the third dimension and the necessity to consider these and similar problems in terms of three space dimensions.

(*h*) *Conclusion*

In all the examples quoted, with the exception of the pebble shape, the third dimension is essential to the problem concerned. Slope is a fundamental factor in the mass-movement of debris on a hillside, but of greater significance in the present context is the volume of material evacuated from the hillside. This volume determines the rate of denudation and, by making quantitative measurements the relative effectiveness of different processes, it can be accurately assessed.

In the circulatory examples discussed briefly the volume of air or water moving vertically clearly plays a deciding role in the effects of the circulation. The occasional failure of water to move up vertically from greater depths is well known off some west coast areas of continents, such as Peru, for its disastrous consequences. Exceptional rainfall, hurricanes, and comparable phenomena, depend essentially on the vertical movement of air. Thus the movement in the third dimension is vital in these fields of enquiry.

9.5 3-Dimensional space and human geography

Although the geographer has become accustomed to using flat maps rather than 3-dimensional models, he should not overlook the fact that all material objects occupy 3-dimensional space and that everything he studies happens in 3-dimensional space. If man's activities on the earth's surface are thought of as happening in layers, then it is feasible that in any given area, several layers are being used simultaneously: examples are mining beneath the ground, a town on the surface, and aircraft using the air above. Non-agricultural productive activities and transportation channels often occur one above another in several layers. In particular, the vertical dimension in mines, urban environments, air transport, and underwater sea activities deserve more consideration from geographers.

When man moves off a horizontal plane complications occur. Greater energy is needed to move things up a gradient than horizontally. The complementary downward movement is not always compensating; often it is difficult and dangerous. Similarly, extra effort is needed to construct a tall building compared with a low one. A massive framework in the lower part is required to support the upper floors, and deep foundations are often needed. Presumably the value of being at a particular site compensates for the high cost of construction, just as a direct but steep railway is more economical in a given context than a much longer one with gentler gradients.

In human geography the vertical dimension seems more likely to need consideration in the study of a small area than in the study of a large area, provided there is some vertical layering of activities. Thus if the U.S.A. is taken as a whole, New York is no more than a flat patch. If the geography of Manhattan Island is being considered, the occurrence of many layers of activity can hardly be overlooked.

Within a 3-dimensional space it is possible for many objects much smaller than the space itself to be accommodated within it and for many communication channels to pass about, all without intersecting (or sharing the same space). In an urban environment, the objects may broadly be classed as

(1) fairly compact, fixed and relatively large blocks, notably buildings.
(2) communication channels, which may be bounded (encased) e.g. underground railway tunnel), unbounded (e.g. air lane for helicopter), semibounded (e.g. road).
(3) Smaller, usually relatively compact, mobile objects, notably people, cars, trains and so on.

The methodical study of fixed and mobile objects in 3-dimensional space presents a new challenge to the geographer in terms both of language and notation, and of techniques. Three co-ordinates, rather than the familiar two, are needed to locate points. Possibly, too, mathematical terms from set theory and topology will offer flexible means of talking about 3-dimensional relationships. The visual representation of 3-dimensional situations is also particularly difficult.

The implications of 3-dimensional spatial relationships for human geography will be discussed briefly, with the help of several pages of diagrams, under the following headings: (*a*) effort, (*b*) diminution of vertical variable as horizontal area increases, (*c*) the urban environment, and (*d*) communication channels.

(*a*) *Effort*

The extra effort needed to move in an environment with slopes rather than over a flat surface has not received much formal attention by geographers. The study of the increased energy (effort) needed to move things up (or down) gradients, additional to friction caused between bodies and the media they are moving through or over, is a subject for physics and engineering. Similarly, the construction of and operation of multi-storey buildings interests the architect more than the geographer. However, some attention should be given to the implications of movement in 3-dimensional space.

Figure 9.20 shows examples of the way man deals with and overcomes slope or gradient. Human pedestrians and to some extent animals, such as pack animals, can negotiate gradients of varying steepness (figure 9.20*a*). It would be interesting to see how speed of movement is reduced for pedestrians as they negotiate steeper and steeper gradients. If there are no obstacles it is customary for people to walk horizontally at about 4–5 feet per second. It therefore requires about 4–5 seconds to walk 20 feet. How long does it take to climb or descend steps joining two floors 20 feet apart vertically? Probably several times as long. An appreciation of this kind of effort is highly relevant both for the study of pedestrian movement in many urban environments and for pedestrian movement in rugged rural environments.

In general, vehicular traffic is even more sensitive to gradients than pedestrian movement (figure 9.20*b*). Vessels moving on water surfaces, as opposed to submarines, which operate in three dimensions, are virtually tied to horizontal movement. In canals locks are costly to build and maintain, and cumbersome to operate. Incline planes, up or down which small vessels can be pulled or lowered, portages, and other devices are of limited practical application. Early steam operated railways were also particularly restricted by gradients, and great efforts were made, as on the early London to Birmingham railway, to minimize these by making cuttings and embankments. More powerful locomotives, together with the help of diesel and electric traction, and devices such as zig-zags and curving tunnels, have considerably reduced the restrictive influence of gradients as railway technology has developed. The cable railway, mine shaft and lift shaft (figure 9.20*d*) may be considered adaptations of the railway to steep slopes and vertical movements. Motor vehicles are less hampered by slope than railway trains. Like the various vehicles running over surfaces or tracks, aircraft require greater effort to ascend than to fly horizontally (figure 9.20*c*). The principle for a commercial air flight, therefore, is to reach the best cruising altitude as soon as convenient and then to fly at that altitude until the point for descent to the destination. Since different aircraft have different cruising heights, they are sorted into layers this way.

When suitable for cultivation because they have a soil cover, steep slopes may nevertheless be difficult to operate and also dangerous to till because of the possibility of soil erosion. In general, terraces to overcome the difficulties and dangers of steep slopes, either may be constructed to give horizontal

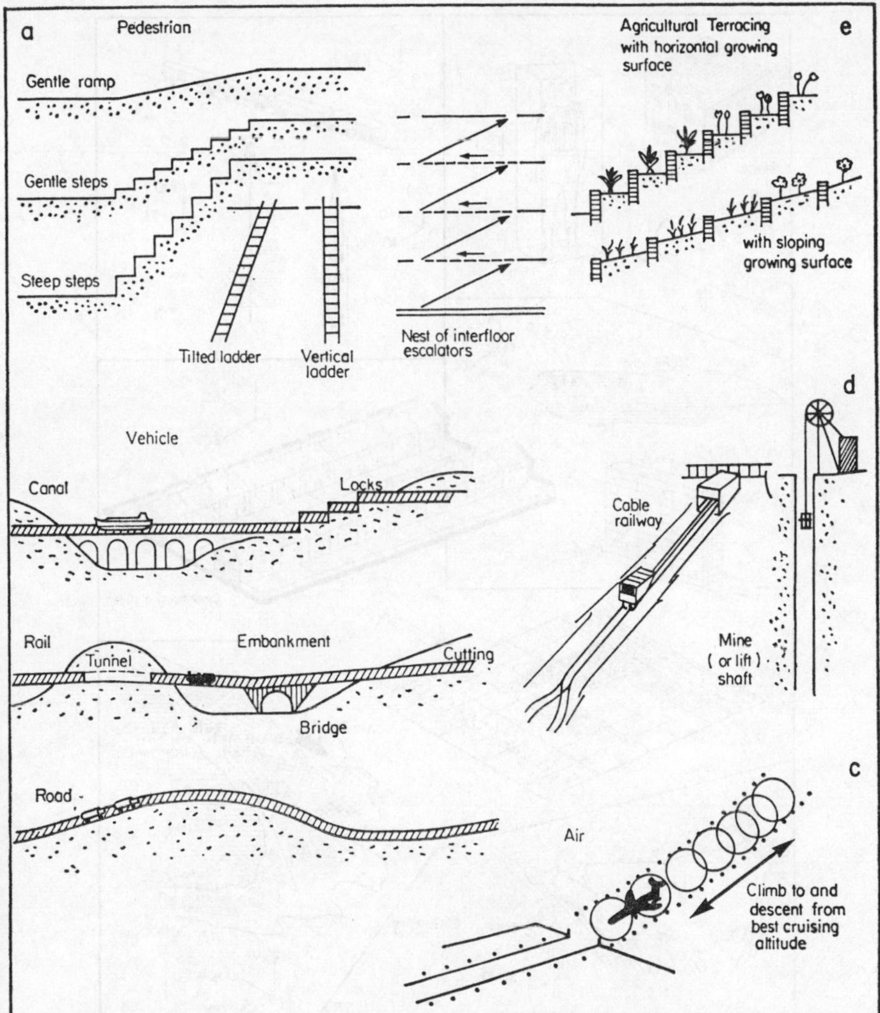

Figure 9.20 Examples of situations in which activities must be considered in a 3-dimensional context. (*a*) Pedestrian movement. (*b*) Vehicle movement along semi-bounded channels. (*c*) Air lane, an unbounded channel. (*d*) Lift. (*e*) Terracing in agriculture.

surfaces for cultivation, or may be built in such a way that they still slope, but slope more gently than the hillsides on which they are made.

(*b*) *The vertical dimension*

The shape and compactness of a 3-dimensional space, or a 3-dimensional object, can be characterized very roughly by the relative lengths (or ratios)

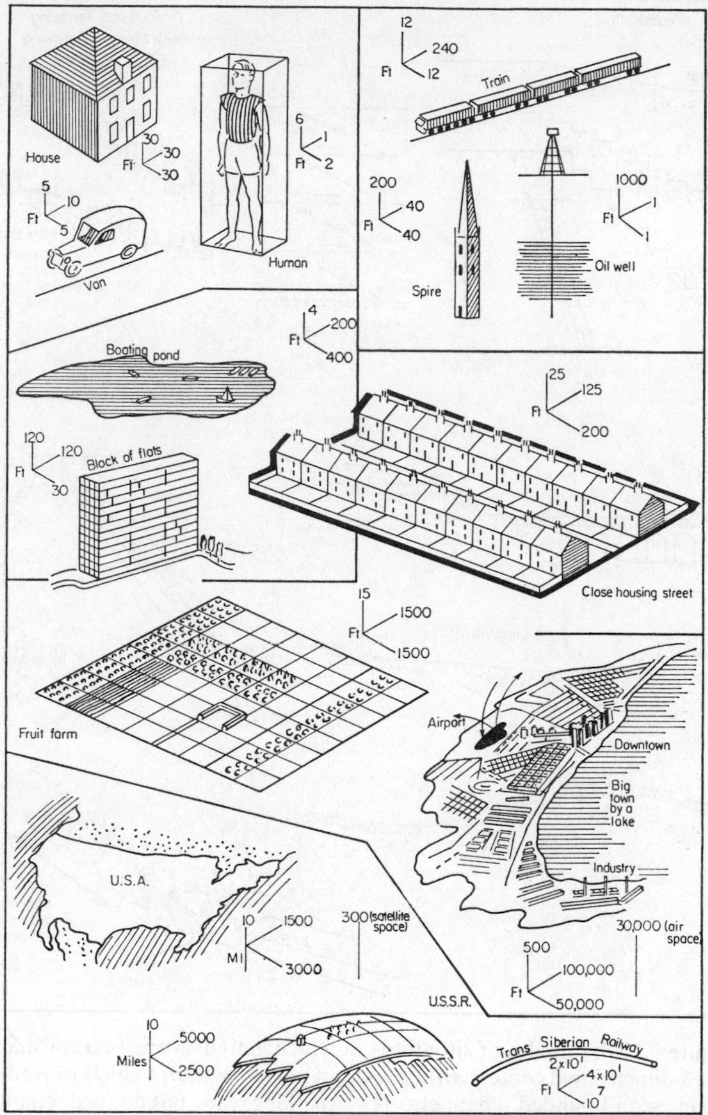

Figure 9.21 A comparison of the lengths of the vertical and two horizontal axes of a number of objects and places. The vertical tends to diminish relatively from the smaller objects at top left to the largest places, lower part. The measurements, which are approximate, are in feet except for the U.S.A. and the U.S.S.R., for which they are in miles. Most of the views are oblique. See also table 9.2 in text.

of three orthogonal axes through it. When the space or object is considered in relation to the earth's surface it may be convenient to make one of the axes always vertical and the other two horizontal. Note, however, that the description of the axes is only relative to the earth's surface.

Figure 9.21 and table 9.2 show, with the help of measurements of the three axes suggested, that the geographer is less likely to be aware of the vertical variable, the larger the objects and spaces he considers. In figure 9.21, each object has a key by the side giving the length of the vertical dimension (height or depth) and two horizontal dimensions.

Many familiar relatively small objects are fairly compact (they have axes that do not differ much). Some, however, have one axis much longer than the others. The long one may be horizontal (train) or vertical (tower, well). Or they may have one axis much shorter than the other two (e.g. the vertical in a shallow pond or lake). The housing area and fruit farm are perhaps the smallest of the spaces and objects in the figure that might be of geographical interest. Already here the horizontal axes are substantially bigger than the vertical. It is possible to compare the vertical with the mean of the two horizontals. In the larger areas, the ratio of the vertical axis to the horizontal depends on whether aircraft space above is taken into account or not, and, with very big areas, satellite space as well. In a large region or on continent scale, even when generous portions of the lithosphere and atmosphere are included in the vertical axis, this is often less than $\frac{1}{100}$ of the horizontal axes. The Trans-Siberian Railway has been added, finally, to illustrate an extremely long but clearly defined space, like a tube curving round the earth's spherical surface as well as making local detours.

The fact that there is not a linear relationship between the volume of an

Table 9.2 3-Dimensional shape

		Absolute scale			Ratio scale[a]		
		V	H	H	V	H	H
House	Feet	30	30	30	1	1	1
Van	,,	5	5	10	1	1	$\frac{1}{3}$
Human	,,	6	1	2	1	$\frac{1}{6}$	$\frac{1}{3}$
Spire	,,	200	40	40	1	$\frac{1}{5}$	$\frac{1}{5}$
Oil well	,,	1000	1	1	1	$\frac{1}{1000}$	$\frac{1}{1000}$
Train	,,	12	12	240	1	1	20
Boating pond	,,	4	200	400	1	50	100
Block of flats	,,	120	30	120	1	$\frac{1}{4}$	1
Close housing	,,	25	125	200	1	5	8
Fruit farm	,,	15	1500	1500	1	100	100
Big town	,,	500	50,000	100,000	1	100	200
U.S.A.	Miles	10	1500	3000	1	150	130
U.S.S.R.	,,	10	2500	5000	1	250	500
Trans-Siberian Railway	Feet	20	40	5,000,000	1	2	250,000

object and its surface area, if its size changes and its shape remains the same, has several interesting implications. The smaller the object is, the greater in proportion is its surface area to its volume. The shape of the object also influences the relationship between volume and surface area. Among regular solids, a sphere has the smallest surface area relative to its volume while a tetrahedron has the largest surface area relative to its volume. This relationship prompted the idea of the tetrahedral theory (see chapter 8, section 4).

Another interesting aspect of the relationship between volume and surface area is the comparison between the size of two very different marine creatures. At one end of the scale the phytoplankton are extremely small, having a very small volume in relation to a comparatively large surface area. This is to enable them to remain floating in the upper layers in which they must remain in order to be able to photosynthesize nutrients by means of the light of the sun. They also increase their surface area by growing excrescences of various sorts and developing intricate shapes. At the other end of the scale are the large whales, the largest of all animals. A large blue whale may be 120 ft long and weigh many tons. It has, however, a relatively small surface area compared with its volume. This is to enable it to maintain its internal temperature in the very cold water in which it lives, as the whale is a warm-blooded mammal. The whales are streamlined and simple in shape to increase the advantage of smaller surface area relative to their volume.

(*c*) *The urban environment*

The urban environment has attracted the attention of many geographers since the second world war. There is much evidence to show that the vertical thickness of a town diminishes from the downtown area towards the periphery as land values fall off. It is feasible, though not very satisfactory, actually to map the number of storeys of buildings, either by different intensities of shading according to height, or by progressively adding shorter and shorter buildings on several comparable maps.

In figure 9.22 it is suggested that many, if not all, of the ingredients of a townscape might be separated into three main categories, buildings (hollow containers), channels for different kinds of vehicle (and for pedestrians) and relatively small, mobile objects such as people, cars, trams. It is interesting to note an analogy between isostasy and buildings, which may be thought of as light containers resting on and sinking into denser material, the ground. They float, like tin cans, on material of varying firmness. (figure 9.23*b*).

The small mobile objects in a town may be located by three Cartesian co-ordinates (see figure 9.23*d*). It is even possible to have addresses of apartments in tall buildings that imply three co-ordinates. In 1961 the address of the British Embassy in Bogota, Colombia, was Carrera 8, Calle No. 15–46, 6th Floor. There is a point of origin at one side of the old town, from which the Carreras and Calles form a co-ordinate grid. Forty-six is the actual door of the embassy on the carrera; it is 46/100 units of distance along the block. The 6th floor represents the vertical axis co-ordinate.

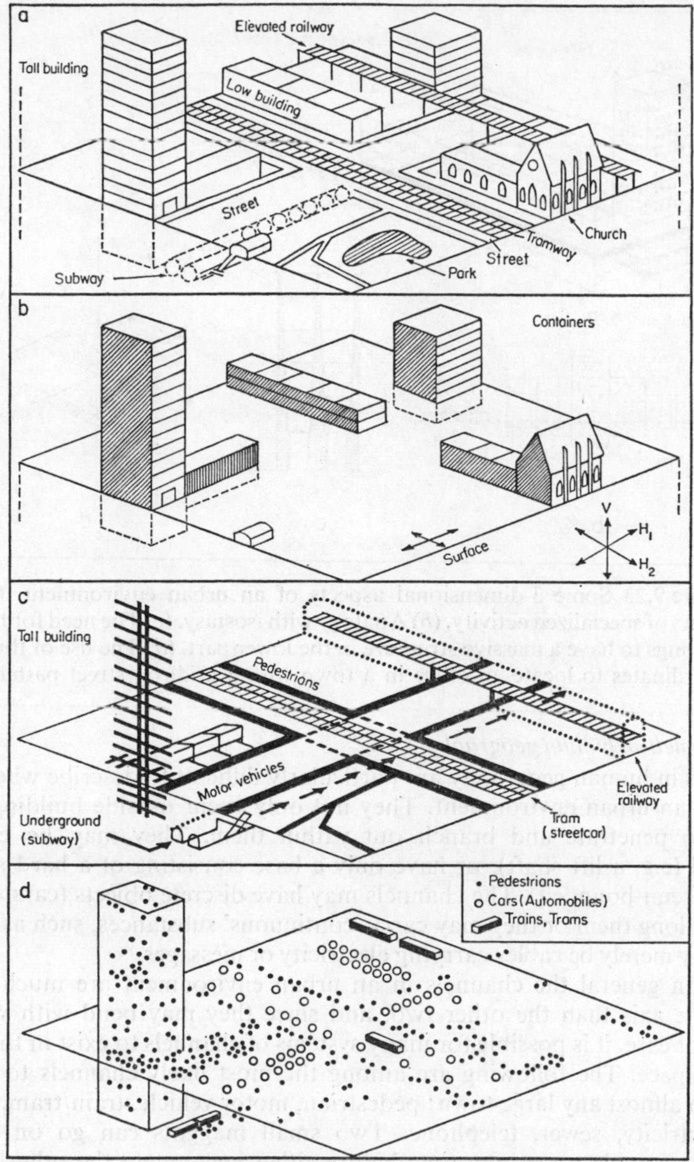

Figure 9.22 Essential ingredients of the urban environment. (*a*) Part of a fictitious town. (*b*) The buildings as hollow containers (wardrobes or chests of drawers). (*c*) The channels designed primarily for movement. Note that some intersect but some are disjoint sets. (*d*) The smaller mobile objects using the channels (*c*) to move between the containers (*b*).

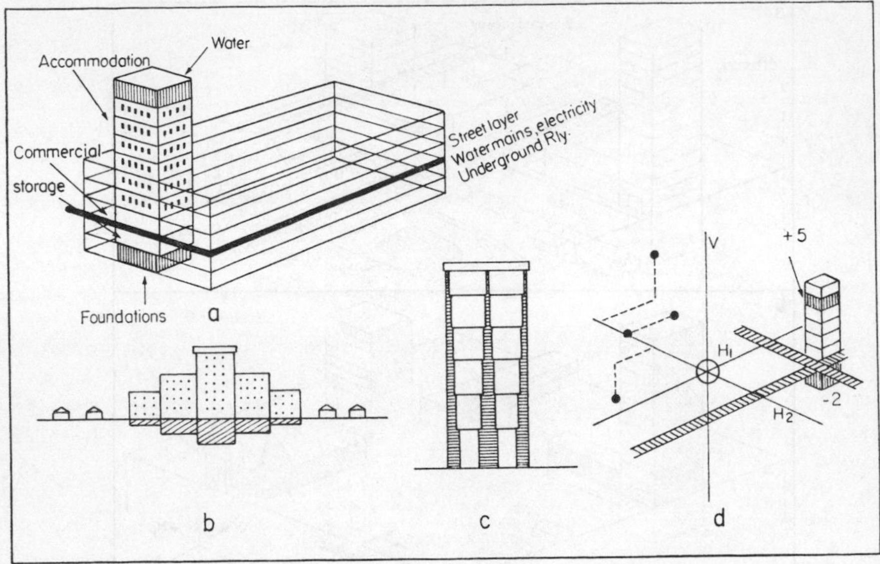

Figure 9.23 Some 3-dimensional aspects of an urban environment. (*a*) Layers of specialized activity. (*b*) Analogy with isostasy. (*c*) The need for tall buildings to have a massive structure in the lower part. (*d*) The use of three co-ordinates to locate an office in a town with a gridiron street pattern.

(*d*) *Channels in human geography*

Channels in human geography are particularly difficult to describe when they occur in an urban environment. They not only occur outside buildings, but they also penetrate and branch out within them. They may be entirely bounded (e.g. a lift shaft), or have only a base consisting of a hard surface or rails (semi-bounded). The channels may have discrete objects (cars, trains) passing along them or they may carry 'continuous' substances, such as water. They may merely be cables carrying electricity or messages.

Since in general the channels in an urban environment are much larger along one axis than the other two, and since they may bend with varying degrees of ease, it is possible for many systems of channels to exist in the same general space. The following are among the most likely channels to be expected in almost any large town: pedestrian, motor vehicle, train/tram, water, gas, electricity, sewer, telephone. Two small maggots can go on boring through an apple for a long time before either encounters the other (figure 9.24*a*). In set language their channels remain disjoint sets. Veins and arteries similarly form two distinct systems in the same body space (figure 9.24*b*).

The satisfactory functioning of a town requires that most of the systems of channels stay disjoint in this sense. If they do intersect, as happens when the pedestrian or tramway channels cross each other or the car channels (figure 9.22*c*) recourse to time is necessary (e.g. traffic signals) to separate

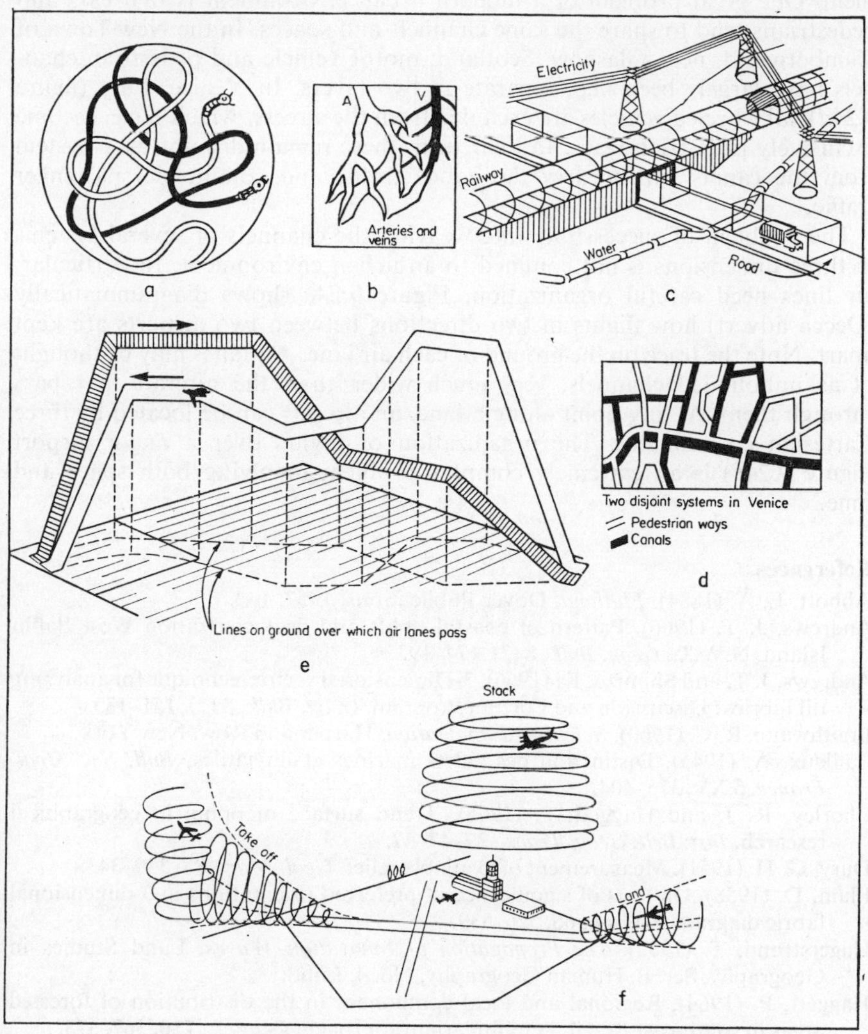

Figure 9.24 Channels sharing 3-dimensional space. (*a*) Two maggots can bore indefinitely through an apple without being aware of each other's existence. (*b*) Arteries and veins intertwine in the same body space but remain disjoint sets. (*c*) Various channels share the same 3-dimensional space but remain disjoint sets. (*d*) The canals and pedestrian ways in Venice, Italy are disjoint systems. (*e*) Air lanes for flights in each direction between two airports are organized to remain disjoint unbounded channels. (*f*) The organization of space above a large airport is a matter of great complexity.

them. One great problem of a modern urban environment is that cars and pedestrians tend to share the same channels and spaces. In the New Town of Cumbernauld, near Glasgow, Scotland, motor vehicle and pedestrian channels have largely been kept separate in two layers. In Venice, Italy (figure 9.24*d*), all wheeled vehicles are excluded from the streets, which have become exclusively pedestrian ways. In their turn, these remain disjoint as a system from the canals which carry the goods traffic and commercial passenger traffic.

The problem of successfully interweaving the channels of several systems in three dimensions is not confined to an urban environment. In particular, air lines need careful organization. Figure 9.24*e* shows diagrammatically (Decca advert) how flights in two directions between two airports are kept apart. Note the trace on the ground of each air lane. Air lanes may be thought of as unbounded channels, very much wider than the aircraft that pass through them. At any point along a lane, an aircraft can be located by three Cartesian co-ordinates. The organization of flights over a major airport (figure 9.24*f*) is an extremely complex problem involving both space and time.

References

Abbott, E. A. (1884). *Flatland*, Dover Publications, 1952, 1–2.

Andrews, J. T. (1966). Pattern of coastal uplift and deglacierization West Baffin Island, N.W.T., *Geog. Bull.*, **8** (2), 174–193.

Andrews, J. T. and Shimizu, K. (1966). 3-Dimensional vector technique for analyzing till fabrics: Discussion and Fortran Program, *Geog. Bull.*, **8** (2), 151–185.

Braithwaite, R. C. (1960). *Scientific Explanation*, Harper and Row, New York.

Cailleux, A. (1945). Distinction des galets marines et fluviatiles, *Bull. Soc. Geol. France*, **5 XV**, 375–404.

Chorley, R. J. and Haggett, P. (1965). Trend surface mapping in geographical research, *Inst. Brit. Geog. Trans.*, **37,** 47–67.

Dury, G. H. (1951). Measurement of available relief, *Geol. Mag.*, **88,** 339–343.

Flinn, D. (1958). On tests of significance of preferred orientations in 3-dimensional fabric diagram, *J. Geol.*, **66,** 526–539.

Hagerstrand, T. (1952). *The Propagation of Innovation Waves*, Lund Studies in Geography, Ser. B. Human Geography, No. 4, Lund.

Haggett, P. (1964). Regional and local component in the distribution of forested areas in south-east Brazil: A multivariate approach, *Geog. J.*, **130,** 367–378.

Hollingworth, S. E. (1931). The glaciation of western Edenside and adjacent areas and the drumlins of Edenside and the Solway Basin, *Quart. J. Geol. Soc.*, **87,** 281–359.

King, C. A. M. (1967). An introduction to trend surface analysis, *Bull. Quantitative Data Geog.*, No. **12,** Nottingham University.

Krumbein, W. C. (1941). Measurement and geological significance of shape and roundness of sedimentary particles, *J. Sed. Petrol.*, **11** (2), 64–72.

Krumbein, W. C. and Graybill, P. A. (1965). *An Introduction to Statistical Models in Geology*, McGraw-Hill, New York.

McCann, S. B. (1966). The main post-glacial raised shoreline of Western Scotland from the Firth of Lorne to Loch Broom, *Inst. Brit. Geog. Trans.*, **39,** 87–99.

Phillips, F. C. (1960). *The Use of Stereographic Protection in Structural Geology*, Arnold, London.

Rapp, A. (1960). Recent development of mountain slopes in Kärkevagge and surroundings, North Scandinavia, *Geog. Ann.*, **42,** 65–200.

Steinmetz, R. (1962). Analysis of vectorial data, *J. Sed. Petrol.*, **32** (4), 801–812.

Strahler, A. N. (1952). Hypsometric (area-altitude) analysis of erosional topography, *Bull. Geol. Soc. Am.*, **63,** 1117–1142.

Swallow, J. C. and Worthington, L. V. (1957). Observations of deep currents in the western North Atlantic, *Nature*, **179,** 1183–1184.

Sweeting, M. M. (1950). Erosion cycles and limestone caverns in the Ingleborough district, *Geog. J.*, **115,** 63–78.

Trotter, F. M. (1929). The tertiary uplift and resultant drainage of the Alston Block and adjacent areas, *Proc. Yorks. Geol. Soc.*, **21,** 161–180.

Whitrow, G. J. (1962). Why physical space has three dimensions, *General Systems, Yearbook Soc. Gen. Systems Res.*, **VII,** 121–129.

Whitten, E. H. T. (1963). A surface-fitting program suitable for testing geological models which involve areally-distributed data, *Office Naval Res. Geog. Branch Tech. Rept. No.* 2, Contrast No. 1228 (26).

10

Geography and time

'Until that time the world seemed to rest, static and fragmentable, on the three axes of its geometry. Now it is a casting from a single mould.

'What makes and classifies a "modern" man . . . is having become capable of seeing in terms not of space and time alone, but also of duration, or — it comes to the same thing — of biological space time; and above all of having become incapable of seeing anything otherwise — anything — *not even himself.*'

Pierre Teilhard de Chardin (1965)

10.1 Introduction

(*a*) *Characteristics of time and space*

This chapter is concerned with space and time. In some ways space and time are analogous but in other ways they differ profoundly. C. D. Broad (1938) has written that the two aspects of time, duration and temporal relations, are very closely interconnected. It is in respect of them that there is a close analogy between time and space. These aspects of time he calls the extensive aspect of temporal facts. Duration is analogous to distance or surface area in space. Both aspects have finite magnitude in time and space respectively. Temporal relations are also analogous to spatial relations, but there is an important difference. A spatial position can be defined by reference to two others on either side. Thus Nottingham lies between Leeds and London. This is not a directed relationship. Three places must be mentioned to express the relationship of betweenness; it is a triadic relation. A temporal position can also be described as between two other events. Thus Wednesday occurs between Monday and Friday in any one week starting on Sunday. This relationship is directed, in that Wednesday must follow Monday and Friday must follow Wednesday. It is not fundamental that Wednesday is between Monday and Friday, but the facts that Wednesday follows Monday and Friday follows Wednesday are the essential ones. The relationship is directed, as it can only go one way. It is enough to specify the relationship of two events to fix their relative position in time. In space this can only be done by reference to a third point or external system of reference. Thus space has no intrinsic direction but time has.

A major difference between space and time is that the latter has a transitory aspect, but space has no such attribute. Time has the characteristics of *pastness*, *presentness* and *futurity*. Every event is continually changing with respect to these three characteristics. An event may be in the remote future, then in the near future, then present, recent past and remote past. There is a danger, according to Broad, in the analogy between events spread out on a time scale and numbers arranged in order of magnitude or objects on a linear space scale. This analogy may suggest that all events must somehow co-exist, but the analogy leaves out one of the very important properties of time — namely, its transitory aspect. Events can, however, begin and end, and objects can begin to exist and cease to exist. Absolute becoming is a reality. Thus the transitory aspect of time is a vital quality and one in which time differs from space. Time may basically be thought of as temporal relationships between events and objects in space.

Time and space pervade all aspects of reality that constitute the data of geography, since everything material must exist in time as well as 3-dimensional space, if space really has three dimensions as is normally assumed. The connexion between space and time is intimate. This is particularly apparent in the theory of relativity, which relates time and space together on the macroscale of the universe and the galaxies. This relationship is seen in the major distance unit of astronomy, the light year, the distance travelled by light in one year. Its definition, therefore, requires units of both space and time.

Another aspect linking time and space is the route dependence of time. Although this property of time is not important in the mesoscale of geographical material it is very significant where high velocities, approaching the speed of light, are encountered. When two bodies are moving at different speeds there is not universal time, as measured on a clock, but each individual body has its own time. This fact allows time to be route dependent, a term which means that times will differ for two bodies according to their movement, relative to each other, provided they are moving at different speeds. Supposing a body is moving uniformly through space and another body is moving at an accelerating speed away from the first. Light signals from the first, made at stated uniform intervals, are received by the second at longer intervals owing to the greater distance that the light has to travel each time. Thus the interval between the receipt of flashes by the second body is greater than the interval at which they are sent out. Time intervals are therefore longer for one body than the other. The interval ratio is K. The value of K depends on the velocity of one body relative to the other. Thus for 186 miles/sec $K = 1.001$ (v is then 0.001 the velocity of light). If v is 0.998 times the velocity of light, $K = 100$.

Because of the private or route dependent qualities of time at high velocities the position of a body must be specified in terms of its own time. Thus four co-ordinates, t, x, y, z are needed to specify the position of the body. The route dependence of time is seen in an interesting way in the aberration of light, first discovered by J. Bradley in 1725. The angle of light from the stars differs slightly from one season to another. This is because the sun is travelling through space and the earth is moving with it. At one season the earth is going in the same direction as the sun; at the other season 6 months later, it is going in the opposite direction. This difference of speed relative to the stars causes slight distortion of the stellar angles.

The fact that time is route dependent and each body has its own time means that events, as viewed from different points, need not necessarily occur in the same order. The so-called light cone illustrates the situation. From any point in space and moment in time a cone extends to the absolute past and absolute future. In between, however, is a zone in which some observers would see one event preceding another, while others would see the order reversed. These are the zones of relative past and future, in which the order of events as seen from different points depends on the route of movement of the points relative to each other and the position at the apex of the light cone (H. Bondi, 1965).

This digression into relativity shows that both time and space take on unusual properties when speeds are high and distances great. Time and position in space become individual attributes of each point considered. It is useful also to bear this in mind when considering geographical data. Each person sees the world from a different viewpoint both in time and space. In other words, while the geographer is attempting to see the 'real' world in an objective way, as it really is, he, and even more, people in general, have their own personal views of the world. Increasing attention is being devoted in geography, partly through the stimulus of studies in psychology, to the way that individuals, particularly influential decision makers, think the world to be. One example is T. F. Saarinen's *Perception of the Drought Hazard in the Great Plains* (1966).

On the much smaller scale of our earth, time and space are intricately linked and often inseparable. The various sections of this chapter draw attention to some of the many relationships between time and space. The sections of the chapter from 4 to 7 inclusive are particularly concerned with the relationships between time and space in geographical data. Even the measurement of time and distance are often carried out in terms of each other. Attention is also drawn in section 9 to the statistical analysis of events which occur on a time scale. Prediction into the future is often the aim of scientific analysis. The extent to which this can be achieved in both physical and human geography is considered in the final section. In this connexion in human geography it is important to distinguish between scientific prediction and prophecy or precognition. The latter poses problems of a different type.

(*b*) *Static and dynamic data.*

In considering geographical data it is important to differentiate between those things that are (1) static, those that (2) change slowly and those that are (3) dynamic. In the first category are features that do not change measurably over the span of a human life or even historical time: for example, the pattern of land and sea, geological structure and major landforms.

In the second category of slowly changing phenomena there are many things: for example, vegetation cover, sea-level changes, landscapes in towns. Movement is not an essential element of these phenomena, but they cannot be considered as static, although normally they can be treated in the same way and by the same techniques as static phenomena. They are slowly changing rather than moving phenomena, in that movement is either not essential or is impossible. Plants cannot move (except within small limits by wind movement) but their distribution can change and their density and height, and so on, as new ones grow and old ones die. In the same way, houses in a town do not move individually but some are pulled down and new ones built so the pattern, although static at any one moment, changes with time. The change in these situations is brought about largely by replacement rather than by movement.

Movement forms an essential element in many situations in the third category. These are the dynamic situations in which time, velocity and acceleration are all vital elements. These dynamic situations can be further subdivided. The criterion for subdivision is whether the changes are, on the one hand, essentially unidirectional or, on the other hand, whether the changes are random in direction or circulatory in character. The unidirectional changes are mostly dependent on the force of gravity for their motivation. Mass movement on slopes, and river flow can be included in this category.

In considering the circulatory dynamic situations, forces that can operate against gravity must be taken into account. Good examples of 3-dimensional circulations are provided by the general atmospheric circulation, the oceanic circulation and the hydrological cycle. In all these an external source of energy counteracts gravity, allowing the circulation to become established and a steady state situation to be set up.

10.2 Measurement of time

(*a*) *Measurement of time and space in terms of each other*

As this chapter is concerned with time and space it is interesting to draw attention briefly to the connexion between time and space in the measurement of these dimensions.

(i) *Space in measuring time.* One of the oldest and simplest methods of measuring time is the hour glass. This method uses the *volume* of sand passing through a small aperture to measure the passage of time. Another old method of measuring time is the sundial. In this instrument the movement of the shadow of the dial, cast by the sun as it apparently moves across the sky, measures the time. Even the common clock records time in angular measurement around a dial in a somewhat similar fashion, thus transferring the time dimension to a spatial one.

(ii) *Time in measuring space.* Time is used to measure distance in several instruments that are increasing in common use. The tellurometer measures distance by timing the passage of waves between the transmitter and receiver. This instrument is becoming increasingly used in geodetic survey work, where long distances must be measured accurately. The echo-sounder, used for measuring depth of water or thickness of ice, is also based on the recording of time taken for waves to travel from the instrument to an interface and be reflected back to the recorder. Position on the earth's surface is determined by latitude and longitude. The latter can only be measured by using time (see chapter 8). This accounts for the inability of early navigators to fix their longitude before the invention of chronometers. The measurement of longitude is based essentially on the difference in time of the passage of the sun across the local meridian and its passage across the meridian of Greenwich or some other fixed meridian. This is the difference between L.M.T. and G.M.T., so that longitude, as measured in hours, must be converted into distance, expressed either in angular terms in degrees or miles east or west of the Greenwich meridian.

(*b*) *Physical methods of measuring time*

The chronology of events is often of significance in physical geography. A number of different techniques have been developed for determining both absolute and relative chronologies in geomorphological studies. In some studies the annual rhythm provides a useful means of measuring time. Dendrochronology measures time according to the number of annual tree rings. This technique covers a limited stretch of time, amounting at a maximum to about 3000 years in the giant sequoias of the west coast of the United States. The annual deposition of coarser and finer sediment in a glacial lake is responsible for the varves that have provided such a good time scale in the elucidation of the deglaciation chronology of Scandinavia and elsewhere. G. de Geer (1912) first used this technique successfully in Sweden where the varve chronology covers nearly 17,000 years from varves laid down in lakes spread out over a distance of almost 1000 km. This long period is made possible by the correlation of the base of the varve sequence in one lake with the top of that in another lake. The correlation depends on the variation of the varve size with climate, which would have been similar for the two lakes.

Another method of providing chronological information is lichenometry. The thalli of lichens grow at a linear rate during the greater part of their life cycle. The maximum size of the lichen thus gives an estimate of the time since the stone on which it grows was first colonized. This dating method can be used to estimate the rate of glacial retreat or the date of moraine formation over a period of about 300 years. The method does not provide an absolute chronology unless the rate of growth of lichen in the area studied can be related to absolute time by some means. In some areas the size of lichens on dated tombstones provides a method of assessing the growth rate. However, so many variables affect the growth rate that absolute dates by this method are not always accurate and each area requires its own growth rate curve (J. B. Bendict, 1967).

One of the most versatile methods of dating geological time is by the use of isotopes. The versatility is due to the great variety in the lengths of the half-lives of naturally radio-active minerals. The average rate of disintegration of radio-active material, containing a large number of nuclei, is the same for any one radio-active substance. Each nucleus, however, disintegrates randomly, in that each nucleus has an equal chance of disintegrating at any given time. The active nuclei gradually disintegrate with time, but as the number of active nuclei decreases steadily the number decaying also decreases. The half-life of the radio-active isotope defines the time when half the active nuclei have disintegrated. In the second half-life period the number of active nuclei is again halved, making the total remaining a quarter of the original number and so on. By the time 10 half-lives have elapsed the nuclei are reduced by a factor just over 1000. For this reason different minerals between them can span the whole of geological time with reasonable accuracy.

One of the longest half-lives, that of thorium-232, is 14,000 million years. Uranium-238, with a half-life of 4500 million years, easily covers the whole of geological time. The radio-active decay of these minerals produces other minerals that decay at a different rate and so on to produce a long radio-active chain. Uranium-238 passes through 14 stages to produce the stable lead-206 isotope. The half-lives of the intervening materials range from 4.5×10^9 years for uranium-238 to 160 microseconds for polonium-214. The results of a study of the ratio of uranium-238 to lead-206 gives rock ages between 50 million years and 4000 million years, which is about the age of the earth's crust. Other mineral isotopes can be used for intermediate ages. For shorter periods up to about 50,000 years, carbon-14 provides the best method of estimating age of organic matter. The half-life of the radio-isotope carbon-14 is 5760 years. Carbon-14 is produced by cosmic rays in the upper atmosphere and is absorbed by growing plants as carbon dioxide, but when they, or the animals which eat them, die, the carbon-14 decays to carbon-12. The proportion of carbon-14 to carbon-12 is thus a measure of time since the organic matter was alive.

(*c*) *Methods of representing time*

The long periods of geological time, which must be counted in millions of years, are difficult to appreciate. A relative time scale can, therefore, be used to illustrate the length of different geological periods. If the age of the earth, which has been estimated to be about 5000 million years, is made equivalent to $3\frac{1}{2}$ days, then on this scale the Palaeozoic period started 10 hours ago, the Mesozoic 3 hours 40 minutes and the Tertiary period 1 hour 10 minutes ago. The Pleistocene period would on this scale have started 1 to $1\frac{1}{2}$ minutes ago and post-glacial time would cover the last 0.6 seconds. On this scale 1 minute is the equivalent of 1 million years. Such a scale helps one to appreciate the recency of the glacial period, and the arrival of man on the earth.

Another method of representing time is shown in figure 10.1. A logarithmic binary scale is used instead of the more familiar logarithmic denary scale. It gives greater emphasis to the shorter periods as each successive division upwards covers double the next lower one. The basic division is one day. On this scale the age of the earth is about 2^{41} or 5,000 million years. The longer periods of distant geological time, which are not accurately known, are given less emphasis than the geographically more relevant period covering 1

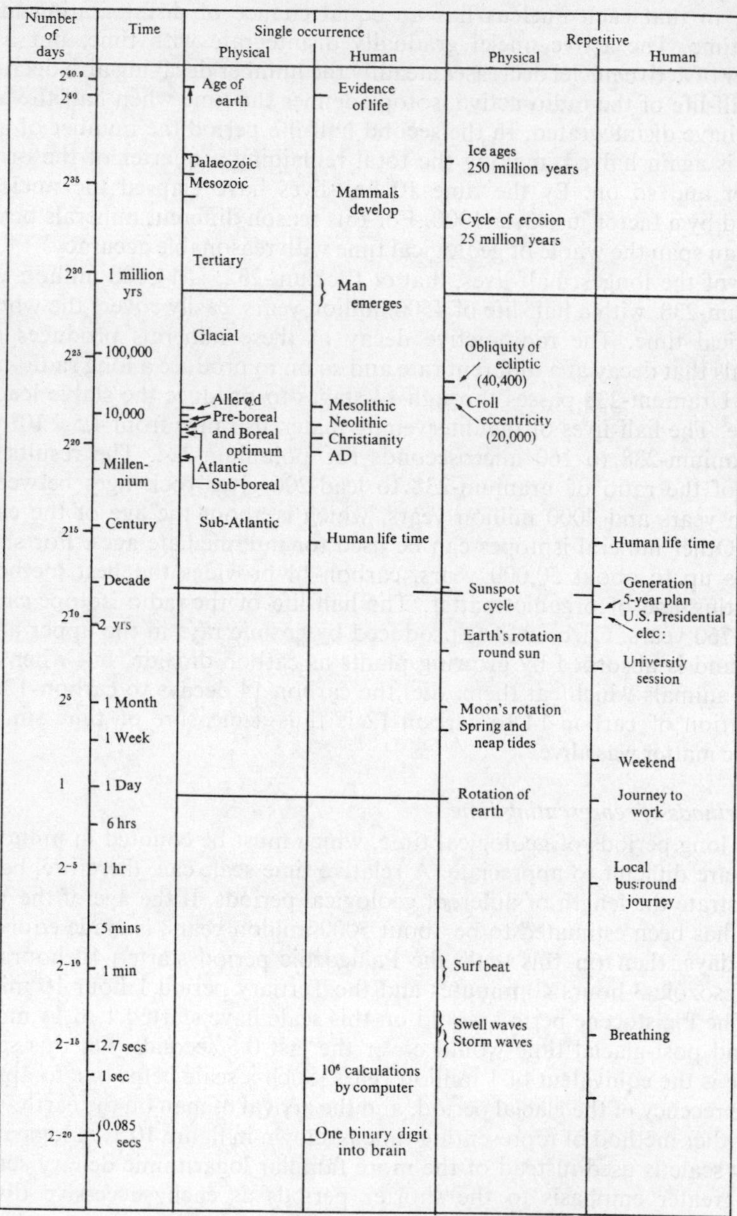

Figure 10.1 A binary time scale based on the unit of one day. The left hand columns show absolute dating for both physical and human events. The right hand columns show cycles in physical and human geography.

minute (about 2^{-10} days) to 1 million years (about 2^{30}). Some absolute dates and a few cycles are shown on figure 10.1. In giving greater space to the recent past, the scale allows for the acceleration of development which is taking place in many aspects of geography, especially in the human field. In the physical sphere, also, more is known of recent events than ones in the distant past, and the scale gives more space to these better known events.

Finite arithmetic (see chapter 2.4) is also often used for the representation of time. The 12 hour or 24 hour clock is based on modulo 12 and 24 respectively and the seven day week uses modulo 7. The familiarity gained by using these different modulos for time provides good examples of this method of counting.

(*d*) *Model time.*

Models are being used increasingly for preliminary research into many problems. Dynamic scale models provide one type in which the time element is important. Models of tidal estuaries, for example, are usually built before major engineering operations are undertaken. These models are dealing with dynamic phenomena, so that it is necessary to maintain dynamic similarity between model and prototype. Froude has developed laws which state that the time and velocity scales of a model must be equal to the square root of the linear scale of the model. One of the difficulties, however, is the fact that the force of gravity is the same for model and prototype in most models. This similarity means that trial and error is often required before correct model scales can be developed and made to simulate known prototype changes. When this has been achieved then known changes can be extrapolated into the future. This is a convenient way of reducing real time so that results of the past, and future expectations, can be speeded up.

10.3 Events on a time scale.

In its simplest form the time scale can be considered as a line on which similar events occur at particular intervals. The events can take place instantaneously or they can occupy a measurable period of time. The events can be compared with objects that exist on a single space dimension. The objects can similarly be considered to have no size and therefore be represented as points on the space dimension line or they can have a measurable length along the space line. In both situations the events or objects can be illustrated graphically by plotting their position and extent on a line.

(*a*) *Discrete events of negligible duration*

A sequence of instantaneous events on a time scale can approximate to three possible patterns. These are regular (usually anti-clustered), random or clustered. Regular events, such as the crossing of the sun over the local meridian, are approximately or exactly equally spaced in time. In this example the periods between successive transits of the sun are not of exactly the same duration owing to the asymmetry of the planetary paths (see equation of time, chapter 8.2 (iii) 2). On the other hand, for convenience, noon by our clocks occurs at precisely regular intervals of 24 hours. Both events can be considered to be regular, however. Random events are those which cannot be predicted as the events have an equal chance of occurring at any moment. Clustered events are those which are biased in their occurrence so that the number of short intervals is greater than random. For example, days with frost cluster in the winter period.

Any given series of events can readily be tested for randomness, regularity or tendency to cluster. The observed set of events can be compared with the expected pattern by use of the chi-square test. Frequencies of observed occurrence are compared with the expected frequencies. The test is similar to that which can be applied to dot distributions in space (see chapter 4, figure 4.11).

(*b*) *Events with significant duration*

Some events are intermittent in character, others are continuous. Thus light and darkness are intermittent and each covers a period of measurable duration. Other events in this category are precipitation or volcanic activity, which may be contrasted with the properties that always exist although they vary with time. Temperature falls into this category. The air always has a temperature, which is a continuous property, measured for convenience at regular intervals. The turning points of temperature, the maxima and minima, can be considered as points of no duration and these points thus fall into one of the categories discussed in (*a*). They also have another aspect worth mentioning. A continually fluctuating value, such as temperature, which is measured at regular intervals, gives a sequence of values, called a time series. The fluctuations of a time series are discussed in section 10.9, where some of the statistical techniques used to analyse time series are mentioned.

An example of a continuously changing time series in the field of human geography is population change. At any one instant in time there is a specific number of people on the earth, though this number is continually changing and for the earth as a whole or even for one country, the changes in number can be smoothed out to form a continuously changing total.

In a small group, however, such as one family, the number may remain constant over a considerable period of time. In another type of small group the situation may occur in which the number remains constant, but the individuals change. The students of one year in a University course or the water in a river provide instances of this. It is another form of constancy, which could be considered as dynamic in character as the individual elements in the system are changing over time but the system may remain uniform. Constancy must not be confused with continuity. The former implies uniformity in quantity and the latter uniformity of presence, but change in quantity may take place.

(*c*) *Reversible and irreversible events*

Some events are reversible, others are not. Material carried away from a beach in a storm, may return again when calmer conditions set in. Glaciers advance and retreat again, sea level falls and rises again, but if a rock falls off a cliff it cannot be replaced naturally. Erosion in a valley cannot be undone. The same principles also apply to human geography up to a point. It is always worth while to consider this aspect of change with time in all fields of enquiry. Irreversible events may often be considered in the context of a closed system and reversible ones are usually associated with an open system (see chapter 11.4*a*).

10.4 Space as evidence for time

Many geomorphological processes act slowly in geological time. It is therefore difficult to observe changes in form in time. In some situations this difficulty can be overcome by substituting space for time. An example of this substitution is R. A. G. Savigear's (1952) study of slope profile development in South Wales. Savigear was studying the progressive modification by subaerial processes of cliffs produced largely by marine erosion. Subaerial erosion could only become the dominant process where marine influence had stopped. In the area studied, marine deposition was spreading along the coast eastwards and thus progressively cutting off the cliff line from the sea. The cliffs that were first isolated from marine action were thus the oldest as far as pure subaerial activity was concerned. The sequence of cliffs from these oldest ones to those still being attacked by the sea thus provided a sequence in space along the coast and these were also arranged in order of age of development by subaerial processes in time. Thus a space sequence could be used as evidence for changes in time. The progressive modification of the sea-cut cliffs could thus be studied at the present time (see figure 10.2).

Archaeological excavations also provide evidence for changes in time, the time dimension being equivalent to the vertical space dimension in this situation. Frequently settlements have been using the same site for a very long period of time, the newer buildings being constructed on the remains of the old. In these situations it is possible, by digging down through the layers of occupation, to go back through time. This has been clearly shown in the

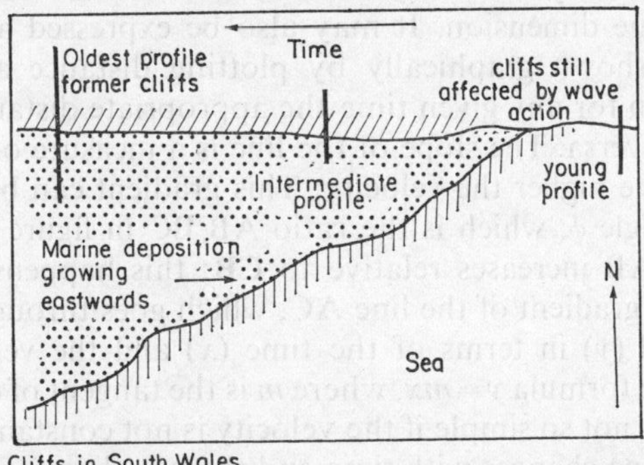

Figure 10.2 Diagram to illustrate how age of cliff profiles can be related to positions in space. The oldest cliff profiles on the left were the first to be isolated from direct marine action. Based on the work of R. A. G. Savigear.

excavations in and around the walls of Jericho. The settlement started a very long time ago and many new additions on top of the older buildings give evidence of the different stages.

In some situations archaeological evidence has a direct bearing on geomorphological processes. V. B. Proudfoot (1965) has pointed out that the upper slope of a prehistoric ditch has a gentler gradient than the lower part. The lower part has been protected from subaerial processes by burial beneath infill while the upper part has weathered back. If the infill includes later archaeological material, then dates for the stages of weathering can be determined. Bronze age relics have been found in Neolithic earthworks. The relics show that 3 to 7 ft of material accumulated during this interval.

Other interesting implications can sometimes arise from archaeological data. It has been suggested that where chalk has been preserved from percolating waters by the building of barrows on its surface, the level of the chalk surface is 15 to 20 in higher than the adjacent unprotected areas. The lowering of the surrounding chalk has taken place during the last 4000 years. Change in elevation can thus be correlated with time and illustrates the importance of solution in lowering chalk surfaces. If these changes are extrapolated back through time then the results are of major geomorphological significance. The chalk surface, at this rate of lowering, must have been lowered about 200 to 400 ft during the Pleistocene and very much more during the Tertiary Period. The rate of lowering is larger than expected, and it shows that lowering must be borne in mind when denudation chronology in chalk areas is being considered.

10.5 Time and distance

(*a*) *Velocity and acceleration*

In this section the movement of objects in relation to space and time is considered. Movement implies velocity, which is expressed as so many units of distance covered in a specified time, miles per hour, or metres per second. Velocity is indicated symbolically as L/T, in terms of one space dimension as well as the time dimension. It may also be expressed as LT^{-1}. Uniform velocity can be shown graphically by plotting distance against time as a straight line. Then for any given time the appropriate distance moved can be read off and vice versa. The slope of the line is a measure of the velocity: the steeper the line the higher the velocity. This gradient can be expressed as the tangent of the angle ϕ, which is the ratio AB/BC in figure 10.3*a*. This value will increase as AB increases relative to CB; this happens when there is a steepening of the gradient of the line AC, which goes through the origin. The distance travelled (y) in terms of the time (x) and the velocity (m) can be obtained from the formula $y = mx$, where m is the tangent of ϕ.

The situation is not so simple if the velocity is not constant. In this instance the angle of the line changes with time and the line therefore becomes curved (figure 10.3*b*). If the speed is increasing (acceleration), then the line steepens with time. If the movement is decelerating, then the line flattens with time. The curve is smooth if the velocity is increasing or decreasing at a constant rate. This is true of a body falling freely under the influence of gravity. This situation arises if the distance covered increases in geometrical proportions as the time increases in arithmetical proportions.

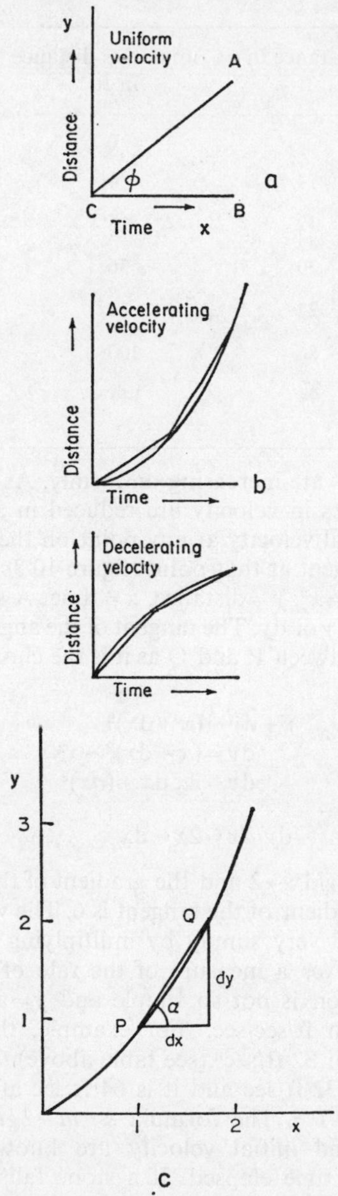

Figure 10.3 Diagrams to illustrate uniform velocity, acceleration and deceleration using graphs with time plotted against distance.

The average velocities for different periods of time can be worked out. The gradients of chords drawn to these points will represent the average velocity for the successive time intervals.

Time in secs	Distance in ft	Cumulative distance in ft	Velocity in ft sec
0		0	
	4	4	4/0.5 = 8
½			
	12	16	12/0.5 = 24
1			
	20	36	20/0.5 = 40
1½			
	28	64	28/0.5 = 56
2			
	36	100	36/0.5 = 72
2½			
	44	144	44/0.5 = 88
3			

These average velocities are increasing uniformly. As the chords are reduced in length so the changes in velocity are reduced in amount until a smooth curve results. The actual velocity at any point on the curve is then given by the gradient of the tangent at that point (figure 10.3*c*). In the example given the curve is given by $y = x^2$, y = distance, x = time. A small change in x of $\mathrm{d}x$ is related to a change in y of $\mathrm{d}y$. The tangent of the angle α, $\mathrm{d}y/\mathrm{d}x$, is a measure of the mean velocity between P and Q as it is the chord of the arc PQ.

If $y = x^2$ then

$$y + \mathrm{d}y = (x + \mathrm{d}x)^2$$
$$\mathrm{d}y = (x + \mathrm{d}x)^2 - x^2$$
$$\therefore \mathrm{d}y = 2x\,\mathrm{d}x + (dx)^2$$

and

$$\mathrm{d}y/dx = 2x + \mathrm{d}x$$

If $x = 1$ then as $\mathrm{d}x \to 0$ $\mathrm{d}y/\mathrm{d}x \to 2$ and the gradient of the chord $\mathrm{d}y/\mathrm{d}x$ becomes 2. If x is 3 then the gradient of the tangent is 6. The velocity at any point on the curve can be found very simply by multiplying the x value by 2. The change of this value gives a measure of the rate of acceleration. In some instances the acceleration is not so simple and $y = \mathrm{a}x^2$ must be used. The acceleration is given in ft/sec/sec. For example, the acceleration due to gravity is 32 ft/sec/sec or 32 ft/sec² (see table above). This means that after 1 second the velocity is 32 ft/sec and it is 64 ft/sec after 2 secs. This can be expressed as $\mathrm{L/T^2}$ or $\mathrm{LT^{-2}}$. The formula $s = ut + \frac{1}{2}gt^2$ can be used to find distance when time and initial velocity are known. u = initial velocity, s = distance and t = the time elapsed. If a stone falls from a cliff to sea in $t = 5$ secs then the height of the cliff is given by $s = ut + \frac{1}{2}gt^2$.

$$s = 0 + \tfrac{1}{2} \times 32 \times 5 \times 5 = 16 \times 25 = 400 \text{ ft.}$$

Velocity is an important aspect of many geomorphological processes such as river flow, while changes of velocity leading to acceleration and deceleration play a very important part in glacier movement. The response of the ice to these changes in velocity has a fundamental effect on the way the ice moves. This affects its behaviour on its bed and its geomorphological effects.

In many geomorphological events the movement is neither continuous and uniform nor is it uniformly accelerating or decelerating. In these situations movement may be intermittent or irregular. Movement along faults provides an example of intermittent activity, but which can in time add up to considerable displacements. The 300 mile movement along the Great Alpine fault in New Zealand is a good example. The western side has moved intermittently north relative to the eastern side. The necessity to shorten telegraph wires that cross the north–south fault from northeast to southwest shows that this movement is still progressing slowly, but probably not continuously (see figure 10.4).

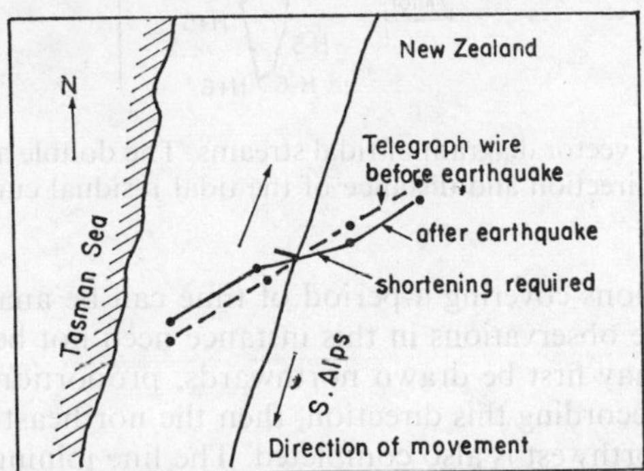

Figure 10.4 A diagrammatic sketch map to show how movement along a fault causes displacement in space. The Alpine fault along the western side of the New Zealand Alps has a large transcurrent movement which is still intermittently in progress. The total movement has been estimated as 300 miles. Refer also to figure 2.20 part *b*.

(*b*) *Vector analysis*

In some situations movement changes with time both in direction and velocity. One of the best examples is the variations of tidal streams through the tidal cycle from high to low tide and back to high tide. The particles of water which follow these complex movements do not necessarily return to their starting point at the end of each cycle. The change of position of the water particle after one cycle is called the resultant or residual and is of great significance. If the water is carrying sediment, the resultant of the movement will determine the net direction of sediment transport.

Vector analysis provides a method whereby the amount and direction of the resultant can be found. Each hourly observation is drawn as a line, the length of line being proportional to the mean velocity for that hour. Each hour is added to the previous one until the end of the last line, that of the twelfth hour, is reached. The length and bearing of the line joining the initial and final point gives the resultant of the flow during the tidal cycle (see figure 10.5).

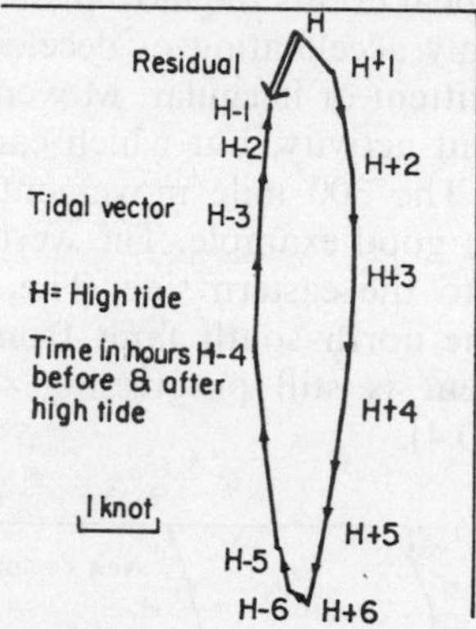

Figure 10.5 A vector diagram of tidal streams. The double arrow indicates the direction and distance of the tidal residual current.

Wind observations covering a period of time can be analysed in a similar way, although the observations in this instance need not be in chronological order. The line may first be drawn northwards, proportional to the number of observations recording this direction, then the northeast line is added and so on until the northwest is also completed. The line joining the first and last points again gives the resultant as shown in figure 10.6. The length of the

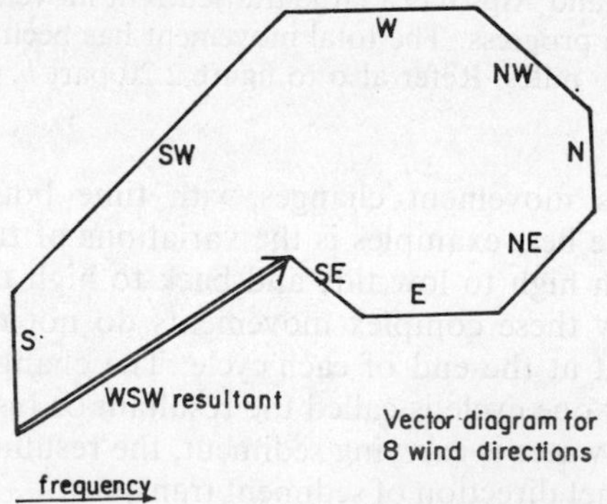

Figure 10.6 A wind vector produced by summing the frequency of wind force records from eight different directions cumulatively. The double arrow indicates the resultant vector, which is from the west-south-west.

lines can be adjusted, by giving greater weight to the stronger winds. Alternatively onshore winds only may be used. It has been shown that the resultant has morphological significance in some areas. Dune orientation in coastal localities can be correlated with the wind resultant.

(*c*) *Changes in velocity in time and space*

It is worth drawing attention to the difference between changes of velocity in time and changes of velocity in space. Change of velocity at a gauging station on a river over a period of time can be compared with the change of velocity between different gauging stations at the same time. The diagram (figure 10.7) illustrates the possible situations. A_1 and A_2 represent the river

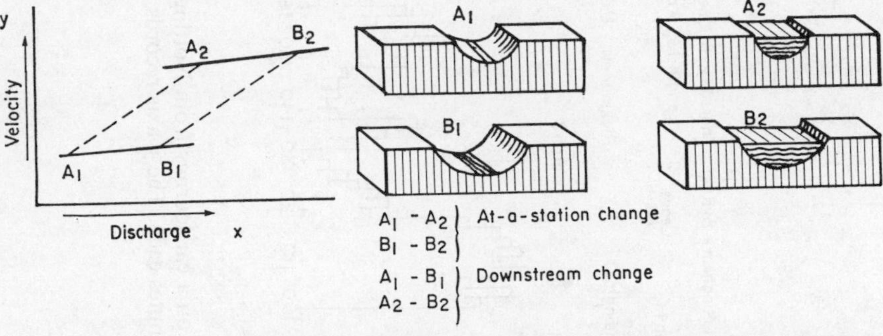

Figure 10.7 The relationship between discharge and velocity with time at-a-station, given by A_1–A_2 and B_1–B_2. The velocity and discharge increase rapidly together. The relationship between discharge and velocity at the same time but at different positions on the stream given by A_1–B_1 and A_2–B_2. The velocity increase is slight with increasing discharge. (After L. B. Leopold, M. G. Wolman and J. P. Miller.)

at the same point upstream at two different times, there being different levels of discharge at the two times, A_2 being greater than A_1. The lower diagrams B_1 and B_2 represent the same river at a station further downstream, B_1 again representing a different time with a lower flow stage than B_2. The discharge (x) and velocity (y) graph illustrates how the velocity changes in these different circumstances. The line A_1 A_2 shows that there is a rapid increase of velocity as the discharge increases near the headwaters of the river. The line B_1 B_2 is nearly parallel and this shows that a similar increase holds also in the lower reaches of the river as discharge increases. The rate of increase of velocity at any one time in a downstream direction is very much less as shown by the much gentler gradient of the lines A_1 B_1 and A_2 B_2 but it is an increase and not a decrease as is often supposed. These two lines are again almost parallel, but the velocity is higher at stages of high discharge, as in situation 2, than at stages of low discharge as in situation 1. These changes of velocity, both in time and space, are important when the action of fluvial processes is being considered.

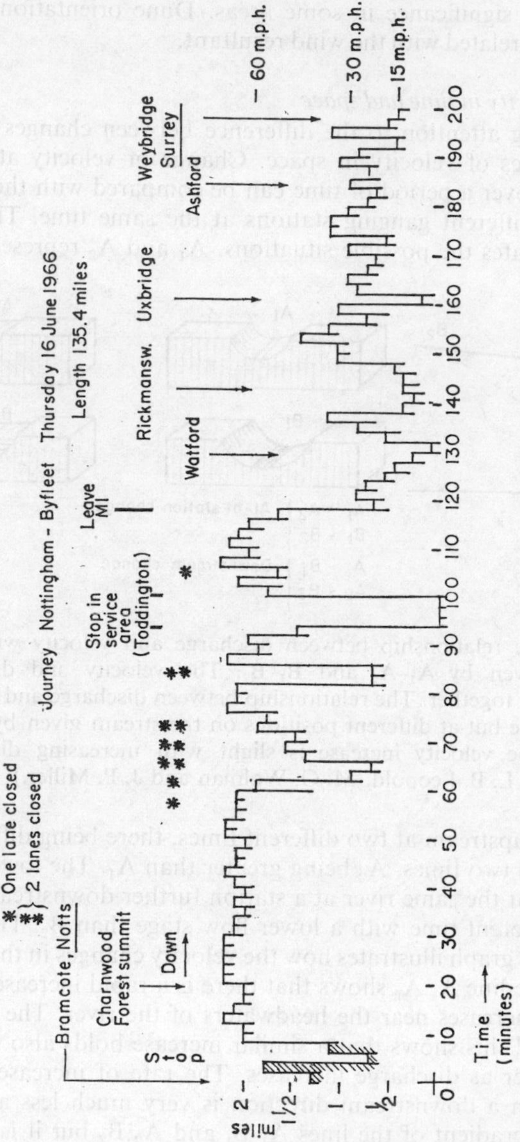

Figure 10.8 Time/distance relationship observed on a car journey from Nottingham to Byfleet (Surrey). The *x*-axis consists of class intervals of 2 minutes each. The *y* axis records the distance covered.

(d) Time and distance on journeys

An example of movement along a line during time in human geography is a journey along a route. Figure 10.8 shows the road journey from Nottingham to Byfleet in Surrey, represented by time on the *x*-axis of a graph and distance covered during given intervals of time on the *y*-axis. Time intervals are held constant, while distance varies. Speed is reflected in the height of the columns on the *y*-axis. These data were collected by reading the mileometer at two-minute intervals. In general, it would seem that the first arrangement emphasizes the variations of distance in fixed time, the second the variation of time over fixed distance. They both bring out delays caused by passage through a town, along a slow road, or travelling behind a slow vehicle. Geographically they help to visually represent mobility or ease of movement at different places along a given route, thus reflecting spatial variations along a line in a special way. Various journeys could be compared in a single diagram.

It is possible and at times visually rewarding to draw a transformed map

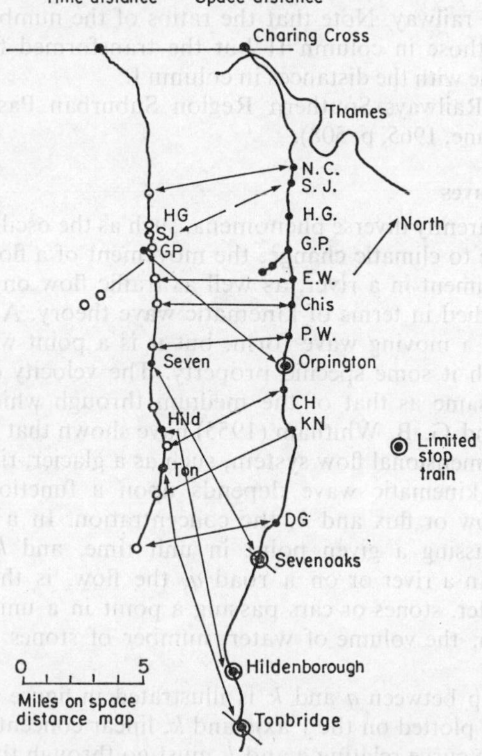

Figure 10.9 Distance from Charing Cross (London) to selected railway stations in southeast London and northwest Kent compared in terms of time distance and space distance, see text.

in which a given set of places is located on the map according to the time taken to reach them, from some centre rather than the distance. A very simple example is illustrated in figure 10.9 and the relevant data are tabulated below. It concerns one particular suburban railway line out of London. The procedure could be repeated for all lines out of London and the results mapped or compared on a graph (see chapter 5).

Table 10.1 contains a list of 16 stations on one of the railways running southeast from London into Kent. Column I shows the rail distance of each station (note that 2 are on a branch line). 24 trains *leaving* London (Charing Cross) between 11.00 and 15.00 hours have been considered. Column II shows the fastest time each station can be reached from Charing Cross. Note for example, that Orpington, though further from Charing Cross than Chislehurst is 'nearer' in time because it is served by a non-stop train. Column III shows the average 'speed' in minutes per mile of the fastest train. The mean for all the stations is 2.3 minutes. Column IV shows the variation about the mean (mean = 100) of the speed of the fastest train to each station. Space distance may be adjusted by multiplying it by the variation index divided by 100. Time distance (column V) may then be used to remap the stations along the railway. Note that the ratios of the numbers in column V are the same as those in column II but the transformed figures are more readily comparable with the distances in column I.
(Source: British Railways Southern Region Suburban Passenger Services, 7 Sept, 1964–13 June, 1965, p. 508).

10.6 Kinematic waves

A number of apparently diverse phenomena, such as the oscillations of glacier fronts in response to climatic change, the movement of a flood down a river and bed load sediment in a river, as well as traffic flow on crowded roads, have all been studied in terms of kinematic wave theory. A kinematic wave is not necessarily a moving wave form, but it is a point which moves and which carries with it some specific property. The velocity of the kinematic wave is not the same as that of the medium through which it is moving. M. J. Lighthill and G. B. Whitham (1955) have shown that kinematic waves can exist in a 1-dimensional flow system, such as a glacier, river or road. The existence of the kinematic wave depends upon a functional relationship between q, the flow or flux and k, the concentration. In a glacier q is the volume of ice passing a given point in unit time, and k is the vertical thickness of ice. In a river or on a road q, the flow, is the transport rate or quantity of water, stones or cars passing a point in a unit time, and k is the concentration, the volume of water, number of stones or cars per unit distance.

The relationship between q and k is illustrated in figure 10.10 in which q (transport rate) is plotted on the y axis and k, linear concentration is plotted on the x-axis. The curve relating q and k must go through the origin because as the linear concentration k tends towards zero the units become very far apart and therefore the transport rate also tends towards zero. On the other

Table 10.1 Stations in Kent

	I Distance in miles	II Fastest time	III Speed minutes per mile	IV Variation about mean speed (=100)	V New time 'distance'	VI Number of trains
Charing Cross	0	0	—	—	—	24
New Cross	5	13	2.6	113	5.65	16
St. Johns	5.5	18	3.3	144	7.9	8
Hither Green	7	17	2.4	104	7.3	16
Grove Park	9	21	2.3	100	9	16
Sundridge Park	10.25	28	2.7	117	12	8
Bromley North	10.5	30	2.9	126	13.2	8
Elmstead Woods	10.25	24	2.3	100	10.25	8
Chislehurst	11.25	26	2.3	100	11.25	8
Petts Wood	12.5	30	2.4	104	13	8
Orpington	13.75	21	1.5	65	8.9	12
Chelsfield	15.25	40	2.6	113	17.2	8
Knockholt	16.5	43	2.6	113	18.7	8
Dunton Green	20.5	49	2.4	104	21.3	8
Sevenoaks	22	31	1.4	61	13.4	16
Hildenborough	27	38	1.4	61	16.5	4
Tonbridge	29.5	42	1.4	61	18.0	8
			36.5			

16 stations — average time: 2.3 minutes.

hand, when the linear concentration is high, the units (stones or cars) become close together and their individual movement is impeded so that a jam occurs. The transport rate is now reduced to zero again. The curve cuts the x-axis at this point also, where q is zero. There must be a point on the curve where

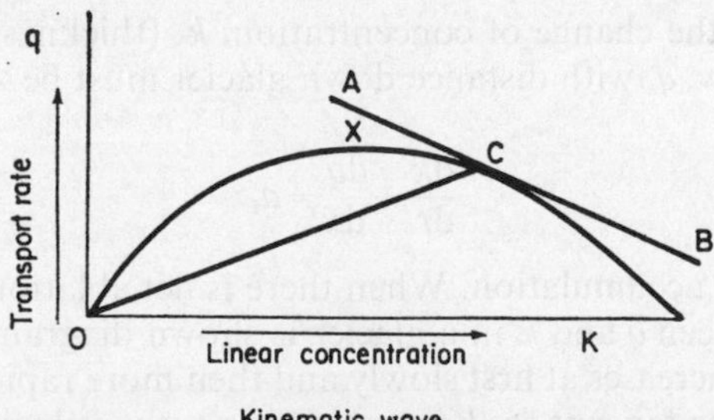

Figure 10.10 Kinematic wave for pebble transport in a stream or cars on a road. q transport rate is plotted against k the linear concentration. (After L. B. Leopold, M. G. Wolman and J. P. Miller.)

the q value reaches a maximum. This is the linear concentration that provides the most efficient and highest transport rate. At this particular value of k, the linear concentration, the transport rate remains constant and the tangent (AB) to the curve is horizontal at this point. The slope of the tangent, AB, to the curve gives the velocity of the kinematic wave. At the point X (figure 10.10) therefore the kinematic wave is stationary. As the linear concentration increases to the right of X, the slope of tangent AB increases, meaning that the kinematic wave increases in velocity and moves downstream or down road as the linear concentration increases. As the linear concentration decreases and the point moves to the left of X the slope of AB is reversed and the velocity of the kinematic wave decelerates and moves upstream or uproad. The velocity of the kinematic wave is independent of the rate of movement of the individual units, stones or cars. The velocity of the individual units is indicated by the slope of the chord OC in the figure. The velocity of each unit gradually decreases as the linear concentration increases, and it becomes zero when the units are jammed where the curve crosses the x-axis with a high linear concentration. Cars or stones stay further apart when they move faster, and when they crowd together they slow down. The speed of stones in a river therefore slows down as the concentration increases, which has been demonstrated experimentally with a single line of stones in a river model. Where stones or cars tend to slow down there will be increased accumulation.

The increased accumulation in a river bed will cause an increase in velocity gradient of the water. The increased velocity gradient will exert a greater fluid force on the stones and some will be moved until a steady state is set up. There is not such an automatic form of alteration with cars as each moves by itself and is not impelled by the water flow as in a river.

In a glacier the steady state may be disturbed by increased accumulation, which affects k, the thickness of the ice (J. F. Nye, 1963). If the accumulation does not change then the equation of continuity must be satisfied by the relation

$$\frac{dk}{dt}+\frac{dq}{dx}=0.$$

This means that the change of concentration, k, (thickness h) with time plus the change in flow, q, with distance down glacier must be zero. When there is net accumulation

$$\frac{dk}{dt}+\frac{dq}{dx}=a,$$

where a is the net accumulation. When there is net ablation a is negative. The relationship between q and k in a glacier is shown diagrammatically in figure 10.11. The flow increases at first slowly and then more rapidly as the thickness increases. At P the tangent to the curve OP is c, the velocity of the kinematic wave. This velocity, c, is given by $c=(dq/dk)x$. It is greater than u, the velocity of the ice. This velocity, u, is given by q/k. The bulge of ice of dimension q can, therefore, pass down the glacier more rapidly than the ice itself, as the slope of

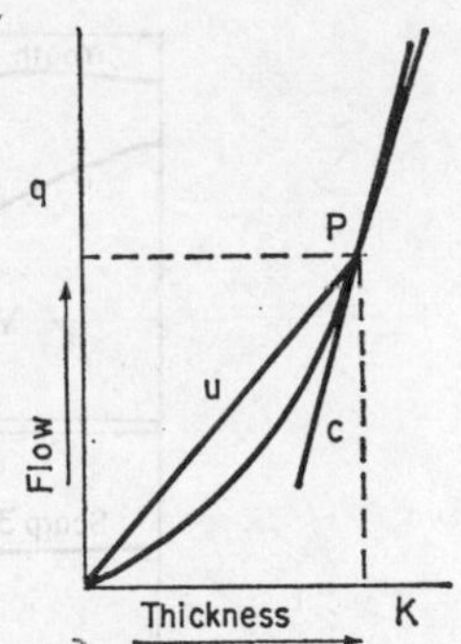

Figure 10.11 Kinematic wave in glacier flow, *u* indicates velocity of the ice and *c* the velocity of the wave at point P, see text. (After Nye.)

c is 3 or 4 times greater than that of *u*, (see figure 10.11). Sometimes if its upper part is extending and its lower part compressing a glacier reacts unstably to changes in accumulation. The instability inherent in the lower part can be stabilized by the arrival of the kinematic wave from up-glacier. The arrival of this kinematic wave at the snout can account for the greater response of the glacier snout to a small change of accumulation along its length. The black Rapids glacier in Alaska showed a sudden surge forward due to the formation of a kinematic wave, as the result of added accumulation. Its snout moved 3 miles in 5 months in 1936–7.

Kinematic waves apply to several types of 1-dimensional flow of interest to geographers. They illustrate well the application of theoretical concepts to very different problems. These problems range from traffic flow on roads to the response of glaciers to changes in climate.

10.7 Time and area

(a) *Physical geography*

In this section some examples of the relationship between time and area will be considered. In geomorphology a large amount of work has been devoted to the study of denudation chronology. Time is involved in this type of study, which aims to elucidate the stages of development of the landscape. Much of the evidence for the study is found in the flatter areas of the surface. Thus the stages of development are found in terms of erosion surfaces. The relationship between the erosion surfaces and the stages of development depend upon the processes operating to form the surfaces. There are at least two schools of thought concerning the processes of denudation. These are the Davisian concepts of peneplain formation by downwearing and the concept of pediplanation by backwearing. According to the Davisian concept a peneplain is formed by the slow lowering of the whole surface until an area of low relief results (figure 10.12*a*). This process does not often operate long enough to produce a single peneplain. Interruptions cause rejuvenation and many partial peneplains are often recognized, rather than one complete one. The several stages of development provide the evidence for the

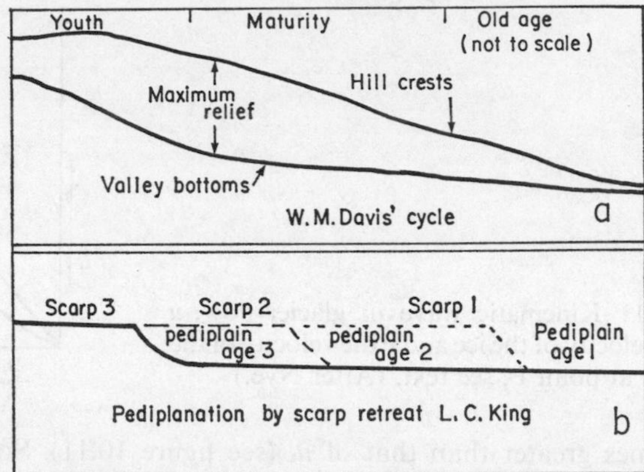

Figure 10.12 Diagrams to illustrate age relationships in W. M. Davis, cycle of erosion and peneplains formed by successive scarp retreat.

denudation chronology. Thus the surfaces in the landscape provide the time element in its development. The surfaces can be fixed absolutely in time if they contain dateable deposits. More often, however, the time scale can only be fixed relatively and not absolutely, according to the height and position of the different surfaces relative to one another. It is normally assumed that the highest are the oldest, but warping must be allowed for in some areas.

The concept of landscape development by pediplanation, advocated by L. C. King, (1962), implies a very different relationship between surface and time. In the peneplanation concept each stage or surface belongs to one time. This relationship does not hold, however, in pedimentation, which is achieved largely by scarp retreat. The growing pediment gradually extends laterally as the scarp is driven back. A surface formed by this process differs in age throughout its extent. It is older at its outer margins and becomes progressively younger as the scarp is approached (see figure 10.12*b*). The whole span of time during which the surface has developed may amount to tens of millions of years. A surface of this type cannot be given one date. Surfaces of similar age can also occur at very different elevations, if a series of scarps are retreating across the landscape.

Many smaller geomorphological features change position with time. The pattern of this change with time can sometimes be studied by plotting suitable variables against time. An example of movement through time is the migration of beach ridges southwards along the southern part of the Lincolnshire coast. This migration can be demonstrated by plotting the landward movement of the ridge crest against time (see figure 10.13). The ridge crests in this area diverge slightly from the coast southwards. Thus if the ridge moves bodily southwards, its crest at any one profile position appears to

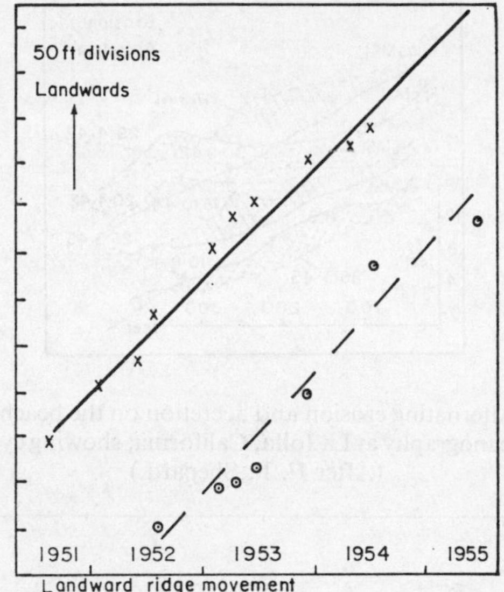

Figure 10.13 The landward movement of ridges on the foreshore of the Lincolnshire coast near Gibraltar Point is plotted against time.

move inland at a very regular rate, as shown on figure 10.13. In interpreting the graph of ridge crest position against time, time itself cannot be causally responsible for the ridge movement. The movement must depend on some other variable that is itself operating regularly through time. In this instance the longshore movement of beach material is probably the process responsible for the migration of the beach ridges southwards. Both the cause and the effect of this migration behave regularly with time in this area, to produce the observed results.

Some small features undergo continual change in extent in response to variations in the processes that form them. A series of beach profiles illustrates clearly the variability of the beach under the action of different waves and other processes. Figure 10.14 shows seasonal changes on a beach in California. It is a dynamic variability which consists of short term changes. There may also be longer term trends in some areas. The shorter and longer time elements can be differentiated and studied by superimposing a large number of profiles taken along one beach section over a period of time. The profiles will vary within certain limits. These limits can be defined by joining all the high points and all the low points by two profiles which form an envelope within which all the variation is contained. The zone between the profiles is called the sweep zone (see figure 10.15). Its width will depend on the mobility of the beach. The difference between sweep zones covering a number of years demonstrates the longer term beach trends. In south Lincolnshire the beach accretion is clearly demonstrated by this method

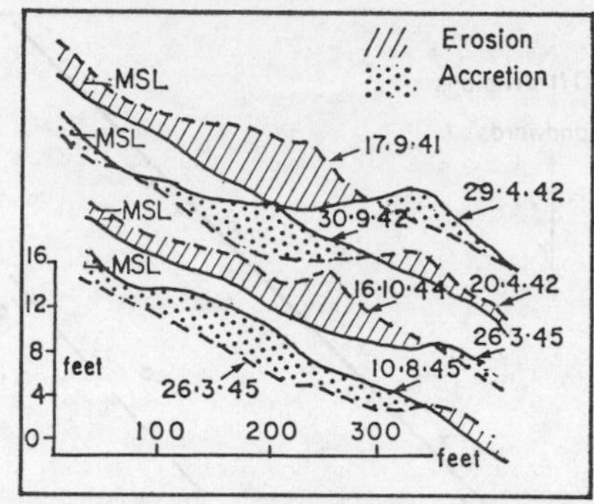

Figure 10.14 Alternating erosion and accretion on the beach at the Scripps Institute of Oceanography at La Jolla, California, showing cyclical changes. (After F. P. Shepard.)

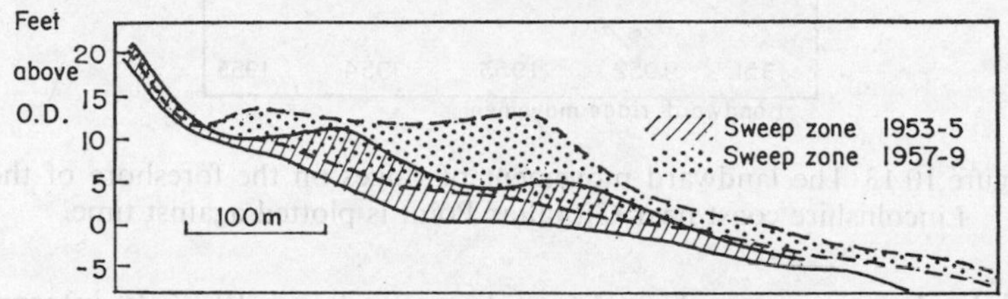

Figure 10.15 The sweep zone of the beach at Derby Avenue, Skegness in Lincolnshire to show the range of variation over the periods specified.

where the sweep zones cover a number of years survey from 1953 to 1966 (figure 10.15). Individual profiles which may be subject to short term variations do not demonstrate the trend of change nearly so clearly.

Another very important relationship between time and area is the diffusion of phenomena in time over area. A good example in physical geography is the spread of waves over the ocean from a storm centre. An interesting study of this type was made by N. F. Barber and F. Ursell (1948) by means of wave recordings at Pendeen in Cornwall. A series of long waves with a period of 18 seconds reached the coast of Pendeen at 19.00 hours on 30 June 1945. The waves reaching the coast became gradually shorter but higher. By 23.00 hours on 2 July the main wave period was 14 seconds. The period of the waves is determined by their velocity and length in deep water. It is, therefore, possible to calculate the distance covered by the waves in a given time if their period and time of arrival is known. The wave paths can be extended backwards in time and space according to their velocity as shown on figure 10.16. The position where the longer and shorter waves meet is the centre from which the waves emanated.

In the example quoted the wave traces converged at a point 2700 miles from Cornwall at 1800 hours on 26 June. Weather charts for this time show a

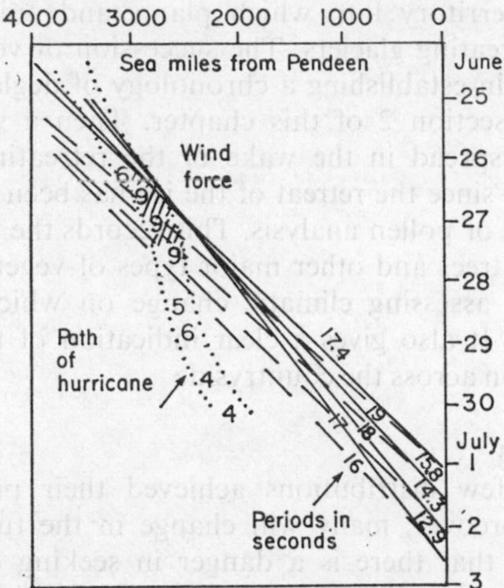

Figure 10.16 The propagation of waves across the Atlantic Ocean to Pendeen generated by a hurricane, 3000 miles away. (After N. F. Barber and F. Ursell.)

hurricane situated off Cape Hatteras about 2500–3000 miles from Cornwall. The waves generated by the force 9 winds blowing at this time in this position then moved out across the Atlantic in all directions. Some of these waves eventually reached Cornwall, the faster, longer waves taking 4 days to make the journey and the slower, shorter waves taking 5½ days to cover the same distance. The diffusion of ocean waves has been traced over 7000 miles in the Pacific Ocean. (It should be noted that the wave trains travel at half the actual wave velocity.)

A very interesting pattern of diffusion in physical geography is the spread of plants and animals. Virgin territory into which plants and animals can spread can be created in a number of ways. New land may be built up by the sea; the colonization of salt marshes and dune ridges illustrates this situation. Volcanic eruptions can destroy all life and leave foundations of readily weathered material onto which plants and animals can spread. When Krakatoa erupted in 1883 all trace of life was destroyed. When the island was first visited in 1886 mosses, algae and ferns were already established. Their spores must have been blown to the island by the wind. Other plants had been carried as seeds across the sea and become established on the shore. Some grasses had already begun to spread inland. Already by 1897 more than 60 species of flowering plants had spread to the island and become established and by 1906 the vegetation formed a closed community. Amongst the animals to return to the island, insects and spiders were the first; they were followed by butterflies and other creatures brought by the wind. Birds followed the insects and they would help to bring the seeds of plants. Small reptiles and mammals could come on driftwood.

Another barren territory into which plants and animals can spread is that left bare by retreating glaciers. The succession of vegetation starts with lichens. Their value in establishing a chronology of deglaciation has already been mentioned in section 2 of this chapter. Then a whole succession of plants and animals spread in the wake of the retreating ice. The gradual change of vegetation since the retreat of the ice has been admirably recorded in the detailed study of pollen analysis. This records the percentage values of the dominant forest trees and other major types of vegetation. It provides a valuable method of assessing climatic change on which the vegetation is basically dependent. It also gives a clear indication of the character of the diffusion of vegetation across the countryside.

(*b*) *Human geography*

It is obvious that few distributions achieved their present spatial form instantaneously. Moreover, many will change in the future. Less obvious, perhaps, is the fact that there is a danger in seeking causal relationships between two distributions when one of them may have had a very different spatial arrangement in the past from now. The purpose of this subsection is to note some of the features of change of distributions during time in human geography, to draw attention to work done, and to discuss difficulties.

The topic under consideration is the diffusion or contraction in space during time of some set of objects or simply of an idea, such as a technique. The space may be a set of points, lines, a 2-dimensional surface, or 3-dimensional. Only 2-dimensional cases will be discussed here, but much of what is said may be applied to other dimensions.

The diffusion may involve a set of physical objects such as people, the potato plant, or disease carrying media; techniques such as coke smelting replacing charcoal smelting; or opinions such as the decision to vote Communist. The spread may start from a point (e.g. a locality where a new mutation appears, or a town), from a line (e.g. settlers moving in from a coast), or from several or many localities, lines or patches simultaneously. The medium over which the spread takes place may be a continuous surface, a network of lines, or no more than points scattered over the surface.

The spread may take place (relatively) slowly, or very rapidly. The diffusion of types of Neolithic implements in Europe presumably took centuries. With the help of modern communications, many ideas spread widely in a matter of a few years or even a few days or hours, in the case of news. The spread may be spontaneous, or it may be consciously fostered (e.g. by advertising). It is possible, also, to study the reverse process, the contraction of an idea, but less work has been done on this than on outward spread.

The whole process of diffusion is complicated by the fact that in the real world many barriers, both physical obstacles such as mountains and rivers, and cultural and other ones, such as language or religious differences, often make the shape of diffusions very distorted. Some barriers act as sieves, allowing only certain things across. Thus the relatively narrow Bering Strait, across which it is thought considerable movement of people and techniques has taken place between Asia and the Americas, lies so far north that most agricultural plants could not be cultivated near it, and presumably therefore could not be taken across by the slow moving peoples of the past.

Barriers may be roughly classified as follows:

PHYSICAL (Water, slope, etc.)	COMPLETE	REBOUNDING
MAN-MADE {Material, Cultural}	PARTIAL {DISCONTINUOUS, SIEVE}	NOT REBOUNDING

They may affect physical objects or human activities.

Some geographers, including particularly T. Hagerstrand (1952, 1965), have attempted to find processes and regularities behind diffusion patterns from empirical studies. Some, also, have attempted to generate diffusion patterns. Random diffusion models (see chapter 11) have used appropriate criteria for determining the distance and direction of diffusion. Obstacles have been placed in the medium over which the diffusion is taking place. Diffusion over parts or all of the earth's surface has also been extensively studied by biologists, archaeologists, and physicists (for example, nuclear fall out in the atmosphere). The reader is referred in particular to papers by W. T. Williams and J. M. Lambert in the *Journal of Ecology*, 1959–1962.

Some examples of diffusion will now be discussed briefly. A relatively early attempt by a geographer to formalize the study of diffusion was made in a paper by T. Hagerstrand (1952). He studied the distribution of car ownership in the extreme southern part of Sweden from the time when cars were first used there. Starting from a time when distribution was almost non-existent, he found that after some years, the index of car ownership per population was much higher in the western part of the area, facing Copenhagen in Denmark, than elsewhere. This difference could not be attributed to areal differences in income. For some time, this imbalance could be noted, but after about two decades, car ownership was comparable throughout the area. Local differences were now related to income, urban/rural and other such differences. In his paper, Hagerstrand made the point that the car idea seems to have spread over his area of study during a period of about two decades, having reached Sweden from Germany via Copenhagen. He pointed out that the idea spread as a wave from west to east through his area. At any given short period during this time, *more* people were obtaining cars in the particular area through which the wave was passing than elsewhere.

This most interesting idea has been followed up by many others, as well as by Hagerstrand himself. E. M. Rogers (1962), has assembled a great deal of material. Information about a specific situation has to be revised as circumstances change. Thus a new idea might spread more rapidly in southern Sweden now than early in this century, but faster then, presumably, than some centuries ago.

There seems considerable evidence that political opinion, voting patterns and allegiance to a political party have often spread out from a few centres. Locally, for example, F. Lägnert (1952), has found this to have happened in part of Sweden. On a larger scale it seems that the present (uneven) distribution of communist voters in Italy and in Japan, to take two examples, is part of an innovation wave. The wave has spread from certain centres, not regularly over time, however, but at partiular periods when conditions were propitious (e.g. 1945–48 in Italy). The Italian communist voters are very much concentrated in north–central Italy (Emilia, Toscana). Their distribution in Italy does not correlate with income, industrialization or other variables popularly associated with communism. It may to some extent be related to land tenure and/or religious considerations; but it looks very much like the product of a successful diffusion of the idea out of Bologna, Firenze and some other towns where the communist idea caught on in the early 1920s. Similarly, Japanese communist voters are heavily concentrated in Tokyo and Osaka, presumably the spearheads of communist propaganda after the second world war, but also the most prosperous parts of Japan.

Other examples of diffusion that have received or may in the future receive the attention of the geographer may be noted:

(i) *World level*: the spread of religions, agricultural and industrial techniques, peoples: for example Christianity, printing, Polynesian peoples over the Pacific. At the IBG Annual Meeting in Sheffield (January 1967), R. G. Ward described the construction of a model to simulate the movement of islanders by sea in order to find how often lost boats would reach islands randomly. This is an excellent example of a situation, often found, where a diffusion process cannot be studied on available evidence alone and must be reconstructed.

(ii) *Regional level*: the cultivation of coffee in São Paulo state, Brazil. This left a 'hollow' distribution behind it, since the wave itself exhausted the land it moved over. Maps of coffee distribution in selected years from 1836 to 1950 shows the diffusion in A. França's (1960) *A marcha do café e as frentes pioneiras*. The spread of a certain kind of woodworm over furniture and wooden building materials has not escaped the notice of the firm of Rentokil, whose product is designed to eliminate the insect. Their map of Britain shows the percentage of houses contaminated by woodworm in different regions.

(iii) *The spread of an idea locally*: of a new gadget through a housing estate, of a new colour for painting doors of houses, farmers taking up new ideas from neighbours.

Figures 10.17 and 10.18 are intended to summarize diagrammatically some of the situations discussed in this subsection.

Figure 10.17a shows the diffusion of objects over a clearly defined compartment during time. The objects enter at a point. At time 0, none have entered. At time 1, there is a concentration of those that have entered by the entrance. By time 3, no more are entering, and those in the compartment have

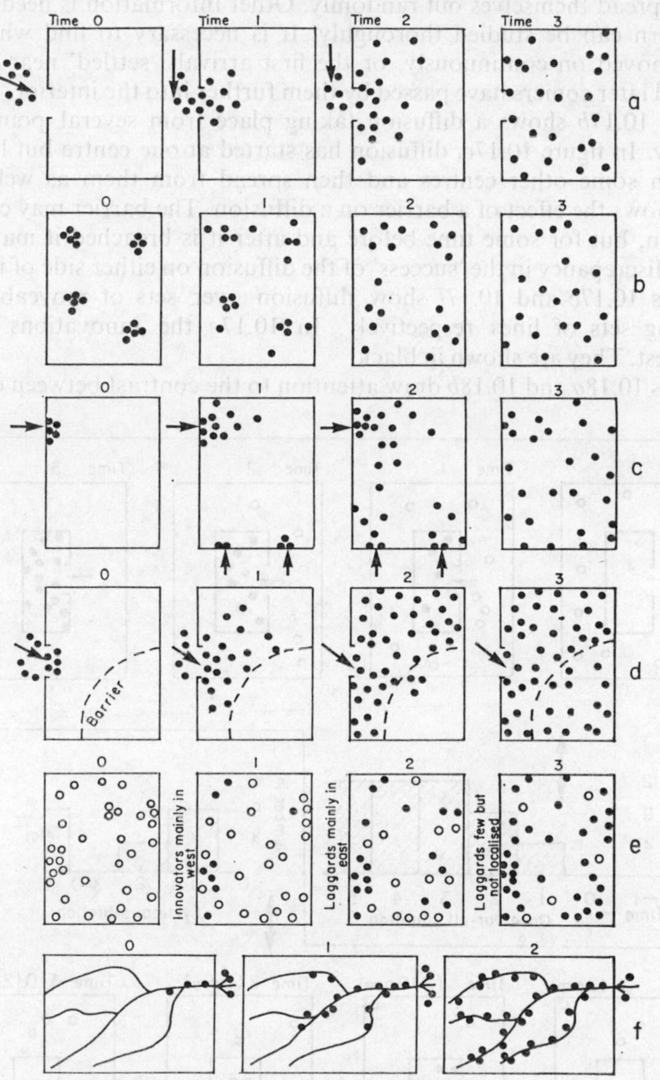

Figure 10.17 Cases of diffusion. (*a*) Diffusion of objects over a patch from one locality. (*b*) Simultaneous diffusion from several localities over a patch. (*c*) Diffusion from one locality, then subsequently from others. (*d*) Diffusion from a locality, hindered temporarily by a line barrier. (*e*) Diffusion spreading from west to east (e.g. Hagerstand's car idea), see text. (*f*) Diffusion of objects along lines (e.g. colonization with settlements exclusively along the lines of penetration, rivers).

already spread themselves out randomly. Other information is needed before the pattern can be studied thoroughly. It is necessary to find whether the objects moved on continuously, or the first arrivals 'settled' near the entry point and later comers have passed by them further into the interior.

Figure 10.17*b* shows a diffusion taking place from several points simultaneously. In figure 10.17*c*, diffusion has started at one centre but later been started in some other centres and then spread from them as well. Figure 10.17*d* shows the effect of a barrier on a diffusion. The barrier may eventually be broken, but for some time before and after it is breached it may cause a marked discrepancy in the 'success' of the diffusion on either side of it.

Figures 10.17*e* and 10.17*f* show diffusion over sets of moveable points and along sets of lines respectively. In 10.17*e* the innovations come in on the west. They are shown in black.

Figures 10.18*a* and 10.18*b* draw attention to the contrast between diffusions

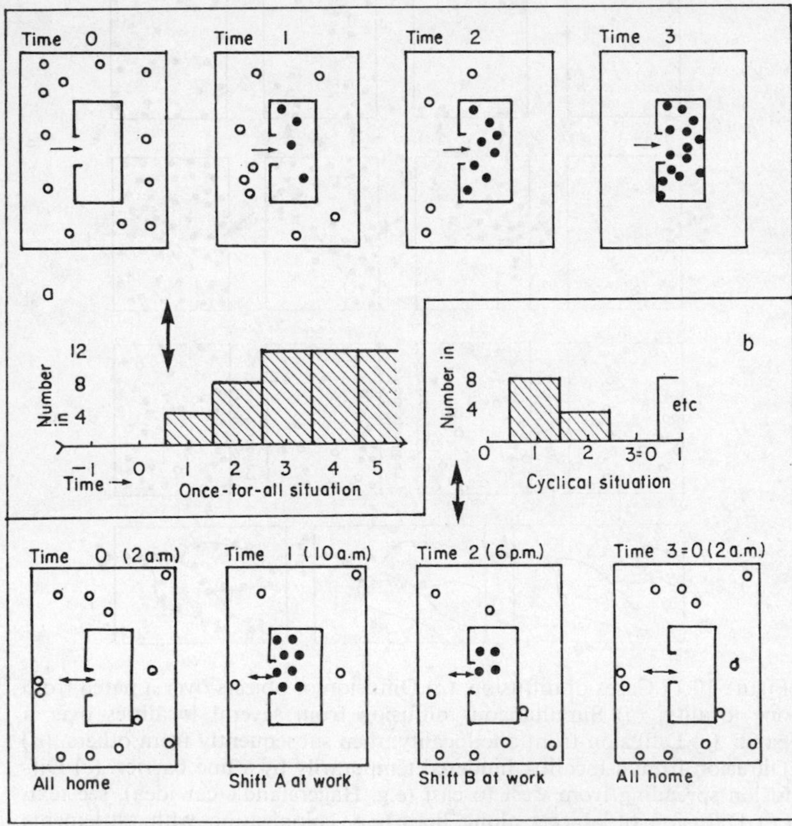

Figure 10.18 Contrast between a once-for-all movement and a cyclical diffusion and convergence. (*a*) Dots all move into area and stay. (*b*) Workers alternate in shifts between home and work.

that take place once-for-all and diffusions which recur cyclically as a process of diffusion and contraction. In the first instance, illustrated by Hagerstrand's car idea in southern Sweden (figure 10.18*a*), there is a time (pre time 0) when the diffusion had not started and perhaps could not have started. There is also a time when the diffusion no longer continues. It might be convenient to consider the period of diffusion as one unit of time (see figure 10.18*a*). Figure 10.18*b* shows workers in a factory or office at time 0. Time 1 catches them travelling home, time 2 at home and time 3 returning. The organization of this commuting pattern resembles finite arithmetic discussed in chapter 2.4, because time 4 would become the same as time 0.

10.8 Cycles in time

Time cycles play an important part in many aspects of geography. Some of these cycles are short term, some long, some are reliable and regular, others are less dependable. Amongst the regular cycles are those associated with the earth as a planet.

(*a*) *Planetary cycles*

The cycles of day and night, the tides and the seasons are very regular although many of these cycles can vary in intensity. Extra high tides can be caused by surges added to the normal tide and the seasons can vary greatly in intensity in many areas. Cold winters can be followed by mild ones, although the seasonal cycle is never lost.

There are other longer term cycles associated with the planetary movements of the earth. Some of these cycles have been used by M. Milankovich

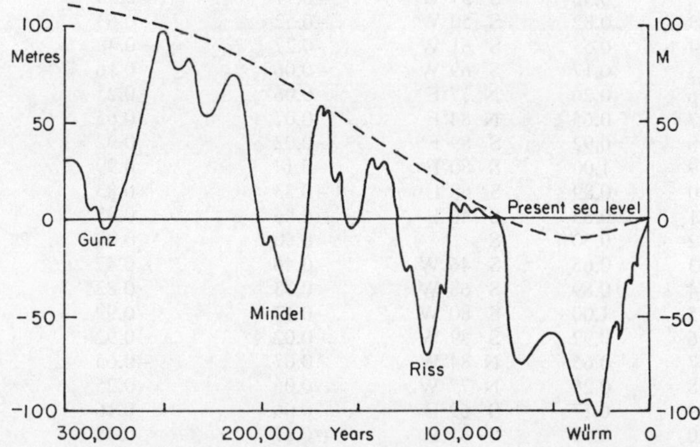

Figure 10.19 Fluctuations of climate and sea-level with time during the Glacial period, showing a trend superimposed on the fluctuations so that each successive lowering of sea-level was greater than the last. (After R. W. Fairbridge.)

(1920) in his attempt to explain the occurrence of alternating glacial and interglacial climatic phases during the Pleistocene. The cyclic pattern of glaciation is shown in connexion with sea level change in figure 10.19. Milankovich takes three main planetary oscillations into account. The first is the obliquity of the ecliptic (see glossary) with a period of 40,400 years. During this cycle the angle of the ecliptic varies from 21° 58′ to 24° 36′. The second is the wandering of the perihelion, with a period of 20,700 years. The third is the eccentricity of the orbit which varies with a period of 91,800 years from 0.00331 to 0.0778. Milankovich combined these three cycles and allowed for their probable effect on the climate. In this way he calculated the probable cycle of cooler and warmer phases. The results of the calculations do bear some resemblance to the glacial and interglacial succession.

(*b*) *Analysis of cycles*

A simple example of harmonic analysis of tidal forces illustrates one method of dealing with cyclic phenomena. The lunar tractive tide producing forces depend on the latitude of the station and the declination of the moon. The forces for one lunar day for 30°N with lunar declination 15°N can be listed as follows:

Table 10.2 Data from Admiralty Manual of Tides

Hour	Intensity	Direction	*N* Component	*E* Component	Hour
1	1.00	S 20°W	−0.94	−0.34	1
2	0.95	S 37°W	−0.77	−0.58	2
3	0.82	S 50°W	−0.52	−0.63	3
4	0.55	S 61°W	−0.27	−0.48	4
5	0.17	S 69°W	−0.06	−0.16	5
6	0.26	N 77°E	0.06	0.25	6
7	0.65	N 84°E	0.07	0.65	7
8	0.92	S 89°E	−0.02	0.92	8
9	1.00	S 80°E	−0.17	0.99	9
10	0.89	S 68°E	−0.33	0.83	10
11	0.65	S 46°E	−0.46	0.47	11
12	0.50	S	−0.50	0.00	12
13	0.65	S 46°W	−0.46	−0.47	13
14	0.89	S 68°W	−0.33	−0.83	14
15	1.00	S 80°W	−0.17	−0.99	15
16	0.92	S 89°W	−0.02	−0.92	16
17	0.65	N 84°W	0.07	−0.65	17
18	0.26	N 77°W	0.06	−0.25	18
19	0.17	S 69°E	−0.06	0.16	19
20	0.55	S 61°E	−0.27	0.48	20
21	0.12	S 50°E	−0.52	0.63	21
22	0.95	S 37°E	−0.77	0.58	22
23	1.00	S 20°E	−0.94	0.34	23
24	1.00	S	−1.00	0.00	24

Table 10.3 Tidal data analysis

Northerly Component	Hrs. 0–11	Hrs. 12–23	Semi-diurnal 0–11	Diurnal 0–11	Easterly component Hrs. 0–11 (*a*)	Hrs. 12–23 (*b*)	$\frac{a+b}{2}$ Semi-diurnal	$\frac{a-b}{2}$ Diurnal
−.35 deducted	−0.65	−0.15	−0.40	−0.25	0.00	0.00	0.00	+0.00
	−0.59	−0.11	−0.35	−0.24	−0.34	−0.47	−0.40	+0.06
	−0.42	0.02	−0.20	−0.22	−0.58	−0.83	−0.70	+0.12
	−0.17	0.18	0.00	−0.18	−0.63	−0.99	−0.81	+0.18
	0.08	0.33	0.20	−0.13	−0.48	−0.92	−0.70	+0.22
	0.29	0.42	0.36	−0.07	−0.16	−0.65	−0.40	+0.24
	0.41	0.41	0.41	0.00	0.25	−0.25	0.00	+0.25
	0.42	0.29	0.36	0.07	0.65	0.16	+0.40	+0.24
	0.33	0.08	0.20	0.13	0.92	0.48	+0.70	+0.22
	0.18	−0.17	0.00	0.18	0.99	0.63	+0.81	+0.18
	0.02	−0.42	−0.20	0.22	0.83	0.58	+0.70	+0.12
	−0.11	−0.19	−0.35	0.24	0.47	0.34	+0.40	+0.06

Tables 10.2 and 10.3 show how the complex curve, shown in figure 10.20*a*, is split first into its northerly and easterly components. Then each of these components in turn is split into their semi-diurnal and diurnal harmonic components, as shown in the two lower tables. The separate and combined curves are shown in graph form in figure 10.20*b*. This method of analysis is suitable for dealing with the simple situation considered, in which there are

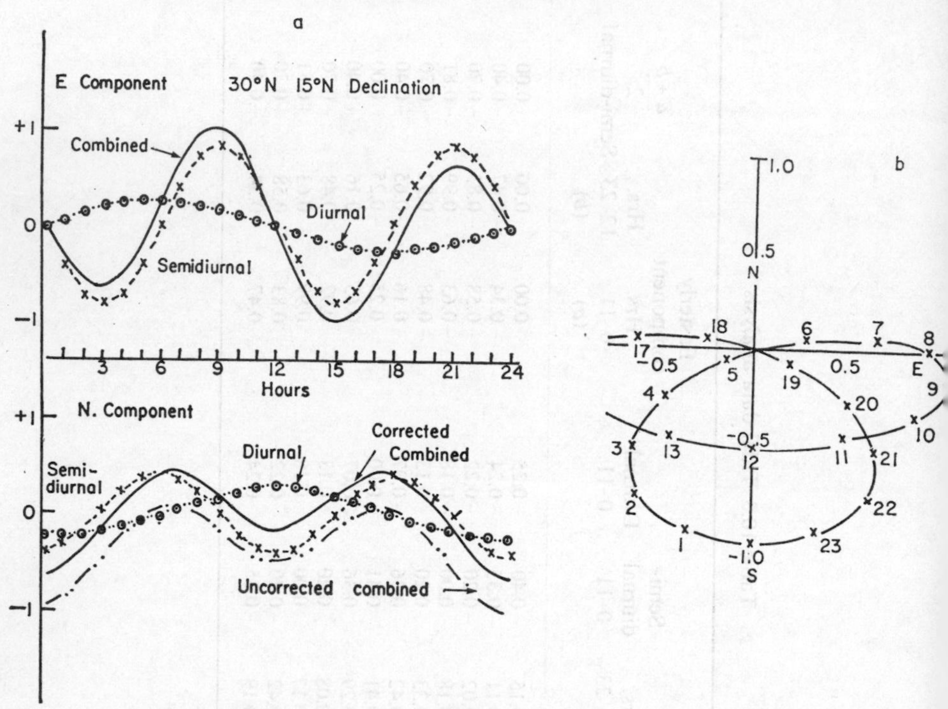

Figure 10.20 The analysis of the tractive tide producing forces for a point at 30°N with a lunar declination of 15°N. The force is divided into its easterly and northerly components. Each of these is divided into simple harmonic diurnal and semi-diurnal components.

only two smooth harmonic constituents. More elaborate methods of analysis are required to deal with the similar problem of analysis of wind wave records. Fourier analysis can be carried out mechanically in order to pick all the constituent periods out of a complex wave record. The type of record that can be analysed in this way is illustrated in figure 10.21, which shows the original wave record and the fourteen component waves from which it is built. The upper diagram illustrates the proportion of energy in each of the wave bands or periods that are incorporated in the complex wave spectrum.

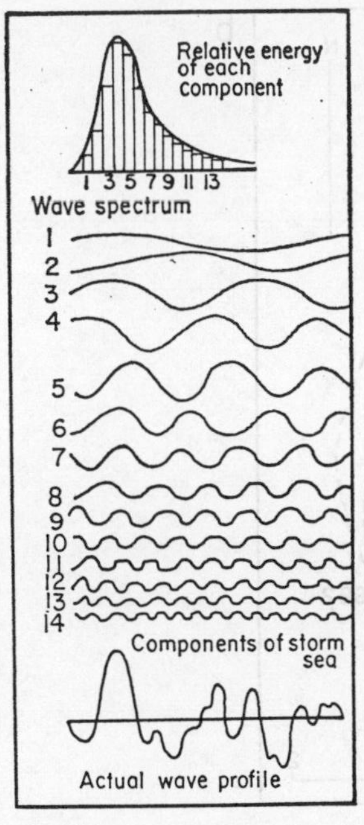

Figure 10.21 An actual wave profile shown at the bottom is divided into its 14 components, to form a wave spectrum. The relative energy in each of these components is shown in the upper diagram.

(c) *Geomorphological cycles*

The best known cycle in geomorphology is the cycle of erosion of W. M. Davis. He postulated, by deductive reasoning, that the landscape passes through the stages of youth, maturity and old age during the cycle of erosion. A cycle of this type implies a closed system in which the available energy in the system is gradually used up as the cycle runs its course. Davis considers that a complete cycle would occupy most of Tertiary time. D. L. Linton (1957), however, considers that in areas of moderate relief at least two cycles could run their course during this period. The whole validity of the cycle concept is based on two assumptions, first, that little erosion is achieved during uplift, and secondly, that a steady state is not set up. Change during the operation of the cycle is assumed to be progressive in the absence of interruptions. Some geomorphologists consider that, under some conditions at least, a steady state of dynamic equilibrium, which resembles an open system, can be set up. This situation would not lead to the progressive change necessary for the cycle concept.

Another example of a much shorter term geomorphological cycle is the growth and decay of Spurn Head south of Holderness. G. de Boer (1964) has shown that the sand and shingle spit goes through a cycle of growth and decay that, in the absence of human interference, lasts about 250 years (see figure 10.22). The cycle is related to the steady coastal erosion which cuts back the land at the root of the spit. This makes the spit progressively more curved and vulnerable to wave attack until it is eventually breached near its

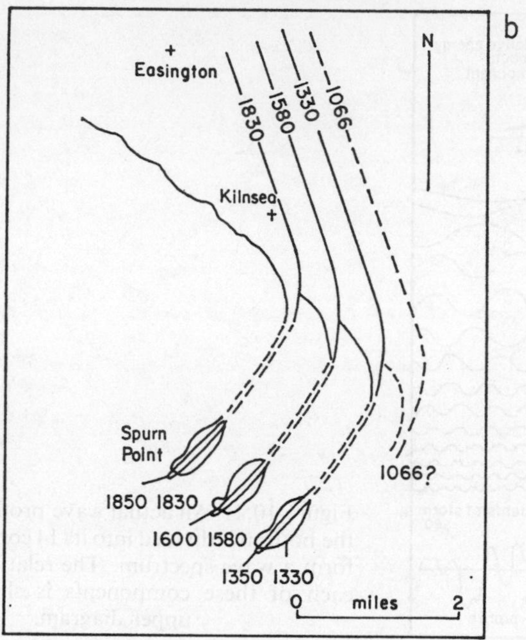

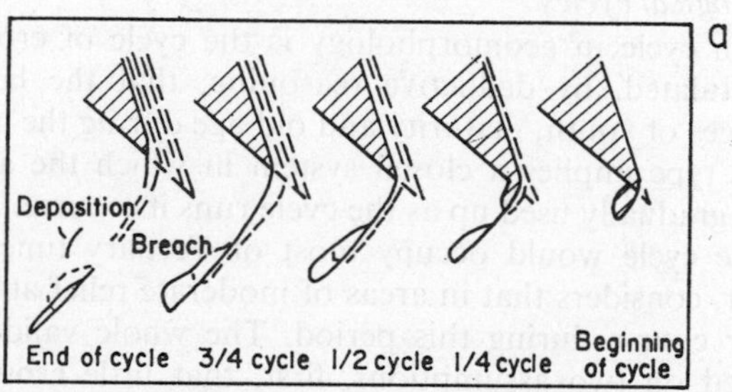

Figure 10.22 The cyclic development of Spurn Head, Yorkshire, is illustrated. The relationship between erosion and accretion is shown in the diagram of the cycle (*a*). Diagram (*b*) shows three recorded positions of Spurn Head at different dates. (After De Boer.)

root. The isolated end of the spit is then transferred across the Humber or offshore and a new spit begins to grow from the angle of the land in a southerly direction. It gets gradually longer until the same process of coastal erosion renders it vulnerable once more. There is historical evidence for at least three different Spurn Head. The last breaching occurred in the middle of the nineteenth century. The normal operation of the cycle, however, was disturbed because the breach was artificially healed. The coast at the root of the spit is now suffering considerable erosion.

(*d*) *Biological cycles*

There is an interesting four year cycle to which many small northern mammals are subject. This cycle is brought about by overcrowding due to the rapid rate of reproduction. The lemming migrations are the best known example. These cyclic movements have been described since the sixteenth century. The lemmings must either move to a new habitat or die of starvation due to overcrowding. The overcrowding produces a type of shock disease which makes the lemmings unnaturally pugnacious and excited and ready to move great distances. When the pressure of numbers is relieved then the lemmings return to their normal quiet character. The cause of the four year cycle may be in part related to a periodicity of this length in the size of the crop of the Norway Spruce cones. These provide an item of diet for the species of birds and mammals that follow this four year cycle. Other plants also seem to operate in phase with the Spruce, such as birch, oak and crowberries. This inter-relationship illustrates the unity of the natural habitat and shows how a cycle in the plant element can have repercussions in the animal element of the biogeography of the area. Both elements show the same cyclic pattern.

There are many more obvious biological cycles dependent upon the animal cycle and the seasons, which affect all plant and animal life in some way. The seasonal rhythm is particularly strongly felt in the higher latitudes where the contrast between winter and summer is great. Equally important is the contrast between wet and dry seasons where these are strongly differentiated as in monsoon areas.

10.9 Time series statistics

A number of statistical tests are particularly applicable to time series. Great care, however, must be used in analysing time series and it is essential that the physical basis of the series be fully understood. A number of methods of dealing with time series will be mentioned. These include (*a*) smoothing time series (running mean), (*b*) calculating trend of time series data, (*c*) testing for cycles in time series (seasonal element), (*d*) testing for randomness in time series, (*e*) correlation test for two time series, (*f*) logistic curve, and (*g*) frequency of occurrence of rare events (Poisson distribution).

(*a*) *Smoothing time series*

Many data that are measured on a time scale show large fluctuations between successive values. For example, precipitation measured on a daily, monthly or annual basis, shows very considerable variation from one value to the next in a climate such as that of eastern U.S.A. or of Britain. The short term variations may be a significant part of the series. On the other hand, the long term variations may be more important in some respects. In the latter situation it is very useful to smooth out the values by preparing a running mean. The running mean includes two or more periods. To calculate a five month running mean, the first five monthly values are averaged, then the second to sixth, the third to seventh and so on through the whole series. This process smooths out the major discontinuities in the series. The example gives the five-month running mean for rainfall for 1964.

Table 10.4 Precipitation data

Month	ins	Total	Mean	
Jan.	0.90			
Feb.	0.92			
Mar.	3.57	9.54	1.91	Jan.–May
Apr.	2.26	11.56	2.31	Feb.–June
May	1.89	13.01	2.60	Mar.–July
June	2.92	11.56	2.31	Apr.–Aug.
July	2.37	10.01	2.00	May–Sep.
Aug.	2.12	9.85	1.97	June–Oct.
Sept.	0.71	8.38	1.68	July–Nov.
Oct.	1.73	9.02	1.80	Aug.–Dec.
Nov.	1.45			
Dec.	3.01			

Nearly all published climatic statistics, especially those concerned with climatic change, are shown as 5, 10 or 40 year running means. This method of showing variation of time series makes the longer term trend of the data more readily discernible.

(b) Calculating the trend of time series

On figure 10.23 the approximate values of mean sea-level are shown biennially for the period from 1940 to 1960 (data modified from R. W. Fairbridge, 1961). The values are also given in table 10.5.

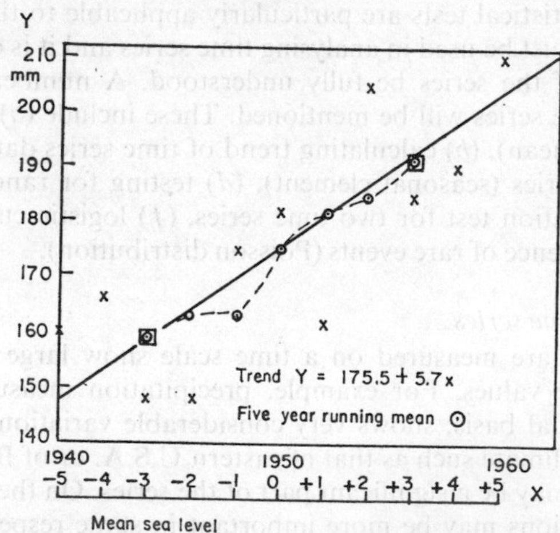

Figure 10.23 Biennial sea-level records are shown by crosses. The circles show the 5 year running mean. The squares indicate the trend as calculated by semi-averages. The line gives the calculated trend of the data and its equation is $Y = 175.5 + 5.7X$.

Table 10.5 Sea-level data

date	*Y* sea level in mm	*X*	X^2	*XY*	5 year running mean
1940	160	−5	25	−800	
1942	166	−4	16	−664	
1944	148	−3	9	−492	159.4
1946	148	−2	4	−296	163.8
1948	175	−1	1	−175	163.0
1950	182	0	0	0	174.8
1952	162	+1	1	+162	181.8
1954	205	+2	4	+410	184.8
1956	185	+3	9	+555	190.4
1958	190	+4	16	+760	
1960	210	+5	25	+1050	
	$\Sigma Y = 1931$	$\Sigma X = 0$	$\Sigma X^2 = 90$	$\Sigma XY = +510$	
	Semi-averages	1940–1948		159.4	
		1952–1960		190.4	

The trend of the data can be calculated either by using the semi-averages or by the method of least squares. The latter is the more exact method and the former the quicker. The semi-averages are found for the first half of the period (in this example 1940–1948) and for the second half (1952–1960). These points are plotted on the graph against 1944 and 1956 respectively and joined to give the trend line. These values are the first and last of the 5-year running means, which are also given in table 10.5.

The least square method of calculating the trend line is also given in the table. The *X* values are entered symmetrically both in the table and on the graph. X^2 is calculated and the sum of X^2 is found. If there are an even number of observations then 0.5 values for *X* must be used to maintain symmetry. The product of *XY* is entered for each value and the algebraic sum calculated. The equation for the trend line which best fits the values is given by $Y = a + bX$, where *a* is the sum of *Y* divided by *n*, the number of values ($a = \Sigma Y/n$) *b* is given by the sum of *XY* divided by the sum of X^2, ($b = \Sigma XY/\Sigma X^2$). In the example given

$$a = \frac{1931}{11} = 175.5 \quad b = \frac{+510}{90} = 5.7$$

therefore $Y = 175.5 + 5.7X$. This trend line is entered on figure 10.23. The line lies very close to that found by joining the semi-average values, and the 5-year running mean values mostly lie close to the trend line as well. The distance of the points from the trend line is a measure of the other elements in the time series; these are seasonal, cyclic and error elements. A method of calculating and removing the seasonal element will be considered.

(*c*) *Testing for cycles* (*seasonal element*) *in time series*
Before testing for cyclic elements it is necessary to remove the trend value. In the example already used this is done as follows. The trend increment for each 2-year period is equal to *b* or 5.7 in the example. The trend component for each 2-yearly observation is:

Table 10.6 Trend of sea-level data

			Trend value	Observed value	Cyclic and error terms
1940	-5×5.7 =	$-28.5+175.5$ =	147.0	160	+13.0
1942	-4×5.7	-22.8	152.7	166	+13.3
1944	-3×5.7	-17.1	158.4	148	−10.4
1946	-2×5.7	-11.4	164.1	145	−19.1
1948	-1×5.7	-5.7	169.8	175	+5.2
1950	0×5.7	0	175.5	182	+6.5
1952	1×5.7	$+5.7$	181.2	162	−19.2
1954	2×5.7	$+11.4$	186.9	205	+18.1
1956	3×5.7	$+17.1$	192.6	185	−7.6
1958	4×5.7	$+22.8$	198.3	190	−8.3
1960	5×5.7	-28.5	204.0	210	+6.0

It will be shown later that this sequence of values (cyclic and error terms) are randomly distributed. Therefore this example is not suitable for cyclic testing. Another example which has a very definite cycle, but no trend, will be considered. There is no need, therefore, to eliminate the trend element. The data

Table 10.7 Tidal height data

1 1st	2 VHST (specific seasonal	3 2nd	4 NT SS	5 3rd	6 NST SS	7 4th	8 NT SS	9 Σ 1 to 4	10 mean 1 to 4
22.6	118	17.3	90	21.1	109	15.9	83	76.9	19.22
23.6	123	16.6	86	21.2	110	15.6	81	77.0	19.22
23.8	124	16.0	84	20.9	109	16.0	84	76.7	19.17
23.4	122	16.0	84	20.3	106	16.7	88	76.4	19.10
22.4	118	16.1	85	20.1	106	17.5	92	76.1	19.00
21.1	111	16.3	86	20.8	110	17.7	93	75.9	18.97
Σ SS −	716		515		650		521		
Typical seasonal $=\dfrac{\Sigma \text{SS}}{n}$									
	119		86		108		87		Total = 400

that are analysed consist of the alternating spring and neap tide heights at high water at Immingham. In the series of values a very high spring tide is followed by a neap tide and this by a less high spring tide and another neap tide. The values then show a four element cycle. This is sometimes called the seasonal effect, as it is very obvious in many climatic or other data in which the four-seasonal influencc is strong. Seasonal influence applies to data in both human and physical geography. The example given is a simple one for illustrative purposes only, although it could show interesting longer term fluctuations in sea-level if continued over a longer period. The data are set out in table 10.7. The ninth column is the sum of the 4 values in the first, third, fifth and seventh columns and the tenth column is their mean. The second, fourth, sixth and eighth columns are the percentage values of each adjacent reading to the left in terms of the mean (col. 10). e.g.

$$\frac{22.6}{19.22}\times 100=118 \quad \text{and} \quad \frac{17.3}{19.22}\times 100=90$$

These values are called specific seasonals. The typical seasonals are found by summing the specific seasonals and dividing by n (6). The values for each entry, adjusted for the seasonal effect, are now found by dividing each value by the appropriate typical seasonal and multiplying by 100. Thus the first row are found as follows:

$$(1)\ \frac{22.6}{119}\times 100=19.00$$

$$(2)\ \frac{17.3}{86}\times 100=20.05$$

$$(3)\ \frac{21.1}{108}\times 100=19.6$$

$$(4)\ \frac{15.9}{87}\times 100=18.3$$

These values are entered in the sixth column of table 10.8 where they may be compared with results obtained by using the four unit running mean method shown in column 5. The four unit mean is used to take account of the four unit seasonal cycle.

The data are entered in the first column of table 10.8. The sum of 4 values is entered opposite the third and the mean of the 4 values is entered in the third column. The deviation between this running mean and each value is entered in the fourth column. These values are collected for each unit of the cycle in table 10.9. They are summed and the mean value found. This mean value for each seasonal unit is then added or subtracted according to its sign to give the corrected value on the 5th column of the previous table. It will be seen that these results closely approximate to those found by the alternative method, which are given in the 6th column of table 10.8.

Table 10.8 Season cycles in tidal data

1 x	2 T	3 Av.	4 Deviation	5 Corrected for seasonal cycle by running mean	6 Corrected for seasonal cycle by typical seasonal
22.6	75.9	18.97	+3.63	18.86	19.0
17.3	77.1	19.27	−1.97	19.97	20.0
21.1	76.9	19.22	+1.88	19.48	19.6
15.9	77.9	19.47	−3.57	18.39	18.3
23.6	77.2	19.30	+4.30	19.96	19.8
16.6	77.3	19.32	−2.72	19.27	19.3
21.2	77.0	19.22	+1.98	19.58	19.6
15.6	77.2	19.30	−3.70	18.09	18.0
23.8	76.6	19.15	+4.65	20.06	20.0
16.0	76.3	19.07	−3.07	18.67	18.4
20.9	76.7	19.17	+1.73	19.28	19.5
16.0	76.3	19.07	−3.07	18.49	18.4
23.4	77.3	19.32	+4.08	19.66	19.6
16.0	75.7	18.92	−2.92	18.67	18.4
20.3	76.4	19.10	+1.20	18.68	18.8
16.7	75.4	18.85	−2.15	19.19	19.2
22.4	75.5	18.87	+3.53	18.66	18.8
16.1	75.3	18.82	−2.72	18.77	18.7
20.1	76.0	19.00	+1.10	18.48	18.6
17.5	74.8	18.70	−1.20	19.99	20.0
21.1	75.0	18.75	+2.35	17.36	17.7
16.3	75.7	18.92	−2.62	18.97	18.9
20.8	75.9	18.97	+1.83	19.18	19.2
17.7	75.8	18.95	−1.25	20.19	20.3

Table 10.9 Tidal data collected for each unit

1	2	3	4
22.6 + 3.63	17.3 − 1.97	21.1 + 1.88	15.9 − 3.57
23.6 + 4.30	16.6 − 2.72	21.2 + 1.98	15.6 − 3.70
23.8 + 4.65	16.0 − 3.07	20.9 + 1.73	16.0 − 3.07
23.4 + 4.08	16.0 − 2.92	20.3 + 1.20	16.7 − 2.15
22.4 + 3.53	16.1 − 2.72	20.1 + 1.10	17.5 − 1.20
21.1 + 2.35	16.3 − 2.60	20.8 + 1.83	17.7 − 1.25
Σ + 22.44	− 16.02	+ 9.72	− 14.94
mean + 3.74	− 2.67	+ 1.62	− 2.49

When the trend and seasonal elements have been eliminated from the time series only the error and perhaps long term cyclic elements remain. The long term cyclic elements are difficult to deal with especially if a number of different periods are present. Great care must be taken if these cycles are extrapolated. This applies also to the use of trends in prediction. They can be very misleading.

(d) Testing for randomness in time series

Two basically very similar tests for randomness in time series will be mentioned. One test depends on the number of runs in the series and the other on the number of turning points. These values are closely related. The runs method will be applied to the data for sea-level already discussed, although it would be desirable to have a longer series of observations or more numerous values. The relevant data after removal of the trend value are as follows (from table 10.6):

Value	Deviation	Run
160	+ 13.0	1
160	+ 13.3	
148	− 10.4	2
145	− 19.1	
175	+ 5.2	3
182	+ 6.5	
162	− 19.2	4
205	+ 18.1	5
185	− 7.6	6
190	− 8.3	
210	+ 6.0	7

There are 7 runs in these values indicated by the alternating + and − signs. There are 6 pluses and 5 minuses, these give $n_1=6$; $n_2=5$; $n=n_1+n_2$; $r=7$. The expected number of runs is $(6+5)/2=5.5$

The value of K is calculated, where $K=\dfrac{(r+\frac{1}{2})-M}{\sigma}$

$$M=\frac{2n_1n_2}{n+1} \qquad \sigma=\sqrt{\left(\frac{2n_1n_2(2n_1n_2-n)}{n^2(n-1)}\right)}$$

$$K=\frac{n(r-\frac{1}{2})-2n_1n_2}{\sqrt{\left(\dfrac{2n_1n_2\,(2n_1n_2-n)}{n-1}\right)}} \qquad K=\frac{11\,(6.5)-60}{\sqrt{\left(\dfrac{60\,(60-11)}{10}\right)}}$$

$$K=\frac{71.5-60}{\sqrt{294}}=\frac{11.5}{17.15}=0.67$$

K gives the number of standard deviations that the observed number of runs differs from the expected value. As the value is less than one the null hypothesis that the series is random is not rejected. Hence its unsuitability for testing for seasonal cyclic elements, because it cannot have a cycle element if it is randomly distributed.

The random turning point method is applied to a series of observations of variation in beach level with time. To be effective this method should not be applied unless there are at least 50 observations. On the first set of values N (number of observations) was 93 and there were 36 turning points. The expected number is given by $E(t)=(2N-4)/3$ $E(t)=(186-4)/3=60.6$. The variance (t) is given by $V(t)=(16N-29)/90=16.2$ $O(t)$ 36. The probability of this occurring is $Z=(36-60.6)/\sqrt{16.2}=6.13$

$$Z=\frac{O(t)-E(t)}{\sqrt{\text{var }(t)}}$$

The Z tables indicate that the probability of this result occurring randomly is very low. Another set of observations gave $O(t)=51$ $N=130$.

$$E(t)=\frac{2N-4}{3}=85.3 \quad V(t)=\sqrt{22.8} \quad Z=\frac{51-85.3}{\sqrt{22.8}}=7.18$$

This set is also not random with a high degree of confidence. These two tests provide a simple means of testing for randomness in a time series.

(*e*) *Correlation test for two time series*

The correlation of time series can be achieved by a simple non parametric method using the chi square test on frequencies of cross difference values. The values are made comparable by taking individual differences and not cumulative values and are arranged chronologically in pairs as shown in table 10.10. The differences between

$$(p_r-q_{r+1}) \quad \text{and} \quad (q_r-p_{r+1})$$

for each pair are entered as sign values in the third and fourth columns and the product of the signs entered for each pair of values in the fifth column. The number of $+s$ and $-s$ are counted separately, omitting the values that include a zero, when $p_r=q_{r+1}$ or $q_r=p_{r+1}$. The values entered are for sand level changes in inches on two adjacent pegs 60 feet apart down the beach at Gibraltar Point from 20 March to 31 December. It is interesting to compare this result with that found for two pegs one 60 feet from the top of the beach and the other 515 feet lower down the beach. There were 44 pairs of observation, of these 14 gave 0, 14 + and 16 −. The value of E is

$$\frac{30}{2}=15. \quad \chi^2=\left(\frac{14-15}{15}\right)^2+\left(\frac{16-15}{15}\right)^2=\frac{1}{15}+\frac{1}{15}=\frac{2}{15}=0.133$$

The null hypothesis in this instance cannot be rejected and there is no correlation between the two series of observations. This suggests that there is no connexion between changes at the top of the beach and those lower down.

This method could also be applied to test the correlation of varve sequences in two adjacent pits. The method of varve analysis requires the matching of the base of one sequence with the top of another to prolong the time scale and this test provides a useful means of accomplishing this. Correlation of deep sea cores can also be done by this method.

Table 10.10 Beach changes (change-level in inches)

Peg 8	Peg 9	$p_r - q_{r+1}$	$q_r - p_{r+1}$	Product	Peg 8	Peg 9	$p_r - q_{r+1}$	$q_r - p_{r+1}$	Product
0	0	−	0	0	−1	+½	−	0	0
0	+1				+½	−½			
0	+½	−	−	+	0	−½	+	0	0
+2½	+½				−½	−½			
−½	−½	0	−	0	+1	0	−	0	0
0	−½				0	−½			
−½	−½	−	+	−	0	−½	−	−	+
−1	0				+½	+½			
+½	+½	+	+	+	+½	+½	0	+	0
−½	0				0	0			
+½	0	0	+	0	0	−1	−	−	+
−½	+½				−1½	+½			
−½	−1	−	0	0	−1½	0	+	+	+
−1	+½				+½	0			
+1	0	+	0	0	0	+½	0	+	0
0	+½				0	0			
+½	−1	0	−	0	−½	0	+	0	0
+½	+½				0	−1			
+2	0	+	+	+	0	+½	+	+	+
−1	−1				0	−½			
+½	0	+	0	0	−1	−3	−	−	+
0	0				+2½	+1			
−½	−1	−	−	+	+½	+4	+	+	+
0	+2½				−½	0			
+½	+1½	0	+	0	−½	−½	0	+	0
0	+½				−1	−½			
+½	+½	+	+	+	−1	−3	+	−	−
−½	0				−2	−7			
+½	0	+	+	+	−1	−1	+	−	−
−1	0				0	−2			
+½	+½	+	+	+	−½	−1	−	−	+
−1	−½				−½	0			
+1½	+½	+	+	+	−1	−½	+	+	+
−½	+½				−1	−2½			
+1	−1	+	−	−	−1½	+2	−	+	−
0	+½				+½	0			
−½	0	+	0	0	−2	−4	−	−	+
0	−1				−1	0			
+1	−½	+	−	−					
+1	0								

No. of 0 16 Observed number
No. of − 6
No. of + 17

Expected number $\dfrac{6+17}{2}=11.5$

$$\chi^2=\frac{(17-11.5)^2}{11.5}+\frac{(6-11.5)^2}{11.5}=\frac{5.5^2}{11.5}+\frac{5.5^2}{11.5}=\frac{60.5}{11.5}=5.27$$

χ^2_{95} for 1 $df=3.84$. Null hypothesis of no correlation is, therefore, rejected with 95 per cent confidence.

Fluctuations of crop yields under different conditions could be treated in this way also. In fact any variable that fluctuates can be used in this test; industrial production, temperatures, precipitation, fish catches could all be tested by this method.

(*f*) *Logistic curve*

Not all series have a linear trend or cyclic elements. One interesting exception is the logistic curve, which often fits organic time series. It has been found to apply to population trends as well as other organic trends such as growth rates of bacteria, insects and other animals. The curve has a lower concave upward portion, during which the numbers are increasing at an accelerating rate. Then a point of inflexion leads to an upper convex up portion during which period the numbers increase at a decelerating rate and approach a steady state of constant size population. The lower part of the curve represents the period in which the rapidly growing population freely occupy the environment. Then the upper part of decelerating growth indicates the period when the land becomes fully occupied. The inflexion point is called the critical point and it separates the accelerating from the decelerating part of curve. At this point there is half of the final population. The inflexion point is denoted by $y_e = \frac{1}{2}k$. The time (x) at which this point is reached is given by $x = (1/a) \log_e b$, a and b and k are constants that depend on the nature of the data. The form of the curve is given by

$$y_e = \frac{k}{1 + be^{-ax}} \quad (e = 2.71828).$$

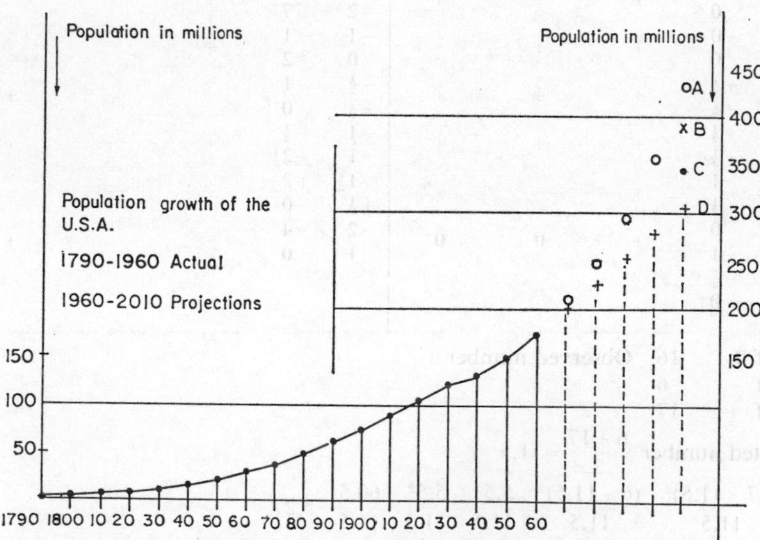

Figure 10.24 Actual and projected population growth of the U.S.A.

The population of the United States seems to have entered the upper part of the curve, the critical point of which was calculated to be March 1914 when the population was 98 million. The value of k should be about 200 million if the value of the constants remain the same until the year 2050 when this value of k is approached. Very often, however, conditions change and then the constants must be modified. Figure 10.24 shows some extrapolations for U.S. population.

(g) Frequency of occurrence of events

Another useful statistical analysis of time data is to work out the expected frequency for events of a given magnitude. This type of analysis can be applied to occurrences such as the frequency of floods of a given magnitude in a river, the occurrence of storm surges in the North Sea, or hurricanes on the coast of the U.S.A. The method of analysis has already been described in chapter 3 when the Poisson distribution was described and an example is given there. It is possible to calculate the expected recurrence interval of a flood of a particular magnitude, and also to indicate the expected magnitude of a flood of a given recurrence interval. Thus a flood with a recurrence interval of 100 years will be expected to have a discharge 2.9 times that of the mean annual flood and one with a recurrence interval of 10 years will have a discharge of 1.7 times the mean annual flood. The mean annual flood is the value obtained by taking the mean of the highest discharge over a number of years. Similarly wave heights can be expressed in terms of the expected frequency of occurrence of waves over a given height in percentage of time. Thus, off New York harbour, waves 27½ ft high are expected to occur for only 0.1 per cent of the time and waves of 33 ft in height will be expected to occur in one observation in five years. These two methods give the expected frequency of occurrence and are illustrated in figure 10.25 for the wave heights. The curves approximate to straight lines when plotted on semilogarithmic paper. This shows that the occurrence of the more extreme events is very rare compared with the events near the mean magnitude. Waves occurring 50 per cent of the time are only 3 ft high or less in an area where the maximum recorded waves are 30 ft high.

10.10 Prediction

(a) Prediction in physical geography

The certainty with which predictions can be made in geography depends upon the knowledge of the cause of the phenomenon to be predicted. Where the cause of an effect is well known the prediction is good. This is a situation in which the dependent variable is closely correlated with the independent variable, which can be readily obtained. The relationship is defined by the regression coefficient, which gives the slope of the regression line. Where the dependent variable is not closely related to a measurable independent variable the correlation is low, and the degree of certainty in prediction is, therefore, poor. This situation implies a lack of knowledge of independent variables to which the dependent variable is related. This is a common situation where

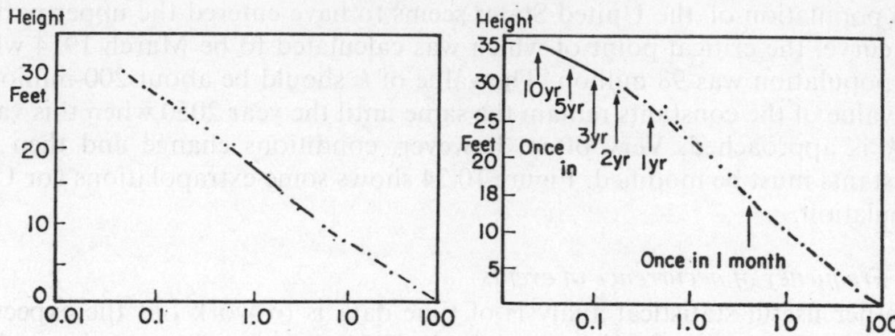

Figure 10.25 The frequency of occurrence of waves of different heights is shown on the left as percentage of time waves exceed various heights. The diagram on the right shows the statistical recurrence interval of waves greater or equal to specified heights.

the independent variables are many, complex and little known or measured This situation is very common in both physical and human geography.

(i) *Harmonic analysis*. One phenomenon that can under normal circumstances be predicted with good accuracy is the height of the tide as this depends upon known and regular movements of the earth, moon and sun. Tidal prediction also illustrates a useful method of analysis, called harmonic analysis. Basically the method consists of calculating the simple harmonic or sine components on which the tide depends and adding these together mechanically. The simple harmonic term can be expressed as $R \cos (nt - k)$, R being the amplitude, n being the speed or number of degrees per hour (mean solar hour), k being the phase-lag and t being the time, in G.M.T. The most important harmonic component is the semidiurnal lunar component, but there are many others of different periods and amplitudes. It is the variation in the periods that makes tidal prediction complex. The harmonic method of tidal prediction was developed by Lord Kelvin. This method is an improvement on the calculation of the individual components hour by hour and the addition of the results. The mechanical means of summing the individual components was developed to save laborious calculation. The equivalent harmonic motions are generated by the revolution of a crank revolving at a regular rate. Many of these motions can be geared together to a main shaft and the harmonic movements of pulleys attached to them can be summed on a revolving drum, as shown diagrammatically in figure 10.26. The tide prediction machine built in 1924 for the Tidal Institute used 26 harmonic components. The maximum number of components used was 37 in a tide-predicting machine made in U.S.A. in 1910. The calculations can now be made more satisfactorily by computer. The accuracy of tidal prediction is great unless random events, such as surges, occur. These can cause local deviations of up to 10 ft from predicted values in the North Sea. This method of tidal prediction is almost the opposite of the Fourier analysis of wave records. In the Fourier process a complex curve is split into its harmonic components. In the case of tidal prediction the complex curve is generated from known harmonic components.

(ii) *Experiment and prediction.* In some situations prediction can be based on experiment. For example, when engineering works such as harbours or river training works are contemplated, experiments are carried out in models to assess the effect of the works. In tidal models, the model is adjusted until known changes can be reproduced, and then the effect of the new works can be predicted with a fair degree of confidence from the results produced in the model.

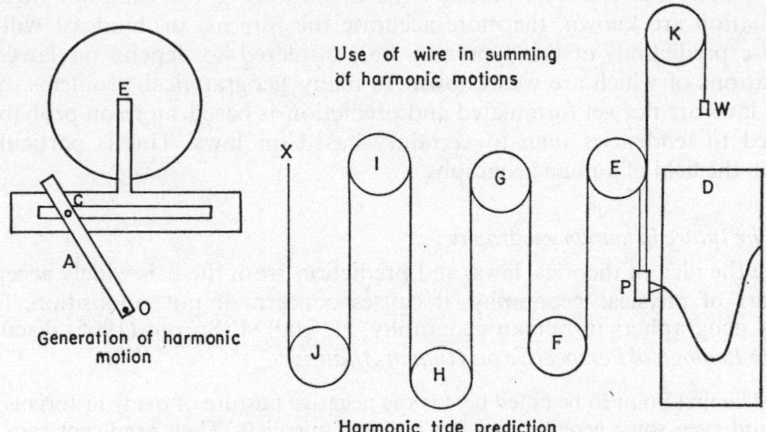

Figure 10.26 Diagram to illustrate the mechanical predictions of tides by summing harmonic motions.

Predictions resulting from full scale experiments under natural conditions are in some ways more satisfactory than model predictions as the scale problem is eliminated. On the other hand, there is the difficulty of experimenting under the whole range of possible natural conditions. Extrapolation from the results of short term experiments can give completely false results if the short period of time used for the experiment is not representative of a long period. An example of this was shown by four weeks' observations of longshore drift off Orfordness in Suffolk, carried out by C. Kidson and A. P. Carr, (1959). The pebbles moved north during these four weeks, in the opposite direction to the long term movement. Had the experiment lasted only for this short period the result would have been very misleading. However, before the end of the experiment more normal conditions were resumed and the pebbles moved rapidly south, which is their long term direction of movement.

(iii) *Extrapolation — forecasting and hindcasting*. Much prediction is done by extrapolation of known trends from the past into the future. Thus weather can be forecast on the basis of the recent past and present situation. Long term forecasting can be done on the basis of finding in past records a situation similar to the present. The development of the past situation may then be expected to be repeated in the present situation. In some instances, it is useful to work backwards in time by hindcasting events that were not recorded at the time by extrapolating backwards from those that were. For example beach changes have been recorded but adequate wave information was not available to interpret the changes. The waves can then be hindcast from data that are available, such as wind observations and other meteorological data.

It is clear that the more precisely the operations of the variables affecting a situation are known, the more accurate the forecast or hindcast will be. Precise predictions of this type may be considered to depend on laws, the operations of which are well known. In many geographical problems, however, laws are not yet formulated and prediction is based more on probability related to tendencies than to certainty based on laws. This is particularly true in the field of human geography.

(*b*) *The future in human geography*

While the idea of theories, laws, and prediction from them is widely accepted as part of physical geography, it causes concern, if not opposition, from many geographers in human geography. H. and M. Sprout (1965) discuss it in *The Ecological Perspective on Human Affairs*:

> 'The first point to be noted here is the negative posture of many historians and even some geographers and political scientists. Their argument runs somewhat as follows: Human behavior is unique; no two milieux are alike; history never repeats; hence it is futile to try to predict events, and particularly futile to predict the specific reactions of particular people; one cannot "pinpoint" the future; therefore, the argument concludes, the student of human affairs should concentrate on historical descriptions that enrich "understanding".'

There are indeed differences between prediction in physical and in human geography. Among them may be noted:

(1) Awareness of the future is an attribute of human beings. Presumably it is also one of many higher animals, though probably it is unconscious. Thus the beaver may have encoded in its brain instructions to make dams without knowing what the outcome will be. Certainly non-organic things are not aware of the future.

(2) Predictions can actually affect the course of events leading up to the particular event or events predicted. In 1960, for example, some influential demographer might have predicted that the population of California would reach 50 million by 2000 AD, resulting in intolerable overcrowding. This prediction could then divert many

prospective migrants elsewhere, thereby upsetting the prediction. The difference is, then, that people can (apparently) make decisions that will affect predictions. It is tempting to think that to some extent the forecasts of Karl Marx have had less chance of coming about because from the communist side they have been artificially stimulated and from the capitalist side, consciously forestalled.

(3) Physical geography is often dealing with a larger population of phenomena than human geography. For this reason it is more easy to be near certain in predictions; but this is not always so. In long range weather forecasting prediction is very inaccurate when based on a 'population' of perhaps only 50 or 100 annual records. However, many of the important man made things on the earth's surface, and many of the makers of big decisions, are unique: Adolf Hitler, Charles de Gaulle, the U.S.A.

An important difference must be noted between precognition and prediction, since many people, perhaps, confuse the two. Precognition is an exact account of something that has happened, is happening, or most often, will happen, by persons not present at the event described by them. Precognition is an attribute apparently belonging only to a few people, though its existence is denied altogether by many people. It involves making a statement about an event that will occur at a given time and place, generally involving one or a few people, often in disastrous circumstances. The probability that such a combination could be achieved by chance often lies far beyond the most cautious confidence levels taken in the testing of hypotheses. Nostradamus, the Provençal physician of the 16th century, wrote many quatrains about future events. Unfortunately he did not date them. His life and writings are discussed by J. Laver (1952). Mrs Jeane Dixon, resident in Washington, D.C., is reputed to have made many remarkable forecasts of events, including the assassination of President Kennedy in Dallas, Texas. Some of her achievements are described by R. Montgomery (1965).

While precognition and prediction should not be separated entirely, they differ basically in that the former often seems to come unexpectedly in a vision, while the latter may be worked out from deliberate and often laborious calculations, following accepted methods such as hypothesis testing and statistical inference. Very broadly, prediction may be thought of as making statements with varying degrees of confidence about one, a few, or many events, occurring in the past, present or future. It is particularly important to appreciate in the words of H. and M. Sprout (1965) 'What one may profitably try to predict, and how to go about it' (p. 175). Moreover, there are various kinds of statement that can be made about the future and varying degrees of certainty.

(1) Some things can be predicted with certainty. Thus in mathematics it is impossible to find an even prime number other than 2 in base 10 numbers. One exception would invalidate the definition.

(2) It is possible to schedule or plan for the future and in this way, make or build it. Indeed all decisions must take effect in the future. On a rail

timetable, for example, a certain train leaving Euston Station, London, each day may be scheduled to reach Manchester at 2.45 p.m. The train may be delayed through unforeseen weather conditions or a breakdown. It may even arrive a few minutes early. It is the intention to arrive at a given time that gives the appearance of certainty to the prediction. In general, at least in a fairly highly organized country, the future for any given person or group of people is roughly mapped out already by arrangements, such as future meeting dates and so on, both private and public. Important events such as congresses and games may be scheduled for many years ahead. Such arrangements may have to be altered, but for the convenience of the community, as few as possible will be changed. Thus a pattern is laid out to be followed. The lives of many people are run on the assumption that the scheduled events will take place.

Both private firms and state undertakings wish to know what the demand will be for their products over the next several years, or even two or three decades, for products ranging from consumer goods to electricity and gas. The motor vehicle firms of Britain use both past figures for this country and figures for the U.S.A. to decide about investment. (*The Times* 'How the Motor Industry Reads the Future', 25 May 1965.)

In 'communist' countries the economic life of whole countries has been planned rigidly. In these cases, consumer demand and sovereignty have been relegated to a very secondary place. One lesson of Soviet planning over the last 50 years has been to avoid being too rigid. A big change in policy (e.g. switching from coal to oil and gas in the late 1950s) was apparently a main cause of the complete abandonment in 1957 of the Five Year Plan, 1956–60. Perhaps big-scale planning is satisfactory only if allowance is made for changes to be made if circumstances themselves change. Both commonsense and long Soviet experience suggest that agriculture is particularly difficult to plan.

(3) Inference can be made about a larger population from an appropriate, unbiased sample from it. Some predictive inference could for example be made about smoking habits of all students in Nottingham University (about 4000) from a sample of, say, 100. Sample studies of farms allow predictions to be made about all the farms in a given area within limits that can be calculated (see chapter 3).

(4) Experience based on empirical studies often allows the future to be predicted satisfactorily. In every day life, pedestrians will cross streets while traffic is stopped by red signals knowing there is a danger risk because from time to time a vehicle will deliberately or inadvertently cross against red. One makes decisions of this kind based on one's own experience or the experience of others, often working out a probability very quickly, as when deciding while driving whether to overtake another vehicle when the oncoming traffic is hidden by a hill or a bend not very far ahead.

In human geography, there is much evidence to build up experience on the basis of the principle of least effort. Because people travel only within certain distances for various kinds of shopping, it is possible to predict that a new shopping centre placed on one side of an urban area would attract more customers from that side of the urban agglomeration than from the opposite side. An analogy borrowed from physics and applied to the distances people, goods and messages travel or are sent is the so-called gravity model, which will be discussed in the next chapter. Similarly it may be assumed that a person making a journey between two places will almost invariably choose the route requiring least effort, or at least the route *in his view* requiring the least effort. This may not necessarily be the shortest in distance; time and/or cost may also be considered.

(5) Predicting general trends. There have been many people in the past who in general terms have foreseen developments decades or centuries ahead. Much admiration must go to Leonardo da Vinci for his detailed description of many devices quite impossible to construct in his own day. On the basis of a large amount of evidence about population growth up to the 1820s, Thomas Malthus (1830) detected: 'The existence of a tendency in mankind to increase, if unchecked, beyond the possibility of an adequate supply of food in a limited territory.'

Many predictions of general trends are based on projections from the past into the future either of sets of data or of technological innovations. Projections of the U.S. population to 2015 are quoted in the *Statistical Abstract of the United States* (p. 8) 1966 (see figure 10.24 and table 10.11). It is an

Table 10.11 U.S. population projections

	1965	2015	Increase %	Increase Absolute
Series A	195 m	475 m	144	280 m
Series B	195 m	425 m	118	230 m
Series C	195 m	370 m	90	175 m
Series D	195 m	322 m	65	127 m

understatement to say that the extremes predicted are very different: an increase ranging from 280 million on the one hand to 127 million on the other. All one can say with confidence is that a decline is not predicted. How useful such information is to planners designing new housing, new factory equipment and so on is doubtful. What would the predictions be in a country with poor past and present data? When general trends are forecast, many limiting conditions are put in the estimates, explicitly or implicitly. Thus the U.S. population predictions quoted above must assume that there will not be a nuclear war in which, say, 100 million Americans are eliminated.

Life insurance companies also work with decades in view ahead. Changes in insurance rates are based on mass observations of patterns of deaths. It is assumed that these will continue. If people suddenly started dying on average ten years earlier or later than had been observed, rates would have to be changed, but as long as dying 'habits' are carefully watched, actuarial data can form the basis for predicting accurately what proportion of a given set of individuals of a certain age group will die within the next so many years.

This can be used to run the affairs and policy of the company on a safe financial basis provided that a large number of people insure themselves. The prediction is still based on probability, not certainty, but probability approaches certainty with 10^5 or 10^6 subscribers.

Although K. Popper (1957) in his *Poverty of Historicism* insists that discoveries, inventions and major technological changes are unpredictable and the future cannot therefore be studied, there have been many stimulating descriptions of a general picture of the world, usually as foreseen some decades ahead. In *Anticipations,* written in 1901 and revised in 1914, H. G. Wells made many interesting points about the development of transportation technology, big cities and so on. He wrongly 'foresaw' (for example) Shrewsbury, Stratford and Exeter as outer suburbs of London. In *The Rise of the Meritocracy 1870–2033*, M. Young (1958) looks back from the 21st century on progress in education and social repercussions thereof. In *Inventing the Future,* D. Gabor (1963) speculates about the continuation of present trends. Another recent piece of prediction is by R. Brech (1963): *Britain 1984: A report prepared for Unilever,* subtitled: *An Experiment in the Economic History of the Future,* and sponsored by Unilever. Among other things, Brech draws on U.S. experience, since in many respects the U.S.A. is now 'ahead' of West Europe. One may assume that if Britain behaves like California, with 5 people per private car now, it can be expected to achieve 3 people per private car in 15 years time. Unilever, obviously, is not the only firm interested in the future.

In his presidential address at the 1964 Southampton meeting of the British Association for the Advancement of Science, Lord Brain developed the theme of thinking ahead. He stressed the need to forecast and prepare for the consequences of accelerating developments in science and technology. He hoped for a 'world brain' to correlate information and extrapolate it on a world scale (*The Times*, 27 August 1964).

References

Barber, N. F. and Ursell, F. (1948). The generation and propagation of ocean waves and swell. Wave periods and velocities, *Phil. Trans. Roy. Soc. A.*, **240,** 527–560.

Benedict, J. B. (1967). Recent glacial history of the Alpine area in the Colorado Front Range, U.S.A. I. Establishing a lichen-growth curve, *J. Glaciol.*, **6** (48), 817–832.

Bondi, H. (1965). *Relativity and Common Sense, A New Approach to Einstein*, The Science Study Series No. 31, Heinemann, London.

Bowden, L. W. (1965). *Diffusion of the Idea to Irrigate, Simulation of the Spread of a New Resources Management Practice in the Colorado Northern High Plains,* University of Chicago, Department of Geography, Research Paper No. **97.**

Brech, R. (1963). *Britain 1984,* Darton, Longman and Todd, London.

Boer, G. de (1964). Spurn Head: Its history and evolution, *Trans. Inst. Brit. Geog.*, **34,** 71–89.

Broad, C. D. (1938). Ostensible Temporality, Reprinted in *Problems of Space and Time* (Ed. J. J. C. Smart), Macmillan, New York, Chapter 35 Examination of McTaggart's Philosophy, **II**, Part 1.
Brown, L. (1965). *Models for Spatial Diffusion Research — A Review*, Research Report, **10**. Department of Geography, Northwestern University, Evanston.
Doodson, A. T. and Warburg, H. D. (1941). *Admiralty Manual of Tides*, H.M.S.O.
Fairbridge, R. W. (1961). Eustatic changes in sea-level, In *Physics and Chemistry of the Earth*, **4,** Pergamon Press, 99–185.
França, A. (1960). *A marcha do café e as frentes pioneiras*, Conselho Nacional de Geografia, Rio de Janeiro.
Gabor, D. (1963). *Inventing the Future*, Secker and Warburg, London, (1964 ed.) Penguin books, Harmondsworth.
Geer, G. de (1912). A geochronology of the last 12,000 years, *C.r.* 11 *Int. Geol. Congr*. (Stockholm 1910), **1,** 241–258.
Granger, C. W. J. (1964). *Spectral Analysis of Economic Time Series*, Princeton University Press.
Hagerstrand, T. (1952). *The Propagation of Innovation Waves*, Lund Studies in Geography, Ser. B., Human Geography, No. 4, Lund, Sweden.
Hagerstrand, T. (1957), *Migration and Area*, In *Migration in Sweden*, Lund Studies in Geography, Ser. B. Human Geography No. 13.
Hagerstrand, T. (1965). A Monte Carlo approach to diffusion, *Archiv. Europ. Sociol.*, **VI,** 43–67.
Kidson, C. and Carr, A. P. (1959). The movement of shingle over the sea bed close inshore, *Geog. J.*, **125,** 380-389.
King, C. A. M. and Barnes, F. A. (1964). Changes in the configuration of the intertidal beach zone of part of the Lincolnshire coast since 1951, *Z. Geomorph.*, Sonderheft. NF **8,** 105*–126*.
King, L. C. (1962). *The Morphology of the Earth*, Oliver and Boyd, Edinburgh, Part B and C, 135–300.
Lägnert, F. (1952). *The Electorate in the Country Districts of Scania 1911–1948*, Lund Studies in Geography, Ser. B. Human Geography No. 5, Lund, 1952.
Laver, J. (1952). *Nostradamus*, Penguin Biography, Harmondsworth.
Lighthill, M. J. and Whitham, G. B. (1955). On kinematic waves. I. Flood movement in long rivers (281–316). II. A theory of traffic flow on long crowded roads (317–345), *Proc. Roy. Soc. London* Ser A., **222,** 281–345.
Linton, D. L. (1957). The everlasting hills, *Advan. Sci.*, **14,** 58–67.
Malthus, T. (1830). *A Summary View of the Principle of Population*, Mentor Books, New York, 1962.
Milankovitch, M. (1920). *Théorie mathématique phénomènes thermiques produits par la radiation solaire*, Paris.
Montgomery, R. (1965). *A Gift of Prophecy, the Phenomenal Jeane Dixon*, Bantam Books, New York.
Nye, J. F. (1963). The response of a glacier to changes in the rate of nourishment and wastage, *Proc. Roy. Soc. London*, Ser. A **275,** 87–112.
Pitts, F. R. (1963). Problems in computer simulation of diffusion, *Papers Proc. Reg. Sci. Assoc.*, **11.**
Popper, K. R. (1957). *The Poverty of Historicism*, Routledge and Kegan Paul, London.
Proudfoot, V. B. (1965). Archaeological evidence for rates of weathering and erosion in Britain, *Geomorph. Sym. Inst. Brit. Geog.*
Rogers, E. M. (1962). *Diffusion of Innovations*, Free Press Glencoe, New York.

Saarinen, T. F. (1966). *Perception of the Drought Hazard on the Great Plains*, University of Chicago, Department of Geography Research Paper No. 106.

Savigear, R. A. G. (1952). Some observations on slope development in South Wales, *Trans. Inst. Brit. Geog.*, **18**, 31–51.

Sprout, H. and M. (1965). *The Ecological Perspective on Human Affairs*, Princeton University Press, Chapter 9.

Teilhard de Chardin, P. (1965). *The Phenomenon of Man*, Fontana Books, 241–2.

Wells, H. G. (1899 and 1914 new edition). *Anticipations of the Reaction of Mechanical and Scientific Progress upon Human Life and Thought*, Chapman and Hall, London.

Williams, W. T. and Lambert, J. M. (1959–62). Papers in *J. Ecol.*, **47–50.**

Young, M. (1958). *The Rise of the Meritocracy*, Thames and Hudson, London (1961 ed. Penguin books, Harmondsworth).

Part 4

Models, theories and organization

11

Models and analogies

'Hume declared, and Mill said much the same thing, that all reasoning whatsoever depends on resemblance or analogy, and the power to recognize it.' *D'Arcy Thompson* (1917)

'But what is genius? . . . William James regarded it as an unusual ability to perceive analogies . . .' *L. M. Milne* (1965)

11.1 Introduction

Many different types of model and analogy have been used in science in general and in geography in particular. Model classification has been suggested by a number of authors, including W. C. Krumbein and F. A. Graybill (1965) and R. J. Chorley (1964). This indicates that there are many possible methods of organizing a description of models. Before considering the order adopted in this chapter, it is useful to draw attention to some of the meanings given to the term model and to differentiate models from theories. A model has been described in a number of ways, some of which imply that model and theory are synonyms. A list of synonyms and characterizations for 'Model' have been assembled by Y. R. Chao (1962). These include the following: A frame of reference (Hockett), archetypical frames of reference (Hockett), a description (Hockett), a theory (Chomsky), analog (Stevens), a proposed method of research (Stevens), a miniature representation of a thing (Oettinger), a representation (Oettinger), an abstraction (Oettinger), a framework in respect to which . . . the subject . . . is described (Harris), a picture of how the system works (Harris), a theory of the structure of something (Harris), a formalized or semi-formalized theory (Braithwaite), theoruncula or theorita (Braithwaite), a psychological crutch (Putnam), an abstraction of a system (Diliberto), an abstraction (Kapp), and a set of constraints (Pask). Braithwaite, in particular, has defined a model as a formalized theory, although he considers that a model is more modest than a theory, and introduces the term 'theoruncula'. Oettinger states that 'The discovery that the replacement of concrete objects by diagrams, and of operations on these objects by corresponding operations on the corresponding diagrams can lead to useful results provides the very justification of models'. This statement can be applied to a number of types of model, including the theoretical model.

The relationship between models and theories is complex. R. J. Chorley (1964) states that 'It is important to differentiate between models and theories.

... A model becomes a theory about the real world only when a segment of the real world has been successfully mapped into it'. Campbell, as discussed by M. B. Hesse (1963), considers that models in some form are essential to the generation and extension of theories. In some instances a model can help a theory to be predictive, which is one of the aims of theories. This applies to certain theoretical and mathematical models, which allow extrapolation from known initial conditions. The alternative view that models are not necessary to the formation of theories has been expressed by Duhem.

One of the main purposes in using models is to provide a simpler situation, or one that is easier to appreciate and study, than the prototype. This simplification is achieved in the theoretical or mathematical model by using symbols that stand in for more complex phenomena and that can be manipulated by the rules of mathematics or logic. In simulation models the situation in the model can be simplified to the required degree by the conditions that are imposed artificially in the model. These models represent a simplification of the real situation, in that the number of possible choices is limited by the conditions set up. For example, in the drainage pattern model to be discussed, the river can only flow in one of the cardinal directions. This limitation is clearly a simplification of the natural situation, in which the possible directions are infinite.

In scale models simplification is also of great importance. In the example of the geological model, the rocks are represented by uniform symbolic material and reflect none of the natural variations or complications. Fossils, for example, are ignored in the model, although they may have played an important part in the original elucidation of the structure. In the scale models of dynamic processes the main advantage of the model is that the situation can be controlled by the simplification of the problem. This is achieved by the control of the variables involved in the process, so that each variable can be studied in turn in a much simpler situation than that occurring in nature. Thus one of the essential benefits derived from using models is the simplification of the problem in one or other of the ways already mentioned.

One of the types of model described by R. J. Chorley (1964) is the simplified model, which he discusses in terms of the abstraction and reduction of the amount of irrelevant data in the situation. The theoretical model is too complex for some purposes in that it attempts to provide a framework for the whole topic under consideration. The mathematical model is rather simpler, but further simplification, by leaving out any extraneous material, leads to the simplified model. Only the most significant aspects of the prototype are preserved in the simplified model, and only those aspects of the model that show some similarity to the prototype are considered.

The use of coloured balls to represent the atomic structure of complex chemicals is an instance of a simplified model. The balls bear little resemblance to the atoms in their external characteristics, but these differences are immaterial and can be ignored. The important aspect of the model is to provide a visual image of a structure in three dimensions, which cannot easily be appreciated by any other means. This is in fact a type of scale model on a very enlarged rather than reduced scale, the latter being the normal type of scale model.

The arrangement of the sections in this chapter is based to a certain extent on increasing abstraction, although it is not possible to fit the different types of models into a specific order on this basis. Another distinction that has been drawn is between models and analogies. Again this distinction is not often clear cut and in some instances analogue models are included in an overall system of model classification. Moreover it is thought that there is a fundamental difference between analogies and models that justifies their separation into a special section. This difference is discussed at the beginning of the section on analogies.

Scale models will be considered first, as in this type of model it is easy to appreciate the correspondence between model and prototype, and the degree of abstraction is minimal. The map may be considered in some respects as a scale model, and the special points in this connexion will be mentioned next. Simulation models, which are then dealt with, are a type of stochastic or probability mathematical model. According to W. C. Krumbein and F. A. Graybill, they should be considered as statistical models. This type of model depends on completely different principles from scale models and is rather more closely linked to the pure mathematical model or deterministic model, as it is termed by Krumbein and Graybill. The essential difference between these two types of mathematical model is that the former is based on probability and the latter on mathematical certainty. The theoretical model is again completely different from either scale or mathematical models. Another term for a theoretical model is a conceptual model. This suggests that this type of model provides a concept or framework that can be used to formulate problems, and to suggest approaches for the solution of these problems. Theoretical models are discussed in chapter 12 as examples of a type of theory.

11.2 Scale models

Scale models, as used in geography, can be of two types. The first type is simply a static replica of natural phenomena. Such a model can be of value for illustrative purposes, but does not provide a means of fundamental research. The second type of model is the working scale model, in which different processes can be reproduced with varying degrees of similarity to the prototype. Models of this type have provided much useful information.

An example of the first type of model is one that shows geological structure in three dimensions in model form. The 3-dimensional aspect of geological structure is often difficult to appreciate, and understanding of the structure can frequently be facilitated by use of models (see figure 9.10*a*). Their construction, however, presupposes a knowledge of the structure. They therefore in no way help to elucidate the structure, but only present it more clearly. Relief models or models of buildings also come into this category. Clearly the scale of the model will determine to some extent how closely it can be made to resemble the prototype and in fact it would rarely be possible or desirable for a complete replica to be made.

The second type of scale model is the more valuable for the purpose of simulating natural processes that are too complex for easy appreciation in the field. One of the problems in working on processes in the study of landforms is the difficulty of controlling the very large number of variables involved. The use of scale models allows these variables to be kept under control. The effect of each one can be determined separately by keeping the others constant during the experiments. If the effect of wave action is being studied, it is possible in a model wave tank to establish the effect of wave steepness or wave length on the gradient of the beach by holding the other variables constant while each is studied in turn.

However this great advantage must be set against the problems of scale. The scale factors in the model and prototype must hold good as far as possible for the geometrical, kinematic and dynamic aspects of the model. It is not possible to make a model that is correct in all these respects because of the model laws that relate these factors. If the model is undistorted geometrically then there will be a direct linear relationship between model and prototype in all three space dimensions and co-ordinates will bear a linear relationship to each other in model and prototype. The linear scale ratio may be defined by the symbol A. By taking into account the forces of gravity and inertia, which are the same in the model and prototype, it can be shown that the time and velocity scales, denoted by B and C, must equal the square-root of the linear scale A. Hence a true scale model in which both geometric and kinematic properties are correct cannot be made. Thus one aspect of the model must be distorted to compensate for the other.

In some instances geometrical similarity is important, as in the study of wave action where wave steepness (the height to length ratio) is an important variable. If the model is made geometrically similar the waves will behave in the same way, but the gradient of the swash slope of the beach will not be comparable. Gradients formed by the smaller model waves will be steeper than the natural beaches formed by waves of similar steepness. In fact a continuous scale of gradient can be established running from wave tank model waves, through small natural waves to large natural waves as shown in figure 11.1. The size of material in a model is another aspect that cannot be directly scaled up. If the natural beach is sand, this to scale would be represented by clay, which behaves quite differently, while shingle on the natural beach would become sand in the model, and this again behaves differently under wave action.

In dealing with tidal models it has been found necessary to introduce some vertical exaggeration into the model in order to be able to reproduce known changes. This type of model is generally made in order to determine the effect of artificial structures on the tidal system. The model is adjusted until known changes can be reproduced. Then the artificial groynes or training works can be added to the model and their effect studied, by extrapolation from known changes. It is often necessary to make the model similar to the

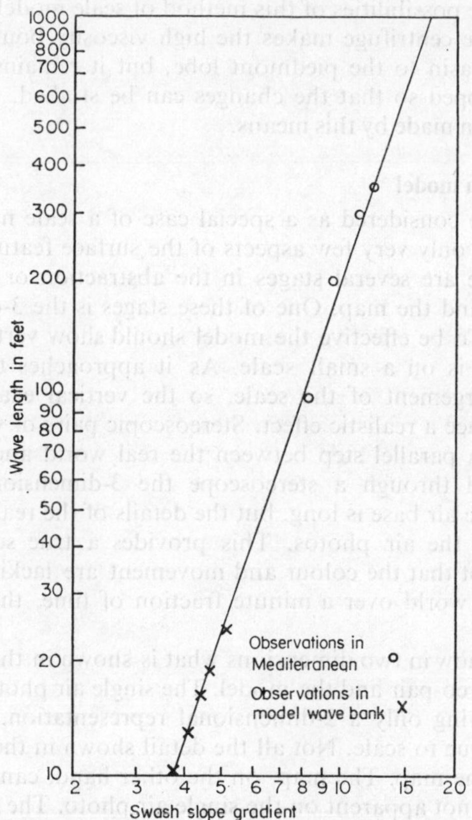

Figure 11.1 Swash slope gradient plotted against wave length. The crosses refer to results obtained in a model wave tank and the circles observations made on the south coast of France in the Mediterranean.

prototype with respect to the materials used. Thus in the Mersey model made by the Hydraulics Research Station at Wallingford, the results of natural water movements could not be accurately simulated until the right amount of salt was added to the sea water in the model to establish the correct density relationships. It is still necessary to use empirical trial and error methods in some models to overcome the problems of scale. Various devices such as the use of less dense materials than those found in nature have been used to help to overcome the problems of scale. Thus sand made of perspex or fine coal has been used successfully in place of the more normal quartz sand.

Another interesting attempt to overcome the control of gravity in model experiments is the use of bouncing putty, a silicone product, in a centrifuge to simulate glacier movement. The work of H. Ramberg (1964) on this problem illustrates the possibilities of this method of scale model study. The speed of whirling of the centrifuge makes the high viscosity bouncing putty move from its upper basin to the piedmont lobe, but it remains stable when the centrifuge is stopped so that the changes can be studied. Realistic moraine patterns have been made by this means.

11.3 The map as a model

The map may be considered as a special case of a scale model. It is a very simplified one as only very few aspects of the surface features can be shown on a map. There are several stages in the abstraction or selection of data between reality and the map. One of these stages is the 3-dimensional relief map or model. To be effective the model should show vertical exaggeration, particularly if it is on a small scale. As it approaches the real world in size, by an enlargement of the scale, so the vertical exaggeration can be reduced to produce a realistic effect. Stereoscopic pairs of vertical air photographs provide a parallel step between the real world and the map. When these are viewed through a stereoscope the 3-dimensional impression is exaggerated if the air base is long, but the details of the real world are shown true to scale on the air photos. This provides a true scale model of the real world, except that the colour and movement are lacking. It records the state of the real world over a minute fraction of time, the exposure of the camera.

A map must show in two dimensions what is shown in three dimensions on the air-photo stereo-pair and the model. The single air photo also has the disadvantage of giving only a 2-dimensional representation. It can, however, show all detail true to scale. Not all the detail shown in the air photo can be reproduced on the map. The map, on the other hand, can show by symbols features that are not apparent on the single air photo. The relief is one of the most important of these and it is usually shown symbolically by contours on the map.

Scale has an important effect on the degree of abstraction or selection of material shown on a map. On a large scale map, such as the 6 in to 1 mile map, buildings and other features can be shown true to scale. However, as the scale of the map diminishes, so the abstraction and selection must increase, until on the very small scale maps all detail must be generalized and simplified and much must be shown symbolically or exaggerated in size. Such a map is a very simplified model of reality, as very many of the features of the real scene have had to be omitted. Nevertheless, it has certain advantages that large scale maps and even reality do not possess. The area covered by the map is large and the relative positions of different objects can be much more readily appreciated on the map than in reality. This applies for example, to the pattern of drainage and transport networks, the distribution of settlements and the character of the outline of the coast and coastal deposits.

Maps which show specific features are even more selective than the topographic map. Any map which shows the distribution of a particular feature, be it rock type or population, is an abstraction of reality, in which a symbol represents the real object. A further stage in the process of abstraction is shown on topological maps where only relative position is maintained and the distances and directions between places are no longer represented correctly. This stage is, however, still a model of the real situation, although it is very abstract and incomplete in many respects (see chapter 2.10.).

11.4 Simulation models

(*a*) *Physical models*

The simulation model bears some relationship to the stochastic mathematical model in that the theory of probability is used in both types of model. In the simulation type of model a specific set of conditions is stated with alternative choices being possible within the stated conditions. Which of the possible alternatives is chosen is then determined purely by chance by using a table of random numbers or similar method to obtain random results. The constraints on the model can be designed to simulate a particular natural situation. By the process of making a series of random choices a model is gradually built up and it can then be compared with the natural prototype in which similar constraints are operating. The closeness of the correspondence between the result of the model, developed by random methods, and the prototype then gives a measure of the randomness of the natural phenomenon.

The use of random numbers in the simulation model requires some justification. It can be related to the attaining of the most probable state. In a closed system there is a certain amount of energy that cannot be renewed once it is used up. In such a system the distribution of energy tends towards the most probable state. This statement is the meaning of the second law of thermodynamics. Thus the system tends towards a state of minimum energy as the energy available is used up through work performed. As this process continues so the system gradually develops towards its final state. The final state is one of maximum entropy. The distribution of energy in the system may be considered as the entropy of the system. As the energy is used up so the entropy is increased. The terms 'entropy' and 'probability' are practically synonymous in this respect, so that as entropy inceases so the state of the system tends to the most probable. This state is where there is the maximum amount of entropy or complete disorder or randomness.

In this state the sum of the probabilities of all possible alternative states is at a maximum, i.e. $p_1+p_2+p_3 \ldots +p_n=1$. The sum of log p is at a maximum when $p_1=p_2=p_3= \ldots p_n$. This means that all possible states are equally probable. If there are 100 possible states then $p=0.01$ and the sum of $\log_e p = -460$. If 50 of the 100 possible states have a probability of 0.005 and the other 50 of 0.015, then the sum of $\log_e p$ is -475. This is not much less than when the probabilities are exactly equal. From this it follows that random characteristics appear to develop fairly rapidly.

The approach to the most probable or final state is rapid at first and so quasi-equilibrium is rapidly set up, although the complete final state may take a long time to achieve. It would be expected, therefore, that natural phenomena, if they are dependent upon this law, should rapidly approach the most likely state in the early stages of development and this state is one of randomness. Hence the value of the random model simulation method, which generates a random system by a process sometimes known as the random walk method.

This method has been applied successfully to drainage basin analysis and this may be taken as an example of the application of the method in physical geography. A drainage basin develops in nature on any surface that is exposed subaerially and where precipitation is sufficient to form a connected system of streams. In natural drainage basins the laws of morphometry, discussed in chapter 13, were established originally by R. E. Horton (1945) and have since been shown to apply to a wide variety of drainage basins in different situations of climate and rock type and stages of geomorphological development. If it can be shown that drainage basins simulated by random walk methods possess the same morphometrical characteristics as natural drainage basins, then there is good evidence to suggest that natural drainage basins are tending towards the most probable state of maximum entropy and randomness.

In order to create a random drainage pattern a square grid is used. An arrow is drawn connecting the centres of two adjacent squares in the direction indicated by a set of random numbers. 1 and 2 can be used to indicate north, 3 and 4 east and so on, numbers 0 and 9 being ignored. If the direction picked from the random number table is the opposite of one already coming into the square, then the next number that shows a different direction is used. Circuits must be prevented by choosing the next possible number that avoids making a circuit. When all the squares have an arrow leading out of them, the isolated pieces of drainage are linked by taking the first possible direction as indicated by the random numbers. The pattern is completed by ensuring that all the streams link and that there is no internal drainage (figure 11.2).

When all the squares are drained by streams that form coherent drainage systems, then the watershed can be added to the diagram. The streams can be ordered and the numbers of streams of each order can be obtained. The number of streams of each order and their mean length can then be plotted against the stream order as in figure 11.3. It is frequently found that the results form straight lines when semi-logarithmic paper is used and that these lines agree closely with those obtained in natural basins. An example of this type of simulation model is shown in the figures 11.2 and 3.

In nature, constraints are imposed by structure and other factors on the development of the system so that the most probable system is not necessarily achieved. The difference between the recorded pattern and the simulated pattern then gives a measure of the effect of the constraints in the particular case under consideration. Constraints can also be placed in the model. Thus the northern component of the streams could be emphasized by allowing 3 or 4 numbers to indicate northern flow while only 2 indicate flow in each of the other directions. Control in other ways can also be introduced into the model.

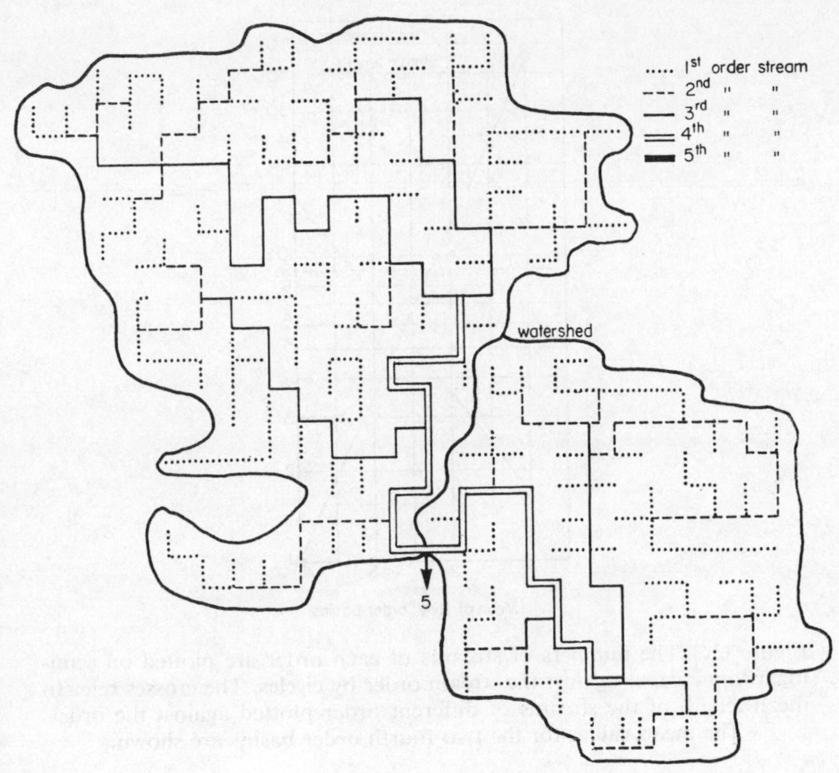

Figure 11.2 Random walk drainage flow in two fourth order streams.

When river profiles are considered instead of the drainage pattern the same general methods can be applied to the development of random walks but rather different principles apply to the model. A river profile is not a closed system that develops towards a static state but an open system. In this situation the most probable state is an isentropic curve. The entropy is distributed equally in this type of curve. In an open system the rate of increase of entropy in the system plus the rate of outflow of entropy is equal to the rate of internal generation of entropy. This rate of internal generation of entropy depends upon the energy gradient towards the base level.

A principle which can be used in the situation described is the principle of least work; nature does not waste energy. The principle of least work can be equated with maximum probability, as it is one way in which maximum probability can be satisfied. If entropy is gained at a uniform rate in each unit length of river, then the resulting curve of the profile is given by $H = ae^{-bx} + C$, a and b being co-efficients, C the base level and H the elevation. This produces the exponential form that is characteristic of most river profiles. This type of curve can be achieved in a random walk model for a

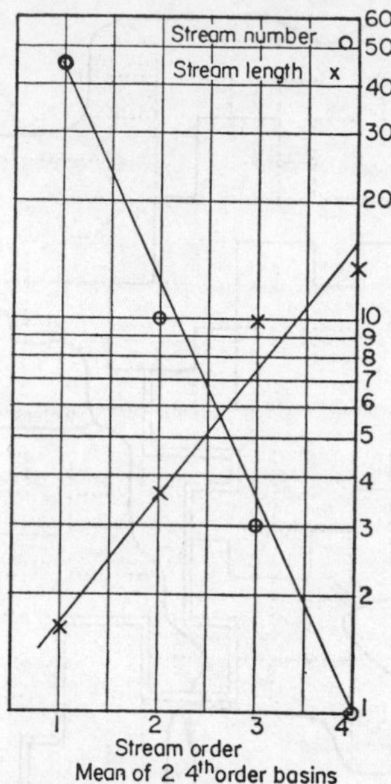

Figure 11.3 The numbers of streams of each order are plotted on semi-logarithmic paper against the stream order by circles. The crosses refer to mean length of the streams of different order plotted against the order. The mean values for the two fourth order basins are shown.

river profile by making the probability of taking a horizontal step rather than a vertical step greater as the elevation decreases. This cannot be done so easily with random numbers but specially designed cards can be drawn randomly.

(*b*) *Simulation models in human geography*

Random numbers have recently been applied fairly widely to simulate distributions, diffusion processes and other features of human geography. Such models may be made simply to provoke thought or as illustrations in the teaching of geography. They may also be useful in research and in applied geography. They have been used in physical sciences in finding how thick a protective shield must be for a nuclear reactor. Imaginary atomic particles are run against this and have various probabilities of passing through, being absorbed or being reflected back. Random numbers are used to determine the actual path of each individual atom. G. A. W. Boehm (1959) illustrates this briefly.

The use of random numbers to generate random distributions and random diffusion processes in models implies usually that something is known about the general behaviour of the people or their decisions being simulated in the model but not about each individual. The individual, therefore, behaves unpredictably. With the help of random simulations, often repeated many times with the help of a computer, it is possible to simulate situations in which very complex probabilities are involved.

The following will be described and discussed briefly: (i) Random distribution of points on a surface. (ii) Random distribution of points on lines. (iii) Political units based on points on a surface. (iv) Random shape. (v) Random diffusion model.

(i) *Random distribution of points (dots) on a surface (figure* 11.4). In this model, every compartment in the grid has an equal chance of receiving any

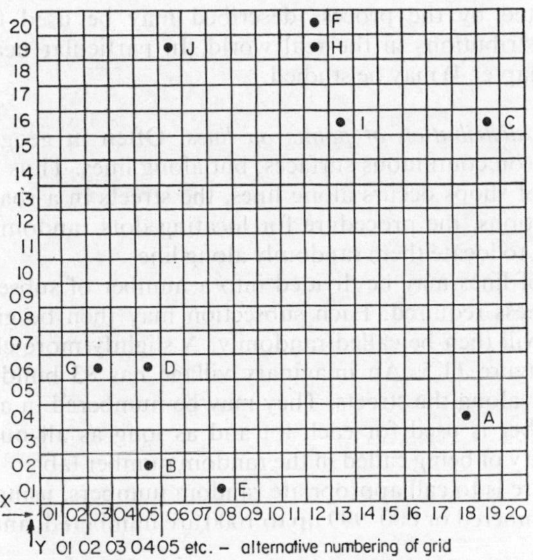

Figure 11.4 A method of constructing a dot distribution on a surface by using pairs of random numbers. This may also be used to provide random sample points (see chapter 3).

point. The sides of the grid may be numbered either, as here, to allow the points to fall in the centres of the compartments of the grid, or to allow them to fall where the lines of the grid cross (see alternative method of numbering grids in figure 11.4). A table of random numbers is required. Numbers will be used in pairs.

Procedure:

Step 1: Choose a pair of random numbers, each with 2 digits or more than 2 digits if required, either from the beginning of a table of random numbers or anywhere else on it.

Step 2: Use the first to locate position on the x (horizontal) axis, the second to locate position on the y (vertical) axis. To economize in the use of numbers, 21–40, 41–60, 61–80, 81–00 may be appropriately adapted to stand in for 01–20, if a 20×20 grid is being used as in this instance.

Step 3: Repeat steps 1 and 2, using new random numbers each time, until enough dots have been located.

Example: (refer to figure 11.4). The following numbers were found in the random number table and have been used to locate the first ten dots: A 18, 44; B 65, 02; C 99, 16; D 25, 86; E 28, 61; F 92, 80; G 03, 26; H 32, 79; I 13, 56; J 86, 19.

Modifications may be made in the size and fineness of the grid and in the number of dots located. The model may be modified in various ways: for example, to preclude the placing of further dots on squares already occupied, on squares contiguous to these, and so on. The random distribution generated by the process described may be used for comparison with actual distributions in the real world. In particular nearest neighbour features (see chapter 4) may be studied.

(ii) *Random distribution of points on lines.* Often in geography dots are distributed not on continuous surfaces, but along lines. Thus the distribution of garages or of shops occurs along lines, the streets in a road system. With some modifications, the procedure for locating dots randomly on a surface can be adapted to locate them randomly along lines.

A network of lines may be divided into a number of subsections with the degree of fineness required. Each subsection may then be given a number. The numbers will then be called randomly. A slightly more elaborate case is illustrated in figure 11.5. An imaginary village has 83 building lots. These face each other along the streets. They may be numbered in any way as long as a new number is used for each lot and as long as all numbers have an equal probability of being called in the random number table.

The procedure is to call appropriate random numbers, using 00–99 if up to 100 lots are numbered or 000–999 up to 1000 are numbered, and so on.

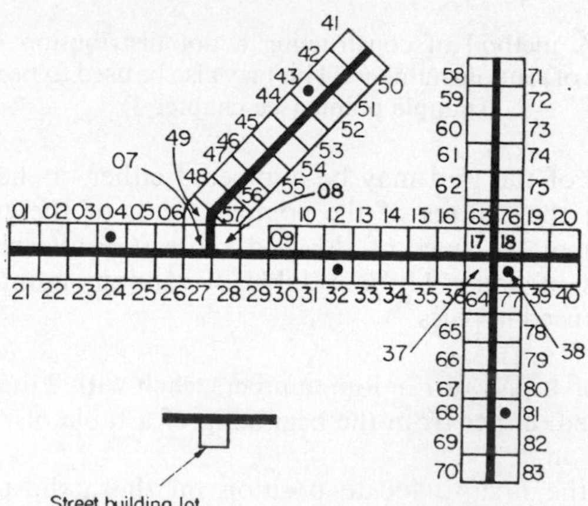

Figure 11.5 A method of constructing a random distribution of dots on a line base.

In the example, the following numbers were called: 38, 32, 92 (rejected), 81, 04, 43 and five lots were thus selected. Their location is shown in the figure.

The study of features such as frequency of occurrence of neighbouring pairs of randomly chosen lots makes it possible to find features of a random distribution of points on lines. Real situations may then be matched against these. In this way it is possible, for example, to test whether shops in a town are more clustered than they would be if randomly distributed.

(iii) *Political units based on points on a surface.* On the basis of a random distribution of dots it is possible to simulate a particular kind of political unit. The model described could be modified in various ways.

Step 1: Produce a random distribution of dots, as described in (i) and shown in figure 11.4. In figure 11.6, the left hand half of the distribution of

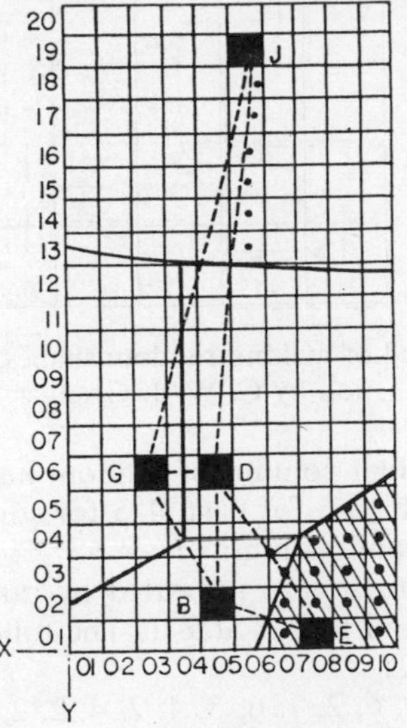

Figure 11.6 A method of building political and other units round an existing or previously constructed (as in 11.4) distribution of dots.

dots has been used (see black squares). These are assumed to be five villages. Their parishes are to be such that any place in the area belongs to the parish of the village nearest to it.

Step 2: Join neighbouring villages by coloured (or dotted) line.

Step 3: Bisect each village–village line and draw a new line at right angles to it. Continue the new line as far as required in each direction. Step 3 has the effect of producing the required parishes. Details must be checked since there appears at times to be some ambiguity.

From the model produced it is possible to study features of political units, particularly area and shape. A rough calculation of area may in fact be quickly made by counting grid squares, whole or over half, falling in each unit. Thus parish E has 15 units of area.

(iv) *Random shape.* The description of shape is very difficult and subjective. The following is a possible method of producing a random shape (see figure 11.7). Use a grid of squares as the base. Select a starting square at the centre of the grid.

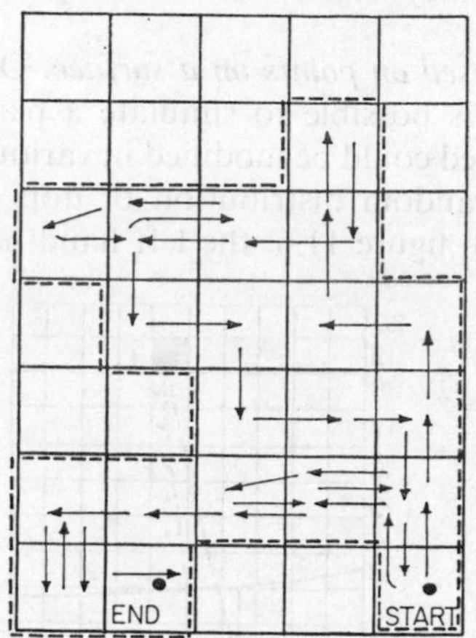

Figure 11.7 A method of making random shapes, suggested in conversation by C. W. J. Granger.

Step 1: Use a single digit column of random numbers to make a random walk, with 0–1 for north, 2–3 for east, 4–5 for south, 6–7 for west (8 and 9 are not used). Find first random number.

Step 2: Draw arrow to square indicated by random number. Make this the new starting point and repeat Step 1. The following digits were used in the example in figure 11.7:

9̸, 1, 8̸, 0, 0, 6, 0, 1, 4, 6, 6, 2, 7, 6, 3, 4, 2, 4, 2, 2, 4, 5, 0, 7, 3, 7, 6, 3, 6, 7, 7, 5, 1, 9̸, 4, 3, 9̸, 9̸. (8 and 9 were rejected /).

(v) *Random diffusion model.* A diffusion model is described here by an example. The reader should be able to develop this along many different lines. The following equipment is necessary. (1) Some base, preferably laid out as a regular grid of squares or hexagons. (2) A diagram for determining the placing of a telling. (3) A table of random numbers. (4) A worksheet to record whether tellings are successful or not. These are illustrated in figure 11.8.

The map shows a fictitious area that might be in India or Africa. Unless otherwise indicated (M), each compartment (square) contains a village. The United Nations (or some such institution) wants to place a number of experts (these are short in supply) in selected villages for a given length of time to teach some new technique (e.g. method of ploughing, new kind of

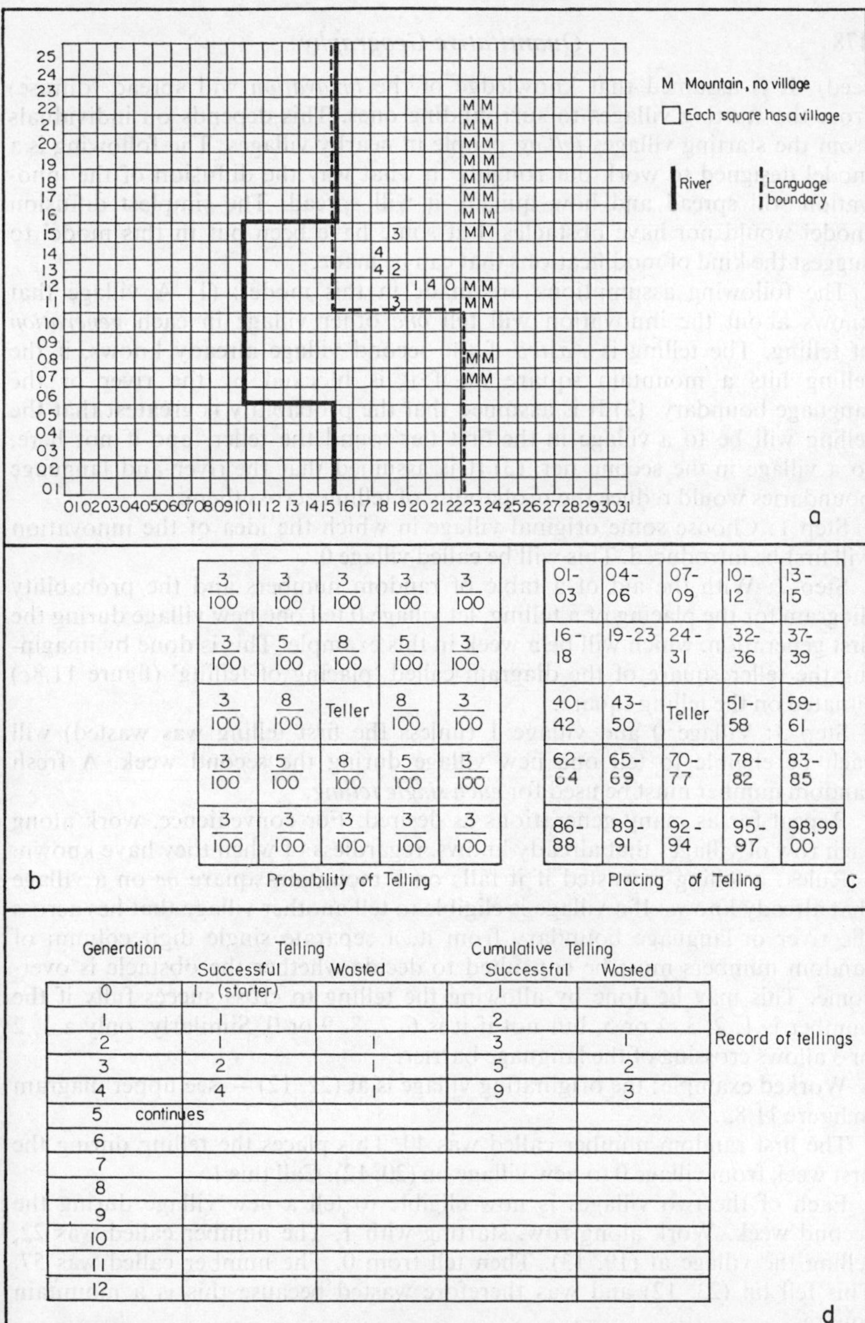

(b) Probability of Telling

3/100	3/100	3/100	3/100	3/100
3/100	5/100	8/100	5/100	3/100
3/100	8/100	Teller	8/100	3/100
3/100	5/100	8/100	5/100	3/100
3/100	3/100	3/100	3/100	3/100

(c) Placing of Telling

01–03	04–06	07–09	10–12	13–15
16–18	19–23	24–31	32–36	37–39
40–42	43–50	Teller	51–58	59–61
62–64	65–69	70–77	78–82	83–85
86–88	89–91	92–94	95–97	98, 99, 00

(d) Record of tellings

Generation	Telling: Successful	Telling: Wasted	Cumulative Telling: Successful	Cumulative Telling: Wasted
0	1 (starter)		1	
1	1		2	
2	1	1	3	1
3	2	1	5	2
4	4	1	9	3
5	continues			
6				
7				
8				
9				
10				
11				
12				

Figure 11.8 A model for simulating the diffusion of an innovation over a surface. Other obstacles could be placed on the surface as required. The first four generations of telling have been done in this example. (*a*) The area over which the diffusion is to take place. (*b*) The probability of telling villages in the first and second tiers away from the teller. (*c*) The corresponding random numbers. (*d*) A method of recording successful and wasted tellings.

seed). It is assumed that knowledge of the *innovation* will spread (diffuse) from the starting villages to surrounding ones. This depends on individuals from the starting villages *telling* people in nearby villages. The following is a model designed to work out roughly in what way the diffusion of the innovation will spread and how quickly it will spread. The simplest diffusion model would not have obstacles, but some have been put in this model to suggest the kind of modifications that can be made.

The following assumptions are made in this model: (1) A village that knows about the innovation will tell *one* other village in each *generation* of telling. The telling is *wasted* if the second village already knows, if the telling hits a mountain square or if it is blocked by the river or the language boundary. (2) It is assumed that the probability is greatest that the telling will be to a village in the first tier round the teller, and if not here, to a village in the second tier. (3) It is assumed that the river and language boundaries would reduce the probability of telling across them.

Step 1: Choose some original village in which the idea of the innovation will first be introduced. This will be called village 0.

Step 2: With the aid of a table of random numbers and the probability diagram for the placing of a telling, let village 0 tell one new village during the first generation, which will be a week in this example. This is done by imagining the teller square of the diagram called 'placing of telling' (figure 11.8*c*) situated on the telling square.

Step 3: Village 0 and village 1 (unless the first telling was wasted) will each be eligible to tell one new village during the second week. A fresh random number must be used for *each single telling*.

Repeat for as many generations as desired. For convenience, work along each row of villages that already knows, regardless of when they have known.

Rules: a telling is wasted if it falls on a mountain square *or* on a village that already knows. If a village is eligible to tell another village that lies across the river or language boundary from it, a separate single digit column of random numbers must be consulted to decide whether the obstacle is overcome. This may be done by allowing the telling to cross successfully if the number is 1, 2, 3, 4 or 5, but not if it is 6, 7, 8, 9 or 0. Similarly, only a 1, 2 or 3 allows crossing of the language barrier.

Worked example: the originating village is at (22, 12) — see upper diagram in figure 11.8*a*.

The first random number called was 40. This places the telling during the first week from village 0 to new village on (20, 12). Call this 1.

Each of the two villages is now eligible to tell a new village during the second week. Work along row, starting with 1. The number called was 22, telling the village at (19, 13). Then tell from 0. The number called was 57. This fell on (23, 12) and was therefore wasted because this is a mountain square.

Each of the three villages is eligible to tell a new village during the third week. Work along rows:

2 calls 09, which hits (19,15).

1 calls 68, which hits (20, 11).

0 calls 88, and is eligible to tell (21,10), across the language boundary, but the calling of a 9 from the single column of digits for boundary crossing disqualifies the telling, which is wasted. Two new villages know, therefore, at the end of the third week.

The diffusion continues as shown by the 4s on the top diagram and for as many generations as desired thereafter. The reader developing his own models on these lines is advised to have sheets of square and hexagonal grids duplicated.

11.5 Mathematical models

(*a*) *Physical*

This type of model is more formal than those discussed so far in that symbols are used to represent variables in the prototype and these are manipulated by the standard techniques of mathematics. This type of model can be divided into two classes according to R. J. Chorley (1964). He differentiates between the deterministic mathematical model and the stochastic mathematical model, the former being based on mathematical certainty and the latter on probability. In most types of problem the mathematical model must be a simplified form of the real prototype. One of the advantages to be gained by using this type of model is that the great complexities of the natural situation can be resolved into the most significant components by ignoring the minor and less essential items. In the natural situation there are normally many unknown variables or at least some whose exact influence is not known. This necessitates the making of assumptions in the mathematical model. If these assumptions are not justified then the model will not behave sufficiently closely to the prototype to provide valuable results and the assumptions will require some modification. The application of this type of model has been made most frequently in physical geography, particularly in problems such as slope development or river action. These processes can be expressed adequately in numerical terms and certain known physical relationships can be used. In this field mathematical models are an extension of models in physics. They are, however, applied to natural conditions in which it is much more difficult to control and to enumerate the variables involved in the process under study. The problems tackled by this method are essentially dynamic problems in which change can be measured, or inferred by reference to examples at different stages of development.

Mathematical models make it possible to deduce from the stated initial conditions, and from the assumptions that are inserted in the model, how the feature modelled will respond. An example of this type of mathematical model approach is J. F. Nye's (1957) work on the movement of a glacier or ice sheet. He makes the basic assumption that glacier ice is a plastic substance and that it obeys the laws associated with the behaviour of plastic substances. In order to set up the model other simplifications must be introduced, for example the geometry of the bed of the ice sheet or glacier must be very drastically simplified from its natural state. Thus an ice cap can be assumed to have a flat base and the ice, assumed to be perfectly plastic, will yield with infinite ease when the critical stress is reached. The stress pattern is shown in figure 11.9 for a glacier. Assumptions must also be made concerning the roughness of the bed.

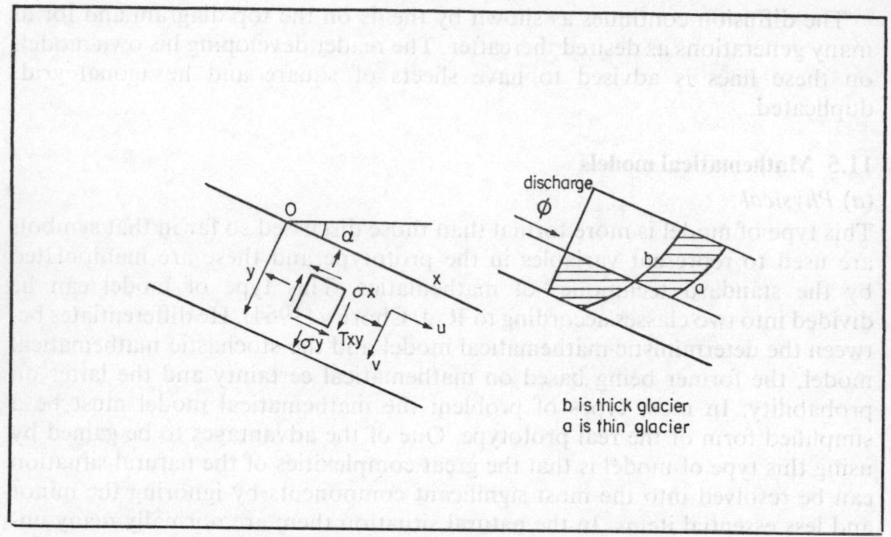

Figure 11.9 Theoretical models of glacier flow. The theoretical stress pattern is shown. (After Nye.)

Differential equations can then be set up. These equations allow the movement of the ice to be calculated by inserting appropriate values for the assumed variables. Different values for such variables as the thickness of the ice, the radius of the ice cap, and the temperature, which are variables that can be measured in reality, can be inserted in the equation. In this way the model can be used to calculate the movement of the ice if the assumptions made are in fact fulfilled. The results of the calculations of the model values then yield the form of the ice cap, which is represented by a parabola. Measurements of the actual ice cap can then be compared with the model results.

The results of such a mathematical model study can be used to provide values of basal ice movement which are difficult to obtain by direct measurement, but which are very important in interpreting the geomorphological effect of ice caps or glaciers. In this situation the problem still remains of assessing the validity of the model results. A model of this type, then, can provide data that can be checked against field observations. If the two results are similar, then confidence in the model enables data to be obtained that cannot be easily measured in the field.

In the stochastic type of model the basis of the model is not a known relationship in physics that can be expressed and worked out by mathematical equations to produce a definite result. In this type of model the results must be expressed in terms of probability based on the use of statistical techniques that are based on the probability theory. A. E. Scheidegger (1961) has reported the work of H. A. Einstein (1950) in this field. He calculates the probability of erosion for any particle on the bed of a river in terms of probability theory. He brings into the equation such variables as the setting velocity of the particle, its density and that of the fluid. The equations yield a value of the bed load in terms of the probability of erosion and the stream velocity. They can then provide a relationship between the stream velocity, which can be readily measured, and the bed load transportation, which is very difficult to measure in the field. However, many variables that enter the equation are difficult to define precisely in quantitative terms. There remains, therefore, some measure of doubt between the two values, and correct numerical values are difficult to achieve.

(*b*) *Mathematical models in human geography*

An example of a model that is mathematical in character and has some bearing on human geography, is the rank size 'rule' as applied to 'geographical' items. It is not the purpose of this subsection to discuss the usefulness or findings of this model but simply to describe with an example how it may be constructed. The model widely applied by G. K. Zipf (1949), is discussed by H. A. Simon (1956). B. J. L. Berry and W. L. Garrison (1958) discuss its application to town sizes. While the rank size rule has largely been used in urban geography, there seems no reason why it should not be applied in political geography (e.g. size of political units within countries) and in economic geography (e.g. size of turnover of firms, data for which are published yearly in *Fortune*).

It has frequently been found empirically, as with incomes in a country, that in a set of observations there will be a few very large values or objects and that as the value decreases, the number of cases increases in a fairly regular way. The following example shows the method of applying the 'rule' to the 50 largest towns of North America, Latin America, the U.S.S.R. and Africa south of the Sahara. The towns are ranked from high to low according to population for each area. To save space, only the complete North American list is given, together with the top five and bottom five for the other three regions. Populations are in thousands (table 11.1). Canadian towns are in italics.

Two procedures may be applied to the data, as shown in figure 11.10. (i) On a graph with rank on the *x*-axis and population on the *y*-axis, plot each town. This is illustrated in the main graph. (ii) On a graph with rank on the *x*-axis and population on the *y*-axis, but with a logarithmic (base 10) scale for both *x* and *y*, plot each town. This is illustrated in the smaller graph.

The first curve shows all towns in similar detail, whereas the second curve stretches out the largest towns but telescopes the smaller ones. The first curve is steeply concave, but the log–log curve should be straight. Both curves allow the inclusion of a town of any size on the *y*-axis and any number of towns on the *x*-axis. However, the goodness of fit in the straight line deterior-

Table 11.1 Complete list of top 50 North American urban agglomerations

Rank		Size	Rank		Size	Rank		Size
1	New York	14,115	21	Kansas City	921	41	Dayton	502
2	Los Angeles	6489	22	Seattle	864	42	Rochester	493
3	Chicago	5959	23	Miami	853	43	*Winnipeg*	476
4	Philadelphia	3635	24	New Orleans	845	44	Akron	458
5	Detroit	3538	25	San Diego	836	45	Albany	455
6	San Francisco	2431	26	Denver	804	46	Sacramento	452
7	Boston	2413	27	*Vancouver*	790	47	Springfield (Mass.)	450
8	*Montreal*	2110	28	Atlanta	768	48	Toledo	438
9	*Toronto*	1824	29	Providence	660	49	*Ottawa*	430
10	Washington	1808	30	Portland (Or.)	652	50	Oklahoma City	429
11	Pittsburgh	1804	31	San Antonio	642		U.S.A. and *Canada*	75,958
12	Cleveland	1785	32	Indianapolis	639			
13	St. Louis	1668	33	Columbus (Oh.)	617			
14	Baltimore	1419	34	Louisville	607			
15	Minneapolis	1377	35	San Jose	603			
16	Milwaukee	1150	36	Phoenix	552			
17	Houston	1140	37	Memphis	545			
18	Buffalo	1054	38	Birmingham	521			
19	Cincinnati	994	39	Norfolk	508			
20	Dallas	932	40	Fort Worth	503			

Top and bottom 5 of lists for other three regions

1	Buenos Aires	6763	Moscow	6296	Johannesburgh	1974
2	Mexico (D.F.)	4871	Leningrad	3498	Cape Town	804
3	Rio de Janeiro	4370	Kiev	1208	Durban	681
4	São Paulo	3872	Baku	1067	Ibadan	600
5	Santiago	1989	Gorky	1025	Addis Ababa	449
6–45 follow						
46	Concepción	232	Zhdanov	321	Yaounde	93
47	Mar del Plata	230	Astrakhan	320	Fort Lamy	91
48	Managua	230	Frunze	312	Mogadiscio	91
49	Bucaramanga	222	Gorlovka	309	Lome	90
50	León	210	Lugansk	306	Abeohuta	84
	Latin America	45,885	U.S.S.R.	37,551	Africa (south of Sahara)	12,609

All population figures in thousands.

ates both at the end where the few large towns occur and at the end where there are very many small settlements.

The curve is of some interest, both for what it shows and for what it implies. In the example in figure 11.10 it is easy both to see anomalies in each of the four main areas and to compare them. It may be noted, for example, that among the ten largest towns, the U.S.S.R. curve drops sharply after Moscow and Leningrad from 6 million–3½ million to several towns between 1½ and 1 million. Latin America's first four towns remain high, however. The

U.S.S.R. and Latin America curves cross at about the 12th town. Even allowing for differences in the total population of the 50 towns of each area, this would not happen if each set of towns fitted the straight line. One thought provoking implication of the straight line fit is that in a system of towns with a clear cut hierarchy of central places of various orders, there should be clustering of towns at certain population levels.

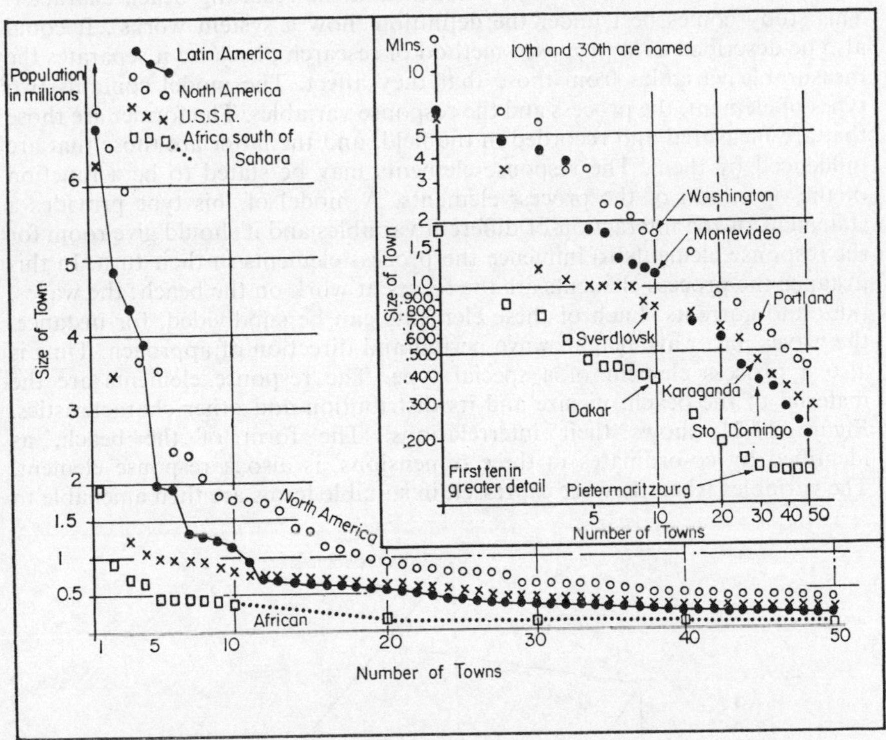

Figure 11.10 The rank size rule applied to the 50 largest urban agglomerations of North America, Latin America, U.S.S.R. and Africa south of the Sahara.

11.6 Theoretical or conceptual models

(a) Introduction

In the list of definitions at the beginning of the chapter the terms 'framework' and 'structure' were introduced. These terms describe well the essence of the theoretical model in one of its meanings. It may be considered to be the broad framework of a particular discipline in which its different parts are linked to provide a general statement of the aims and subject matter of the whole discipline. This type of unifying framework or broad theoretical model has been lacking in geography, and the dichotomy between physical and human geography has often not been effectively bridged. Attempts have been made in this book, however, to show that these two aspects can quite often be linked both by the methods used to study them and also in their general characteristics.

An example of a theoretical model in physical geography of a more limited scope is the conceptual model set up by W. C. Krumbein (1963) to study the movement of sand on a beach and the resulting beach character. This study comes best under the definition 'how a system works'. It could also be described as 'a proposed method of research', because it separates the measurable variables from those that they affect. The model contains two types of element, the process and the response variables. The former are those that are measured and recorded in the field, and the latter are those that are influenced by them. The response elements may be stated to be a function of the operation of the process elements. A model of this type provides a statement of the interaction of different variables and it should give room for the response elements to influence the process elements in their turn. In this instance the process elements are the forces at work on the beach: the waves, tides and currents. Each of these elements can be subdivided, for instance, the waves into wave height, wave period and direction of approach. Time is also a process element of a special type. The response elements are the material of the beach, its size and its distribution and other characteristics. Figure 11.11 shows their interrelations. The form of the beach, as identified by co-ordinates in three dimensions, is also a response element. The variables when they are expressed in suitable terms are then amenable to

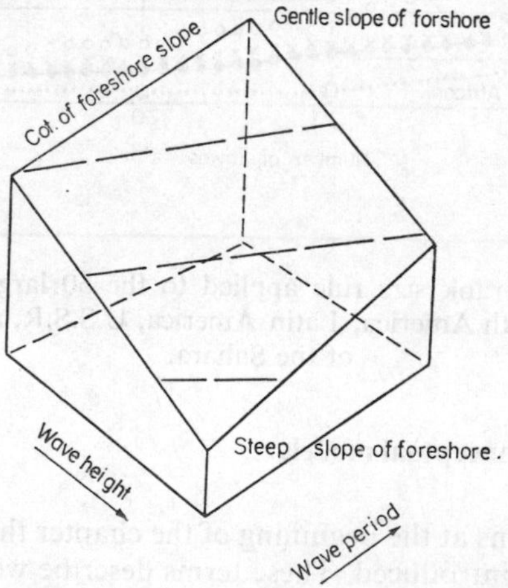

Figure 11.11 Conceptual beach model to show effect of wave period and wave height on beach foreshore gradients. For any given wave heights the slope gets gentler as the wave period increases (see also figure 12.1). For any given period the slope gets steeper as the wave height increases. (After Krumbein.)

statistical analysis by appropriate methods. The results of the analysis provide a model of the relationships between the response and process elements in this field of study. Further theoretical models are considered in chapter 12.

(*b*) *Spacing*

One of the most persistent preoccupations in human geography is about the actual and the theoretical arrangement in space of man-made objects, particularly settlements. Although the question is usually directed towards spacing on a 2-dimensional surface, the same concept applies to the spacing of objects on a line (e.g. stations along a railway) and in 3-dimensional space.

Spacing is connected with a branch of mathematics called packing theory, which studies the packing of things in various dimensions. The need to space or pack things in a good way on the earth's surface as well as in boxes, honeycombs and so on, is related to the principle of making optimum use of space available. The optimum or at least a good arrangement of objects in space is connected with the so-called principle of least effort associated particularly with the name of G. K. Zipf (1949). Thus, for example, the best spacing of service centres in an area implies that the people using them will in total spend the least amount of effort necessary in doing so. In general, the success or otherwise of a particular spacing of objects depends on people or things (or both) moving to and from them. Time, therefore, is also a prerequisite for spacing to become operationally meaningful.

Spacing is an idea that must have preoccupied man ever since he placed dwellings together in a village. The arrangement of plants in fields at equal intervals along rows is clearly a spacing problem, as is the distribution of market towns in many parts of Europe in the Middle Ages, and other parts of the world before this. Thomas More considered the spacing of his towns in Utopia (first published 1516) to be worth mention:

> 'There are fifty-four splendid big towns on the island, all with the same language, laws, customs and institutions. They're all built on the same plan, and, so far as the sites will allow, they all look exactly alike. The minimum distance between towns is twenty-four miles, and the maximum, nor more than a day's walk.'

One is left full of admiration at the distance Utopians could walk.

It is proposed in this subsection to show by examples how some kinds of spacing are superior to others, rather than to go into elaborate theory either from a mathematical point of view, or a geographical one. For more advanced development of hierarchies and other related topics, the reader is referred to various papers in *Proceedings of the IGU Symposium in Urban Geography*, held in 1960, edited by K. Norborg (1962), especially papers by W. Christaller (p. 3) and M. F. Dacey (p. 55). The subject is also approached from various angles by P. Haggett (1965), especially in parts of chapters 4 and 5.

Figure 11.12*b–f* shows five fictitious areas all equal, and each divided into 81 smaller equal sized squares. The total area is 81 units and each small square is 1 unit of area. Each of the five areas has been subdivided into nine farms. Each farm consists of a farm plot (one unit of area) and eight fields (total 8

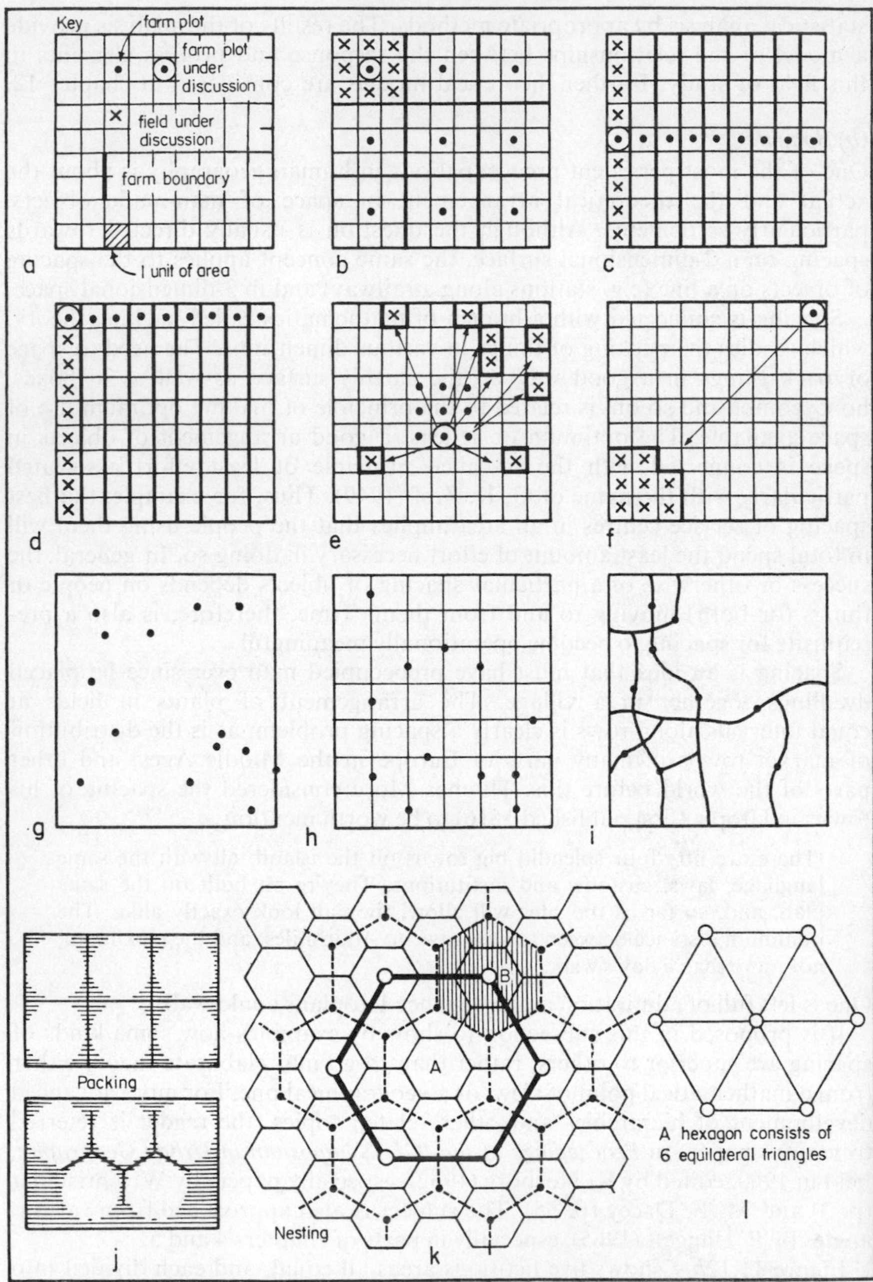

Figure 11.12 Examples of the spacing of dots and the nesting of hexagons (*a*)–(*f*) Farms and fields in various arrangements. (*g*)–(*i*) Dots on a continuous surface. (*h*) A regular network. (*i*) An irregular network. (*j*) Packing. (*k*) Nesting of hexagons. (*l*) Hexagon as six equilateral triangles.

units of area). The example illustrates three interconnected matters, the spacing of farms, the centring of farm buildings within farm areas, and the shape of the farms themselves.

Case (b) shows 9 square farms. In general hexagonal compartments would be somewhat more compact, but in this case there is little to choose because the number of compartments involved is small. Case (b) therefore represents a satisfactory arrangement in the situation being considered.

Cases (c) and (d) have greatly elongated farm areas. Such a situation could occur where three bands of soil of varying fertility are shared fairly among them. In case (e) the farms are fragmented and one of the nine farms has its eight fields distributed as shown. These were located with random numbers. In case (f) the farm is actually away from the land it cultivates, which itself forms a compact unit of 8 fields.

Figure 11.12 is included primarily to show the relationship of the spacing of farms to total effort expended by farmers to visit and work their land. It is assumed that any farmer visits each of his eight fields with equal frequency. It has been found empirically in many cases that the more distant land of a farm or a village community tends either to be neglected or to have uses that require relatively little labour. This theme is well documented by M. Chisholm (1962).

It is possible to compare numerically the effort required for the farmers in the given examples. The side of each small square is one unit of distance. The units of distance to cover a journey from the farmstead to each of the eight fields is calculated as follows:

Case (b) $4 \text{ units} + 4 \times \sqrt{2} \text{ units} = \text{approx. } 11.6$

Case (c) $2\,(1+2+3+4) \text{ units} = 20$

Case (d) $1+2+3+4+5+6+7+8=36$

Case (e) $5.5+5+4.5+3.5+5.5+3+3=30$

Case (f) approximately 80

If the possibility of the farmer visiting more than one field on a journey is ruled out, then the regular dispersal of farms in Case (b) gives the farmers less travelling distance than any other case and shows the spacing of farms that gives the least amount of total effort to the farmers. In case (f), the inclusion of the other 8 farms and their respective fields would bring the average distance down somewhat since not all the farms could be as far from their fields as the one illustrated.

With regard to the positioning of each farm plot in its farm area, the farms in both cases (b) and (c) are well centred, whereas the farm in (d) is not. Varying shapes of farm have also emerged from the example: fairly compact in case (b), elongated in (c) and (d) and fragmented in different ways in (e) and (f). Even this simple example suggests that case (b), the most anti-clustered type of spacing in the cases shown, might be expected to occur spontaneously in farming areas and to be aimed at by planners in areas of planned settlement.

Spacing will now be considered from a more general viewpoint. Spacing requires firstly a space and secondly objects on it. The space may be assumed to be uniform in a simple model, but it may be varied if required. The number of objects ranges from one to potentially many millions. One object is a special case since it is not 'competing' with other objects for a place. At the other extreme, the geographer does not usually in practice deal with the spacing of more than several hundred discrete objects in a given study. In two dimensions, the space is usually thought of as a continuous surface, upon which objects may occupy any point (figure 11.12*g*). The space could, however, be a regular or irregular network on a surface rather than the whole surface (see figures 11.12*h–i*).

While in the farm example the arrangement of farms at the centres of squares produced the farm spacing giving least effort to the farmers, the spacing of points on a surface at the centres of hexagons is the most advantageous way, if the number of points exceeds about 30 and the surface is bounded. This can be shown experimentally with some tins (cans) in a box, or with coins on a table top. If the number of coins that can be fitted into a given rectangular frame is small, it may be possible to pack more in the frame if the centre of each coin is placed at the centre of a square rather than of a hexagon. This is shown in figure 11.12*j*,

The equilateral triangle, square and hexagon are the only polygons with all sides of equal length that will cover a surface without overlapping or leaving residual spaces between. Of the three, the hexagon is the most compact, and most closely approaches the circle, the most compact of all figures on a 2-dimensional surface. A further property of a network of hexagons is that nesting may take place. On a given network of hexagons, a coarser network with only a third as many hexagons, each three times as large, can be fitted perfectly. The second network will be 'tilted' somewhat in relation to the first. These features of hexagons are illustrated in figure 11.12*k*. In the nesting diagram there is a basic network of small hexagons. Letter A stands at the centre of one of them. A second network, with broken line sides, rests on this. One hexagon in this network, centring on B, has been shaded. Its area can be seen to be three times that of the smaller hexagons because it contains the small hexagon centring on B together with $\frac{1}{3}$ of the area of each six contiguous small hexagons. A third network is also shown. One complete hexagon in this (with thick sides and open circle nodes) centres on A. It is three times larger than the hexagons in the middle level and nine times larger than those in the small level. Its orientation is similar to that of the basic network of smallest hexagons.

Central place theory, involving both the spacing and the hierarchical functional role of settlements, and also accompanying communication networks, is related to the geometrical properties of the hexagonal system illustrated. There have been attempts, as by W. Isard (1956), to remove some of the rigidity of the Euclidean geometrical system by suggesting that away from large concentrations of population, density would diminish, the distance between centres increase, and the hexagons be transformed in a topological fashion by an appropriate stretching of the network.

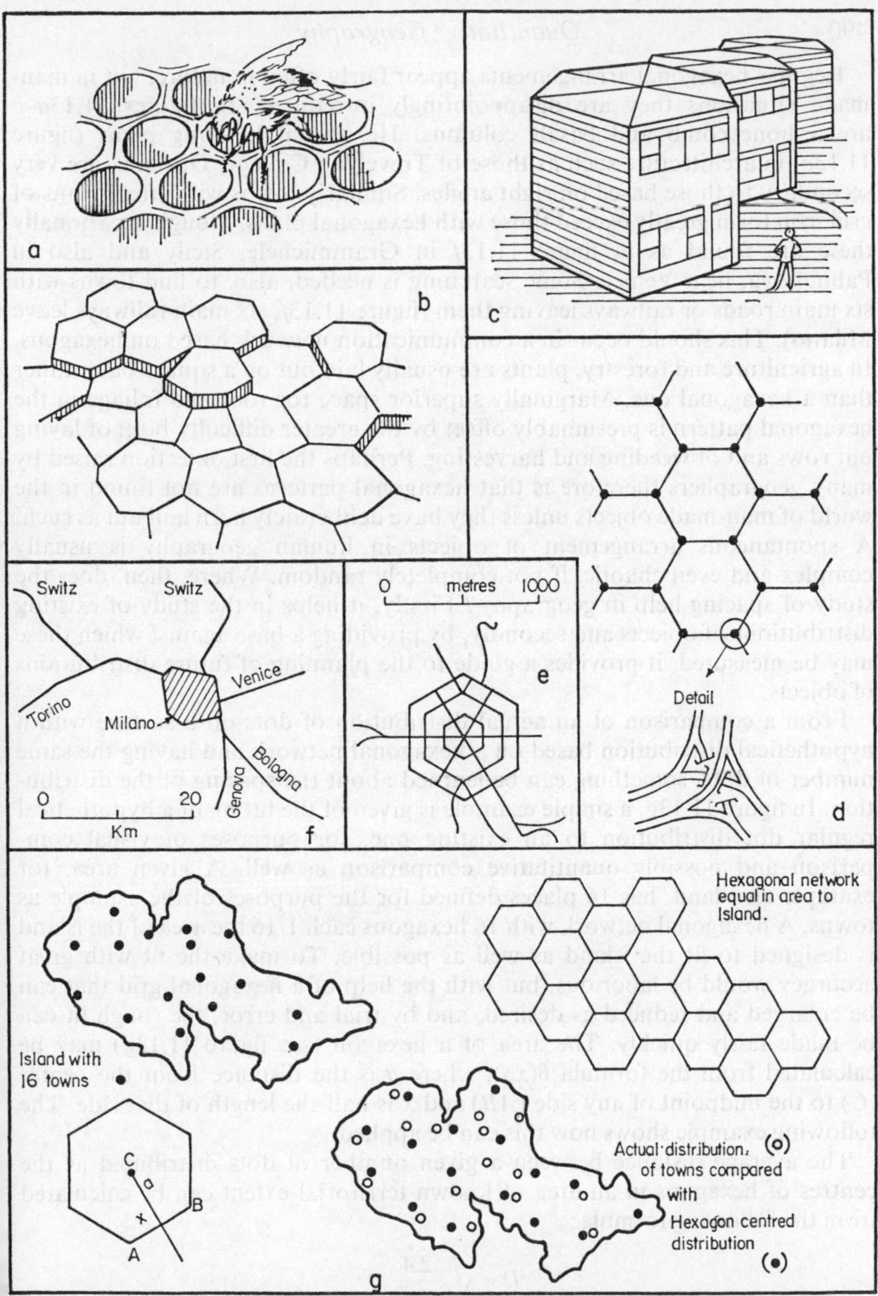

Figure 11.13 The hexagon and its occurrence and application. (*a*) (*b*) in nature. (*a*) Honeycomb. (*b*) Surface eroded basalt columns, Giant's Causeway, Antrim, Northern Ireland. Note that the columns are not regular in section. (*c*) Hexagonal 2-storey houses. (*d*) Hexagonal street layout suggested in *Traffic in Towns* (p. 43). (*e*) Selected streets of Grammichele planned 17th century village, Sicily. (*f*) Six principal railways leaving Milano, North Italy. (*g*) Actual distribution of dots compared with theoretical distribution (same number of dots as hexagon centres).

Regular hexagonal arrangements appear fairly often in nature, but in man-made situations they are disappointingly infrequent. In figures 11.13*a–b* are a honeycomb and basalt columns. Hexagonal building plans (figure 11.13*c*) in architecture such as those of Trevelyan College, Durham, are very secondary to those based on right angles. Similarly towns with street plans of grid iron form greatly exceed those with hexagonal plans, though occasionally these are found as in figure 11.13*f* in Grammichele, Sicily and also in Palmanova, near Venice. Some searching is needed, also, to find towns with six main roads or railways leaving them (figure 11.13*f*, six main railways leave Milano). This should occur in a communication network based on hexagons. In agriculture and forestry, plants are usually laid out on a square base rather than a hexagonal one. Marginally superior space for root and foliage in the hexagonal pattern is presumably offset by the greater difficulty both of laying out rows and of weeding and harvesting. Perhaps the first objection raised by many geographers therefore is that hexagonal patterns are not found in the world of man-made objects unless they have deliberately been laid out as such. A spontaneous arrangement of objects in human geography is usually complex and even chaotic, if not completely random. Where, then, does the study of spacing help in geography? Firstly, it helps in the study of existing distributions of objects and secondly, by providing a base against which these may be measured, it provides a guide to the planning of future distributions of objects.

From a comparison of an actual distribution of dots on a surface with a hypothetical distribution based on a hexagonal network and having the same number of dots, something can be learned about the spacing of the distribution. In figure 11.13*g*, a simple example is given of the fitting of a hypothetical regular dot distribution to an existing one, for purposes of visual comparison and possibly quantitative comparison as well. A given area, for example an island, has 16 places defined for the purposes of the example as towns. A hexagonal network with 16 hexagons each 1/16 the area of the island is designed to fit the island as well as possible. To make the fit with great accuracy would be laborious, but with the help of a hexagonal grid that can be enlarged and reduced as desired, and by trial and error, the rough fit can be made fairly quickly. The area of a hexagon (see figure 11.13*g*) may be calculated from the formula $6(xa)$, where a is the distance from the centre (C) to the midpoint of any side (AB) and x is half the length of that side. The following example shows how this can be applied.

The average distance between a given number of dots distributed as the centres of hexagons in an area of known territorial extent can be calculated from the following formula:

$$D = \sqrt{\frac{2A}{n\sqrt{3}}}$$

D = the average distance between neighbouring dots.
A = the total area of the piece of territory under consideration.
n = the number of dots.

Areal and linear measurements must be in the same units. Example: an urbanized area of 150 sq miles in extent has 500,000 houses. What is the distance between each house and its six nearest neighbours if all the houses are placed at the centres of hexagons in a network?

$A = 150, n = 500{,}000$

$$D = \sqrt{\frac{2 \times 150}{500{,}000\sqrt{3}}} = \sqrt{\frac{300}{850{,}000}} = \sqrt{\frac{1}{2833}}$$

$$\simeq \tfrac{1}{53} \text{ mile} \simeq 33 \text{ yards}$$

What kind of spacing problems are planners considering now? Many manufacturing and service activities are concerned with the problem of organizing branches within a given area, such as a country. How many does a given firm need, and where should it locate them? The problem is sometimes referred to as a 'warehousing' problem. But it involves the deployment of such diverse things as factories with a regional rather than a national market, sales offices, hospitals, schools, churches, petrol stations and even airports. Even the deployment of letter boxes, fire hydrants and lamp posts, is a spacing problem, usually on a line network. Unfortunately for the geographer, the space on which his n objects may be located is usually so varied that it is absurd to think of every point on it having equal eligibility for the location of the objects to be spaced.

To conclude this subsection, an interesting example from Paris will be briefly referred to.* During 1966–80 it is planned to build 142 new churches in Greater Paris. It has been assumed on the basis of sociological research that town dwellers will not frequent a church unless it is within about 600 metres of their home. In many suburbs of Paris this requirement is not fulfilled. Moreover, it has been decided that 1000 people is the maximum desirable capacity for any one church. The total population of Greater Paris and the proportion of this that are regular churchgoers, together with certain other information, have made it possible to plan in total, rather than to continue putting new churches haphazardly as before. T. Lakshmanan (1964) outlined a model for the location of retail hubs in the Baltimore region. Traffic patterns and retail expenditure were taken into consideration.

(c) *Centre and periphery*

In this subsection a special situation will be considered, that of the best location of one thing in an area. What is under consideration might be thought of as a question of a position of command or of accessibility. It is a special situation of spacing, in which interest is not focused on the relationship of many dots to each other, as was the primary consideration in the previous subsection, but on the relationship of one dot to the space, line or more often a 2-dimensional space, on which it is situated. The object is assumed to have some relationship to the whole space in which it is situated.

The relationship will usually be in the form of journeys between the object

* *Times* — 29 April, 1964.

in question and so many other places. Most of the journeys will in fact occur on networks rather than over a continuous surface. Moreover the more technologically advanced and therefore 'mobile' the community being studied, the longer and/or more frequent will the journeys tend to be. The fact that frequency of journeys between places tends to diminish as distance increases will be overlooked at this stage. The theory behind the command position model and the techniques for studying it are closely related to those discussed in chapter 4 under the running mean in area. In general, too, the best location is the centre of minimum aggregate travel.

If the space in which an object or dot is located is itself uniform, and if the object to be located has some sort of relationship to all of the space under consideration, then in very general terms, a central location is more advantageous than a peripheral one. Centre and periphery may be defined for the purposes of a particular study as an inner area and a residue (for example, the innermost 25 per cent or 50 per cent) or the two terms may be used more loosely. It is also possible to think of an area as divided not into two discrete parts, but as a continuous surface with points at varying distances between the centre (mean centre of area) and the remotest point on the periphery. The advantage of a central position over a peripheral one is often retained even when the base under consideration does not have things evenly spread over it.

The points made so far will now be illustrated with some examples in figure 11.14. In chess all squares are of equal value, but those in the central part of the board are recognized by most players to be important in the sense that command of them often leads to victory. One important reason for this seems to be that the number of squares controlled by four of the five pieces (excluding pawns), the king, queen, bishop and knight, but not the rook, is greater on the central squares and tends to diminish somewhat irregularly towards the periphery. The knight's move is illustrated in figures 11.14*a*. At position (1) the knight can move to any of eight squares, at (2), only to four squares and in the corner, at (3), only to two. He could even do better than this from off the board, if this were allowed in the game, as from (4), like an offshore pirate radio station near a big population centre. In chess, of the five pieces, only the rook controls the same number of squares (as long as there are no intervening pieces) from any place on the board.

Figure 11.4*b* shows that Madrid, capital of Spain, has within an arbitrarily chosen radius of 250 miles, a much larger percentage of the total national area than do either Barcelona or Seville. The position of command is illustrated diagrammatically in the next three figures (11.14*c*). The circle on the right has a radius of one unit of distance. From the centre, D commands the whole circle. F, on the periphery, has considerably less than half of the circle within a distance of 1 unit from it, while even E, in a half way position, only extends over about $\frac{2}{3}$. A network of hexagons forms a useful basis for studying and experimenting with both imaginary and real situations. If the distance between the centres of adjoining hexagons is considered to be 1 unit of distance, then it can be seen from figure 11.14*c* that the centres of all the

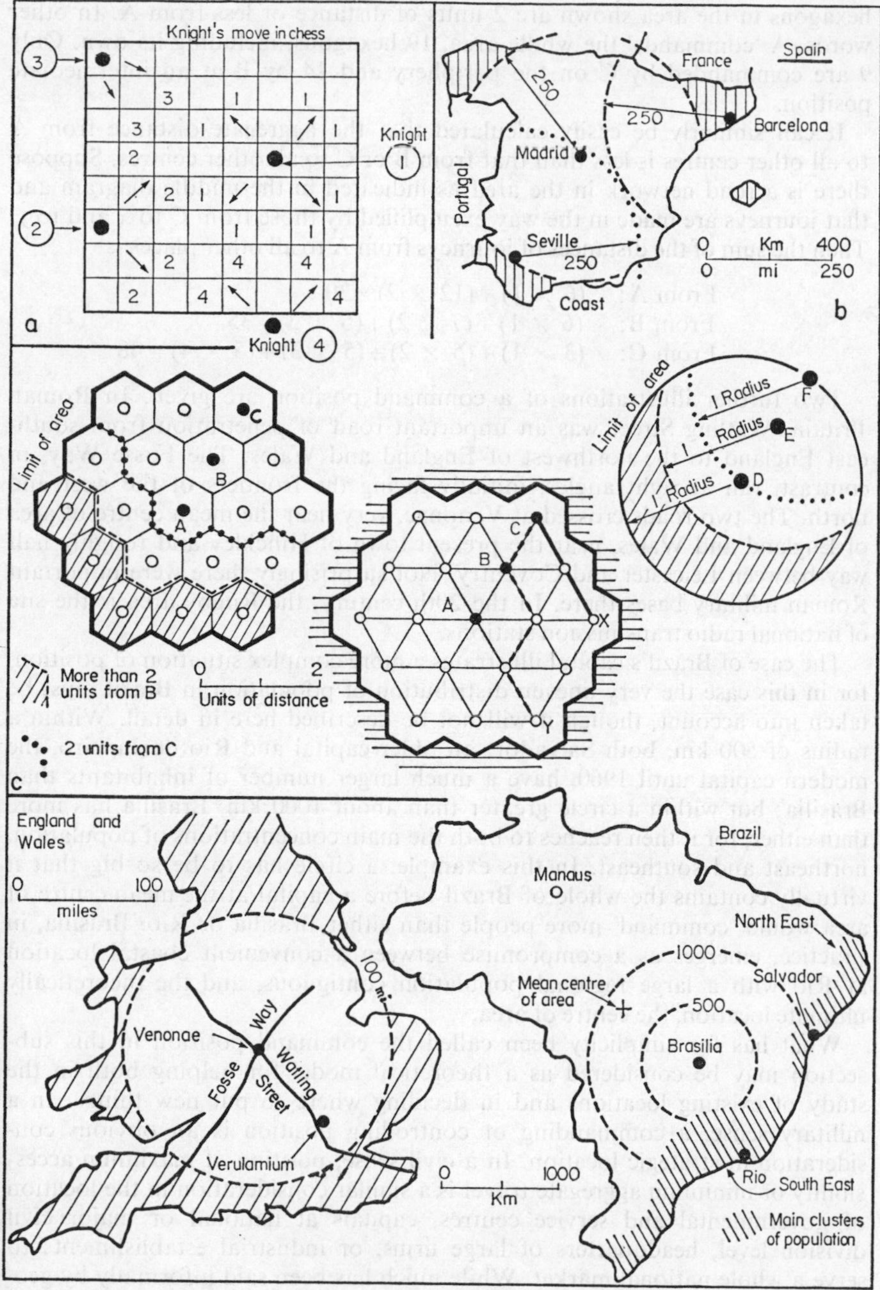

Figure 11.14 Positions of command. (*a*) Squares controlled by the knight in chess. (*b*) The location of Madrid, Spain. (*c*) Comparison of areas commanded from different positions. (*d*) Commanding position in Roman Britain. (*e*) The position of Brasilia, Brazil.

hexagons in the area shown are 2 units of distance or less from A. In other words, A 'commands' the whole area, 19 hexagons, including its own. Only 9 are commanded by C on the periphery and 14 by B at an intermediate position.

It can similarly be easily calculated that the aggregate distance from A to all other centres is less than that from B or C to all other centres. Suppose there is a road network in the area as indicated in the middle diagram and that journeys are made in the way exemplified by those from C to *x* and to *y*. Then the sum of the distances of journeys from A to all other places is:

From A: $(6 \times 1)+(12 \times 2)=30$
From B: $(6 \times 1)+(7 \times 2)+(5 \times 3)=35$
From C: $(3 \times 1)+(5 \times 2)+(5 \times 3)+(5 \times 4)=48$

Two further illustrations of a command position are given. In Roman Britain, Watling Street was an important road of penetration from south-east England to the northwest of England and Wales. The Fosse Way, in contrast, ran at right angles, initially facing the frontiers of the west and north. The two roads crossed at Venonae, very near the mean centre of area of England and Wales, near the present town of Hinckley and roughly half way between Leicester and Coventry. Not surprisingly there were important Roman military bases there. In the 20th century, the Rugby area is the site of national radio transmission stations.

The case of Brazil's capital illustrates a more complex situation of position, for in this case the very uneven distribution of population in Brazil must be taken into account, though it will not be described here in detail. Within a radius of 500 km, both Salvador, an older capital and Rio de Janeiro, the modern capital until 1960, have a much larger number of inhabitants than Brasilia; but within a circle greater than about 1000 km, Brasilia has more than either, for it then reaches to both the main concentrations of population, northeast and southeast. In this example, a circle has to be so big that it virtually contains the whole of Brazil before a capital at the mean centre of area would 'command' more people than either Brasilia or Rio. Brasilia, in practice, emerges as a compromise between a convenient coastal location at Rio with a large regional population contiguous, and the theoretically ultimate location, the centre of area.

What has for simplicity been called the command position in this sub-section may be considered as a theoretical model for helping both in the study of existing locations and in deciding where to put new things. In a military sense, a commanding of controlling position is an obvious consideration in strategic location. In a civil sense, position of maximum accessibility or minimum aggregate travel is a similar consideration in the location of governmental and service centres, capitals at national or major civil division level, headquarters of large firms, or industrial establishments to serve a whole national market. While much has been said informally by geographers about advantages and disadvantages of central and peripheral locations, relatively little has been done to measure with any precision the effect position might have. In their respective national areas, Northern Ireland, Brittany, Sicily and Sardinia, and the Soviet Far East all hold peripheral positions. Each, too, has its particular development problems. To what extent these are incurable outcomes of peripheral location it is not easy to say.

Some fundamental features of position are not widely appreciated. It surely causes surprise to find for instance, that the improvement of transportation links between peripheral areas and the centre of a country may strengthen the centre at the expense of the periphery. One reason may be that because of the greater usefulness of good routes, such as motor ways, between centre and periphery than between periphery and periphery, the radial routes will be built first. Moreover, even when all peripheral routes are built as in the diagram in figure 11.15*a*, each of the six peripheral regions still only has half as much 'good' route as the central region. Such a model could be applied to the transportation systems of various countries.

(*d*) *The accumulation of journeys in the centre of an area*

It emerges both from theoretical considerations and empirical situations that in an area that is integrated in the sense that journeys take place between different parts of it, the central part will have more journeys passing through it than the periphery. This assumption is based on the existence of a uniform surface. Moreover, it does not take into account journeys from places within the region to places outside it.

In figure 11.15*b* there are seven regional centres, I–VII. Centre I is central, while the other six occupy peripheral locations. One journey is shown from each centre to every other. The journeys are intended to be straight line ones. There are 21 journeys altogether. Three of these, II–V III–VI and IV–VII, pass through I (they by-pass it in the diagram). I is the only centre through which journeys not starting or ending in it pass. Moreover, it can be seen visually (and actually measured) that the density of journeys per unit of area falls off from the centre (I).

In figure 11.15*c* (left hand circle), a circular area has an inner or central half and an outer or peripheral half. There are four places, Cs and Ps respectively, in each of the two halves. The following features of journeys between these places may be noted:

(1) Journeys between any pair of Cs always occur within the inner half.
(2) Journeys between any C and any P always occur partly in the inner half, partly in the outer half.
(3) Some journeys between pairs of Ps occur only in the outer half, but some *cross* the inner half.

The inner half therefore gets used by some journeys originating and ending in the outer half, while the reverse never occurs. If inner (25 per cent) middle (next 25 per cent) and outer (50 per cent) areas are considered (as in the right hand circle) the situation above occurs in principle, though in a more complicated fashion, with, for example M–N journeys sometimes occurring only in the middle ring, but sometimes passing through the inner area.

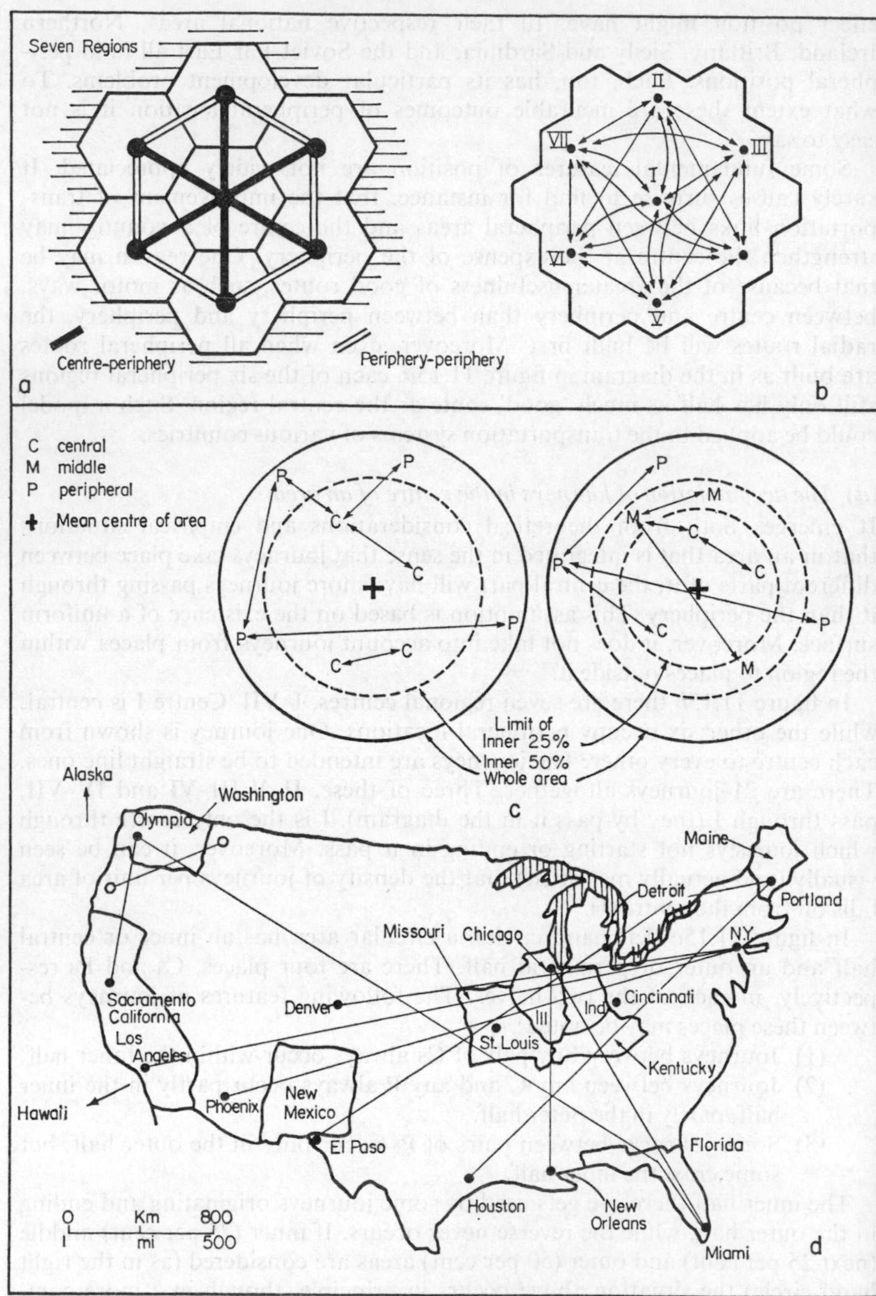

Figure 11.15 The concentration of journeys in the central part of a region. (*a*) Centre and periphery. (*b*) All pairs of journeys between 7 places. (*c*) Selected journeys related to centre and periphery. (*d*) Selected journeys in the U.S.A.

From the example of the circles, a continuous surface with an infinite number of points, or at least a very large number, may be visualized. It may be thought that the circle case is special or exceptional. In fact, it does illustrate satisfactorily what happens in general. Indeed, if the circle is transformed topologically to an indented shape like Britain, then a little thought will show that many land journeys, though not of course air journeys, between places on the periphery, will be forced into the central part and out again.

The theoretical model outlined above might be called the journey pile-up model. Some features can be illustrated briefly with reference to the U.S.A. and some of its 50 states. The selected straight line journeys shown on the map in figure 11.15*d* may be thought of either as air routes or as land journeys roughly following highway routes between pairs of places. It must be the experience of many Americans that at some time or other they have passed through such central states as Missouri and Illinois on journeys between other states. Few, on the other hand, will have travelled *through* Maine, Florida or Washington when travelling between two other states, except perhaps on a complicated tourist route. Alaska and Hawaii are the extreme cases of states in which journeys between pairs of U.S. states will only originate or end, not pass through.

What has been said about the U.S.A. is true also for a relatively small area, such as a town, where through traffic may cause physical congestion. In a larger area there will not be an obvious pile-up of traffic. There may, however, be a greater development of facilities for servicing rail or road vehicles, for catering and so on. Some manufacturing activities could conceivably be attracted to busier central parts of an area. These would seem at first sight minor advantages, but it should not be overlooked that the central area must tend to become better known to more people in the total area in question, though this may be offset for some peripheral areas by other attractions. The concept of journey pile-up is closely related to that of command position of the centre. Northern Ireland, Brittany, Sicily and Sardinia, even the Soviet Far East, may miss something not only on account of their peripheral positions but because they are backwaters or dead ends through which no one else's journeys pass.

11.7 Analogies in physical geography

The essential difference between models and analogies is that in the former either smaller or large replicas of the original or symbols are used to stand in for the prototype features. In the latter the features compared are quite dissimilar in nearly all respects except those that constitute the positive analogy. To clarify this distinction the example of a model glacier may be cited. This is a scale model in which the prototype is represented on a smaller scale and, although different material may be substituted for the real material, another type of feature is not being compared with a glacier. In some models simplification leads to the elimination of certain aspects of the situation but not the substitution of one object for another or the comparison of one thing with another. On a map symbols stand in for particular natural features, but they are in no way compared with reality; they merely represent it.

In analogies one feature or operation is compared with another often apparently dissimilar feature or process. The value of analogies lies in the ability to appreciate the similarities which constitute the analogy. The two objects or ideas that are compared consist of many aspects that differ and some aspects that are similar in some respect. Those aspects that differ are called the negative analogy and the other features that do not concern the problem under consideration or which are non-comparable or irrelevant are the neutral analogy and can be ignored. It is the positive analogy which is of significance in using the method of analogy, or analogue models, for research purposes.

Analogy plays a large part in general systems theory, which is an interdisciplinary method of study. It attempts to build up theoretical models that can apply to many disciplines. In order to achieve this aim the common ground or positive analogy of different subjects must be sought and classified. Most subjects have some common ground that can be exploited in this way. K. Boulding (1956) shows that most disciplines have populations that are added to and subtracted from by birth and decay. Most phenomena show some interaction between the individual and its environment, a theme that was very popular in deterministic geography. Growth is a phenomenon that occurs in nearly all disciplines and it is of universal importance.

Boulding has drawn attention to the different levels on which theoretical ideas and analogies can be sought between different fields of interest. Firstly, there is the level of frameworks which provides the broad pattern in which details can be fitted. Thus in cartography, the geodetic survey provides the framework into which detailed mapping can be fitted.

The second level is that of clockworks, which are simple dynamic systems, like the solar system on a large scale, or a machine on a smaller scale. This level includes much of physics and can be classed as a stochastic dynamic system.

The third is the level of the thermostat, which is a system in stable equilibrium and applies both to physics and to biology. The process of homeostatis (see glossary) whereby animals and man can maintain a steady state of temperature and other factors is an example of this level. This state requires the feedback of information in order to maintain the steady state. This might even be applied to the regulation of the economy of a country with the help of its politico-administrative 'machinery'.

The fourth level is that of the open-system. It consists of a self-maintaining structure than can be either living or not. As already mentioned earlier in this chapter, a river is an open system, in which there is a through-put of energy and material in the form of water and sediment. The same principle can apply also to the cell, which can reproduce and maintain itself. All living organisms are open systems, and only recently has physics advanced into the open system field according to L. van Bertalanffy (1956).

The open system state is necessary to allow advance. It has already been shown how a closed system leads to a levelling down, when all the energy is spread out evenly and the low temperatures would cause the heat death of the world. The living world of organisms, however, shows the opposite tendency; it is tending to develop greater specialization and heterogeneity and more complex organization. Increasingly it is moving through the levels under discussion.

The fifth level shows a step further up this path; it is the genetic-societal level. Plants show a division of labour among the cells; there are root, leaves and seeds. The sixth level is the animal level, with increasing awareness and greater communication. The seventh level is the human level. The individual human possesses self consciousness and uses symbolism, for example in language, and is aware of the past and future as well as the present. The eighth level is that of human organizations which incorporate all the former levels and whose complexity is very greatly increased. The subject matter of human geography, which belongs in this level, thus lies well up the scale levels, nearly at its most complex. The ninth and final level is that of transcendental systems. These are the ultimates and absolutes about which very little is known.

Empirical knowledge is lacking at nearly all the levels. The second level is probably the best known. Boulding considers that theoretical models are completely lacking beyond the fourth level, in that living organisms cannot be reproduced. However, it is by analogy from one system and one discipline to another that progress is likely to be made, and the unity of the operation of systems in these different spheres demonstrated.

Some examples of the analogies that have provided valuable ideas in the field of geography will be discussed briefly. James Hutton who lived from 1726 to 1797, and laid the foundations of modern geomorphology was one of the first to appreciate the value of analogy in geomorphology. He had a wide scientific background, having studied chemistry, medicine and agriculture through which he came to geology and geomorphology. His medical training and familiarity with the circulation of the blood led him to appreciate, by analogy, the circulation of matter in the growth and decay of landscapes. He was the author of the theory of uniformitarianism set up in opposition to catastrophism, which at that date was the rival school of thought. Thus he supported the open, as opposed to the closed, system, summarizing his views in the words 'no vestige of a beginning, no prospect of an end'. He himself used analogy in his major work, published in 1795, 'A theory of the earth with proofs and illustrations', in which he referred to the physiology of the earth.

The idea of circulation can be applied to many other spheres as well as that suggested by Hutton. The circulation of the atmosphere can be used as an analogy for the circulation of the ocean waters, although in this comparison there are important negative analogies as well as positive ones. The circulation of water through the hydrological cycle can be compared with the circulation through the body, although again there are negative as well as positive analogies involved.

Analogies can be of several types. In one type one physical feature can be used as an analogy to elucidate another less well understood phenomenon. In another type a physical concept can be used in fields other than that in which it was first developed. Physical processes can also be used as analogies for processes of human geography. Alternatively biological processes may give ideas concerning physical processes.

As an example of a process in physical geography providing an analogy for another physical process, the work of W. V. Lewis (1960) on the formation of corrie glaciers may be cited. He was familiar with the rotational shear slips that take place on the south coast of England near Folkestone. Here the chalk is underlain by clay. Water seeps through into the clay and causes it to become weaker as it absorbs the water. At the toe of the slope, marine erosion renders the slope unstable, and when the weight of the material exceeds the strength of the weakening rock, a slump takes place along a circular arc. This causes a back slope to develop on the slumped block and the strata are tipped up so that they dip into the hill, and the ground at the base of the slope is pushed up (see figure 11.16*a*). The form of a corrie glacier bears a close similarity to that of the rotational shear slump in that both slip on approximately circular arcs. This similarity is enhanced by the stratification

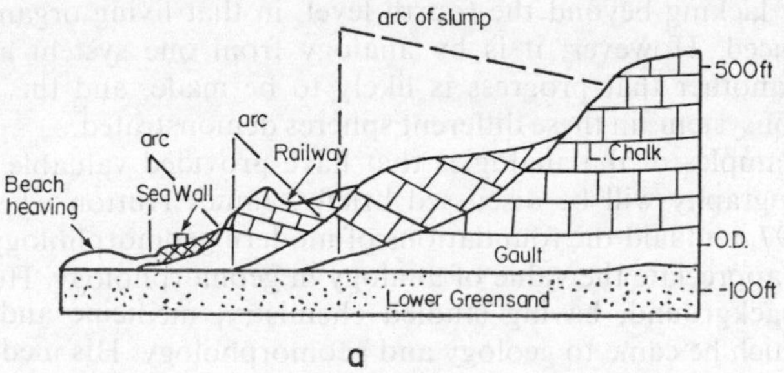

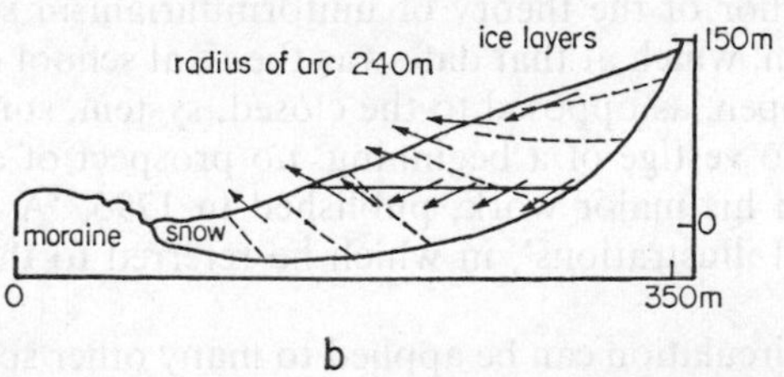

Figure 11.16 The upper part of the figure shows (*a*) a section through the rotational shear slump at Folkstone, Kent, showing the circular arc along which movement took place. The lower part of the figure (*b*) shows the movement of ice in a small corrie in Skauthoebreen in Norway. Its base is an almost circular arc. The ice layers are shown to rotate to dip up-glacier near the snout (partly after J. G. McCall).

of the layers of ice in the corrie glacier. It was this factor that first led Lewis to suggest that corrie glaciers might be rotating on their beds. This rotation would cause the individual annual ice layers to tilt up towards the snout of the glacier and associated debris would emerge from the ice with an up-glacier dip. Detailed survey work on Skauthoebreen (see figure 11.16*b*) in Norway has confirmed this suggestion that corrie glacier movement has a large element of pure rotation although this is not the only type of movement. In a glacier, the instability is produced both by removal of mass from the bottom of the slope by ablation and by addition of snowfall at the top of the slope. These changes increase the gradient until the strength of the ice is less than the stress applied to it. In this analogy the type of movement is basically similar and the resultant form has similarities, particularly in the circular slip arc. The materials involved, however, are very different.

Another analogy of a similar type is the use of the deformation of an ice sheet to simulate or act as a model for the deformation of the earth's crust. An ice-sheet deforms at a measurable rate and the properties of ice on which this deformation depends can be measured in the laboratory. The strength of the rocks of the earth's crust can also be measured in the laboratory. The processes of earth movement, however, are so slow that they are not measurable under normal circumstances, particularly in the case of movement by slow flow rather than rapid adjustment in earthquakes. The movement of the ice can be used as an analogy to study the movement of the crust. Suitable changes must be made in the scales of time and strength. Much fruitful work in glaciology has come from the analogies that can be drawn between the behaviour of metals and ice under stress.

R. J. Chorley (1959) has drawn attention to an interesting analogy between the shape of birds' eggs and the shape of drumlins. At first sight there appears to be little positive analogy between these two objects. Nevertheless interesting comparisons can be made. Both things are streamlined in shape and this is their most important similarity. Drumlins are streamlined by the ice flowing over them, while eggs are streamlined to facilitate the process of laying them. It is recognized by biologists that when small birds lay relatively large eggs, the eggs tend to be more elongated in shape, but that larger birds laying relatively smaller eggs produce eggs of rounder shape. This is because it is easier for larger birds to lay smaller eggs and they need not be so strongly streamlined as a result. In drumlins also, where the pressure of the ice forming the drumlin is greater, as in the small bird laying the large egg, the form of the drumlin is considerably more elongated. This pattern of drumlins is clearly shown in the elongation of those in the Eden valley in Northwest England. Around Carlisle and Appleby, which were situated near basal ice-sheds, the ice pressure was not very great and the drumlins have a low value of elongation. West of Carlisle where the ice was moving rapidly westwards the pressure was great and the drumlins become very much more elongated. Thus the positive analogy in this instance is the form relative to the pressure that produces it.

A good example of an analogy of a physical type transferred to various different fields is that of the application of the idea of kinematic waves to a wide variety of problems. The kinematic wave theory was first developed by M. J. Lighthill and G. B. Whitham (1955) (see chapter 10), who showed that this type of wave could be applied to 1-dimensional flow phenomena. They also showed that these waves differ significantly from the classical wave motion in dynamic systems. This wave theory can be used to study problems where there is a relation between the flow, or quantity passing a point in unit time, and the concentration, or the quantity per unit distance. It can be applied to a wide variety of phenomena that have the necessary properties in common. A study has been made by this method of the movement of small steps over a crystal face during crystal growth. The passage of a flood wave down a river has also been studied, and the movement of a surge through a glacier resulting from variations in climate causing variations in the amount of accumulation and ablation. This latter aspect of the problem has been studied by J. F. Nye (1960). Lighthill and Whitham themselves have applied this method of analysis to the movement of traffic on crowded roads. Not all these examples are strictly geographical. Nevertheless as a method of analysis the concept can be applied to geographical problems where these fulfil the conditions that create the positive analogy.

There are other examples in which physical geographical processes can be used as analogies for certain aspects of human geography. R. J. Chorley quotes the examples of W. Bunge's study of the shift of highways in comparison with the shift of rivers. W. L. Garrison (1960) has developed an 'ice-cap analogy' of city growth. In this type of analogy, it is important to differentiate the positive from the negative part of the analogy.

There is another point which should be borne in mind when using analogies to attempt to understand some processes more fully. In some instances the analogy is such that the feature used for the comparison is less understood, or at least no more understood, than the feature with which it is being compared. Thus it could be argued that no more is known about the formation of ice sheets and their movement than the earth movements that they are supposed to clarify by analogy. Despite these drawbacks, however, the use of analogy is a powerful tool in the study of many phenomena.

11.8 Analogies in human geography

A number of analogies have been made at different times between situations in human geography and in other disciplines, particularly physics, but also biology. Since the Second World War, borrowing from physics appears to have increased greatly. Knowledge in physics about interaction both of relatively few masses (e.g. the planets in the solar system) and of enormously many particles (e.g. atomic particles) has been used by geographers to throw light on such aspects of geography as the frequency of inter-city transactions and the behaviour of motor vehicles in restricted spaces. Various methods of studying traffic flow are discussed by R. Herman and K. Gardels (1963). They note other analogies with physics such as (p. 43): 'Multilane flow is described by an equation resembling the Boltzmann equation for a gas'. Political geography has been embarrassed* if not enriched, by the temptation to talk of the state in terms of a biological organism and to make analogies between features and behaviour patterns of higher organisms and of political units.

This section is only an introduction to some of the possibilities and drawbacks of analogue models in human geography. It includes (*a*) the gravity model (*b*) other models from physics together with the organismic model for organized territorial spaces.

(*a*) *Gravity model*

The so-called gravity model in geography is borrowed from physics. It is based on the observation that the force of attraction of two bodies is proportional to the product of their masses, divided by the distance between them, often but not always squared. In human geography, commonsense suggests that all other things being equal, the number of transactions, such as journeys or telephone calls made, will be greater between two nearby places than between two places far apart. In reality of course, there might be an insuperable physical, administrative or linguistic barrier between two nearby places and each might have more transactions with other places further away.

The gravity model can be introduced with a simple example. Suppose there are three small settlements, A, B and C, each with ten inhabitants. A is one unit of distance from B, while C is two units of distance from B (figure 11.17*a*). The gravity model would suggest that *all other things being equal* transactions would be more numerous between B and A than between B and C, because B is nearer to A than it is to C and all three settlements have the *same* number of people. If T is the number of *transactions* between *any two places i* and *j*, and the number of transactions expected in a given time span diminishes as *distance* increases, then the formula

$$T = \frac{(\text{Population of } i) \times (\text{Population of } j)}{\text{Distance between } i \text{ and } j}$$

can be used to ascertain an index relating to the number of transactions.

The formula can be written more concisely

$$T_{ij} = \frac{P_i P_j}{d_{ij}}$$

P meaning population, represents 'mass' (by analogy). In the example in figure 11.17*a* then,

$$T_{\text{AB}} = \frac{10 \text{ (the population of A)} \times 10 \text{ (the population of B)}}{1 \text{ (the distance A to B)}} = 100$$

$$T_{\text{BC}} = \frac{10 \times 10}{2} = 50$$

* In particular, German geopoliticians justified the expansionist policy of Nazi Germany in the 1930s on the grounds (among others), that a powerful organism is justified to expand at the expense of weaker ones.

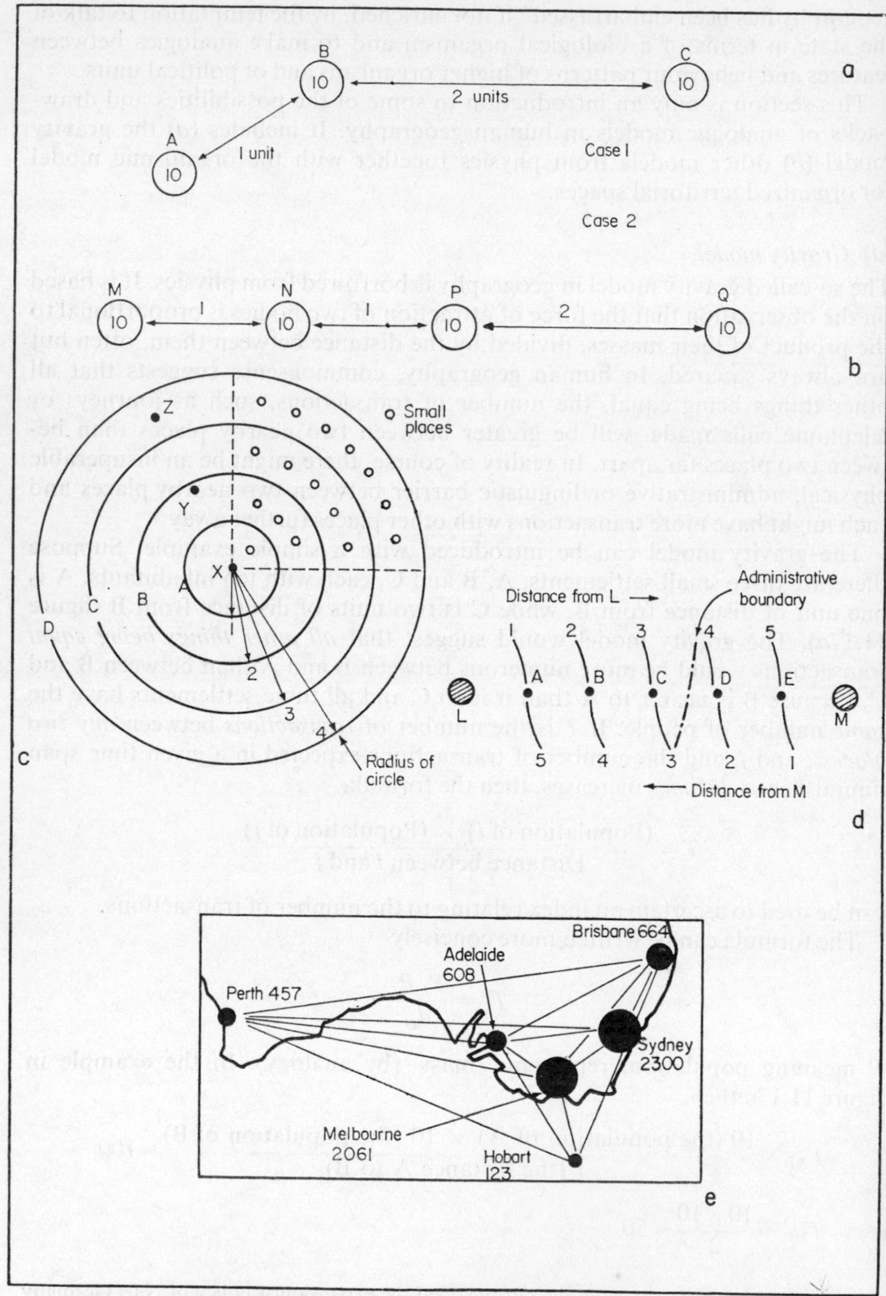

Figure 11.17 Simple aspects of the gravity model (see text). (*a*) Places of equal size at different distances. (*b*) Special case of (*a*) on a straight line. (*c*) Transactions between a central place and many localities spread over a surrounding area. (*d*) Diagram illustrating the contrast between gradual diminution of transactions with distance in the gravity model and instant cut-off by an administrative boundary. (*e*) The six state capitals of Australia.

because the distance B–C is double the distance B–A. So far the model is very simple, for all it suggests is that the transactions diminish in a linear fashion as the distance grows.

It must be made clear at this stage that the actual value of T (100 and 50 in this case) is only relative. It does not tell how many transactions will be made over a given period between places. It merely suggests that for every 100 transactions between A and B, 50 would be expected between B and C during the same length of time.

Various refinements may be added to the model: (1) the population of the two or more places in which one is interested will rarely be identical. This does not affect the formula itself. Note also that the 'mass' of the places need not be assessed by number of inhabitants; it could be income, retail sales, motor vehicles, or some other criterion. But one should be consistent in a given exercise in using the same criterion. (2) Distance need not be measured in terms of spatial distance (miles). Time taken to travel, or cost of travel (e.g. of transporting goods) may often be more meaningful. (3) In some circumstances, places with massive populations not far apart from each other will give a very big, inconvenient T value. Thus Greater London (about 10^7) and Greater Birmingham (abont 2×10^6) are about 100 miles apart. Their T value would be 2×10^{11}. A reducer constant k may be placed in the formula to bring down the actual T number to wieldy proportions

$$T_{ij} = k\frac{P_i\, P_j}{d_{ij}}$$

If k were made to be 10^{-7} in this case, then the result would be 2×10^4. k may also serve as a converter changing, for example, miles to kilometres. (4) Of more direct relevance to the results is the treatment of distance between places. As the diminution of transactions with increase in distance may not be linear, an exponent is needed for d. So far, the distance d has not been modified. In the gravitational model in physics, d^2 is used, or d^3 if lunar tractive forces are used for tide generation. This gives a sharper reduction in transactions with distance. Thus in the example (figure 11.17*a*), if

$$T_{AB} = \frac{P_A\, P_B}{d_{AB}^2} \quad \text{the result is still 100}$$

but
$$T_{BC} = \frac{P_B\, P_C}{d_{BC}^2} = \frac{10 \times 10}{2^2} = 25$$

In other words, the ratio AB to BC has dropped from 2 : 1 to 4 : 1 when d is squared. It is of course possible to have any exponent of d, but d^2 has been found often to fit real world experience in human geography fairly well.

Figure 11.17*b* shows four places, M, N, P and Q, along a straight line. The following table compares T values between all different pairs of the four places

Distance units		Value for d	Value for d^2
1	M–N, N–P	100	100
2	M–P, P–Q	50	25
3	N–Q	33.3	11
4	M–Q	25	6.25

Like most of the models in human geography discussed in this chapter, the gravity model may be used, preferably with some caution, both to study existing transactions between places, and to help in the location of new places. The applicability of the gravity model to journeys was appreciated by W. J. Reilley (1931). Its application is discussed by G. K. Zipf (1949). M. Chisholm (1962) gives examples in rural geography. An interesting application of the gravity model to the study of references to England and to the American colonies in the American press in the 18th century is described by R. Merritt (1964).

Finally, some dangers and drawbacks may be noted.

(1) It would be a remarkable coincidence if d^2 should be exactly the best exponential to fit the reducing effect of distance on all transactions.

(2) In a physical model, intervening masses might not be expected to reduce interaction between places on either side. In human geography, however, the presence of N between M and P in figure 11.17*b* might be expected to reduce transactions between M–P compared with P–Q, although the model would give these the same T value. The influence of the intervening settlement might, however, merely be to increase travelling time from M–P compared with P–Q, in which case distance might be replaced by travelling time for the d in the formula.

(3) The gravity model serves primarily to predict transactions between relatively small but densely populated places, such as nucleated settlements, which can be treated as points. The study of transactions between a central place and a surrounding area, say a town and its countryside, has also been the subject of much study, starting a century and a half ago with J. H. von Thünen (1826) if not before. It has been noted empirically that in agricultural areas with farmers living in nucleated settlements but using land up to several miles from the settlement, the land nearest the settlement tends to be visited more often and used more intensively than land on the periphery, even though the land further away may be more fertile. Thus in figure 11.17*c* a farmer at X would visit a field at Y more often than one at Z. The influence of distance is at work. It might even be assumed that the farmer would pay four times as many visits to Y as to Z, since in the illustration, Y is $1\frac{3}{4}$ units of distance from X while Z is $3\frac{1}{2}$ units from X.

When, however, two bands or rings at different distances from X are considered, rather than two points or small patches, then the ratio of transactions changes. Suppose in the simplest case that the spacing of small places, represented by small open circles in the figure, is uniform, then the number of small places in each ring will be proportional to the area of the ring. The following points may be noted in figure 11.17*c*: there are four areas, the circular figure enclosed by circles A–B, B–C and C–D. Each circle is one unit radius further from centre X than the preceding one. Thus if the first circle is $\pi \times 1^2$ in area, the second is $\pi \times 2^2$ or four times as large and so on. Thus the ratio of the innermost circle to the three bands is as follows:

Within		
A	1	
A–B	$4-1=3$	
B–C	$9-4=5$	
C–D	$16-9=7$	and so on.

For every 3 small places in ring A–B, 7 may be expected in C–D. Therefore although the distance from small places in A–B to X is only roughly half the distance from small places in C–D from X, and the number of transactions per small place therefore only about 0.25 times as much (d^2), there are more than twice as many places in C–D as in A–B.

(4) Serious reservations must be placed on the gravity model with regard to the actual measurement of distance. Cost and/or time of movement from place to place must be taken into account.

(5) In cases where a countryside is divided by line boundaries into administrative units and people travel to a given centre and to that centre only for a particular purpose like education or health services, the gravity model is superfluous in assessing the distance over which a centre might be expected to exert its influence. Consider figure 11.17*d*, L and M are two major centres with equal populations. A, B, C, D and E are smaller centres, all with the same population. Their distances from L and M are shown. The number of transactions between the smaller centres and L and M respectively would be proportional to distance, or the value actually shown, in a case where d is used, or to the square root of the distance if d^2 is used. In the d^2 case, then, A would have L, 25, i.e. 5^2, times as many transactions as it would with M. The model does however allow for the occasional transaction between A and M. If however there is an administrative boundary across the area between C and D, then A, B and C might conceivably always transact with L for certain purposes such as education, and *never* with M.

The reader is invited to complete the data in table 11.2 for transactions between the six state capitals of Australia (see figure 11.17*e*) applying a reducer constant. The reducer constant in this case is 1/1,000,000 because population is given in thousands. A new matrix can be prepared with expected transactions. Each pairwise transaction can then be expressed as a percentage of the total expected. Frequency of air services could then be compared with the expected transactions.

For example, the expected transactions between Sydney and Melbourne are:

$$T_{\mathrm{SM}} = \frac{P_{\mathrm{S}} \times P_{\mathrm{M}}}{d_{\mathrm{SM}}^2} = \frac{2300 \times 2061}{420^2} = \frac{4{,}740{,}000}{176{,}000} = \text{approximately } 27$$

(*b*) *Other analogies from physics*

This subsection discusses briefly some other analogies borrowed from physics.

Table 11.2 Straight line distance in miles[a]

1964 population in thousands		Sydney	Melbourne	Brisbane	Adelaide	Perth	Hobart
2300	Sydney	0	420	470	700	2040	490
2061	Melbourne		0	760	410	1710	360
664	Brisbane			0	970	2250	960
608	Adelaide				0	1330	730
457	Perth					0	1900
123	Hobart						0

[a] 1964 estimated population of urban agglomeration. Source: *United Nations Demographic Yearbook* 1965, p. 154.

(1) Mechanical analogue models have been used to study the location of industries. The location of certain kinds of industrial establishment, often those in which the raw materials and/or finished products are costly to transport, has suggested analogies with mechanical weights and pulleys.

(2) Physics and historical processes. In *Science and History*, E. Cuneo (1963) writes (pp. 4–5)

'As an instrument for the interpretation of the histories of nations, empires, civilizations and peoples, nothing could be simpler than the basic formula of kinetic energy. This formula is emphasized because the energies of nations determine their relation to each other. As a mere illustration that science is applicable to history, this basic formula is used:

$KE = \frac{1}{2}MV^2$. Energy equals one-half the mass times the velocity squared.'

Later, (p. 22) he illustrates the idea:

'The ratio of men-in-numbers to each other — population — does not measure the relative energy of their respective nations. That is, ten thousand Australian bushmen (sic) do not generate as much as ten thousand Europeans.' Cuneo attributes this assumption to the fact that the European is more mobile.

Cuneo also refers to the First Law of Thermodynamics, the Law of the Conservation of Energy (p. 210): 'Whole peoples are seldom destroyed, still less the ideas of culture. Accordingly, except in rare cases of complete annihilation of the conquered, the *Law of the Conservation of Energy* is applicable to history. That is, the surviving folk will combine in a new political identity.'

(3) The Second Law of Thermodynamics has also been considered by physical geographers, or at least the concept of a process leading towards randomization connected with it. In 'The Nature of Thermodynamics'

P. W. Bridgman (1961) expresses the widespread uneasiness of physicists about thermodynamics (p. 3): 'In some of his most exalted moments the physicist ascribes to the first and second laws of thermodynamics the most far reaching generality he has ever achieved in formulating laws of nature, while on other grayer mornings he depreciates the insight attainable by thermodynamic analysis, extols the superior virtues of statistical mechanics and kinetic theory. . . . (p. 4). Many physicists, I suspect, feel about the principle of relativity, much as they feel about thermodynamics — they do not see why the thing should work; there is a lack of perspicuousness about it that leaves them uncomfortable.'

Possibly, however, an analogy, unconscious rather than conscious with thermodynamics is what influences many politicians and administrators who consider that population in a country should be as evenly spread as possible. Just as the particles of sugar, once dissolved and stirred, distribute themselves through a cup of coffee, so population should fill up a space and leave no vacua (see figure 11.18*a*). At some time or other this has certainly been a consideration, often a policy, of governments not only in the six territorially most extensive countries of the world, the U.S.S.R., Canada, China, Brazil, the U.S.A. and Australia, but even in many smaller ones. Urbanization is by analogy an important example of the way in which man concentrates, thus moving towards a less entropic state than previously. Presumably such a trend is advantageous. Yet even in such small countries as Britain and France there is alarm at the continuing flow of people to a few centres.

(4) A simpler analogy between population migration and physics is that suggested by a magnet (see figure 11.18*b*). As centripetal forces 'increase' with the accumulation of non-agricultural activities in central places, the attractive force of certain localities or relatively small areas spreads out more and more widely. People may be likened to iron filings and the central places to magnets. As income prospects rise in the central places relative to those elsewhere, so the strength of the magnets grows, and iron filings are attracted from greater and greater distances. The iron filings themselves could be modified by making some (the laggards) heavier and more difficult to 'drag in' than others. In passing, it may be noted that the attraction of the magnets really works only along the transportation network, a line set. If the magnets are suddenly removed (switched off) the iron filings presumably remain concentrated. It needs new magnets to attract them back to where they were, or elsewhere. The attractive forces represented by the magnet analogy are presumably higher income, better employment prospects, superior amenities, and so on.

(5) 'Soap films, which always assume a shape with the least possible area, illustrate a concern of differential geometry: surfaces of least area bounded by curves in space. Differential geometry is also applicable to problems of map making and curvature of surfaces'. So runs a caption beneath a photograph of a soap bubble in a paper by M. Kline (1964). W. Bunge (1962) has a tantalizingly short section on the feasibility of using soap bubbles to find networks that form 'the shortest set of lines connecting

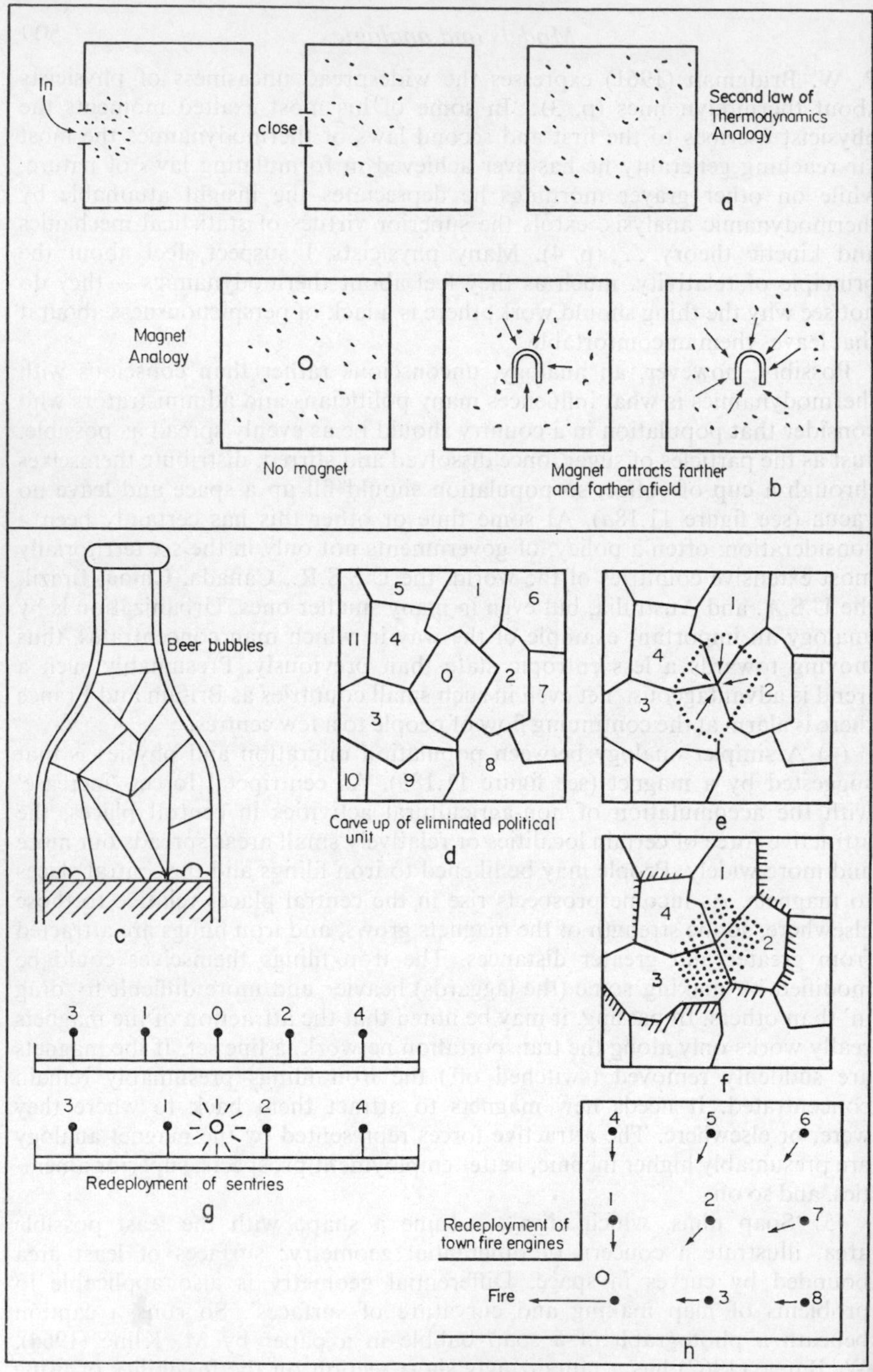

Figure 11.18 Some analogies from physics. (*a*) Second law of thermodynamics. (*b*) Magnet and population migration. (*c*) Beer bubbles. (*d–f*) the carving up of a political unit. (*e*) With adjustments only in adjoining units. (*f*) With adjustments over whole area. (*g*) Readjustment of sentries on a wall. (*h*) Readjustment of deployment of town fire engines when one is called to a fire.

all points to all others'. A classic work on the subject is *Soap Bubbles* by C. V. Boys (republished 1960).

The reader may have noticed that when part of the contents of a beer bottle are emptied and the bottle is then replaced in a standing position, bubbles will often form. These are sketched in figure 11.18*c*. As they gradually dry up, one film after another breaks. After each film breaks, a sliding readjustment of the remaining films takes place. The enclosed 3-dimensional compartments change shape as the films move to new minimum area positions. Sometimes what is a difficult concept to explain in words may quickly be grasped by an analogy. It has been customary since the unification of Italy in the 1860s to create new provinces from time to time round some particular town that has gained quickly in prominence (for example Varese in North Italy, Latina in the Peninsula). Similarly, in the U.S.S.R., new oblasts have been created and then removed from time to time. Whether a new unit is created or an existing one eliminated, the change usually involves alterations only in contiguous provinces. By analogy with the beer bubbles, but eliminating one dimension, it might theoretically be best to spread the effect of the change throughout the network. Such an 'infilling' situation is illustrated by an imaginary example in figures 11.18*d–f*.

Even in a 1-dimensional situation, such a possibility is not only conceivable, but common experience. Suppose a bus stop on a town street is suspended. It would be convenient to bring towards the vacuum at least the two neighbouring stops. Or suppose a wall is guarded at intervals by sentries and one sentry is eliminated (figure 11.18*g*). Theoretically all the sentries should move towards the gap. Similarly in figure 11.18*h* the fire engines should redeploy themselves after one has become engaged in fighting a fire.

While the beer bubble analogy may not have more than marginal significance for human geography, it does show how some basic spatial idea can be illustrated by an example of common experience and then intuitively used to link in a very basic way a number of otherwise apparently disconnected situations. The analogy need not, of course, be taken from physics.

11.9 Operational gaming models

(*a*) *Introduction*

Since the term model is clearly a very broad one, it seems convenient to include a particular kind of game in this chapter. Perhaps operational games represent such a fundamentally new approach at least to the teaching of human geography that they should have more space than one section to do them justice.

R. D. Duke (1965) uses the term 'Operational gaming', which is not the same as game theory, though the two things are connected. Game theory discussed in chapter 14, is a special, recent, branch of mathematics that studies the play, strategy and outcome of games. Most games are so complex that only a few simple ones have actually been worked out for all outcomes. Operational gaming, on the other hand, is a general, old, widely practiced method of learning about the real world. An operational game is a simulation model that depends basically upon human decision, and therefore, on skill rather than chance for success. Here then, physical and human geography part company, though physical features may be passive players in operational games.

War games have been used for a very long time, both for instructional purposes and to test alternative war plans. The use of games for training businessmen and urban planners, on the other hand, has largely been developed since the second world war. Even more recently there have been signs in geography that the usefulness of games is at least being considered.

R. L. Meier (1965) writes:

> 'The word simulation has taken on a more precise meaning in the last decade. A simulation now refers to a working model of a system in the real world. Just as a map is ordinarily used to illustrate relationships in space, a simulation produces a pattern of interactions in another variable, usually time.'

Operational games are used to give persons in training some experience of what they will encounter in the real world, to get the feel of and understand better decision making, and to experiment in a kind of laboratory, in which wrong decisions can do no harm. One drawback of the social sciences is that experiments of a well regulated kind are often inconvenient and costly to run, owing to the complexity of the situations being studied (e.g. a large urban agglomeration), and to the fact that the experiment may be affected because it involves people who know they are part of the experiment. With reference to the study of cities, Meier (1965) writes:

> 'But it turns out that it is extraordinarily difficult to synthesize a realistic city — much too difficult to be worth our effort . . . Stated in more technical language: the community is a non-zerosum game between players with assymetric roles. Our job is to find rules that compress this game to about one-millionth of its normal "real World" dimensions so that the basic structure can be understood within hours or days.'

A feature of most games is that 'real time' is greatly reduced. It is convenient in more complex gaming simulations also to have a computer to process decisions and produce outcomes that can then be studied and compared.

R. D. Duke (1965) notes the importance of systems theory in urban simulation. He also notes that, in contrast to game theory, (p. 296) 'there is no theory of optimal solution that can be derived from operational gaming. Rather the technique serves as a learning environment, a substitute for experience'.

M. Beloff (1967) is less enthusiastic about the use of games in some circumstances. Talking of a collection of papers from the Peace Research Centre of the University of Lancaster, he says: 'The first paper reports on a "Simulation" of the Vietnam conflict . . . Apart from the indecency of making an intellectual game out of large-scale human suffering, the idea that what Ho Chi Minh or Mao Tse-Tung are likely to do in certain circumstances can be discovered by graduate students playing about with mathematical formulae is breathtaking in its absurdity'.

There is a considerable literature on games as seen from various points of view, though little about their usefulness in geography. E. Berne (1964) looks at games from the viewpoint of the psychologist. R. Caillois (1962) makes an interesting basic fourfold classification of games into: Competition, Chance, Simulation and Vertigo.

(*b*) *Games in geography*

What kinds of game might be useful to a geographer? A tentative outline of some of the ways games may be classified is essential before the question can be answered. The following features appear important:

(i) *The degree of chance and skill.* Some games depend entirely on chance. Gambling games and games in which the throw of a dice is the only thing deciding an outcome are examples. Other games depend entirely on skill. Draughts and chess are examples, though for good players, the starter has a slight advantage. Many games, including for example bridge, include some chance (the hand of cards received) and some skill (the bidding and play). When chance and skill are mixed, the more skilful player will win after many turns, though a poorer player may win in the short run simply through receiving a sequence of good hands.

In most of the games to be mentioned in this section, chance has either been eliminated or has been given a limited role compared with that given to human decision making. In contrast, the diffusion model described earlier in this chapter was run entirely with random numbers, though to a set of rules designed to give it a general direction.

(ii) *Spatial or not.* A game may or may not be played on a space. Thus most card games do not need a board. The geographer may expect to find games played on a space of greatest potential use. The space may be 1-dimensional, like snakes and ladders, and Monopoly, both of which, however, depend on the throw of a dice for movement in space. The space may also be 3-dimensional; there is a game similar to draughts in 3 dimensions. Noughts and crosses (tick tack toe) may also be played in three dimensions. Most spatial games, however, take place on two–dimensions: draughts and chess, Oriental Go, many battle games, such as D-Day and Stalingrad.* Usually the space is divided into square or hexagonal compartments and is not therefore continuous. It is difficult to imagine a game in which time is not essential in some way, so time will be considered a further essential. Like space in games, time is usually greatly telescoped (e.g. two years in Stalingrad can be played through in a few hours) and is also divided into sections, as for example months in Stalingrad. Actual military manoeuvres, however, in which real equipment is used, are often 'played' in real time.

* D-Day and Stalingrad are two examples of the many war games now produced by the Avalon Hill Company. D-Day is played over the whole surface of France and adjoining countries, Stalingrad on European U.S.S.R.

(iii) *Degree of abstraction.* Games vary greatly in the amount of abstraction they contain compared with similar situations in the real world. At one extreme, life itself has been described as a game, since society operates within a system of accepted rules, albeit very complex and very loose ones. At the other extreme, many card games are based on highly complex arrangements of numbers and probabilities. A 'live' military manoeuvre is an example of a game with a minimum of abstraction. A game of chess is highly abstract, but the pieces still convey a greater feeling of reality than the indifferent pieces of draughts. A model railway or road system lies somewhere between the extremes.

(iv) *Number of players.* The number of persons playing a game and actually making decisions in it can vary. A game with no players is difficult to visualize, but many games are designed for one player. The opponent will be chance (card patience games), the 'decisions' which are decided with the help of a dice, random numbers and so on. A game with two or more players almost inevitably involves some form of conflict. A two-person game, however, is a special case, for whenever there are more than two independent players (unless teamed in pairs or more) then collusion is possible between any (proper) subset of them. In Diplomacy each of seven players represents a European power in 1900. The game need not be played with the formation of alliances, but this does often happen.

(v) *Patterns of play.* Games played on a 2-dimensional surface often take one (sometimes a combination) of several distinct forms:

(1) Getting from one place to another: e.g. Ludo, Halma, Buccaneer, Great Britain rail race, Dispatcher.
(2) Gaining control of a space or part of a space, e.g. noughts and crosses, Oriental Go, Hex (one version),* Tangle, Stalingrad, D-Day.
(3) Achieving some kind of contiguity: e.g. dominoes, Hex (one version).
(4) Elimination of pieces, often with advantage in control of space: e.g. draughts, chess, Dover patrol, Waterloo, Gettysburg, Diplomacy, Stalingrad, D-Day.

It is possible that some very basic situations and problems in human geography resemble the patterns of play in the above mentioned spatial games. Control of space through good positioning, the transference of objects from place to place, the best deployment of things in space, conflict or competition, routing, are all familiar. They form much of the content of new techniques and models in human geography. Can, then, existing commercial entertainment games be used in the training of geographers, and can even more useful games be designed for the needs of geographers? It would be easy to leave the reader with the few references so far available on this subject to decide for himself. However, it seems more opportune to add an example

* One version of Hex involves competing for control of a space. Another consists of placing hexagons contiguous to one another, with points for correctly matching edges.

of two simple games to illustrate how geographical games might be expected to work. It is interesting to note that there is nothing very new in the idea. In 1787 a game 'for young learners of geography' was printed by Robert Sayer and Co. (London). It represents a tour of Europe.

(*c*) *Simple examples*

While many commercial games, both old and new, no doubt teach something about spatial situations, they often lack even a minimum of reality to enable the players to connect their experience in the game with their experience in the real world. There are, however, some excellent commercial games that have been designed primarily for entertainment, but also with instructional undertones. Monopoly appeared some decades ago though it is more financial than geographical in what it teaches. Diplomacy is worth playing for its lessons in diplomatic geography. The recent games of the Avalon Hill Company, however, are very promising. Among these the reader is recommended to try D-Day, Stalingrad, Waterloo, Gettysburg, Dispatcher.

Some games have been designed by geographers, and others have been projected, if not perfected. H. Stafford has helped to design a game called Metfab for the U.S. Geography High School Project. This is a role playing game for a group of several pupils, who represent a board chosen to consider the location, in one out of 25 possible major U.S. cities, of a metal manufacturing plant. Each player has particular interests (e.g. sales manager, personnel manager) and certain information. The president co-ordinates the decisions. The game has been tried and experience has shown that the playing of it brings in many considerations of industrial location. An attempt has also been made to develop a game for political geography in which an imaginary boundary dispute in Maine between the U.S.A. and Canada is settled by role playing pupils.

The following two games are selected from among ten designed by G. A. Smith and J. P. Cole (1967). They illustrate the possibilities of using games to teach geographical situations and problems to school children of various ages, and perhaps even to university students, with both imaginary and real places. It is hoped particularly to develop regional geography games, firstly to play through what has actually happened (e.g. the American West, inter-war U.S.S.R.), then to play into the future (e.g. the next motorways in Britain, the development of Siberia).* It might be useful also to have available for teaching purposes one or more fictitious regional laboratories. A mid-latitude island, St. Lucius, could be designed with certain features and certain information. Players could develop it without any bias or pre-conceived ideas about it.

The section concludes with a description of two very simple games that have already been tried on children.

* The same principle applies to the use of scale models to elucidate hydraulic engineering problems. The scales and material of the model are adjusted until it can successfully reproduce known changes that have occurred. Then the model can be used to extrapolate changes into the future. The probable effect of engineering works can be tested in this way.

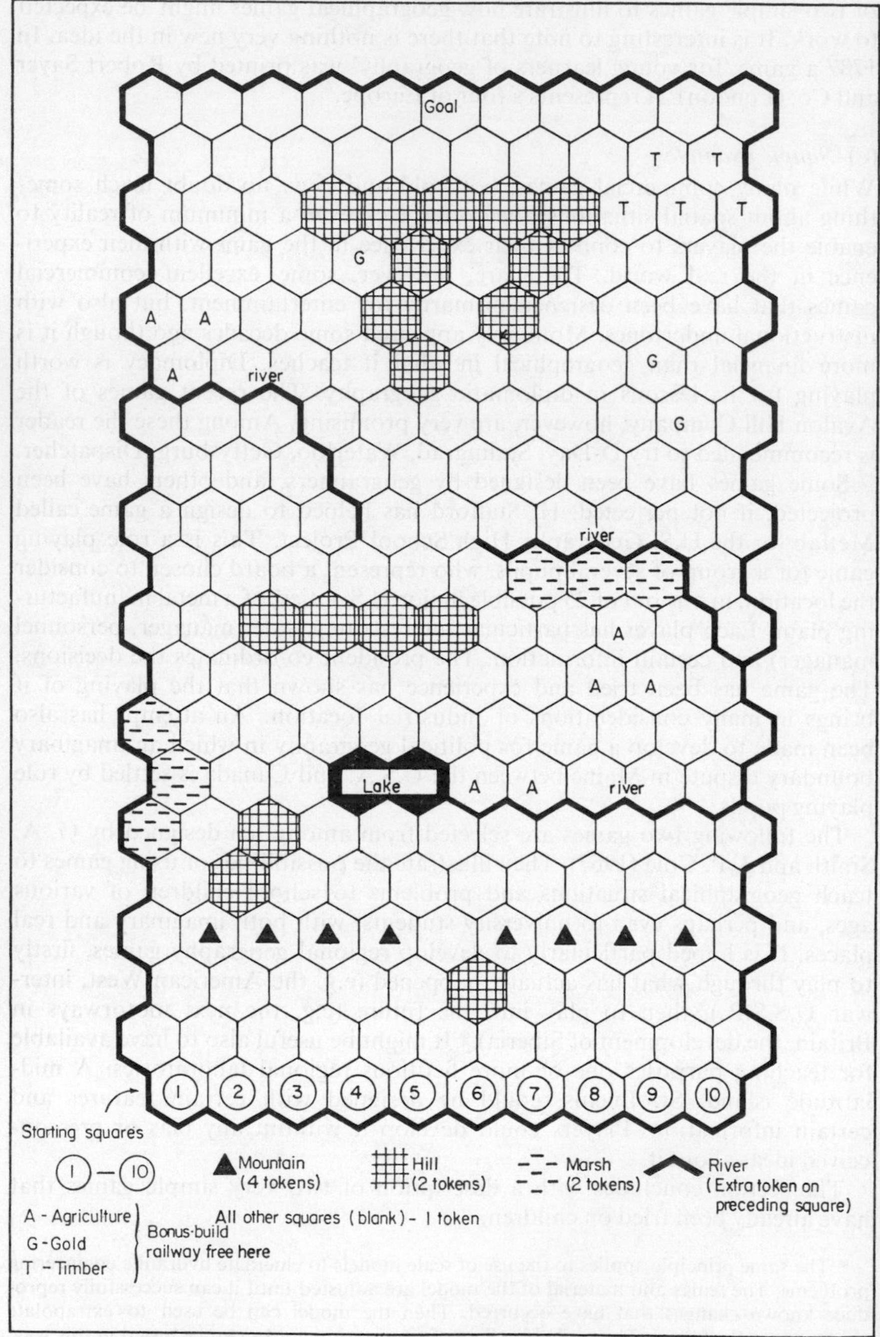

Figure 11.19 Railway building game. The cost of building a railway across the area from one of the starting squares to the GOAL is counted in tokens. Each player places one token at a time in turn. Some squares require more than one token, some are free. Most require one token.

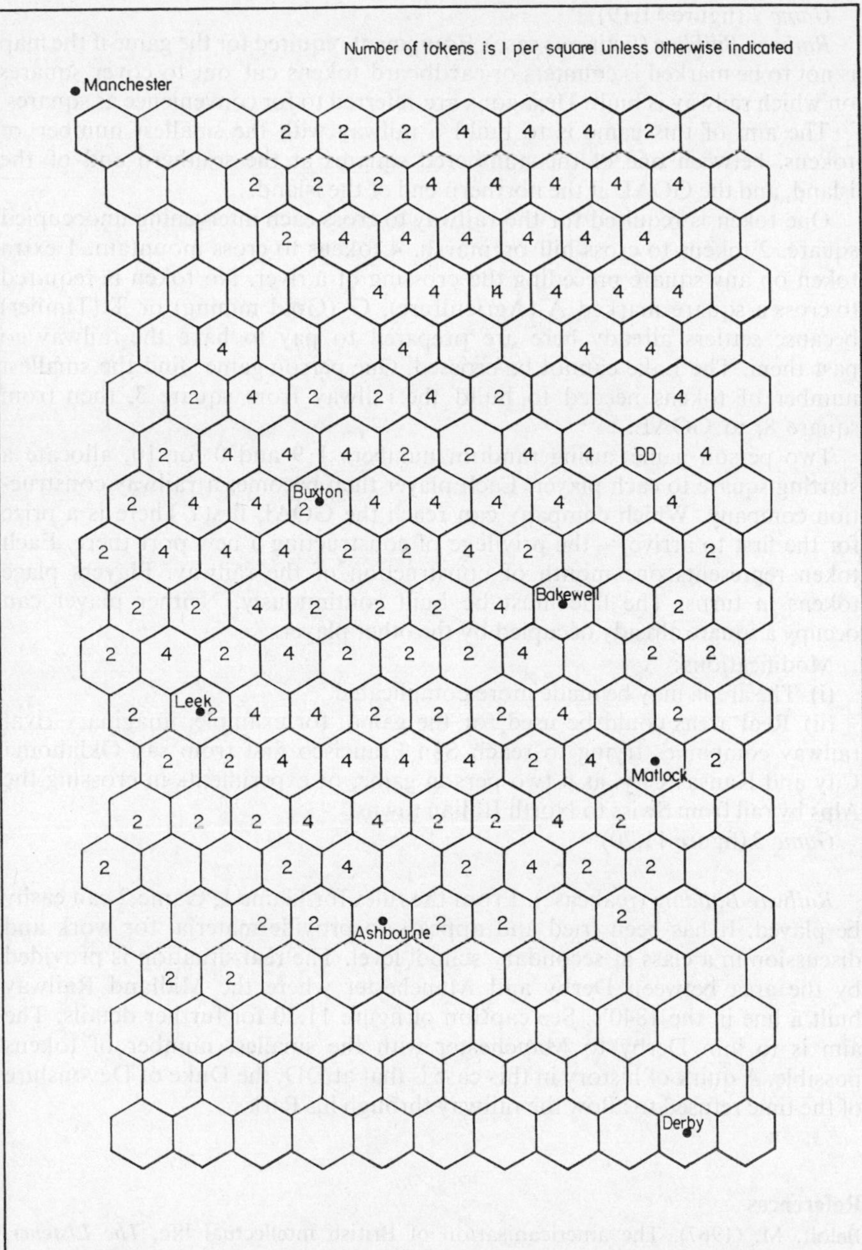

Figure 11.20 The building of the Midland Railway between Derby and Manchester around 1840. Very roughly, the blank squares, requiring one token, are relatively easy to build on, the (2) squares are hilly, the (4) squares particularly rugged. In practice, the squares DD occupied by Chatsworth House, were precluded from having a railway by the Duke of Devonshire.

Game 1 (figure 11.19)

Railway Building (fictitious case). Equipment required for the game if the map is not to be marked is counters or cardboard tokens cut out to cover squares on which railway is built. Hexagons are referred to for convenience as squares.

The aim of this game is to build a railway with the smallest number of tokens, between one of the numbered squares at the southern end of the island, and the GOAL at the northern end of the island.

One token is required for the railway to cross each intervening unoccupied square. 2 tokens to cross hill or marsh. 4 tokens to cross mountain. 1 extra token on any square preceding the crossing of a river. No token is required to cross a square marked A (Agriculture), G (Gold mining) or T (Timber) because settlers already here are prepared to pay to have the railway go past them. The Lake cannot be crossed. One person game: find the smallest number of tokens needed to build the railway from square 3, then from square 8, to GOAL.

Two person game: using random numbers 1–9 and 0 for 10, allocate a starting square to each player. Each player then becomes a railway construction company. Which company can reach the GOAL first? There is a prize for the first to arrive — the privilege of constructing a new port there. Each token represents one month of construction of the railway. Players place tokens in turns. The line must be built continuously. Neither player can occupy a square already occupied by the other player.

Modifications:

(i) The arena may be made more complicated.

(ii) Real areas could be used for the game: for example, imaginary rival railway companies trying to reach San Francisco first from say Oklahoma City and Kansas City, as a two person game, or experiments in crossing the Alps by rail from Swiss to North Italian towns.

Game 2 (figure 11.20)

Railway Building (real case). From the rules for Game 1, Game 2 can easily be played. It has been tried and appears to provide material for work and discussion in a class at secondary school level. The real situation is provided by the area between Derby and Manchester where the Midland Railway built a line in the 1840's. See caption of figure 11.20 for further details. The aim is to join Derby to Manchester with the smallest number of tokens possible. A quirk of history in this case is that at DD, the Duke of Devonshire of the time refused to allow the railway through his Park.

References

Beloff, M. (1967). The americanisation of British intellectual life, *The Listener*, 14th Sept. 1967, 327–328.

Berne, E. (1964). *Games People Play*, Grove Press, New York.

Berry, B. J. L. and Garrison, W. L. (1958). Alternate explanations of urban rank-size relationships, *Ann. Assoc. Am. Geog.*, **48,** 83–91.

Berry, B. J. L., Simmons, J. W. and Tennant, R. J. (1963). Urban population densities: structure and change, *Geog. Rev.*, **53,** No. 3, 389–405.

Bertalanffy, L. Von (1956). General system theory, *Yearbook Soc. Advan. Gen. Systems Theory*, **1,** 1–10.

Blair, B. J. and Biss, T. H. (1967). The measurement of shape in geography: an appraisal of methods and techniques, *Bull. Quantitative Data Geog.*, **11,** Nottingham University.

Boehm, G. A. W. (159). *The New World of Mathematics*, Faber and Faber, London, 80–81.

Boulding, K. (1956). General system theory — The skeleton of science, *Yearbook Soc. Advan. Gen. Systems Theory*, 1-17.

Boys, C. V. (1902). *Soap Bubbles*, Republished 1960, Heinemann, London.

Bridgman, P. W. (1961). *The Nature of Thermodynamics*, Harper, New York.

Bunge, W. (1962). *Theoretical Geography*, Lund, 183–4.

Caillois, R. (1962). *Man, Play and Games (translated from French)*, Thames and Hudson, London.

Chao, Y. R. (1962). Models in linguistics and models in general, 558–566. Logic, methodology and philosophy of science, *Proc. 1960 Internat. Congr.* (Ed. by E. Nagel, P. Suppes and A. Tarski), Stanford University Press.

Chisholm, M. (1962). *Rural Settlement and Land Use*, Hutchinson University Library, London.

Chisholm, M. (1967). General systems theory and geography, *Trans. Inst. Brit. Geog.*, **42,** 45–52.

Chorley, R. J. (1959).The shape of drumlins, *J. Glaciol.*, 3 (25), 339–344.

Chorley, R. J. (1960). Geomorphology and general systems theory, *U.S. Geol. Surv. Prof. Paper*, 500–B, Washington, B.1–B.10.

Chorley, R. J. (1964). Geography and analogue theory, *Ann. Assoc. Am. Geog.*, **54,** 127–137.

Cuneo, E. (1963). *Science and History*, Cassell, London.

Duke, R. D. (1964). *Gaming — Simulation in Urban Research*, Institute for Community Development and Services, Michigan State University.

Duke, R. D. (1965), *Gaming Urban Systems*, in *Planning 1965*, American Society of Planning Officials, Chicago, 294, 297.

Einstein, H. A. (1965). The bed-load function for sediment transportation in open channel flows, *U.S. Dept. Agric. Soil Cons. Tech. Bull.*, No. 1026.

Garrison, W. L. (1960). Notes on the simulation of urban growth and development, *Dept. Geog. Univ. Washington, Discuss. Paper*, **34.**

Gould, P. (1964). A note on research into the diffusion of development, *J. Modern African Studies*, **2,** No. 1, 123–125.

Guetzkow, H. (Ed.) (1962). *Simulation in Social Science: Readings*, Prentice Hall, New Jersey.

Haggett, P. (1965). *Locational Analysis in Human Geography*, Arnold, London.

Herman, R. and Gardels, K. (1963). Vehicular traffic flow, *Sci. Am.*, **209,** No. 6, 35–43.

Hesse, M. B. (1963). *Models and Analogies in Science*, Newman History and Philosophy of Science Series, Sheed and Ward, London and New York.

Horton, R. E. (1945). Erosional development of streams and their drainage basins; hydrophysical approach to quantitative morphology, *Bull. Geol. Soc. Am.*, **56,** 275–370.

Isard, W. (1956). *Location and Space Economy: A General Theory Relating to Industrial Location, Market Areas, Land Use and Urban Structure*, New York.

Isard, W. and Smith, T. E. (1967). Location gāmes: with applications to classic location problems, *Papers Reg. Sci. Assoc.*, **19**, 45–80.

Katz, E. (1963). Traditions of research on the diffusion of innovation, *Am. Sociol. Rev.*, **28**, 237–257.

King, L. J. (1961). A multivariate analysis of the spacing of urban settlements in the United States, *Ann. Assoc. Am. Geog.*, **51**, No. 2, 222–233.

Kline, M. (1964). Geometry, *Sci. Am.*, **211**, No. 3, 60–69.

Krumbein, W. C. (1963). A geological process-response model for analysis of beach phenomena, *Beach Erosion Board Ann. Bull.*, **17**, 1–15.

Krumbein, W. C. and Graybill, F. A. (1965). *An Introduction to Statistical Models in Geology*, McGraw-Hill, New York.

Lakshmanan, T. R. (1964). Analysis of market potential for a set of urban retail centres, Paper presented at the Urban Symposium (Nottingham), 20*th Internat. Geog. Congr.*

Lewis, W. V. (Ed.) (1960). Norwegian cirque glaciers, *Roy. Geog. Soc. Res. Ser.* No. **4**, Chapt. IX, 97–100 and J. G. McCall, Flow characteristics of a cirque glacier and their effect on glacial structure and formation, Chapt. V, 39–62.

Losch, A. (1954). *The Economics of Location*, New Haven.

Meier, R. L. (1965). *The Gaming Simulation in Urban Planning*, in *Planning 1965*, American Society of Planning Officials, Chicago, 287 and 289.

Merritt, R. L. (1964). Distance and interaction among political communities. *Yearbook Soc. Gen. Systems Res.*, **IX**, 255–263.

Morrill, R. and Pitts, F. R. (1967). Marriage, migration and the mean information field: a study in uniqueness and generality, *Ann. Assoc. Am. Geog.*, **57**, No. 2, 401–422.

Norborg, K. (1962), *Proc. Internat. Geog. Union Symp. Urban Geography, Lund 1960*, Lund Studies in Geography, Ser. B. Human Geography No. 24.

Nye, J. F. (1957). The distribution of stress and velocity in glaciers and ice-sheets, *Proc. Roy. Soc. London*, A **239**, 113–133.

Olsson, G. (1966). Central place systems, spatial interaction, and stochastic processes, *Reg. Sci. Assoc. Papers*, **1967**, pp. 13–45.

Ramberg, H. (1964). Note on model studies of folding of moraines in piedmont glaciers, *J. Glaciol.*, **5**, 207–218.

Reilley, W. J. (1931). *The Law of Retail Gravitation*, Reilley, New York.

Scheidegger, A. E. (1961). *Theoretical Geomorphology*, Springer-Verlag, Berlin.

Simon, H. A. (1956). On a class of skew distribution functions, *Biometrika*, **42**, 425.

Smith, G. A. and Cole, J. P. (1967). Geographical games, *Bull. Quantitative Data for Geographers*, No. 7, Nottingham University.

Thompson, D'Arcy (1961). *On Growth and Form*, Abridged edition (Ed. J. T. Bonner), Cambridge University Press.

Thunen, J. H. von (1826). *Der isolierte Staat in Beziehung auf Landwirtschaft und National ökonomie*, Rostock 1826.

Zipf, G. K. (1949). *Human Behavior and the Principle of Least Effort*, 1965 ed. Hafner, New York.

12

Theories, tendencies and laws

> 'One branch after another of chemistry, physics and cosmology has merged in the majestic river as it approaches the estuary — to be swallowed up by the ocean, lose its identity, and evaporate into the clouds; the final act of the great vanishing process, and the beginning one hopes of a new cycle . . . It seems that the more universal the "laws" which we discover, the more elusive they become, and that the ultimate consummation of all rivers of knowledge is in the cloud of unknowing.'
>
> *A. Koestler* (1964)

12.1 Introduction and definitions

The aim of every scientific discipline is to describe and explain the phenomena that come within its field of study and to establish inter-relationships between the phenomena and also between them and their environment. The explanation of phenomena requires the setting up of theories and their testing by observation or experiment. Tendencies or laws may be established from the results of these tests. In this chapter various types of theories, tendencies and laws are defined and illustrated by reference to geographical examples. Nearly all the examples discussed apply to physical geography. This does not mean that there are no theories, tendencies and laws in human geography, but that they are of a rather different character.

Some theoretical models concerned with spacing and inter-relationships in human geography are referred to as theories, if not laws, by some geographers. Thus central place theory has some of the features both of a theoretical model and a working hypothesis. Economic 'laws' have also been applied to spatial situations concerning the location of economic activities. Nevertheless, it is more difficult to formulate theories, tendencies and laws in human geography than in physical geography. The reason is certainly partly due to the fact that individual human decisions cannot be predicted. It seems also to arise from the fact that so many variables are involved in human geography that one or even a viably small number can rarely be isolated and used for prediction.

Theories can be built up using empirical observations and the theories then in turn lead to the recognition of tendencies and laws. A theory or hypothesis can also be used to develop suitable techniques of analysis with which to test the theory. If the test is successful then the theory may become a tendency or law.

The terms 'theories', 'tendencies' and 'laws' have several meanings in scientific writing. Definitions of these terms are given before the discussion of examples.

Definitions

(*A*) *Theory*

(1) Theory — the nature of a topic or subject and its relations with other disciplines.
(2) Theoretical model — a method of approach or framework. (This type of theory overlaps almost completely with theoretical models, discussed in chapter 11.)
(3) A working hypothesis that is amenable to testing.
(4) An apparently reasonable hypothesis, but one that is *not* subject to testing.

(*B*) *Tendency*

(1) A commonsense relationship.
(2) A qualitative impression of a relationship.
(3) A hypothesis that has been tested and found significant at a fairly low confidence limit (90 per cent for example).

(*C*) *Law*

(1) A statement in mathematical logic.
(2) A mathematical relationship.
(3) A theory that has been exhaustively tested and not rejected.
(4) An empirical relationship found by observation or experiment.
(5) A legal law or regulation.
(6) A concept or principle.

12.2 Theories

Theory Type 1

The first type of theory is concerned with the philosophical character of the discipline. It can be applied to a whole field of knowledge, such as a theory of geography, or to a more restricted topic, for example communication theory, which is interdisciplinary. In this type of theory not only the nature of the subject itself is examined but also its contacts with other subjects. The study of this type of theory becomes largely philosophical. R. Hartshorne's *Perspective on the Nature of Geography* (1959) is perhaps the best known theory of this type in geography. This type of theory is not discussed further here.

Theory Type 2. Theoretical Model

The second type of theory, the theoretical model, is more practical and less philosphical. It has already been considered in the last chapter, in which examples of theoretical models, mainly in human geography, were described. In this sense 'theory' can include the method of approach to the subject. The theoretical ideas can be set up as a framework or model with which real situations may be compared. The essential element of this type of theory is the setting up of a conceptual or theoretical situation rather than the study of the real situation. The latter only enters later when reality is compared with the theory.

An initial state must first be assumed in a theoretical model of this type, and specified steps are then carried through to lead to a theoretical final state. A type 2 theory can be devised to study the operation of processes through time. The theory need not, however, be dynamic as in the example to be considered. A synchronic theoretical model can be set up, as in the human geography examples considered in chapter 11.

A basic theory must first be conceived before the model can be set up. It is on the basis of the theory that the steps to be followed are worked out and the initial state is specified. When the theoretical model leads to the real observed final state, then it may be argued that the conditions and steps in the model may also have occurred in the real situation. In this way it should be possible to discover the relative effectiveness of different processes which may be simulated in the model in turn.

An account of the effect of a falling and rising sea-level on a steep land surface will exemplify a simple theoretical model. It is sometimes assumed that erosion surfaces are formed by wave erosion when sea-level is falling intermittently. The processes whereby the erosion surfaces are cut is often not adequately explored. The theoretical model sets out to establish the effect of wave erosion on a steep coast with a falling sea-level. The effect of a rising sea-level is also investigated. The assumptions made in the model are the following: (1) the waves erode only to a given depth, the wave base, which is of the order of 40 ft. (2) The waves reduce the surface to a minimum slope of about 1 in 100. They can no longer erode below this gradient. (3) The initial coast slopes more steeply than 1 in 100. (4) The steps by which sea-level falls are less than the wave base. Figure 12.1 shows the results of the model. In the case of the falling sea-level, if the fall is continuous the resulting slope is parallel to the original one and no gently sloping erosion surface would form. If the fall is intermittent small steps would occur. Only with a slowly rising sea level can waves form a wide, gently sloping erosion surface. The presence in reality of wide surfaces, which have probably been formed with a falling sea-level, suggests that other marine processes besides wave action are also involved (C. A. M. King, 1963).

Many much more elaborate theoretical models have been worked out for slope formation. The work of A. Penck (1953), J. P. Bakker and J. W. N. Le Heux (1952) and A. Young (1963) are notable in this respect. The theoretical development of the slope under specified conditions has been studied by sequential steps by all these geomorphologists.

Theory Type 3

The third meaning of theory is perhaps the most common. A theory in this sense is an idea or hypothesis which has been tested and not discarded. The

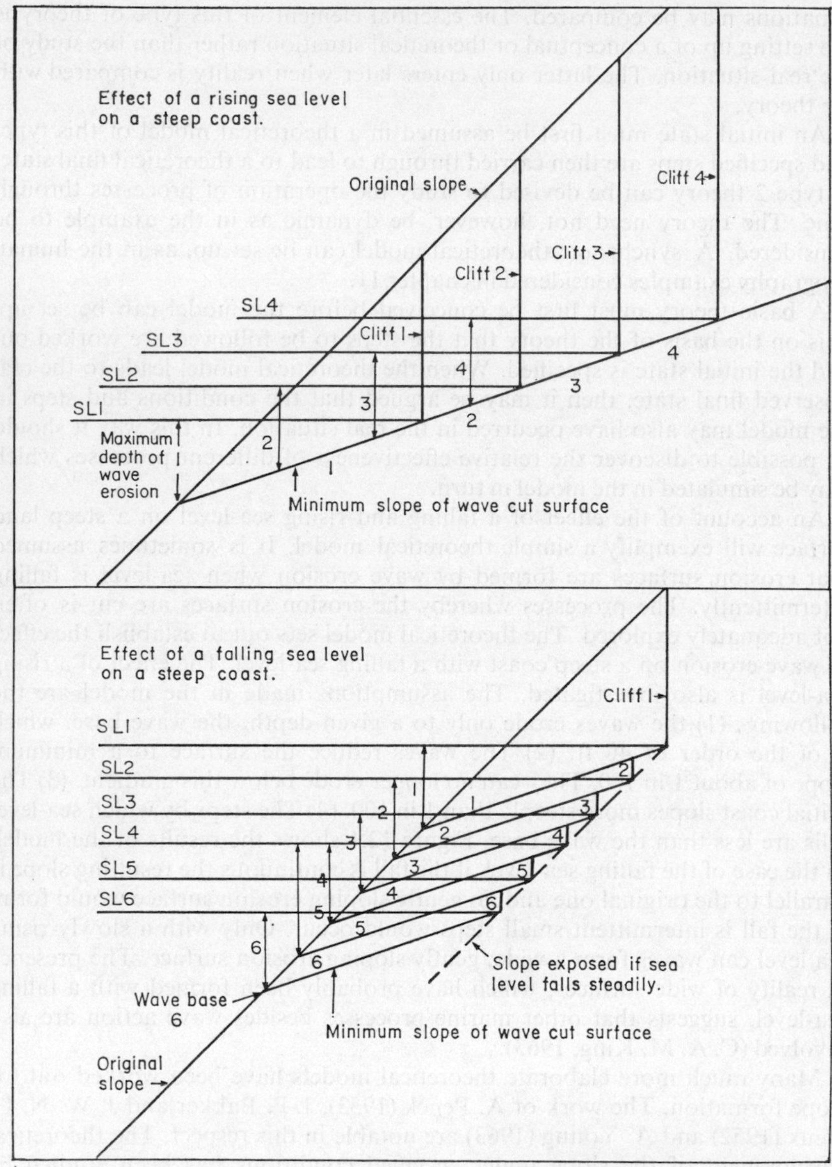

Figure 12.1 Diagram to illustrate theoretical model of the erosional effect of wave action with a rising and a falling sea-level.

essential point which differentiates this type of theory from the fourth type is that it is subject to testing and may therefore lead to the formulation of a tendency or law. The third type of theory often emerges from the method of multiple hypotheses in scientific investigation. As many alternative hypotheses as possible are explored and each is tested against the available data and that which best explains the data then becomes the working hypothesis or theory. There is always the possibility that subsequent testing on new data will lead to the rejection or modification of the theory.

The theory of continental drift is a good example of theoretical thinking. The idea was first elaborated by A. Wegener in 1912. It was an attempt to find an explanation for certain observations and other rather puzzling facts. One observation that led him towards the theory was the apparent jig-saw-like fit of the continents on either side of the Atlantic Ocean. However, this was only circumstantial evidence, supporting his main reasons for proposing the theory.

Wegener was primarily a meteorologist and was interested in palaeoclimatic data. The distribution of the tillite deposits, indicating past glacial conditions during the Permo-Carboniferous period, was particularly significant in this respect. These deposits were found in great thicknesses in South Africa and they also occurred in South America, India and possibly Australia. There is also evidence in Antarctica of temperate vegetation at this time. Wegener reasoned that the wide distribution of the areas of tillite deposition, which now lie in relatively low latitudes, could best be accounted for if the areas had once been closer together and also in higher latitudes. He therefore suggested that all the southern continents had once been grouped together around the south pole which then lay a short distance off the present position of South Africa. Having once formulated the idea. Wegener was able to find much other evidence from geological, botanical and zoological* sources to support his theory. The geological evidence was particularly strong as rocks and structural lines were found to match very closely on both sides of the Atlantic Ocean.

When Wegener first proposed the theory, much of the evidence on which it now rests was hidden beneath the sea, and the theory met considerable criticism, as so many new and revolutionary ideas usually do. One of the main points made by the critics was that no force capable of producing this vast movement could be envisaged. The forces suggested by Wegener himself were manifestly incapable of producing the movements that he proposed. Another point raised by the critics was that there was no reason why the primeval continent Pangea should have suddenly broken up in the Mesozoic period and drifted apart to form the present continents. The strongest criticism was therefore largely negative in character.

Ever since the theory was proposed arguments both for and against it have been discussed. Data have been reinterpreted, and alternative hypotheses explored. In the early 1950's opinion amongst experts was divided about half in favour of it and half opposed to it.

* The idea cited in chapter 13.5*b*(iii) is of interest in this respect.

One of the difficulties of a theory of this type is to find evidence that will either prove or disprove it conclusively. In the 1928 edition of his book A. Wegener gives evidence to show that the longitude difference between Europe and North America has increased by an amount considerably larger than the probable error of the observations.

Much new evidence has since been collected in favour of the theory. This information is largely concerned with geophysical data and oceanographic observations. Heat flow studies, seismic and gravity studies, both on land and at sea, have all provided valuable new evidence. The studies of palaeomagnetic data have, however, provided the most convincing new evidence in favour of the theory. These studies have been carried out on rocks which received their magnetism at the time of their formation. Their original magnetism was related to the magnetic field at the time of their formation. From this evidence their original latitude can be determined. If the rocks now occur in a different latitude then it may be assumed that the rocks have been moved relative to the position of the poles. There are many problems in the interpretation of palaeomagnetic data but the majority of workers in this field have come to the conclusion that the evidence indicates a movement of the continents both relative to each other and relative to the poles, (P. M. S. Blackett, 1961). Palaeoclimatic evidence also supports palaeomagnetic interpretations of polar wandering.

A modified version of the theory of continental drift has been put forward by J. T. Wilson (1963) and R. A. Dietz (1961). This version of the theory incorporates much new evidence, with particular emphasis on that found in the ocean basins. It is suggested that the oceanic floor is spreading outwards from the mid-oceanic ridges causing the continents to drift further from the central ridge. The force which causes this outward movement is suggested to be convection currents within the earth's mantle. The currents appear to move up under the median ridges, along beneath the oceans away from the ridges and down under the continental borders.

The evidence of the form of the median ridge supports this theory. There is a rift valley in the crest of the ridge and its high heat flow values also supports this suggestion. Both the Atlantic and Indian Oceans have well developed median ridges with rift valleys associated with recent and current volcanic activity. The alternating strips of differing polarity found in the magnetic survey of the ridge borderland are of particular interest in this respect. The polarity of the earth's magnetic field changes fairly frequently and rapidly. A strip of new igneous material emerging at the centre of the ridge received the polarity existing when it solidified. As it moved out from the centre it was replaced by more material from below and by the time this younger material solidified the polarity had changed and this material showed the reverse polarity. The continuation of the process has led to alternating strips of north and south polarity in the rocks adjacent to the median ridges. It has been suggested that this outward movement takes place at the rate of a few centimetres per annum. The complex patterns of the island arcs and the great transcurrent faults around the Pacific Basin can also be interpreted to fit into the theory. These arcs occur in the zones at the edge of the continent where the convection currents are turning downwards, and the pressures are compressive rather than tensional as in the central ridges.

The weight of evidence assembled in favour of the theory of continental drift is now so great that most people accept the theory. The measurably growing rift of the Red Sea provides a small scale example of how continental drift can work, but there still remain many problems concerning the precise mechanism of the process and its world wide pattern of operation. In time, however, as geophysical techniques improve and more data are assembled the exact operation of the process should become known. This is because the process is still operating and data are available to prove the theory correct. In time the theory will be accepted in the same way as the theory of the ice age, which was so hotly debated in the middle of the last century. This theory is now accepted by all and taken for granted.

Theory Type 4

The fourth type of theory differs from the third in that it is not amenable to testing. As the theory cannot be tested conclusively, it cannot become a tendency or law, even though it does not disagree with any of the known facts. The inability to test the theory is often due to the loss or removal of evidence that would be required to test it. As a result, this fourth type of theory is often concerned with past events rather than current processes or situations, which can be observed. An example of a theory of this type is that which attempts to explain drainage patterns. Several different and conflicting hypotheses have been put forward to account for the pattern of certain river systems in terms of their development through time. Although a great deal has been written about the Appalachians, the knowledge of their geomorphological development is in some respects only at the stage of the fourth type of theory. In this respect the development of drainage in this area is still subject to uncertainty as is the drainage development in Britain.

The most striking feature concerning part of the Appalachian region is the very strong structural control on relief and minor streams, but equally important is the disregard of major streams for the structural control. Much evidence for drainage adjustment remains, but probably some of the vital clues have been lost. This loss would account for the lack of agreement concerning the stages of development and the large number of theories put forward to explain the major drainage pattern. They probably originally flowed northwest but they now flow southeast.

W. M. Davis in 1909 put forward the first theory to account for the drainage pattern. (See figure 12.2). He believed that drainage flowed northwest in the Permian but that uplift of the Appalachian ridges across the streams allowed swift southeast streams to dismember and reverse much of the original drainage pattern. The lower portions of the southeast flowing streams he thought were younger, flowing first on the emerging Cretaceous cover and later being superimposed on older rocks with complex structures.

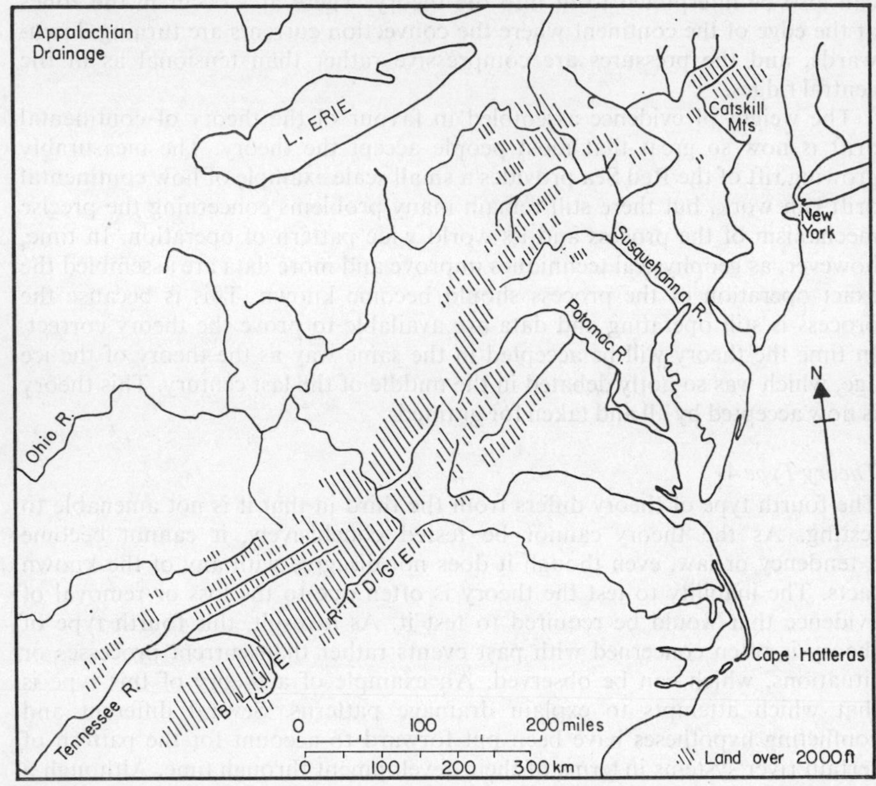

Figure 12.2 Map to show the pattern of Appalachian drainage in relation to major relief.

D. W. Johnson (1931) put forward the theory of superimposed drainage. He assumed that the whole area was peneplained in the pre-Cretaceous period and that the present drainage was initiated on an uplifted cover of Cretaceous rocks. The streams were considered to flow southeast on the emerging sea floor of Cretaceous strata. The lack of adjustment to structure of the main streams is accounted for by epigenesis on to the older folded rocks beneath. There is, however, no evidence of the former widespread Cretaceous cover. These rocks, he supposed, were removed by the erosion which created the Schooley peneplain, during the Tertiary.

H. D. Thompson (1939) considered that progressive river capture could account for the drainage reversal. The original watershed he supposed to lie about the Blue Ridge, but the steeper shorter slope to the Atlantic allowed the watershed to be pushed northwestwards. The distance the watershed moved is thought to be 80 to 100 miles in the northern Appalachians, at a rate of 7/10 miles in 1 million years. The streams cut across the ridges at their weakest points, an observation which supports this theory.

Various other possibilities have been suggested (see W. D. Thornbury, 1954, pp. 230–242, for detailed references). The number of proposed theories is sufficient evidence that an answer to this problem is probably impossible in that the theories cannot be tested, owing to removal of the vital evidence. The similarity of the situation with regard to theories of drainage development in this area and in Britain is striking.

12.3 Tendencies

One of the reasons why tendencies rather than laws are common in physical geography is the principle of indeterminacy. This concept, according to L. B. Leopold and W. B. Langbein (1963), has long been recognized in physics, but it has only recently been applied in the field of geomorphology.

Indeterminacy

There are two concepts involved in the principle of indeterminacy. The first depends on the fact that many laws or relationships amongst natural phenomena contain several variables. The second depends on the interaction of mutually opposing forces.

The first concept shows that the same result can be achieved if the variables contributing take on different values, by altering in different directions. A simple example will illustrate the principle. The discharge of a stream Q, depends on the breadth b, the depth d and the mean velocity of flow v. Thus $Q = bdv$, and this equation can be satisfied in several ways if

	$b=4$,	$d=2$,	$v=6$	then	$Q=48$
or	$b=2$,	$d=2$,	$v=12$	then	$Q=48$
or	$b=8$,	$d=1$,	$v=6$	then	$Q=48$

A large number of possible conditions of the three variables can satisfy the equation and give the same value of Q. No one variable is uniquely determined. The variables are, therefore, indeterminate and can yet give a specific value for Q. In this first concept of indeterminacy it is the variables that are indeterminate.

The second concept considers the effect of mutually opposing forces, and the interaction between them is in this case indeterminate. When one force is working, its effect is limited by the opposing one. This results in a range of possible states within which equilibrium conditions must lie. This range of variation is too large to enable precise unique predictions to be made for individual instances. Prediction can only be made in terms of mean values. The range of variation, which is dependent upon the character of the opposing forces, determines the variance.

An example of this concept can be seen in the variability of any one beach gradient. The slope is dependent upon the wave characteristics and the opposing effects of the swash and backwash. When the swash is constructive and is steepening the gradient by deposition, then gravity increases as the slope steepens. This increases the power of the backwash until equilibrium is achieved. The reverse applies when the swash is destructive and material is carried seaward and the slope reduced. This reduces the effect of gravity, and less material is moved so the destructive effect of the swash is limited. A balance between the opposing forces gives a range within which the beach gradient varies. The precise value cannot be determined unless all the conditions operating are known. Only a value within the range of indeterminacy and a mean can be stated.

The important point concerning the principle of indeterminacy is that, within certain specified limits, geomorphological phenomena will vary in value. All rills on a valley side for instance will not be equally deep. This variability in natural land forms is an inherent characteristic, quite different from measurement errors and differences due to dissimilar controls. In this type of situation probabilistic methods are most suitable as these methods imply inherent variability in the data. This approach is more likely to lead to the discovery of tendencies than laws. Mean values and measures of variance or dispersion are arrived at rather than precise or unique values applicable to individual instances.

Owing to the inherent variability of many phenomena it will always be easier to measure and record them than to explain and predict them. Because of indeterminacy a probabilistic statement is more likely to approximate to reality than a statement based on laws of mechanics when the situation is complex. Hence tendencies are more likely to be common and realistic than laws in physical geography.

Tendency type 1

One definition of the word 'tendency' is to describe a relationship based on every day experience, a relationship that may not have been systematically tested but which appears reasonable. This type of tendency is of limited interest.

Tendency type 2

A second use of the word is to refer to a quantitative observation or imprecise statement concerning the relationship between two variables or the character of a feature formed by a particular process. Some of the second type of tendency may provide suggestions that are worth pursuing to the third type, which is of considerably greater importance.

There are many examples in physical geography of the second type of tendency. Some of these were suggested before much quantitative work had been undertaken. The statement that glaciated valleys tend to be U-shaped, while stream valleys tend to be V-shaped, or that slopes tend to flatten with time are examples. A map could be examined and the statement made that slopes facing northwest tend to be steeper than those facing southwest, or that most corries face northeast. Similarly, the study of maps combined with observations in the field might suggest that cultivation and settlement extends higher up the south facing than the north facing slopes of the Alps. All these apparent tendencies could be further investigated quantitatively and the examples quoted have in fact been treated in this way.

Tendency type 3

The third type of tendency refers to quantitative relationships that have been tested, but in which the relationship established is only significant at a fairly low level and the scatter of values is considerable. A tendency expressed in these terms is a probabilistic statement dependent upon statistical analysis. This type of statement is common in geography owing to the complexity of the variables on which so many relationships depend. Human geography at least is dealing with the higher levels of organization (see chapter 13). Even in physical geography most problems are complex and the variables numerous. Thus it is difficult to arrive at simple and valid relationships that can be expressed with the very high degree of confidence necessary to establish a law.

The third type of tendency is the most important in physical geography. An example is the relationship between beach gradient and material size. A beach of coarse material tends to be steeper than one of fine material. A shingle beach is very definitely much steeper than a fine sand beach. If material size were the only variable that affected beach gradient, then all beaches formed of the same type of material could be expected to have the same gradient. This is not so because any one beach which is always composed of the same material, varies considerably in steepness. The relationship is only a tendency, partly because of the principle of indeterminacy for the reasons already mentioned.

The slope of any beach at any moment in time depends not only on the size of material but also on other variables. The two relevant variables are the wave length and wave steepness, both of which affect the relative effectiveness and character of the swash and backwash. The directions of their effect can be demonstrated in a model wave tank, in which the other variables can be held constant. In these conditions very high coefficients of correlation can be obtained between the slope and the individual variables that influence it (see figure 12.3). In natural conditions, however, the variables cannot be controlled and it is the combined effect of all three that determines the slope at any one time. There is a considerable scatter of points relating gradient to the two most changeable variables on which it depends, the wave length and steepness. The range of variability depends on the range of wave sizes experienced.

A rather similar tendency has been demonstrated by C. C. Mason and R. L. Folk (1958) in the difference in sorting (standard deviation) of dune and beach sand. The dune sands showed a statistically significant tendency, at the 95 per cent level, to be better sorted than the beach sands, although their mean size was not significantly different. The scatter of points is again quite considerable.

These tendencies are statistical ones that can be expressed in terms of probability and confidence levels. The probabilities are such that the results are statistically significant. Usually in geomorphological work the 95 per cent level of significance is chosen as a minimum level of significance. The resulting

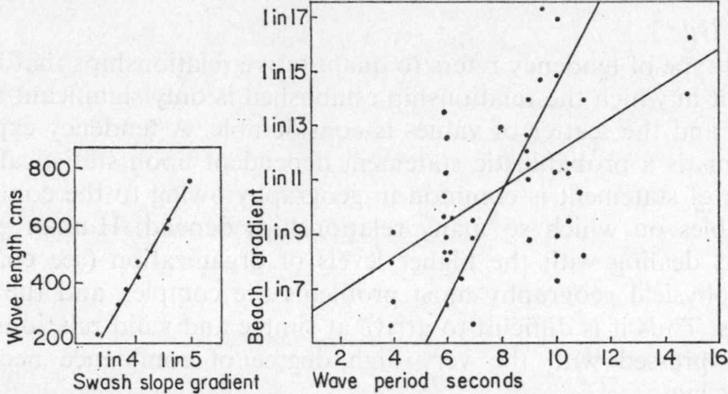

Figure 12.3 The left hand graph shows a swash slope gradient from results obtained under controlled conditions of a model wave tank. The beach gradient is highly correlated with wave length. The right hand graph shows results obtained under natural conditions in which the scatter of data is much greater as the variables cannot be controlled. (Marsden Bay, Co. Durham.)

relationships should be classed as tendencies, not laws. It is also due to the principle of indeterminacy and errors of measurement, sampling and possible omission of essential variables, that cannot easily be measured or identified.

12.4 Laws

The vagueness of the term 'law' in science has been recognized by E. Nagel (1961) who states that ' "scientific law" is not a technical term that has been rigidly defined in each of the sciences'. The term 'law' can be used in a number of senses.

Law type 1

Perhaps the simplest form of law is that applied to a statement of mathematical logic. The laws of Euclidean geometry, which follow logically from certain fundamental postulates, fall into this category of law. If the postulates are accepted as true, then the laws follow logically and cannot be false.

Law type 2

A mathematical relationship between two scales of measurement is another simple type of law, that can be stated simply and categorically. The relationship is a matter of definition. These laws are predictive, in that prediction, which will be fulfilled, can be made by them.

Some of the precise mathematical relationships, belonging to the second type of law, can be used to assist the analysis of geomorphological problems. Snell's law will be considered as an example. This law is derived from the theory of wave motion and can be stated in the form

$$\frac{\sin \alpha_2}{\sin \alpha_1} = \frac{c_2}{c_1}$$

α is the angle of the wave front with the shore and c is the wave velocity. The suffixes refer to two different depths, 1 and 2. This relationship is based on the effect of depth, d, on wave velocity as given by

$$c^2 = \frac{gL}{2\pi} \tanh \frac{2\pi d}{L}, \text{ where } L \text{ is the wavelength.}$$

Its importance in geomorphology is based on its application to the refraction of waves in the sea. Wave refraction diagrams are drawn on the basis of this law. It can be used to construct wave rays where the bottom relief and wave characteristics are known. The law leads to another which states that wave crests approximate to the shape of the depth contours. These laws can, therefore, be used as a step in the analysis of coastal features, where these depend on the direction of approach of refracted waves and the variations of intensity of wave attack due to refraction. Wave refraction greatly affects both the direction of wave approach and the intensity of wave attack, particularly in areas subject to the action of long swells, which are greatly refracted before they eventually reach the coast.

Snell's law influences only a relatively small field of study. The more important laws in the second category are those of greatest generalization that influence the widest field. The great natural laws of motion, enunciated by Newton in 1686, are the most wide ranging in their application and form the basis of dynamics. The first law concerns inertia, and states that every body will continue in a state of rest or of uniform motion in a straight line unless compelled by the action of external forces to change that state. The second law states that force is equal to mass times acceleration. The third states that every action has an equal and opposite reaction. These laws are of the highest generalization and apply to all situations that come within the field of geography. The laws were based on astronomical calculations and cannot be demonstrated by direct experiment. Their justification is demonstrated in the very good agreement between the predicted and actual time of astronomical events, such as eclipses. The minor inaccuracies noted have resulted in the discovery of new planets, whose presence had not previously been allowed for in the calculations. It is in their ability to be used to predict events that these laws, and the law of gravitation, which was developed from them, can be called genuine laws.

The laws of thermodynamics are rather similar in character and are valid over all ranges of values of interest to the geographer. The second law of thermodynamics has already been referred to in connexion with the development of closed and open systems (see chapter 11.7). Analogies have already been drawn between these laws and the law of gravity, on the one hand, and situations in human geography, on the other (see chapter 11.8).

Law type 3

The third type of law is probabilistic rather than predictive. It is probably the most important type of law from the geographical point of view. M. Scriven (1959) has stated that 'it is virtually impossible to find a simple example of something that is normally called a law in science which can be precisely formulated in non-probability terms, I think it certain that all scientific explanations, with quite negligible exceptions are "statistical" or better, "probabilistic" '. This type of statistical law can develop out of a theory that has been exhaustively tested and not rejected. A law of this type should have a higher level of statistical significance or probability than a tendency, but it is difficult to set a precise limit. The law is usually based on a study of a sample rather than the whole population. Statistical allowance must be made for this by relating the results to the population as a whole. Often in geographical work the 95 per cent confidence level is accepted to indicate a significant relationship, but this degree of significance would not justify the formulation of a law. The 99 or 99.9 per cent level of confidence might be suggested as the limit defining a law. Even then the relationship could be classed as a law only if the exceptions could be accounted for. Laws of this type are more common in physics than in the social sciences. The laws in physics hold because they apply only to ideal, controlled conditions that can be precisely stated, but which are frequently not fulfilled in the real world. Laws in human geography, on the other hand, must be statistical as the conditions cannot be controlled or even stated theoretically as they can in theoretical physical work. Physicists can control their experimental work sufficiently closely to approximate to ideal conditions (H. M. Blalock, 1961, p. 17).

The law of gravity applies to all processes operating on the earth's surface and it therefore forms the basis of rather less general laws that relate to processes operating in physical geography. G. K. Gilbert (1877) was one of the earliest geomorphologists to formulate laws of type 3. The laws that he has put forward depend fundamentally on the laws of gravity and motion. One of them is the law of declivity (slope in a river profile) which states that declivity bears an inverse relation to quantity of water, other things being equal. This law is seen to be true in that the steeper reaches of rivers are nearly always in the upper reaches where the volume of flow is less.

The law of divides is another of Gilbert's laws, stated as 'the nearer the watershed or divide the steeper the slope'. This law may be stated in a slightly modified form, in which it is often referred to as the law of unequal slopes. In this form it states that the watershed will move away from the steeper slope, thus equalizing slopes on either side of the watershed. This is because gravity has a greater effect on the steeper slopes and erosion is more effective on them as a result. River capture is an effective process in this respect. There are many examples of river capture that support the law of unequal slopes. In U.S.A. the classical area where this type of capture has taken place is in the Catskill Mountains, where streams flowing down the steep east-facing scarp slope have captured some of the headward drainage of the west-flowing dip stream, thus causing the watershed to move west. The diversion of the Upper Conway to the Irish Sea and the capture of the headwaters of the east-flowing river Ure by the steeper, west-flowing Eden and Lune, are further examples in Wales and northwest England respectively.

Law type 4

In the fourth type of law it is not necessary to set up a theory or hypothesis before the law is established by testing the theory. The fourth type of law is an empirical relationship found by observation or experiment. If a high degree of correlation is found between two variables they will be related by a law, although there may be no causal relationship between them. H. M. Blalock (1961) has pointed out that it is difficult to establish a connexion between a theoretical causal law, derived by theory, and the results of experiment and observation. It may be shown that B always follows A, but it is difficult to prove experimentally that B is caused by A. Once an empirical law has been set up, a theory can be developed to explain it. In testing the theory further the law may be extended to greater generalization.

An interesting set of laws, based on map measurement supplemented by field work, was first put forward by R. E. Horton in 1945. These are now often referred to as the laws of morphometry. Horton showed that certain relationships exist between parameters used to describe the characteristics of drainage basins. The first relationship is the law of stream numbers, the second the law of stream lengths and the third the law of drainage areas. Horton showed that all these three parameters bear a simple geometrical relationship to stream order. If the values for each parameter are plotted on semi-logarithmic paper against the stream order the line joining the points will nearly always be a straight line. This relationship has been shown to apply very widely and where there are exceptions these can usually be accounted for.

Some indication of the possible reason for these relationships has already been discussed in chapter 11.4 in connexion with simulation models. It was shown that the random walk model of drainage basin parameters agreed closely with the relationship found in natural drainage basins. These natural relationships, therefore, as expressed in the laws of morphometry, appear to be based on the second law of thermodynamics, which is one of the fundamental natural laws. This illustrates the situation where a law or relationship is found empirically, and the reason for the relationship only becomes clear after the empirical law has been established and shown to be valid over a considerable range of conditions.

The flow law of ice, will be mentioned as an example of an empirical law of type 4 that has been found by experiment. This law was derived from laboratory experimental data by J. W. Glen (1958) and has since been tested by observation in the field in glacier tunnels. The law can be expressed as an equation $\epsilon = k\sigma^n$, where ϵ is the strain rate, σ is the stress and k and n are constants. The power relationship was found to hold in the laboratory under a reasonable range of temperatures and pressures. The constants k and n must be adjusted for different temperatures because ice moves more readily at higher temperatures close to freezing point. A law of this type, derived from experimental physics, provides a necessary value for substitution in equations developed to describe the way in which ice flows. The law provides basic information concerning the way in which polycrystalline ice reacts to stresses upon it. This information can then be incorporated in theories of glacier flow under different stress conditions.

Law type 5

The fifth type of law differs from the others in being presumptive and not predictive. This is the legal law or regulation. Such laws, unlike any of the others, can be broken and yet remain laws. They have, in common with the first two types of law, an independence of probability statements. Their influence is widely felt in human geography but they have no effect on physical geography except through the operation of human interference. They cannot be described as a 'scientific' type of law.

Law type 6

There are certain other terms that are used particularly in physical geography which apply to broad generalizations of wide application. The terms 'concept' and 'principle' are often used, as in the concept of uniformitarianism. This is a concept or principle which runs throughout the whole field of geomorphological processes and development. It is a statement of greater generalization but lesser precision than most of the statements to which the terms law or tendency are applied. Similarly, the 'law' or 'principle' of least effort applicable to many human activities may be considered here.

Geomorphology as a whole has developed relatively few generalizations that are defined by the term 'law'. Such generalizations are not, however, entirely lacking, although they are more often called concepts or principles. L. C. King (1953) has used the term 'canon' to describe statements concerning landscape evolution that fall into the category defined as law type 6. His first canon states that 'landscape is a function of process, slope and structure. The relative importance of these is indicated by their order'. This generalization is based on the concept of W. M. Davis. Many of L. C. King's canons are more specific and less general. They are, therefore, less like the general laws applicable in all fields. In fact many of them appear to be tendencies rather than laws as the term 'may' is used instead of 'is'. In number 20, for instance, 'Gullying may appear upon pediments where laminar flow of water is changed to linear flow'. W. D. Thornbury (1954) uses the term 'concept' to describe his major generalizations. He states that 'landscape evolution is sequential, having distinctive characteristics at successive stages', a statement also derived from the work of W. M. Davis. Another of his concepts is that geomorphological processes leave their imprint on the landscape and develop their own characteristic landscape. This concept is the basis of much recent work in climatic geomorphology.

Conclusion

A number of different types of theories, tendencies and laws have been considered. These lead in turn to greater degrees of certainty in the relationships established (see figure 12.4).

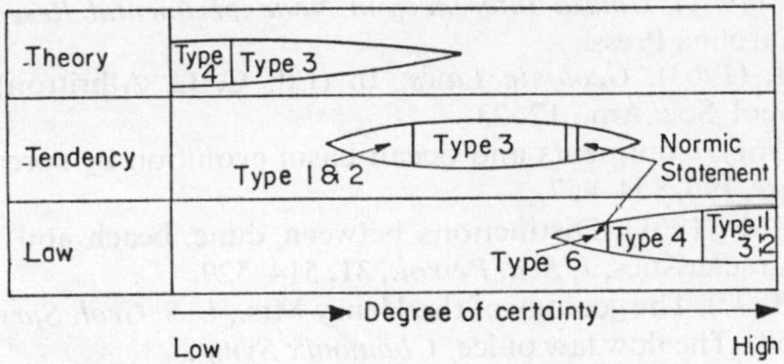

Figure 12.4 Diagram to illustrate degree of certainty of theories, tendencies and laws.

All these stages of certainty must be based on sound observation and experimentation. W. M. Bradley (1963) has suggested that concepts or principles come somewhere about halfway between the primary observations and the formulation of laws. The diagram in figure 12.5 shows that much of

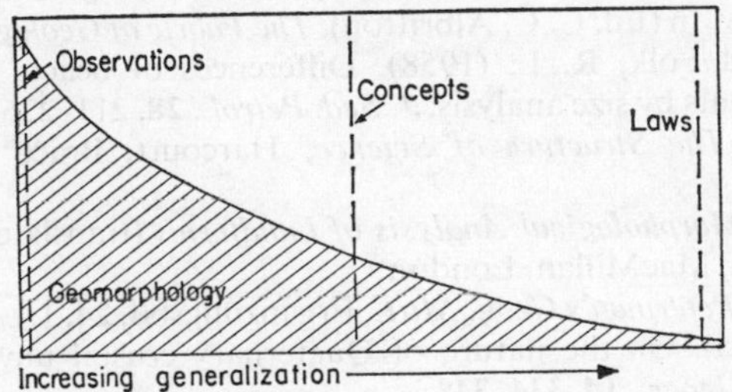

Figure 12.5 The relationship between observation, concepts and laws in terms of increasing generalization. (After Bradley.)

the work of geomorphologists is still concerned with the primary stage of observation. As yet there are few concepts and still fewer laws. It is in this direction that generalization increases. It is necessary to use the laws of physics, chemistry and biology in order to proceed to the formulation of laws in geomorphology. In geomorphology the variables are so complex and their inter-relationships so little understood that the formulation of generalizations is not easy. When the operation of the processes are more clearly known, the laws will follow. In human geography at the present time the formulation of general laws seems even more remote.

References

Bakker, J. P. and Le Heux, J. W. N. (1952). A remarkable new geomorphological law, *Proc. Koninkl. Akand. Weknshop*, Amsterdam, B **55** (4), 399–571.

Blackett, P. M. S. (1961). Comparison of ancient climates with the ancient latitudes deduced from rock magnetic measurements, *Proc. Roy. Soc., London*, A **263**, 1–30.

Blalock, H. M. (1961). *Causal Inferences in Nonexperimental Research*, University of North Carolina Press.

Bradley, W. H. (1963). *Geologic Laws*. In (Ed. C. C. Albritton) *The Fabric of Geology*, Geol. Soc. Am., 12–23.

Dietz, R. A. (1961). Continents and ocean basin evolution by spreading of the sea floor, *Nature*, **190**, 854–857.

Friedmann, G. M. (1961). Distinctions between dune, beach and river sand from textural characteristics, *J. Sed. Petrol.*, **31**, 514–529.

Gilbert, G. K. (1877). The geology of the Henry Mts., *U.S. Geol. Surv.*

Glen, J. W. (1958). The flow law of ice, *Chamonix Symp.*

Hartshorne, R. (1959). *Perspective on the Nature of Geography*, Rand McNally, Chicago.

Horton, R. E. (1945). Erosional development of streams and their drainage basins, hydrophysical approach to quantitative morphology, *Bull. Geol. Soc. Am.*, **56**, 275–370.

King, C. A. M. (1963). Some problems concerning marine planations and the formation of erosion surfaces, *Trans. Papers Inst. Brit. Geog.*, **33**, 29–43.

King, L. C. (1953). Canons of landscape evolution, *Bull. Geol. Soc. Am.*, **64** (7), 721–752.

Leopold, L. B. and Langbein, W. B. (1963). *Association and indeterminacy in geomorphology*. In (Ed. C. C. Albritton), *The Fabric of Geology*, 184–192.

Mason, C. C. and Folk, R. L. (1958). Differences of beach, dune and aeolian flat environments by size analysis, *J. Sed. Petrol.*, **28**, 211–226.

Nagel, E. (1961). *The Structure of Science*, Harcourt, Brace and World, New York, xiii.

Penck, A. (1953). *Morphological Analysis of Landforms* (translated by H. Czech and K. C. Boswell), MacMillan, London.

Richter, E. (1901). *Peterman's Geog. Mitt.*, Erganzungsbd, **24,** 1.

Sauramo, M. (1955). On the nature of Quaternary crustal unwarping in Fennoscandia, *Aeta Geogr.*, **14,** 334–348.

Scriven, M. (1958). *Definitions, explanations and theories*. In (H. Feigl, *et al.* Eds.) *Minnesota Studies in the Philosophy of Science*, **2,** University of Minnesota Press, Minneapolis, XV.

Scriven, M. (1959). *Truisms as The Grounds for Historical Explanations*. In (Ed. P. Gardiner), *Theories of History*, Free Press of Glencoe, Illinois, IX.

Thornbury, W. D. (1954). *Principles of Geomorphology*, Wiley, New York.

Wilson, J. T. (1963). Continental drift, *Sci. Am., reprint*, 868.

Young, A. (1963). Deductive models of slope evolution, *Nach. Akand. Wissen. Göttingen*, II, **5,** 45–66.

13

Networks and classification

'They are called wise who put things in their right order.'

St Thomas Aquinas

13.1 Graphs or networks

Figure 13.1 shows the development of a network. It is used to introduce some aspects of graph theory. Suppose that E, F, G and H are four football teams, Everham, Fulton, Gatesfoot and Hoole. At a given stage in the football season, some teams may already have played each other, others not. In figure 13.1, cases 1–8, the four teams are represented by nodes, E, F, G and H. Eight different situations are illustrated. In the first, no games have been played. In the second, E and F have played each other. An arc is drawn joining the nodes E and F to record this fact. In the table below, 1 is recorded for the E–F game, zero for all the unplayed games. As the number of games increases, so the number of arcs in the network increases. In cases 3 and 4, each team has played at least one other team. Four other situations are illustrated in cases 5–8. Note that case 1, where no games have been played, is called a *null graph*. Case 8, in which all games have been played, is called a *complete* (or *universal*) *graph*. An intermediate stage is a *connected* graph, where every node is connected to the network but does not necessarily have direct connexions with every other. Note also how a check can be kept on the arcs in the graph by recording them in the tables below.

The following features may be noted (n = the number of nodes being considered):

(1) For every node to be linked to at least one other, $n/2$ arcs are necessary, where n is the number of nodes. When there is an odd number of nodes, $(n+1)/2$ arcs are needed.

(2) For all nodes to be linked to one another, at least $n-1$ arcs are necessary, when n is the number of nodes.

(3) For each node to have a link with every other node without the linking arc passing through an intervening node, $n(n-1)/2$ arcs are necessary, where n is the number of nodes. $n(n-1)/2$ is also the number of compartments in half of the table, less the diagonal. In the example there are 4 nodes. Therefore $4(4-1)/2=6$ arcs. They are illustrated in case 8.

(4) In figure 13.1, the four nodes, E, F, G and H are linked in seven

different ways (cases 2–8) while they are not linked at all in case 1. The 8 situations shown are only a small proportion of all possible ones. In this example, the 6 possible arcs may be present or absent in 2^6 or 64 different combinations. Thus if x is the number of arcs, then 2^x combinations of arcs are possible. Therefore from the number of nodes in a net-

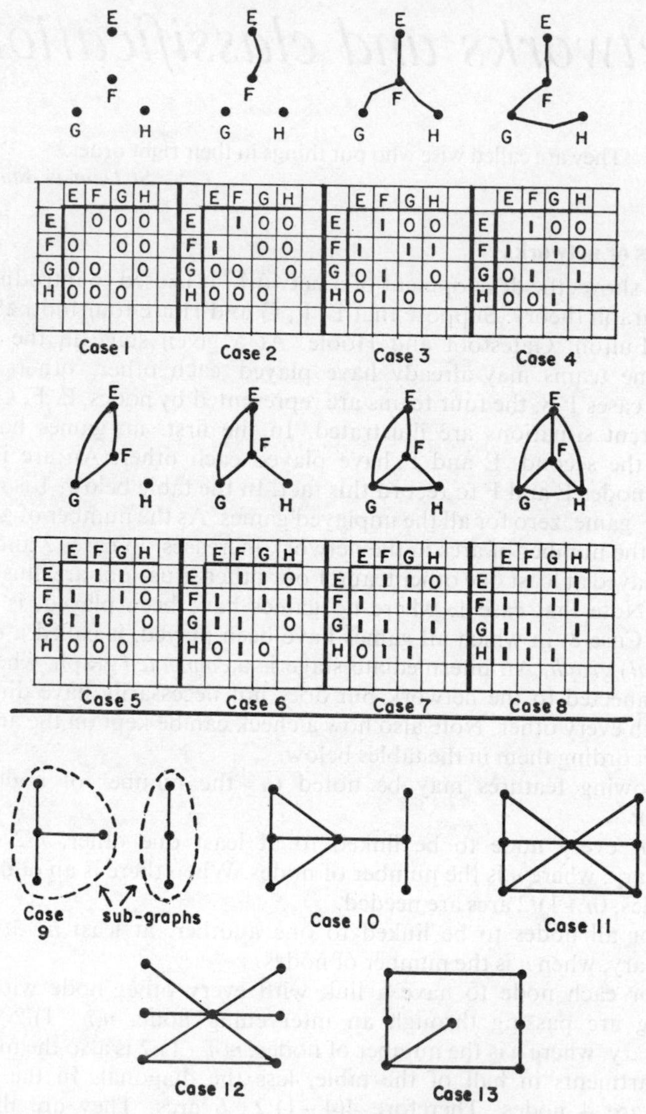

Figure 13.1 Examples of graphs. Cases 1–8 illustrate the description of a simple network. Cases 9–13 contrast different combinations of arcs.

work the number of possible combinations of arcs may be derived directly from the following formula: $2^{n(n-1)/2}$ where n is the number of nodes. Some further features of graphs must also be noted.

(5) When only one *route* is possible on a graph between any pair of nodes, the graph is called a *tree* graph (case 1 in figure 13.2). Thus in case 1, the only possible route between A and X passes along the following *arcs*: AZ, ZY, YX. In case 2, there are three possible routes between E and F: EF, EGF, EGHF. But case 2 also has some *branches* (e.g. HJ). In case 2 it is possible to travel from E back to E without going along any arc twice. EGHFE is called a *circuit*. It will be noted that in case 2 there are also regions (marked r). In case 3 there are no branches. In a graph without branches,nodes must be at least 3-nodes. The formula described in chapter 2.11, $a+2=n+r$ $(8+2=5+5)$ is true for such a graph. In the special situation in case 4, two nodes are joined by three different arcs. The above formula holds true here. New complications which cannot be discussed here arise however when more than one arc (or channel) links to nodes. More important, if more nodes in case 3 are joined, a time comes when some arcs cross (e.g. join N–K (round outside) then L–O). This situation is dealt with next.

(6) Graphs confined to a 2-dimensional surface are called *planar* graphs. Cases 1–4 in figure 13.2 are all planar graphs. In the non-planar graph (case 8) in figure 13.2, however, the arcs between R and S and between V and Q cross. On a planar graph they would intersect (share the same crossing point in common). On a non-planar graph they pass without intersecting. This is because one of the arcs is permitted to pass for a short distance into the third space dimension. The examples of road junctions shown in the lower part of figure 13.2 illustrate two kinds of junction at the crossing of two dual carriageways (north–south and east–west). The non-planar junction allows traffic in the main directions to flow continuously, while the changing traffic filters in without crossing. Traffic lights are desirable for a busy planar junction.

The situation may be seen in a more diagrammatic form in figure 13.2, lowest left, as putting four inflow nodes in a one to three correspondence with combinations of three outflow nodes.

(7) The number of possible journeys on a well linked graph becomes very large as the number of nodes grows. Thus in figure 13.2, case 5, the number of possible journeys from A to Z combining visits to L, M and/or N or none is 8, derived from the formula 2^n, n being the three intermediate points.

(8) From the point of view of the geographer, topological descriptions of networks should be supplemented by distance information where necessary. Thus, for example, in cases 6 and 7, two graphs are equivalent topologically, but case 6 has a greater proportion of its system in branch lines, which are less advantageous in general than other lines. The completeness of the network could be further described by measuring direct and actual distances between all or between selected nodes and

then comparing the ratios in the two cases. Direct and actual distances are discussed in section 13.6.

13.2 Types of network

Networks vary greatly in many ways such as their size, connectivity, and dimensions. Some networks such as streams and railways have relatively

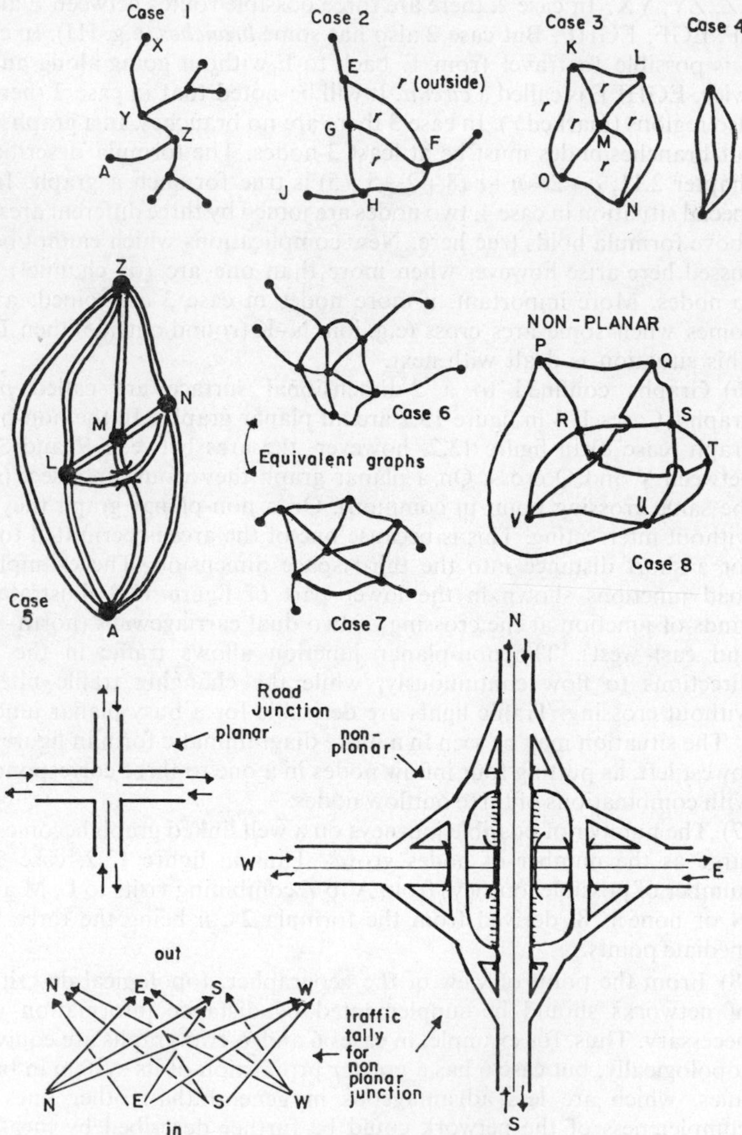

Figure 13.2 Types of graphs discussed in the text.

fixed channels on the ground. Networks describing relationships, as between entities in an administrative hierarchy or elements in a chemical substance, are conceptual lines that may not exist at all in a material form but which help one to conceive relationships.* The organization and classification of material in a discipline is also closely connected with networks.

The study of networks is covered in particular by graph theory. The term graph used here should be distinguished from the more familiar graph on which co-ordinate positions are plotted. Graph in the new sense, and network, are not entirely interchangeable, but they are treated as similar here. In graph theory the nodes and arcs are the main ingredients. Each arc (lines) should be thought of as a discrete feature rather than as an infinite set of points. However new nodes can be placed at will along arcs.

Networks may be classified on the basis of several criteria or combinations of these. A rather mixed selection of criteria is listed below and then discussed in this section. At least the following questions should be asked about a network in any attempt to classify it:

(1) How many nodes (2, 3, several, many)?
(2) How many arcs, and other descriptive indices derived from node/arc relationships?
(3) Is the network abstract or material? Does it have channels? If material, is it dynamic or static (do things move along the channels or not)?
(4) Is it directed or not (one way or two way along the arcs)?
(5) What sort of correspondences are there between the nodes?
(6) What sort of distances are involved? Or time periods if the arcs are proportional to duration of time?
(7) What is the orientation of the network?

Further features should be borne in mind including: how many dimensions (the present chapter deals with two) and how many regions (related to (1) and (2)). If material (3), do the channels carry material things and if so, are they discrete objects (e.g. cars) or continuous (e.g. water, gas)?

(1) *The number of nodes*. At least two nodes are required before a graph can be created. The possible complexity of a network grows rapidly as the number of nodes grows. For example, only 1 arc is needed to join 2 nodes, 3 to join each pair of 3 nodes and 6 to join each pair of 4 nodes. But 45 arcs are needed to join each pair of 10 nodes.

(2) *The number of arcs*. A graph does not necessarily have all its nodes joined by arcs. There are various ways of measuring and describing the completeness of a graph. For example in figure 13.1, case 3, there are only branches, no closed regions. In case 6 there is a region and also a branch. In case 8 there are regions but no branches. In a network with more nodes it is possible to express the number of regions as a ratio of the number of branches. The higher the proportion of regions, the more developed the network. Note that in transportation networks, where distance must be considered as well as the fact of having a direct link,

* The relationships here are not the causal kind measured in terms of correlation indices.

the ratio described, and others of the same kind, must be treated only as part of the study of networks.

(3) *Real or imaginary*. Is the graph representing a real network with channels or is it imaginary? Concrete or abstract, material or conceptual? If it is real, do objects move along the arcs between the nodes? The function of a railway network is to carry trains, material objects, whereas a telephone network carries messages; the messages themselves are not material objects, at least during their passage through the system.

(4) *Directed*. Is the graph directed or not? Are the relationships between pairs of nodes one-way or two-way? In a stream system, water particles flow only in one direction, while in a railway system, trains usually go in both directions, even on a single track line.

(5) What sort of *correspondences* are there between nodes? In an administrative hierarchy, down which orders pass, and which usually is in tree form, there is a one to many correspondence from each level down to the one below it. In a railway system there are usually many to many correspondences because it is desirable to be able to make a journey from any station to any other should the need arise. The introduction of the idea of correspondences is only a guide. On the whole, a similar description would be given to networks by correspondences as by describing the arrangement of arcs.

(6) *Distance*. In many networks of a material kind, spatial distance between nodes is of importance. Two systems might have the same number of regions and branches, but in one the length of branches might be much greater in proportion to the length of other arcs, than in the other. Two imaginary topologically equivalent graphs are shown in figure 13.1, cases 6 and 7. Case 6 has much larger branch arcs than case 7.

In general, spatial distance between nodes should be taken into account particularly when items being moved along the network are heavy, bulky, perishable or in some other way inconvenient to move. Even the cost of transmitting messages grows with distance. In networks of this kind, it is obviously desirable to have arcs running between nodes as directly as possible. A rough measure of the efficiency of the network could therefore be given by the ratio of direct to actual distance between nodes. Kansky (1963) refers to this as degree of circuity (p. 32).

(7) *Orientation*. The term orientation is used in preference to direction, which has already been used, to describe the relationship of a network to compass points.

It would be useful to have some accepted notation and set of symbols to portray the various features of a network. To introduce one here would be more confusing than useful. It is more profitable to express some familiar networks in the light of the features listed. Eleven networks are illustrated in figure 13.3 and their features are noted in the chart (figure 13.4).

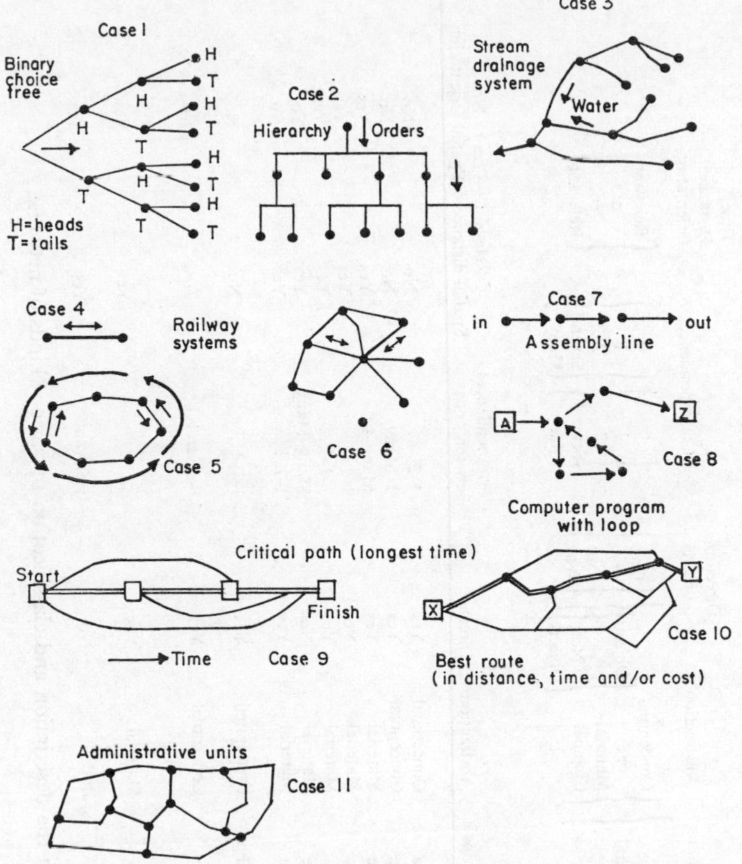

Figure 13.3. The application of graphs in various fields.

13.3 Stream networks

The stream drainage pattern is one of the most conspicuous patterns on a physical or topographic map. It consists of streams organized into particular patterns. These patterns can be analysed in a number of different ways. This section aims to draw attention to some aspects of stream network organization and show how they can be used in geomorphological analysis. A genetic type of stream classification differentiates consequent, subsequent and other streams according to their order of development and relationship to structure. The relationship between stream patterns and structural patterns provides valuable evidence. Those sections of the streams that are adjusted to the structure, so that the two patterns are in harmony, can be differentiated from those parts of the streams where the two patterns are not related, and

Dimensions	Nodes	How many arcs	Regions	Abstraction		Direction on arcs	Correspondences		Time, Distance, Orientation
One Two Three Time Other	Two Several Many	One Several Many More (than nodes)	None Several Many	Conceptual or Material Channel	Objects No Objects	One or Two Way	0 1 2 Several Many	0 1 2 Several Many	Relevant or Not Relevant

Case	Nodes	Arcs	Regions	Abstraction	Directed	Correspondences between nodes	Relevant — Yes or No: Spatial distance	Relevant — Yes or No: Spatial Orientation
1 Binary tree	Many	as nodes	None	Conceptual	Yes	1–2	No	No
2 Hierarchy	Many	as nodes	None	Conceptual	Yes	1–Many	No	No
3 Streams	Many	as nodes	None	Material	Yes	Many–1	Yes	Yes
4 Rail	Two	One	None	Material	No	1–1	Yes	Yes
5 Rail	Many	as nodes	Two	Material	Yes	1–1–1 . . .	Yes	Yes
6 Rail	Many	More	Some	Material	No	Many–Many	Yes	Yes
7 Assembly line	Several	as nodes	None	Material	Yes	1–1–1	Yes	No
8 Computer program	Several or many	as nodes	Possibly	Conceptual	Yes	1–1–1	No	No
9 Critical path analysis	Several or many	More	Some	Conceptual	Yes	1–several–1	*Time*	No
10 Best route								
11 Admin. units	Several or many	More	Some	Material	Yes	?	Yes	Yes
		$(a+2=n+r)$		Both	No	Irrelevant	Yes	Yes

Figure 13.4 Chart to aid in the description and classification of various kinds of network.

the drainage is discordant. The latter situation implies epigenesis or antecedence, more frequently the former. These patterns are concerned with the layout of the streams in a horizontal plane as shown on a topographic map. However, more abstract relationships in stream organization have been recognized.

The laws of morphometry are concerned with these relationships. They were discussed originally by R. E. Horton (1945) and have been mentioned in chapter 11.4 and in relation to simulation models. They are concerned with the organization of drainage networks. The drainage basin is the framework in which the streams develop. The relationships that have been established are concerned with the numbers and lengths of different orders and other variables. Several significant relationships have been established by this type of analysis. Density of drainage, given by the length of stream in relation to the basin area, and the degree of integration, given by the bifurcation ratio, provide useful information concerning the degree of development of the drainage pattern. The bifurcation ratio is found by dividing the number of first order streams by the number of second order streams.

An example of the application of the results of this method of analysing

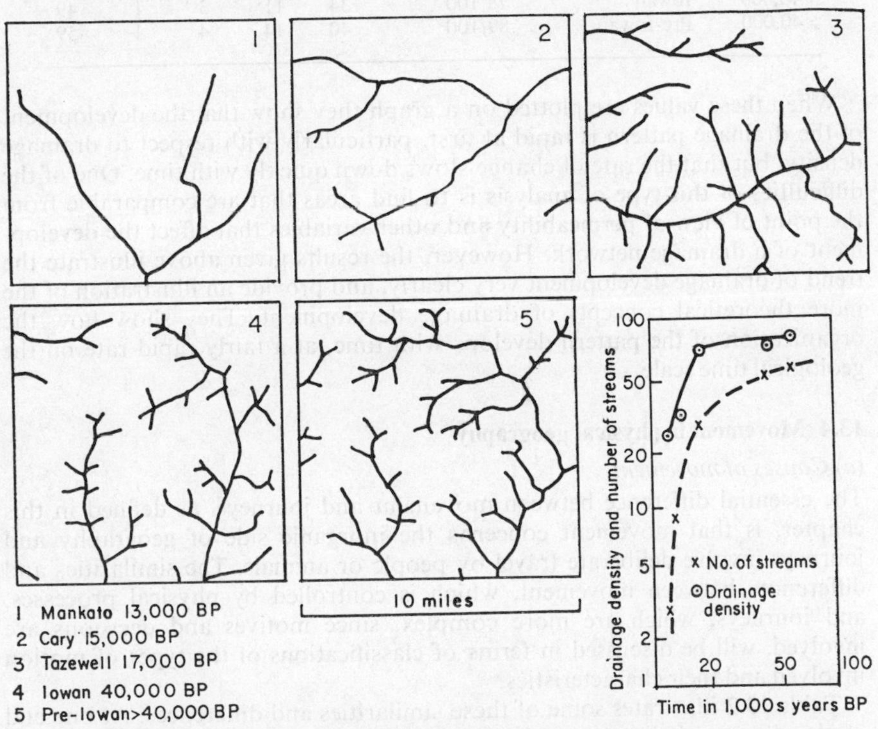

Figure 13.5 The progressive development of a drainage pattern on tills of increasing age. The increase of stream numbers and drainage density with time is shown in the graph. (After Leopold, Wolman and Miller.)

drainage patterns will be mentioned. The study of the drainage pattern on glacial tills of the various ages illustrates the gradual integration of drainage networks with time. This increasing integration and organization is shown by the increasing number of second and higher order streams and the greater length of streams relative to the area of the drainage basin. The number of streams of various orders have been calculated by L. B. Leopold, M. G. Wolman and J. P. Miller (1964) (pp. 423–6) from maps of drainage patterns given by R. V. Ruhe (1952), which are shown in figure 13.5. The values of drainage density have also been worked out as shown in table 13.1:

Table 13.1 Stream density data

Year B.P.	Glaciation	Drainage density	Number of streams 1st	2nd	3rd	4th	Total
13,000	Mankato	25/100	2	1	0	0	3
15,000	Cary	31/100	6	2	1	0	9
17,000	Tazewell	77/100	22	6	2	0	30
40,000	Iowan	78/100	34	11	3	1	49
>40,000	Pre-Iowan	89/100	40	14	4	1	59

When these values are plotted on a graph they show that the development of the drainage pattern is rapid at first, particularly with respect to drainage density, but that the rate of change slows down quickly with time. One of the difficulties of this type of analysis is to find areas that are comparable from the point of view of permeability and other variables that affect the development of a drainage network. However, the results given above illustrate the trend of drainage development very clearly, and provide an illustration of the more theoretical concepts of drainage development. They show how the organization of the pattern develops with time, at a fairly rapid rate on the geological time scale.

13.4 Movement in physical geography

(*a*) *Causes of movement*

The essential difference between movement and journeys, as defined in this chapter, is that movement concerns the inorganic side of geography and journeys involve deliberate travel by people or animals. The similarities and differences between movement, which is controlled by physical processes, and journeys, which are more complex, since motives and decisions are involved, will be discussed in terms of classifications of the types of motion involved and their characteristics.

Table 13.2 illustrates some of these similarities and differences. At one end of the diagram it is suggested that energy is the force which provides the power to move. On the physical side the force of gravity, which always acts downwards towards the centre of the earth, must be balanced by a force

Table 13.2 Types of movement

	Energy	Load	Type of route / Topological	Length	Channel Type	Orientation pattern reduced to a horizontal plane.
Physical Movement Inorganic	Applies mainly to units consisting of many linked, inseparable particles to form continuous bodies, e.g. rivers, glaciers, but also applies to soil, etc.		Cycles common for moving media involving change of state, e.g. air + water (atmosphere) Mainly one way movement for sediment, except for reversal by endogenetic forces			
	Exogenetic and endogenetic processes — energy of *uplift* by natural processes only, by forces counteracting gravity. *Gravity* is fundamental downward force Physical laws only dependent on energy imparted to wind and water No motive	Water {Vapour (gas), Liquid, Ice (solid)} Sediment Heat		Depends upon landform and relief	Narrow (sharply defined) Broad (sharply defined) Ribbon Diffuse	Patterns, e.g. ocean currents, drainage networks.
Organic Human and Animal Journeys	Motive + Physical laws + Controllable energy up and down based on food and fuel Applies mainly to individual discrete units — men, animals or vehicles acting separately	Goods People	Efficiency of network may be assessed in terms of arcs, nodes	May be considered in terms of distance, time or cost	Bounded (e.g. tunnel) Partly bounded (e.g. road) Unbounded (e.g. air lane)	Orientation in itself not usually of interest

which lifts matter upwards. The upward force comes from two major sources. The sun provides the external source of energy that is directly or indirectly responsible for the upward movement of air and water especially. The internal source of energy is derived from processes, such as radioactive heating, taking place inside the earth. These processes can produce uplift of the earth's crust in several ways that need not be elaborated here. These processes, which counteract gravity and which allow gravity to operate by raising matter, obey physical laws. No motive on the part of the moving matter enters the problem. This simplifies the physical side of the problem to a considerable degree. On the human side the fact that journeys have motives complicates the problem. Nevertheless, like physical movement, human journeys and animal journeys besides requiring motive also require energy.

In the field of physical geography movements in the vertical sense depend on the balance between the forces that are pulling downwards and the counteracting upward forces. The latter depend on the availability of energy and the former depend on gravity. Unless some upward movement is imparted to a system the movement will always be downwards under the force of gravity. The upward force can be made manifest by the operation of a number of different processes. The most massive upward transfer of energy against gravity is probably the convectional and other activity in the atmosphere which carries air and moisture, in the form of vapour, upwards. From the point of view of biogeography another important upward force is the capillary action that takes place within the soil moisture zone and allows plants to obtain water from a greater depth than would otherwise be available to them. This movement depends on the molecular attraction within the pore spaces and is most effective when these are fairly small. Another method of overcoming gravity is by hydrostatic pressure, which allows subglacial melt-water streams to flow uphill for a distance that is dependent on the elevation of the outflowing stream in relation to the intake. A similar force allows water to rise above the surface in some artesian wells. The fountain in Trafalgar Square used to demonstrate this process until the water table in the neighbourhood of London was lowered below the level of the fountain. All these forces allow water to rise upwards against the force of gravity, convection being the most effective of them.

(b) Hydrological cycle

The hydrological cycle illustrates clearly the consideration of upward and downward forces operating to produce a complex circulatory pattern, in which matter moves in a complicated way to complete a circuit of shorter or longer distance and duration. This cycle has been briefly mentioned in chapter 11 as an example of an analogy with the circulation of the blood, and the energy that provides the upward or positive side of the circulation has already been alluded to in this chapter.

The diagram, figure 13.6, indicates some of the many routes by which water can complete the hydrological cycle, which provides the water on the earth's surface without which life on earth would be impossible. The distribu-

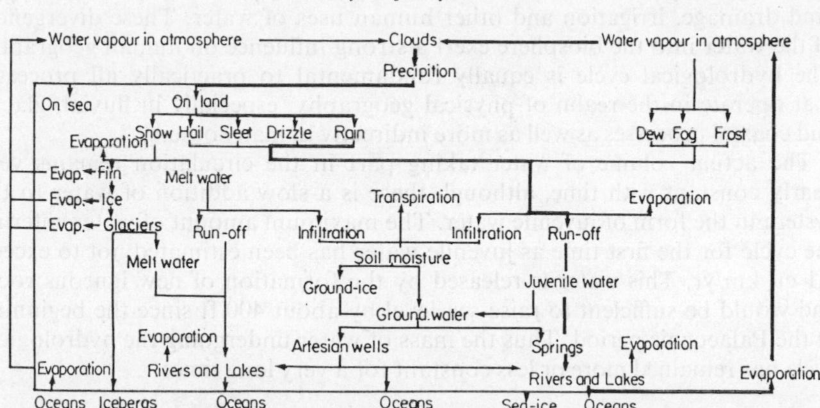

Figure 13.6 A flow diagram to illustrate the hydrological cycle. The path carrying the greatest amount of water over the land surface is suggested by the thickest line.

tion of water on the earth, derived from the great reservoir provided by the oceans, is determined by the upward part of the cycle. The upward movement of water in the form of vapour is dependent upon the processes of evaporation from water surfaces of which the sea is the most important, and by transpiration from plants. The vapour is condensed into water droplets by cloud formation in the atmosphere. These positive processes are determined by the energy provided by the heat of the sun, which gives rise to convection. More indirectly the energy causes frontogenesis, which leads to the condensation of water vapour to form clouds and then precipitation.

The cycle can be completed by a short route if the precipitation falls directly back into the ocean. The part of the cycle in which precipitation moves over or through the earth is much more important geographically. This movement can be very slow as, for example, when the precipitation falls as snow and remains locked up as ice for a long period. Alternatively, the precipitation may fall as rain and run off directly along a river back into the sea, making a rapid circuit. Some of the falling water pursues a slower course to the sea within the ground, seeping down as groundwater and eventually reappearing on the surface through a spring or perhaps being pumped to the surface in a well. There are a very great number of possible routes that any one particle of water can take within the framework of the cycle. Water may fall on to the land surface as rain, then run off into a basin of inland drainage and be re-evaporated into the atmosphere and so complete a circuit without returning to the ocean.

Of great significance to all forms of life is the water which passes through the soil into the plants and through them directly or indirectly into food, as well as the water that is used directly for drinking and other purposes. The hydrological cycle is thus intimately connected with problems of water supply, land drainage, irrigation and other human uses of water. These divergences of the water into the biosphere exert a strong influence on human geography. The hydrological cycle is equally fundamental to practically all processes that operate in the realm of physical geography, especially in fluvial, glacial and coastal processes as well as more indirectly in many others.

The actual volume of water taking part in the circulation remains very nearly constant with time, although there is a slow addition of water to the system in the form of juvenile water. The maximum amount of water entering the cycle for the first time as juvenile water has been estimated not to exceed 0.1 cu km/yr. This water is released by the formation of new igneous rock, and would be sufficient to raise sea level by about 400 ft since the beginning of the Palaeozoic period. Thus the mass of water undergoing the hydrological cycle has remained more or less constant for a very long time.

(*c*) *Character of movement*

The external forces acting on the earth's shell from outside have been stressed, but the internal ones must not be ignored. If they were not operative the earth would long ago have been reduced to a uniform level. It is the external forces, however, that play the major part in producing the most common and familiar movements in the physical world. These forces involve the movement of air and water, to which they impart energy, both potential and kinetic. Thus the movement involves a large number of minute units that link to form large masses of air or water, or more contained but continuous flow, as in a river or glacier, where the water is transformed into its liquid or solid state.

There is a contrast between the form of physical movement, where the moving media flow continuously, and human movement. In human journeys the moving media are often isolated, discrete units. They are the men, vehicles or animals that move individually. Their movements are, however, not independent of one another, as traffic jams illustrate very clearly. In this type of problem, concerning group phenomena, the same methods can be applied to both physical movement (e.g. of flood waves) and human traffic problems, as discussed in chapter 10 where kinematic waves are commented upon.

There are several other ways in which physical movement and human journeys can be linked. Some of these will be considered. The physical side of the problem will be dealt with first and then it will be shown that human journeys can be treated similarly, but some require different and/or more complex treatment. These aspects are as follows: (i) Load. (ii) Type of route — a topological approach. (iii) Length of route — Euclidean. (iv) Type of channel for movement or travel — Euclidean. (v) Pattern of routes — orientation reduced to horizontal plane — Euclidean.

In each of these aspects some similarity will be found, showing that the two aspects of geography may be used to throw light on each other.

(i) *Load*. The load in physical movement is carried mainly downwards under the force of gravity. There are minor exceptions, such as wind blown climbing dunes and glaciers carrying drift upwards, but the general balance of load carrying must be downwards under the force of gravity. The load is largely composed of sediment carried along on the moving belt of a river or glacier or by gravity down a slope. The volume of load is partly a function of the volume of discharge. There is no close parallel in journeys, although the goods carried can be compared with the sediment load of a river.

A parallel may be seen in the fact that in both aspects, physical and human, both concrete objects and intangible phenomena are carried. The most important intangible property that is carried by the physical movements is heat. The spreading of heat more uniformly over the earth's surface by air and water movements is very important in accounting for the physical character of the earth. The difference of climate at similar latitudes on either side of the Atlantic indicates the importance of this type of movement. It has its counterpart in the flow of intangible ideas and messages in the human sphere.

(ii) *Type of route*. Movement in the physical sphere is usually unidirectional. It may well be intermittent but it is rarely reversing. This is not true of movement under wave action where flow may be reversing, zig-zag or very complex. Wave energy provides a force that can counteract gravity.

Where no material load is concerned the movement is often cyclic, frequently involving changes of state in water. The hydrological cycle which is the most important and complex type of movement involved in the physical sphere has been considered already. It has similarities with some of the complex patterns of journeys by men.

(iii) *Length of route*. Fairly clear cut divisions between types of journeys and their length can be made in human journeys, where the method of transport depends partly on the length of the journey. This division is not apparent in physical geography. Where the transport of load in natural conditions is concerned the length of its movement depends mainly upon the relief and the size of the land or seabed over which it is moving. The distance the load moves at any one time may be long, with continuous movement. This applies to the suspension load in active rivers and load carried by a glacier, which moves continuously. Alternatively it may be intermittent, for example where rivers can only move their load at times of high flow or with turbidity current flow in the ocean. In the latter case a state of instability must be built up or triggered off by some disturbance and then frictionless flow, in the form of auto-suspension, allows movement over a long distance and on very low gradients. An equivalent land process is seen in the nuées ardentes of explosive volcanic activity. In these intermittent types of movement the recurrence interval depends upon many factors and so does the distance the load travels in any one phase of movement. Most human journeys and transport have more in common with this type of intermittent movement.

(iv) *Channel type*. Channels may be narrowed and deep (river gorge), semi-circular (some rivers) broad and deep (valley glaciers) or broad and shallow (rivers on outwash plains or some ocean currents like the Cromwell current). The Cromwell current stretches horizontally 2 degrees either side of the Equator but is only 200m in depth. Some channels are diffuse without clear limits such as those followed by air masses or the more diffuse ocean currents such as the West Wind Drift in the Southern Ocean. Generally as the channel becomes more diffuse the velocity is reduced but the volume carried through the cross section increases. These variables may balance to produce approximate equality of discharge although this is by no means always so.

The more clear-cut and well defined the channel is, the more definite will be the network formed by the channel and its association with other channels of similar character with which it may merge or from which it may diverge.

(v) *Orientation pattern*. The best type of network of movement channels that occurs in physical geography is the drainage network. This is a branching pattern if followed upstream or a linking pattern if followed downstream. There are exceptions in braided channels and delta distributions in which the pattern becomes complex or the reverse of normal. The normal drainage pattern has been considered earlier in chapter 11 in connexion with simulation models and is further discussed in section 13.4*b* of this chapter, in dealing with the analysis of networks. Well-defined patterns of a circulatory type also occur in ocean currents. Both these patterns have their analogues in human geography, although there are important negative as well as positive analogies. The unidirectional flow in the physical movement channels must be contrasted with the bidirectional movement along a road network. The flow in a unidirectional stream system is controlled by gravity, which acts along the vertical component of the slope, thus the direction of slope determines the direction of flow where gravity is the dominant motive force.

In the circulatory patterns, however, gravity, in the form of pressure gradients, is counterbalanced by the Coriolis Force. The balance between the pressure gradient forces and the Coriolis Force determines the direction of geostrophic flow in the circulation of both ocean currents and air flows (winds). The gradient winds also take into account the curvature of the isobars and thus are influenced also by centrifugal force. The direction of flow, being a resultant of several forces, has a strong positive analogy in human transport patterns in which several forces play a part in determining the pattern. Gravity plays a part in the selection of route to avoid hills, and the disadvantage of greater distance may be outweighed by the easier gradients. Both attractive and repelling forces also help to fashion the human transport pattern, which forms a network of routes.

13.5 Movement in biogeography

The movement of animals comes somewhere between the inorganic aspect of movement and the purposeful human journeys. For this reason a classification of animal movement provides a useful link. The classes of animal movement also provide ideas in considering distribution as affected by movement. L. Reid (1962) (pp. 229–252) gives a classification of animal movement as follows: (*a*) Movement of the habitat. (*b*) Movements within the habitat.

(i) Drift.
(ii) Movements connected with food and predators.
(iii) Movements for breeding purposes.
(iv) Migration caused by overcrowding.

(a) Movements of the habitat

The basic dichotomy stresses very clearly the importance of the total environment of any animal species. Each species has adapted itself over a long period of time to live in its own biological community. The community in turn depends on the physical character of the environment and the ecological system evolves towards a climax state in which both plants and animals can make best use of the natural properties of the physical or inorganic framework. This framework depends to a considerable degree on the climate, which controls the major vegetation regions. However, as the climate changes so the whole environment shifts and the ecological system connected with it moves too. This type of movement belongs to the first category and it exerts a strong influence on the distribution of plant and animal species. Category (a) can affect both plants and animals though it is a very slow process.

(b) Movements within the habitat

The second type of movement is restricted to animals, as plants are fixed. The reason for this is that for animals movement is essential. They need food because, unlike plants, they cannot make it for themselves. Therefore animals must move while plants, which make their own food from soil chemicals, can remain fixed. In the same way movement is essential for humans to survive. Animals are, however, only free to move within their habitat, which provides their essential needs. It is on the basis of these needs that the second category is mainly subdivided, although this is not true of the first group, as this covers the involuntary movement of creatures that cannot themselves resist the movement.

(i) *Drift*. A good example of the first group, the drifters, is the zooplankton in the oceans. These minute creatures must move with the currents, but some of them have developed a system of vertical migration that enables them to change their environment to a certain extent.

(ii) *Movements connected with food and predators*. In the second group food provides an obvious purpose for movement and so does escape from predators. There is a very great range in the distances involved in these movements.

(iii) *Movements for breeding purposes.* The third group, dependent on reproduction, suggests some interesting relationships between environment and distribution. Many animals have fixed breeding places, such as the cod of the North Sea. Other animals, particularly birds, cover very great distances to breed. The Arctic tern is an extreme case; it breeds in high northern latitudes in the northern summer, but spends the northern winter in high southern latitudes. Wilson's petrel reverses this pattern, travelling from its breeding grounds in the Antarctic to winter in Scotland or Labrador in the northern summer. It is possible that light stimulates these birds to make the best of both hemispheres.

The eels that breed in the Sargasso Sea are another familiar example. They live their adult lives in European rivers and move 3000 miles across the Atlantic to spawn. An interesting suggestion to account for this movement is related to continental drift (chapter 12). It is suggested that they first started to breed not far off the coasts of western Europe but as America moved west their breeding ground gradually receded from the rivers in which they lived. The movement was, however, so slow that they did not notice the gradually increasing length of their journeys until it is now 3000 miles long. Such a theory cannot be proved and belongs to category 4 (see chapter 12) but it is possible that this suggestion could both support the theory of continental drift as well as explaining the long journey of the eels to breed.

(iv) *Migration caused by overcrowding.* The fourth category of overcrowding as a cause of movement in animals has obvious similarities with human migrations. One of the best known examples is the mass migration of lemmings in Scandinavia. They reach such large numbers about every four years that they must either migrate or starve. In doing the former many perish but those that survive can carry on in better circumstances. These migrations are associated with cyclical phenomena amongst other members of the ecological system. These phenomena affect both plant and animal members of the community.

In the biosphere movements become more complex to analyse as the plants and animals become more elaborate in their organization. The small phytoplankton move only with the currents in which they live. The larger zooplankton have rather more freedom of movement, and, although they must drift helplessly in a horizontal plane, they are capable of changing their environment by vertical migration. This movement is in some way connected with their tendency to avoid light. It may be compared with the minor movements associated with many growing plants, whose position is fixed by their roots, but which have an ability to turn towards the light by some complex photosynthetic enzyme process. As animals become more complex so their movements become greater and depend on instinct and group activity. There is still much room for investigation of the methods used by migrating animals to find their way about and concerning the constraints that help to determine their movements. Some of these constraints are physical, such as land and sea barriers or wind and current systems, while others must be related to the physiology of the animals.

13.6 Transportation and communication networks

(*a*) *Networks and journeys*

Networks and journeys are so intimately associated that it is difficult to consider either alone. The main purpose of this section is to outline ways of describing spatial features of networks in human geography, but some more general questions will be discussed first.

Very broadly, transportation and communication between places can be achieved within one of the following three spaces: a clearly defined or bounded channel, a roughly defined or unbounded channel, or in a diffused way along no particular channel at all. Pipelines and railways (note its tunnels) are examples of the first kind. Air lanes and shipping lanes are examples of the second kind of channel. They are very wide (*and* deep for aircraft) compared with the vehicles using them. The broadcasting of information by radio and other such media is usually diffuse in a spatial sense and exemplifies the third kind. Only the first medium, the bounded channel, produces a set of 'lines' with a ratio of one horizontal dimension very high compared with the second horizontal and the vertical. This can be described in terms of arcs and nodes.

The following questions may be asked about networks and the answers help to classify them. They are primarily intended for the systems with channels, but to some extent are relevant to those with unbounded channels.

(1) Does the system carry vehicles, materials, or information? If vehicles, are the vehicles designed to carry passengers, materials, or both?

(2) If vehicles are used, are they controlled by a person within or are they automatically controlled?

(3) What means of propulsion are used — animal or machine? If machine, is the power produced within the vehicle or gathered from outside?

(4) Is the network directed or two-way? For example, a system of channels carrying water for irrigation by gravity is directed. At any given time, a channel must be carrying things in one direction only.

(5) Are the items moving on the network continuous (e.g. gas, oil, water, coal suspended in water) or are they discrete (e.g. trains, cars).

(6) What kind of lengths (distances) are involved in the network? Here it is interesting to think of the kinds of length journeys are. The following (in km) are possibilities:

within room (0.001), within public building (0.01), within vicinity (e.g. street) (0.1), within neighbourhood (1.0), county (10), state (100), country (1000), world (10,000), to moon (400,000).

The definitions have been fitted round a denary base kilometre measure in this example. Perhaps a different base (e.g. octal) would be an improvement.

(7) What is the orientation of the network? This may not often be relevant. However, it is possible to consider this by finding if there are preferred orientations of arcs in the network.

(8) The general layout of the network may be described in verbal terms like: gridiron, radial, spider's web, tree like and so on. This kind of description is unsatisfactory, however, for at least two reasons. Firstly, it is imprecise and unsuitable for comparison among several networks. Secondly, it is possible that the visual appearance of the layout is only a superficial feature and that underlying features that can be described in terms of topological ratios or detour indices cover more meaningful aspects of the network and describe better whether it is successful and efficient.

The above list of items is not exhaustive, but it covers some of the main features of a transportation network. The following aspects may be measured:

(1) The quality or class of the channels (e.g. the class of road, number of tracks of railway, diameter of pipeline).
(2) The volume of traffic or material using different parts of the network in a given period.
(3) Various ratios involving the number of arcs and nodes. Topological (or graph theoretical) descriptions, briefly discussed below ignore distance and are therefore more relevant to messages and electricity, which travel very rapidly, than to bulky fragile and/or perishable materials.
(4) Measurements of length expressed in various ways. The following may be considered: length of arcs per population or per area, volume of goods carried per length of arc, proportion of length of 'dead end' arcs to total length of arcs, the ratio between direct distance and actual distance between appropriate places. The last measure will be discussed in subsection 13.6*c*.

While the above measurements are straightforward to make, they do not, however, seem to produce good variables for purposes of correlating transportation networks with other variables. This difficulty will be discussed in section 15.3, under problems of regional geography. It may be noted that the layout of transportation networks may be affected by existing features in an area (e.g. desirability of serving large towns, densely peopled farming areas) but may itself also affect the presence of new features.

Various indices are used to describe the connectivity (completeness) of a graph, or lack of it. Some examples are described. Further description and application of these will be found in K. J. Kansky's *Structure of Transportation Networks* (1963).

(1) The cyclomatic number, μ (mu), is the number of arcs less the number of nodes plus the number of subgraphs (p).

$$\mu = a - n + p$$

In figure 13.1, cases 9–11 there are 7 nodes.

Case 9 has 5 arcs and two subgraphs, thus

$$\mu = 5 - 7 + 2 = 0$$

Case 10 $\quad \mu = 8 - 7 + 1 = 2$

Case 11 $\quad \mu = 11 - 7 + 1 = 5$

The more complete the network, the higher its cyclomatic number. However, networks with differing layout may have the same cyclomatic number. Compare cases 12 and 13 with case 9.

(2) An illustration of a ratio index for describing a graph is the alpha index. For non-planar graphs it is as follows

$$\text{alpha} = \frac{a - n + p}{\frac{n(n-1)}{2} - (n-1)}$$

In figure 3.1, case 6

$$\frac{4 - 4 + 1}{6 - 3} = \frac{1}{3} = 0.33$$

Case 7

$$\frac{5 - 4 + 1}{3} = \frac{2}{3} = 0.67$$

Case 8

$$\frac{6 - 4 + 1}{3} = \frac{3}{3} = 1$$

Note that the alpha index uses the μ number $(a - n + p)$ as its numerator. When μ is equal to or less than 0, the alpha index is 0. When μ is equal to or greater than 1, the index is

$$0 < \text{alpha} \leqslant 1$$

Although the arcs, nodes and where relevant the number of subgraphs are counted in whole numbers (integers), the alpha index is a ratio of two integers and is therefore a real number. There are various other descriptive indices of networks, some ratios, some not. Those described give an idea of their nature and scope.

(b) Topological maps and networks

The use of topological maps in route networks is so widely found both in advertising and in route information displays, as at appropriate places on the Underground of London, that it is not necessary to elaborate this aspect of the application of topology. In figure 13.7*a* a topological route map used by United Counties to show express coach services between London and the Midlands is compared with the topographical layout, which retains correct scale and orientation (after rotation). It is interesting to note that in detail the topological transformation has been made incorrectly. Thus Lo (Loughborough) and Da (Daventry) are not correctly located in relation to the

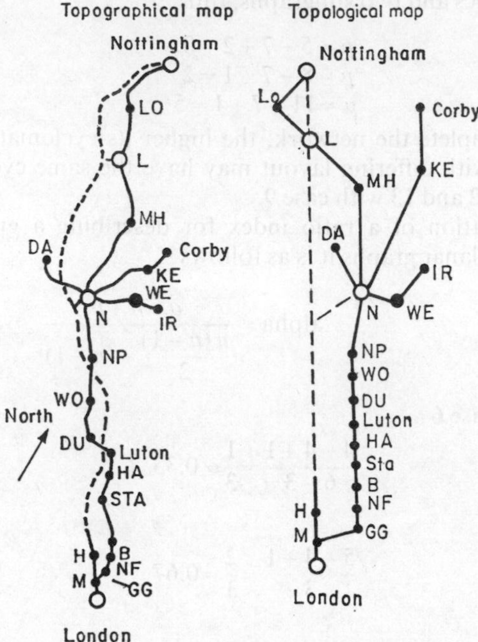

Figure 13.7*a* Comparison of topographical and topological maps of a 'bus company system.

Motorway (dashed line). But this does not matter to the person finding his or her route on the map.

Some topological terms could be used both to describe the rail network of New Zealand as it is in figure 13.7*b*, or after transformation. The network has only three regions. The construction of some quite short stretches of line would greatly reduce journeys between certain places.

(*c*) *Direct and actual distance*

The success and efficiency of a transportation network can to some extent be measured by comparing the direct distance between pairs of nodes and the shortest actual distance between them on a network. The calculation assumes firstly that people always desire to make the journey between two places with the least amount of effort, and secondly that the shortest in terms of distance is the one involving least effort. The procedures outlined in this section are based on this assumption. In reality, there may occasionally be instances when the least effort journey between two places is not chosen, and also when a longer journey than necessary is made in error. It is also possible that the shortest distance on the network is not the one involving least effort. Thus a longer but less hilly road may be quicker in time and/or more economical in cost than the shortest. Such a situation is illustrated in figure 13.8*a*, which will be elaborated below.

In general, few nodes on a transportation network are connected by straight line-arcs. Moreover, since a Euclidean straight line is one that becomes a point if looked at end on, such a concept is of little meaning on a spherical surface since it would not be commercially or technically possible to link distant places by a straight line channel through the earth; W. Warntz (1967) elaborates this idea. Direct distance in this section implies, therefore, a straight line for a small area but a great circle for a larger area.

A numerical detour index may be obtained by expressing the shortest

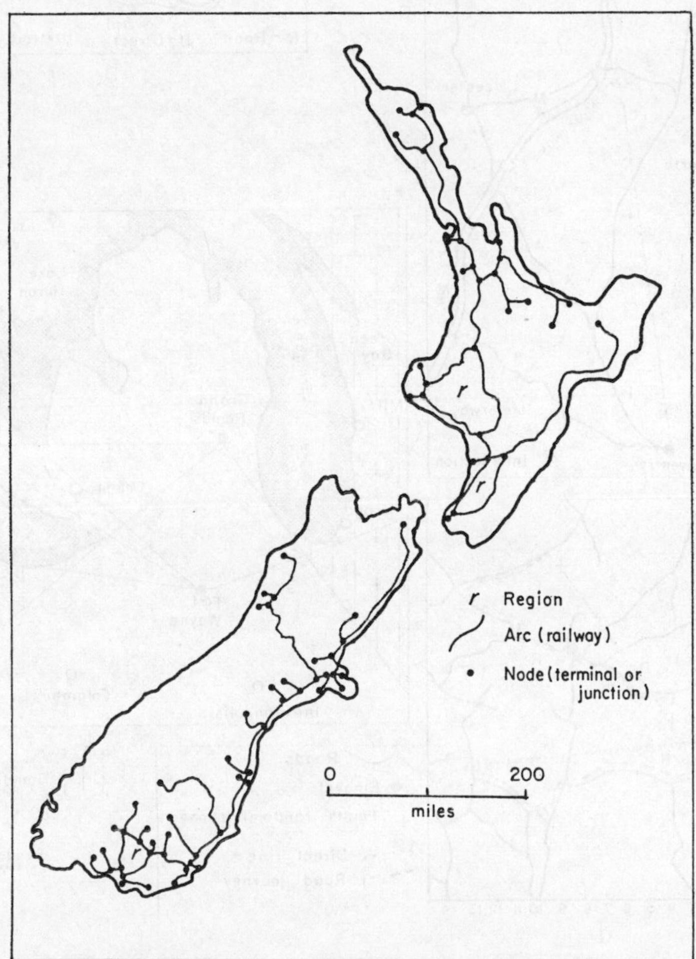

Figure 13.7*b* The railways of New Zealand (Source: Ed. A. M. Mackintosh *Descriptive Atlas of New Zealand*, 1959).

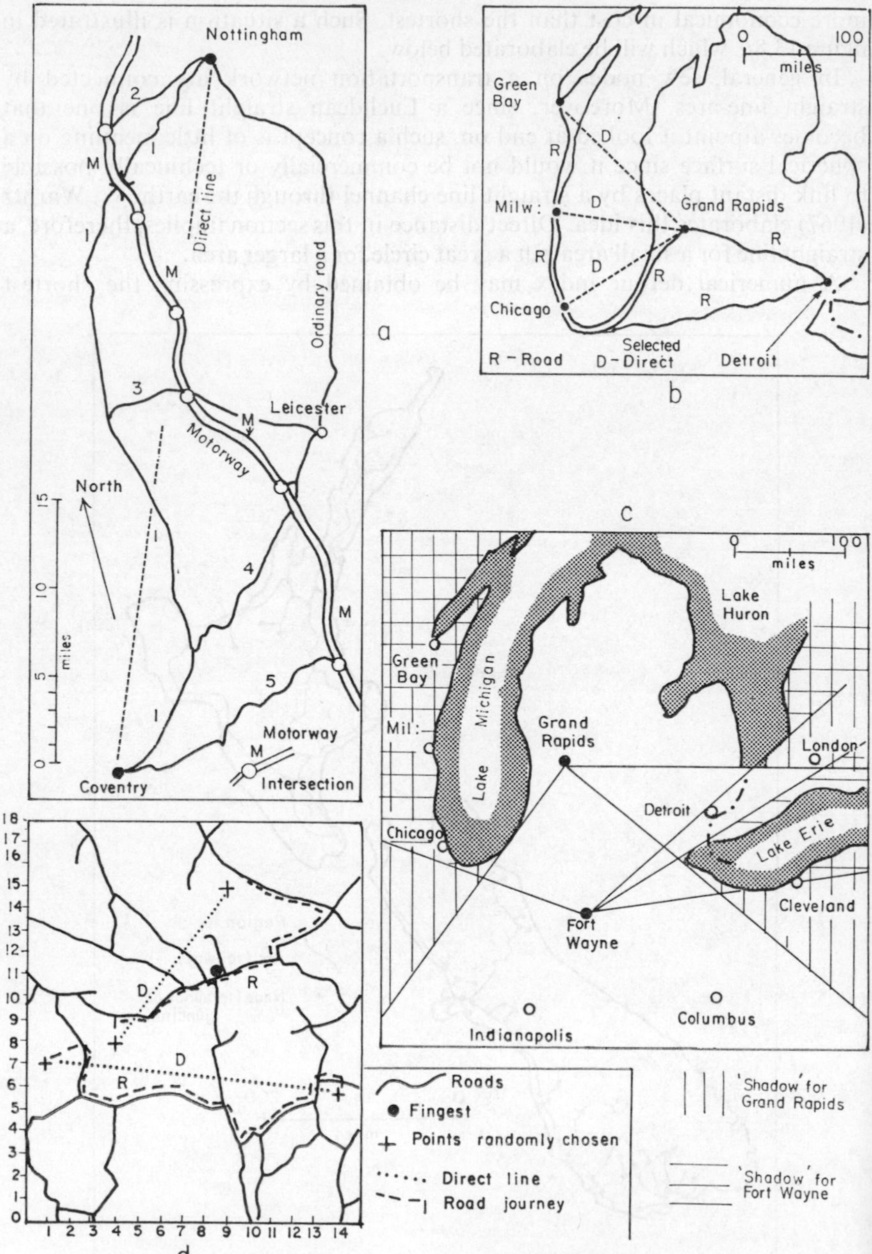

Figure 13.8 The comparison of direct and of actual road distances between places. (*a*) The alternative routes between Nottingham and Coventry. (*b*, *c*) Journeys between places in Michigan and Illinois. (*d*) Journeys between selected pairs of places in rural Buckinghamshire, England.

distance between a given pair of nodes on a network as a ratio of the direct distance (as defined above) between the same two nodes. The two distances must be measured in the same units. The calculation may be made as follows:

$$\frac{\text{Shortest distance on network}}{\text{Direct distance}} \times 100$$

The $\times 100$ eliminates decimals. The calculation is an approximate one and further precision would be spurious. The detour index gives a numerical ratio (direct = 100) that serves for comparison between areas of different scale. Thus the detour index on a journey from Grand Rapids to Green Bay (see figure 13.8*b*) may be identical with one over a winding road between two villages, or by sea (shortest feasible sea lane) between two ports round an intervening continent.

The above calculation may be made for a single journey. It may also be made at various times in a given network in order to characterize the total network as well as individual journeys. Two methods may be suggested: the first considers all pairs of journeys between selected (usually important) nodes on the network; the second considers journeys between randomly selected pairs of places (either nodes on the network or points anywhere within the area served by the network). The methods will be described with examples:

(i) The single journey is exemplified in figure 13.8*a* by the journey between Nottingham and Coventry. Since the opening of the Motorway (M on map) several alternative routes have become possible. The following will be considered:

Table 13.3 Routes between Nottingham and Coventry

	Miles	Detour index
Nottingham–Coventry direct	41	
I By route 1	47	114
II By 2, M, 3, 1	55	134
III By 2, M, 4, 1	54	132
IV By 2, M, 5	60	146

The original usual route (I) has a detour index of only 114 (direct = 100 units). The biggest detour is route IV, with an index of 146. As already noted, however, it may be more attractive in terms of time, consumption of fuel and/or relaxed driving, to cover a greater distance. Roughly, it is possible for a passenger car to average 30 m.p.h. on the ordinary roads of the area, but to average 60 m.p.h. on the motorway. To calculate the relative time required in this example it is sufficient to multiply ordinary road distances to find the expected journey time in minutes (table 13.4). The following calculation shows that the increased distance of route II over route I is not compensated for by the 17 miles of motorway used. On the other hand, route III takes about 10 minutes less than I and II, although it is longer than I and nearly equal to II in distance. Finally, IV is longer than II in distance but no shorter in time. It is a matter of opinion which is the least effort route in this example.

Table 13.4 Distances

	Distances in miles		Times in minutes		
	Ordinary road	Motorway	Ordinary road × 2	Motorway	Total
I	47	0	94	0	94
II	38	17	76	17	93
III	29	25	58	25	83
IV	24	36	48	36	84

(ii) Journeys between all pairs of five selected places (nodes) are illustrated in the example of highway journeys in the Great Lakes area, U.S.A., in figures 13.8*b–c*. The relevant measurements are recorded in the three matrices below. Note that the matrices in table 13.5 are square and symmetric.

Table 13.5 Distances in U.S.A.

	I					II					III				
From / To	D	GR	C	M	GB	D	GR	C	M	GB	D	GR	C	M	GB
Detroit	0	153	275	364	476	0	134	231	245	284	0	114	119	149	168
Grand Rapids	153	0	178	269	381	134	0	128	114	160	114	0	139	236	238
Chicago	275	178	0	91	203	231	128	0	87	188	119	139	0	105	108
Milwaukee	364	269	91	0	112	245	114	87	0	100	149	236	105	0	112
Green Bay	476	381	203	112	0	284	160	188	100	0	168	238	108	112	0
	1268	981	747	836	1172	894	536	634	546	732					

I contains the road distances between pairs of places as recorded in a U.S. road atlas (Rand McNally Road Atlas, 42nd Annual Edition). From the previous example it will be clear that there might be several other equally attractive alternatives.

II contains the direct distances between the pairs of places.

III contains the detour indices between the places. For example:

$$\text{Detroit–Green Bay}\ \frac{476}{284}\times 100 = 168.$$

The following numerical information may be extracted from the matrices:

From III, the index of detour for each journey.

From the sums of columns (or rows) in I, the total distance from each place to all others (e.g. contrast Chicago 747, Detroit 1268).

From the sums of columns in I divided by the sums of columns in II, the overall detour index for each place.
From the grand total (sum of sums of columns) for matrix I divided by the grand total (sum of sums of columns) for matrix II, the overall detour index for the network in question.

The example given contains too few nodes for the results to be of much interest, but it serves to illustrate the method. J. P. Cole (1965) has used the method to bring out contrasts in road systems of Latin American countries, particularly Peru.

(iii) The same procedures can be applied to distances between pairs of randomly chosen places in an area. One method of choosing places is suggested in figure 13.8*d*. The area is part of rural Buckinghamshire. The roads are only suitable for small motor vehicles. Two journeys have been selected randomly. They are between (4, 8) and (9, 15), and between (1, 7) and (14, 6). The dotted lines (D) show the direct distances, the broken lines (R) the road journeys. Usually the journey end points will not fall on a road. They must therefore be 'joined' to the nearest road for the road journey to be calculated. It would also be possible to produce random pairs of journeys between actual nodes (e.g. towns, crossroads) on a network by numbering these and then calling them randomly in pairs.

Calculations (unpublished) by students in the Geography Department, Nottingham University, suggest that big differences in overall detour indices occur for rural areas in various parts of Britain.

District	*Detour index*	*District*	*Detour index*
Bath area (Somerset)	117	Sleaford (Lincs)	142
East Anglia	119	Skegness area (Lincs)	154
Falkirk (Scotland)	120	York	158
Hampshire	120	Kings Lynn (Norfolk)	163
Isle of Wight	125	Mid-Wales (Aberdovey)	219
Nottingham/Sheffield	127		
Stirling (Scotland)	135		

The detours in a network may be present for the following reasons:

(1) Only a continuous surface (e.g. a town square, a football pitch, the sea), can accommodate direct journeys between any pair of points. Some road networks achieve a high degree of directness, however, at least between more important places.

(2) A detour is often the result of avoidance of an obstacle that would be extremely costly if not technically impossible to cross directly (e.g. Grand Rapids to Milwaukee, Green Bay).

(3) There might be a detour between places X and Z to take in Y on the way. Arcs X–Y and Y–Z may be direct.

(4) There may be no link at all between two important nearby places.

Carefully calculated overall detour indices should help to characterize transportation networks better than verbal statements and in a way complementary to the topological descriptions of graph theory. The following are among the questions they may help to answer:

(1) How strong are physical influences (particularly slope, water surfaces) in an area, and how strong is its level of economic development in affecting the overall detour index?

(2) Do important places such as national capitals have better detour indices than other places because the most direct routes in the network tend to converge upon them?

(3) Does the detour index tend to be greater for short journeys than for long ones?

(4) What will the effect on the overall detour index be of the closing of an arc in the network and, similarly, of the opening of a new arc? The opening of the Suez Canal (1869) and its closure in 1956–7 and again in 1967 have affected world shipping routes. Many large modern rail and road bridges and tunnels (e.g. Verrazano Narrows over the entrance to New York harbour, the Mont Blanc and St. Bernard tunnels in the Alps) have had local and even regional repercussions in transportation networks.

13.7 Networks that form compartments

All the networks that have been discussed in this chapter so far, whether conceptual or material, may be thought of as having things moving along the arcs. An important type of network in geography, that forming a grid, regular or irregular, differs fundamentally in that its arcs form barriers and delimit containers, the regions. Sentries can be visualized marching round town walls, and parishioners used to walk round and beat the boundaries of their parishes. But the main function is not to facilitate movement along arcs, but to prevent it across them. Thus it is the compartments themselves and the things that may and may not move *across* the boundaries between the compartments that are of interest. Such networks will for convenience be referred to as grids in this section. The following types may be suggested:

(1) Layout. This may be regular or irregular. Regular here implies that once some part of the grid has been seen the pattern of the rest can be predicted.

(2) Abstract or material. This question is not easy to decide. At one extreme is a conceptual grid such as that put down on a map to facilitate the location of places on the map. The grid is not on the ground; it need not even be marked on the map, so long as the frame of the map has appropriate co-ordinate scales. At the other extreme is the physical barrier: a fence, hedge, wall, 'iron curtain'. In between are boundaries on the ground that are marked at intervals or merely agreed upon but not actually indicated at all on the ground.

Whatever the layout and degree of material manifestation of grids, however, they differ from most other networks discussed in that they rarely have loose end arcs and terminator nodes. In this section grids will for convenience be discussed under two heads. (*a*) Conceptual grids. (*b*) Region forming grids.

(a) Conceptual grids

It is often useful in geography to create a grid on a map for such purposes as the collecting of data or the organization of a spatial sample. Existing administrative grids may be used, but they are rarely regular. Basically there are three figures that cover a surface without overlapping or leaving gaps between, the rectangle in figure 13.9*c*, *f* (of which the square is a subset), the equilateral triangle (13.9*k*) and the hexagon (13.9*g–i*, *k–m*). Other regular polygons do not pack perfectly. Note that a hexagon can be regarded as being composed of six equilateral triangles (13.9*k*). Grids of squares and of hexagons are the two most widely used to cover a surface with regular, equal area and resonably compact units. The hexagon is considerably more com-

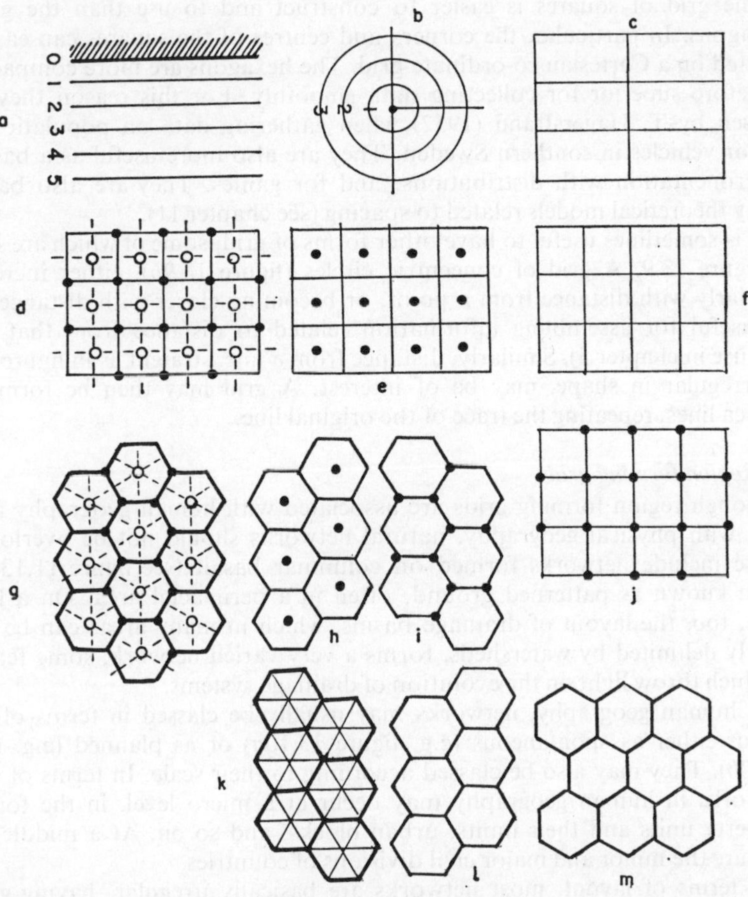

Figure 13.9 Examples of grids applicable in geography for the collection of data over area, the operation of models and other purposes. Letters are referred to in text.

pact than the square. Note that the grid of squares has 4-nodes except at its periphery (13.9*d*), whereas the grid with hexagons has only 3-nodes. Another difference is that a grid of squares is the same if rotated 90 degrees whereas a grid of hexagons is different in appearance (13.9*l m*), if rotated 90 degrees.

If a node is placed at the centre of each square in a grid of squares (13.9*e*) and the new nodes are joined by arcs, a new grid of squares will be created identical with the existing one (13.9*d*). Its arcs will bisect and be bisected by the arcs of the existing grid. If a node is placed at the centre of each hexagon in a grid of hexagons (13.9*h*) and the new nodes are joined by arcs (13.9*g*), a new grid of equilateral triangles will be created, each *half* the area of the existing hexagons. Its arcs will bisect and be bisected by the areas of the hexagonal grid.

The grid of squares is easier to construct and to use than the grid of hexagons. In particular, the corners and centres of the squares can easily be located on a Cartesian co-ordinate grid. The hexagons are more compact and therefore superior for collecting data smoothly. For this reason they were chosen by T. Hagerstrand (1952) when gathering data on population and motor vehicles in southern Sweden. They are also more useful as a basis for experimentation with distributions, and for games. They are also basic to many theoretical models related to spacing (see chapter 11).

It is sometimes useful to have other forms of grid, some of which are shown in figure 13.9. A grid of concentric circles (figure 13.9*b*), either increasing regularly with distance from a point, or becoming closer with distance, may be useful for assembling information related to distance from that point (see use in chapter 5). Similarly, distance from a line, straight as in figure 13.9*a* or irregular in shape, may be of interest. A grid may then be formed of replica lines, repeating the trace of the original line.

(*b*) *Region forming grids*

Although region forming grids are associated with human geography rather than with physical geography, natural networks should not be overlooked. These include networks formed on columnar basalt (see figure 11.13*b*), or those known as patterned ground, often in a periglacial area. On a larger scale, too, the layout of drainage basins, which in many areas can be quite clearly delimited by watersheds, forms a very varied network, some features of which throw light on the evolution of drainage systems.

In human geography, networks may usually be classed in terms of their layout either as spontaneous (e.g. figure 13.10*a*) or as planned (e.g. figure 12.10*b*). They may also be classed according to their scale. In terms of scale, networks in human geography may occur at a micro level, in the form of property units and their limits, urban blocks, and so on. At a middle level they are the minor and major civil divisions of countries.

In terms of layout, most networks are basically irregular, having grown up and been modified over considerable periods of time. In some parts of the world, however, basically regular networks have been put down on the surface during some brief period of time. They often have minor irregularities

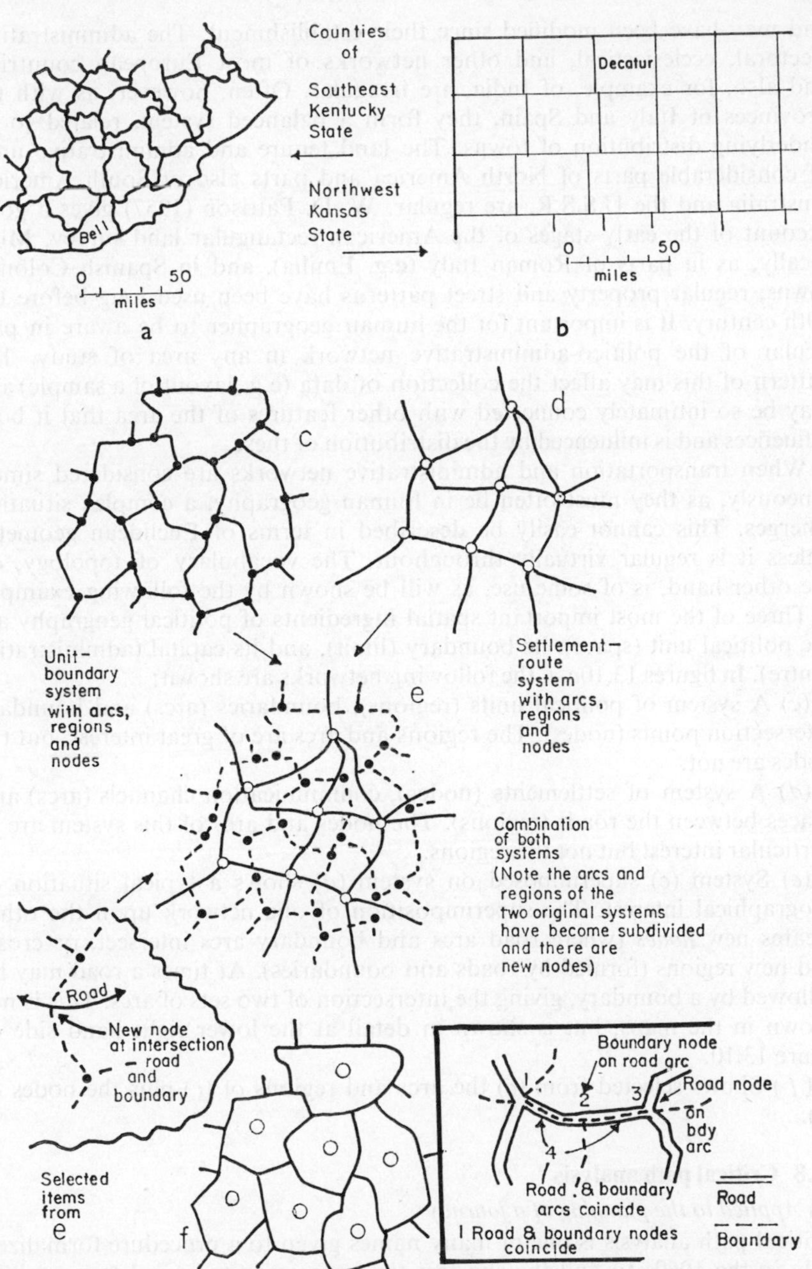

Figure 13.10 Examples of politico–administrative grids. (*a*) Spontaneous. (*b*) Planned. (*c*)–(*f*) Some of the chief ingredients of political geography described in terms of topology and the intersection of systems of nodes, arcs, and regions.

and may have been modified since their establishment. The administrative, electoral, ecclesiastical, and other networks of most European countries, and also, for example, of India, are irregular. Often, however, as with the provinces of Italy and Spain, they form a balanced system, related to an underlying distribution of towns. The land tenure and administrative units of considerable parts of North America and parts also of South America, Australia and the U.S.S.R. are regular. W. D. Pattison (1957) gives a good account of the early stages of the American rectangular land survey. More locally, as in parts of Roman Italy (e.g. Emilia), and in Spanish Colonial towns, regular property and street patterns have been used long before the 19th century. It is important for the human geographer to be aware in particular of the politico-administrative network in any area of study. The pattern of this may affect the collection of data (e.g. layout of a sample) and may be so intimately connected with other features of the area that it both influences and is influenced by the distribution of these.

When transportation and administrative networks are considered simultaneously, as they must often be in human geography, a complex situation emerges. This cannot easily be described in terms of Euclidean geometry unless it is regular virtually throughout. The vocabulary of topology, on the other hand, is of some use, as will be shown by the following example.

Three of the most important spatial ingredients of political geography are the political unit (space), its boundary (limit), and its capital (administrative centre). In figures 13.10*c–f*, the following networks are shown:

(*c*) A system of political units (regions), boundaries (arcs) and boundary intersection points (nodes). The regions and arcs are of great interest, but the nodes are not.

(*d*) A system of settlements (nodes), communication channels (arcs) and spaces between the roads (regions). The nodes and arcs of this system are of particular interest but not the regions.

(*e*) System (*c*) superimposed on system (*d*) shows a typical situation of geographical interest. The superimposition of one network upon the other creates new *nodes* (where road arcs and boundary arcs intersect, or cross) and new regions (formed by roads and boundaries). At times a road may be followed by a boundary, giving the intersection of two sets of arcs. This is not shown in the maps, but is shown in detail at the lower right hand side of figure 13.10.

(*f*) Shows selected from (*e*) the arcs and regions of (*c*) plus the nodes of (*d*).

13.8 Critical path analysis

(*a*) *Applied to the planning of a journey*

Critical path analysis is one of many names given to a procedure formalized only in the 1960s to find the shortest time necessary to complete a project. The builders of Stonehenge, Salisbury Cathedral and the Empire State building presumably used some comparable procedure. Basically, critical path analysis helps to sort out which operations in a project cannot be started before others have been completed, and which are 'free'. It has become vital in planning such projects as missile launching and space shots, since a very large number of parts have to be provided and assembled in as short a time as possible. It has been widely used already in less spectacular projects such as launching a new detergent, building a power station, and so on. The finding of the critical or longest duration path is vital since if the duration of the project is to be made shorter, it is useless to cut down the time required on any operation unless it is *in* the critical path. Getting up and making breakfast, cooking Sunday dinner, or making a cup of tea are commonly used as examples to illustrate critical path analysis. Here the example is a journey. The fundamental features of critical path analysis are that it is a *one way* network, that the links are measured in units of time, and that the critical path is the *longest* path in time through the network. The two principal symbols used are directed lines representing activities or jobs, and circles representing the completion of these. In figure 13.11*a*, the heavier arrow represents the critical path.

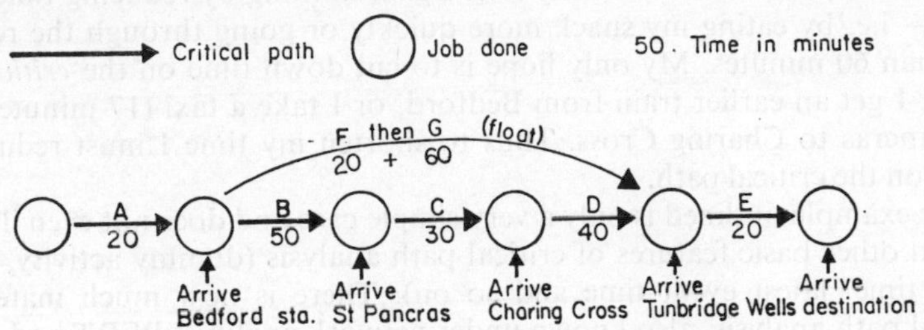

Figure 13.11*a* A diagram to illustrate the layout of a critical path analysis chart.

The following problem will now be discussed: I must travel (urgently) from Bedford to Tunbridge Wells. It is not convenient to go by road (London traffic) so I shall travel by train. The times required are as follows, table 13.6:

Table 13.6 Data for critical path analysis

Job[a]		Time in mins
A	Home to Bedford Station and wait on station	20
B	Train to St. Pancras	50
C	Tube to Charing Cross	30
D	Train to Tunbridge Wells	40
E	Tunbridge Wells Station to Appointment	20
F	Eat lunch (have packed sandwiches)	20
G	Read report	60

[a] Small jobs such as buying ticket have been included in bigger ones.

To start with I do not say how soon it is hoped that I will arrive at my appointment in Tunbridge Wells. The five pieces of travel *must follow* one another in the sequence shown. I can (though I need not) eat my lunch *and* read the report while travelling. I prefer to eat my lunch before tackling the report, and to do both during the two train journeys or crossing London, rather than on my way to and from the stations.

A detail that could be built in to the model is that I would not want to eat my snack on the underground so I therefore should complete it before arriving at St. Pancras. I do not therefore have to start eating my snack until I am within 20 minutes of St. Pancras. The 20 minute snack can 'float' in the 50 minute train journey. This still leaves $30+40=70$ minutes crossing London and travelling to Tunbridge Wells in which the 60 minutes of reading the report can float.

Now supposing I find that there is only an interval of 20 minutes between arrival at St. Pancras and the departure of the train I would like to catch at Charing Cross and that if I miss this, then the delay would make my arrival at the destination inconveniently late. I gain nothing by reducing time on the float — i.e. by eating my snack more quickly or going through the report in less than 60 minutes. My only hope is to cut down time on the *critical path.* Either I get an earlier train from Bedford, or I take a taxi (17 minutes) from St. Pancras to Charing Cross. Thus to shorten my time I must reduce time spent on the critical path.

The example outlined is only a very simple case and does not even illustrate certain other basic features of critical path analysis (dummy activity, earliest event time, latest event time and so on). There is now much material on critical path analysis, also known under network analysis, PERT and various other labels. Examples are *Critical Path Analysis* by the British Productivity Council (undated) and a paper by L. J. Rawle (1964).

Some additional aspects and applications of critical path analysis may be noted briefly. Critical path analysis is somewhat comparable to the problem of finding a shortest route between places but in the routing problem the 'critical route' is the *shortest* (in length and presumably in time) whereas in critical path analysis it is the *longest* (in duration).

Critical path analysis is of potential interest to the geographer firstly to assist in setting out a complicated piece of work such as the preparation of a thesis, report, even a book, and secondly, in studying the feasibility of specific projects on the ground, and of planning in general. It might also provide a new approach to studying the course of history, since certain events could not happen before the completion of others. The motor car of the present could not be made before the invention of the internal combustion engine using petrol. This in turn had to await the invention of the sparking plug, and so on.

(b) Critical path analysis in geomorphology

It is sometimes important to establish the sequence of events in geomorphological problems. The events leave their mark on the landscape and by studying the landforms it is often possible to establish the sequence of past events. In a geomorphological type of analysis it is the sequence of events that is important, not, as in human examples, the saving of time. The sequence of deglaciation in western Baffin Island can be taken as an example of the application of the principle. Only two of the very complex events that took place as the ice retreated will be mentioned. One of the earlier events for which there is evidence is the spreading of the ice cap from Baffin Island out over Foxe Basin [(1), figure 13.11*b*], until the dispersal centre was situated far to the southwest of its original position. From the Foxe Basin centre ice

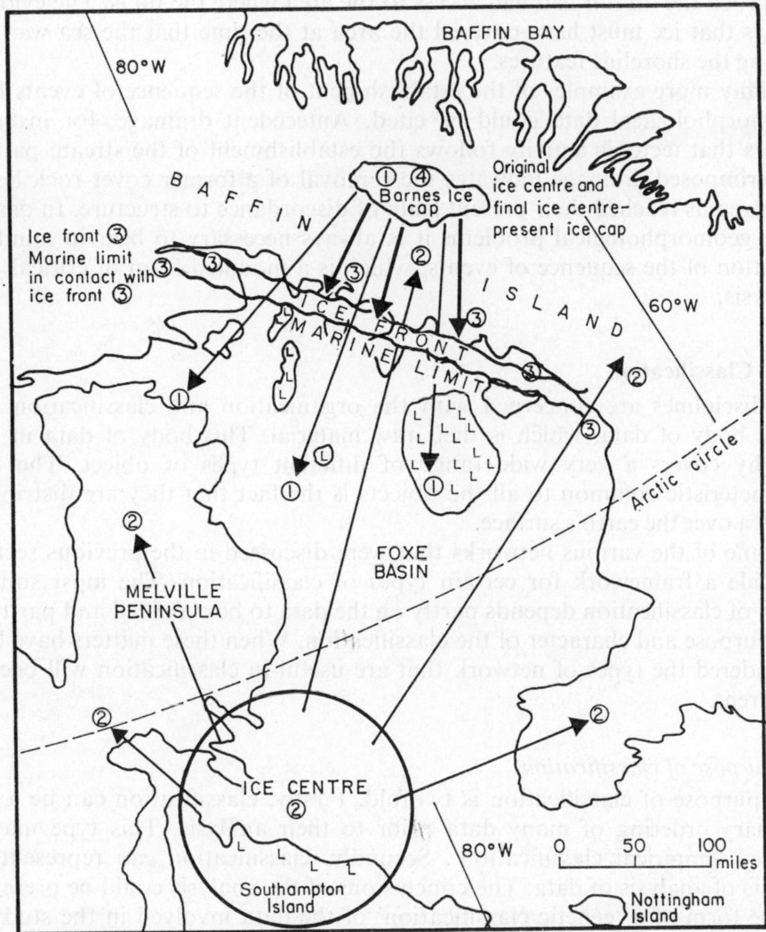

Figure 13.11*b* Map of Foxe Basin and part of Baffin Island to illustrate the sequence of events concerning some of the more recent glacial events and their relation to sea-level.

passed northeastwards across the coast [(2), figure 13.11]. The northeastward movement is indicated by the presence of limestone erratics, carried inland from the coast. The movement must have preceded the marine limit [(3), figure 13.11*b*], because it necessitates ice over the limestone outcrops, erratics from which were carried to the northeast. A little later the ice front was situated in a position near the present coast [(3), figure 13.11*b*]. The contemporaneity of this stage with the marine limit (3) can be established by the study of the contact zone. The zone shows in one place undisturbed till at a lower level than the marine limit shoreline features. The sea would have disturbed the till if it had had access to the area where the till is. The evidence shows that ice must have covered the area at the time that the sea was producing the shoreline features.

Many more examples of the establishment of the sequence of events from geomorphological data could be cited. Antecedent drainage, for instance, shows that tectonic activity follows the establishment of the stream pattern. Superimposed drainage indicates the removal of a former cover rock before the streams reached their present state of discordance to structure. In dealing with geomorphological problems it is always necessary to bear in mind the question of the sequence of events, which is a modified form of critical path analysis.

13.9 Classification

All disciplines are concerned with the organization and classification of a large body of data, which is their raw material. This body of data in geography covers a very wide range of different types of object. The only characteristic common to all the objects is the fact that they are distributed in area over the earth's surface.

Some of the various networks that were discussed in the previous section, provide a framework for certain types of classification. The most suitable form of classification depends partly on the data to be classified and partly on the purpose and character of the classification. When these matters have been considered the types of network that are useful in classification will become apparent.

(*a*) *Purpose of classification*

The purpose of classification is twofold. Firstly, classification can be a preliminary ordering of many data prior to their analysis. This type may be termed 'empirical classification'. Secondly, classification can represent the results of analysis of data. The conclusions of the analysis could be presented in the form of a 'genetic classification' of the data involved in the study. A good example of the difference of approach to classication, based on these two major divisions, is the classification of landforms by mapping. There is the empirical form of mapping, normally carried out in the field. This is not dependent on any preconceived ideas concerning the development of the landforms to be mapped. The map, which is called a morphological map.

only records as accurately and meaningfully as possible the actual landform types visible in the field. Breaks and changes of slope and other relevant information are plotted. The map can then be used as a basis for further study of the landforms.

The morphological map may be contrasted with a geomorphological map on which is recorded the genetic character of the landforms. In order to produce a geomorphological map, it is necessary first to have studied the individual forms and to be able to recognize them in the field. They must be assigned to their correct group in the classification into which the features are subdivided. The basis for these two methods of classifying and recording landforms is therefore quite different, one coming at the beginning of the study and the other forming the conclusion to it.

Classification implies the recognition of pattern and organization in the data being considered. As the level of organization increases so does the complexity of the classification that is needed to describe the data adequately. The scale of organization suggested by K. Boulding (1956) and mentioned in chapter 11 is useful in this respect. He gives nine levels of organization. Concerning the structure of systems he writes: 'One advantage of exhibiting a hierarchy of systems in this way is that it gives us some idea of the present gaps in both theoretical and empirical knowledge. Adequate theoretical models extend up to about the fourth level, and not much beyond. Empirical knowledge is deficient at practically all levels. Thus at the level of the static structure, fairly adequate descriptive models are available for geography, chemistry, geology, anatomy and descriptive social sciences. Even at the simplest level, however, the problem of the adequate description of complex structures is still far from solved. The theory of indexing and cataloguing, for instance, is only in its infancy. Librarians are fairly good at cataloguing books, chemists have begun to catalogue structural formulae, and anthropologists have begun to catalogue culture traits. The cataloguing of events, ideas, theories, statistics, and empirical data has hardly begun.' This was written more than a decade ago but there is still room for much more classification of data at both stages of enquiry. It has been suggested that geography lies far up the scale of levels of structure and complexity and hence classification in it must give rise to many problems. Biological classification or taxonomy used to be based on the similarity between organisms. The greater number of similar characteristics that the groups have in common, the better the classification will be. With the increasing availability of computers it is now possible to use a large number of correlated characters to define biological groups. In numerical classification all the features used are given equal weight. One of the essential foundations of the method is that similarity of character must be defined numerically. The organisms compared must have in common all features measured.

Measurement can be on any of the four scales of measurement, nominal, ordinal, interval and ratio. The data are arranged in a matrix and the similarities between all possible pairs computed. This can be done in a number of ways. Units will cluster according to their degree of similarity. A similarity matrix can be constructed and the values can be so arranged that similar ones are arranged together as triangles along the diagonal of the matrix. The triangles of similar values then indicate the separate classes into which the units can best be fitted.

The units may also be clustered into classes of similar characteristics by tree linkage diagrams. The closest links provide the species, where the similarity coefficient is high. A limiting value can provide the cutoff point for species and a lower value of similarity can limit the genus. There are three possible methods of forming the links between units. First, single linkage starts with the shortest distance between any two units and then takes in other units in order of magnitude of the distance between them. Second, in average linkage, each unit joins a cluster if the average distance between it and the centre of gravity of the cluster is less than any other distance. Third, complete linkage forms connexions only when the relationships between the unit to be joined and already linked clusters are all at a minimum distance for a particular clustering cycle.

Computers have now made it possible to use automatic sensing and recording devices. Optical scanners which can identify and classify outlines are particularly relevant in geography. A fairly simple device has been suggested for this purpose by R. R. Sokal (1966). A random mask was made from 25 cards each with 25 randomly chosen holes punched in it. An outline was overlaid in turn with all the cards and the holes showing the black outline were coded 1 and those blank 0. There were 625 holes and the score for each outline was compared with that of another outline. The values of the scores would enable the two outlines to be classified relative to each other, and when enough outlines had been tested, each unit could be classified into the appropriate group by this means. The method provides an example of automatic classification.

One of the problems of classification in geography is concerned with objects that are unique. All objects are unique in some respect and cannot be classified in this respect in that each unique object would require a separate class. There are, however, very few objects, if any, that have absolutely no attribute in common with other objects, and as soon as an object can be shown to share an attribute, it can become a member of the class of objects possessing that particular attribute. The earth might be considered to be a unique object and in many respects this is indeed true, as it lies at a unique distance from the sun and is the home of the human race. On the other hand, the earth is one of a series of planets that move round the sun and thus it can be classified as a planet. Paris is unique, yet it is a member of the class of capital cities.

Any unique object is rarely incapable of being subdivided. It is the unique combination of its several elements which give it its unique character. The Atlantic Ocean is a unique feature on the earth's surface. However, in its component parts it is similar to the other ocean basins and can therefore be classed with them. It has deep water, currents, a median ridge, abyssal plains and other features that together give it its oceanic character. This aspect of classification may be called classification by component properties. It is dependent on the splitting up or analysis of the unitary object into its component parts, the presence of some of which at least are essential in order to assign the object to its correct class.

The character of the component parts, rather than their mere presence, can be used to make a deeper and often more valuable classification. In the example of the Atlantic Ocean the nature of its median ridge, the character of its marginal shelves and coastal patterns, can be used to assign the ocean to the genetic class of rift ocean. This type of classification falls into the second major group while the other is of the first type.

In order to arrive at the most valuable type of classification it is necessary to study unique features in order to make generalizations which are the basis of classifications. By analysing unique features at all levels of subdivision it is possible to arrive at a basis for meaningful classifications. Thus it is necessary to identify those aspects of unique features that they have in common with other unique features. In this way a classification can be built up. In classifying mountains, for instance, the element that all mountains have in common is their elevation, which may be set at an arbitrary level, such as 2000 ft. Then those mountains that have high relative relief as well as absolute relief can be differentiated. Alternatively, in a deeper classification, the character of the structure could be used as a basis for further subdivision. The study of many individual mountain ranges has shown that they can be classified as fold mountains, volcanoes or horst mountains, etc. Unless the individual unique mountains had been studied it would not have been possible to arrive at these criteria for useful classification on a genetic basis.

It will be apparent that nearly all classifications are based on a series of successive subdivisions, which results in a pattern of nesting subsets. Each subset lower down the scale provides a more specific feature and a lower level of generalization. In some instances the hierarchy of subsets is fixed by the relationship between them. The hierarchy can best be represented on a vertical axis in this instance. In other instances the subsets represent differences of type which bear no fixed relationship to other subsets at an equal level of subdivision. Two examples will illustrate this difference and the character of nesting subset types of classification.

The geological time scale illustrates the first type where the hierarchy must maintain a fixed order, based on relative age. The major divisions are into eras, such as Palaeozoic, which is subdivided into periods, such as the Carboniferous. This again is subdivided into epochs, for example the Yoredale Beds and Carboniferous limestone. Further subdivision is into stages, such as the Main Limestone, and zones, such as the *Cyrtina septosa* band in the Great Scar Limestone. Each of these stages has a unique place in the time scale. Although different names are given to rocks from different areas, the common factor is the time of deposition.

An example of the second type of nesting subset classification, which can best be expressed in the pattern of a family tree, is the classification of glacial deposits.

Table 13.7 Glacial deposits classified

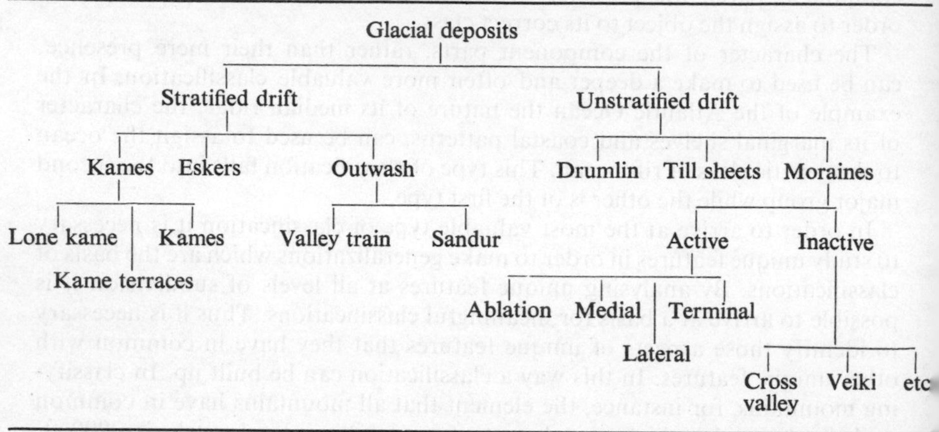

Table 13.7 is not complete, but it illustrates the point that none of the groups is uniquely determined by any property, in the way that the geological time scale is fixed by its age relations. Thus the order of the disjoint sets is not fixed, as it was in the previous example. Subdivision can go on until each

Table 13.8 Classification

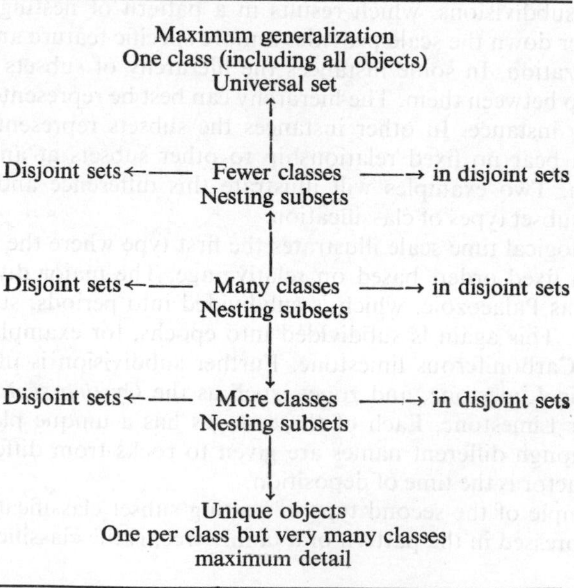

group consists of only one unique item, which is the ultimate stage beyond which classification cannot go unless the basis for subdivision is changed.

There is a limit to the value of continued subdivision in any one discipline, although the degree of subdivision that is warranted varies with different subjects. Thus geographers are usually only interested in whole plants or plant assemblages, whereas botanists would start at this point and carry the subdivision much further. The chemist and physicist go even further and may not be satisfied until they have split the atom into its component elementary particles in their classifications. There is a danger in geography that the line of useful subdivision, and hence depth of enquiry, may at times be crossed and then the study is no longer geographical. In many studies of a geographical character it is, however, not easy to define this line distinctly.

A great many classifications consist of two elements. Firstly, there are the nesting subsets that can be considered to lie on the vertical axis in a family tree network diagram, where each succeeding lower level consists of subdivisions of the one above. Generalization increases in an upward direction. Secondly, disjoint subsets form the classes on the horizontal axis at any one level of classification (table 13.8). This pattern applies to the classification of plants or animals. The various coastal classifications that have been proposed also illustrate this type. H. Valentin's (1952) may be taken as an example:

Table 13.9 Coastal classifications

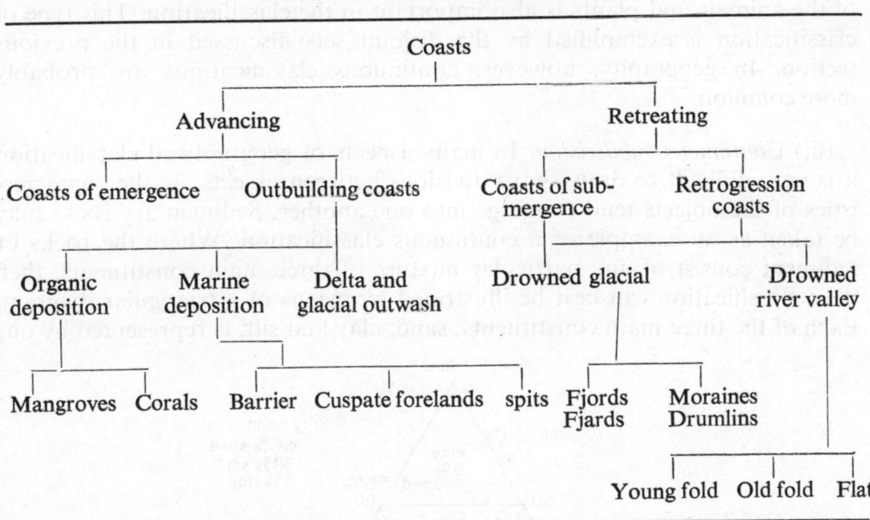

Both aspects are shown in this genetic classification, which is illustrated diagrammatically in table 13.9.

(*b*) *The basis for subdivision*

The value of any classification depends very much on the selection of appropriate subdivisions. The basis of subdivision depends on the character and purpose of the classification. A few of the large number of possible bases of subdivision that can be used in geographical classification will next be mentioned.

(i) *Arbitrary subdivision.* Where the features that are to be classified have a continuous range of character then it may be necessary to make arbitrary subdivisions. The classification of slopes may be made on a basis of their gradient. The categories could consist of steep slopes, medium slopes and gentle slopes. In order to divide slopes into the categories, it would be necessary to set arbitrary limits. Steep slopes could be designated as those over 25 degrees, and gentle slopes those less than 5 degrees, all others falling into the medium category. Although these divisions may appear arbitrary it is possible to fix limits that have some physical significance. These may be where a change in the dominant process at some critical slope angle takes place. Creep may change to rill wash when the slope exceeds 5 degrees. The values given to the classification boundaries will be relative and one set of divisions may be applicable to one area but unsuitable to another.

(ii) *Discrete subdivision.* When different and distinguishable objects are being classified then the divisions can be arranged so that each object falls into one of the available categories. This applies, for example, to the classification of plants and animals by species. (The subdivisions are made on the basis that the animals within them can reproduce.) The physical character of the animals and plants is also important in the classification. This type of classification is exemplified by the disjoint sets discussed in the previous section. In geography, however, continuous classifications are probably more common.

(iii) *Continuous subdivision.* In many aspects of geographical classification it is very difficult to draw sharp divisions between objects, as the characteristics of the objects tend to merge into one another. Sedimentary rocks may be taken as an example of a continuous classification. Where the rocks or sediment consist of any particular mixture of three main constituents, then their classification can best be illustrated by means of a triangular diagram. Each of the three main constituents, sand, clay and silt, is represented by one

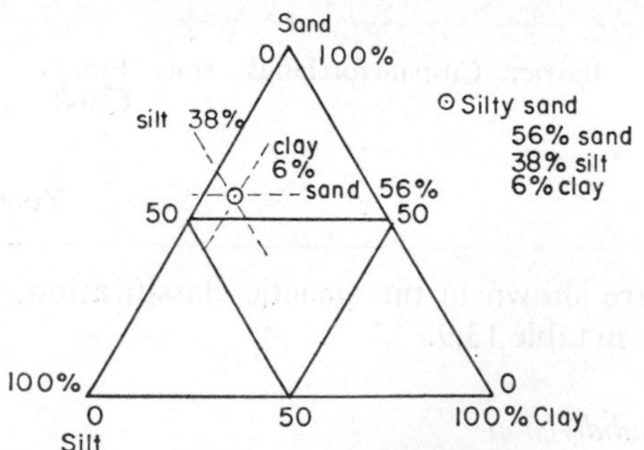

Figure 13.12 A triangular diagram to illustrate the classification of sediment in three categories.

corner of figure 13.12. The percentage of each can readily be obtained from such a triangular graph. The resulting rock could be called a sandy clay, or a clayey sand, or a silty clay, or clayey silt, according to the proportions of each material. Another example of a continuous classification frequently used in biogeography is the Munsell colour chart for the classification of soil by colour. In this chart the colours merge and their intensity shades into very finely divided categories.

(iv) *Dual continuous subdivision*. Some classifications are based on two properties. A good example of this is the classification of morphogenetic regions based on a combination of temperature and precipitation. In this type of classification hot-dry, hot-wet, cold-dry, cold-wet conditions are identified by their representative morphogenetic type. The boundaries between the categories merge in this classification, which can best be illustrated on a 2-dimensional graph where one property is plotted against the other, as shown in figure 13.13.

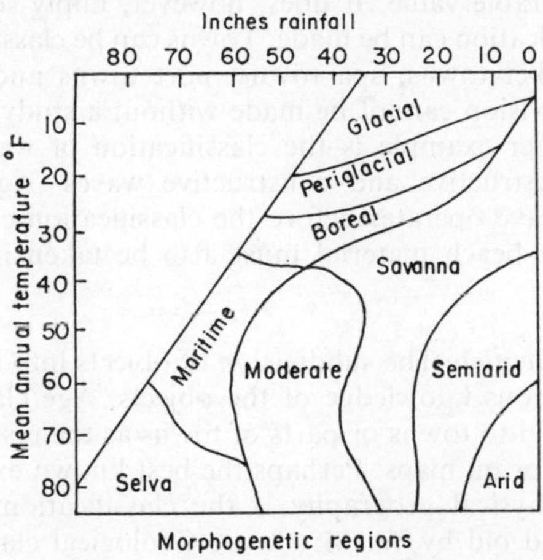

Figure 13.13. A dual continuous classification based on rainfall and temperature. The limits of various morphogenetic regions are suggested.

(v) *Dual discrete*. A good example of this type of classification is the simple division of igneous rocks by chemical content on one axis, and mode of formation on the other axis.

Mode of formation	Chemical character		
	Acid	Intermediate	Basic
Volcanic	Rhyolite	Andesite	Basalt
Hypabyssal	Quartz Porphyry	Felsite	Dolerite
Plutonic	Granite	Syenite	Gabbro

(vi) *Complex classification*. An example of a complex classification is C. W. Thornthwaite's (1933) climatic classification. Climate is itself made up of many elements that vary throughout the seasons, with altitude, aspects, distance from the sea and many other variables. To express these variations simply is very difficult and the resulting classification is very liable to become complex. Thornthwaite uses the potential evapo-transpiration (*e*) which is a function of temperature. The humidity index and aridity index take account of precipitation and form the primary basis of his classification $I(m)$. Subdivisions are on the basis of seasonal concentration. Each climate requires 4 figures to classify it, the first two giving the appropriate $I(m)$ and e value, the second two providing the seasonal element for each group. Thus a number of variables are included in the classification.

The first six types of subdivision are based on different criteria from those still to be mentioned and the types may cut across each other.

(vii) *Functional divisions*. The use of functional divisions in a classification can be of considerable value. It does, however, imply some analysis of the data before classification can be made. Towns can be classified on a functional basis, giving market towns, spa towns, port towns and many other categories. This subdivision cannot be made without a study of the character of each town. Another example is the classification of waves into functional types, such as destructive and constructive waves. Again it is necessary to know how a wave operates before the classification can be made and its effect on different beach material must also be taken into account in this instance.

(viii) *Age classification*. The subdivision of objects into an age classification also requires previous knowledge of the objects. Age classification can, for example, be applied to towns or parts of towns as the result of a study of the towns in the field or on maps. Perhaps the best known example of classification by age in physical geography is the classification of landforms into young, mature and old by W. M. Davis. Geological classifications are also based on age.

(ix) *Genetic classification*. In many fields of physical geography a genetic classification provides a useful summary of the analysis of a wide variety of data. It constitutes the conclusion of the study and hence falls into the second type of classification. An example of a very early genetic classification is that of stream types into divisions such as consequent, subsequent and obsequent, etc. This classification implied a knowledge of the origin of the stream, reflected in its character in relation to the structure. The classifications of coasts put forward by D. W. Johnson (1919) and F. P. Shepard (1948) are also genetic in character although they are very different in the essential elements of their dichotomies.

All these methods of subdivision can be applied to a wide variety of different types of object, which may be concrete or intangible, or even abstract ideas (see classification of theories and laws, etc. in chapter 12), although classification is normally applied to the more concrete objects.

Any individual group of features can usually be classified by a variety of types of subdivision. One example will be considered to illustrate this aspect of classification. Glaciers and ice-sheets can usefully be classified in three ways. They can be classified according to their temperature regime. This is a thermal classification, which divides them into (1) cold or polar glaciers and (2) temperate glaciers. The former are frozen to their beds while the latter are at their pressure melting point. Various other subdivisions are also possible on this basis, as suggested by H. W. Ahlmann (1948). The temperature classification is very important from the point of view of the geomorphological activity of glaciers.

The second basis of classification is on the dynamic activity of the glaciers. On this basis they can be subdivided into three groups, these being respectively (1) active, (2) passive or inactive, and thirdly, (3) dead glaciers. These three types will behave differently from the geomorphological point of view, and they will produce very different deposits.

The third type of glacier classification is into morphological groups. This type of division requires least previous knowledge of the ice-mass concerned as subdivision can be made on the basis of good maps or aerial photos. The divisions in this classification are basically in order of size, ranging from niche glaciers and corrie glaciers at one end to the large continental ice-sheets at the other end.

Niche glacier
Corrie glacier
Valley glacier { normal { outlet type; firn field (Alpine type) }; regenerated }
Piedmont glacier
Mountain ice cap
Continental ice sheet

This example shows that different classifications can all yield valuable data concerning the operation and character of the objects that are classified.

13.10 Linkage analysis

Linkage analysis, also referred to as cluster analysis, is a technique that is being used increasingly in a number of disciplines, including biology and psychology. It requires many calculations and has therefore benefited greatly from the advent of computers. Linkage analysis is designed to take a given set of things and to reduce them step by step into a smaller number of classes. The most similar pair are 'collapsed' into one class, the next most similar are then joined, and so on, down to a desired number of classes. Many variables can be taken into consideration.

The idea of linkage analysis will be illustrated first with a very simple one-variable situation. There are four people, aged 6, 10, 45 and 47. These can be reduced to three classes: (1) one 6 year old, (2) one 10 year old, and (3) two averaging 46 years old. A further reduction to two classes (1) with two members averaging 8 years old and (2) two averaging 46 years old, loses somewhat more information but retains the basic contrast between younger and older members. Information about the variable, age, may be thought of as distributed on a line scale. The criterion for 'collapsing' the first pair is greatest similarity, represented by shortest distance on the line scale. Similarly, the next pair to be collapsed are the next nearest on the line, 6 and 10. Had the ages been 4, 6, 10 and 45, then the 'younger' class would have contained three members, the 'older' class still only one, after two steps. With ages 14, 26, 34, 45, classes are less clearcut.

It would be possible to consider the original four people not only on the basis of their age, but also on the basis of their height or some other variable. Thus with the following information it becomes more difficult to see at a

Table 13.10 Data table for linkage

Individual	Age (years)	Height (inches)
A	6	43
B	10	55
C	45	68
D	47	61

glance which pair out of A, B, C and D, are most similar, because age and height do not correlate closely. A scatter diagram can be drawn, and the position of each person established on this according to the two variables studied. Distance apart on the diagram can then be measured. The values for age must, however, be made comparable with the values for height, because if this is not done, one variable may command more distance on the scatter diagram than the other and bias the assessment of distance apart and therefore of similarity. Clearly, however, when many things are to be classified, not just four, and many variables are to be considered, not just one or two, it is impossible to represent the relevant information graphically and a mathematical method must be adopted.

Linkage analysis has not been used widely by geographers, but B. J. L. Berry (1961) has shown with a worked example the possibility of using linkage analysis to reduce a given number of regions to a smaller number. Considerably earlier, M. J. Hagood and others (1941) pointed the way to this kind of work. R. L. Kaesler and M. N. McElroy (1966) discuss 'Classification of subsurface localities of the Reagan sandstone (Upper Cambrian) of Central and Northwest Kansas'. They use 17 characters (or variables) for 80 localities (wells), and produce a *dendrogram* from these. R. R. Sokal (1966), gives a very clear outline of problems of classification and illustrates his article with examples from biology. The *linkage tree* of Berry, *dendrogram* of Kaesler and McElroy and the *phenogram* of Sokal are basically the same.

The remainder of this section describes briefly the results of the application of linkage analysis to information about the 50 states of the U.S.A. The first five factors derived for these from the 25 variables already tabulated and discussed in chapter 6 are used as the 'variables' for linkage analysis. Each of the 50 states has a position on five co-ordinates, determined by its individual weighting on each of the 5 factors, which in this example contain about 80 per cent of the total variation among 25 variables.

The basic role of linkage analysis in regional geography is to detect homogeneous regions. Since the states in the U.S.A., like major civil divisions in general, are non-intersecting compartments which together cover the whole universal area under consideration, the regions produced by linkage analysis similarly cover the whole area. If, however, only standard metropolitan areas,

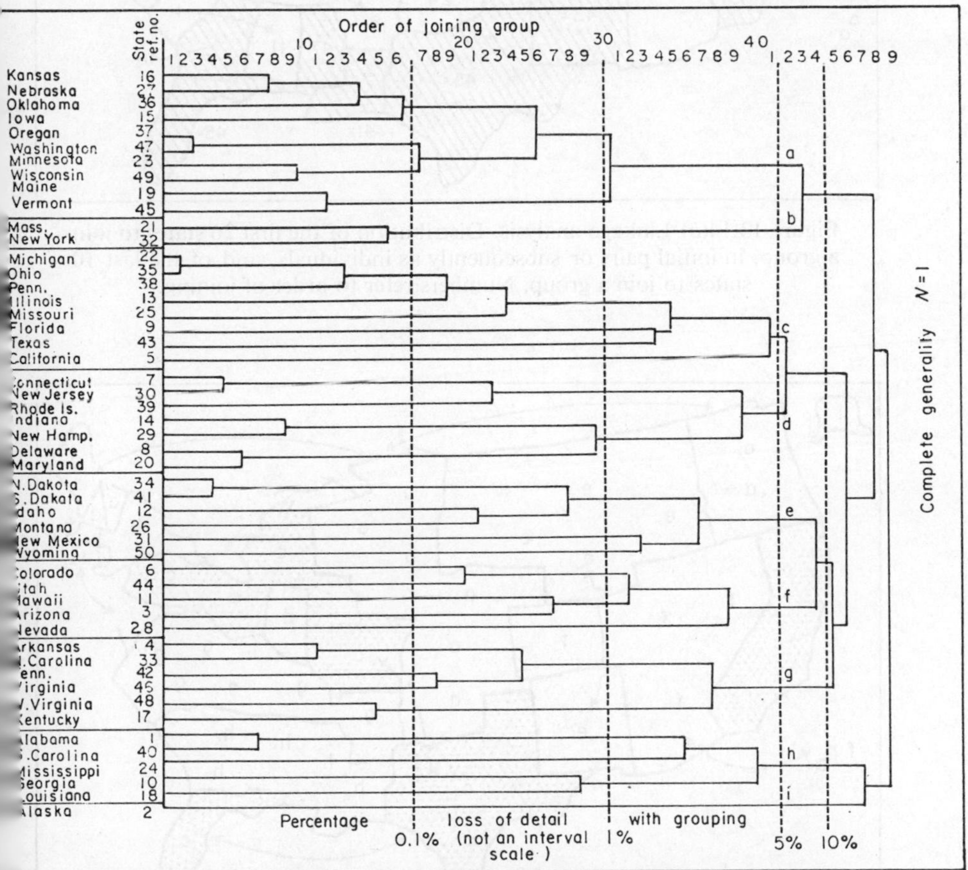

Figure 13.14(*a*) Linkage analysis. Linkage tree for 50 states of the U.S.A. classified on the basis of 5 factors derived from 25 variables tabulated in chapter 6.4, vertical axis, the 50 states; horizontal axis, top scale, the order in which the 50 states are reduced to one unit. (*Note*: the scale used is an ordinal one. Usually a logarithmic scale is used). Bottom scale, percentage of detail lost at selected stages of grouping. See also table in text for further detail.

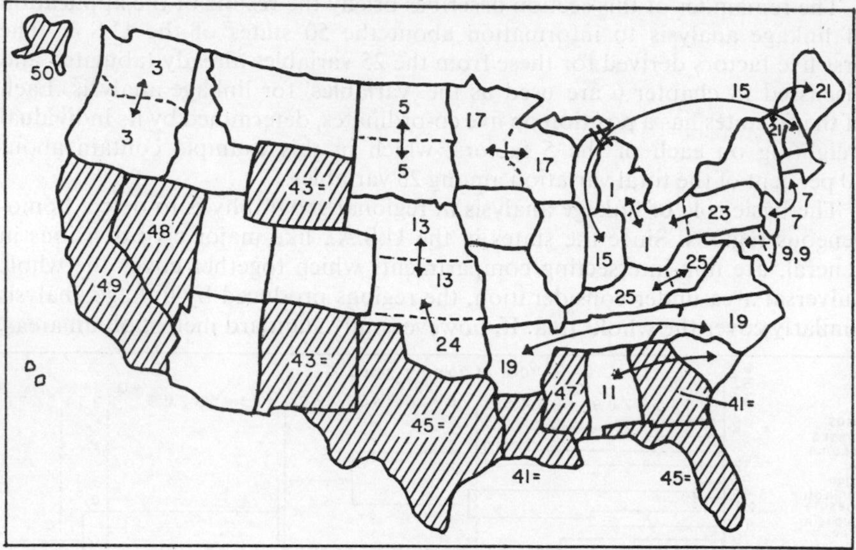

Figure 13.14(*b*) Linkage analysis. Distribution of the first 26 states to join a group, in initial pairs or subsequently as individuals, and of the last 10 states to join a group. Numbers refer to order of joining.

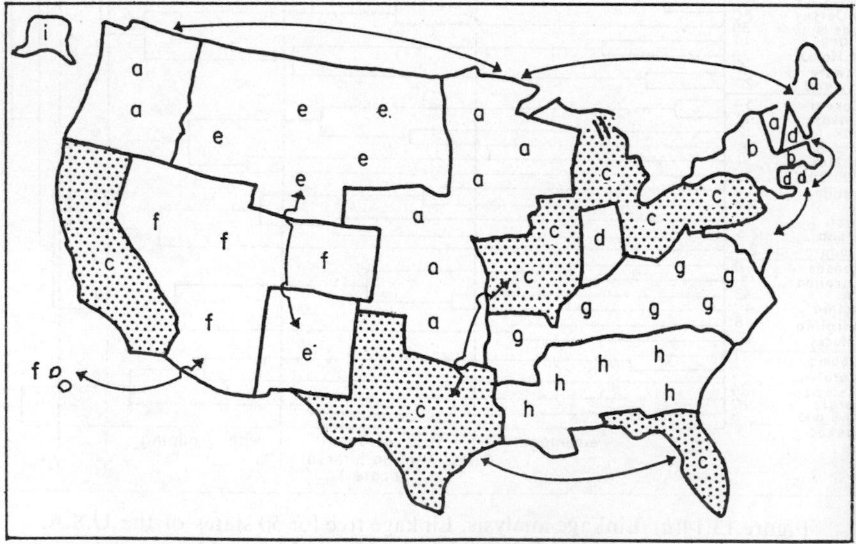

Figure 13.14(*c*) Linkage analysis. Nine groups (regions) remaining after grouping achieving only 5 per cent loss of detail.

or urban districts, were taken, then the whole surface of the country would not be contained in regions.

In the present example, the fifty states are reduced to a progressively smaller number of regions until a desired number is reached. The desired number might be for instance ten planning regions, or it might be the number at which only 10 per cent of the detail was lost. Since the procedure collapses the more similar units first, if many of the states are similar to each other it is possible to reduce the number appreciably while losing only a small amount of information. Thus the 50 states were reduced to 20 regions with the loss of only about 1 per cent of detail, and to 10 with the loss of only about 5 per cent. It is possible to add a constraint to the computer program to allow collapsing only of contiguous states. This was not done in the present exercise.

Figures 13.14*a–c* show the linkage tree for the 50 states of the U.S.A. and some of the early and late joiners. The computer program used gives each successive pair of units to be joined. Some manual sorting is then required to produce the final tree. The tree shows reduction from 50 states with 'perfect detail' (Berry 1961) to one, the U.S.A. itself, with 'complete generality'. Michigan and Ohio emerge on the verdict of the 5 factors as the most similar ('closest') pair of states, Oregon and Washington, the next pair and so on. Alaska, California, Texas and Florida are among the last to join. It should be noted that at any stage except the first, the 'next' state to join may join an existing pair or bigger group of states, rather than another single state like itself. Pairs of groups also become joined as the process continues.

This section does no more than introduce the question of taxonomy. P. Macnaughton-Smith (1965) gives a concise introduction to the general question of classifying individuals and may usefully be referred to by the reader for his text, mathematical techniques and bibliography. In particular, he points out that it is as reasonable to proceed divisively, starting with an initial set and splitting it, as it is to proceed agglomeratively, as was done in the example in this section.

References

Ahlmann, H. W. (1948). Glaciological research on the North Atlantic coasts, *Roy. Geog. Soc. Res. Paper*, **1**.

Berge, C. (1958). *The Theory of Graphs*, Methuen, London.

Berry, B. J. L. (1958). A note concerning methods of classification. *Ann. Assoc. Am. Geog.*, **48**, 300–303.

Berry, B. J. L. (1961). A method for deriving multi-factor uniform regions, *Przeglad Geograficzny*, **XXXIII** Z. 2, 263–282.

Boyce, R. R. and Clark, W. A. V. (1964). The concept of shape in geography, *Geog. Rev.*, **LIV**, No. 4, 561–572.

Baye, Y. (1965). *Routing Methods*, Lund (Sweden) Studies in Geography, Ser. C. General and Mathematical Geography No. 5.

British Productivity Council (undated) *Critical Path Analysis: Eighteen Case Histories*, BPC Vintry House, London.

Cole, J. P. (1965). *Latin America: An Economic and Social Geography*, Butterworths, London, 124–127.

Grigg, D. (1965). The logic of regional systems, *Ann. Assoc. Am. Geog.*, **55,** No. 3.

Hagerstrand, T. (1952). *The Propagation of Innovation Waves,* Lund (Sweden) Studies in Geography. Ser. B. Human Geography No. 4.

Hagood, M. J., Danilevsky, N. and Benn, C. O. (June 25, 1941), *An Examination of the Use of Factor Analysis in the Problem of Subregional Delineation.* Paper read before the Conference on the Analysis of Social and Economic Data. North Carolina State College, University of North Carolina.

Harary, F., Norman, R. and Cartwright, D. (1965). *Structural Models, An Introduction to the Theory of Directed Graphs,* Wiley, New York.

Horton, R. E. (1945). Erosional development of streams and their drainage basins, hydrophysical approach to quantitative morphology, *Bull. Geol. Soc. Am.*, **56,** 275–370.

Johnson, D. W. (1919). *Shore Processes and Shoreline Development,* Wiley, New York.

Kaesler, R. L. and McElroy, M. N. (1966). In (D. F. Merriam, Ed.) *Computer, Applications in The Earth Sciences: Colloquium on Classification Procedures* Computer Contribution 7, State Geological Survey, University of Kansas Lawrence, 42–47.

Kansky, K. J. (1963). *Structure of Transportation Networks,* University of Chicago Department of Geography, Research Paper 84.

Leopold, L. B., Wolman, M. G. and Miller, J. P. (1964). *Fluvial Processes in Geomorphology,* Freeman, San Francisco.

Macnaughton-Smith, P. (1965). Home Office Studies in the Causes of Delinquency and the Treatment of Offenders. 6, *Some Statistical and Other Numerical Techniques for Classifying Individuals,* H.M.S.O., London, 1965.

Ore, O. (1963). *Graphs and Their Uses,* Random House, Yale University.

Pattison, W. D. (1957). *Beginnings of the American Rectangular Land Survey System,* 1784–1800, Dept. of Geography Research Paper No. 50. University of Chicago, Chicago, Illinois.

Rawle, L. J. (1964). The Right Order of Things: Network Analysis, *Progr.*, The Unilever Quarterly, **3,** 128.

Reid, L. (1962). *The Sociology of Nature,* Penguin, Harmondsworth.

Ruhe, R. V. (1952). Topographic discontinuities of the Des Moines lobe. *Am. J. Sci.*, **250,** 45–56.

Shepard, F. P. (1948). *Submarine Geology,* Harper, New York.

Sneath, P. H. A. (1967). Trend-Surface analysis of transformation grids, *J. Zool.*, London, **151,** 65–122.

Sokal, R. R. and Sneath, P. H. A. (1963). *Principles of Numerical Taxonomy,* Freeman, San Francisco.

Sokal, R. R. (1966). Numerical Taxonomy, *Sci. Am.*, **215,** No. 6, 106–116.

Taaffe, E. J. (1958). A Map Analysis of United States Airline Competition, *J. Air Law Comm.*, **25,** Part 1, Part 2.

Taaffe, E. J., Morrill, R. L. and Gould, P. R. (1963). Transport Expansion in Underdeveloped Countries: A comparative analysis, *Geog. Rev.*, **53,** No. 4, 503–529.

Thornthwaite, C. W. (1933). The climates of the earth, *Geog. Rev.*, **23,** 433–439.

Valentin, H. (1952). Die Küste der Erde, *Petermann's Geog. Mitt.*, Ergänzungsheft, **246.**

Warntz, W. (1967). Global science and the tyranny of space, *Papers Reg. Sci. Assoc.*, **19,** 7–19.

14

Some applications of geography

'Hitherto it was the mission of philosophers to interpret the world; now it is our business to change it.' *Karl Marx*

'On the other hand, there are sciences that seem to have "frozen" at the descriptive stage, sciences about which one rarely hears a good word from representatives of applied fields. Among such sciences is geography, which clearly does not occupy a vanguard position in world science . . .

'This question is far from academic. It is just such a synthesis that is needed first of all, and urgently, in the applied fields. Only deep, comprehensive study of the geographic environment insures that it will be properly taken into account in economic activity.'

V. A. Anuchin (*Moscow*, 1965) *in* '*A Sad Tale about Geography*'

14.1 Introduction

Geography is not a vocational subject. Nevertheless it is the hope of geographers, both teachers and students of the subject, that their subject should be of practical value and have some application. Often they are frustrated to find that geographers are not consulted when decisions of a geographical nature are made, either by public bodies or by private firms. Sometimes they find suggestions made by geographers being ignored, or recommendations that should be made by geographers being made by others. It should not be overlooked, however, that while a student of geography may be studying purely academic themes, there may be some less tangible practical value simply in the training. He may, therefore, be equipped to think in a special way unique to geography, and thus be able to analyse certain complex situations during his subsequent career.

This chapter outlines some of the more obviously practical or applied forms of geography. It contains examples of things geographers have already done and of things they could usefully do. Success in terms of application and practical value would be measured by some people in terms of the amount of work done by geographers and used as a basis for decisions by others. Here physical and human geography become very close, for often the two have to be considered side by side. A plan to irrigate a new area, to build a new bridge or to plant a forest involves decisions made by people about physical conditions as well as the future arrangement and/or movement of human beings. Man is usually either modifying or conserving physical patterns

Some of the fields in which the geographer has traditionally made a practical contribution, such as weather study and cartography, have become extremely specialized. Other fields have been invaded by outside disciplines. Economists have become keenly interested in the spatial distribution of economic activity, as the flourishing Regional Science Association clearly shows. J. R. Meyer (1967) provides a good introduction to 'regional economics'. Similarly, architects have become more aware of spatial distributions on a middle level as well as a micro level, considering the layout of towns, regional planning and then national planning.

In a less obvious sense, geography has potentially an important part to play in the occupation of leisure time. As a way of thinking and of seeing the world, it seems particularly relevant at a time when leisure is becoming more widely available, and broadcasting and the press are making people more and more aware of the world region. Geography can help to produce informed citizens able to form political opinions and to view world affairs with interest and intelligence. Travel, both connected with work, and for leisure (now increasingly possible) can be planned and used more fully. It seems, then, that school geography might be thought of as having a practical application in enabling people to gain increased enjoyment from their leisure time.

14.2 Gathering, storing and retrieving data

(*a*) *Remote sensors*

The advent of aerial photography and the increasing availability of aerial photographs of different parts of the world in the interwar period rapidly opened for geographers a new source of information hitherto unobtainable. New views of archaeology, geomorphology and vegetation were obtained. What has become available in this way has been minute compared with the material now being produced by means of remote sensors in aircraft and spacecraft. At the Regional Conference of the International Geographical Union in Mexico City in August 1966, Dr. A. C. Gerlach outlined some of the possibilities of new techniques, and made it clear that in his view, geography and remote sensing have much to offer each other.

Earth orbital spacecraft (satellites) will soon be able to carry payloads of many tons, including complicated instruments for transmitting and processing data. There have been many developments in the field of remote sensors, instruments that can detect and identify the properties of an object without being in contact with it. The eye and the camera are familiar examples of remote sensors. New devices are increasing the range of the part of the electronic spectrum than can be perceived. Very sensitive photographic equipment, infrared sensors, radar sensors and laser beams are examples of such instruments, which can be installed both in aircraft and in satellites. Temperature, mass and magnetic field are among properties that can be perceived by appropriate sensors at great distances.

A. Gerlach (1966) writes in his paper: 'Application studies of remote sensor data in such fields as agriculture, cartography, forestry, geology, hydrology, meteorology and oceanography have already revealed new possibilities for the rapid and accurate assessment of surface features and materials and helped in searching for highly mineralized areas, oil reserves and water resources'. It is hoped that many other physical fields can be assisted in the future. 'Basic, and practically synoptic daylight surveys could be made of the entire earth every 24 hours as it rotates under polar orbited satellites, and periodic surveys could be scheduled to discern annual, seasonal, diurnal, and even more frequent changes of earth phenomena'.

In very general terms, human geography will be less immediately affected by the impact of remote sensors, but (p. 93): 'People are emitters of electromagnetic radiation in the thermal band, and it may become possible for them to be counted and mapped by thermal means in the same way as one might take a radioautograph of radioactive tracers'.

So great is the potential amount of data, that it (p. 93) 'would exceed the capacity of people to process it in this century, even with modern electronic computers'. G. C. Ewing (1965) gives many examples of the way in which remote sensor data from satellites has been contributing to oceanography. A. Morrison and others (1965) show how information from satellites is interpreted for the land surface in Southern Africa.

(b) Work in Canada and Sweden

The Canada Land Inventory based in Ottawa and developed by IBM and the Canadian Government serves as an example of the way in which a very large amount of information will be gathered, stored and retrieved. The main aim of the project is to bring together large amounts of data on resources to provide a basis for sound decisions on the best utilization of land. Over the last 40 years, some 30,000 maps have been compiled with land use information about roughly 1 million square miles or $\frac{1}{3}$ of the area of Canada. Information in map form is difficult to retrieve and co-ordinate on account of the large number of maps. The kind of questions to be answered concern the suitability or best use of different pieces of land (e.g. for farming, forest, wild life reserves, recreation) and the best sites for new enterprises (e.g. a pulp and paper industry).

An automatic scanning device is being developed to convert map information into binary data for computer processing. Each map is placed round a drum. The map only has black boundary lines (8/1000 of an inch) on a white opaque plastic sheet. Each fibre optic channel covers a four-mile square area. If at least half of the area under the channel is black when the pulse is received, a 1 bit is generated, if not, a 0. One and zero for each pulse are recorded in magnetic tape as a byte (=8 bits). A 16 ft square map, with 18 million bytes, can be scanned in less than 11 minutes. Once in digitized form the map information can be rapidly sorted and handled in other ways by the computer (see chapter 2).

A similar project is being developed in Sweden. It is proposed to record the Cartesian co-ordinate position of each locality for which census information is collected. It will then be possible to know the location of each place in

Sweden relative to every other. Thus the correct location will be available for a very large amount of information. Experiments have already shown the feasibility of drawing rapidly on this kind of information with the help of computers, and producing maps showing, for example, the number of people within given radii of various places (see chapter 4, running mean idea). Such information is useful in helping to decide where it is practicable to locate establishments such as hospitals and schools, that require a given number of users within a certain distance to be viable or economical in terms of scale. The scope of the project was outlined by T. Hagerstrand (1967) at the Meeting of the Institute of British Geographers, held at Sheffield in January 1967.

(*c*) *Soviet planning aata*

Many countries have used or now use economic planning regions as the basis for some decisions of national importance. There was great interest in the U.S.A. in the 1930s in planning regions, and a long report by the National Resources Committee (1935) went into the regional concept and the regionalization of the U.S.A. at great length. Currently the U.K. and France have introduced planning regions. This subsection outlines three forms for storing data about the economic activities of a country. The U.S.S.R. is used as an example since it is perhaps the country that has been most preoccupied with planning regions in the last few decades. The 19 economic planning regions used since the early 1960s are considered. They vary greatly both in area and in population, which detracts from their usefulness to some extent. This aspect will not, however, be considered here.

As a basis for drawing up their major plans, Soviet planners have generally used information about branches (sector planning) of the economy and about regions (territorial planning) of the national area. Emphasis has shifted between these over the last 40 years. Very broadly, three types of data matrix can be used for storing and handling data. In the first, the input-output matrix, rows and columns both represent branches; in the second, rows represent regions and columns represent branches (or vice versa); in the third, both rows and columns represent regions. To illustrate the three kinds of matrix, a small portion of each kind is illustrated.

(i) *Branch–branch matrix* (table 14.1). This matrix is of more *direct* importance to economists than to geographers since it contains no spatial information. A very small part of the 1959 Soviet interbranch (or interindustry) transactions matrix with 73 items listed and 73×73 transactions is shown for 8 selected branches. Other information, such as input coefficients, may also be recorded in such a matrix. The reader will find a reduced version of the 1959 Soviet input-output table in English in *New Direction in the Soviet Economy*, Part II A — Economic Performance, U. S. Government Printing Office, Washington 1966. The U.S. equivalent matrix is discussed at a reasonably elementary level by W. W. Leontief (1965) in the *Scientific American*.

The matrix in table 14.1 is a tiny portion of a 73 × 73 (=5329) (square) matrix, which is not symmetric. Each of the 73 branches occupies a row, which shows the value of its *transactions*, outgoing, if any, to itself and to each of the other 72 branches. Each of the 73 branches also occupies a column, which shows the value of its transactions, incoming, if any, from itself and each of the other 72 branches. For example, much timber (item 50) goes into furniture (item 51) but virtually nothing returns. On the other hand, there is a two way flow between metals (item 1) and coal (item 5). Finally, there are no transactions either way between sugar and furniture.

Table 14.1 Branch to branch matrix

From \ To	1 Metals	5 Coal	6 Oil	12 Energy	50 Timber	51 Furniture	63 Sugar	73 Other
1 Metals	3122	32	1	5	8	2	5	2
·								
5 Coal	584	841	—	797	10	—	47	3
6 Oil	3	—	2	31	—	—	—	—
·								
12 Energy	332	139	46	4	33	—	1	3
·								
50 Timber	19	49	1	1	607	248	2	2
51 Furniture	2	1	—	—	—	—	—	—
·								
63 Sugar	—	—	—	—	—	—	329	—
·								
73 Other	177	—	—	—	—	—	53	—

Values in millions of roubles — under ½ million roubles.
Source: *Narodnoye khozyaystvo SSSR v* 1960 *godu, statisticheskiy yezhegodnik*, Moskva (1961), pp. 104–143.

(ii) *Region–branch matrix* (table 14.2). This matrix shows the share each region (of a given set of regions) has of each branch of the economy. The table shows a small portion of a possible 19 row × n column matrix showing the regional distribution of selected branches of industry. Matrix A, with six columns, shows absolute production figures for a given year period (mostly 1963). Production is expressed in different units for different branches.

It is, therefore, convenient to express the production for each branch for the U.S.S.R. as a percentage or in per thousands of the national total. This has been done for the first two branches in Matrix B. Since the regions vary greatly in population (extremes are Centre (2), 26.4 million, Moldavia (19) 3.2 million), it is also convenient to express production in per capita terms. Further modifications of the data are possible, such as ranking of the regions (see chapter 6) and the calculation of an index of concentration.

Table 14.2 Region to branch matrix

	Matrix A						Matrix B	
	Meat[a]	Milk[a]	Timber[b]	Electricity $\times 10^9$ kwh	Cement[a]	Cotton textiles[c]	Meat	Milk
1 Northwest	270	2,653	91	22	3,256	222	27	43
2 Central	848	6,569	31	37	6,114	4,906	83	107
3 Volga-Vyatka	314	2,342	33	11	961	111	31	38
.								
.								
.								
10 Far East	144	917	23	7	1,692	9	14	15
11 Donets-Dnepr	906	4,955	1	59	6,570	16	89	81
.								
.								
.								
17 Kazakhstan	665	2,834	2	15	3,425	20	65	46
18 Belorussia	430	3,210	8	6	1,266	7	42	53
19 Moldavia	144	584	0	1	445	2	14	10
USSR	10,195	61,248	370	412	61,018	6,619	1000	1000

[a] Thous. of tons
[b] Millions of cu. metres
[c] Millions of metres

Source: From tables 1 and 5, pages 301 and 306 in *Geography of the USSR*, J. P. Cole, Penguin, 1967.

(iii) *Region–region matrix* (table 14.3). A third kind of matrix can be compiled to show region–region transactions. In the example, interregional movement of coal is recorded. For simplicity here, a 1 is recorded if coal goes from one region to another, a 0 if it does not. More elaborate information can be given, smaller, more numerous regions can be put in the matrix, and distances between regions calculated so that observed interregional flows may be compared with expected flows according to the gravity model (see chapter 11) or other models.

Table 14.3 Region to region matrix

From \ To	1 Northwest	2 Central	3 Volga-Vyatka	10 Far East	11 Donets-Dnepr	17 Kazakhstan	18 Belorussia	19 Moldavia
1 Northwest	1	1	1	0	0	0	1	0
2 Central	0	1	0	0	0	0	0	0
3 Volga-Vyatka	0	0	0	0	0	0	0	0
⋮								
10 Far East	0	0	0	1	0	0	0	0
11 Donets-Dnepr	1	1	1	0	1	0	1	1
⋮								
17 Kazakhstan	0	1	1	0	0	1	0	0
18 Belorussia	0	0	0	0	0	0	0	0
19 Moldavia	0	0	0	0	0	0	0	0

Source: Based on information in *Ekonomicheskaya geografiya transporta SSSR*. M. I. Galitskiy and others, Moskva 1965, especially p. 81. Note that the matrix is square (19 × 19 in complete form), but not symmetric, since a two-way flow between each region must be allowed for. The principal diagonal may be used to record within region flows. Where it has a 0 (e.g. Moldavia (19)), this means that the region itself produces no coal. Thus for some purposes the rows for non-producing regions (3, 18 and 19 in this example) are superfluous.

(*d*) *Rapid reference for the United Kingdom*

In addition to the sophisticated data banks outlined in the previous sections, it would be useful to have works of reference on the same lines but more limited in scale. So-called geographical data are available in maps and atlases in abundance. It is not easy, however, to find a large number of relevant facts about a particular area since information about it may be dispersed through many maps. Data are also available listed alphabetically in gazetteers though the places, usually towns, countries or states, may not be the places of interest. Rapid reference tables might find many users and cover many situations. The kind of table under consideration will be outlined with an example from the United Kingdom.

Places are constantly being considered and chosen for the location of new factories, service establishments, transport terminals and so on. It would be useful to be able to eliminate quickly many unsuitable places and to focus attention on a few likely ones. The United Kingdom could be divided into a number of regions, combinations where necessary of the 100 or so counties.

Information could be tabulated for each of these. Each region could then be subdivided into ten smaller units, combinations and adaptations of minor civil division units. Information could then be tabulated for these. A finer level still might also be provided.

In a most original talk on the B.B.C., D. Rigby Childs (1965), indicated the need for many different levels of region: 'What I have been doing is to devise a spatial scale within which all other features are systematically compared. When I came to apply this method to Britain's conditions my starting point was to work out a range of significant size of areas as a clothing manufacturer will produce different ranges of sizes. I found I wanted about 50 different sizes in order to produce an adequate range to span the range of physical environment from a single dwelling up to the whole island'.

One user might be an industralist considering locations for a new light manufacturing establishment. He might want to be in a region with a port, within 100 miles of London, with at least 20 per cent of the national market in a radius of 70 miles and an international airport less than 2 hours away by road. Reference to the table would quickly enable him to eliminate perhaps 30 out of 40 of the regions. He could then refer to the appropriate tables for subregions and find which subregions within the 10 main regions fulfilled more local requirements, which might be suitable housing available for staff, a golf course or a football team within ten miles and so on.

14.3 Applying knowledge of physical geography to solve problems

(*a*) *Coasts*

A knowledge of the processes operating along the coast is of considerable value in dealing with problems associated with coast erosion and coastal amenities, such as beach preservation in a good state and the removal of beach material for construction purposes. Much of this work is carried out by engineers, including the U.S. Army Coastal Engineering Research Council, which has taken over from the Beach Erosion Board, and the coastal division of the Hydraulics Research Station in Britain. These organizations have the facilities for experimental work both in the field and in model wave tanks. Nevertheless the geographer's special point of view can add considerably to the solving of many of these problems. An appreciation of the interaction of a large number of factors operating together and influencing a considerable stretch of coast line is necessary. There have been examples in the past of coast defences built to protect one area, without adequate study of the effect of the structures in adjacent areas. The defences have caused even more serious erosion to occur in areas adjacent to that protected. This has occurred along the coast of Holderness in Yorkshire.

An interesting experiment in a new method of beach building is being carried out at Bournemouth where the authorities are worried by the loss of beach material. This experiment is an attempt to build up the level of the beach by planting artificial seaweed made out of an artificial fabric in a belt about 320 to 480 ft out from the promenade over a distance of 500 ft along the coast in a water depth that was originally about 9 to 10 ft deep. Surveys have been carried out both before and after the planting of the artificial seaweed and these show that there has been a considerable build up of sand at the landward edge of the strip of material but the sand has been eroded from the lower part of the strip and in the offshore zone. Sand has also accumulated on the down wave or east side of the strip. The maximum gain of sand amounts to over 6 ft to the east and inland of the strip.

To account for these changes the action of the waves and the effect of the seaweed on this action must be taken into account. The seaweed reduces the height of the waves without affecting their length and thus may be expected to increase their constructive activity before they break, which was at a point inland of the strip at first. This accretion of sand in turn affects the breaking position of the waves, which break in water which is about 4/3 of their breaker height in depth. It has been noted that the waves now break further offshore in this zone. This has had the result of causing a break-point bar to develop, particularly when large, high and destructive waves are breaking on the developing bar and has resulted in a loss of sand nearer the promenade and the formation of a deep trough landward of the new accumulation. Thus by changing the character of the waves the whole pattern of beach development has been altered. The down-wave spread of the bar feature can be accounted for by the eastern longshore movement of beach material in this area. This takes place most effectively in the breaker zone under some conditions. The character of the wave refraction will also have been altered by the changing bottom configuration and this will change the angle of wave approach and the effect of the waves on the beach.

An interpretation of the results of these artificially induced changes must take into account many different aspects of coastal processes over a wide area. The fetch from which the waves come is important. The adjacent coast, which supplies the material moved along the coast from the west by longshore transport under both wave and tidal action, must also be considered.

(*b*) *Harbour maintenance problems*

A knowledge of the operation of coastal processes is useful in an assessment of the best method of dealing with specific problems of harbour maintenance. An example to illustrate this application of geographical knowledge to the solving of practical problems is concerned with the Tees estuary (figure 14.1*a*). The particular problem was concerned with the effect of deepening a channel and reclaiming land within the estuary, consisting of the Seal Sands. Before the effects of the proposals to deepen Seaton Channel and reclaim the Seal Sands could be suggested, the nature of the estuary had to be investigated.

Many observations of the tidal flow in the river were available and measurements of waves and currents in Tees Bay north of the river mouth had been made. By use of these data it was possible to establish the way in which the water and sediment in the estuary were moving. The effect of the waves resulted in a southward movement of sand along the open coast and a movement of sand to the southwest across the North Gare Sands into the

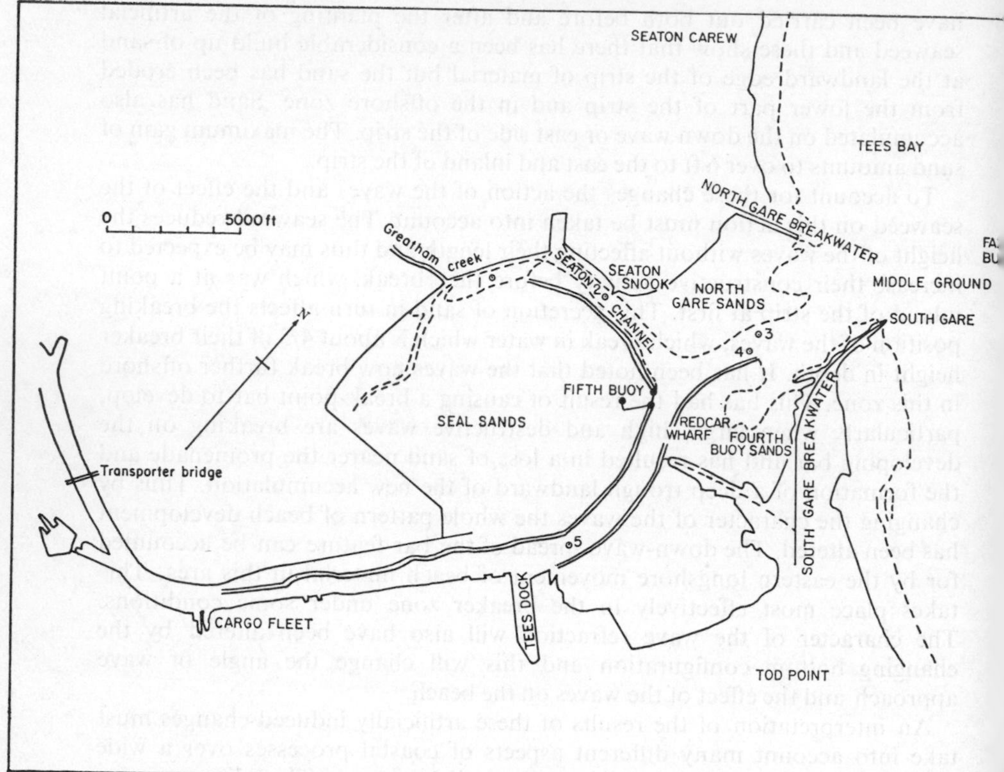

Figure 14.1 (*a*) A map of the Tees estuary. The figures in the channel refer to the spring tide vectors illustrated below.

estuary. This movement is due to the longer fetch to the north as the frequency of waves coming from north and south of normal to the coast is about equal. The waves, however, are only of significance in the outer estuary and in the inner part the tide is of much greater significance.

The relative duration of the ebb and flood tide is of particular significance in the movement of sediment. Thus a study of the tidal residuals based on the construction of vector diagrams (see figure 14.1*b*) provides useful evidence of the relative strength of the flood and ebb tide in different parts of the channels. It is worth noting that in the Seaton Channel there is an ebb residual, owing to the addition of water passing over the Seal Sand from the main river when the tide is high. This water accentuates the ebb flow and scours the Seaton Channel. This scouring action would, therefore, be reduced if the Sands were reclaimed. On the other hand the material that does accumulate in the west end of the Seaton Channel is largely derived from the river as it is fine silt and this material would be prevented from reaching the channel. In the main river the flood stream is stronger than the ebb flow at

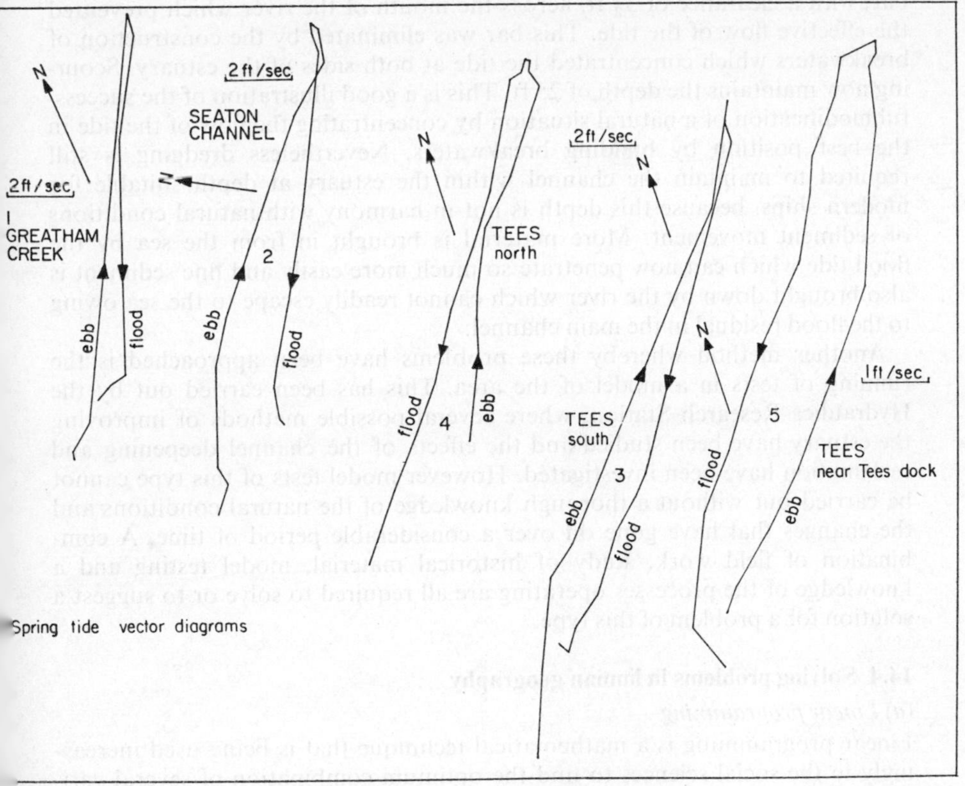

Figure 14.1 (*b*) The spring tide vectors.

the bottom of the channel where the sediment is moving. To account for this fact, when the volume of the ebb flow, which is augmented by the river flow, is greater than that of the flood flow, the salinity stratification in the estuary must be considered. The inflowing flood stream consists of dense salt water from the sea while the outflowing ebb stream consists of fresher river water. The flood stream is, therefore, concentrated on the bottom of the channel and continues to flow upstream while the seaward flowing ebb current has already started to flow in the upper layers of water. The material that enters the estuary is derived, therefore, largely from the sea and is brought into the area by the flood tide. It is this material as well as fine river sediment that necessitates the dredging that is required to maintain the channels.

The channels in the Tees are now artificially maintained by dredging and by other training works which confine the channel. In its natural state, before 1841, the river was much shallower and followed shifting channels. Sand from the sea side was however, partially excluded by a very shallow bar, with a clearance of $3\frac{1}{2}$ ft, across the mouth of the river which prevented the effective flow of the tide. This bar was eliminated by the construction of breakwaters which concentrated the tide at both sides of the estuary. Scouring now maintains the depth of 25 ft. This is a good illustration of the successful modification of a natural situation by concentrating the flow of the tide in the best position by building breakwaters. Nevertheless dredging is still required to maintain the channel within the estuary at depth suitable for modern ships, because this depth is not in harmony with natural conditions of sediment movement. More material is brought in from the sea by the flood tide which can now penetrate so much more easily and fine sediment is also brought down by the river which cannot readily escape to the sea owing to the flood residual in the main channel.

Another method whereby these problems have been approached is the running of tests in a model of the area. This has been carried out by the Hydraulics Research Station, where several possible methods of improving the estuary have been studied and the effects of the channel deepening and reclamation have been investigated. However model tests of this type cannot be carried out without a thorough knowledge of the natural conditions and the changes that have gone on over a considerable period of time. A combination of field work, study of historical material, model testing and a knowledge of the processes operating are all required to solve or to suggest a solution for a problem of this type.

14.4 Solving problems in human geography

(*a*) *Linear programming*

Linear programming is a mathematical technique that is being used increasingly in the social sciences to find the optimum combination of several variables or processes. The method can be adapted to assist in the solution of certain geographical problems, in which a spatial aspect enters the data. A graphical solution to problems of linear programming is easiest to understand. The method is based on a series of linear equations which intersect each other. The points at which they intersect are used to calculate the optimum solution to the problem in hand.

A few simple examples will illustrate best how the method works. The first problem concerns the siting of a plant which requires 30 tons of sand and 20 tons of gravel per day. There are two quarries 40 miles apart from which gravel and sand can be obtained. Quarry x produces 1 ton of sand for every 1 ton of gravel. Quarry y produces 3 tons of sand for every $\frac{2}{3}$ ton gravel. The problem is to find the best site for the plant so that there is no surplus sand or gravel. The data are set out as follows:

x produces 1 ton sand, 1 ton gravel.
y produces 3 tons sand, $\frac{2}{3}$ ton gravel.
x and y are 40 miles apart.
30 tons sand and 20 tons gravel are required.

The equation giving the required quantity for the sand is given by $x+3y=30$.

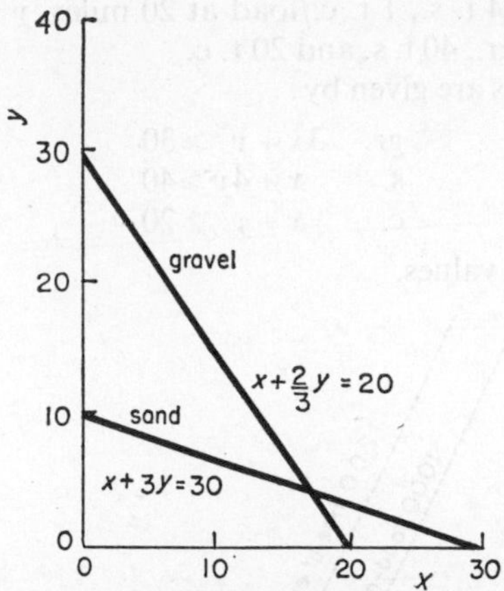

Figure 14.2 Graph to illustrate amount of sand and gravel obtainable from two quarries.

This line is drawn on the graph (figure 14.2). The equation for the gravel is $x+\frac{2}{3}y=20$ and this line is also drawn. The appropriate amount of sand would be obtained all along the sand line $x+3y=30$, and similarly for the gravel. The point at which these two lines cross must, therefore, indicate the proportions required from the two quarries to fulfil exactly the needs of the plant and *leave no surplus*. These values are nearly at $x=17$ and $y=4.4$. These values will yield the desired amount because quarry x provides 17 tons of sand and 17 tons of gravel. Quarry y provides 3×4.4 or 13 tons of sand and $\frac{2}{3}\times4.4$ or 3 tons of gravel, making a total of $17+13=30$ tons of sand and $17+3$ or 20 tons of gravel. Quarry x therefore supplies a total of 34 tons and y of 16 tons, making the total requirement of 50 tons. If the material is supplied in loads of 2 tons 17 loads must come from x and 8 loads from quarry y. The site must be chosen so that the load-miles from the two quarries to the site are equal. If a is the distance from quarry x and b is the distance from quarry y then $17a=8b$. The quarries are 40 miles apart so $a+b=40$, solving these simultaneous equations $a=12.8$ miles and b is 27.2 miles. The plant is therefore sited 12.8 miles from quarry x and 27.2 miles from quarry y. $12.8+17=27.2\times8=218$ load-miles from each quarry to the site.

The second example concerns two quarries in till. Quarry A which is 40 miles from the site under consideration produces 3 tons gravel, 1 ton sand and 1 ton clay per load. Quarry B, which is in a sandy till, is 20 miles from the site and produces 1 ton gravel, 4 tons sand and 1 ton clay per load. The requirement of material at the site consists of 30 tons gravel, 40 tons sand and 20 tons clay. The problem is to reduce transport, in terms of load miles to a minimum by obtaining the optimum proportion of material from the two quarries. The data are set out:

Quarry A — 3 t. gr., 1 t. s., 1 t. c./load at 40 miles, x is load miles from A.

Quarry B — 1 t. gr., 4 t. s., 1 t. c./load at 20 miles, y is load miles from B. Requirements are 30 t. gr., 40 t. s. and 20 t. c.

The linear inequalities are given by:

$$\begin{array}{ll} \text{gr.} & 3x+y \geqslant 30 \\ \text{s.} & x+4y \geqslant 40 \\ \text{c.} & x+y \geqslant 20 \end{array}$$

$40x+20y$ gives distance values.

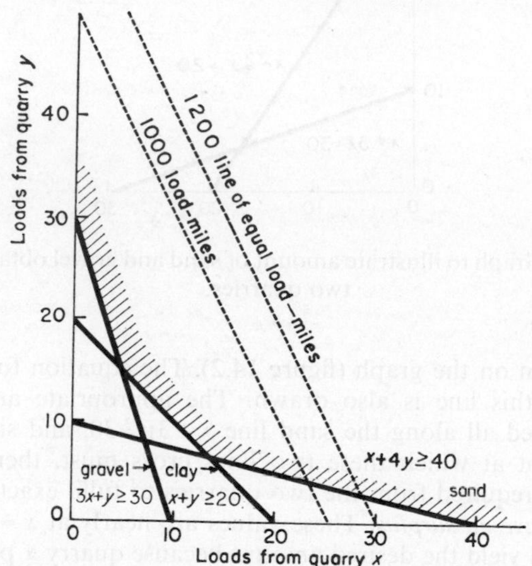

Figure 14.3 A problem of linear programming to calculate optimum amount of sand, gravel and clay obtainable from two quarries. The x-axis represents number of loads from quarry x and the y-axis number of loads from quarry y. Two lines of equal load-miles are shown.

These inequalities are drawn in the graph, figure 14.3. There are four points where the inequalities cut each other or the x-and y-axes. These points are as follows:

$x=0$	$y=30$	$0+600=600$ load miles
$x=5$	$y=15$	$200+300=500$ load miles
$x=13.5$	$y=6.7$	$540+134=674$ load miles
$x=40$	$y=0$	$1600+0=1600$ load miles

$40x$ and $20y$ are substituted into the equations to obtain the load mile values for the four points. The minimum value is 500 for $x=5$, $y=15$. This point, therefore, represents the optimum proportions. The loads would therefore be made up as follows:

From quarry A	15 t. gravel, 5 t. sand, 5 t. clay = 5 loads
From quarry B	15 t. gravel, 60 t. sand, 15 t. clay = 15 loads
	30 t. gravel, 65 t. sand, 20 t. clay

The required amount is exact for gravel and clay, which is easily verified by noting that it is the gravel and clay lines that cross at $x=5$, $y=15$. There is, however, a surplus at 25 tons of sand. It is impossible to obtain the correct amount of all these materials as the 3 inequalities do not meet at a point. The necessary conditions are fulfilled throughout the shaded part of the graph. Lines of equal load milcs are entered to show how these decrease in value towards the axis of the graph.

A third example shows how this method can also be applied usefully in physical geography. The problem is to find the optimum conditions for, and amount of ice formation in, different soil types. The data are as follows: soil x is fine, it contains 10 per cent water and freezes at 5 cm/hr. Soil y is coarse, it contains 30 per cent water and freezes at 4 cm/hr. The maximum time available for freezing is 8 hours and the total soil thickness is 50 cm. x represents the thickness of frozen fine soil and y of frozen coarse soil. No extra water is available. A number of inequalities can be set up: $x \geqslant 0$, $x \leqslant 40$ cm, $y \geqslant 0$, $y \leqslant 32$ cm, $x+y \leqslant 50$ cm. The amount of ice depends upon the water present, therefore, $10x+30y$ must be substituted in the equations The portion of the graph bounded by the inequalities and the axes contains the necessary point. The points where the inequalities cut each other or the axes are as follows (see figure 14.4.)

$$\begin{aligned} x &= 0, & y &= 32; & 10x+30y &= 960 \\ x &= 18, & y &= 32; & 10x+30y &= 180+960=1140 \\ x &= 40, & y &= 12; & 10x+30y &= 400+360= \ \ 760 \\ x &= 50, & y &= 0; & 10x+30y &= 500 \end{aligned}$$

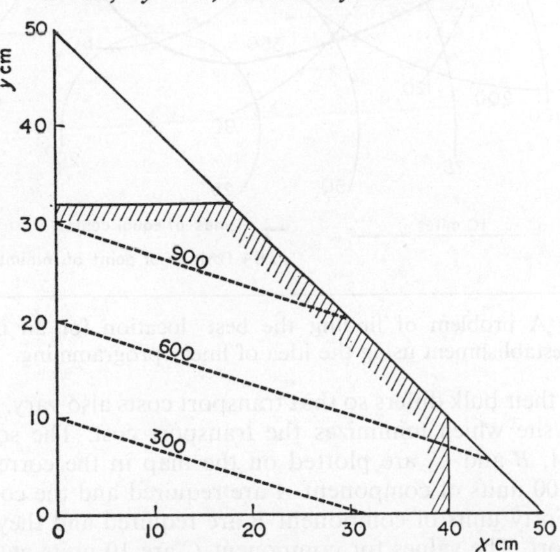

Figure 14.4 A problem of linear programming, x represents the depth of penetration of frost in soil which is fine, and y the penetration of frost in soil which is coarser. The values on the sloping lines refer to volume of ice.

The optimum thicknesses are, therefore, 18 cm of fine soil and 32 cm of coarse soil. Lines of equal ice formation are entered on the graph. The values in this example are not very realistic, but it does illustrate a possible method of calculating the optimum conditions for ice formation, which would be associated with maximum heave and periglacial sorting for instance.

A modified method of linear programming can be adapted to solve certain locational problems. A simple example of this method is illustrated in figure 14.5. In this example a factory requires three components in different

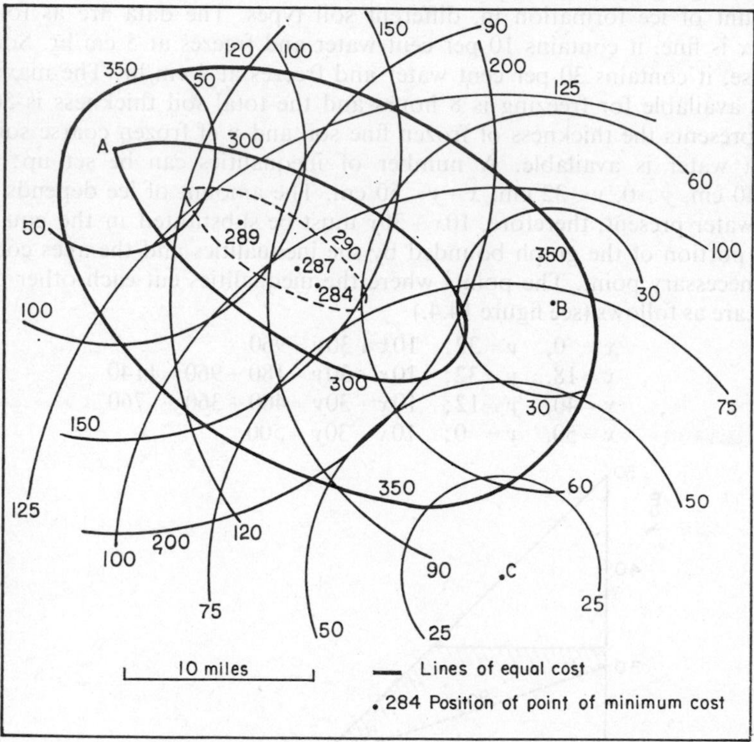

Figure 14.5 A problem of finding the best location for an industrial establishment using the idea of linear programming.

quantities and their bulk differs so that transport costs also vary. The problem is to locate a site which minimizes the transport cost. The sources of the components, *A*, *B* and *C*, are plotted on the map in the correct places. In this example 100 units of component *A* are required and the cost per unit is 2/- per mile. Forty units of component *B* are required and they cost 3/- per mile to transport. The values for component *C* are 10 units at 10/- per mile per unit. The total cost of component *A* is then £10/mile, *B* £6/mile and *C* £5/mile. A scale is drawn representing the cost of each component and circles are drawn around each point for the appropriate cost. On a scale of 5 miles to 1 inch, 1 inch radius represents £50 for component *A*, £30 for component *B* and £25 for component *C*. Radii increasing by one inch are drawn as multiples of these values as shown on figure 14.5. It is then possible at any point to add up the total cost for the three components' transport. The required point is where this value reaches a minimum. In the example shown this point has a value of £284 and is situated almost half way between *A* and *B*. The cost of components is not the only factor to be considered in a location problem so that it may be useful to delimit a wider zone in which the cost is not much greater. In this example the isopleths for £290, £300 and £350 have been drawn in. It can be seen that there is quite a large area in which the cost varies relatively little, but beyond the £300 isopleth the rate of increase is much greater, particularly to the northwest of *A*. The £300 isopleth is pulled right over to *A* because this component is the most expensive to transport.

This method can be adapted to solve many similar problems, one of which is discussed by D. M. Smith (1966). In his example he is concerned with balance between market area, labour supply and raw material supply, but the basis of the problem is similar.

(*b*) *Game theory*

This brief section is intended only to give some idea of the purpose and basis of game theory and to illustrate by means of simple examples a geographical application of the theory. While connected with operational games, game theory has different aims. Operational games are intended to give experience of real world situations.

Game theory is applicable to situations in which there is an element of conflict and choice. It is, therefore, relevant to human geography rather than to physical geography, although the physical element of a situation may be one of the players in the game. This applies to the example discussed. Game theory grew from a study of games in which strategy was an essential element, the opposing players each attempting to devise the optimum strategy to enable them to reap the highest reward. In some of the games and situations in which game theory is applicable chance also plays a part, but games of pure chance are not susceptible to game theoretical analysis in the same way as games of skill, such as chess.

One of the essential elements of game theory is the classification of games, and other situations amenable to game theory, into different types. Each different type can be analysed by the correct application of game theory. The basic method of analysis is to consider the various possible choices open to the rival players and to tabulate the results of each possible choice. This analysis can most easily be achieved in simple games by forming a game matrix, though in practice, only the simplest games can be contained in matrices of a viable size. The matrix sets out the results of all possible moves by each player. It therefore shows which moves will benefit which player. In working out the solution of the game, the game theorist assumes that the players will act 'rationally'. This means that they will try to optimize their own profit (or payoff as it is called technically) without being either altruistic or vindictive. In order for a situation to be amenable to game theory analysis the players must be able to distinguish the relative merits of certain courses of action or choices and to know that a certain outcome will follow their choice.

Within this framework games fall into a number of different classes: (i) in some games there is 'perfect information'. This means that each player knows what choice the other has made as the game proceeds. Chess is a game of perfect information, but card games are usually not, as one player does not know what cards the other has. In games of imperfect information chance often plays a part. (ii) Some games are referred to as zero sum games. This means that what one player wins the other loses, so that there is no net profit or loss when the game ends. Non zero sum games are not limited in this way. (iii) Games can also be divided into negotiated or non-negotiated games, according to the presence or absence of co-operation between the players. Two players may agree to a certain choice which provides the maximum joint payoff and then, on the next move, come into conflict over the division of the payoff between them. (iv) The number of players can be two or more than two.

Games may also be classified according to the nature of the solution that is possible. Some games have a dominating strategy open to either one or both players. Some, on the other hand, have no dominating strategy for either player. A dominating strategy is one which leads to a payoff at least as good as that provided by any other move, regardless of the subsequent move made by the opposing player. A rational player will always choose a dominant strategy move. There are other situations in which one player makes his choice on the assumption that his opponent will act rationally. In this situation the result will be mutually the best in view of the fact that both players act rationally. This is called the minimax solution and occurs in games which have saddle points in the game matrix. A saddlepoint is defined as the cell in the matrix in which the value is at the same time the row maximum and column minimum, or vice versa. The example below shows a saddle point:

Table 14.4 Payoff table

	B_1	B_2	B_3
A_1	-2	2	-6
A_2	①	5	4
A_3	-1	-3	2

Values give *A*'s payoff. *B*'s payoff is reverse of *A*'s.

A can choose A_1, A_2 or A_3. He will clearly choose A_2 which is his dominating strategy, in that whichever *B* chooses, *A* will get the maximum payoff. *B* will choose B_1 as it is the best for *B*, knowing as *B* does that *A* will choose the rational move A_2 ($-1 > -5$ or -4 for *B*).

Saddle points do not necessarily have to be associated with dominating strategy, as in the last example. If, however, a saddle point exists without a dominating strategy for either player, this point is still the rational one to choose so that both players get the best on the assumption that the other is rational.

Table 14.5 Payoff table

	B_1	B_2	B_3
A_1	−2	6	−7
A_2	①	5	4
A_3	−1	−3	8

If A chooses A_2 he cannot do worse than get 1. B will choose B_1 on this assumption and will only lose 1. A_2 B_1 is the saddle point.

The mathematical solution of games without saddle points is very complex except for the simplest case of a 2×2 game matrix.

Table 14.6 Game matrix

	B_1	B_2
A_1	a	b
A_2	c	d

If there is no saddle point then $a \geqslant d > b \geqslant c$. In other words the largest values must lie in one diagonal and the smallest in the other. The optimum solution for A is given by

$$\frac{d-c}{(a+d)-(b+c)}$$

and for B

$$\frac{d-b}{(a+d)-(b+c)}$$

This formula provides a solution for a 2×2 game with no saddle point.

A geographical example illustrates the application of this method of solving a 2×2 game. The players are the farmer and the weather. The weather can be hot or cold. If it is hot the farmer gets more profit from growing a large percentage of wheat. If cold he would do better with oats. His profits can be set out in the 2×2 game matrix.

Table 14.7 Game matrix data

	Hot	Cold
Wheat	a 80	b 45
Oats	c 25	d 70

However, as the farmer does not know what the weather will be, he decides to plant that proportion of crops which will ensure that, if the worst weather occurs over a number of years, he will get the maximum profit possible under these conditions. This is the optimum solution for the farmer. It is given by

$$\frac{d-c}{(a+d)-(b+c)}.$$

The values for a, b, c and d are 80, 45, 25, 70 as set out in the game matrix. The value is

$$\frac{70-25}{150-70}=\frac{45}{80}\times 100=56.3 \text{ per cent.}$$

The farmer should plant 56.3 per cent wheat and 43.7 per cent oats.

By solving the game matrix vertically it is possible to calculate the proportion of hot and cold years that would minimize the farmer's return, i.e. using the formula

$$\frac{d-b}{(a+d)-(b+c)}.$$

The result gives 69.8 per cent hot years and 31.2 per cent cold years. It is this situation, therefore, that the farmer is considering in calculating what proportion of crops to plant.

If the situation is more complex so that the choices and possibilities give a larger game matrix then an approximation to the exact solution may be gained by repeated random trials when the choices of one of the players in the game are dictated by chance. Examples of the results of this method, using geographical data, illustrate some of the potentialities of this solution to a game theory problem. The players are again the farmer and the weather. The latter being unpredictable is considered random. Four types of weather are possible and the farmer grows five crops, each of which is influenced differently by the weather. He wishes to optimize his profits, which are set out for the different crops and conditions in the game matrix as follows:

Table 14.8 Optimization data

	1, 2 Warm Wet	3, 4 Warm Dry	5, 6 Cold Wet	7, 8 Cold Dry	(Total)
Barley	4	4	4	4	(16)
Grass	8	4	2	2	(16)
Oats	1	5	7	3	(16)
Rye	0	4	1	11	(16)
Wheat	3	5	5	3	(16)

Each weather type has an equal frequency as shown in the two random numbers allocated to each. Random numbers are used to indicate the succession of weather conditions. The profits obtained on each crop are added cumulatively. A count is kept of the number of times each crop for any one year shows a maximum accumulated profit. The results of a series of several short random runs are shown and also larger cumulative runs. The first few steps are shown:

Table 14.9 Agricultural data

	CD	WD	WW	CW	CD	Score
Barley	4	8	12	16*	20	$\frac{1}{5}$
Grass	2	6	14	16*	18	$\frac{1}{5}$
Oats	3	8	9	16*	19	$\frac{1}{5}$
Rye	11*	15*	15*	16*	27*	$4\frac{1}{5}$
Wheat	3	8	11	16*	19	$\frac{1}{5}$

After 15 steps the score was B $\frac{11}{15}$; G $6\frac{11}{15}$; O $\frac{9}{10}$; R $5\frac{11}{15}$; W $\frac{9}{10}$.

The following figures record the results of different trials, using a varying number of iterations:

Table 14.10 Agricultural data

Number of iterations		Separate trials	Cumulative to 50		
	15	15	17 →	32 →	50
Barley	$\frac{11}{15}$	0	0	$11\frac{1}{2}$	$11\frac{1}{2}$
Grass	$6\frac{11}{15}$	5	$8\frac{1}{2}$	$8\frac{1}{3}$	$8\frac{1}{3}$
Oats	$\frac{9}{10}$	9	0	0	$2\frac{1}{3}$
Rye	$5\frac{11}{15}$	0	$8\frac{1}{2}$	$10\frac{1}{3}$	$25\frac{5}{6}$
Wheat	$\frac{9}{10}$	1	0	$1\frac{5}{6}$	$2\frac{1}{6}$

The number of times the different weather types were recorded are: Warm Wet 22; Warm Dry 24; Cold Wet 14; Cold Dry 22.

The relative lack of Cold Wet years explains the poor yields from oats in particular and wheat also. The weather does not affect the barley crop, but it appears that a greater profit is obtained from a crop that does very well under some conditions and poorly under others than from one that does not vary from year to year. This same trend also emerged from another test using different profit values and giving warm dry and cold wet years greater weight. The game matrix and the result of the first 30 iterations are shown in the table:

Table 14.11 Agricultural data

	1, 2 Warm Wet	3, 4, 5 Warm Dry	6, 7, 8 Cold Wet	9, 0 Cold Dry	Results
Barley	3	4	4	4	0
Grass	6	3	5	2	0
Oats	2	1	7	6	$11\frac{1}{2}$
Sugar beet	5	8	2	1	$18\frac{1}{2}$
Wheat	4	6	3	2	0

The high yields of sugar beet and oats in the conditions that occur most frequently is the cause of the high profitability on these two crops. The results suggest that the farmer would do well to concentrate on sugar beet and oats, as one or other of these crops will do well whatever the weather. A short series of trials may lead to varying results owing to the occasional concentration of similar random numbers together. A long series of trials will approximate to the mean profits to be expected over a long period, and provide a theoretical framework to help the farmer optimize his profits. How in reality the method is applied by the farmer depends on whether or not he knows the proportions and therefore probability of each type of season. In an area of older settlement he presumably would know. In a new area he might assume equal proportions but later modify his crop mix.

One of the problems of game theory is that in all games except the very simplest the matrix of the game becomes impossibly complex owing to the very large number of possible strategies. Nevertheless game theory can show that under certain conditions optimal strategies do exist and in some real situations the choices open, as in the case of the farmer, may be fairly small. Thus game theory can be applied more readily to real life situations, such as military strategy, business problems and similar decisions, than to parlour games. In situations with a saddle point the outcome is easily determined if the game is a zero-sum type. In non zero-sum situations the theory cannot usually be applied successfully to real life situations.

Further material on game theory is provided by A. Rapoport (1966) and by S. Vajda (1963). P. R. Gould (1963) gives examples of its application in geographical situations.

(c) Locating a motorway

While from a geographical point of view it is not necessarily any more difficult to find the best location for a very large costly item or establishment than for a small one, it is obviously financially more important to find a suitable one for the costly item than for the small one. The placing of a new community hall in a village or small town may affect the hall's usefulness, and if not suitable cause inconvenience. But the suitable locating of one large nuclear power station or international airport could save more in terms of cost and effort than perhaps hundreds of small decisions. This subsection considers a few of the spatial aspects of the location of a motorway in south-west England. Other considerations could readily be substituted, as appropriate, to make the approach suitable for any location problem.

For the sake of this example it has been decided to allocate enough funds to construct approximately 80 miles of motorway (limited access highway) in the area west of a line joining Bristol and Weymouth in figure 14.6. The problem is to locate it where it will be most useful. A great deal of information is available already on maps and in other forms about terrain conditions, e.g. Dartmoor upland, rivers difficult to cross, existing roads, built-up areas and so on. There is also some information about traffic flows. It is assumed that the motorway will be of use at least for several decades and that traffic will continue to grow in volume for some time. Special features of the region studied are its peninsular nature and the fact that its traffic builds up appreciably in the summer period through holiday visitors from other regions of Britain.

It must be decided firstly whether the motorway will be continuous for 80 miles or whether it will form two or more discrete stretches. The second question is to decide the course of the motorway. The motorway locating problem was given to geography students at Nottingham University. Figure 14.6*c* lower map shows the choices of 13 students. Two out of 13 split the 80 miles. In discussion of the problem, the four following approaches were among several suggested:

(1) Base the decision on information collected from traffic censuses at appropriate points. The busiest traffic flows could be used to decide the best course. This approach might be called empirical or pragmatic.

(2) Base the decision on a theoretical model that could predict the heaviest traffic flows. For example, 'desire lines' in figure 14.6 (middle map) join traffic generating centres. Only 11 have been used, three (3, 6 and 9) for larger towns, the remainder for areas (e.g. 1 for west Cornwall). A greater number of traffic generating centres could be used, and the thickness of the desire lines made proportional in width to the volume of traffic expected according to the gravity model (see chapter 11). Such desire lines have been widely used to give a rough idea of journey frequencies. They were used as early as the 1830s in a map of pre-railway Ireland available in the British Museum Library, and more recently, for example in *Traffic in Towns* (1963), and in a Soviet publication, *Ekonomicheskiye svyazi i transport* (1963). If they do nothing else, the desire lines show where it is superfluous to hold censuses of traffic.

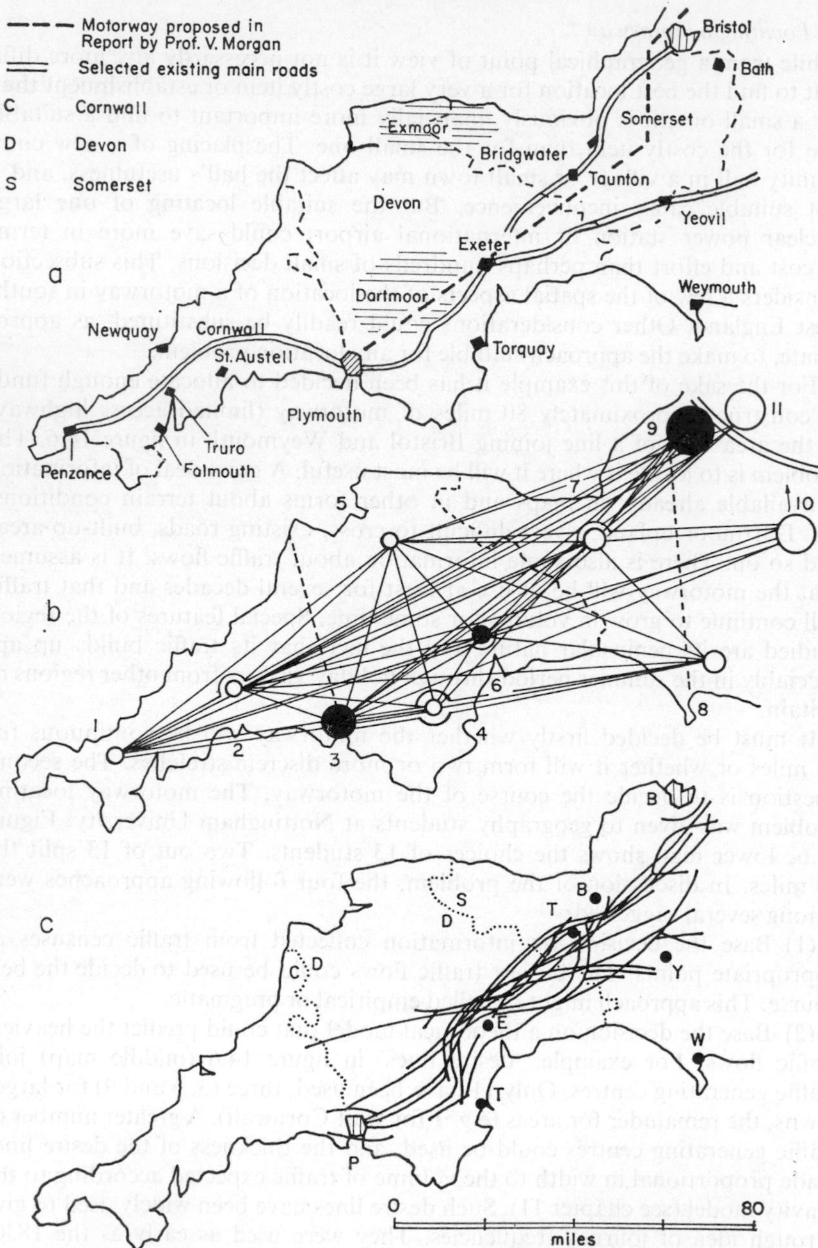

Figure 14.6 The problem of finding the best location for 80 miles of motorway in Southwest England. (*a*) Selected features of the area and route proposed by Professor V. Morgan. (*b*) Desire lines for journeys between 11 selected places or areas. (*c*) Routes suggested independently by 13 students in tutorial exercises, Geography Department, Nottingham University.

The problem is also related to the accumulation of journeys and the advantage of centre over periphery discussed in chapter 11.

(3) A comparative approach might also be a possibility, though perhaps not very fruitful in this example. Where else in the world are there motorways running into a peninsula? Two obvious examples, both of larger areas than southwest England, are Florida, U.S.A., and Peninsular Italy (Autostrada del Sole). As stressed in chapter 5, comparison of a few things is difficult in geography. Some rotation, dilation and topological transformation could be helpful.

(4) A consensus of informed opinion might be a valuable short cut to the best location of the motorway. A team of preferably disinterested traffic engineers, economists, geographers and others might independently give their views and the results could be superimposed. If there was general agreement, it might be sufficient to take the course recommended. If not, the problem would be a more complex one. The agreement of the 13 students in the lower map is close.

The course of a motorway proposed by Prof. V. Morgan of Swansea University in a report for the Roads Campaign Council (*Ford Bulletin*, Nov. 5, 1965, p. 7) is shown on figure 14.6*a*.

(*d*) *Improving local government and electoral districts*

Most countries of the world have two if not more levels of political administrative units (regional and local government). In some countries (e.g. Brazil, Italy, U.S.S.R.) the units are constantly under review for change. In other countries (e.g. the U.K., Sweden) change has tended to come during certain periods of reform. In certain countries (e.g. Switzerland, U.S.A.) change is very difficult to introduce at all. Since underlying distributions of population, economic activity and transportation media are now changing rapidly in many parts of the world, local government areas may quickly become unsatisfactory. There is, however, surprisingly little guide as to the spatial form a network of local government units might best have to fit the needs of a given distribution of population and other features.

The following are among the many questions the geographer might consider about a given region:

(1) How many people should each subdivision ideally have?
(2) What territorial extent should it ideally have?
(3) How important is compactness?

The first two questions will be briefly considered.

(1) Theoretically it is convenient for administrative units at the same level in a hierarchy to have roughly the same number of inhabitants, or at least, not to fall below a certain minimum. This is for the provision of services of appropriate scales. In figure 14.7, cases 1 and 3 show systems of units in which the number of inhabitants is the same in each unit.

(2) Theoretically it is convenient for administrative units at the same level in a hierarchy to be roughly equal in area, or at least not to exceed a certain maximum extent. This is for the provision of services over the area (e.g. fire

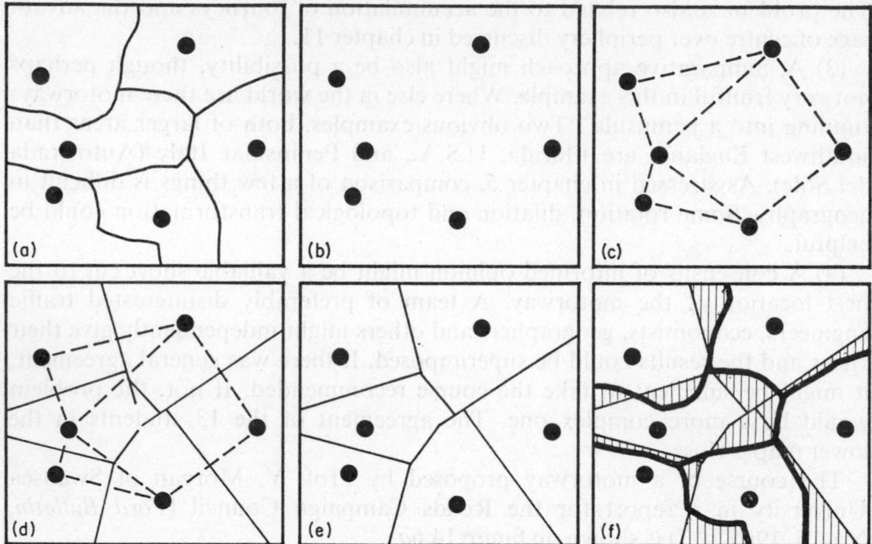

Figure 14.7 A device for assessing how closely a system of administrative or comparable units fits the system of centres it is serving.

service, police). In figure 14.7, case 2 shows a system of units in which the area of each unit is equal.

(3) Theoretically it is desirable for areal units to be arranged about the administrative centres in such a way that any place falls under the administration of the centre nearest to it. Figure 14.8 shows how approximation to

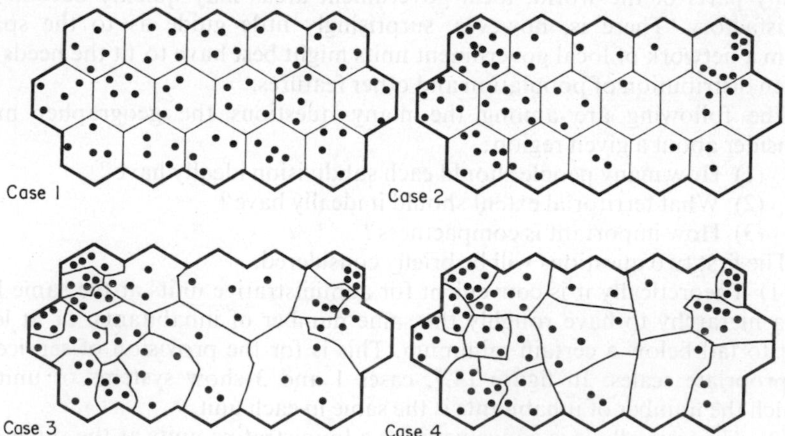

Figure 14.8 Four cases of fitting a system of units to a distribution of people.

such an ideal situation is actually achieved in an area with six centres. The shaded portions, arrived at by the steps shown in *a–e*, show the anomalous areas.

The ideal situation, in which units are equal in area *and* have the same number of inhabitants, is illustrated in figure 14.7, case 1. This is highly unlikely in reality since any sizable region usually has some clusters of population. Given an uneven distribution of population, the political units often have a compromise fit, illustrated in figure 14.7, case 4, in which units that are smaller in area tend to have more people than large units. Examples of countries in which this tends to occur are Mexico, Venezuela, Sweden (see figure 14.9*a*) and the U.S.S.R. In some countries (e.g. France, Spain, Poland),

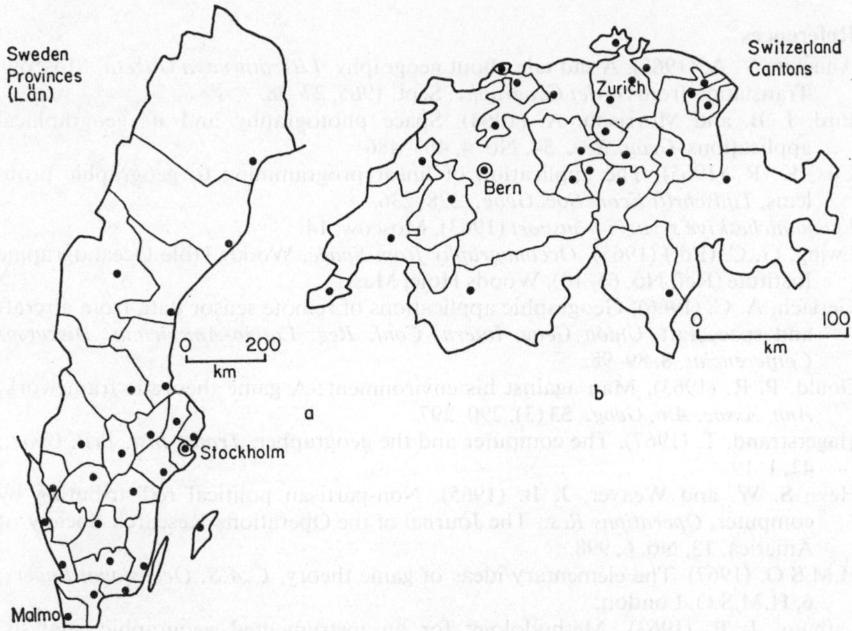

Figure 14.9 A comparison of the major civil divisions of (*a*) Sweden and (*b*) Switzerland.

emphasis tends to be on preserving equal areas. In extreme cases (counties of Iowa, Kansas), this has clearly been a matter of deliberate policy when the units were mapped out prior to extensive settlement. In certain countries, notably Switzerland (figure 14.9*b*) and Great Britain, the cantons and counties respectively still reflect complex feudal land ownership arrangements and population distributions of several centuries ago.

Clearly area and number of inhabitants are only two of the aspects of local government that a geographer could study with a view to contributing to a theory of political units. This subsection merely draws attention to the kind of questions a geographer might help to resolve.

The reform of electoral districts involves a somewhat different set of problems from that for local government. Often it is desirable to create districts that contain as near as possible the same number of voters. In areas where population is moving in or out in large numbers, this is very difficult. Moreover, it is desirable, also, to have a reasonable mix of voters in terms of income, 'race' where relevant, and so on. The term gerrymandering has long been used in the U.S.A. to refer to the 'unfair' apportionment of voters to districts. R. C. Silva (1965) gives an interesting general view of the problem in 'Reapportionment and Redistricting'. S. W. Hess and J. B. Weaver (1965) discuss a method of obtaining a balanced mix of voters in electoral districts.

References

Anuchin, V. A. (1965). A sad tale about geography, *Literaturnaya Gazeta*, Moscow. Translation from *Soviet Geography*, Sept. 1965, 27–28.

Bird, J. B. and Morrison, A. (1964). Space photography and its geographical applications, *Geog. Rev.*, **54,** No. 4, 463–486.

Cox, K. R. (1965). The application of linear programming to geographic problems, *Tijdschrift Econ. Soc. Geog.*, 228–236.

Ekonomicheskiye svyazi i transport (1963), Moscow, 14.

Ewing, G. C. (Ed.) (1965). *Oceanography from Space*, Woods Hole Oceanographic Institute (Ref. No. 65–10), Woods Hole, Mass.

Gerlach, A. C. (1966), Geographic applications of remote sensor data from aircraft and spacecraft, *Unión Geog. Intern. Conf. Reg. Latino-Americana, Discursos Conferencias*, **8,** 89–98.

Gould, P. R. (1963). Man against his environment: A game theoretic framework, *Ann. Assoc. Am. Geog.*, **53** (3), 290–297.

Hagerstrand, T. (1967). The computer and the geographer, *Trans. Inst. Brit. Geog.*, **42,** 1–19.

Hess, S. W. and Weaver, J. B. (1965). Non-partisan political redistributing by computer, *Operations Res.*, The Journal of the Operations Research Society of America, **13,** No. 6, 998.

H.M.S.O. (1967). The elementary ideas of game theory, *C.A.S. Occasional Papers*, **6,** H.M.S.O. London.

Latham, J. P. (1963). Methodology for an instrumented geographic analysis, *Ann. Assoc. Am. Geog.*, **53,** No. 2, 194–209.

Leontief, W. W. (1965). The structure of the U.S. economy, *Sci. Am.*, **212,** No. 4.

Meyer, J. R. (1967). *Regional Economics*: A Survey, in *Surveys of Economic Theory*, **II,** Survey VIII, 240–271. Macmillan, London: St. Martin's Press, New York.

Morrison, A., Bird, J. B. and Lundgren, M. (1965). *Atlas of Interpreted Photography of Southern Africa*, From Tiros Satellites I to VIII, Prepared by McGill University, Montreal for National Aeronautics and Space Administration.

National Resources Committee (1935). *Regional Factors in National Planning and Development*, U.S. Govt. Printing Office, Washington.

Rapoport, A. (1966). *Two Person Game Theory*, University of Michigan Press, Ann Arbor.

Rigby Childs, D. (1965). A new approach to regional planning, *Listener*, **73,** No. 1884, 667–669.

Silva, R. C. (1965). Reapportionment and redistricting, *Sci. Am.*, **213**, No. 5, 20–27.

Simpson, R. B. (1966). Radar, a geographic tool, *Ann. Assoc. Am. Geog.*, **56**, No. 1, 80–96.

Smith, D. M. (1966). A theoretical framework for geographical studies of industrial locations, *Econ. Geog.* **42** (2), 95–113.

Traffic in Towns (1963). H.M.S.O., London, 92.

Vajda, S. (1963). *An Introduction to Linear Programming and the Theory of Games*, Methuen, London.

Young, O. R. (1968). *Systems of Political Science*, Prentice-Hall, New Jersey.

15

Organization

'Geography, coming late in the paradigm race, has the compensating advantage that it can study at leisure the "take-off" paradigms of other sciences. There is good reason to think that those subjects which have modelled their forms on mathematics and physics have climbed considerably more rapidly than those which have attempted to build internal or ideographic structures.'

R. J. Chorley and P. Haggett (1967)

15.1 An approach to geomorphology

Before the approaches to a field of study can be considered it is necessary to have a clear idea of the aims of the study. In geomorphology the main aim is to understand the shape of the earth's surface. In order to achieve this, the character of the surface must be known and described, an operation that may be aided by classification of the first main type. It is also necessary to understand the processes that have been and are modifying the landforms that have been described. As landscape is a developing and changing entity it is also necessary to appreciate the changes that have taken place in the past, often under different controlling factors, to gain a full understanding of the present landscape. There are, therefore, many interlinking aspects of the study that must be woven together to provide a satisfying approach to the subject.

At its boundaries geomorphology impinges on other sciences, such as geology, geophysics, pedology, physics, chemistry and several others. One of the problems of the study of geomorphology is to know when the study is getting outside the proper field into another discipline. For example, to understand glacial landforms it is necessary to know something about the behaviour of ice, and this study is the proper domain of the physicist. However geomorphologists must make use of the physicists findings if their results are to be of maximum value.

Geomorphology can be approached by a number of paths, none of which alone provides a fully satisfactory study. These approaches are indicated on the diagram (figure 15.1) which is in the form of a circulation or network. The separate branches must be linked to arrive at the final aim. This is represented by regional geomorphology on the diagram, as this involves all the methods of approach and has an areal aspect that is fundamental to geography of which geomorphology is a part. However there is not yet

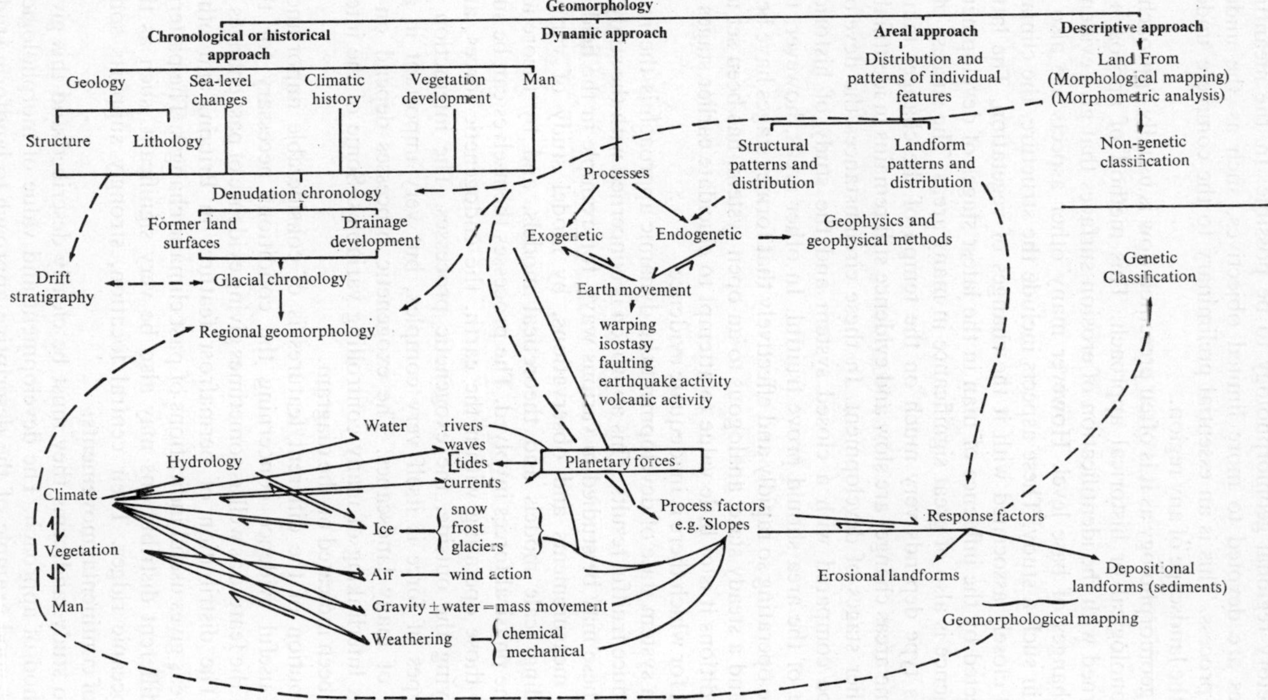

Figure 15.1 A diagrammatic plan of various possible approaches in the study of geomorphology. The arrows indicate possible links. Where this is an interaction between two elements the length of the arrows indicates the relative importance of each direction.

sufficient known concerning some of the other methods of approach for a fully satisfactory regional geomorphology to be possible. In the meantime many studies are devoted to more limited objectives, such as the understanding of process. This is an essential preliminary to the complete understanding of the landscape of any region.

Regional geomorphology as it is often presented now is usually approached via the chronological or historical approach. This method of approach is often concerned with the identification of erosion surfaces that give evidence of former changes of base level. However many other aspects are also of importance in such a study. These aspects include the structure, the climatic history, and closely associated with it the changes of vegetation. The latter are also affected by the influence of man in the latest stages of development. Glacial influence is also of great significance in many areas. The success of a study of this type depends very much on the tempo of the changes taking place. In some areas changes are slow and evidence still remains in identifiable form of earlier stages of development. In these circumstances the development may be compared with a closed system and the study of historical development of the area should prove fruitful. In other areas, however, the processes are operating so rapidly and effectively that former stages have been obliterated and a steady state, analogous to an open system, has been set up. In these conditions it is of little value to attempt to elucidate earlier stages of development for which there is inadequate evidence.

In an open system state of development the dynamic approach is the most likely to produce fruitful results. This approach is concerned with the study of processes. These may be studied in various ways, for example in the field by quantitative measurements and observations, by model study of various types, including scale models and theoretical models, and by theoretical analysis of the physical forces involved. The processes themselves can be subdivided into those operating within the earth, the endogenetic forces, and those modifying the outside, the exogenetic processes. The interaction of these two types of force is itself very complex, but very important in an appreciation of many landscapes. The exogenetic processes depend on a very complex interlinking of many controlling variables. Some of the interactions have been indicated on the diagram.

The distribution of the different features is of considerable importance. It can give useful evidence concerning the conditions necessary for the formation of the feature as well as sometimes giving evidence of past changes of conditions. The distribution of permafrost features in Britain, and other temperate areas gives useful indications of past climatic changes. The patterns formed by different distributions may also be very significant, such as the pattern of oceanic ridges. Their central location, strongly supports some recent ideas of continental movements.

In order to study landforms they must be clearly described and this gives another method of approach. The development and value of morphological mapping is a good example of the descriptive approach to landform study, which is only a preliminary step, but nevertheless a valuable one, in the final understanding of the features concerned. It is particularly applicable to the study of slopes which are difficult to describe quantitatively on an areal basis.

15.2 An operational framework for political geography

The last word has certainly not been said on the content, aims and organization of political geography, although there have been several attempts to define it (e.g. R. Hartshorne (1954), S. B. Jones (1954), W. A. D. Jackson (1958, 1964), H. and M. Sprout (1965)). The present section is no more than another attempt to outline the content of political geography. It is included as an example of the way in which some organization and a conceptual framework are necessary before work can have aims and direction. The present scheme is a framework into which work already done can be placed and through which new tasks can be detected.

Politics is one of three useful general terms widely accepted and employed to cover many (but not all) human activities. The other two are economics and sociology. The three overlap (intersect) to a considerable extent. Politics is concerned among other things with the study of government (theory, institutions, machinery, laws, representation), with conflict (non-military and military) between powers, decision making, and the system of politico-administrative units that cover most of the earth's land surface. Perhaps the critical and most outstanding role of politicians is that of forming policy about the spending of public funds (e.g. taxes, rates). Geography deals among other things with variations and relationships on the earth's surface. Political geography is the intersection of politics and geography.

The present outline assumes that political geography is concerned with three basic ingredients, space, people and institutions. Naturally time is essential also before a political system can function. *Space* at present is the vicinity of the earth's surface, appropriate parts of the lithosphere, hydrosphere and atmosphere. *People* are some 3×10^9 in number. Almost every member of the world's total population is a resident of some political unit. *Institutions* are created for purposes of organizing political activities.

For the purposes of public administration, the earth's land surface (almost all), together with parts of the sea surface, is divided and subdivided into compartments, politico-administrative units (these will be referred to hereafter as political units).

A political unit will first be considered in isolation. The unit will generally have the following features (ingredients). These are illustrated in figure 15.2*a*.

As a system the political unit functions very roughly over time as sketched in figure 15.2*b*.

1. *Space*. On a map this will appear as a 2-dimensional surface. In reality it is a 3-dimensional space.
2. *Boundary*. On a map this will appear as a line. In reality it is a 2-dimensional surface, open at the bottom towards the centre of the earth and the top towards outer space.
3. *People*. These will be residents or citizens of the unit. They will often

be associated with the unit by some document (identity card, passport). In general they will make some contribution (financial, work) to the unit as a whole.

4. *Capital.* Almost always a unit has only one capital. This is the seat of government and it often contains also various institutions connected with the government.

5. *Communications.* Ideally every citizen will be accessible to the government of the unit through one or more channels. Communications

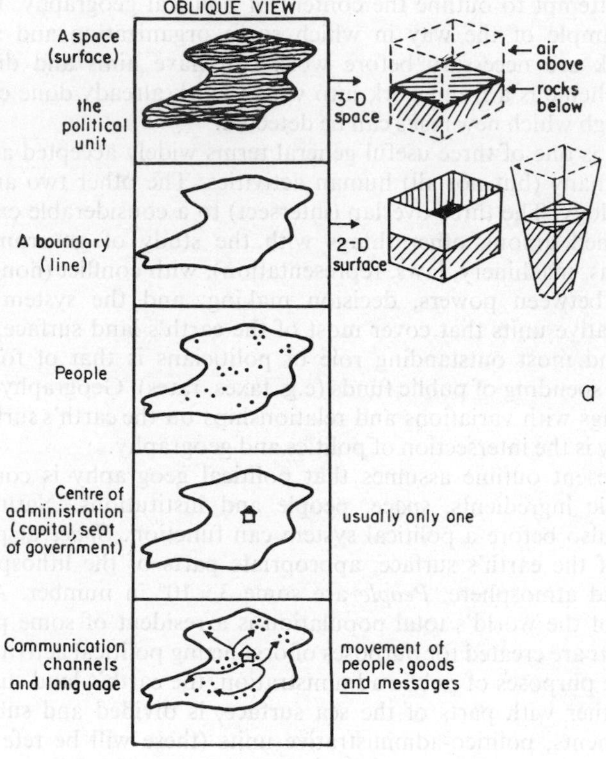

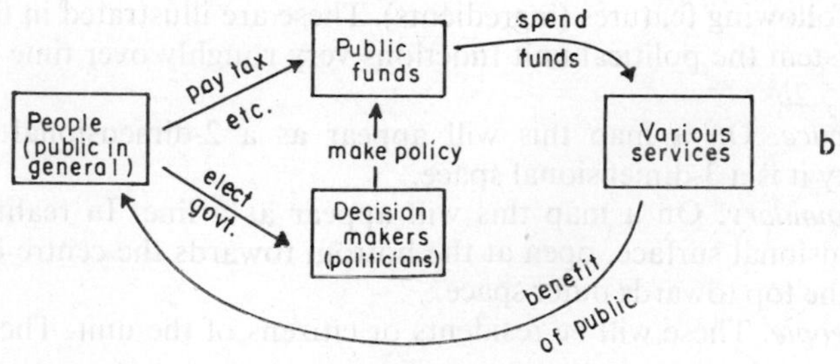

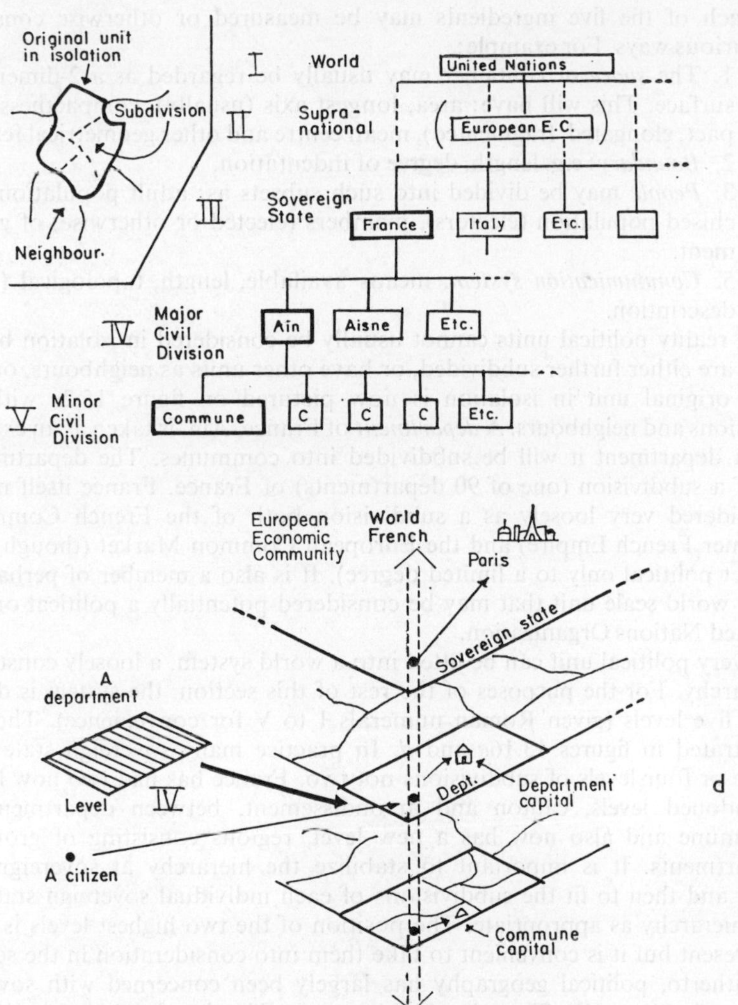

Figure 15.2 Political geography. (*a*) Ingredients of a unit in isolation. (*b*) The organization of a political unit. (*c*) World politico–administrative hierarchy. (*d*) The position of an individual person in the world politico–administrative system.

here are thought of in the broadest sense of movement not only of goods and people but also of information. For the purposes of the diagrams that follow, however, roads (paths) will be used to represent typical communications.

Each of the five ingredients may be measured or otherwise considered in various ways. For example:

1. The *space* or container may usually be regarded as a 2-dimensional surface. This will have: area, longest axis (usually), compactness (compact, elongated, fragmented), mean centre and other geometrical features.
2. *Boundary*: e.g. length, degree of indentation.
3. *People* may be divided into such subsets as: adult population, franchised population (electors), members (elected or otherwise) of government.
5. *Communication system*, means available, length, topological (graph) description.

In reality political units cannot usually be considered in isolation because they are either further subdivided, or have other units as neighbours, or both. The original unit in isolation is now pictured on figure 15.2*c* with subdivisions and neighbours. A *department* of France, Ain, is taken as an example. As a department it will be subdivided into communes. The department is itself a subdivision (one of 90 departments) of France. France itself may be considered very loosely as a subdivision both of the French Community (former French Empire) and the European Common Market (though this is as yet political only to a limited degree). It is also a member of perhaps the only world scale unit that may be considered potentially a political one, the United Nations Organization.

Every political unit can be fitted into a world system, a loosely constructed hierarchy. For the purposes of the rest of this section, the system is divided into five levels (given Roman numerals I to V for convenience). These are illustrated in figures 13.16*c* and *d*. In practice many sovereign states have three or four levels of subdivisions not two. France has had two now largely abandoned levels, canton and arrondissement, between department and commune and also now has a new level, regions consisting of groups of departments. It is important to stabilize the hierarchy at sovereign state level and then to fit the subdivisions of each individual sovereign state into the hierarchy as appropriate. The position of the two highest levels is vague at present but it is convenient to take them into consideration in the scheme.

Hitherto, political geography has largely been concerned with sovereign state units (level III). There has been some justification for this, since it is the most vital for world affairs, the level at which most of the momentous decisions are made and perhaps the level on which most people feel greatest allegiance politically. Since there are only about 120 sizeable sovereign states and since these vary enormously in *size* (while having the same *status*) it is difficult to make generalizations about them. At level IV there are at least several thousand units and at level V, perhaps a million. Thus France (level III) for example, has 90 departments (level IV) and about 35,000 communes (level V).

The following points about the world hierarchy of politico-administrative units should be borne in mind:

At any one of the three levels III, IV and V there is very little overlapping of units at *that* level. Similarly, at any of the three levels in question there is rarely any residual space between units. Usually there is an exact fit. Units are therefore disjoint area or patch sets; only their boundaries intersect. In this respect political geography differs fundamentally from and is simpler than urban and economic geography, in which urban service areas, market areas and so on may overlap extensively or be separated by large gaps. In general the power of a government extends exactly to the boundary of its political unit. At sovereign state level, however, this statement should perhaps be reconsidered, since power may diminish towards the periphery of the unit and at the same time extend beyond it in some form.

It may be assumed that in general a person (citizen) resides in and pays taxes in *one unit and only one unit* at any level in the politico-administrative hierarchy. This important assumption is illustrated in figure 15.2*d*.

Now that the original political unit in isolation has been considered as part of a system, the original list of 5 ingredients must be extended. The following may be added: 6 Status (level in hierarchy), 7 Subdivisions, 8 Neighbours, and others.

Up to a point, each of the eight features listed so far has an equivalent in political science. Thus for example the capital of the geographer is the seat of government to the political scientist. However the features are so obvious that there should be no misunderstanding between political scientists and geographers as to the general meaning of the terms that refer to them. The same features also appear in economics (and economic geography) and sociology (and social geography). It is important to appreciate that different branches of human geography often share the same spaces and objects in the real world but look at them in different ways.

The eight features of a political unit are suggested in the simple map in figure 15.3*a*. Other features may be added.

Four further important features of the politico-administrative hierarchy of the world must now be mentioned.

(1) Often the system of units in a given area is not suitable for some particular project. In figure 15.3*d–f*, lower part, four situations are illustrated. In situation A, only a small subset of the total population of the political unit is involved in a particular project. The small area could for example be a blight area being rehabilitated with public funds from the whole unit. In situation B, exactly the whole set of people in the unit is involved in a project. The project could be modernization of schools or fire services in the unit. In situation C, part of the population in the political unit is involved in a project, together with part of the population of one neighbouring unit (or several): a project such as the Tennessee Valley Authority, occupying a drainage basin that covers parts of several states in the U.S.A. In situation D, funds from one unit may be spent entirely on a project completely outside the unit. This seems most likely to happen at sovereign state level (III). Thus one country might spend its own public funds to construct military bases on the territory of other countries.

Only in situation B of the above examples does the existing political

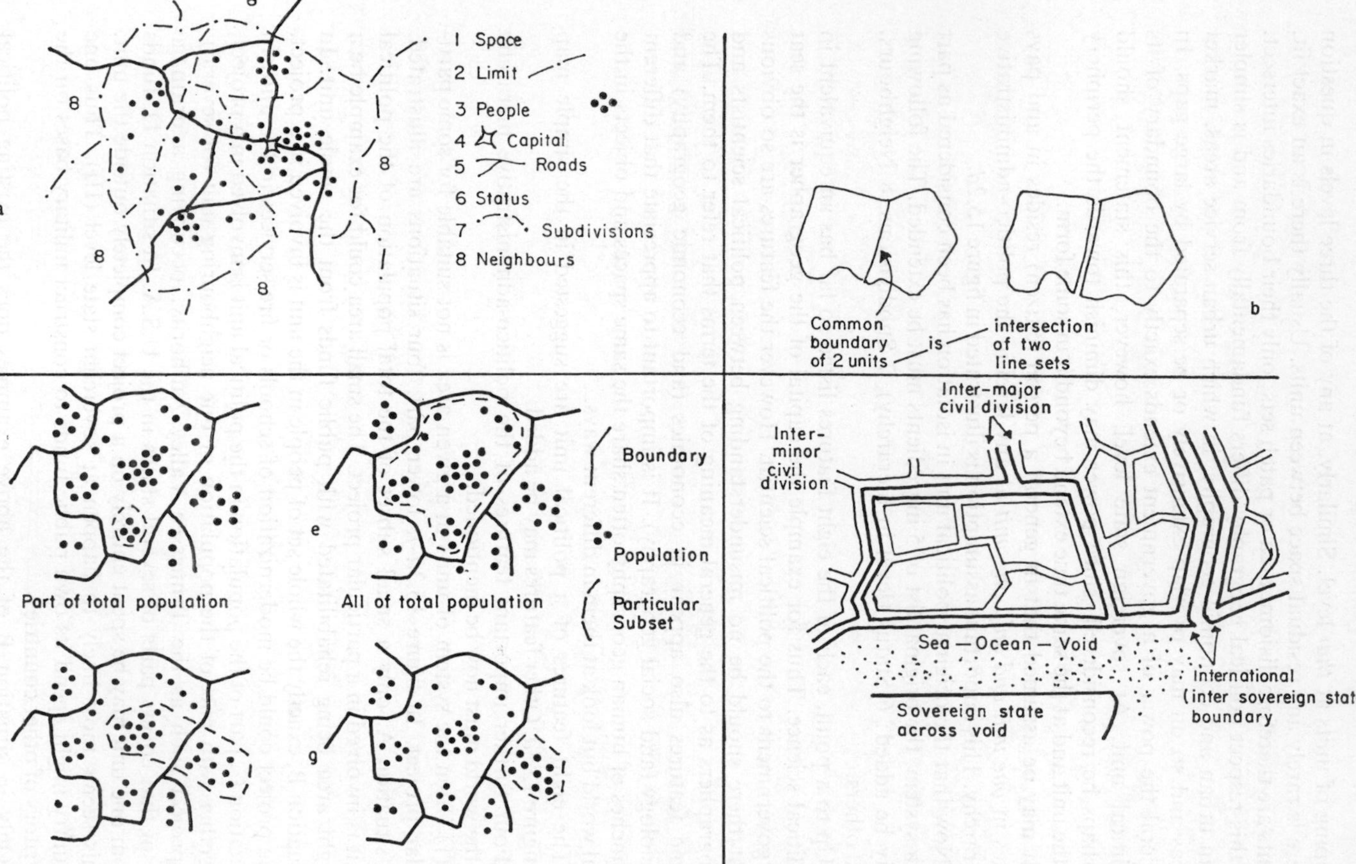

Figure 15.3 Features of political units. (*a*) Eight basic ingredients. (*b* and *c*) Boundaries at different levels. (*d–g*) Lack of

unit serve exactly the requirements of a particular kind or phase of public spending. It is important to bear in mind, therefore, that political geography is complicated by the fact that the spending of public funds may be carried out in areas that are only parts of units or in areas that cut across several units.

(2) When all or several of the features of a system of political units are put on a map, considerable visual confusion may occur. This question was discussed in section 13.7*b* and illustrated in figure 13.10*c–f*.

(3) A further complication of the politico-administrative hierarchy is almost invariably overlooked on maps and therefore often forgotten, namely that unless a given boundary faces a void (the ocean, perhaps an ice cap, a desert) it is really two boundaries. It is the intersection of the boundary lines of the units on each side of it. Moreover, unless it is a boundary between a pair of units at the lowest level in the political administrative hierarchy, it will really be four boundaries, or even six. Thus any particular stretch of the international boundary between France and Italy separates France from Italy, some particular French *department* from an Italian *provincia*, and some particular French *commune* from its adjoining Italian *commune*. An imaginary example is illustrated in figure 15.3*b* and *c*. More attention might be given by political geographers to the *order* of a boundary. Stream systems and road systems among others, are classified in this way. In general the barrier effect of a given boundary is as strong as the barrier established by the highest level of units on either side. Thus in the French–Italian example, regulations for movement between the two sovereign states override regulations (if any) for movement between units at major and minor civil division level. It is important to note that political boundaries may be imaginary lines rather than material barriers on the earth's surface.

(4) Sovereign states and major civil divisions (e.g. U.S. states) are often used by geographers and economists as the basis for mapping distributions and for making correlations. However unless the network is reasonably uniform at the level being used, the system is very unsatisfactory. Moreover it is important not to confuse politico-geographical features of the system with other information that is collected in the units of the system for convenience. Four possible situations are shown in figure 13.17 *d–g*.

The following programme is suggested for the study of political geography. It is designed to provide a place for any work done on the subject and at the same time to point out the major gaps and the lines that might usefully be developed. Three kinds of investigation are suggested as the *first stage*. The first stage covers the spatial, geometrical aspects of political geography. The second stage brings in human ingredients of the political scene, the decision makers of governments and the electors. How do these people view the system of political units in which they find themselves, how do they use the system and why do they find it desirable to change the spatial layout from time to time?

The following three lines of work are suggested for the first stage:

Descriptive material must be assembled and made accessible about such

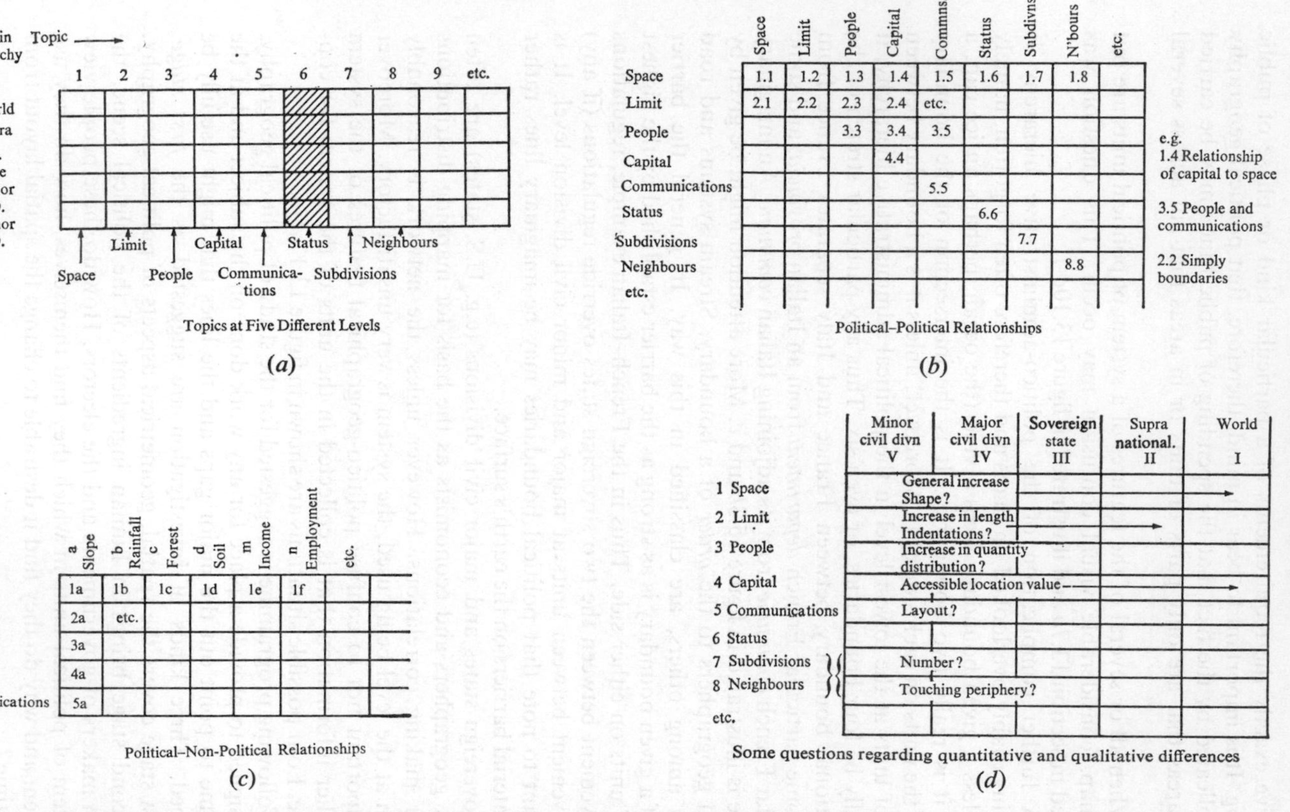

Figure 15.4 The organization of work in political geography. (*a*) Relations between levels and individual topics or ingredients. (*b*) Relationships between political geographical ingredients (*within* political geography). (*c*) Relationships between political geographical and other geographical topics or ingredients (*between* political geography and other branches). (*d*) Worksheet for checking which systematic aspects remain the same and which change at different levels in the hierarchy

single features of political units as comparative size, shape, communication networks, number of subdivisions and so on. Figure 15.4*a* shows how each topic can be considered at each of five levels.

Relationships between these should be studied. *Pairwise* and *multiple* correlation techniques should be used. These should be applied both to political–political relationships, as in figure 15.4*b* (e.g. size and compactness of a political unit) and political–non-political relationships, as in figure 15.4*c* (e.g. compactness and slope). Comparative methods should be used to deal with the relatively few but large and important units at sovereign state level. Correlation techniques, using sampling where appropriate, should be used for the much larger number of small units at lower levels. From these *inductive* studies, some generalizations should be possible.

Models for political units should be developed. These would consider the theoretically desirable deployment of people, size of units, communications, and so on in a political system. The findings of this *deductive* approach could then be compared with those of the inductive approach. Some of the questions that might be asked are discussed in chapter 14.4*d*.

The material gathered and conclusions that emerge from the work suggested above should somehow be integrated and used to trace the ways in which individual political units and whole systems of units change through time.

In the second stage of the study, the following kinds of question might usefully be asked, once something had been established about the actual spatial layout of the system.

How do the politicians view the system in which they operate? What are their viewpoints? Do the viewpoints of decision makers vary greatly?

How do the people who elect the politicians (the key decision makers), themselves see the people they elect and the system in which they live?

How is the spending of public funds actually distributed (or deployed) spatially? Firstly, what shares do different branches (e.g. education, roads, health services) receive? Secondly, how is spending distributed territorially? Evenly according to people, or unevenly? To what extent are funds used for services (e.g. police, fire services), to underpin private activities (e.g. to attract industrialists), on defence spending (largely sovereign state level), on long term activities (e.g. research)?

To what extent does politics in general, and do individual politicians in particular, stay the same at different levels in the politico-administrative hierarchy, and to what extent do they differ? An initial study of the problems could be made with the help of the diagram in figure 15.4*d*.

There are many possible approaches to political geography or to important parts of it. Some of these are now noted briefly. The approaches will be divided into two broad kinds, conceptual and applied. It must be appreciated that the approaches are not considered to be mutually exclusive.

Conceptual:

1. *Ideal — model.* Plato (The Republic) and Thomas More (Utopia), were largely concerned about non-spatial aspects of their ideal political units. A non-spatial approach has generally been characteristic of

thinkers who have conceived ideal political, social and economic systems.

2. *Political approach*, the study of the subject by political scientists. This approach often omits or gives slight attention to spatial aspects of politics.

3. *Psychological approach.* This focuses on the way in which individuals see the world in which they live (e.g. W. Kirk, 1963). There is a growing interest currently in the U.S.A. in the way decision makers view the real world.

4. *Historical — evolutionary.* Obviously this approach emphasizes changes of political systems and units over time. G. Vico in the 18th century suggested three stages in the life of a political system. O. Spengler and A. Toynbee have applied evolutionary concepts to whole cultures rather than to individual political units. W. W. Rostow (1960) has recently proposed stages of economic growth in countries. These are of considerable relevance to political geography at least at sovereign state level.

5. *Geographical–spatial.* There are many studies by geographers, both regional and systematic. Regional studies focus on the micro-level (say V–IV), the meso-level (say IV–III–II) and the macro-level (world affairs at III–I). H. Mackinder (1904) some decades ago, S. Cohen (1964) lately, are two exponents.

6. *Organismic.* A biological approach seeking analogies between organisms and political units. German political scientists brought this approach into disrepute in the interwar period. One exponent was K. Haushofer.

7. *Physics.* An approach through analogies from physics. Some of these were discussed in chapter 11.

8. *General systems approach.* General systems (see in particular various Yearbooks) offers both a vocabulary and concepts. Some of the terms and concepts have been applied to political science in papers in General Systems Yearbooks.

Applied:

1. From research in political geography, can patterns be discerned and useful generalizations and recommendations be offered to practising politicians to help them in decision making, and to reformers of political units (e.g. local government reform) to help them to reshape the spatial arrangement of systems?

2. Can systems of electoral divisions be made to eliminate gerrymandering (deliberate or accidental) so that election results are not biased in favour of some political party or group at the expense of others?

3. What light can political geography throw on conflicts (e.g. R. J. Rummel, 1963).

4. Can game theory and operational gaming be applied in the teaching of political geography and the training of politicians? (e.g. A. Rapoport, 1966).

15.3 Regional geography

Regional geography has lately been attacked both for the descriptive, disconnected, mechanical, inventory nature of much of its content, and for the dubious nature of the limits of many regions used. Some geographers have tried to put more purpose into it by stressing themes (e.g. poverty in South Italy, oil in Venezuela, voting patterns in Iowa). Others have condemned it completely, or have simply ignored it; some such people, perhaps, have never been fortunate enough to become well acquainted with a region anyway. The main purpose of this section is to suggest approaches to regional geography for those readers who are interested in it. Many of the techniques and ideas in preceding chapters will be referred to.

If the whole of the earth's shell is under consideration, then a single world region does not need defining. Any other region, however, will only be a subset of the whole shell. Whether it is called a region, area, space, geomer or some other label does not affect the subset itself. The question of how to arrive at 'acceptable' regions has however caused considerable discussion and argument in geography. A region may be very closely defined on the basis of a single criterion, such as a watershed, an administrative boundary, land and sea, or a contour. In places two or even several criteria coincide. Often, however, a region is a 'fuzzy' set.

Regions need not be delimited by lines on a continuous surface. They can be built up from mosaics of smaller regions into relatively homogeneous bigger ones. There is a very large number of possible combinations even of 100 small regions: 2^{100} or about 10^{32} (see chapter 2).

The problem of defining regions will not be further discussed here. Reference should be made to J. R. McDonald (1966). Once a space or region has been chosen by the geographer, what is he to do with it? The rest of this section sets out some of the steps that might be followed. Points will be exemplified by the U.S.S.R. For simplicity, the region will be considered in two dimensions rather than three.

Is the region relatively *open or closed* to outside influences? In general, the smaller the region in question, the more exposed it is likely to be, if only because the length of its boundary is roughly related to the square root of its area. In the U.S.S.R. there is relatively little movement of people and trade across the limits of the region compared with within it.

Subdivisions. The region must be subdivided for the collection of information. Though the subdivision is no longer apparent in many maps (e.g. most atlas maps), some basic grid, whether regular, or based on irregular administrative divisions, will usually have been employed at some stage for collecting data. Even data distributed on dot locations are finite, though sometimes very large, in number. In other words, some breakdown of the region under consideration is almost essential before its regional geography can be studied quantitatively.

Variables. The spatial distribution of the objects in the region, whether they belong to the lithosphere, hydrosphere, atmosphere or biosphere, may be measured in many ways. The measurements are often referred to as variables, but also as parameters and as spatial distributions. In considering the nature of variables, the following points should be borne in mind:

(1) Some variables measure a whole 'population', some only a subset of a whole population. Thus all arable land may be under consideration, or only arable land under wheat or rye. Wheat may further be subdivided into spring wheat and winter wheat. There are many rather loose hierarchies of variables.

(2) Variables may be expressed as absolute quantities or as ratios. Thus total number of people may be considered, or persons per unit of area. Total wheat area may be considered, or wheat area as a percentage of total arable area or even of total land area. Theoretically, any variable can be expressed as a ratio of any other (e.g. motor vehicles per 1000 people, motor vehicles per 10 miles of road, motor vehicles per 1000 acres of wheat).

(3) Variables behave differently in their distribution over area. Some show a continuous trend in some direction (e.g. cold increases towards the northeast of the U.S.S.R.), some show gradual fluctuations (e.g. precipitation in the U.S.S.R.) and some show rapid changes over area (e.g. distribution of population in many areas). The *appearance* of spatial variation is affected by the fineness of the grid from which information is collected over an area. The finer the module, the more varied the distribution may appear. The actual trends can be studied with the help of trend surface analysis (see chapter 9) while it has also been suggested by W. Tobler (1967) that spectral analysis could be adapted to consider changes in spatial distribution through time.

(4) Which variables? Once the nature and drawbacks of different kinds of variables have been given careful consideration, the question of which to use for a particular regional study must be decided. Naturally, of course, there may be serious limits to the choice on account of lack of availability of data. Figure 15.5 shows a list of 19 items (two spaces have been left vacant either side of population, No. 11) that are familiar in regional geography. It should be noted firstly that each could be greatly subdivided or otherwise elaborated (e.g. vegetation into tree, shrub and herbaceous, tree into coniferous, broadleaf tree; individual species); secondly, that some important features are not considered (e.g. age of rock, educational level of population); and thirdly, that the order of the 19 items selected for the example is one out of 19 (!) or approximately 2×10^{16} ways in which they could have been arranged. The items listed all reflect some object or objects of the four 'spheres', the actual objects in the regional space under consideration. Some regions, especially smaller ones, might entirely lack or have in only negligible quantities some of the things listed: for example, an urban region very little soil or vegetation, Antarctica no permanent people and therefore none of items 13–21 in the table.

The selection of variables for the description of a region and for explaining relationships within it is one of personal choice. An analogy

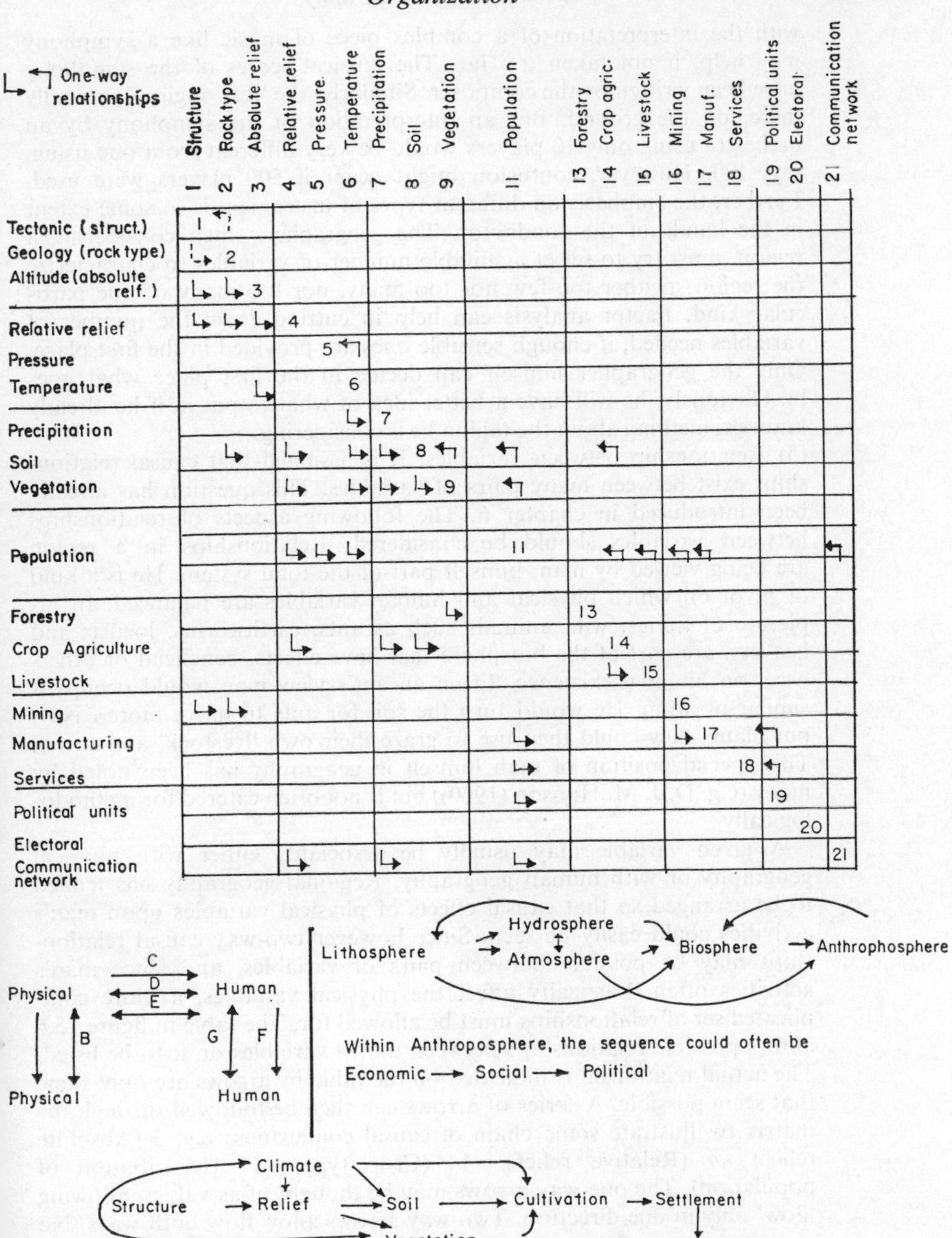

Figure 15.5 Matrix for following possible relationships within physical geography, within human geography, and between the two.

with the interpretation of a complex piece of music like a symphony may help, if not taken too far. The musical scores of the symphony have been written by the composer. Similarly, the 'real' region is actually there, on the ground. But an interpretation of the symphony by an orchestra using only 10 players would be very different from one using, say, 100. However, confusion might occur if 500 players were used. Further, the emphasis on different types of instrument is to some extent in the hands of the conductor. The geographer, when considering a region, must try to select a suitable number of variables to characterize the region: neither too few nor too many, nor too many of one particular kind. Factor analysis can help in cutting down the number of variables needed, if enough sensible ones are provided in the first place. Only the geographer himself can decide in the first place what goes in. Obviously, he will have a better idea of what to put in if he already knows something about the region he is considering.

(5) Relationship between variables. It is assumed that causal relationships exist between many pairs of variables. This question has already been introduced in chapter 6. The following aspects of relationships between variables should be considered: Relationships in a region are being viewed by man, himself part of the total system. He is a kind of pivot on which physical and human variables are balanced. In his picture of things, wild animals such as ants, earthworms, locusts and beavers, are part of the biosphere that have effects, beneficial or otherwise, on his own existence. From an ant's view man would occupy a similar position. He would turn the soil for ants to move more easily, put plants they could then use to graze their own livestock, and so on. This special position of man himself in geography has been noted by many (e.g. D. J. M. Hooson (1960)) but is not often catered for methodologically.

A given variable may usually be associated either with physical geography or with human geography. Regional geography has tended to be arranged so that causal effects of physical variables upon man's activities could easily be seen. Since however two-way causal relationships may be possible between pairs of variables, and since man's activities often drastically affect the physical variables, a more complicated set of relationships must be allowed for. The table in figure 15.6 allows possible relationships between the 19 variables on it to be listed. The actual relationships indicated on the table by arrows are only some that seem possible. A series of arrows can then be followed through the matrix to illustrate some chain of causal connexions: e.g. 3 (Absolute relief) → 4 (Relative relief) → 14 (Crop type) → 11 (Distribution of population). The one-way arrows may be thought of as valves, allowing 'flow' only in one direction. Two-way arrows allow flow both ways. No arrow means no flow at all. Examples of pairwise relationships within and between physical and human variables were discussed in chapter 6.1.

Time is a prerequisite for causal relationship to occur. Man can only change the nature of soil by cultivating if allowed some span of time. Moreover, some distributions change more easily over a given period of time than others do (e.g. weather conditions compared with structural features). It is useful to have in mind a scale for the possible rate of change of distributions over time. This might be built into the time chart in chapter 10, figure 10.1.

Further aspects of relationships between variables that need consideration are: firstly, what are the implications of the circularity arising from the fact that in geography (as in economics) there are apparently many two-way causal relationships? Secondly, do some pairs of variables always correlate highly (e.g. high temperature and low pressure, high per capita income and high number of motor vehicles in circulation per 1000 people)? Are there others that correlate differently in different regions?

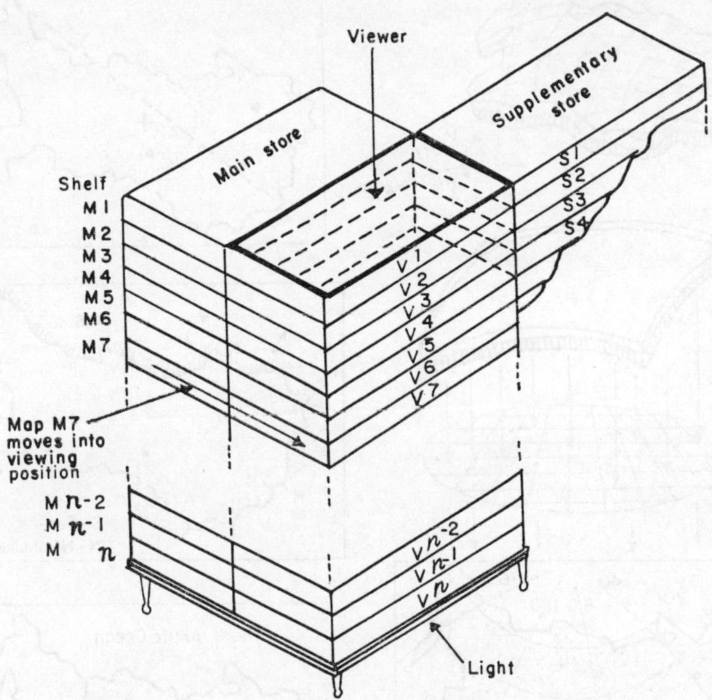

Figure 15.6 An imaginary cabinet for facilitating the visual comparison of distribution maps drawn on transparent paper. Each shelf in store, M_1, $M_2 \ldots M_n$ contains one main map. These may be drawn forward into the viewer in combinations as required. Supplementary maps, as of a main distribution at different times in the past, may also be pulled in. There are approximately 10^8 or 100,000,000 possible combinations of $n=27$ main maps. Allowing 10 seconds to scan each combination, a person would take many decades to view every combination.

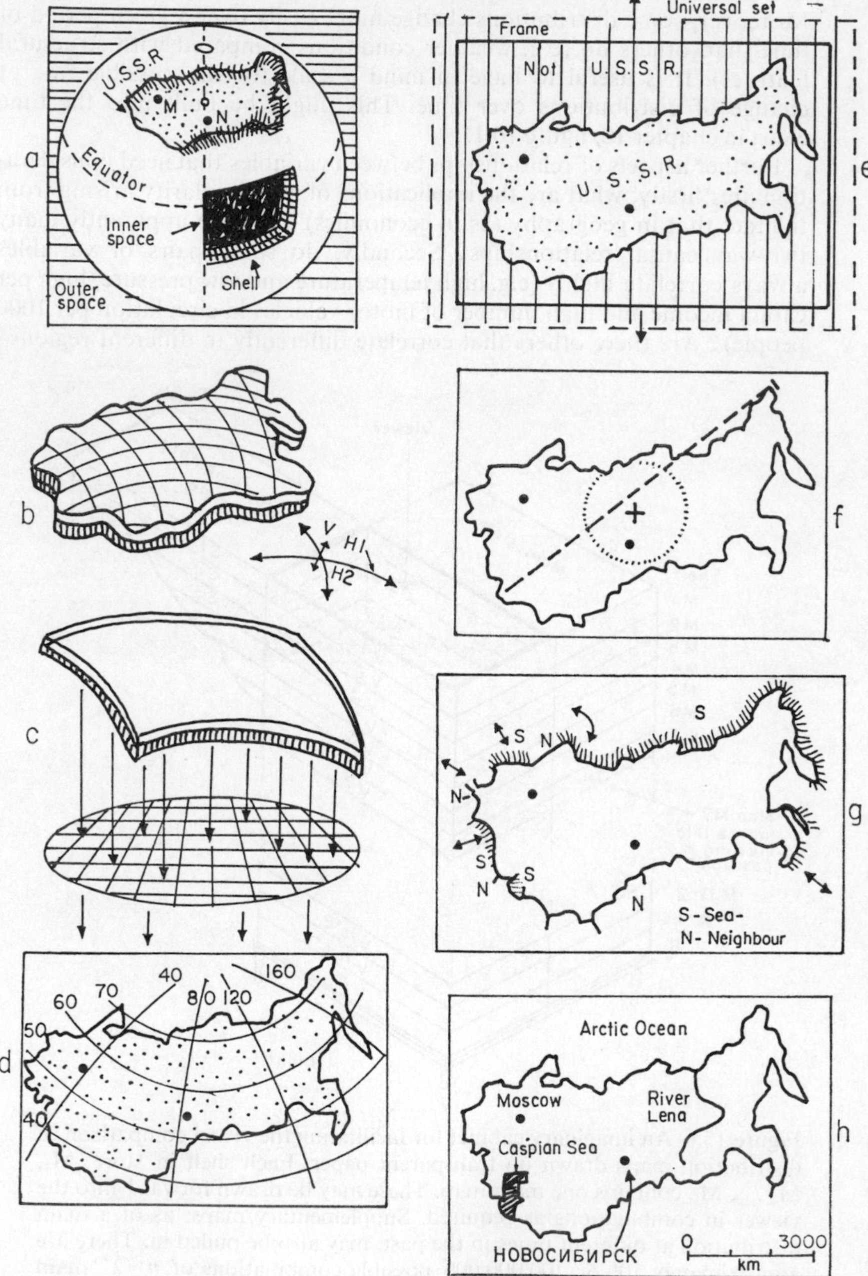
a
U.S.S.R
M
N
Equator
Inner space
Shell
Outer space
Universal set
Frame
Not
U.S.S.R.
U.S.S.R.
e
b
V
H1
H2
f
c
S
N
S
N
S
S
N
S
N
g
S - Sea -
N - Neighbour
d
40
160
60
70
80
120
50
40
Arctic Ocean
Moscow
River Lena
Caspian Sea
НОВОСИБИРСК
0
3000
km
h

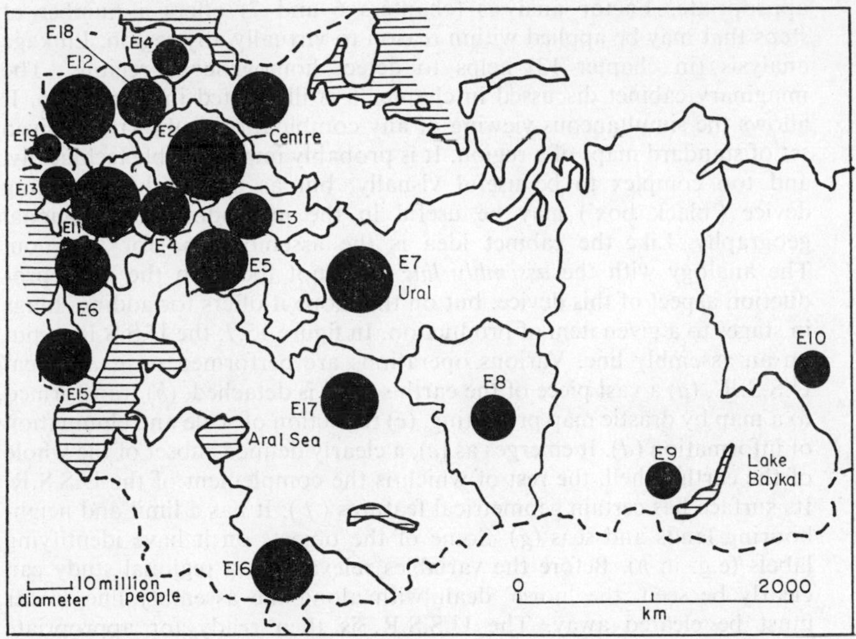

Figure 15.7 (*a–h*) The U.S.S.R. on an assembly line (i) shows the 19 economic regions, which form the first good level of compartments for obtaining data. They are as follows (E = economic region):

E1	Northwest	E9	East Siberia	E15	Transcaucasia
E2	Centre	E10	Far East	E16	Central Asia
E3	Volga-Vyatka	E11	Donets-Dnepr	E17	Kazakhstan
E4	Central Blackearth	E12	Southwest	E18	Belorussia
E5	Volga	E13	South	E19	Moldavia
E6	North Caucasus	E14	Baltic		
E7	Ural				
E8	West Siberia				

(e.g. forest on steep slopes in one area, but on gentle slopes or level ground in another, communist vote correlating highly with industrial population in one country, but highly with agricultural population in another). Thirdly, are there some leading link variables (e.g. absolute relief, soil type, urban population), that do a large amount of work in characterizing regions because they tend to gather round themselves clusters of variables? (6) Devices for organizing regional geography. From what has been said so far, it is clear that some kinds of scheme for organizing regional geography would be useful. Some are suggested and are illustrated where appropriate. Factor analysis (chapters 6 and 7) offers a number of steps that may be applied within reason to virtually any region. Linkage analysis (in chapter 13) helps to detect homogeneous regions. The imaginary cabinet discussed in chapter 6 is illustrated in figure 15.6. It allows the simultaneous viewing of any combination (subset) of a given set of standard maps of a region. It is probably impracticable technically, and too complex to be useful visually, but as a thought provoking device ('black box') may be useful in the methodology of regional geography. Like the cabinet idea is the assembly line for a region. The analogy with the *assembly line* does not focus on the mass production aspect of this device, but on the scope it offers for adding things in stages to a given item of production. In figure 15.7, the U.S.S.R. is put on an assembly line. Various operations are performed on it: the real U.S.S.R., (*a*) a vast piece of the earth's shell is detached, (*b*) transformed to a map by drastic map projecting, (*c*) reduction of scale and elimination of information (*d*). It emerges as (*e*), a clearly defined subset of the whole of the earth's shell, the rest of which is the complement of the U.S.S.R. Its surface has certain geometrical features (*f*). It has a limit and neighbouring lands and seas (*g*). Some of the objects on it have identifying labels (e.g. in *h*). Before the variables relevant to a regional study can clearly be seen, the 'noise' dealt with along the assembly line so far must be cleared away. The U.S.S.R. is then ready for appropriate variables to be added, which may be done with the help of some set of administrative compartments (figure 15.9), for which data are available: the nineteen economic regions of the early 1960s. Sets of real objects will then be transformed to symbols, usually represented by dots, lines or patches (figure 15.8), and mapped on the region as required. Time might also be taken into account. The compilation of an atlas is very much like an assembly line, but the continuity might be stressed more and the feeling of integration kept to the fore. With the help of

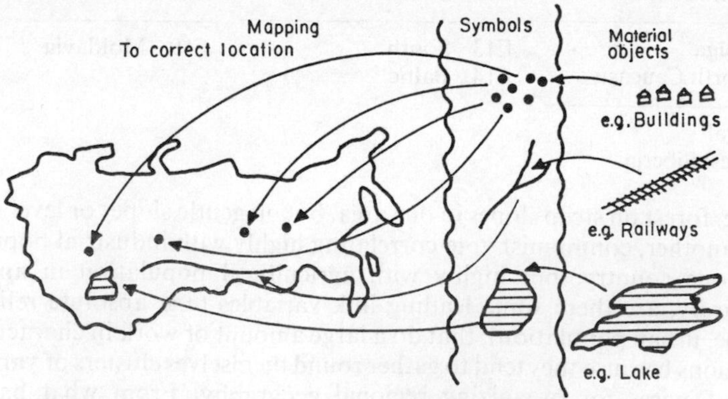

Figure 15.8 The objects from the real U.S.S.R. selected and mapped as symbols in the form of dots, lines and patches.

general systems concepts and vocabulary it might be feasible to consider regions as very complex systems. An attempt (unpublished) has been made by R. W. Kates (Clark University, 1966) to picture in diagram form the relationship of the hydrologic cycle, with its various subsystems (e.g. the atmospheric, the oceanic) to a total region. In figure 15.9 the chain of regional relationships has been reconsidered as a kind of system. The idea of a system seems to help in conceiving the integrated nature of phenomena in a region. An imaginary region, a kind of laboratory, might be set up as an operational game for experimentation. It would be impossible, for various reasons, to take an actual part of the earth's surface. But the model island of St. Lucius, somewhere in the Pacific Ocean, might prove useful. Availability of information would be no problem here. Unlike the static Utopia, the features of which were fixed by its designer, St. Lucius would be dynamic. It could be played through by different geographers, who settle it and develop it in model time and then compare their results. Such a game in fact exists and is used for an urban area, Metropolis, described by R. D. Duke (1964). Ultimately, the problem of assembling, storing and processing a very large amount of data makes any single map or matrix inadequate. The data for factor analysis uses a 2-dimensional matrix, with areas along one side and variables along the other. To represent the relationships between the areas, however, two dimensions are needed. A third dimension would then have to be added to the matrix for variables, a fourth, conceivably, for time, and others, perhaps, for other considerations hitherto ignored or underemphasized in regional geography.

5.4 Conclusion

Geography may be compared with a building. When a large and intricate building is being constructed, detailed plans are prepared and it is built up brick by brick each fitting into its appropriate place. Geography in its present state is like a building that lacks an architect and a plan. There are plenty of bricks and stones. There are even some elaborate small sections of the building complete, but there is no overall framework into which these stones and other parts may be fitted. It is being put together in several styles and it lacks the massive solidarity and unity of a great building. Individual geographers at times may pursue their own paths. As a result a general approach tends to be lacking.

This situation is aggravated by lack of communication between the several aspects within the discipline, although some attempts have been made recently to remedy this. The mixed nature of the contents of many journals and other professional publications adds to the feeling of diversity and lack of purpose and overall plan. Another equally great danger is the lack of communication between workers at different levels of the subject. It is hoped that this book, explaining some of the new geography and some of the older geography in new terms, has with its rather lengthy descriptions of methods and ideas,

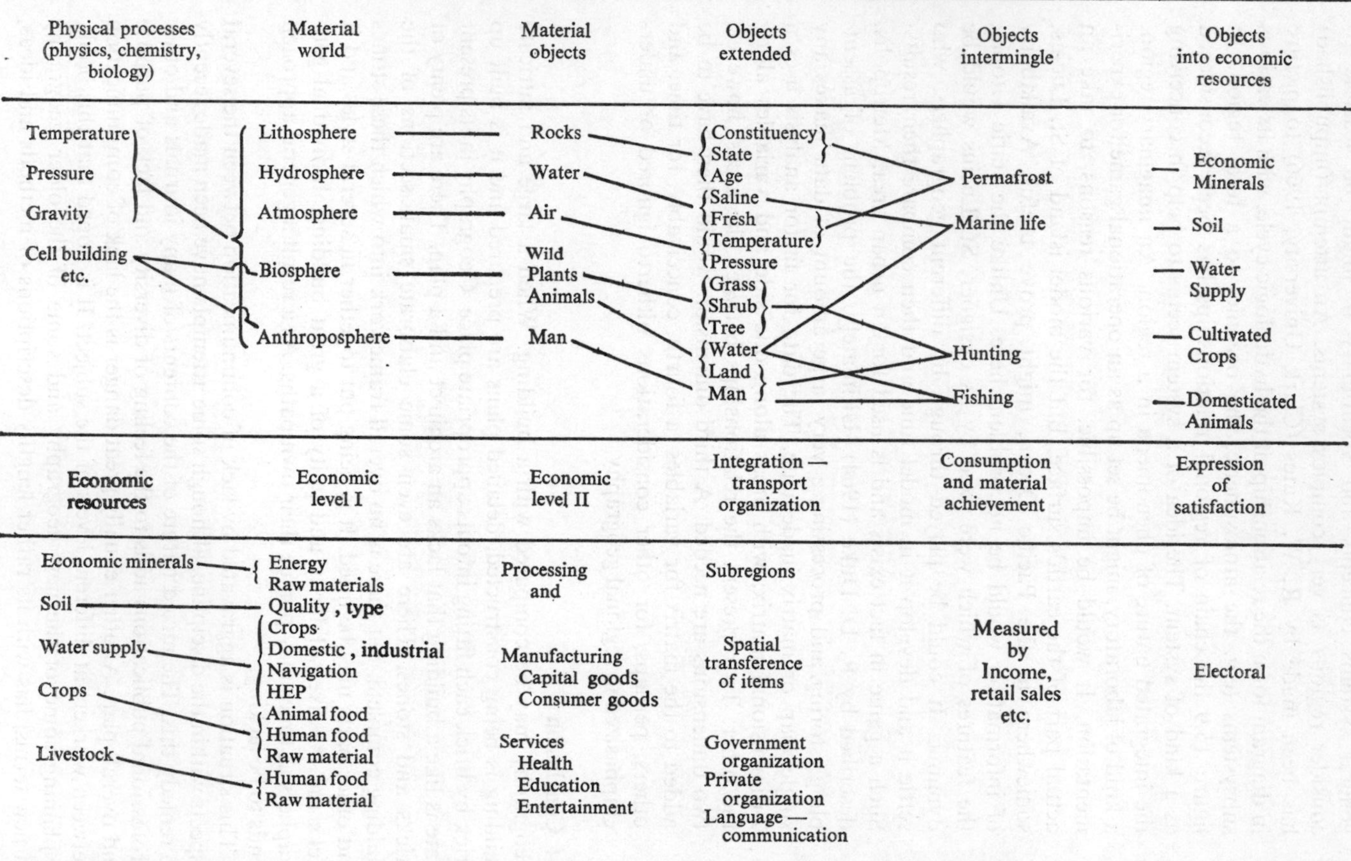

Figure 15.9 Network of relationships to be taken into account in a regional system.

made some of these accessible to geographers and to non-geographers working at many different levels.

Perhaps the greatest need of geography is a conceptual framework which will unite its many very good building bricks into a single edifice. E. A. Ackerman (1958) remarks: 'Without a conceptual framework, the vast numbers of data which can be accumulated about space content on the earth's surface would have minimum meaning.'

The analogy of the building may be carried a step further by referring to figure 15.10. This 3-dimensional model could be considered as a plan for a

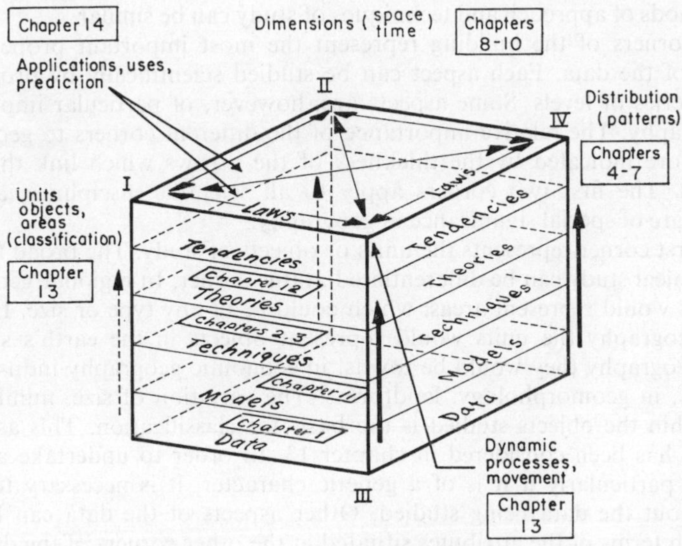

Figure 15.10 A theoretical model of geography, showing how the different chapters in the book are related to each other. The arrows indicate direction and strength of relationships between the different aspects.

square modern building which consists of six stories or levels. Each of these levels represents a stage in the analysis of geographical data. The levels follow the steps in the scientific method discussed in section 1.5 of chapter 1.

In general it is necessary to progress by the stages indicated in the diagram, although at times it may be necessary to return to the beginning. If, for example, a theory does not prove satisfactory, then it is necessary to start again, perhaps even with new data and certainly with a new model and modified techniques.

The arrangement of the levels applies to many scientific disciplines. It is, however, the subject matter and particularly the way in which this is organized, which differentiates geography from other disciplines. The corners of figure 15.10 represent the methods of organization and approaches to the geographical data which are the subject matter of geography. The emphasis is on methods of approach and general characteristics of the data rather than on the data itself. In this respect the diagram is similar to the more recent matrix approach suggested by B. J. L. Berry (1964) and also by R. J. Chorley and P. Haggett (1967). Their vectors are concerned with topological and geometrical form on the one hand and with locational relativity on the other. Because the diagram (figure 15.10) is based mainly on properties and methods of approach it can accommodate all the branches of geography within the scheme. The plan, therefore, suggests that there is no essential dichotomy between physical and human geography. The nature of the data will differ from one branch to the other, but the properties of the data are similar and the methods of approach and techniques of study can be similar.

The corners of the building represent the most important properties or aspects of the data. Each aspect can be studied scientifically by proceeding up the series of levels. Some aspects are, however, of particular importance to geography. The relative importance of the different corners to geography is therefore indicated by the thickness of the arrows which link the levels upwards. The first two corners apply to all scientific disciplines, and the last two are of special significance in geography.

The first corner represents the units or objects of study. The broad fields of geographical study can be differentiated at this corner. In regional geography the units would represent areas, which could be of any type or size. In systematic geography the units would represent objects in the earth's shell. In urban geography they would be towns, in economic geography industries or products, in geomorphology, landforms. The variation of size, number and type within the objects studied is the basis for classification. This aspect of the data has been considered in chapter 13. In order to undertake a classification, particularly if it is of a genetic character, it is necessary to know more about the data being studied. Other aspects of the data can be considered in terms of the attributes situated at the other corners of the diagram.

The second corner of the diagram is concerned with the dimensions of the objects. These are fundamental characteristics of all objects, which exist in space and time. Sometimes it is possible to simplify the description of objects by omitting some of their real dimensions. It has already been pointed out in chapter 1.3 and in chapter 9, that 3-dimensional objects can, for purposes of study, often be regarded as points, lines or surfaces. In other situations it is essential that the 3 space dimensional character of objects be taken into account. Chapter 9 was devoted to a discussion of some of these situations. Time is often regarded as a dimension. Studies in which time is an essential element fit into the framework at this corner. Some of these studies were considered in chapter 10.

Those studies in which time is important link very closely to the third corner of the diagram, in this corner the dynamic situations are considered. Dynamic processes by their nature imply movement and change through time. Dynamic processes apply to all fields of geography. In geomorphology the study of process is claiming a great deal of attention. Movement and change are also essential elements of all other branches of geography, and change forms the basis of historical geography. Some of these dynamic situations were discussed in chapter 13.

The final corner is perhaps the most important in geography. It is concerned with the distribution of the units and, thus, represents their spatial aspect. It can represent the distribution of one set of data in one area, or more complex situations can be considered with many variables distributed in a number of areas. The various possibilities are considered in chapters 4 to 7. Distributions may give rise to patterns, and it is at this position in the framework that a study of patterns can be placed.

At all levels of the diagram the corners can be interlinked with each other. These links are indicated by the two headed arrows on the top of the diagram, although they run throughout all the levels. Thus data can be described at any corner of the diagram in terms that bring in all the other corners, or produce an effect on them. Models and techniques are available for all corners of the diagram. The level of models was discussed in chapter 11. Classification models were considered in chapter 13, and these apply at the first corner but must depend upon data at the other corners. Scale models, which deal primarily with data at the third corner, depend very greatly on scale, which is a matter of dimensions, relevant to the second corner. Diffusion models in human geography study movement at the third corner, and this affects the distribution of the data at the fourth corner. It is not necessary to describe all the different levels and the interconnexions between their corners. The interlocking of the different levels and the methods of approach cannot readily be broken down into isolated parts. The system forms one solid whole, which encompasses the study of geography and shows how its separate parts fit together.

The progress upwards through the building requires the use of all the different approaches, leading finally to the formulation of geographical laws and prediction. If this book has assisted in suggesting paths through this very complex maze of possible routes then its purpose will have been fulfilled.

References

Ackerman, E. A. (1958). *Geography as a Fundamental Research Discipline*, University of Chicago, Department of Geography, Research Paper No. 53, p. 14.

Berry, B. J. L. (1964). Approaches to regional analysis: a synthesis, *Ann. Assoc. Am. Geog.*, **54**, 2–11.

Chorley, R. J. and Haggett, P. (Eds.) (1967). *Models in Geography*, Methuen, London.

Cohen, S. B. (1964). *Geography and Politics in a Divided World*, Methuen, London.

Duke, R. D. (1964). *Gaming-Simulation in Urban Research*, Institute for Community Development and Services, Michigan State University, East Lansing, Michigan.

Gerlach, A. C. (1964). New approaches to the geography of the United States, *Ann. Assoc. Am. Geog.*, **54**, No. 1, 1–23. Papers by B. J. L. Berry, E. H. Hammond and G. T. Kimble.

Hartshorne, R. (1954). *What is Political Geography?* In (Eds. P. E. James and C. F. Jones) *American Geography Inventory and Prospect*, Syracuse University Press.

Hooson, D. J. M. (1960). The distribution of population as the essential geographical expression, *Canadian Geog.*, **17,** 10–20.

Jackson, W. A. D. (1958). Whither political geography?, *Ann. Assoc. Am. Geog.*, **48,** No. 2, 178–183.

Jackson, W. A. D. (Ed.) (1964). *Politics and Geographical Relationships*, Prentice-Hall, Englewood Cliffs. Subtitle *Readings on the Nature of Geography* (contains several key passages on concepts and methodology by different authors).

Jones, S. B. (1954). A unified field theory of political geography, *Ann. Assoc. Am. Geog.*, **44,** No. 2, 111–123.

Kirk, W. (1963). Problems of geography, *Geog.*, **48,** 357–371.

Mackinder, H. (1904). The geographical pivot of history, *Geog. J.*, **23,** No. 4.

McDonald, J. R. (1966). The region: Its conception, design and limitations. *Ann. Assoc. Am. Geog.*, **56,** No. 3, 516–528.

Rapoport, A. (1966). *Two Person Game Theory*, Ann Arbor, University of Michigan Press (concludes with a discussion of the usefulness of game theory).

Rostow, W. W. (1960). *The Stages of Economic Growth*, Cambridge University Press.

Rummel, R. J. (1963). Dimensions of conflict behaviour within and between nations, *Yearbook Soc. Gen. Systems Res.*, **8,** 1.

Sprout, H. and Sprout, M. (1965). *The Ecological Perspective on Human Affairs*, Princeton University Press.

Tobler, W. (1967). *Spectral Analysis of Spatial Series* (duplicated paper). Department of Geography, University of Michigan.

Glossary

* Means see elsewhere in glossary. (1), (2) etc. are distinct meanings of words in glossary. Reference is also made to extended definitions in text where appropriate. Some mathematical definitions are perhaps not rigorous enough to satisfy a mathematician.

Abscissa The distance of a point along the horizontal *x*-axis from the vertical *y*-axis in a frame of reference such as a graph with two axes at right angles to one another.

Abstract(ion) Used in two senses.
(1) Extraction of the essence of an idea, to reduce to a minimum.
(2) The opposite of material.

Accessibility Used fairly loosely to refer to ease of movement between places in area. Related to distance. More particularly connected with studies in urban and economic geography.

Accuracy An accurate method of measuring is one which is both precise* and unbiased. The results of many values of one measurement cluster closely about the true mean value.

Actuarial data Data related to births and deaths; therefore a part of all demographic data. Collected by actuaries or in censuses and drawn on, for example, by life insurance companies.

Additivity Variables under investigation must have no mutual effect or interaction on each other. If this is so then the data have the property of additivity.

Adiabatic Changes in the temperature, pressure and volume of air which take place without transfer of heat. These changes normally take place by vertical displacement of air. The dry adiabatic lapse rate is 5.4°F per 1000 ft in ascent and descent. The saturated adiabatic lapse rate is less owing to latent heat of condensation.

Affine geometry Allows the transformations of Euclidean* geometry and also projection by parallel rays to a plane that can be tilted.

Algorithm(-ism) Any particular set of steps for performing a computation. For example, addition is taught in several possible ways to western school children.

Aliquot An exact divisor, e.g. 2 and 3 are aliquot parts of 6.

Amphidromic system A tidal system in which the tide rotates anticlockwise around a nodal point in the Northern Hemisphere and vice versa in the Southern Hemisphere. It is caused by the effect of the coriolis force* on a standing oscillation (stationary wave). The tidal range increases outwards from the nodal point and the tidal streams rotate in the same sense as the whole system.

Analogy(-gue) Similarity between two apparently unlike objects or processes. A powerful method of scientific reasoning is by the recognition of similarities between understood and unexplained objects or processes. The *positive* analogy includes all like aspects, the *negative* analogy all unlike aspects and the *neutral* analogy all irrelevant aspects (chapter 11).

Anomaly In a general sense, the occurrence of an event, object or observation out of step or accordance either with those near it or with some expected* observation. For example, temperatures in Western Europe are higher than those generally found at similar latitudes elsewhere and are therefore anomalous. In geography, trend surface analysis (chapter 9) is a method of bringing out anomalies in spatial distributions.

Antecedence Used of drainage in areas where localized uplift has not disrupted a stream, which has cut down faster than the uplift across its path, e.g. Rhine gorge through Eifel District, R. Arun in Himalayas.

Antipodes(-al) Points on the earth's surface at opposite ends of a diameter passing through the centre of the earth.

A posteriori statement Statements or predictions made on the basis of existing experience. For example, long range weather forecasting may be based on the study of past weather. Life insurance premiums are based on observed actuarial* data. Statements made on an empirical basis. They may be revised on the basis of new evidence. Thus it was found in postwar British elections that the Labour Party was under-represented in Parliament because of the arrangement of constituencies. This statement was revised on the evidence of the 1964 and 1965 elections.

A priori statement Statements that can be proved to be true or false by reason not by experience. Contrast with *a posteriori*.

Arc (1) Any part of a curve. For example, a circle may be divided into two arcs of equal length, semicircles.
(2) In topology, any line joining two nodes.* The line need not be part of a curve (chapter 2.11).

Area (1) General term used also in geography to refer to some fairly clearly defined space. Often interchangeable with region.*
(2) Precise term of measurement of the extent of a portion of a plane surface or of a solid such as a sphere, provided the portion has a clearly marked boundary. Usually measured in square units.

Array A regular arrangement of information, particularly of numbers, as in a matrix* (chapter 2.8).

Astrolabe An instrument developed in 1585 to calculate latitude and longitude. It is based on an equatorial stereographic projection, on top of which there is a moveable arm or pointer.

At-a-station Observations made at one point on a river by gauging discharge. These observations provide information of changes of flow with time. They should be differentiated from observations made at one time at many places on the river. (The latter are downstream observations.)

Attribute (noun) Often in statistics retained for a feature that is confined to the nominal* scale (e.g. *black* stones and *white* stones, *male* or *female*) or to yes/no scale (e.g. a village has or does not have a post office).

Average Often synonymous with arithmetic mean, but also used loosely to indicate any central tendency (see chapter 3). Semi-average: the averages of the first half and of the last half of a set of observations, used to measure trend.*

Axis (1) Earth's Axis. The diameter around which the earth rotates.
(2) Two lines at right angles on a graph represent the axes of the graph. Coordinates are measured from them.
(3) The lines which determine the dimensions of a geometrical figure. In three

dimensions there are the long (*a*) axis, the intermediate or (*b*) axis and the short (*c*) axis, representing maximum length, breadth and thickness of the figure.

Base (1) General term frequently found in geography (e.g. energy base of a country).
(2) In surveying, an initial length measurement of a survey.
(3) In numbers, the number of digits* used in a positional method of counting. The familiar denary* method is base ten because it uses a zero plus nine digits; binary* is base two because it uses a zero plus one digit. Note that the bases are being defined here in the language of denary or base ten.

Bias (1) A biased set of values is not centred on the true mean of the measurement.
(2) A biased sample is one that is not random in that all members of the population have not an equal chance of being selected.

Binary A positional method of counting (see chapter 2.5) using only two digits,* zero and one (0 and 1). This is the minimum number of digits for a positional method.

Binomial formula The expansion of $(a+b)^n$ (see chapter 3).

Bit An abbreviation of *binary digit*, a term used in information theory and computing, referring to *one* binary* digit (i.e. 0 or 1). It is possible to reduce for example speech to bits and thereby measure the speed at which information can be handled in a simple, standard form.

Boolean algebra A system of notation (called after George Boole, 1815–1864) using set theory and binary computation. It is introduced in chapter 2.3.

Boundary (1) Mathematical: a line forming the limit of a 2-dimensional surface or a surface forming the limit of a 3-dimensional figure.
(2) Geographical: a line forming the limit of a unit of area, such as a clearly defined administrative unit. Contrast frontier for a zone between two units of area.

Breakdown The detail available in data. For example, there may be a regional breakdown of certain data into counties in the U.S.A., or into individual types of employment in the U.S. economy.

Cardinal numbers (1) The numbers 0, 1, 2, 3 …, or the positive integers* or whole numbers. Cardinal numbers are used to count aggregates. A set of objects may be counted with the cardinal numbers if each can be put in a one-one correspondence with the objects. Contrast ordinal numbers: 3 sheep, 3rd sheep. Zero belongs to the counting numbers according to some but not all sources consulted.
(2) Cardinal points are N, S, E and W directions on a compass.

Cartesian co-ordinates The name of the French mathematician Descartes is often given to a system of two co-ordinate axes (*x*-axis and *y*-axis) perpendicular to each other. Also co-ordinate or analytic geometry. With this approach, geometry can be dealt with in terms of algebra.

Cartography The preparation of maps of all types but not including the basic surveying for topographic survey maps.

Case A very general word used also technically in a statistical sense for each member of a set of objects or observations. For example, in a study of 30 towns, each of the 30 towns would be a case. Not to be confused with case study.

Catastrophism A school of thought which considered that the basic elements of the landscape were formed by sudden catastrophic occurrences, (e.g. Noachian deluge). See uniformitarianism,* the opposing school of thought.

Causality A relation assumed among two objects or events or sets of objects or events whereby one is explained in terms of the other. Causal relationships, the detection of which is the concern of much scientific work, are often studied by statistical techniques.

Cell In addition to its meanings in biology and physics, a cell may be used to refer to each unit of area, usually clearly defined, in a grid on a surface: for example the communes of France. Also cell in a matrix.

Central place theory Now a vast branch of urban geography dealing with the spacing of settlements and more particularly the study of the hierarchy of larger centres with increasingly specialized functions. Owes much of its initial impetus to W. Christaller.

Central tendency measures Averages of various types. These are calculated to describe or summarize in one term a whole group of data. Discussed in chapter 3. Five kinds are: Arithmetic Mean, Geometric Mean, Harmonic Mean, Median, Mode.

Centre of gravity The point through which the line of action of the weight of a body always passes, however the body may be placed. Used in human geography to denote the point of balance of a distribution, centre of gravity (= mean centre) of an area either of a continuous surface or of a dot distribution placed on a 'weightless' area.

Centrifuge(al) An instrument which revolves at a high speed so that normal accelerations very much greater than the acceleration of gravity are obtained. Centrifugal force is the outward force induced by rotation. Centrifugal force balances gravitational attraction to keep bodies (e.g. earth and moon, revolving round each other).

Centripetal force The force which causes a body to depart from a straight line and travel in a curved path. In a rotating circular motion the force is directed into the centre of the circle. It is the gravitational interaction between 2 bodies (e.g. earth and satellite).

Channel A greatly elongated 3-dimensional object usually bounded only on the bottom and sides. Extended to any means of communication, along which material objects and even information may pass. Discussed in chapter 13.

Chord Mathematical: a line connecting the extremities of an arc* (e.g. any straight line with a circle that cuts the circumference of the circle at two points).

Chronological Referring to matters of time (e.g. events mentioned in chronological order are events mentioned in order of occurrence).

Chorological Referring to matters of space. Not used so frequently as chronological, but offering the possibility of chorometrics as a term for quantitative geography.

Circle The line traced out by a point which moves through 360 degrees at a fixed distance (the radius) from a fixed point. *Great* circle on the globe is one whose centre is at the centre of the globe (e.g. equator or meridian). *Small* circle is one whose centre is not at the centre of the globe (e.g. any parallel of latitude except equator).

Circuity A term used by some to refer to the degree of detour required on a journey between two places in excess of the direct distance.

Circular measure Measurements made in radians.*

Circumference Ordinarily refers to the line that limits the area of a circle. In the description of a network, such as the transportation network of a town, circumferential routes are those running in the general direction of the circumference as rings, rather than radial lines.

Coefficient (1) of reliability — an index or measure of reliability;
(2) of correlation — an index or measure giving a precise value to the relationship between two or more variables;
(3) a constant in an equation.

Column A term in matrix algebra (see chapter 2.8) referring to the entries arrayed vertically (as opposed to rows, arrayed horizontally).

Combinations The number of possible arrangements of a given number (n) of things (elements) regardless of their sequence, 0 at a time, 1 at a time, and so on, e.g. ABC: ϕ (none), A. B, C, AB, AC, BC, ABC (but not BA as well as AB).

Communication theory Also information theory, is a mathematical approach to a very broad, general and now extensive field including speech, signals, interference (noise), and media for transferring information from one person or machine to another. Origins attributed to C. E. Shannon in 1948 paper *A Mathematical Theory of Communication.*

Compactness In geography, particularly one of the most important aspects of the shape of a 2-dimensional figure. If a circle is considered the most compact shape possible, then other shapes may be given appropriate indices of diminishing compactness for elongation, fragmentation and so on.

Complex A situation in which many variables must be considered simultaneously.

Complex number A combination of a real number and a real number multiplied by the square root of minus 1. For example $a+bi$, where a and b are real numbers (see chapter 2.5) and i is $\sqrt{-1}$.

Component, principal A group of closely related or co-varying variables. Also basically the same as a *factor*, in factor analysis, though different in interpretation.

Computer (1) A general term for any device used for making arithmetic calculations.

(2) More particularly applied now to a very high speed electronic computer (see chapter 2.7).

Concave A curve, facing upwards, which is decreasing continuously in angle from the horizontal downwards. A concave slope steepens upwards.

Concept An abstract idea concerning a particular object or situation.

Confidence limit The proportion of times a given outcome may be expected to occur by chance in statistical analysis. Ninety-five per cent confidence limit implies that the given outcome will only be expected to occur by chance 1 in 20 times, 99 per cent confidence 1 in 100 times, etc. See also significance level.*

Cone, constant of The ratio of the angle at the apex of a developed (unwrapped) cone to 360°. The constant is proportional to the sine of the latitude at which the cone touches the sphere. It varies from 1 for a plane tangent at the pole to 0 for a cylinder.

Conical projections are based on a cone tangent to the sphere along a line of latitude.

Connectivity In particular, the degree of interlinking of nodes by arcs in a graph or network. The more links, the stronger the connectivity.

Consensus Recourse to a body of informed opinion about some matter to be decided. For example, one method of predicting when such developments as a manned landing may be expected on Mars is to consult and pool the opinions of an appropriate body of experts. A political election may be regarded as a very broadly based consensus of opinion.

Constant A value in an equation which does not vary. A reducer constant is a value by which all values are multiplied to yield more manageable figures. Constancy — a value which remains the same.

Contiguity The state of being in contact. For example, the states of Nevada and California share a boundary and are therefore contiguous.

Contingency table A table in the form of a matrix* with rows and columns, designed to discover whether two criteria of classification are independent of one another, for example, sex and favourite colour. The table is important in setting up the chi-square test.

Continuous An attribute which is always present and which varies by very small increments, which can be smoothed to make a series without abrupt breaks. (See discrete.*)

Convection current A current induced by instability, usually caused by a less dense (and often hotter) medium lying below a denser (and often cooler) medium. The less dense medium has a tendency to rise.

Convex A curve, facing upwards, which is increasing continuously in angle from the horizontal downwards. A convex slope flattens upwards.

Co-ordinates Lines of reference by means of which the positions of points in a plane relative to one another may be described. Two axes, the abscissa* and ordinate* perpendicular to one another are used.

Coriolis force A force which deflects objects moving on the earth. It is caused by the earth's rotation. The deflexion is to the right of the path of movement in the N. Hemisphere and to the left in S. Hemisphere. It is proportional to $2v\omega \sin \phi$, where v is velocity of object, ω is angular velocity of the earth's rotation and ϕ is latitude.

Correlation The degree of relationship between pairs of variables.

(1) Coefficient — see coefficient.*

(2) Multiple — the correlation between two or more independent variables and one dependent variable.

(3) Negative — the relationship in which one variable increases as the other decreases.

(4) Partial — the correlation between two variables when a third, which is related to both, is controlled.

(5) Positive — the relationship when two variables increase together.

(6) Product-Moment — a parametric test for correlation between two variables expressed on ratio or interval scale.

(7) Spearman Rank — a nonparametric test for correlation between two variables expressed on rank or ordinal scale.

Correspondence A fundamental concept in set theory. In a one–one (also one to one) correspondence, the members of two sets may be exactly paired. Ideally the books borrowed from a library and the cards in the librarian's desk; or the sheep in a field and the counting numbers. One–two, one–many correspondences are also possible.

Covariance The relationship between two variables, given by $\Sigma\, xy/n$, where x and y are the differences between each variate and the mean of its set, and n is the number of pairs.

Critical path analysis Discussed in chapter 13. One term for a procedure used increasingly in industry and other fields to assess the time needed to complete a project. Usually in a project some steps cannot be started before others are completed, whereas some can be carried out simultaneously and are free to float. Critical path analysis finds the longest sequence of steps or path in the project. Here, only, savings in time can shorten the duration of the project.

Cubic surface The third degree trend surface analysis surface in which the formula for the surface is given by

$$Z=\beta_0+\beta_1X+\beta_2Y+\beta_3X^2+\beta_4XY+\beta_5Y^2+\beta_6X^3+\beta_7X^2Y+\beta_8XY^2+\beta_9Y^3$$

This surface has a double curvature in two dimensions, X and Y. Z represents the third dimension.

Cumulative curve A frequency curve in which each group is added to the preceding one until the total number of observations are included, often adding to 100 per cent.

Cycle A sequence of events which returns to its starting point. A biological cycle is usually represented by alternating increases and decreases of numbers. The hydrological cycle is the circulation of water through the atmosphere, earth and ocean.

Cyclomatic number A term used in graph* theory to refer to an index μ, which is equal to the number of arcs or edges minus the number of nodes or vertices plus the number of subgraphs. (Scc chapter 13.)

Cylindrical A group of projections based on a cylinder that touches the globe along the equator in the normal case, a meridian in the transverse case and any other great circle in the oblique case.

Decimal (1) Decimal number system is the familiar number system using ten digits,* 0, 1, 2, 3, 4, 5, 6, 7, 8, 9. Recommend the term denary for this use.
(2) Decimal point is the starting point, to the right of which decimal fractions (e.g. 7/10) may be written without a denominator.

Declination The angular distance of the sun or moon north or south of the equator measured along the meridian. The declination of the sun varies between 23° 27′ N and S and the moon between 28° 35′ N and S.

Declivity Another term for slope.

Deductive Deductive reasoning proceeds from the general to the particular, from the theoretical to the specific. Contrast inductive reasoning, which builds on empirical evidence.

Degrees of freedom (D.F. df) 'The number of degrees of freedom is equal to the number of quantities which are unknown, minus the number of independent equations linking these unknowns.' (Blalock) The maximum number of variates* that can be freely assigned before all the rest are completely determined.

Denary An alternative term for decimal when referring to base* ten numbers.

Density The mass* of a unit volume of a substance is known as its density. *Drainage density* is calculated by dividing the area of the basin by the total stream length in it. *Population density* is defined as numbers of people per unit area.

Descriptive Account of what something is like or what does happen or what is done. (See prescriptive,* proscriptive* and predictive.*)

Desire lines Lines, usually straight, representing the *expected* or hypothetical flow of transactions based on some assumption or model between places on a map. Contrast flow lines in economic mapping drawn proportional in width to *observed* transactions, and often following precise routes.

Determinism The concept of the inevitability of occurrence of some set of events or relationships due to the influence of e.g. physical conditions (environmental determinism), economic patterns (economic determinism).

Deviation The difference between an observed and an expected or calculated value. (See residual.*) Also used in the sense of a variation from the normal or the average.

Diagonal (1) In geometry, a line segment joining any two vertices of a polygon when the two vertices are not adjacent.
(2) The entries running from top left to bottom right in a square matrix* are called the principal diagonal.

Diastrophic Processes of earth movement and rock deformation, many of which cause changes in relative position, both vertical and horizontal, of portions of the earth's crust.*

Dichotomy Division into two, especially in classification.

Diffusion A general term that has become associated in a more technical sense with the spread of (e.g. an idea or innovation, or of a set of people from a given point or set of points usually over a surface but also over the lines of a network).

Digit A basic counting unit represented by a symbol in a number system. In binary* the digits are 0 and 1, in duodecimal (base* 12), the digits are 0, 1, 2, 3, 4, 5, 6, 7, 8, 9, *t*, *e*. *t* and *e* are 10 and 11 respectively in denary.

Dilation A term popularly associated with expansion, but in Euclidean geometry applicable both to the enlargement and reduction of a figure (e.g. a square, which is still a square, regardless of change of size by the transformation of dilation). (See chapter 2.9.)

Dimension There are three generally recognized space dimensions, but in multivariate analysis a multi-dimensional space is conceived. *Dimensionless* refers to values that can be expressed without linear measurements, (e.g. slope or wave steepness, L/L, in which the dimensions cancel. Time can be used as a dimension in this respect, e.g. Reynolds number has

$$\frac{ML^{-1}T^{-1}}{ML^{-1}T^{-1}} = 0 \text{ dimensions.}$$

Dip The angle of inclination of bedded strata; true dip is maximum angle of inclination, apparent dip is any smaller angle of dip. True dip is measured at right angles to the strike.*

Dipole magnet A magnet with two poles, one being positive and one negative, like a simple bar magnet.

Discrete A separate entity, opposed to continuous.*

Disjoint In set theory, two sets (or subsets) are disjoint if they have no members (elements*) in common (e.g. in a set of people, those under 25 and those over 40 are disjoint sets).

Disparate Things that are so unlike that they have no basis for comparison are disparate.

Dispersion The spread of a set of observations or objects (usually) about some central point or place. To describe the dispersion or dispersal of a distribution is to consider only this particular aspect of a distribution, on a line, on a surface or in a space.

Distance (1) Spatial distance, measured in units of length which are usually conceived in terms of existing known distances (e.g. foot, furlong, kilometre (1/40,000 of the length of the equator)).

(2) Time distance, also referred to as length, but measured in time units, usually conceived in terms of phenomena experienced (e.g. day–night). The term *duration* may be preferred to length or distance when time is considered.

(3) Other distance (e.g. social distance: distance on a social class scale, income distance).

Distribution To the geographer, usually a distribution of points (dots) on a surface. More generally, a distribution may occur in any number of dimensions from one upwards. It may be events on a time scale or other observations (e.g. examination marks) on a scale (see chapters 1 and 3). Important descriptive terms for distributions are clustered, random, anti-clustered; regular and irregular (see chapter 4).

Division A general term, also used technically to refer to a compartment of administrative or other function. For convenience politico-administrative subdivisions of countries may be referred to as major and minor civil divisions.

Dynamic A situation in which movement is an essential element, as opposed to static. Refers to the operation of processes that cause change through movement.

Earth A planet moving round the sun.

Core, the central part of the earth occupying approximately half its radius and having some properties of a liquid.

Crust, the outermost layer of rocks above the Mohorovičić* discontinuity.

Mantle, the part of the earth between the crust and the core.

Shell, the part of the earth of interest to geography, extending from the upper boundary of the mantle to the atmosphere, a thickness of about 50–100 miles. Also soil (e.g. Blackearth).

Eccentricity The amount by which a point is distant from the centre of a circle. The sun is not exactly in the centre of the earth's orbit. Also means not quite circular, i.e. earth's orbit is not a true circle.

Ecliptic The plane of the earth's orbit around the sun. It is set at approximately 23° 27′ to the earth's equatorial plane. The apparent path of the sun in the celestial sphere.

Ecology The mutual relations between organisms and their environment (S.O.E.D.).

Edge Technical term in 2-dimensional topology* used by some mathematicians for the line joining two vertices* or nodes.* Similar to arc* but the two are sometimes given distinct meanings.

Eigenvector and eigenvalue An eigenvector in mathematics is one which does not change direction when a mathematical transformation is performed upon it. V is an eigenvector if $V^\star = \lambda V$, λ is called the eigenvalue, $V^\star$ is the value after the transformation and V the value before it. Eigenvalues in factor analysis are the characteristic or latent roots of the correlation matrix. There are therefore as many eigenvalues as variables in the matrix. The eigenvalues determine the values of the factors, which are in fact the eigenvalues of the correlation matrix. They may be thought of roughly as the length of the vector which passes through a scatter of points in multidimensional space in such a way as to reduce to a minimum the distance between the points and the vector.

Element Technical term in set theory for the members of a set. A given set, finite or infinite, consists of so many elements, which must be clearly defined (e.g. the letters *R, E, N* and *O* in *Reno*).

Ellipse An oval figure which is produced by cutting a cone at any angle between the base and the side of the cone. An ellipse has a major or a minor axis. Its area is twice that of the circle whose diameter is the minor axis.

Ellipsoid is a 3-dimensional figure in which all plane sections through the short axis are ellipses and planes through the long axis are circles or ellipses.

The formula for an ellipse is $\frac{x^2}{a^2}+\frac{y^2}{b^2}=1$

x and y are distances along the horizontal and vertical axes; a and b are major and minor semi axes.

Empirical Based on experiment or observation rather than on theory.

Endogenetic Forces acting internally within the earth, such as volcanic and orogenic (diastrophic*) forces.

Energy Energy in a system is its ability to do work.

Kinetic energy is the energy a body possesses by virtue of its movement and it is the amount of work a body can perform in coming to rest.

Potential energy is energy due to position, usually energy due to height. Energy is also used to imply a source of power (e.g. coal, steam, water-power, etc.) with which work can be done. For purposes of comparison of production of different kinds of energy, some equivalent, such as coal equivalent, may be used. Energy gradient in a stream is usually approximated by water-surface slope.

Entropy In thermodynamics the ratio H/T where H is heat and T temperature. H/T remains constant on lines of constant entropy (isentropic lines). In more general terms 'entropy' is used to denote the change from organization to disorganization or randomness in a closed system.* The greater the randomness the higher the entropy.

Equation of time The difference between mean time and apparent time. It is caused by the obliquity of the ecliptic* and the ellipticity of the earth's orbit. It reaches a maximum of 16 minutes.

Equilateral triangle A triangle having all 3 sides equal. Its 3 angles are then also all equal.

Equilibrium A body is in equilibrium when the resultant of the forces acting on it is zero. Stable equilibrium is defined as the condition when, if the body is moved slightly, it returns to the same place. In unstable equilibrium it will move further away. *Quasi-equilibrium* refers to a state of near equilibrium. Used in an economic sense to imply general balance and stability.

Error Used in statistics in two senses. Error type 1 is when a hypothesis being tested is in fact true, but the sample used implies that it is false. Error type 2 is when the hypothesis being tested is false, but it is concluded from the sample that it is true.

Euclidean geometry In general, the geometry in the works of the Greek geometer, Eukleides. In particular, a geometry that studies points, lines and planes. Euclidean geometry defines parallel lines as lines that never meet (the fifth postulate). Non-Euclidean geometry does not.

Even There is some scope for confusion here over even as (1) *even* number (any integer* exactly divisible by 2) and (2) *even* distribution, usually implying regular or predictable distribution. The opposite of even (1) is odd, of even (2) is uneven.

Exclusive, mutually Two events are said to be mutually exclusive if the occurrence of one precludes the occurrence of the other (e.g. in cards, a 5 may be drawn randomly from the pack). This event precludes the drawing of any other number on that draw. However, the suit that the 5 belongs to may or may not be hearts. In set theory, mutually exclusive subsets are disjoint.

Exogenetic Forces and processes acting externally on the surface of the earth.

Expansion Used in mathematics to denote the conversion of an equation into a series of many terms. A series is called *convergent* if its terms when added approach a finite value. See series* and exponential.*

Expected Popularly, something that it is anticipated will happen. In statistics and the testing of models it refers to a set of values that would occur if the model were working perfectly. The observed values from the real world may then be matched with the expected. In this sense, the expected is not necessarily anticipated.

Experimental (1) Something undergoing trial.
(2) Something based on the result of an experiment (e.g. experimental evidence) as opposed to something based on intuition or guesswork.

Explicit Something expressed or made clear in a statement, as opposed to something implicit, understood but not openly expressed.

Exponent A value, n, which indicates the power to which any number, a, is to be raised, written a^n: e.g. $10^3 = 10 \times 10 \times 10 = 1000$; $e^{1.5}$.

Extrapolation The extension of curves beyond the limit of observations, by continuing the trend shown in the observed portion.

Factor Two uses of the term not to be confused.
(1) A cause. Often one variable supposedly causing the variations in another set of observations.
(2) A family or cluster of co-varying variables, as in factor analysis.

Feedback A term now widely used, though in several senses, in general systems and in computing. An important meaning is the ability of an organism or a machine to adjust itself or correct itself as it proceeds through time. The feedback is the information given to the control of the organism or machine to regulate itself.

Fetch The distance over which the wind blows in a constant direction when it is generating waves.

Finite arithmetic Also modulo arithmetic or clock arithmetic (see chapter 2.4). Arithmetical system functioning with only a finite set of numbers, e.g. 1, 2, 3, 0 (e.g. calculations on a twelve-hour clock are based on finite arithmetic modulo 12).

Flatness A measure of 3-dimensional shape. A. Cailleux's formula is

$$\frac{a+b}{2c} \times 100 \text{ (see axis*)}$$

Fourier analysis A mathematical technique used to split up a complex curve into its harmonic (sinusoidal) parts. Used in the analysis of wave records, see wave spectrum.*

Fraction In general, a quotient of two quantities (e.g. $\frac{3}{8}$), the upper part of which is called the *numerator* and the lower part the *denominator*. The line is an instruction to divide the numerator (dividend) by the denominator (divisor).

Frequency Used in statistics to define a particular type of distribution of variates in a sample, see Binomial,* Normal,* Gaussian,* Poisson* frequency distributions.

Frontogenesis The processes in the atmosphere by which meteorological fronts are developed.

Function In addition to meaning in a general sense the purpose of a thing, function in mathematics refers to relationship implying that a change in one variable depends on a change in another. The second is then a function of the first. It can be expressed as $y=f(x)$: y is a function of or depends on x. Each value of x must correspond to just one value of y.

Game theory Recent mathematical theory that studies the strategies and outcomes of games in a formal way. Associated terms are: zero sum *game*, non-zero sum *game*, negotiated *game*, non-negotiated *game*, dominating strategy.

Gaussian curve see *Normal curve* after K. F. Gauss (1777–1855).

General systems The Society for General Systems (also produces a Yearbook) seeks to find common ground between disciplines. In particular, the concept of a system is common to many disciplines.

Generalization Increased generalization denotes a law or tendency that includes an increasing range of phenomena within its terms. The laws of gravity and motion are very general laws of a high degree of generalization. The opposite of particular.

Generation Two distinct uses sometimes become confused.
(1) To generate something (e.g. waves).
(2) A span of time, defined as required in terms of its duration.

Generic Concerning the classification of things (from genus).

Genetic Concerning the origin of things; Genesis.

Geodesy The science which aims to measure the figure of the earth. A geodetic survey is a very precise survey covering a considerable area of the earth's surface.

Geoid Earth shaped.

Geostrophic The current or wind which results from the effect of the Coriolis force* on the pressure gradient force. It results in movement parallel to the isobars and proportional to their distance apart.

Geothermal Geothermal power is derived from heat within the earth usually in the form of steam. Used for generating electricity in Italy, New Zealand and Iceland.

Gerrymandering The arrangements of districts for a specific purpose, particularly elections, in such a way that some controlling criterion such as population equality is distorted in the interest of some person or party. Called after Governor E. Gerry of Massachusetts, 1812.

Grade The division of a circle into 400 equal angles gives 400 grades. 400 grades therefore equals 360 degrees.

Graph (1) A continuous plane surface upon which points, lines and smaller surfaces may be depicted in correct relative positions in relation to co-ordinates* perpendicular to each other.

(2) A network on a 2-dimensional surface in topology. Graph theory studies the number, arrangement and proportions of nodes and arcs in a network. The following terms may be found: null graph, complete graph, universal graph, connected graph, tree graph, planar graph, non-planar graph.

Gravity The force which attracts two bodies towards each other. It depends on their respective masses, m, m^1 and their distance apart, r, and is proportional to mm^1/r^2. Used as an analogy in human geography, called gravity model.

Grid A network, usually, though not always, conceptual and often regular (e.g. composed of squares, hexagons). Also special uses (e.g. electricity grid, national grid or rectangular co-ordinates used by the British Ordnance Survey for reference to places in Britain).

Habitat Natural home of plants or animals.

Half-life The time it takes half the radio-active elements in a substance to change to another form.

Harmonic analysis The process of splitting a complex curve into its harmonic or sine constituents, which will differ in period and amplitude. (See Fourier analysis* and wave spectrum.*)

Heterogeneous A number of elements of diverse character. Composed of diverse elements.

Heuristic Heuristic conveys the idea of solving problems by trial and error rather than according to a preconceived model. The criterion of success is successful progress towards the final result required especially if achieved within specified time or financial limits. This contrasts with a set procedure for solving a problem, as algorithmic.*

Hexagon A 6-sided polygon. Mostly a *regular* 6 sided polygon is what is implied by hexagon in geography.

Hierarchy A set of items (e.g. persons, districts) classified by rank or order. Often represented visually by a tree diagram.

Histogram A diagram based on two axes perpendicular to each other with the horizontal axis divided into classes (e.g. age groups of people) and a column raised above each and measuring frequency (e.g. number of people in each class) proportionally on the vertical axis. Popularly called a graph.

Homeostatis The ability of a living organism to maintain a balance by adjusting itself to changes in its environment (e.g. maintenance of body temperature at a given level). The idea of self-regulation.

Homogeneous(-eity) Composed of like elements.

Hydrostatic Hydrostatic pressure enables water to flow up hill for some distance in an enclosed channel when the outlet of the channel is below the point of entry.

Hypothesis An idea which can be tested to establish its truth. A working hypothesis is one which has not been disproved, but which has not yet been substantiated.

Hypsometric A hypsometric curve is one in which area is plotted against height. The hypsometric integral is a measure of the area below the curve, when it has been reduced to dimensionless units varying between 0 and 1.

Icosahedron A geometric solid having 20 faces. The application of a regular icosahedron is discussed in chapter 8.

Imaginary number Properly a *pure* imaginary number, is the square root of a negative number (e.g. $\sqrt{-3}$).

Implicit Implied but not openly expressed (contrast explicit*). The use of implicit as without questioning (e.g. to accept implicitly) is to be deplored.

Indeterminacy The degree of variation to which most natural phenomena are susceptible. Also implies that when several attributes are concerned in one measure, the value of this measure can be obtained by variations between the attributes, e.g. River discharge depends on the attributes of river width, depth and velocity, and these can vary to give the same discharge if one increases as the other decreases (see chapter 12.3).

Individual weighting A term used in factor* analysis to refer to the position of an individual case on a factor. Also factor score.

Inductive A process of reasoning leading to general laws from (many) particular instances (contrast *deductive**).

Inequality An equation in the terms equal to or more than, or alternatively, equal to or less than, indicated by the symbols $x \geqslant y$ and $x \leqslant y$.

Infinite In a mathematical sense, infinite is defined in set vocabulary. Thus an infinite set is one that can be put in a one-to-one correspondence with a subset of itself. The term infinite should not be used to describe very large but finite numbers.

Infinitesimal An infinitely small quantity, the inverse or reciprocal of an infinite quantity, $1/\infty$.

Inflection point The point at which a curve changes curvature (i.e. from concave* to convex* or vice versa).

Input Various uses in computing, particularly information or data transferred from external storage to internal storage. In economics, input/output models deal for example with the empirical analysis of production.

Integer A whole number (positive, zero, or negative). An integer always contains the unit (one) an exact number of times.

Integral The symbol $\int$ used in calculus to denote integration, or summation. The limits are indicated at the top and bottom of the symbol. Integration of a drainage pattern implies the joining of streams by capture to increase the stream orders but to decrease the stream numbers.

Interaction Used in analysis of variance to indicate the interaction or effect of two sets of observations on each other (see additivity*).

Interpolation The insertion of values between measured values. Used to denote the drawing of contours with reference to a series of spot heights for example.

Interval (1) The range between two extremes, *a* and *b*, over which a variable, *x*, can take any real number value.

(2) Interval data in statistics contrasts with ordinal data in having real number values, rather than integer placings as in ordinal data.

Inversion Turning upside down or more generally, any reversal of expected order or position (as when temperature increases with altitude).

Irrational number A number that does not terminate or recur (e.g. π is an irrational number but 0.500 and 3.333 recurring are *rational*). Rational and irrational numbers together form the set of real numbers.

Isobase A line joining points of equal uplift on a raised shoreline.

Isopleth A line (usually on a map) joining points of equal quantity of any variable.

Isostatic Equal standing. Balance of density at all points achieved by smaller volume where the mass* is greater, and greater volume where mass is less (e.g. roots of mountains in which light material extends to greater depth beneath higher land). Isostatic recovery is uplift of land surface following removal of mass by retreat of ice.

Iteration Repetition of a procedure. For example the repetition of a group of computer instructions for as long as required.

Juvenile water Water which enters the hydrological cycle for the first time, coming directly from magma within the earth.

Kinematic wave A kinematic wave consists of a concentration of units in a unidirectional flow system. The concentration may move through the units making up the system as a wave. The velocity of the wave may therefore be greater than the velocity of the individual units.

Kurtosis The fourth moment measure about the mean. It is a measure of the peakedness of a curve. $\mu_4 = E[x - E(x)]^4 E(x)$ is the expectation of the random variable x. Platykurtic is flatter than the normal curve. Leptokurtic is more peaked than the normal curve.

Least effort A very general but also somewhat vague principle taken as guiding much of human activity. In a simple example, it may be assumed that in making a journey between two places a person will choose the one requiring least effort (in terms of energy, time, cost, and so on).

Lemniscate loop A curve which has a stream-lined tapering form. The equation of the loop is $\rho^2 = l \cos k\theta$, where $k = (l^2\pi)/4A$ and expresses the elongation of the lemniscate loop. Where l is length of long axis and A the area of figure.

Light year The distance travelled by light in one year. Light travels at 186,000 miles per sec.

Linear programming A method developed in the last three decades for finding the best solution from among every solution of a system of linear inequalities. (See chapter 14.) Applied to optimize the delivery of war materials from various U.S. ports to various U.K. ports in World War 2.

Loading Term in factor analysis referring to the degree to which a given variable is attached to a factor,* in the sense of co-varying with the family of variables that most contribute to that factor.

Logarithm See Mathematical and Statistical signs.

Logistic curve A time series curve shows an accelerating growth rate followed by a decelerating growth rate leading to a saturation level. The point of inflection is called the critical point. The equation of the curve is $Y_e = k/(1 + be^{-ax})$. a, b and k are constants determined by the data and e is 2.71828.... k is the size of the population at saturation level. The critical point occurs at $x = 1/a \log_e b$ when $Y_e = \frac{1}{2}k$. This pattern of growth often applies to population change.

Lorenz curve A curve designed to show to what extent a given distribution is uneven compared with an even distribution. See chapter 5.

Loxodrome A straight line on a Mercator projection is a line of constant bearing and is called a loxodrome or rhumb line.*

Magnetic Field A region around a magnet is called its magnetic field. Lines of magnetic force indicate the pattern within the field. A magnetic line of force is a line drawn in the magnetic field so that the tangent to it at any point is in the direction of the magnetic force at that point.

Magnitude, order of An order of magnitude is the number of digits before or after the decimal point. Values 10–99 are one order of magnitude less than values 100–999.

Mantissa The decimal part of a logarithm (base 10) (e.g. in the logarithm 2.469, .469 is the mantissa).

Mapping (1) An important mathematical term referring to the operation of establishing correspondences between points in two regions. The word *transformation* is very similar in meaning. The points in the second region are called the images of the points in the first region.
(2) In geography, the transference of selected objects in the real world to symbols (dots, lines, etc.) on a surface (e.g. piece of paper), usually with the aim of keeping scale and shape correct. Other kinds of map (e.g. topological maps, are however possible).

Mass The mass of a body represents the quantity of matter in it and is measured in units of weight. It should not be confused with volume or density.*

Matrix An ordered array of numbers on which certain operations may be performed (e.g. multiplication). A matrix is itself considered to be a special kind of number. See chapter 2.8 for a discussion of features of a matrix.
The following kinds of matrix may be encountered: Identity *m.*, transpose *m.*, inverse *m.*, square *m.*, symmetric *m.*, row *m.*, column *m.*, correlation *m.*, data *m.*, input/output *m.*

Mean The *arithmetic* mean is the value found by summing all values and dividing by the number of values (see average*). The *geometric* mean is the nth root of the product of all the values, where n is the number of values. The *harmonic* mean is given by $M_n = \dfrac{n}{\frac{1}{x_1}+\frac{1}{x_2}\cdots\frac{1}{x_n}}$

Mechanistic A view of the world that assumes that the behaviour of all biological phenomena may be explained in mechanico-chemical terms.

Median The median is the value that is central in an ordered series of values, having an equal number of values above and below it. On a cumulative frequency distribution it is the value of the 50th percentile.

Mesoscopic Also meso-scale. A term suggested by analogy with mesolithic meaning middle (stone), to refer to the visual world as opposed to the microscopic and macroscopic worlds, viewed properly only with the help of the microscope or telescope.

Metachronic Similar to diachronic, of a different age in different parts (e.g. of a raised shoreline). See synchronic* which is the opposite.

Metric (1) Measurement system founded on the base ten number system.
(2) Measuring in general, e.g. metric geometry, econometrics.

Minimax A value which is at the same time a maximum in a row of a matrix and a minimum in the column which intersects at the value concerned. Also called saddlepoint in game theory.

Mode The highest value in a series of grouped data is the modal group or class. The class (or classes) for which the frequency is greatest.

Models Representation of a situation or object often simplified. Used in very many different ways. *Mathematical* model, *theoretical* model, *scale* model, *analogue* model. Discussed in chapter 11.1.

Module One use refers to the fineness or otherwise of a grid (e.g. for collecting data).

Modulo Also modulus, the constant base of a variable function.

Modulus See *Modulo*. Also may be an instruction to ignore a plus or minus sign (i.e. sign irrelevant in a given context).

Mohorovičić discontinuity Also called Moho or M-discontinuity. The level at which primary seismic waves suddenly accelerate downwards to 8.1 km per sec. The boundary is usually taken to indicate the upper limit of the mantle,* separating it from the crust* above. It may be a change of state boundary not a change of material.

Molecular attraction Very important in ground water hydrology. The attraction of rock surfaces for molecules of water and the attraction of the molecules of water for one another. In fine grained rocks or sediment much water is held by molecular attraction.

Moment measures Used in statistics to describe the character of a distribution curve. The first moment measure is the mean.* The second is the standard deviation.* The third is the skewness.* The fourth is the kurtosis.*

Monadnock A residual area of high ground of limited extent above a peneplain.

Morphometry The measurement of the characteristics of a drainage basin from maps or photos. Includes ordering streams, measuring stream length, bifurcation ratio, drainage density.*

Mosaic The difference between a mosaic made of discrete parts (e.g. tiles) and a painting, forms a useful analogy for contrasting a surface divided into a number of discrete compartments (e.g. fields, provinces) with one that is continuous.

Multi-variate analysis The study of the relationship between many variables rather than between only two. Factor* analysis is one kind.

Nearest neighbour Concept and techniques associated with the relationship of dots to one another in a 2-dimensional distribution of dots (see chapter 4).

Nesting A mathematical term used for example to describe the way in which networks of hexagons of different module* can fit one into another, or to describe sampling within designated subdivisions of network.

Network (1) A system of lines linking a given set of points. A network may be material (channels in a stream or transportation network), or conceptual (see chapter 13).

(2) Network analysis refers specifically to a system of planning a number of steps or operations in a good order.

Node A term in topology referring to a point or dot (of no dimensions).

Noise Recently accepted term to refer to interference as in the transmission of messages and also to excessive detail or irrelevant information on a map distracting from the visual assessment of important things.

Nominal A classification of cases on the basis of verbal description rather than numerical values, whether ordinal* or interval* (e.g. male and female; red, brown and black stones).

Normal (1) At right angles to, or perpendicular to, or orthogonal* to.

(2) A normal probability distribution curve, which is bell shaped, also called Gaussian curve. Most observations lie near the mean of the values at the apex of the curve. The equation of the curve is

$$Y = \frac{N}{\sigma\sqrt{2\pi}} e^{-(X-m)^2/2\sigma^2}$$

m = mean; σ = standard deviation; Y is frequency with which any X occurs.

Null hypothesis The null hypothesis may for example state that the relationship between two variables is random. This may then be accepted or rejected with so much confidence* as a result of an appropriate test.

Numeracy A concept similar to literacy and now widely used to cover the ability to appreciate the meaning of numbers and quantities in particular and mathematics in general.

Numerical Described or defined by a number. Note the difference between a *number*, a concept of quantity, and a *numeral*, a symbol or group of symbols used to represent the number. Any number can be represented by an infinite number of different numerals (e.g. 2, base ten, is one zero (written 10) in binary).

Observe (1) *Observation* Used sometimes in statistics to refer to a case.*
(2) *Observed* Actual data referring to phenomena in the material world (e.g. rainfall data, livestock data are observed values). These may then be compared with some expected* set of values.

Operational gaming A term recently accepted to refer to games that are played for educational purposes, possibly also for research purposes, and in which role playing is often a vital part.

Operations research The application of analytic methods derived from mathematics to the solving of operational problems.

Optimum The best result that can be obtained. To optimize is to so arrange the situation that the best result is arrived at.

Ordinal The arrangement according to order or rank (e.g. population size of towns), see chapter 3. Ordinal data have cases ranked, 1st, 2nd, 3rd and so on. When there is a tie (e.g. equal 5th) then a value of 5.5 is given to the two equal cases, and the ranking is resumed at the 7th.

Ordinate The vertical or y axis in a graph, in contrast to the horizontal or x axis, the abscissa.*

Orthogonal At right angles to, used (e.g. of lines at right angles to wave crest drawn to determine distribution of wave energy on a coast).

Orthomorphism The property of correct shape in map projections. It applies strictly only to points. It is achieved when the meridians and parallels cross at right angles and the scale is equally exaggerated or reduced in all directions from any point on the projection.

Outcrop The area over which a particular geological stratum is exposed on the surface. Sometimes used to denote the line of junctions between two different strata on the surface.

Packing theory Packing theory is a branch of mathematics with some relevance to geography since it is concerned with the arrangement of things in spaces.

Paleomagnetism Magnetism induced in a rock when it was first extruded, solidified or deposited sometime in the past. The magnetic field of the time was fossilized in the rock and can be measured relative to the present magnetic field after making allowance for rock structure. The magnetism of old rocks.

Parabola The mathematical function for a parabola is $y=x^2$. The curve passes through the origin of the graph and increases steadily in gradient away from the axis on both sides, the value on the y-axis increasing as the square of the value on the x-axis.

Parameter The term parameter in statistics should be applied to the true fixed values of the mean, standard deviation, etc. of the entire population and not to a sample of the population. It is often used loosely as more or less synonymous with variable.*

Pascal's triangle Pascal's triangle gives the coefficients for the expansion of $(x+y)^n$. Thus e.g. $(x+y)^2 = 1x^2 + 2xy + 1y^2$
e.g. $(x+y)^4 = 1x^4 + 4x^3y + 6x^2y^2 + 4xy^3 + 1y^4$
the triangle is built up summing the values to the immediate right and left above each new entry thus

```
        1  2  1
      1  3   3  1
    1  4   6   4  1
  1  5  10  10  5  1
1  6  15  20  15  6  1
```

It is the basis of the symmetrical binominal distribution. (Blaise Pascale — 1623 to 1662.)

Payoff table A table (also matrix) showing the rewards in terms of profit or loss of various outcomes of a game or of all possible acts or decisions in for example a business problem.

Periphery Strictly speaking, a boundary* line or surface. It is also used to refer to the outlying portions of an area such as a political unit.

Per mil Per thousand (short form $^0/_{00}$), a useful way of expressing data with adequate precision without the decimal place that would be necessary in per cent. Used, for example, for salinity measurements in oceanography.

Permutations The number of different orders in which a set of things may be arranged (see chapter 2.11) e.g. ABC, ACB, BAC, BCA, CAB, CBA. It is also possible to have permutations of subsets of ABC (e.g. AC, CA).

Phenomenon (plural *-mena*) Refers to things recognized by the senses; therefore the use of phenomenal for unusual, exceptional or large is to be deplored.

Plankton That which is made to wander or drift. Minute creatures of the marine environment most of whom are carried passively by the currents. *Phytoplankton* are plants and *zooplankton* are animals. Some zooplankton have the ability to move vertically by their own power, but none move far horizontally except by passive movement.

Plottable error The minimum distance that can be shown on a map to scale. The distance depends on the scale of the map and is based on the assumption that a line on the map is 1/100 in thick. For a scale of 1 in to 100 ft the plottable error is 1 ft and for 1 in to 1 mile it is 52.8 ft.

Point (1) A location without dimensions, a pre-requisite for the construction of Euclidean geometry.
(2) Decimal point.
(3) A small object represented on a map by a dot. The use of dot rather than point is recommended. (See chapter 4.)

Poisson distribution The Poisson distribution is one in which the probability of an event occurring (p) is very small compared with the probability that it will not occur (q). p is very much smaller than q. (In a normal curve $p = q = \frac{1}{2}$). The Poisson distribution should be used if $np < 5$. It is a very asymmetrical distribution curve. If z is the expected number of occurrences the Poisson distribution is given by

$$e^{-z}\left(1 + z + \frac{z^2}{2!} + \frac{z^3}{3!} \dots\right)$$

Polarity The direction toward which a magnetic compass points. At present the earth has a north polarity, but the polarity changes at intervals.

Polygon A surface bounded by three or more straight lines.

Polyhedron (plural, *-drons*, *-dra*) A solid bounded by planes. There are five possible regular polyhedra, tetrahedron (4 faces), cube (6 faces), octahedron (8 faces), dodecahedron (12 faces) and icosahedron (20 faces).

Polynomial A polynomial algebraic expression with more than one term. For example a quadratic or higher order equation has terms such as

$$Y = a_0 + a_1 X + a_2 X^2 \ldots$$

A polynomial can also be used for two variables in the form of

$$Y = a_{00} + a_{10}U + a_{01}V + a_{20}U^2 + a_{11}UV + a_{02}V^2 \cdots$$

Population (1) General and geographical usage: people.
(2) Statistical population, virtually any set of things under consideration (e.g. people, drumlins, light bulbs).

Positional notation Number system in which the value of a symbol varies according to its position in a set of symbols. Contrast non-positional Roman III with denary 111, in which each one is ten times more than the one to the right of it.

Power (1) Statistical use in the expression power of a test, which can be expressed as $(1-\beta)$ where β is the probability of committing a type 2 error.*
(2) A power function is one which contains an exponent* and can be written as $Y = aX^b$. It can also be written in the form $\text{Log } Y = \log a + b \times \log X$.

Pragmatic An approach to things that considers only the practical value of what is done.

Precision The closeness with which observations of one value cluster together. It depends on the measuring technique (see accuracy*).

Predictive Stating what will happen or what will be done (see prescriptive,* proscriptive* and descriptive,* see also chapter 10).

Prescriptive Stating what ought to happen or what ought to be done (see descriptive,* predictive* and proscriptive*).

Principal components Similar to factors in factor analysis; clusters of variables in multi-variate analysis.

Principle A broad concept or idea which is at a fairly high level of generalization.

Probability (1) Experience of the occurrence of some event expressed for future reference in terms of the likelihood of it happening.
(2) A subset of events that could occur out of a total population of events, conventionally expressed as a proportion of 1 (e.g. the mathematical probability of getting a head when tossing a coin is $\frac{1}{2}$ or 0.5).

Program Spelt this way when referring to a set of instructions telling an electronic computer what to do with a set of data.

Projective Geometry A kind of geometry concerned with projection from a point source.

Propagation The spread or dissemination of objects or ideas in area (e.g. the spreading of waves from their generating area over the sea). Can be abstract or material.

Proscriptive Stating what ought not to happen or what ought not to be done (see descriptive,* predictive* and prescriptive*).

Prototype The real thing, usually used in opposition to a model of the real thing.

Quadratic surface The second degree of surface fitted in trend surface analysis. The equation for the quadratic surface is in the form given under polynomial function* with 2 variables. The variables being the X and Y or U and V axes of the surface. The surface has one change of curvature in each direction.

Quantitative Very general term usually used in contrast to qualitative and associated with data that can be handled in ordinal (ranked), interval or ratio form as opposed to data that cannot be put in a numerical scale. A quantitative difference and a qualitative difference between two things or sets of things are also two contrasting concepts.

Quartile A distribution may be divided into four equal quarters. These are referred to as quartiles.

Radial From radius, a straight line joining the centre of a circle to any point on its circumference. Radial is commonly used in the description of networks of a certain kind with reference to lines extending out from a point. Contrast circumferential.*

Radian A radian is equal to 57.3°. The use of radians to express angles is referred to as circular measure. The angle subtended at the centre of a circle by an arc equal in length to the radius of the circle is one radian.

Random Implies the equal chance of any of a number of events occurring (e.g. in a table of random digits, each space has an *equal* chance (probability) of 1/10 of obtaining one of the digits 0, 1, 2 ... 9). In fact, it obtains one of these. In a random sample, every item in a population* has an equal chance of being selected.

Ranking Arranging in (descending) order according to some criterion such as performance in an examination, size and so on. Ordinal data are thus produced.

Ratio The relationship of one quantity to another of the same kind (e.g. the ratio of 12 to 4 is 3). The sign for a ratio is a colon (e.g. 12 : 4, 1 : 63,360). Rational* numbers (see chapter 2.5) are related to ratio, not to reason (a different sense of rational). The need to go beyond whole numbers (integers) arises from the need to express ratios (e.g. 14 : 4 is 3.5).

Real (1) The term real world is widely used (e.g. in statistics, philosophy). Rapoport defines reality as 'an assumed carrier of an invariance', an invariance being something which remains constant as other things change. The real world may be thought of as shared with someone with whom one is communicating.

(2) To appreciate the wide variety of meanings given to real in different contexts it is interesting to consider possible opposites: real/unreal, real/model, real/conceptual, real/theoretical. At times real is similar to empirical, material, physical, concrete.

(3) Real numbers are those quantities that can be represented on a number line. There is a real number for every point in the line.

Reciprocal In mathematics a reciprocal of one expression is that which multiplied by the first expression equals 1 [e.g. if first expression is 4, its reciprocal is $\frac{1}{4}$ ($\frac{1}{4} \times 4 = 1$)].

Rectilinear Forming a straight line or bounded by straight lines. Rectilinear tidal streams flow in two opposite directions.

Recurring decimal Also repeating decimal, a non-terminating decimal which however repeats an initial digit or pattern of digits (e.g. 0.3333, 0.176292929). Recurring decimals belong to the rational* numbers, a subset of all real numbers.

Reflexion A general term used specifically in Euclidean geometry for a particular transformation involving flipping a figure over on some axis (e.g. horizontal axis) to produce a mirror image (see chapter 2.9).

Refraction (*wave*) The bending of wave crests when depth influences their velocity. This process operates in the sea when the depth is less than half the wave length and causes waves to turn more nearly parallel to the depth contours.

Region (1) A term in geography with a long history of wrangling and a mystique, referring often to an area defined by a particular criterion or set of criteria.

(2) In mathematics, a general term for a clearly bounded space (e.g. in Euclidean

geometry, the surface within the circumference of a circle. In topology, any shaped figure).

Regression The fitting of a line through a scatter of points in such a way that the sum of the squares of the distance between the points and the line is reduced to a minimum. The regression coefficient* gives the slope of the regression line and can be used to predict the relationship between the two variables.

Relief absolute relief, the height above sea level. *Relative* relief, the difference in height between the hill tops and the valley bottoms.

Replica line A line drawn parallel (as far as possible) to another line (not necessarily straight) (e.g. the Panama Canal Zone boundary follows a course parallel to the Canal at a distance of some miles either side of it for much of its length).

Resection The method in surveying by which an unsurveyed point, at which the surveyor is situated, is fixed by reference to three visible, fixed points.

Residual The difference between an observed and a computed value; see also deviation.*

Resultant The distance and direction between the first and last point of a vector analysis, especially used in connection with resultant tidal currents or wind resultant.

Rhumb line See loxodrome.

Riemannian geometry Geometry can be classed in three ways according to the kind of surface on which it takes place.

(1) Euclidean* is on a plane surface.
(2) Riemannian on a curved surface.
(3) Lobachevskian on a saddle shaped surface.

The contrast between the three comes in the behaviour of angles and parallel lines. In Riemannian geometry there are no parallel lines, and the angles of a triangle always exceed 180°. In Lobachevskian geometry, an infinite number of lines can be drawn through any point parallel to another line, while the angles of a triangle always add up to less than 180°.

Right triangle Also right angled triangle, one in which one of the angles is formed by two lines perpendicular to each other.

Rotated factors Rotation of factors is a procedure in factor analysis designed to reexpress factors more strongly once they have been calculated by standard methods.

Rotation A transformation in geometry whereby a plane is turned as many degrees as required about a point fixed.

Routing In transportation economics and geography, refers to various ways of organizing journeys between places.

Row An array of numbers arranged horizontally (strictly left to right) across a page rather than vertically (a column) (e.g. a row vector is a matrix* consisting of a single row of figures).

Running mean A procedure designed to smooth a series of numbers, useful particularly if adjoining numbers fluctuate considerably. Each number is in turn converted to the mean of the sum of itself plus its two nearest neighbours, four nearest neighbours and so on as required (e.g. in 19, 28, 10, *14*, 33, 16, 9, the 14 would become $(10+14+33) \div 3 = 19$).

Sample A subset of the total population. Systematic sample — grid pattern. Random sampling — every member of the population has an equal chance of being selected. Multi-stage or nested sampling, some randomly chosen areas are again randomly sampled. Stratified sampling, several different strata are sampled in proportion to their area of outcrop.

Scalar In matrix algebra, a single number that is used to multiply every number in the matrix (see chapter 2.8).

Scientific (1) Popularly refers to disciplines that study natural or physical as opposed to man-made phenomena.

(2) Refers to an approach to study, generally regardless of the nature of the thing studied. An exercise in strict objective enquiry. Assumptions are used, hypotheses tested and conclusions drawn up in a methodical way.

(3) Science as opposed to Arts is a useful if confusing way of subdividing disciplines for purposes of educational organization.

Seismic A study of earthquake waves in the earth is called seismology. Seismic waves are those generated by movement within the earth and are of three main types.

Sensor, remote Any device that picks up, records and transmits information without actually being in physical contact with the source of the information (e.g. the human eye, a radar scanner). See chapter 14.

Series A succession of quantities, each of which is derived from the quantity preceding it. A mathematical term for an equation with an infinite number of terms. A *convergent* series is one which approaches a finite value as the number of terms approaches infinity, a *divergent* series approaches infinity as the number of terms approach infinity. An example is

$$\sin x = x - \frac{x^3}{3!} + \frac{x^5}{5!} - \frac{x^7}{7!} \cdots \frac{x^n \sin \frac{n\pi}{2}}{n!}$$

where x is in radians.

Set A definable collection of things. These are the elements or members of the set A set may contain any number of elements from 0 to an infinite quantity. The elements may be people, grains of sand, integers, or points on a plane. See chapter 2.2.

Shape The form of a 2- or 3-dimensional figure (e.g. of a county, a stone, a building). Irregular shape cannot easily be characterized by a single word or numerical value. Different features of shape such as compactness, elongation, indentation of the boundary, are often studied.

Shell (earth's) See earth.*

Significance level In statistics a hypothesis is accepted if the calculated probability exceeds a given value α, which is called the significance level. The result is significant if α is <0.05, i.e. 95 per cent confidence limit and highly significant if α is <0.01, 99 per cent confidence limit.*,

Simplification The smoothing of an intricate outline which becomes necessary as the scale of a map is reduced.

Simulation A technique used in certain models often based on the laws of chance, sometimes called Monte Carlo models, to represent a simplified version of a situation in the real world.

Sinusoidal A curve defined by the sine of the angle. It is a smooth harmonic curve. It is at 0 at the origin, rises to $+1$ at $\pi/2$ (90°), and falls to the x axis at π (180°) and to -1 at $3/2\pi$ 270°, returning to 0 at 2π (or 360°).

Skewness The third moment measure, which defines the asymmetry of a distribution curve. This can be expressed in terms of the difference between the mean and the median.

$\mu_3 = E[x - E(x)]^3$. $E(x)$ is the expectation of the random variable x.

Source of deep water The sources of deep water in the ocean are those areas where water can *sink* to the lowest layers of the ocean because of an increase in density. There are two major ones, situated in the North Atlantic near Greenland and the South Atlantic in the Weddell Sea.

Space Associated with the first three dimensions, space defies definition except in a naive way such as something essential for objects to exist in. Since geography is concerned with the arrangement and relationship of things in space and the word spatial is now commonly found, it is important for the geographer to conceive spaces on the earth's shell and not confine space to 'outer space'.

Spectrum A wave spectrum consists of all the individual waves which are present in the sea at any one place. The waves are of different heights and periods (see Fourier analysis*).

Sphere atmosphere The envelope of air which surrounds the earth. *Biosphere* the living beings which inhabit the earth's shell. *Hydrosphere* the water in the earth's shell including water vapour in the atmosphere, surface and ground water on land and the oceans. *Lithosphere* the rocks of the earth's shell. *Stratosphere* the upper atmosphere. *Troposphere* the lower part of the atmosphere below the tropopause.

Spherical excess The amount by which the angles of a spherical triangle exceed 180°. *Spherical triangle* a triangle on the surface of a sphere, the sides and angles of which are both expressed in angular units.

Sphericity The shape of a 3-dimensional object expressed in terms of its long (a), intermediate (b) and short (c) axes. Sphericity (s) $= \sqrt[3]{(bc/a^2)}$.

Spheroid, oblate The earth is shaped approximately as a slightly flattened or oblate sphere. Its equatorial diameter is rather longer than its polar diameter owing to its speed of rotation.

Standard deviation The second moment measure. The standard deviation indicates the spread of the values on either side of the mean in a frequency distribution. $66\frac{2}{3}$ per cent of the variates in the sample would be expected to lie within one standard deviation of the mean and 95 per cent within two standard deviations of the mean. It is a measure of dispersion. It is given by $\sqrt{(\Sigma x^2/N)}$ where x is difference between each value and the mean and N is number in sample. An estimate of the population standard deviation can be obtained from the sample standard deviation.

Stereographic A projection invented in 150 B.C. in which angles are preserved correctly and the projection is orthomorphic.* It is a true projection in that the lines of latitude and longitude are projected from one end of a diameter into a plane tangent at the other end.

Stereoscopic A pair of vertical aerial photographs, taken from slightly different positions, can be fused to give an impression of relief. These are a stereoscopic pair.

Stochastic A method of reaching a solution with the help of probability, which enables confidence intervals to be stated.

Strain The response of an object to the force acting upon it (stress*). Strain rate is the speed at which the object responds to the force.

Stress The force acting on an object.

Strike A horizontal line at right angles to the true dip* of bedded geological strata. It is a contour line on a geological stratum and is straight when the dip is uniform in direction.

Subscript Auxiliary symbols (numerical or letter) to specify the location or relationship of values (e.g. in a series, in a matrix). For example a_{23} is the number in the second row and third column of a matrix.

Sum of squares The sum of squares is used in many statistical tests, especially analysis of variance. It represents the sum of the squared differences between the individual values and the mean of the values.

Summation An instruction to add or the operation of adding two or more numbers or quantities. The symbol for this instruction is a capital sigma Σ.

Superimposition A drainage pattern which was initiated at a higher level on an upper series of strata, to which it was conformable. It was then let down by erosion onto an older discordant series of strata below. The term *epigenesis* is also used to refer to this situation.

Superscript Usually an instruction in numerical or letter form written above a given number or letter to raise its power (e.g. a^2, 5^2).

Surface (1) A concept in mathematics referring to an infinite set of points making a 2-dimensional space (of no 'thickness').

(2) A material object such as the earth's surface, meaning shell and therefore 3-dimensional albeit 'thin'. Interchangeable to some extent with area, patch.

Surveying, plane Surveying is the technique of measuring and recording the position and heights of places on a fairly large scale so that the curvature of the earth may be ignored.

Spherical surveying is carried out on a large area by accurate triangulation and allowance must be made for the curvature of the earth, and its major and minor axes.

Sweep zone The vertical envelope within which a beach profile varies over a period of time.

Synchronic A feature of the same age throughout, formed at one time. Opposed to diachronic and metachronic.*

System 'A system is a set of objects together with relationships between the objects and between their attributes' (*General Systems*, Vol. I, 1956, p. 18). Various other definitions are also available. They stress the compound nature, wholeness, operational aspects, integration of a system as a thing made up of many elements or parts.

Systematic A term often used in geography for branches that cover topics such as climate, plants, manufacturing, as opposed to portions of the earth's surface, regions.

T/ϕ graph The symbols refer to temperature and entropy in the atmosphere. It is a graph on which upper air observations are plotted and includes lines of equal temperature, water vapour content, pressure, dry and saturated adiabatic* lapse rates. Isentropic lines, along which entropy* is constant, are shown also.

Tail A statistical term for the tapering end of a distribution, furthest from the mean. A one tailed test only considers one end of the distribution, a two tailed test considers both ends of the distribution.

Tangent (1) A straight line touching a circle at right angles to the radius, or a straight line touching a curve perpendicular to the radius of curvature at the point of contact.

(2) The trigonometric ratio of perpendicular over base or opposite side over adjacent side to the angle used. (See Mathematical Signs.)

Taxonomy The science of classification, of particular importance in the biological sciences, especially in relation to the general laws and principles of classification.

Telling A term used in diffusion* models to describe the operation of passing on information about an idea, innovation, etc.

Tellurometer An instrument which emits micro-waves which are recorded at a receiver. The time of wave travel gives a measure of the distance between the two points to a high degree of precision. Used now in geodetic survey for measuring distances 30 to 50 miles long.

Temporal Used in this book to refer to things connected with time in contrast to space (i.e. temporal and spatial, not temporal and spiritual).

Territory (1) Generally referred to things happening in area, space, e.g. territorial extent, territorial (=regional) planning (as opposed to branch (=systematic) planning).
(2) A special kind of politico-administrative unit, usually thinly populated and controlled or subsidized by a central government.

Tetrahedron A polyhedron with four faces.

Thermodynamics The study of the movement induced by changes in heat. The first law of thermodynamics is the law of the conservation of energy. The second law of thermodynamics states that heat cannot be transferred from a colder to a warmer body without an outside source of energy.

Till fabric The arrangement of particles within a till or glacial deposit.

Times series A series of successive observations of the same phenomenon over a period of time.

Topographical Variously used in geography with reference to maps showing relief features and/or such objects as vegetation and buildings. The term might more usefully be employed to stand in contrast to topological.

Topology A branch of geometry concerned with the preservation of certain relationships (e.g. relative position) without concern over distance or direction (see chapter 2.10). Called the rubber sheet geometry.

Tractive force The force which produces the tides on the earth's surface. It is directed along the surface tangentially and is given by $\frac{3}{2} g \frac{M}{E} \frac{e^3}{r^3} \sin 2C$ for the lunar tractive forces. M and E are the masses of the moon and earth respectively. e and r are radius of the earth and the distance between earth and moon respectively and C is the latitude. The forces vary with position and the declination* of the moon.

Transactions A special use is for the exchange of people, goods or messages between places in the study, empirical or more often theoretical, of the interaction of places (e.g. towns).

Transcurrent A wrench fault or tear fault along which the movement is in the horizontal plane.

Transformation See mapping.*

Translation In addition to its linguistic connotations, translation is a transformation (see mapping*) involving the movement of every point in a plane, a given distance in a given direction.

Transpiration Moisture given off to the atmosphere from vegetation. Potential evapo-transpiration is the total amount of water vapour that would be lost by both evaporation and transpiration combined, assuming a permanent supply of water is available.

Trend The general direction in which a series of observations are increasing or decreasing, used especially of trend in time series and trend surface analysis.

Trend surface analysis The fitting of linear, quadratic*, cubic* or higher degree polynomial* surfaces to data distributed in area by means of least sum of squares* regression.* The result produces a plane or higher surface in three dimensions.

Trochoidal The curve traced out by a point within a circle which is rolled along a straight line.

Uniformitarianism The view that processes acting on the earth at present are more or less similar to those which have been acting throughout geological time (see the opposite view — catastrophism*).

Variable One of the items which enters into a situation that is being studied. A dependent variable is one which is influenced by an independent variable. A variable can take on a wide range of individual values.

Variance The variance is the square of the standard deviation.* In factor analysis the *common* variance is that part of the total variance which correlates with other variables. *Specific* variance is that part of the total variance which does not correlate with any other variable. *Error* variance is chance variance due to sampling errors, measurement errors.

Variate One member of a set of values for one variable. An individual observation.

Vector (1) A matrix with either a single row or a single column.
(2) In geometry, a line, directed and proportional in length to scalar magnitude. Vectors are not confined to lines on planes but may exist in many dimensions (see eigenvector*).

Vernier scale An optical device using a moveable scale to enable more precise observations to be made on certain instruments such as abney levels, sextants, some theodolites, barometers. It depends on the principle that it is easier to see coincidences between two lines than to judge their distance apart.

Vertex (1) In Euclidean geometry the intersections of lines on the plane.
(2) In topology, nodes,* points in 2-dimensional topology.

Vertical exaggeration The ratio of the vertical scale on a profile to the horizontal scale which is usually drawn as a smaller scale than the vertical.

Water table The surface in three dimensions which divides a permanently saturated zone below from an unsaturated zone above, within the ground.

Zenithal A group of projections based on a plane which is tangent to the globe in the normal case at a pole in the equatorial case at any point on the equator and in the oblique case at any other point on the surface of the globe.

Mathematical and statistical signs

Commonly used signs

$a+b$ usually means add b to a (but not in Boolean algebra)
$a-b$ usually means subtract b from a
$a \times b$ or $a \,.\, b$ or ab means multiply a by b
$a \div b$ or $\frac{a}{b}$ or a/b means divide a (numerator) by b (denominator)

Note: a dot can be the operator *and* in Boolean algebra, a decimal point, the instruction to multiply or sometimes *such that* in set notation

$=$ equals, e.g. $2+3=5$
$\triangleq \simeq \doteqdot$ approximates to
$\neq$ is not equal to
$a>b$: a is greater than b
$a<b$: a is less than b
$a \geqslant b$: a is greater than or equal to b
$a \leqslant b$: a is less than or equal to b

e.g. $0<a<2$ a is greater than 0 but less than 2
$0 \leqslant a \leqslant 2$ a is greater than or equal to 0, less than or equal to 2

Order of arithmetical operations: (1) multiply, (2) divide, (3) add, (4) subtract
Brackets give instructions if conventional order is to be modified, e.g. if $a=4$, $b=2$, $c=3$

$$a(b+c)=4(2+3)=4(5)=20$$

whereas
$$ab+c=4\times 2+3=8+3=11$$

Miscellaneous signs and symbols

$\rightarrow$ becomes, tends to
∞ infinity
$\Rightarrow$ implies
$\pm$ plus or minus, as used to give range of values either side of a given number
$:$ means *is to* (a ratio)
$|x|$ modulus x, absolute value of x, the negative or positive sign is to be ignored
% per cent, ‰ per mil (per thousand)
π pi, the ratio of the diameter of a circle to its circumference. pi can be derived from the convergent series

$$\frac{\pi}{4}=1-\frac{1}{3}+\frac{1}{5}-\frac{1}{7}+\frac{1}{9}-\frac{1}{11}\cdots$$

Matrices (see chapter 2.8)

Matrix $\begin{bmatrix} 2 & 1 \\ 5 & 3 \\ 6 & 4 \\ 7 & 9 \end{bmatrix}$ 4×2 matrix, see figure 2.12 and 2.15*a* and *b*

Identity matrix $\begin{bmatrix} 1 & 0 & 0 \\ 0 & 1 & 0 \\ 0 & 0 & 1 \end{bmatrix}$ Column vector $\begin{bmatrix} 1 \\ 2 \\ 3 \end{bmatrix}$ Row vector $[1 \quad 2 \quad 3]$

$[\mathrm{a} \quad \mathrm{b}]^{-1}$ matrix inversion

Ordered pairs (1, 2). Ordered triples (1, 2, 3) to give positions in 2 or 3 dimensions respectively.

Set Notation

$\{\ \}$	encloses members of a set
$\in$	is an element of
ϕ	empty set
$\mathscr{U}$ or $\mathscr{E}$ or I	universal set
$\supset$ (or $\supseteq$)	is a subset of, e.g. $A \supset B$: B is a subset of A
$\bar{A}$ A'	complement of set A
$\cup$	union of sets (or)
$\cap$	intersection of sets (and)
$A \cap B = \phi$	sets A and B are disjoint
$A = B$	sets A and B are equal
$\Leftrightarrow$	one to one correspondence (equivalence)
$\mid$	such that
N	the set of natural numbers
Z	the set of integers
Q	the set of rational numbers
R	the set of real numbers

Subscripts (generally 'labels')

$x_1, x_2, x_3 \ldots x_i, x_j, \ldots x_n$ (or some other letter instead of x) refer to values for a set of individual cases (observations). i (and if necessary j) refers conventionally to some member of the set without any particular member being specified. n is the 'last' member of the set and is equal to the total number in it.

The instruction Σ, which means sum (add) all the values (numbers) that follow, is often used with subscripted letters. Suppose the following data are under consideration

x_1	43	
x_2	29	Any or all of the five
x_3	36	may be referred to as the *i*th
x_4	32	x_5 is the *n*th
$x_{5\,(n)}$	40	

The instruction: $\sum_{i=1}^{i=n} x_i$ means take all the values (x_i) of x between x_1 and x_n (including these) and add them.

The instruction $\sum_{i=1}^{i=3} x_i$ means add only x_1, x_2 and x_3

$$\Sigma x_i \quad \text{or} \quad \sum^{n} x_i \quad \text{or} \quad \sum_{i=1}^{n} x_i \quad \text{all mean the same as} \quad \sum_{i=1}^{i=n} x_i$$

Sometimes it is desired to calculate or express the sum of the sums of two or more columns of numbers. This situation may be illustrated with the case of a matrix. A matrix has m rows and n columns

	1	2	j	n
1	a_{11}	a_{12}	a_{1j}	a_{1n}
2	a_{21}	a_{22}	a_{2j}	a_{2n}
i	a_{i1}	a_{i2}	a_{ij}	a_{in}
m	a_{m1}	a_{m2}	a_{mj}	a_{mn}

The sum of the values in the first column will be $\sum_{i=1}^{i=m} a_{i1}$

The sum of the values in the second column will be $\sum_{i=1}^{i=m} a_{i2}$

The sum of the sums of all the columns becomes $\sum_{j=1}^{j=n} \sum_{i=1}^{i=m} a_{ij}$

Exponents

$\sqrt{\ }$ square root e.g. $\sqrt{4}=2$

$\sqrt[3]{\ }$ cube root e.g. $\sqrt[3]{27}=3$

$$y^1=y$$

$$y^2=y\times y \qquad \text{e.g. } 3\times 3=9$$

$$y^{-1}=\frac{1}{y} \qquad \text{e.g. } 10^{-1}=\frac{1}{10} \text{ or } 0.1$$

$$y^{-2}=\frac{1}{y^2} \qquad \text{e.g. } 10^{-2}=\frac{1}{10^2}=\frac{1}{100}=0.01$$

$$y^{\frac{1}{2}}=\sqrt{y} \qquad \text{e.g. } 9^{\frac{1}{2}}=\sqrt{9}=3$$

$$y^{-\frac{1}{2}}=\frac{1}{\sqrt{y}} \qquad \text{e.g. } 9^{-\frac{1}{2}}=\frac{1}{\sqrt{9}}=\frac{1}{3}$$

$$y^{p/q}=\sqrt[q]{y^p} \qquad \text{e.g. } 27^{2/3}=\sqrt[3]{27^2}=\sqrt[3]{729}=9$$

Logarithms (Short Log). Various bases may be used for logarithms

Base 10 logarithms are commonly used

Log_{10}	10^0	log 0.0
Log_{10}	10^1	log 1.0
Log_{10}	10^2	log 2.0

The number before the point is the characteristic of the log.

The characteristic depends upon the number of digits in the number

e.g. 356	is log 2.5514
35.6	is log 1.5514
3.56	is log 0.5514
0.356	is log $\bar{1}.5514$
0.0356	is log $\bar{2}.5514$

The numeral to right of the point is the mantissa of the log.

Other bases can be used, e.g. Natural or Napierian logarithms have the base e, which is given by

$$e = 1 + \frac{1}{1} + \frac{1}{2!} + \frac{1}{3!} + \ldots = 2.7182818$$

Examples of the use of logarithms to work out exponents.

$$y^{4.11} = \text{Log } y \times 4.11 \text{ antilogged}$$

if $y = 232$ then

$$232^{4.11} = 2.3655 \times 4.11 = \text{Log } 9.722205$$

Number is 5,274,700,000 approx.

$$y^{-2.4} = \frac{1}{y^{2.4}}$$

if $y = 12$ then $y^{-2.4} = \dfrac{1}{\text{Log } 12 \times 2.4} = \dfrac{1}{1.0792 \times 2.4} = \dfrac{1}{2.59008} = \dfrac{1}{389.5} = .00257$

Trigonometry uses ratios in right-angled triangles

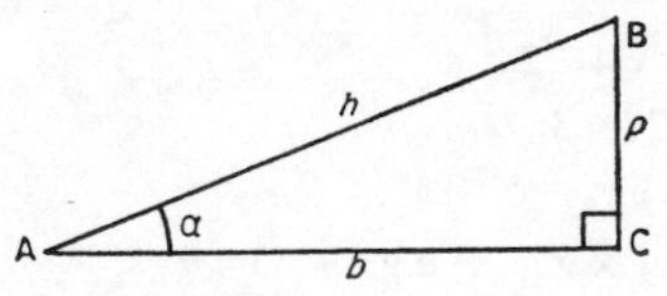

sin (sine) $\alpha = p/h$ or BC/AB   cosec (cosecant) $\alpha = h/p$ AB/BC
cos (cosine) $\alpha = b/h$ or AC/AB   sec (secant) $\alpha = h/b$ AB/AC
tan (tangent) $\alpha = p/b$ or BC/AC   cot (cotangent) $\alpha = b/p$ AC/BC

e.g., if h is 5 and b is 4 and p is 3

$$\tan \alpha = \frac{3}{4} = 0.75 \qquad \alpha = 36°\ 50' \text{ (to the nearest } 10')$$

$$\sin \alpha = \frac{3}{5} = 0.60 \qquad \alpha = 36°\ 50'$$

$$\cos \alpha = \frac{4}{5} = 0.80 \qquad \alpha = 36°\ 50'$$

$$\cos 60° = \sin 30° = 0.5 \quad \tan 45° = 1.0$$

Factorial

! or (!) means factorial, e.g. $3! = 3 \times 2 \times 1$ $4! = 4 \times 3 \times 2 \times 1$

Combinations and Permutations

$$P\binom{n}{r} \text{ or } {}^nP_r = n!/(n-r)! = \frac{n!}{(n-r)!} \quad \text{Permutations}$$

$$C\binom{n}{r} \text{ or } {}^nC_r = \binom{n}{r} = n!/[r!(n-r)!] \quad = \frac{n!}{r!(n-r)!} \quad \text{Combinations}$$

where n is number of objects taken r at a time.

When

$$r=2 \quad n!/[r!(n-r)!] = \frac{n(n-1)(n-2)\ldots 2\times 1}{2\times 1(n-2)\ldots 2\times 1} = \frac{n(n-1)}{2}$$

$$n=5 \quad r=2 \quad 5!/[2!(5-2)!] = \frac{5\times 4\times \not{3}\times \not{2}\times \not{1}}{2\,(\not{3}\times \not{2}\times \not{1})} = \frac{20}{2} = 10$$

Statistics

$$\text{Normal curve} \quad y = \frac{1}{\sigma\sqrt{2\pi}}\, e^{-(x-\bar{x})^2/2\sigma^2}$$

Z is the distance from the centre of the normal curve such that when $Z=1$ it is one standard deviation away, $Z=2$, two standard deviations away and so on

X any variate; x deviation of any variate from the arithmetic mean of the set

$\overline{X}$ arithmetic mean of set of variates, alternatively m

μ population mean (where population is very large)

σ standard deviation of population

$\hat{\sigma}$ estimate of standard deviation of population

s, S standard deviation of sample

α_2 skewness, also μ_3

β_2 kurtosis, also μ_4

DF df degrees of freedom

O or o observed

E or e expected

χ^2 chi square

ν degree of freedom for F ratio, which is ratio of variances

$\nu_1\nu_2$ degrees of freedom of denominator and numerator respectively

SS sum of squares total

SSE within sample sum of squares

SST between sample sum of squares

LSD least significant difference

p P probability, e.g. p (ONE) could be the probability of obtaining a 1 on the throw of a dice

V the coefficient of variance

ρ rho Spearman rank correlation coefficient

d commonly used for difference

n, N number

r_{xy} product moment correlation of x and y

b_{xy} regression coefficient of x on y, b_{yx} the regression of y on x. These give the angles of the regression lines as tangents

a gives intercept as one axis when the value of the other is zero

Random digits

These digits may be used singly, in pairs, or in larger groups, as required. They are mainly relevant to chapter 11. The authors are grateful to Mr. M. J. McCullagh for writing a computer program to produce them.

9 1 4 1 3 4 7 5 7 8 0 7 3 2 4 4 4 7 0 1 1 4 0 2 7 0 4 4 0 1 2 4 4 5 5 9 9 0 3 3
2 1 3 4 7 3 7 6 8 9 8 8 5 8 4 4 6 2 3 8 3 4 1 3 2 4 0 3 5 5 2 4 0 7 6 2 1 3 2 0
7 2 3 0 9 0 7 2 4 3 1 9 2 3 7 3 3 7 1 9 5 3 1 1 4 2 8 4 4 2 8 4 6 7 2 4 5 2 2 3
4 4 4 7 2 1 2 1 3 3 2 2 2 9 2 6 2 4 3 4 7 2 1 3 4 8 5 6 1 4 2 0 1 2 2 8 4 6 3 0
5 3 2 4 3 0 3 0 4 6 1 8 3 7 8 3 4 3 1 8 5 6 0 4 1 4 1 5 9 7 7 8 6 4 3 3 8 0 8 9
8 7 2 0 3 7 0 1 2 7 1 3 9 8 0 8 5 8 7 9 7 1 7 4 0 3 0 8 4 9 6 1 4 3 8 6 1 6 8 4
8 6 7 2 0 2 9 1 4 5 3 9 7 1 4 9 6 8 5 5 5 9 2 9 7 5 6 8 2 2 7 2 4 6 3 4 6 6 7 7
5 9 7 7 7 4 2 6 6 1 5 9 9 5 1 2 1 4 1 9 5 4 9 1 7 2 4 6 9 9 4 7 3 1 5 7 4 6 3 3
3 7 0 6 9 5 6 5 0 3 4 1 0 9 2 0 9 9 5 1 6 1 5 8 6 0 0 3 8 8 1 8 3 8 0 4 3 2 0 3
4 3 3 2 5 3 3 7 5 0 1 2 1 3 7 8 5 3 4 6 4 0 2 3 8 8 4 8 9 2 9 0 3 5 4 7 9 2 1 5
3 4 9 6 0 7 5 8 3 0 7 5 7 6 8 9 0 7 8 3 3 9 2 0 5 6 7 7 0 9 5 2 9 4 2 1 7 4 0 2
8 2 9 1 2 4 5 2 1 0 9 4 8 5 9 0 4 9 7 9 9 6 0 7 0 6 4 8 9 9 7 4 8 3 7 9 6 5 9 4
0 5 6 7 0 6 4 0 5 9 0 6 9 0 8 3 0 4 4 3 0 2 0 6 1 0 5 9 1 6 8 4 2 0 0 4 9 0 2 5
4 2 9 2 2 5 3 6 7 2 0 5 9 5 5 5 6 1 3 9 5 3 0 2 1 8 9 7 6 8 8 8 7 9 5 9 9 9 6 6
0 5 5 1 7 5 6 7 9 6 6 2 7 7 6 7 7 2 4 4 9 1 1 0 6 8 7 6 1 7 8 7 8 7 6 2 9 9 4 3
1 2 6 1 4 5 4 3 4 9 1 1 4 8 6 8 5 8 9 3 7 7 6 0 2 0 6 1 9 8 2 5 4 6 6 0 8 0 5 8
8 0 2 5 7 1 4 6 5 8 0 6 0 1 8 2 4 2 1 4 9 6 4 8 4 9 8 1 4 5 2 7 7 5 6 6 7 7 5 0
1 2 6 7 7 4 7 9 9 1 3 7 2 5 0 8 3 4 9 9 6 0 3 3 9 3 4 0 8 1 9 4 5 9 4 5 5 7 5 9
3 6 8 7 2 9 8 2 3 0 3 6 2 6 6 2 0 3 0 7 1 2 1 2 3 9 2 9 1 4 4 6 2 2 2 1 3 8 9 5
3 4 6 3 2 1 9 4 5 6 1 7 5 0 9 6 4 6 5 6 9 5 1 9 9 2 2 4 6 3 6 8 3 2 3 4 7 8 1 3
0 2 7 5 1 9 7 5 3 6 8 5 2 7 2 1 0 8 4 2 2 1 3 8 3 5 5 9 9 0 1 7 5 3 9 2 3 0 2 1
9 1 9 7 1 8 3 8 3 7 8 1 1 9 7 6 9 6 3 1 4 2 3 4 3 9 0 0 4 0 7 4 4 3 1 4 8 2 0 2
4 2 8 2 4 8 7 1 0 6 7 1 0 5 0 6 6 5 5 6 3 8 0 8 1 4 7 8 4 1 6 5 9 7 1 5 2 3 5 2
1 6 9 4 6 1 4 7 1 0 1 5 4 0 2 6 7 7 6 7 9 1 4 7 4 0 8 2 3 5 2 1 7 3 3 3 3 2 6 4
0 9 8 1 1 3 0 1 4 1 7 4 7 3 6 0 3 7 1 5 0 8 3 1 1 1 0 2 3 9 9 1 5 1 3 4 1 0 0 2
9 3 8 0 4 6 5 6 7 5 6 4 0 4 3 5 9 7 6 6 5 9 8 3 2 9 6 8 1 6 8 0 3 1 1 0 4 2 5 7
4 8 6 3 5 0 6 6 5 1 7 6 3 8 7 2 1 9 6 0 0 2 3 4 3 7 2 9 0 6 1 5 0 9 3 1 0 0 6 3
5 9 2 7 7 0 7 6 7 0 1 3 1 9 9 4 1 9 9 4 7 9 4 8 1 1 0 7 2 4 6 0 2 3 2 7 2 8 0 4
6 1 5 4 1 7 7 0 6 6 6 4 4 1 7 2 5 1 3 2 3 0 1 7 5 0 7 1 0 1 4 9 4 2 5 4 2 0 3 6
9 0 4 2 7 3 8 0 5 6 8 4 8 2 8 3 8 8 2 2 3 9 9 1 8 3 4 6 8 6 8 8 3 5 8 3 0 9 5 4
7 4 3 0 0 1 7 2 9 7 4 0 1 8 4 9 4 9 9 6 8 8 1 7 2 1 3 4 4 1 9 5 1 8 0 6 4 1 8 7
8 2 7 9 0 4 0 5 1 9 8 3 4 7 3 8 9 6 9 0 5 1 6 0 9 4 4 8 0 4 9 6 8 4 0 8 9 5 1 6
1 6 1 7 0 7 7 4 6 9 3 5 1 6 0 6 6 7 7 5 8 8 6 4 9 6 7 5 2 6 0 6 0 4 0 2 6 0 1 4
1 4 4 6 6 9 0 6 5 7 5 5 3 6 8 1 4 2 7 7 1 2 1 6 5 5 5 5 7 3 0 2 4 7 4 2 5 6 1 5
2 4 1 5 6 8 7 8 8 5 8 8 6 4 4 6 0 4 6 2 9 3 9 4 3 8 4 9 7 4 2 7 1 5 5 9 8 6 3 4
5 0 0 7 5 4 0 5 6 7 1 5 1 3 8 0 6 6 8 9 1 2 5 5 8 6 3 7 9 6 6 5 1 2 9 2 4 3 9 3
3 5 4 7 3 5 1 9 5 5 8 8 6 8 1 8 6 5 5 2 1 5 6 7 1 0 5 5 4 1 6 9 3 6 5 6 4 8 6 7
6 6 2 0 0 8 3 1 1 2 2 2 5 8 5 6 8 0 8 0 1 7 8 3 2 3 9 2 9 7 0 8 6 5 1 0 1 7 6 4
1 3 3 9 6 5 7 1 3 9 4 6 4 3 0 5 4 7 8 5 6 2 7 4 0 7 0 1 1 5 2 4 4 1 5 4 2 1 2 5
0 1 6 1 2 4 8 9 3 3 3 1 4 8 6 7 0 8 0 6 5 0 0 6 0 2 2 6 6 3 8 4 2 3 0 4 8 9 0 5
4 4 8 2 4 3 5 4 3 0 6 9 5 8 4 9 4 0 0 7 7 8 6 2 7 5 5 8 5 0 0 2 7 5 6 9 0 4 1 8
8 1 6 9 3 9 5 0 6 5 4 3 0 8 8 1 3 8 9 2 7 5 7 0 8 2 6 3 1 3 6 7 7 2 4 8 1 5 6 9
2 0 4 7 7 1 9 6 4 6 1 5 2 3 9 8 9 4 9 9 7 7 6 5 4 3 2 1 4 2 3 1 1 6 4 1 5 3 7 4
1 5 8 6 5 5 5 0 7 5 3 7 6 1 5 7 9 1 6 1 6 4 0 1 1 1 6 5 7 8 4 7 7 9 7 7 1 9 3 0

Author Index

Subject Index

In this index *italic* numbers are used to denote reference to figures while bold face numbers denote the major reference to an entry.